AF575453

SAP® Cloud ALM

Matthias Friedrich, Fred Kindler, Marcel Kirsten, Daniel Kloppich,
Mathias List, Bert Lorenz, Lars Teuber

SAP® Cloud ALM

Application Lifecycle Management

Editor Meagan White
Acquisitions Editor Hareem Shafi
German Edition Editor Nicole Hohmann
Copyeditor Kim Hannel
Cover Design Silke Braun
Photo Credit iStockphoto: 1003629220/© IakovKalinin
Layout Design Vera Brauner
Production Kelly O'Callaghan
Typesetting SatzPro, Germany
Printed and bound in the United States of America, on paper from sustainable sources

ISBN 978-1-4932-2595-8

1st edition 2024
1st German edition published 2023 by Rheinwerk Verlag, Bonn, Germany

Library of Congress Cataloging-in-Publication Data
Names: Friedrich, Matthias, author.
Title: SAP cloud ALM : application lifecycle management / by Matthias Friedrich, Fred Kindler, Marcel Kirsten, Daniel Kloppich, Mathias List, Bert Lorenz, and Lars Teuber.
Description: 1st edition. | Bonn ; Boston : Rheinwerk Publishing, 2024. | Includes index.
Identifiers: LCCN 2024015379 | ISBN 9781493225958 (hardcover) | ISBN 9781493225965 (ebook)
Subjects: LCSH: Cloud computing--Computer programs. | Software maintenance--Computer programs. | Agile software development. | Computer programming--Management.
Classification: LCC QA76.585 .F754 2024 | DDC 004.67/82--dc23/eng/20240409
LC record available at https://lccn.loc.gov/2024015379

Contents at a Glance

Contents

Part II Explore

Part VI Extend and Integrate

Preface

The wealth of foreign experience acquired from books is called erudition. One's own experience is wisdom.
—Gotthold Ephraim Lessing (1729-1781)

Cloud technology has grown enormously in importance in recent years, as companies are increasingly reliant on flexible and scalable IT solutions to keep pace with fast-moving market conditions. With *SAP Cloud ALM* (*ALM* stands for application lifecycle management), SAP has developed a suite for application lifecycle management that is not only a cloud product itself, but is specifically designed for managing applications in the cloud—including in SAP S/4HANA. It offers companies a range of functions, including project management, quality assurance, test management, and operational support. In this book, we will show you how to successfully manage and optimize your applications in the cloud and on premise using SAP Cloud ALM.

Goal of This Book

The point of this book is to help you successfully manage your applications in the cloud with SAP Cloud ALM. Our aim is to provide a little support in your daily work to all those who also have to deal with IT and SAP systems on a daily basis. Above all, administrators, IT project managers, test managers, IT landscape architects, and IT landscape owners—as well as consultants—should feel inspired by this book. We give you a comprehensive introduction to the functions of SAP Cloud ALM and share best practices and tips for successfully implementing and managing your applications in the cloud. Our approach is neither a product introduction nor a complete implementation guide; that would simply not be feasible for a cloud product in a printed version. Instead, we will introduce you to SAP Cloud ALM and familiarize you by using example scenarios.

Structure of This Book

The book is divided into six major parts, with a total of 22 chapters. As the planning and development phase of an application is particularly complex, the focus of this book is initially on the build phase. The first four parts are therefore based on the SAP Activate phases for implementing cloud applications (prepare, explore, realize, and deploy) and deal with the topics relevant to each of these phases. Part V deals with important operational topics. Part VI looks at the options for extending SAP Cloud ALM or integrating it with other ALM tools. We will repeatedly refer to a project example that you will become familiar with in Chapter 1 and that we will primarily use to illustrate the implementation of the individual tools. The aim of this approach is to enable you to decide which of the example scenarios shown make sense for use in your landscape.

Before we get into the project work, we start in **Chapter 1** with a rough overview of the SAP Cloud ALM product and the SAP Activate methodology. You will learn about the individual areas of SAP's new ALM product and how these areas can be integrated into the ALM process. We will also give you insight into important SAP Activate concepts and terms.

Part I of this book covers the prepare phase of an SAP implementation project with SAP Cloud ALM. This phase is crucial for the success of the project, as it forms the basis for all subsequent phases. The chapters in this part cover the following tasks within the prepare phase:

- A new SAP cloud tenant is created in **Chapter 2**. The *tenant* is the area in the cloud in which the project is carried out.
- **Chapter 3** sets up the environment for the project. This includes the provision of resources such as servers, storage, and network.
- In **Chapter 4**, the systems involved in the project are connected to the transport system. The *transport system* is the system that manages changes to the systems.
- **Chapter 5** describes the agile methodology to be used for the project. An *agile methodology* is an approach to project implementation that is based on the needs of the customer.
- **Chapter 6** defines the guidelines and conventions that should apply to the project. These guidelines and conventions help to improve the quality of the project and promote communication between the project participants.
- In **Chapter 7**, the project is set up in SAP Cloud ALM. We will go into the creation of projects, teams, and tasks.
- In **Chapter 8**, you will learn how the project participants are prepared for the project in training sessions, workshops, and meetings.

We move on to the next phase in Part II of the book. The explore phase is about understanding and documenting the company's requirements for the project. We will address the following four key topics in detail in the individual chapters:

- **Chapter 9** explains how you can use SAP processes for your project. Processes help you structure and document your requirements.
- **Chapter 10** explains how you can create and manage requirements for your project. Requirements form the basis for all other project activities.
- **Chapter 11** introduces fit-to-standard and fit-gap workshops, in which the SAP processes are compared to the company's requirements. These workshops help you ensure that the project meets the requirements.

- **Chapter 12** describes how to plan and execute tests for your project. Tests help you to ensure that the project meets the requirements.

Part III of the book deals with the realize phase of an SAP implementation project. In this phase, the company's requirements for the project are implemented. **Chapter 13** through **Chapter 15** explain the tasks that are carried out in this phase:

- **Chapter 13** describes how you can create user stories for your project. User stories are the smallest units of work in an agile project.
- **Chapter 14** explains how you can create test cases for your project. Test cases help you ensure that the requirements are met.
- In **Chapter 15**, you will learn how to use SAP Cloud ALM to perform tests for your project.

In Part IV of this book, we look at the final build phase of an implementation project, the deploy phase:

- In **Chapter 16**, the changes to the systems are planned and prepared. This includes the creation of change requests, test planning, and test execution.
- In **Chapter 17**, the changes to the systems are implemented and tested. You will learn how changes are made available in the systems, how tests are carried out, and how changes are released for production.

In Part V, we come to the run part:

- In **Chapter 18**, we will provide an overview of SAP Cloud ALM for operations, with a focus on business service management and incident management.
- In **Chapter 19**, we will show you how to implement a suitable monitoring system in SAP Cloud ALM.

Finally, in Part VI, we introduce you to the extension options of SAP Cloud ALM:

- In **Chapter 20**, we start with the basics. You will learn which requirements must be met in order to implement extensions.
- In **Chapter 21**, we go into more detail about the extensions that are already available.
- In **Chapter 22**, we present practical examples of extensions.

So that you know at all times which phase you are in within the book, the individual subprocesses of the phases are shown in Figure 1 with the respective chapter number. The diagram represents a complete cycle for the integration of an application into your business processes. In practice, details of the processes described in this book may differ from our approach depending on your SAP Cloud ALM use case. As you can see, Part V and Part VI are not shown in this diagram, as the topics covered in Chapter 18 through Chapter 22 are outside the SAP Activate methodology.

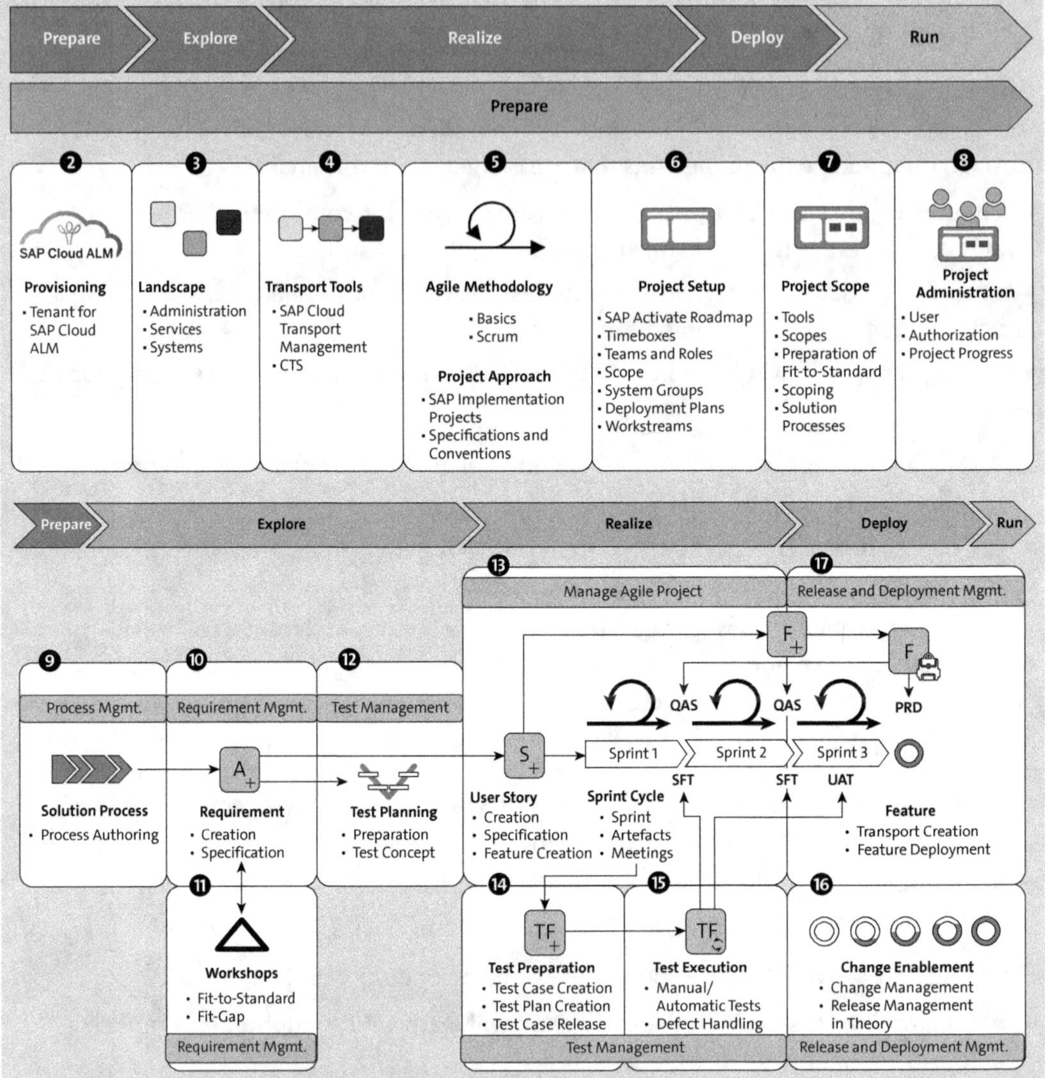

Figure 1 Schematic Representation of the Content of This Book

Highlighted information boxes in this book contain content that is worth knowing and helpful but is somewhat outside of the actual explanation. We have marked the boxes with symbols so that you can immediately classify the information:

[»] ■ In boxes marked with this symbol, you will find information on *further topics* or important content that you should remember.

[!] ■ This symbol draws your attention to *special features* that you should be aware of. It also warns you of common errors or problems that may occur.

[+] ■ This symbol indicates *tips* and *advice* from professional practice that provide practical recommendations that can make your work easier.

Acknowledgments

This book incorporates the wealth of experience and extensive expertise of many contributors—both authors and coauthors—who have invested their time, energy, and patience to make this book a valuable resource for you. But we could not have done this without the people who supported and encouraged us during the writing of this book: our families and loved ones.

If we are writing about an SAP product here, of course it wouldn't work without the support of the manufacturer. We would also like to thank SAP SE, and in particular the SAP Cloud ALM product development team, who have generously supported us in the creation of this book. Thanks to their support, we had the opportunity to look behind the scenes of SAP Cloud ALM and gain valuable insights into sometimes difficult topics. We appreciate their open cooperation and the tireless efforts of all our colleagues in project development.

Special thanks also go to Rheinwerk Publishing, who critically reviewed the manuscript and provided valuable feedback.

Last but not least, we would like to thank our colleagues who actively supported and encouraged us during the creation of this book. Their advice, support, and expertise have made a significant contribution to the success of this book.

We hope you find this book useful and that it helps you to deepen your knowledge of SAP Cloud ALM. Thank you for your support—and have fun and good luck with our book!

Matthias Friedrich, Fred Kindler, Marcel Kirsten, Daniel Kloppich,
Mathias List, Bert Lorenz, and **Lars Teuber**

Acknowledgments

Chapter 1
Introduction to SAP Cloud ALM and SAP Activate

This introductory chapter is intended to give you an overview of the features and areas of SAP Cloud ALM and show the connection with SAP Activate. SAP Activate is a central component of SAP Cloud ALM for the planning, implementation, and introduction of an SAP solution. You will get to know SAP Activate as a framework and by its individual components. At the same time, the integration of these components in SAP Cloud ALM will be explained.

The digital transformation has changed the way companies do business today. New technologies such as cloud computing, big data, and artificial intelligence enable companies to respond to changes faster and more agilely, thus securing competitive advantages. In the SAP environment, SAP Cloud ALM plays an important role in the implementation of these technologies. SAP Cloud ALM is a cloud-based application that helps companies achieve the full performance of their applications in the cloud and optimize their business processes. This solution from SAP offers a variety of functions to monitor the performance of applications, diagnose and resolve problems, and minimize risks. In addition, it enables close collaboration between different departments within a company. To reap the benefits of SAP Cloud ALM, it is important that companies have an effective implementation strategy. The *SAP Activate* methodology offers such a proven implementation strategy: It is based on an agile approach and provides a clear framework for planning, implementing, and monitoring projects.

In this chapter, we will take a detailed look at SAP Cloud ALM and SAP Activate. We will look at the theoretical foundations and concepts behind them, consider their practical application, and outline the benefits and challenges of implementing them.

By understanding SAP Cloud ALM and SAP Activate, you can optimize your implementation strategies and make your business processes more effective. You will be able to take advantage of cloud computing technology to optimize your business processes and increase your competitiveness. In the following sections, we will look at the functions and features of SAP Cloud ALM and SAP Activate in detail and highlight their impact on businesses.

1.1 Introduction to SAP Cloud ALM

With SAP Cloud ALM, SAP is offering a new proprietary tool to support your ALM process for cloud applications and the current SAP S/4HANA suite. This modified tool is primarily aimed at the new and evolving challenges of managing cloud applications within a company's own business processes. There are fundamental differences in how SAP Cloud ALM executes ALM processes for applications as well as in how you implement it, compared to SAP Solution Manager. In the following sections, we will explain the idea and, above all, the methodology behind the tool and differentiate SAP Cloud ALM from other SAP ALM products—specifically, SAP Solution Manager and SAP Focused Run.

For more than a decade, SAP has been offering a tool for managing and mapping the entire ALM cycle: *SAP Solution Manager*. Many customers have invested a great deal of time and effort in SAP Solution Manager and its integration into their IT processes. In recent years, SAP Solution Manager has been expanded, particularly in the areas of operation (monitoring of technical components), individual change management, and support for project planning of migration projects. Both its strength and its weakness is that it is designed as a monolithic standalone solution for underlying company systems that are local, on-premise system landscapes. In the meantime, applications in IT landscapes are increasingly being provided according to the *software as a service (SaaS)* model. With this model, you as a company are no longer responsible for the quality and stability of the application, but merely consume the IT service. This change in handling applications naturally requires a change in handling application lifecycle management in your IT processes. To close this gap, SAP developed the SAP Cloud ALM solution for managing ALM processes. It is primarily intended to meet the changed requirements for application management and support the management of cloud applications.

The strategy today and in the coming years is to gradually introduce SAP Cloud ALM and support you as a customer while pursuing parallel support of the tools in line with the current challenges of heterogeneous system landscapes. SAP Solution Manager will continue to be supported until at least 2027. The end of maintenance for SAP Solution Manager is part of SAP's strategy to replace SAP Business Suite 7 solutions with the next generation of SAP solutions.

To simplify this migration, SAP has decided to migrate its customers to the SAP S/4HANA platform, which is based on the latest technology and offers more powerful functions. SAP Solution Manager, on the other hand, is based on the older SAP Business Suite 7 platform and does not support some of the new functions of SAP S/4HANA; in addition, some functions such as customer relationship management (CRM), on which the SAP Solution Manager product is based, are no longer available. It is therefore part of SAP's strategy to also phase out support for SAP Solution Manager with SAP Business Suite 7 in 2027, though SAP Solution Manager can continue to be used until 2030 as part of extended maintenance. This will be a decisive advantage for migrations, as SAP

Solution Manager is the only tool that supports SAP Business Suite 7 and SAP S/4HANA in the same manner.

In Figure 1.1, you can see an overview of the maintenance strategy moving forward for SAP's ALM tools.

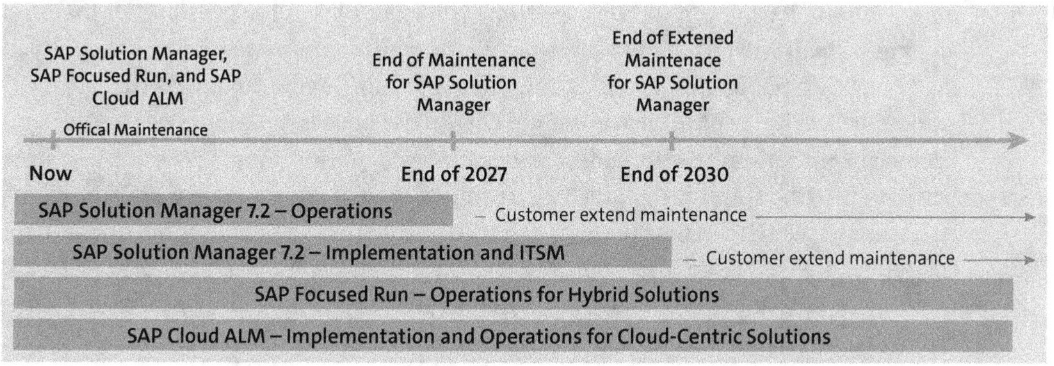

Figure 1.1 Planned Support of SAP Solution Manager as an ALM Tool

1.1.1 Features of SAP Cloud ALM

SAP Cloud ALM is primarily designed for the management of cloud applications, where it unfolds its full-service potential. In on-premise system landscapes or in environments that are provided according to the *infrastructure as a service (IaaS)* model, e.g., in a private cloud, the SAP Cloud ALM service offering can currently only be used in part of the system. In the future, however, it should be possible to map the entirety of the SAP S/4HANA suite. The current objective is to close the gap between the existing tools and to be able to fully support customers with public cloud applications (SaaS). You can decide whether the use of SAP Cloud ALM offers advantages within the ALM cycle based on your own IT strategy and the technologies used in your IT landscape.

As already described, there are different challenges when operating cloud applications in your own business processes than when operating on-premise system landscapes. SAP Cloud ALM is designed to support customers with the challenges that arise when cloud applications are no longer operated independently. However, this also means that SAP Cloud ALM is not intended for customers with a high proportion of on-premise solutions—especially from the old SAP Business Suite 7 product generation—because the previous patterns of responsibility for the management of on-premise system landscapes cannot be mapped there. But a direct comparison of SAP Solution Manager and SAP Cloud ALM in terms of their functional scope is difficult. SAP is also not planning to completely replace SAP Solution Manager with SAP Cloud ALM. Rather, parts of SAP Solution Manager will be retained that will make sense for future environments; anything that will no longer be needed will not be migrated.

Another special feature of SAP Cloud ALM is that it is a native public cloud application and is therefore subject to the same technical mechanism as any other SaaS product: provision as a service. SAP Cloud ALM is an open software solution based on SAP standards. This means that it can be supplemented and optimized with your own extensions, e.g., within the SAP Business Technology Platform (SAP BTP) or with third-party extensions. In Part VI of this book, we show you examples of the possibilities that such an extension in SAP Cloud ALM can open up for you as a customer. Another key advantage of using a public cloud application is its operation—you can access the latest version of the software solution at any time and do not have to use your own resources to keep the solution up-to-date. In comparison, SAP Solution Manager requires a great deal of effort to stay up-to-date. This includes regular maintenance and individual adjustments, which can take place up to twice a year and can interrupt the operation of the tool as well as tie up important resources for updates and maintenance. With SAP Cloud ALM, this effort is almost eliminated as SAP, as a service provider, incorporates changes directly and makes them available in your environment with minimal delay.

In addition to the classic operational issues, the use of an SaaS service also offers further advantages in the functional area. All change requests, adaptations, and enhancements resulting from the experience of the various customer projects come together in SAP Cloud ALM, and everyone always has access to the same level of development and information. There are no longer any individual solutions in the product, but a range of potentially individual solutions to which you automatically have access.

One advantage of SAP Cloud ALM, and a contrast to competitor products, is that the solution is operated entirely by SAP, which provides an additional boost in stability and availability. Within SAP Cloud ALM, you have no additional effort due to individual coding and extensions and can align and design the tests specifically to your own requirements. As the provider, SAP already provides comprehensive technical security for the application. This greatly reduces the direct costs of operating your ALM solution landscape compared to a product such as SAP Solution Manager. Every customer with an SAP cloud product or a product from the SAP S/4HANA product family in their SAP Enterprise Support license agreement receives an SAP Cloud ALM tenant that they can call up from SAP at any time. For more information on the steps required to test the functions of SAP Cloud ALM, see Chapter 2.

SAP currently offers a free usage option for SAP Cloud ALM, though it has a limit on the amount of storage space you can use. The amount has risen steadily in recent years; you can currently (as of 2024) have 8 GB of storage space at their disposal; if they require more, they can purchase it. SAP Cloud ALM can be used in full with all available functions in all options, whether free of charge or for a fee. In contrast, the operating costs of SAP Solution Manager are comparatively high, and there are also the costs for maintaining and optimizing your own landscape.

1.1.2 Differentiating Between SAP Cloud ALM from Other ALM Tools

SAP Cloud ALM is an application lifecycle management tool designed primarily to help you manage and optimize your cloud applications and processes. It offers a comprehensive solution that covers the entire lifecycle of an application—from planning and development to operation and continuous improvement. Compared to other ALM tools on the market, SAP Cloud ALM differs in the following areas:

- **Integration into SAP system landscapes**
  SAP Cloud ALM is specifically tailored to the needs of companies already using or planning to use SAP solutions. It offers seamless integration with existing SAP systems, applications, and services, which is a significant advantage over other ALM tools that may not offer the same depth of integration.
- **Cloud-native architecture**
  Unlike some older ALM tools that were originally developed for use in on-premise environments, SAP Cloud ALM is designed from the ground up as a cloud-native solution. This enables better scalability, flexibility, and agility to manage modern cloud applications and infrastructures.
- **Focus on SaaS applications**
  SAP Cloud ALM is designed to address the unique challenges of managing SaaS applications. This focus sets it apart from other ALM tools that may not offer the same depth of support and functionality.
- **SAP-specific best practices and process support**
  SAP Cloud ALM offers a variety of best practices and process support that are specifically tailored to SAP solutions and technologies. This enables companies using SAP software to make their ALM activities more effective and efficient.
- **Limited applicability outside of SAP environments**
  One of the drawbacks of SAP Cloud ALM compared to other ALM tools is that it may be less useful for organizations that are not using or planning to use SAP software. Because SAP Cloud ALM is heavily focused on SAP technologies and integrations, there may be gaps in support for non-SAP technologies and applications.

As already mentioned, SAP Cloud ALM is therefore particularly useful in landscapes with SAP applications and clearly sets itself apart from the competition.

SAP itself is currently involved in ALM management with three tools on the market:

- SAP Cloud ALM
- SAP Solution Manager
- SAP Focused Run

From today's perspective, each of these three products is useful—which product is the right one depends above all on how the IT landscapes are structured and what requirements the company has. The more complex the system landscape and requirements are, the more difficult it is to decide on a tool to support the ALM process. It may even be necessary to use more than one tool to take advantage of individual benefits. In the following section, we offer you some further differentiation options to help you make the best tool selection for you.

One of the most important differences between the three SAP ALM tools is the *target group*. Even if this distinction, as shown in Figure 1.2, can only apply to the current situation, it should clarify the focus of the three tools. SAP Cloud ALM is targeted primarily at companies that run their SAP applications in a cloud or plan to do so in the future. An example of companies that can benefit from SAP Cloud ALM are enterprises that have already implemented their SAP application almost completely in a cloud environment using SaaS applications or are planning to do so (see the right-hand path in Figure 1.2). SAP Solution Manager, on the other hand, is a much more comprehensive solution aimed at companies that require a centralized management platform (see the left-hand path in Figure 1.2) to manage their SAP applications. An example of a company that could benefit from SAP Solution Manager is a larger organization with multiple SAP systems, especially SAP Business Suite 7 products, in a heterogeneous IT landscape that includes both on-premise and cloud applications. The third product, SAP Focused Run, is primarily aimed at companies with a very large variety and, above all, many SAP applications in their IT environment. The use of SAP Focused Run is recommended for 300 system instances and predominantly in the on-premise area.

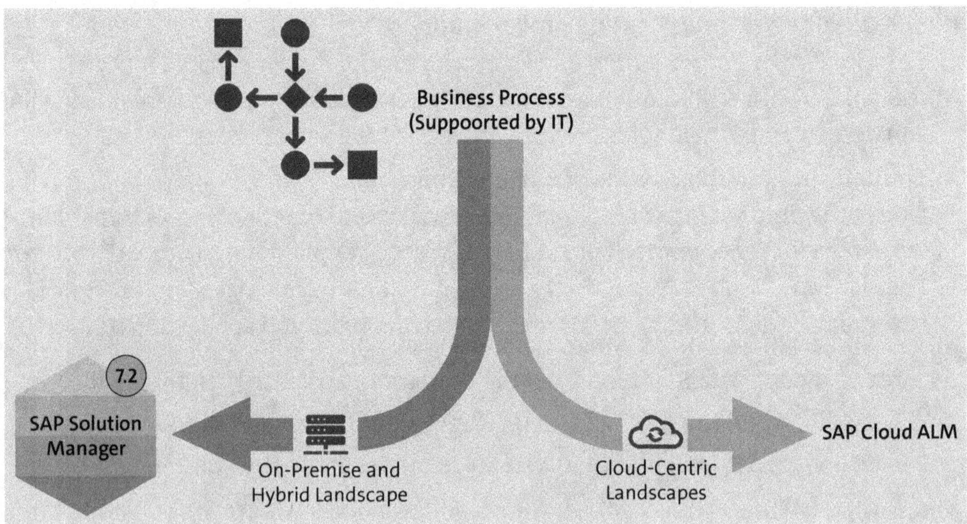

Figure 1.2 SAP's ALM Tools and Their Focus

Another feature for differentiation is the availability and installation of the tools. SAP Cloud ALM is offered by SAP as a pure SaaS solution. SAP Solution Manager, on the other hand, is provided as both an SaaS and an on-premise solution. Companies can choose whether they want to operate the solution in their own cloud, through SAP or other service providers, or on their own servers. SAP Focused Run is only available as an on-premise solution. Both SAP Solution Manager and SAP Focused Run require installation and subsequent configuration—in contrast to SAP Cloud ALM, where this is not necessary.

There are also clear differences in terms of the range of functions offered by the three tools: SAP Cloud ALM offers functions to support the entire application lifecycle, including project management, application management, test management, and change management. One example of the use of SAP Cloud ALM is support for the implementation of a new SAP application in the cloud, including the planning, execution, and monitoring of tests. SAP Solution Manager offers a much broader range of functions to support the entire application lifecycle, including project management, monitoring, incident management, change management, and test management. SAP Solution Manager is typically used in large IT landscapes due to this centralized approach and the use of proprietary process mapping and landscape information (solution documentation). In contrast, SAP Focused Run offers targeted functions for monitoring, troubleshooting, and managing SAP applications in on-premise environments and only supports the ALM phase of operations.

A final aspect that distinguishes the three tools from one another is the ability to integrate applications. Although there are differences due to the architecture and technology, the options for integrating other applications are similar for all three tools. SAP Cloud ALM primarily offers integration options with other cloud applications such as SAP Integration Suite (part of SAP BTP), SAP SuccessFactors, or SAP Ariba, while SAP Solution Manager enables classic integration with SAP BusinessObjects Business Intelligence or SAP Process Integration. Due to the cloud-native technology, integration into SAP Cloud ALM using SAP Integration Suite is significantly easier and more individual for companies than with the other two ALM solutions.

In summary, SAP Cloud ALM differs from other products on the market primarily due to its cloud-centric approach. But it depends on your specific requirements and the characteristics of your IT infrastructure as to which solution is best suited for you. Examples of the use of these solutions can vary depending on the industry, size, and IT landscape of the company.

1.2 SAP Cloud ALM Areas

SAP Cloud ALM is an integrated solution for managing the entire lifecycle of applications that are developed, deployed, and operated on SAP cloud platforms. This section explains how SAP Cloud ALM is linked to the various phases of the ALM process and how the solution can be ideally used.

In the implementation phase, it is particularly important to translate the requirements into possible applications. SAP Cloud ALM supports this process with a variety of functions, applications, and integration options:

- **Process management**
  Process management in SAP Cloud ALM forms the basis for every implementation project. The project scope to be implemented is defined based on the SAP best practice processes provided. In addition to process documentation and the process diagrams provided, SAP Cloud ALM offers the option of defining your own processes and describing them in line with requirements.
- **Requirements management**
  SAP Cloud ALM makes it possible to record, track, and prioritize requirements. This enables companies to ensure that the applications developed meet the business requirements.
- **Project management**
  With SAP Cloud ALM, project management can monitor the progress of projects, manage resources, and assign tasks. The solution also offers real-time dashboards and reports to support decision-making.
- **Development and testing**
  SAP Cloud ALM offers integration options for transport tools to enable controlled change management. It also supports test management by integrating test automation tools such as test automation in SAP S/4HANA Cloud Public Edition.
- **Continuous integration and deployment**
  SAP Cloud ALM enables automated build and deployment processes that shorten development cycles and improve the quality of applications.

The supporting functions and tools in the operating phase are just as extensive. The two main components here are *monitoring* and *root cause analysis*. These two functions contain all the tools required for the operation of cloud applications and SAP S/4HANA applications. Basically, the focus here is not limited to the operation of the cloud application alone but is rather about interpreting the overall process of your company and finding out how the cloud applications contribute to its execution.

The service area of SAP Cloud ALM includes the interaction with customers and end users to offer the highest possible customer satisfaction and an optimal user experience. It is important that you as the customer and SAP as the service provider have a shared overview of the entire service delivery process. One of the core functions of SAP Cloud ALM for service is the introduction of a collaboration platform that makes it possible to share results, tasks, and status messages with everyone involved in the service delivery process, thus offering increased transparency about the status and history of each service delivery.

In the following section, we will take a closer look at the areas of SAP Cloud ALM to provide you with the necessary basics and understanding for the following chapters of the book. SAP Cloud ALM is currently divided into three subareas. Starting with the overall ALM process, there is the large area of *implementation*, which is all about supporting the planning and deployment of applications in your landscape. The second area is *operations*. To support the optimization phase, SAP Cloud ALM also offers the third area, *service*. Here we will address the various processes, technologies, and best practices that can ultimately be helpful and necessary for the successful implementation, operation, and service provision of cloud applications. In the following chapters, we will focus primarily on the areas of implementation and operation, as they will be your key to success.

1.2.1 SAP Cloud ALM for Implementation

The implementation area of SAP Cloud ALM, *SAP Cloud ALM for implementation*, can be divided into two areas. The first is strategic planning. In this phase, the goals and requirements of the organization that are to be achieved in the cloud are determined, including:

- **Analysis of the business requirements**
  To make optimal use of cloud applications and services, it is important to understand and prioritize the organization's business requirements and goals.
- **Integration and migration**
  A thorough analysis of existing IT systems and applications is required to enable seamless integration and migration to the cloud.
- **Compliance and security**
  It is important to consider the applicable compliance and security guidelines to minimize the risks when implementing cloud applications.

The next phase, and the second area of implementation, is the development and implementation of the application. The focus here (especially for cloud applications) is on the different technologies, development environments, and various development methods that are supported by SAP Cloud ALM and primarily applied using best practices such as continuous integration/continuous delivery (CI/CD) to ensure the quality and efficiency of the development process. In addition to application development, *infrastructure management* is also crucial. Cloud infrastructure management includes the provision, scaling, and monitoring of resources. Infrastructure as code (IaC) and automated provisioning processes should be used to improve efficiency and control. To ensure that cloud applications are error-free and perform well, a comprehensive testing and quality assurance process is required prior to implementation. This includes both manual and automated tests, performance tests, and security tests. The final step is deployment and orchestration. Effective management of application provisioning and the use of orchestration tools such as SAP BTP, Cloud Foundry environment, and Kubernetes (in a future step) facilitate the management of complex cloud applications and services.

1.2.2 SAP Cloud ALM for Operations

The *SAP Cloud ALM for operations* area is crucial for ensuring the smooth operation of cloud applications and services, optimizing their performance, and controlling their quality. The most important aspects of the operations area are:

- **Monitoring and management of applications, infrastructure, and resources**
  This includes the monitoring of key performance indicators, network traffic, and resource utilization as well as the management of capacities and their scaling.
- **Troubleshooting and support**
  Quickly identifying and resolving issues is critical to minimizing downtime and ensuring user satisfaction. This includes root cause analysis, troubleshooting, and corrective actions for proactive troubleshooting.
- **Continuous improvement and innovation**
  In the operations area of SAP Cloud ALM, it is important to drive continuous improvement and innovation. This includes identifying potential for improvement, introducing new technologies and solutions, and constantly optimizing processes and resources.

In summary, operations in SAP Cloud ALM are central to ensuring the successful operation of cloud applications and services. By effectively monitoring and managing applications, infrastructure, and resources, ensuring security and compliance, rapid troubleshooting and support, and continuous improvement and innovation, operations in SAP Cloud ALM can help optimize the efficiency, performance, and cost-effectiveness of cloud solutions.

1.2.3 SAP Cloud ALM for Service

SAP Cloud ALM for service is crucial for the successful provision and continuous improvement of cloud services and applications. But because SAP Cloud ALM for service is currently still under development and is primarily aimed at increasing transparency in service delivery between SAP and its customers, and the focus of this book is primarily on presenting the implementation options for you as a customer, we will not go into this topic in detail in the rest of the book. SAP Cloud ALM for service is sufficiently described in this section.

The cloud architecture of SAP Cloud ALM enables real-time collaboration during the entire service delivery process. The *Service Delivery Center* is the central access point to all information, providing a shared overview of the status of service delivery and its history. The Service Result app provides easy access to the details of the analysis and the service results. The digitized format of the service results makes it possible to highlight issues according to access rights and information needs. The Issues and Actions Management app presents the results of service delivery as action-oriented tasks. This makes it easy to respond to the service outcome, assign tasks to the right employees, and monitor progress and status.

1.3 SAP Activate

In 2015, the process model for on-premise projects *Accelerated SAP* (ASAP, in the latest version 8) was harmonized and merged with the methodology for cloud projects, *SAP Launch* (see Figure 1.3). The name for this new process model is *SAP Activate*. It applies to all SAP solutions and makes use of the agility methodology.

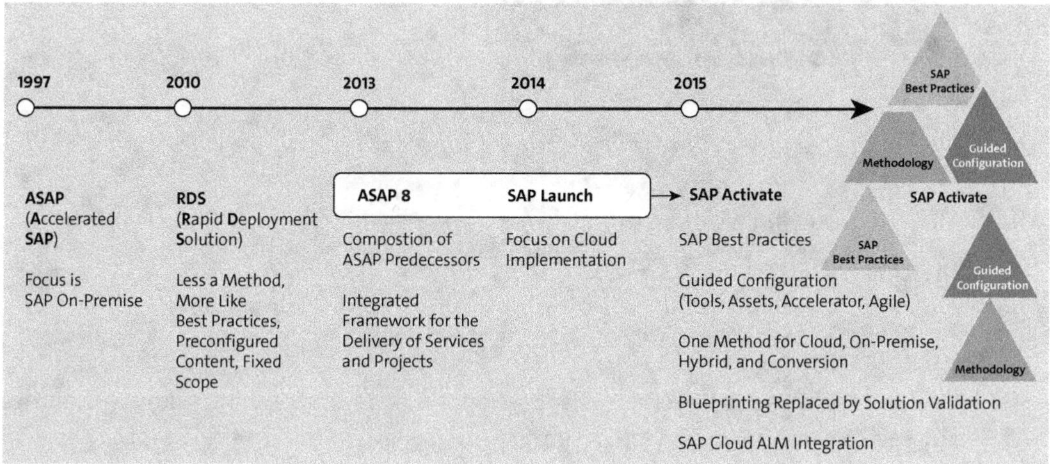

Figure 1.3 Evolution of SAP Activate

The SAP Activate methodology was specifically developed to support SAP customers during implementation and upgrade. SAP Activate accelerates and improves complex solution implementation projects by providing clear guidelines and roadmaps.

[«]

SAP Community for SAP Activate

Before we come to the consideration of SAP Activate as a framework, we would like to point out the existing SAP Community for SAP Activate, where you can get the latest information on the topic and ask questions yourself: *https://community.sap.com/topics/activate*.

1.3.1 SAP Activate as a Framework

As already shown in Figure 1.3, customers have access to various templates and tools for carrying out implementations within SAP Activate. Another tool that supports the entire application lifecycle is SAP Cloud ALM. Together, they can be regarded as the SAP

Activate framework. This book explains the implementation of SAP S/4HANA Cloud using SAP Cloud ALM. The SAP Activate framework is described in Figure 1.4 for an SAP S/4HANA Cloud solution.

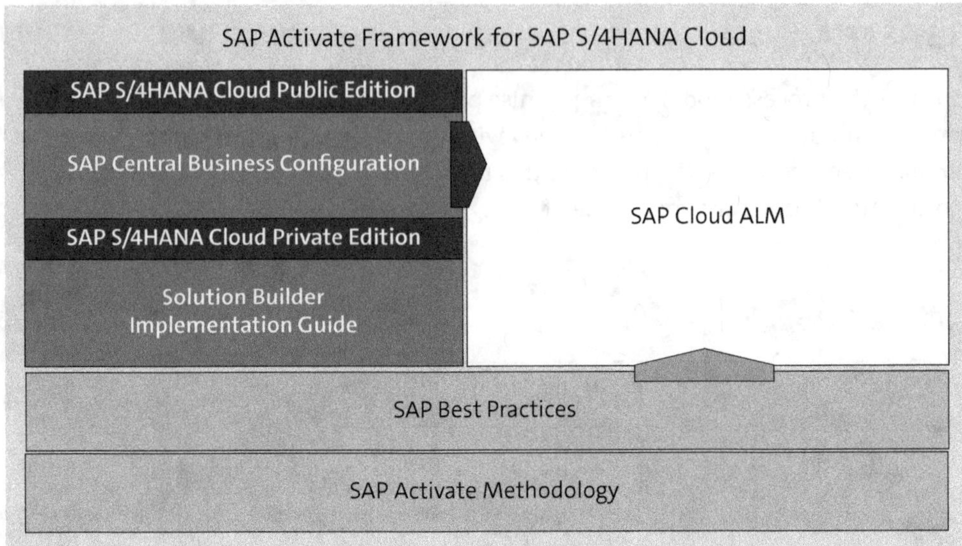

Figure 1.4 SAP Activate Framework

The *SAP Best Practices* are the basis for preparing and implementing an SAP solution and the *SAP Activate methodology*. Both form the basis with the templates they provide but are also part of the projects in SAP Cloud ALM. Chapter 6 describes how these template contents are assigned. Another component of the framework is the *guided configurations* such as *SAP Central Business Configuration* for SAP S/4HANA Cloud Public Edition and the *SAP solution builder tool* for SAP S/4HANA Cloud Private Edition. Both are used in the project and are used for the initial configuration of the SAP S/4HANA Cloud solution. SAP S/4HANA Cloud Public Edition is configured by selecting the preconfigured standard business processes and customer-specific settings.

1.3.2 SAP Activate Methodology

The SAP Activate methodology describes a procedure for implementing a specific SAP solution based on SAP best practices processes and with the help of a roadmap. The roadmap itself is divided into different project phases and provides tasks for the implementation of an SAP solution, subdivided according to deliverables. These tasks are managed in predefined workstreams and are assigned to the responsible processors and executed during the project. In addition, there are accelerators in the form of templates, examples, and tools that help implement tasks and deliverable results more quickly.

Deliverable Results (Deliverables)

The *deliverable results* are a compilation of related tasks that are assigned to a project phase and a specific working group. Templates provided by SAP are intended to help speed up the implementation of the tasks. An example from the on-premise roadmap: The Project Team Enablement deliverable must be carried out in the prepare phase. It contains several roadmap tasks that are assigned to the responsible agents from the Customer Team Enablement working group during processing.

Working Groups (Workstreams)

A *working group* is a cross-project phase compilation of deliverable results that are grouped into different areas. Examples include project management, data management, tests, and solution adoption.

You can find the roadmaps for implementation in the *SAP Roadmap Viewer* (see *https://go.support.sap.com/roadmapviewer*). Figure 1.5 shows the project phases along the top and the working groups on the left. The deliverable results are listed within the project phases.

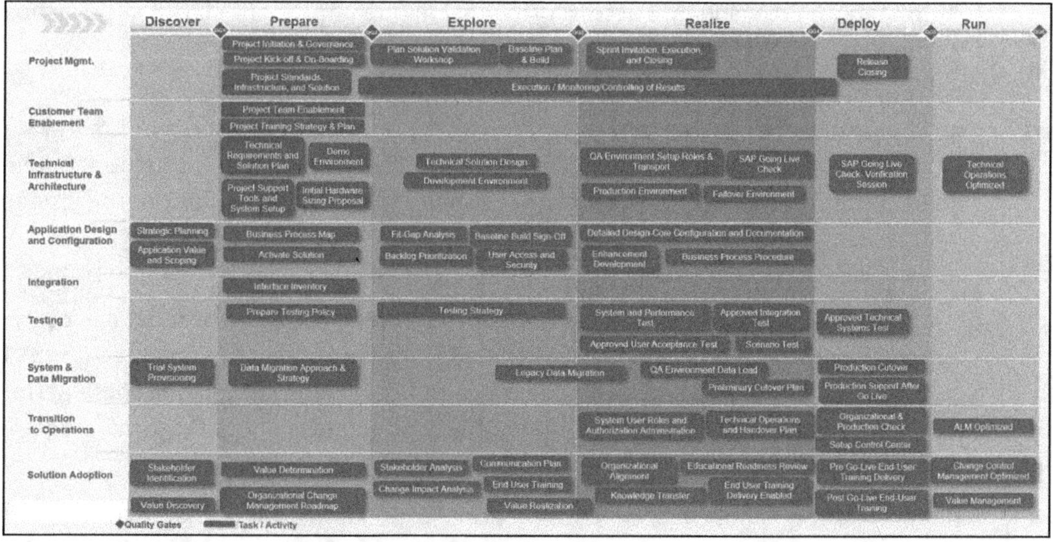

Figure 1.5 Roadmap for SAP Business Suite and On-Premise Solutions (Source: SAP)

Following is an overview of the project phases of the roadmap. You can find a more detailed insight into this in the SAP Roadmap Viewer or in the book *SAP Activate—Project Management for SAP S/4HANA and SAP S/4HANA Cloud* by Sven Denecken et al. (SAP PRESS, 2022). SAP Activate includes the following six project phases:

- **Discovery phase (discover)**
  The first phase is about discovering the possibilities of the selected SAP solution and understanding the business value for the business context to be covered. A test environment of this SAP solution is used to gain initial experience with its use and to identify necessary business processes.
- **Preparation phase (prepare)**
  This phase is used to plan and prepare the project. SAP Cloud ALM is also used here. The project is defined and set up in SAP Cloud ALM; specially defined roadmap tasks from the Cloud ALM setup task source are available for this purpose. The onboarding of project members also takes place in this phase. The project landscape is defined and integrated into SAP Cloud ALM for management.
- **Exploration phase (explore)**
  The third phase of the project serves to validate and confirm the business processes to be implemented with the help of fit-to-standards workshops. This is based on business requirements that are to be fulfilled. Identified configurations and enhancements are added to the backlog.
- **Realization phase (realize)**
  In this project phase, the implementation of the requirements from the backlog begins. This book describes this implementation using an agile approach. The configurations are made, and necessary extensions are implemented. The settings must meet the *definition-of-done criteria* and are tested sufficiently.
- **Commissioning phase (deploy)**
  The penultimate project phase is used to set up the production environment. Settings are summarized as a release and made available for the production system. User training ensures the organization's ability to use the SAP solution. The productive SAP solution is handed over to support operations.
- **Phase of ongoing operation (run)**
  This phase marks the end of the implementation project. Here it must be ensured that the productive solution functions within the performance parameters specified for it. SAP Cloud ALM offers a wide range of options for monitoring these performance parameters. Improvements to the SAP solution should also be regularly derived and implemented from this phase.

1.3.3 SAP Best Practices Explorer and SAP Signavio Process Navigator

Since 2015, *SAP Best Practices Explorer* has been a service that provided detailed information about SAP implementations. The service has a broad target group: SAP prospects, customers, and partners.

In November 2022, SAP announced that this service would be discontinued as of June 30, 2023. One reason was the changing requirements of target groups, including the trends toward the transformation to the cloud, the simplification of integrations, the growing demand for a broader country-specific scope, the acquisition of *SAP Signavio*, and the development of *business process as a service (BPaaS)* offerings with new business opportunities.

SAP is responding to the changed requirements with the release of a new service offering, which has been available since June 2023. This new service is *SAP Signavio Process Navigator*, part of the SAP Signavio product family. Although this is not the successor to SAP Best Practices Explorer, it has adopted content from it, albeit with a new data model; this includes process information, documentation, and accelerators.

Despite the new name, SAP Best Practices will continue to exist. Access is via login as an S-user or a P-user; anonymous access is not planned. The new service has been integrated into the SAP customer portal *SAP for Me* in the application lifecycle management (ALM) area and can be accessed free of charge at the URL *https://me.sap.com/processnavigator*. SAP for Me offers a consolidated view of the customer-specific SAP solution portfolio and aggregates information in a central location. You can find helpful tips and first steps with SAP Signavio Process Navigator at *https://help.sap.com/docs/cloud-alm/getting-started-process-navigator/overview*.

SAP Signavio Process Navigator is not part of SAP Cloud ALM but was developed on the same technology stack. Customers using SAP Cloud ALM, (e.g., as part of the RISE with SAP offering) automatically receive SAP Best Practices for the implementation. This process for publishing SAP Best Practices has been in productive use since the fourth quarter of 2022.

In the SAP Signavio Process Navigator environment, there are related or similar services and applications that we would like to briefly discuss here. First, *SAP Enterprise Architecture Reference Library* (SAP EA Reference Library) is an internal SAP tool for creating business and solution architectures that provides SAP Signavio Process Navigator with basic process structures. SAP Enterprise Architecture Reference Library is based on *SAP Enterprise Architecture Methodology* (SAP EA Methodology) and supports a process-centered data model, which in turn supports SAP's BPaaS vision.

The *One Process Acceleration Layer* and its navigation tool *SAP Signavio Process Explorer* offer similar functions to the new SAP Signavio Process Navigator service, as both services use the same data source. However, compared to SAP Signavio Process Navigator, SAP Signavio Process Explorer offers broader access to other functions, such as process mining. SAP Signavio Process Navigator, on the other hand, offers deeper and vertical access to processes and deployment details. You can find more details on the differentiation between the two tools at *http://s-prs.co/v589700*.

In this book, the solution scenario content is activated in scopes (see Chapter 6, Section 6.4), and the solution processes are defined in the scopes (see Chapter 7, Section 7.4).

There they are available for the fit-to-standard workshops for the purpose of confirmation or requirements determination (see Chapter 11, Section 11.2, and Chapter 10, Section 10.2.1).

1.3.4 SAP Central Business Configuration

SAP Central Business Configuration is a central configuration environment designed to enable initial and ongoing configurations of SAP cloud solutions. It is currently used to configure the SAP S/4HANA Cloud Public Edition solution. SAP Central Business Configuration structures your project into phases and activities, which are displayed in the **Project Experience** workspace. In addition, you are guided through all configuration-related steps of your implementation in a meaningful sequence. In the Define Scope app (see Figure 1.6), you can first define the solution scope of your implementation and put it together yourself. This procedure is similar to the procedure in SAP Cloud ALM, as the processes to be implemented (process scoping) also serve as the basis for implementation projects. You can find more information on this in Chapter 7.

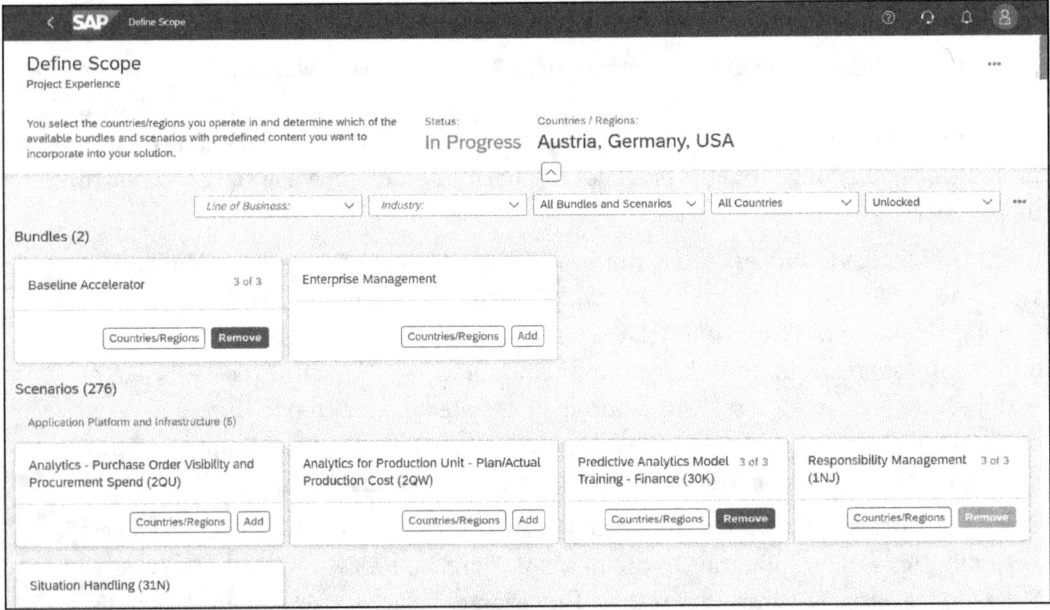

Figure 1.6 Definition of the Configuration Scope

Then create your organizational structure in the Set Up Organizational Structure app area (see Figure 1.7). As soon as changes or additions are made to your organization, these are implemented again via the configuration view shown.

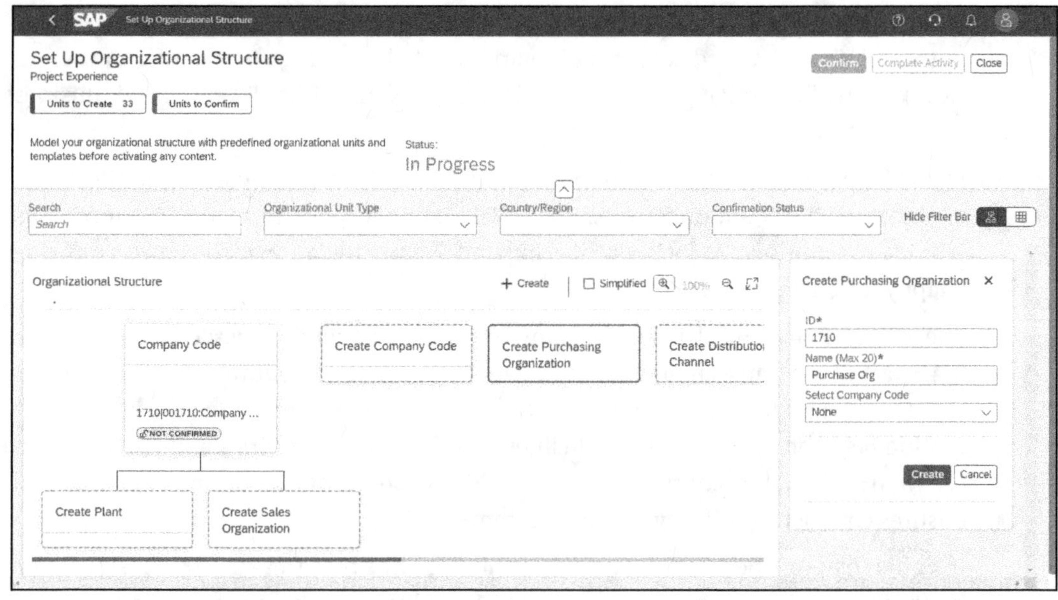

Figure 1.7 Layout of the Organizational Structure

Once you have defined your solution scope and created your organizational structure, SAP Central Business Configuration displays a list of all configuration activities that are relevant for your implementation. You can find this overview in the Project Experience app on the **Configuration Activities** tab (see Figure 1.8).

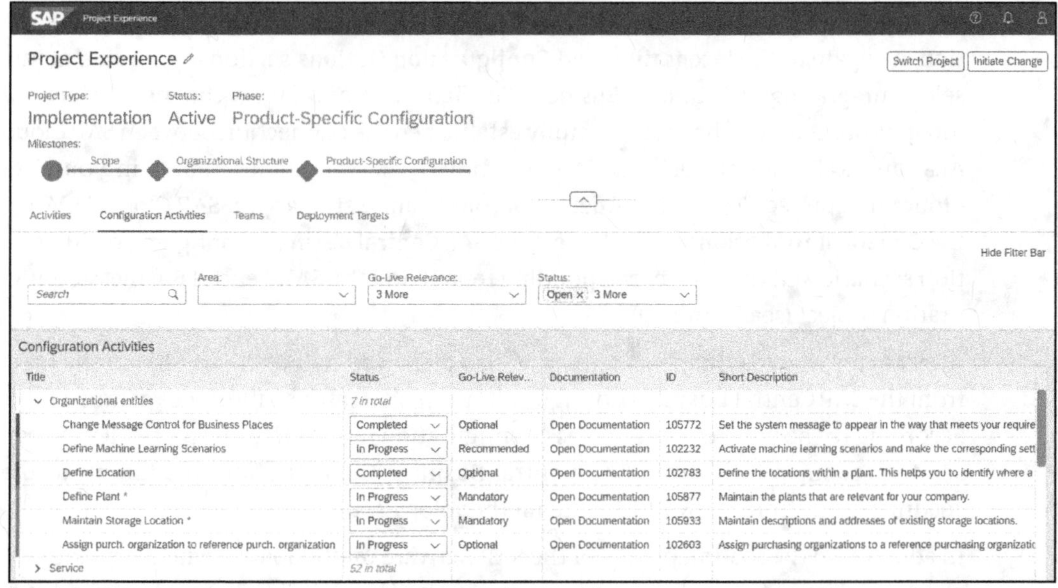

Figure 1.8 Carrying Out the Configuration

Via the respective configuration activity, you can seamlessly jump to the corresponding SAP S/4HANA Cloud configuration and make the settings. The configuration carried out within SAP S/4HANA Cloud is synchronized in SAP Central Business Configuration.

Even after your SAP cloud solution has gone live, SAP Central Business Configuration allows you to carry out ongoing activities. For example, you can add new countries and regions or additional solution scope, expand the organizational structure, or further adapt your business processes.

The use of SAP Central Business Configuration is generally prescribed during SAP S/4HANA Cloud implementation and maintenance. The procedure it describes has several advantages during the configuration phase. One such advantage is the close integration between the SAP Central Business Configuration activities and the associated SAP S/4HANA Cloud configuration activities for adopting the defined settings. This ensures complete feasibility of the requirement.

It would therefore be a great advantage to also provide these SAP Central Business Configuration activities within SAP Cloud ALM. In addition to the existing project tasks, the necessary configuration tasks would then also be displayed. These configuration tasks would be assigned to the responsible project members or the team for implementation. This in turn would help to significantly increase the transparency of the progress of the entire project.

You will find the necessary configuration steps via SAP Help Portal. First open the URL *https://help.sap.com/docs/cloud-alm* and select the **Use** tab on the page. From here you can open the **Setup and Administration Guide** documentation. Within the documentation, navigate to the **Integration and Configuration Options** section. In the side menu, select **Integrating SAP Central Business Configuration** and follow the instructions for integration. Once you have successfully established the connection between SAP Cloud ALM and SAP Central Business Configuration, the assignment is made in your SAP Cloud ALM project. To do this, open the Projects and Setup app in SAP Cloud ALM. On the **General information** tab, first select the **SAP Central Business Configuration** integration scenario and then the corresponding tenant and the SAP Central Business Configuration project (see Figure 1.9).

After saving, the roadmap tasks are now also displayed in SAP Cloud ALM in the tasks from the SAP Central Business Configuration app. You can also filter the roadmap tasks according to the SAP Central Business Configuration task source. In Figure 1.10 you can see the roadmap tasks from SAP Central Business Configuration in addition to the roadmap tasks from the SAP Activate methodology source.

In addition to the roadmap tasks of the SAP Activate methodology, which provide procedural guidance for the implementation of SAP S/4HANA Cloud, the additional roadmap tasks for the specific business configuration of SAP S/4HANA Cloud Public Edition are provided due to the integration with SAP Central Business Configuration. You can see an example of the detailed view of a roadmap task from the SAP Central Business Configuration source in Figure 1.11.

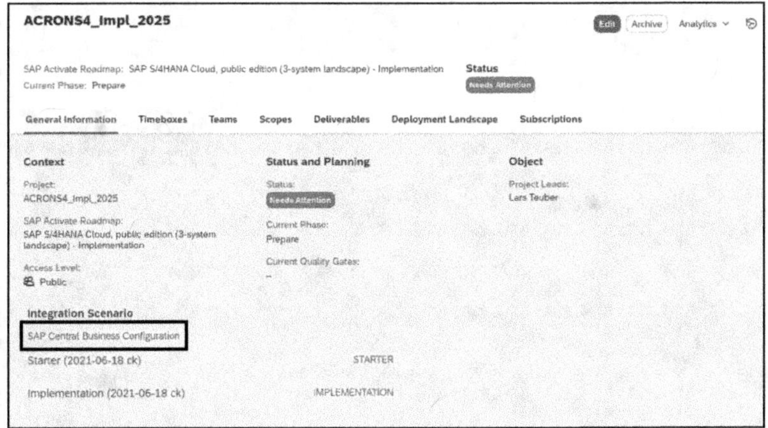

Figure 1.9 Integration of SAP Central Business Configuration in SAP Cloud ALM

In addition to the assignment of responsibilities, such as the editor or team, further subtasks can be created and their priority defined here. A timebox can also be assigned (see Chapter 6, Section 6.2). The roadmap task also contains the link to jump to SAP Central Business Configuration.

Figure 1.10 Roadmap Tasks from SAP Activate Extended by SAP Central Business Configuration

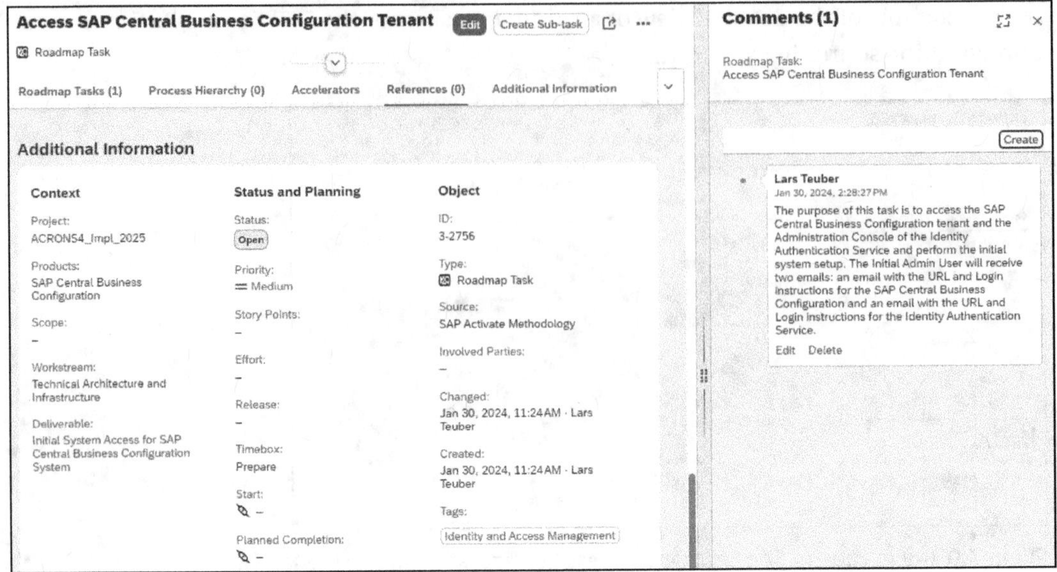

Figure 1.11 Roadmap Task with Reference to SAP Central Business Configuration

1.3.5 SAP Activate in Connection with SAP Cloud ALM

As already explained, the phases of SAP Activate form the basis for the structure of this book. The main idea was to use this introduction to show you, the reader, how closely SAP Cloud ALM and SAP Activate are interlinked during implementation and operation. The roadmap tasks from SAP Activate describe the activities that need to be implemented next to enable a successful project. In SAP Cloud ALM, we use these roadmap tasks to delegate responsibilities and to log and monitor progress, also in relation to the project phase. The contents of these roadmap tasks are often described in general terms with activities. This is where the practical relevance of the book comes in and shows examples of the implementation of these activities using the SAP Cloud ALM functions.

Another advantage of dividing the book into project phases is that it is easier to navigate to the desired chapter. This means that it can be used as a reference book at any time to find out which SAP Cloud ALM-related activities should be considered in the current project phase.

1.4 Our Practical Example

Now we will present the real-world example that serves as the central reference for detailing the various phases, challenges, and recommendations related to the implementation and use of SAP Cloud ALM throughout the book. This example allows us to provide you with a better understanding of the real-world use cases, effective deployment options, and benefits of SAP Cloud ALM by applying the relevant concepts and strategies to concrete situations and business requirements.

By using a hands-on scenario, we can not only teach theoretical concepts and fundamentals, but also provide a practical perspective that will help you understand the complexity and dynamics of deploying SAP Cloud ALM in a real business environment. You will have the opportunity to understand the methods, techniques, and best practices presented in the context of a realistic project and how they can help increase the efficiency, effectiveness, and quality of the implementation and operation of SAP solutions. The detailed examination of this real-world example will help you better apply the concepts and recommendations discussed throughout the book to your own business needs and challenges. It will also help you gain a comprehensive understanding of the opportunities, benefits, and challenges of SAP Cloud ALM and make informed decisions regarding the planning, implementation, and optimization of your own SAP solution landscape.

Our example company operates an outdated non-SAP ERP system in the logistics area, which is used in three different business units. The company is now planning to switch to SAP S/4HANA in a greenfield approach and operate this system in a private cloud environment. The aim is to deliver global functions to two plants in Dresden and Madrid in the form of rollouts using a template approach. Madrid represents the country entities Spain and Portugal, while Dresden covers Germany and Austria.

To meet plant- and country-specific requirements, individual template components may need to be adapted and new process and functional areas defined and developed. To extend and optimize the functions of SAP S/4HANA, some customer-specific applications and extensions are developed with SAP Business Technology Platform (SAP BTP). SAP Integration Suite is used to implement interfaces for the exchange of information between the various systems. SAP Ariba will be used to manage and collaborate with suppliers. As the company is also planning to build new plants, it wants to standardize its personnel management processes. The cloud solution SAP SuccessFactors is being considered for this.

To efficiently manage the implementation of the various SAP solutions and ensure smooth operations later, the company decided to rely on the SAP Cloud ALM suite. This solution allows you to manage the entire lifecycle of the SAP solutions used, from planning and implementation through to operation and continuous optimization.

In this book, we will use this example scenario to explain the various phases of implementation and operation in more detail, as well as the challenges that can arise when introducing SAP Cloud ALM. The aim is to give you a comprehensive understanding of the possibilities and benefits of SAP Cloud ALM.

1.5 Summary

In this chapter, we have given you an introduction to important topics such as SAP Activate and the areas of SAP Cloud ALM that you will encounter again and again in the other chapters of this book. We have also presented our practical example, which we use as the basis for the examples shown in the book. Read on to further understand SAP Cloud ALM.

PART I
Prepare

Chapter 2
Provisioning of the SAP Cloud ALM Tenant

This chapter explains the prerequisites of and how to provision the SAP Cloud ALM tenant, as well as some further preparatory activities you should consider before handing over SAP Cloud ALM to the project team. Possible additional costs that could be incurred during the project or during operation due to the use of SAP Cloud ALM are also highlighted.

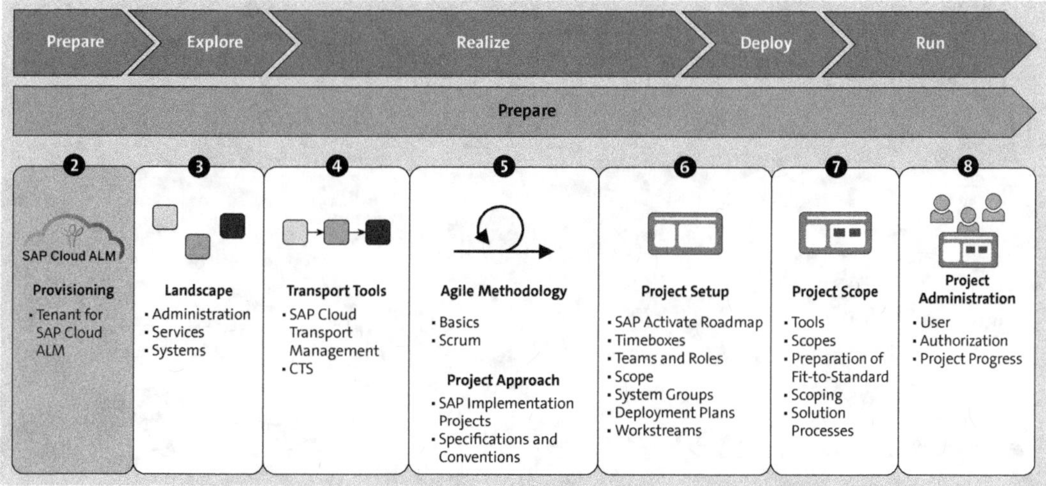

For the preparation of your project, known as the prepare phase, you first need an SAP Cloud ALM tenant. Find out if an active SAP Cloud ALM tenant is already available. If not, you must begin the process for requesting the SAP Cloud ALM tenant. This chapter explains the necessary steps.

The term *tenant* figuratively refers to an apartment that is inhabited by one specific customer. The usage rights for SAP Cloud ALM are included in the SAP cloud service subscriptions and SAP Enterprise Support; further information on the rights of use can be found in this book in Section 2.1.1. The costs for hardware, operation, and data transfer per month are already included in the basic provisioning of the tenant, so customers do not incur any costs for this initial configuration. You can find more information on this in the SAP Help Portal within the online documentation *Setup and Administration for SAP Cloud ALM* in the **Fair Use** section at *https://help.sap.com/docs/cloud-alm/setup-administration/fair-use*. Additional costs may be incurred if more data storage is required (see Section 2.4). Please note that only one productive SAP Cloud ALM tenant is provided by SAP at this time, though the possibility of ordering additional tenants is planned by SAP. A development or quality assurance tenant, e.g. for piloting, training, and testing purposes, is also not yet available.

In Figure 2.1, you can see the entire setup procedure for SAP Cloud ALM. The setup is carried out by different user groups. Users can be created in several steps. For example, the administration and project management users require access at the very beginning of the project. The project management user is responsible for creating the project structure, while the administration must ensure communication with the managed SAP solutions (services, systems, or business services) and set up interfaces to possible third-party tools. At the start of the project, the project members are added via the project activity called *onboarding*.

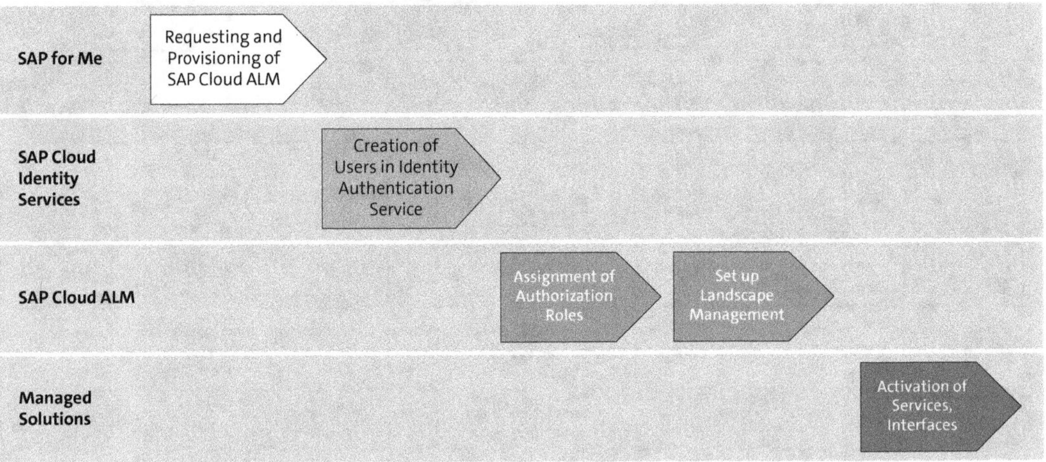

Figure 2.1 SAP Cloud ALM Setup Procedure

2.1 Prerequisites

Before starting the provisioning process, it must be ensured that an *entitlement* (i.e., a usage authorization) for SAP Cloud ALM exists. Your IT management department should be able to provide information about the current SAP contract or the availability of cloud-based solutions. The provisioning process is carried out by the cloud administrator. In the following sections, you will find more detailed information regarding the usage privileges of SAP Cloud ALM, the necessary authorizations, and the entry point for the cloud administrator to request SAP Cloud ALM.

2.1.1 Scope of Usage Rights

A major advantage of SAP Cloud ALM is that SAP customers can use the solution without any additional license fees. For example, customers of the *RISE with SAP* service automatically receive SAP Cloud ALM in a complete package in addition to SAP S/4HANA Cloud. Organizations with the following contracts and license models are also allowed to use SAP Cloud ALM:

- SAP Enterprise Support, cloud editions
- SAP Enterprise Support with an on-premise maintenance agreement/contract
- SAP ActiveAttention
- SAP MaxAttention

On-premise SAP solutions are allowed to be managed with SAP Cloud ALM if they are covered by the customer's SAP support contract.

In addition to the contractual aspects, the following additional usage conditions have to be considered:

- Use of SAP Cloud ALM for customer-specific, internal ALM processes only
- No automation for productive processes with test automation of SAP Cloud ALM
- A baseline of 8 GB SAP HANA memory
- A baseline of 8 GB monthly outbound API data transfer

You can find current information on the usage rights of SAP Cloud ALM on the SAP Support Portal: *https://support.sap.com/usage-rights*.

If the basic configuration is exceeded due to extensive use, fees may be incurred (see Section 2.4).

2.1.2 Authorizations for the Cloud Administrator

An S-user or SAP Universal ID with an SAP customer number is a prerequisite for the cloud administrator to be able to carry out the provisioning process.

Applying for the SAP Cloud ALM tenant begins via the online portal *SAP for Me*. The portal, which you can access at *https://me.sap.com/*, provides you an overview of your entire SAP product portfolio and allows you to set up, operate, extend, and expand your SAP products. The **Edit Cloud Data** authorization role is required for selecting SAP Cloud ALM. In addition, the cloud administrator needs administrative authorizations in the SAP Business Technology Platform (SAP BTP) cockpit and must be defined within the *Identity Authentication tenant*.

[»]

Further Information on Authorizations

For details on the required authorizations, please refer to the documentation in the SAP Help Portal (*https://help.sap.com/docs/cloud-alm/setup-administration/required-setup*) as well as in SAP Note 3248116.

After SAP Cloud ALM has been provisioned, it is a good idea to authorize more cloud administrators for the new SAP Cloud ALM account in SAP BTP. We explain this in detail in Section 2.3.2.

2.2 Provisioning Process

The starting point for the SAP Cloud ALM provisioning process can be found in SAP for Me in the **Provisioning** menu of the **Systems & Provisioning** dashboard. The two sections **Products Available for Provisioning** and **Provisioning Status** are displayed in this menu. If the prerequisites described in Section 2.1.1 are fulfilled and no entry for the SAP Cloud ALM product is visible in the **Products Available for Provisioning** area, the provisioning has already taken place. You should find the entry for the SAP Cloud ALM product in the **Provisioning Status** area, and the status should already be set to **Provided.** In this case, you can skip the following explanations.

Alternatively, the provisioning process can also be started via the SAP Support Portal (see *https://support.sap.com/en/alm/sap-cloud-alm.html*) by clicking on the **Request** button (see Figure 2.2).

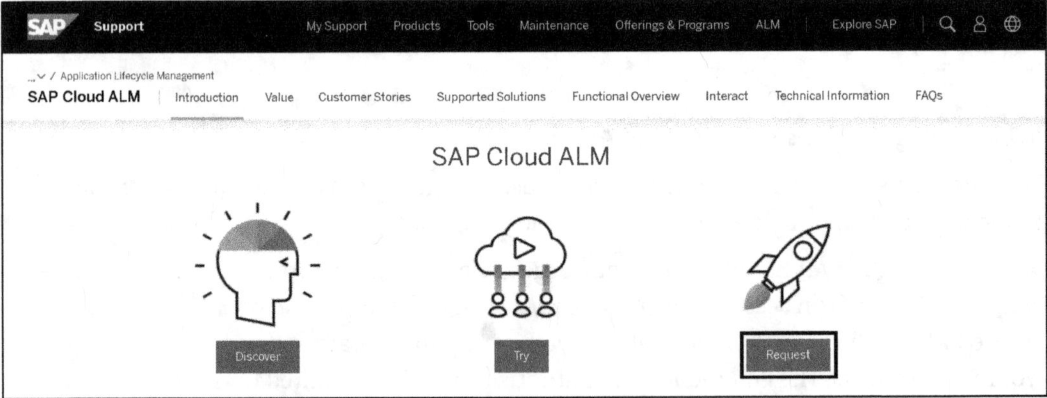

Figure 2.2 Provisioning of SAP Cloud ALM via the SAP Support Portal

To start the provisioning process, navigate to the **Products Available for Provisioning** area in SAP for Me and select the line for **SAP Cloud ALM.** On the right-hand side, you will see the **Start Provisioning** action, as shown in Figure 2.3.

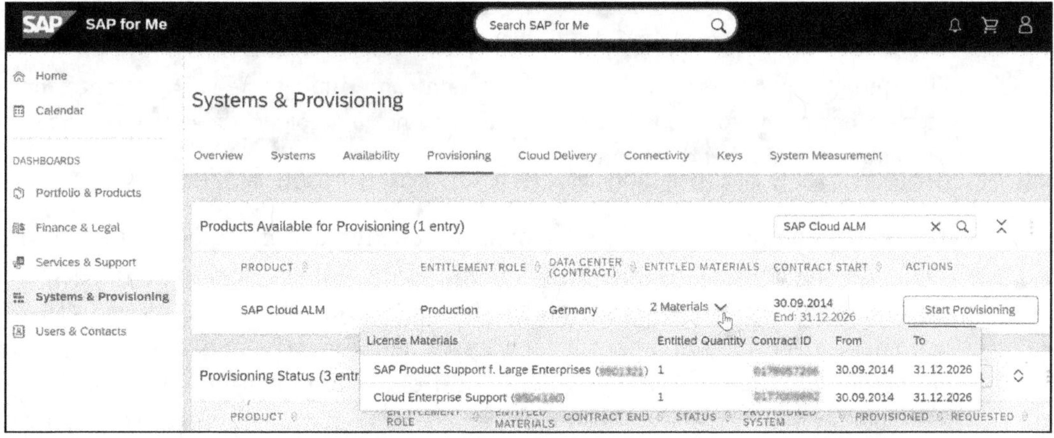

Figure 2.3 Starting the Provisioning Process for SAP Cloud ALM

A suitable data center is automatically assigned in the following dialog box. Next, you will be asked to enter a new and unique name for the subdomain to create the SAP Cloud ALM subaccount in SAP BTP. The subdomain entered here will also be part of the URL that is used to access SAP Cloud ALM. It is therefore recommended to use the suffix "cloudalm" for better recognition (see Figure 2.4). In companies with several customer IDs and the technical possibility of provisioning multiple productive SAP Cloud ALM tenants, it is important to ensure that these SAP Cloud ALM subdomains are subdivided appropriately.

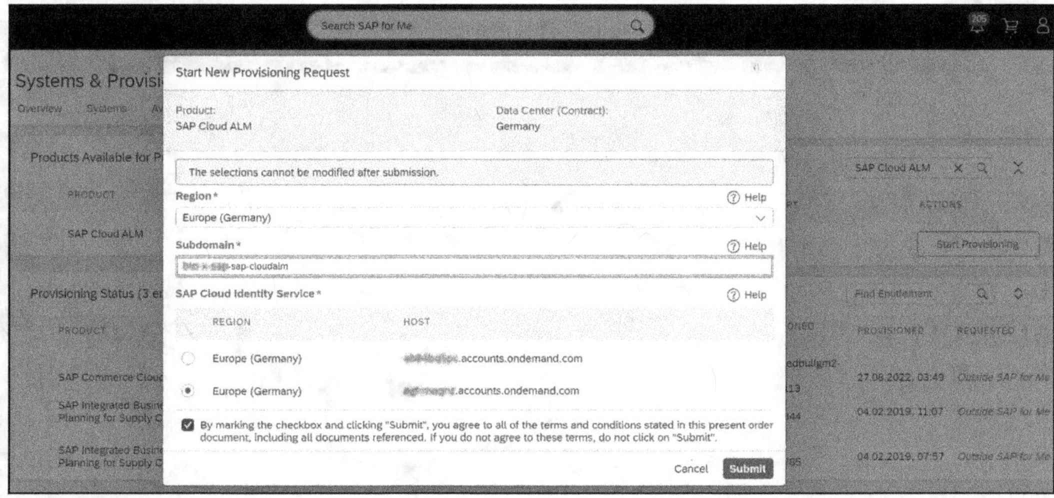

Figure 2.4 Define Region, Subdomain Names, and SAP Cloud Identity Service

[!]

Name of the Subdomain Can No Longer Be Changed

Please note that the name for the subdomain can no longer be changed once the tenant has been provisioned. It is also currently not possible to move the SAP cloud application to another data center.

The last entry represents the selection of the productive *Identity Authentication service tenant*. If several productive Identity Authentication tenants are displayed, select the one that is already in use for the productive SAP cloud solution, e.g., SAP S/4HANA Cloud. You can get an overview of all Identity Authentication tenants at *https://iamtenants.accounts.cloud.sap*. The migration of user data from Identity Authentication to SAP Cloud ALM is currently not supported, so users can only be created manually in SAP Cloud ALM. If necessary, the Identity Authentication assignment for SAP Cloud ALM can be subsequently changed using SAP Note 3020352.

Finally, you need to consent to the general terms and conditions by selecting the corresponding checkbox. Click the **Submit** button to submit all of the settings. The requested SAP Cloud ALM tenant then takes on the provisioning status of **Request Submitted** (see Figure 2.5).

And just that quickly, SAP Cloud ALM is fully ready for use. In addition to the **Provided** status displayed in green, the responsible person on the SAP cloud administration team also receives an email notification with the access information, including the unique SAP Cloud ALM tenant URL.

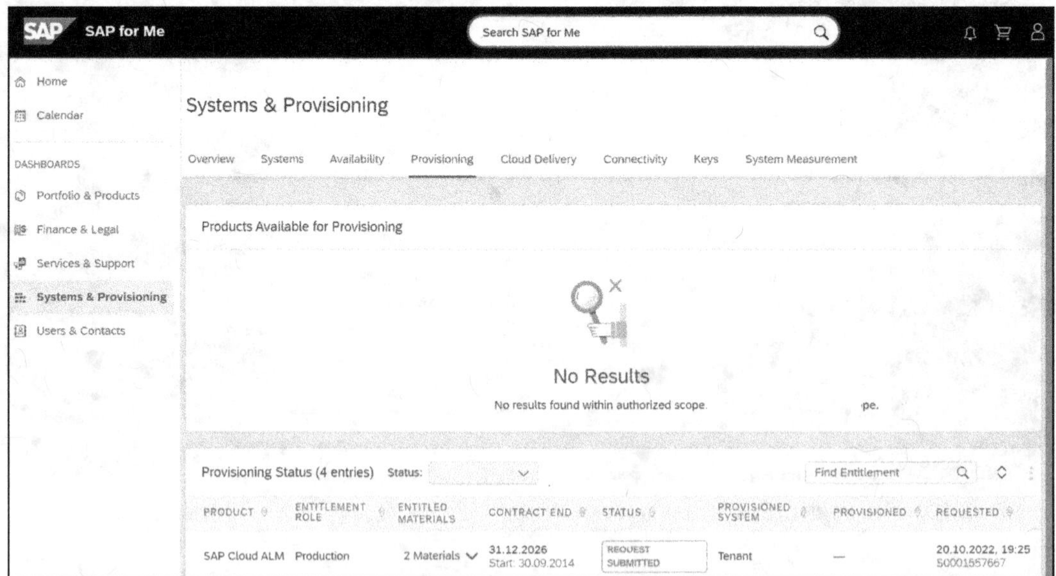

Figure 2.5 Request Submitted Provisioning Status

2.3 Further Preparation Steps

Once you have completed the onboarding process (see Chapter 8, Section 8.1), the URL for accessing the SAP Cloud ALM tenant is sent to each user via an email with the following text: **Welcome! You are now ready to access SAP Cloud ALM**. By clicking on the URL in the email, users are directed to the SAP Cloud ALM logon page (see Figure 2.6).

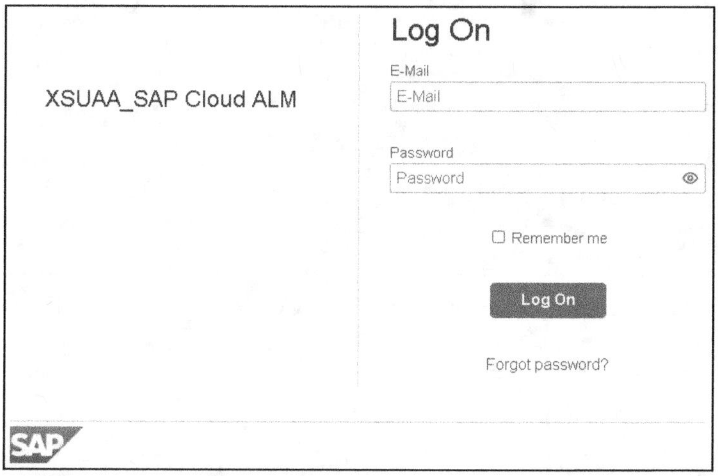

Figure 2.6 Logon Page for SAP Cloud ALM

By default, the logon title and application name are mentioned as **XSUSAA_SAP Cloud ALM**. To improve the user experience, the name and logo can be customized according to your company profile. We outline this procedure in the following section.

2.3.1 Customize Logon Title and Logo

The settings for the logon title and the associated logo can be found in SAP Cloud Identity Services. After logging in, navigate to the **Applications & Resources** menu and click on the **Applications** submenu. Next, select the initial application name **XSUSS_SAP Cloud ALM** within the **Bundled Applications** area (see Figure 2.7).

Select the application name and switch to edit mode using the **Edit** button. You can now enter the desired name in the **Display Name** field and save the settings (see Figure 2.8).

You can change the associated logo in the same submenu via the **Branding and Layout** tab. There, select the **Logo** setting and use **Browse** to upload the image from your local storage location. The maximum image size of *300 × 100* pixels must be observed (see Figure 2.9).

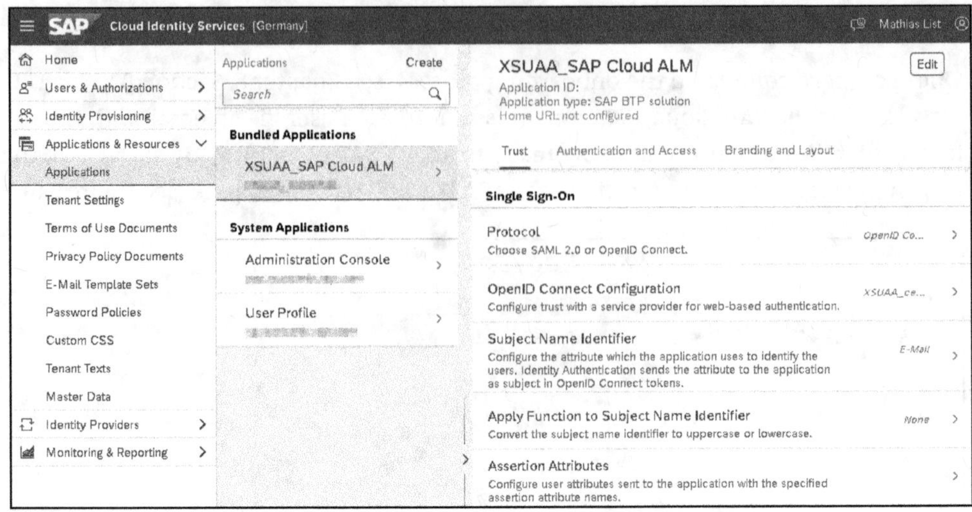

Figure 2.7 Settings for SAP Cloud ALM in SAP Cloud Identity Services

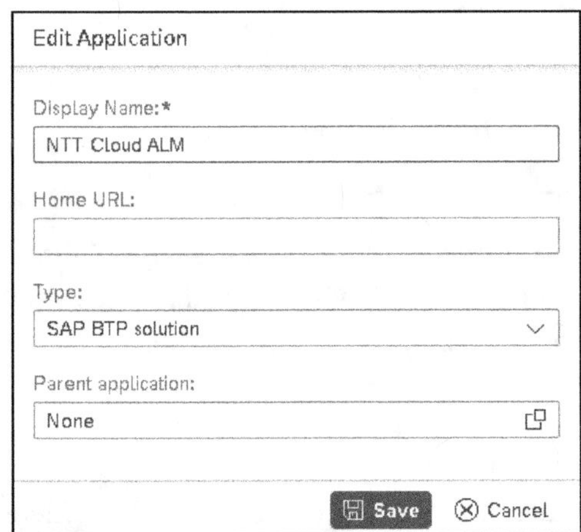

Figure 2.8 Customize Application Name for SAP Cloud ALM

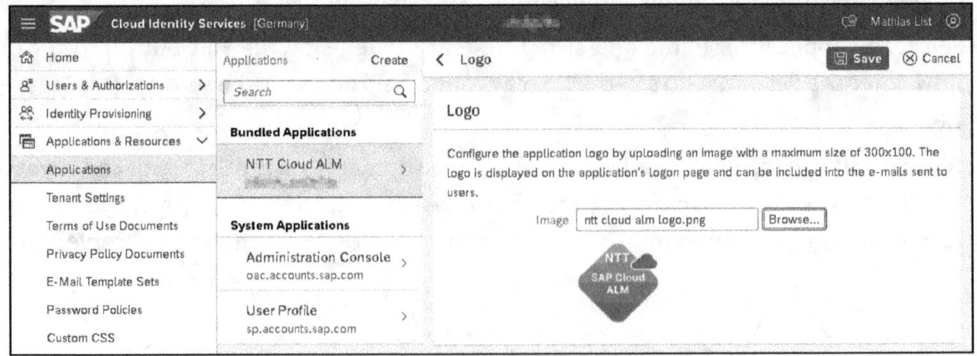

Figure 2.9 Customize Logon Logo for SAP Cloud ALM

Confirm the new logo using the **Save** button to activate it for the next calls of the URL for SAP Cloud ALM. The result should look like Figure 2.10.

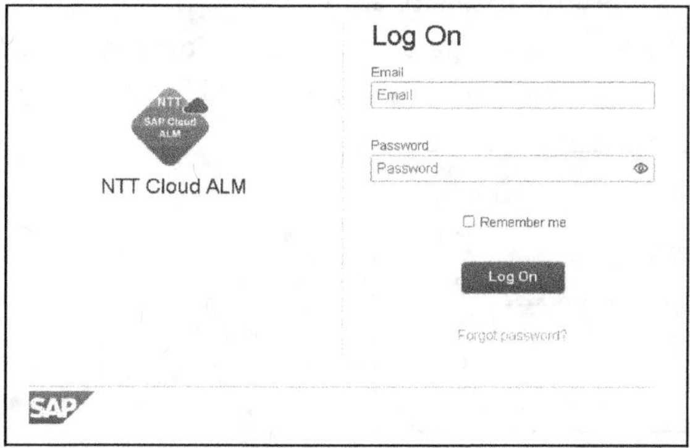

Figure 2.10 Customized Logon Page for SAP Cloud ALM

2.3.2 Create Cloud and Project Administrators

Once the provisioning of the SAP Cloud ALM tenant has been completed, it is advantageous to enable additional administrators to manage it. That access includes the following authorizations:

- Administrative access to the productive Identity Authentication tenant used by the SAP cloud tenant
- SAP BTP administrative access to the global account and the subaccount containing SAP Cloud ALM and the role of Org Manager of the subaccount
- Global Administrator and User Administrator in SAP Cloud ALM

The detailed steps for setting up the first two authorizations can be found in SAP Note 3248116. Further information on the required authorizations can be found on SAP Help Portal at *https://help.sap.com/docs/cloud-alm/setup-administration/required-setup*.

To set up the roles for Global Administrator and User Administrator, navigate to the **Administration** area in SAP Cloud ALM and open the User Management app (see Figure 2.11).

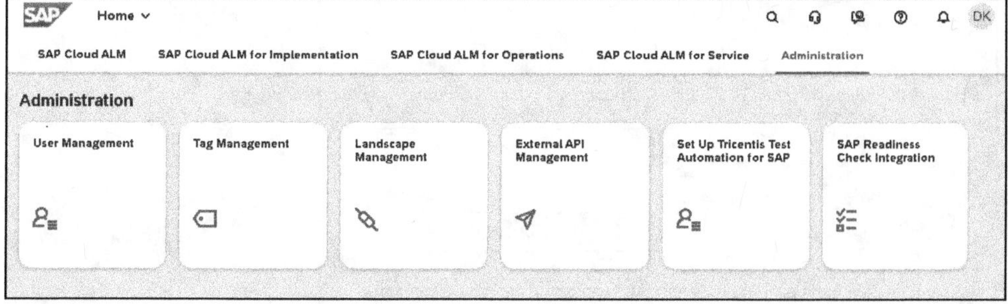

Figure 2.11 Administration Area of SAP Cloud ALM

In the user administration overview, click the **Users** button in the upper right area. The **Users** menu opens, where you can assign authorization roles to existing users and create new ones. Select the user who will be authorized for administration. The view for role assignment opens on the right-hand side. Navigate to the **Cross Topics** area and assign the roles **Global Administrator** and **User Administrator**, as shown in Figure 2.12.

Figure 2.12 Administrative Roles for the Management of SAP Cloud ALM

The Global Administrator role consists of the single roles `x_calm_GlobalAdministrator` and `FlexOperator`. It allows the user to create and manage all SAP Cloud ALM objects, with the exception of personal or confidential data. The User Administrator consists of the single roles `x_uam_UserAdministrator` and `FlexOperator`. With this role, you can create and manage users and assign and approve authorizations.

In addition to the administration of SAP Cloud ALM, necessary services must also be activated in the managed SAP solutions, and possible interfaces to other applications must be set up. For this purpose, additional authorizations for the cloud administration team have to be taken into account.

In addition to the administration team for the technical management of SAP Cloud ALM, the project manager should be equipped with the role of Project Administrator before onboarding the project members. This role is required for creating the project structure within SAP Cloud ALM. It also enables the assignment of users to the respective project teams. To authorize the person managing the project as the Project Administrator, the role is assigned in the **Implementation** area in the **Project Management** field, as shown in Figure 2.13.

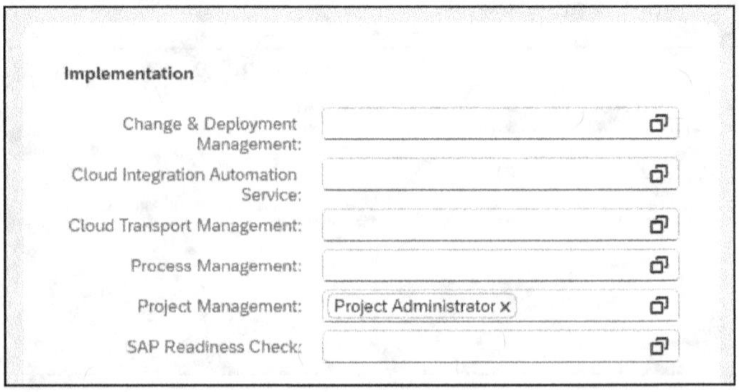

Figure 2.13 Add Role for Project Administration in SAP Cloud ALM

The **Project Administrator** role consists of the single role `imp_pjm_ProjectAdministrator`. With this authorization, it is possible to perform the entire project setup. You can find an introduction to this in Chapter 8.

2.3.3 Planning the Onboarding and Training Concept

The next step in preparing the project within SAP Cloud ALM is the onboarding—that is, creating users with the corresponding authorizations and planning the necessary user training.

[«]

User Training

User training ensures that the project member is trained in the use of SAP Cloud ALM in relation to their role within the project.

For the subsequent creation of users, it is helpful to create an *authorization matrix* to keep track of which user groups require which kinds of roles. If there are several planned and active projects available, access restrictions at project level must also be taken into account. A detailed user overview in which the user name, the user ID, the e-mail address, and the reference to the user group as well as the respective project are stored helps quite a lot. A differentiation or documentation according to user types such as employees, SAP, partners, and others is also an option but has no functional influence within the application. Further information on user creation can be found in Chapter 8, Section 8.1.2.

With regard to the user training, you should determine which user accounts should be used to access SAP Cloud ALM. There are two possible options:

- The project members use their own personalized users for the training.
- Separate role-based exercise users with the corresponding roles are provided.

With option 2, additional, non-personalized e-mail addresses must be used. The concept of role-based training users is employed, for example, in the SAP Cloud ALM public demo tenant (refer to *https://support.sap.com/en/alm/demo-systems/cloud-alm-demo-system.html*). In addition to the users, a training project is also created, which ideally follows a predefined, suitable project structure.

[!]

Training Is Only Possible on Productive Tenant

Please note that the training can currently only take place on your productive SAP Cloud ALM tenant. After the training, the project can be archived, or you can change the access level.

2.4 Additional Costs

As mentioned in Section 2.1.1, it may become necessary in the course of using SAP Cloud ALM to purchase additional data storage. In principle, each tenant is provided with a basic configuration of 8 GB SAP HANA memory. It should also be noted that in addition to the memory capacity, the outbound API data transfer has to be considered as well. Using and connecting your own extensions via the provided APIs and the associated data exchange or an intensive use of the monitoring functions of many solutions can result in higher outgoing data transfer rates. Regarding the SAP HANA memory, you will have to expect to pay around $328 per month for additional 4 GB space extension, which corresponds to one package. The billing takes place monthly on the basis of the actual consumption. Assuming a permanent additional space extension of 4 GB, you should expect total annual costs of about $4,000. The extension is provided via the *SAP Cloud ALM, memory extension* service, which is part of the *Cloud Platform Enterprise Agreement (CPEA)* (see Figure 2.14). Detailed information on this can be found in the service catalog of *SAP Discovery Center* at *https://discovery-center.cloud.sap/* by entering the name of the service in the search bar. If you use more than the memory provided by default, SAP will inform you that overutilization has been detected and will prompt you to activate the CPEA memory extension service.

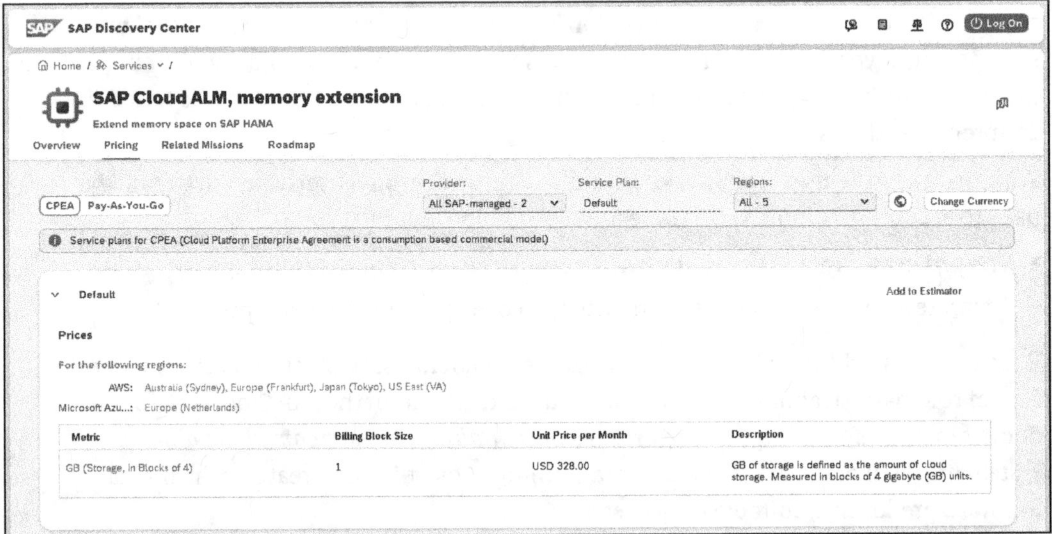

Figure 2.14 SAP Cloud ALM, Memory Extension

The example in Figure 2.15 shows how a customer receives a message at the end of April informing them to activate the CPEA memory extension service. For the month of May, in this example, 13 GB of space is required. This means that two blocks with costs totaling $656 are incurred. In June, some data housekeeping has been performed. Thus, a total of 10 GB of space is required, resulting in additional costs of $328.

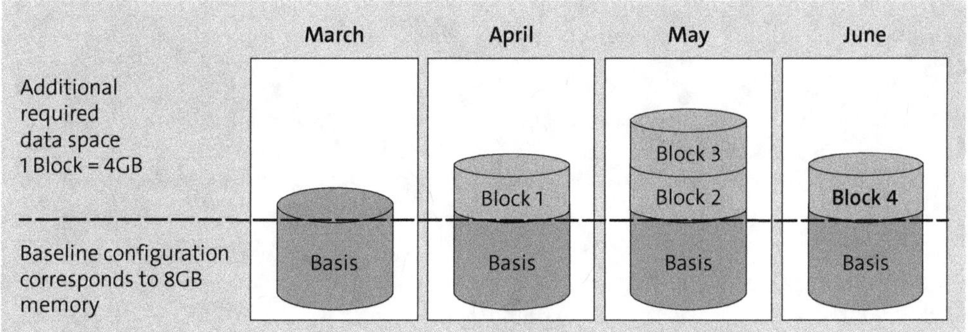

Figure 2.15 Example of Data Requirements by Month

2.5 Summary

In this chapter, we explained the prerequisites and the necessary steps for provisioning SAP Cloud ALM in detail. You now have the opportunity to log on to the application to get familiar with SAP Cloud ALM and gain your first impressions. In the next chapter, we will look at the underlying project landscape. We will explain what information must exist for the landscape of our example scenario for the upcoming project and for later operation.

Chapter 3
Management of the Project Landscape

When you come to the discover phase, you have already defined the project landscape. This landscape usually contains various SAP solutions that have been selected based on the business requirements. Certain information must be available so that the landscape can be used in SAP Cloud ALM at the start of the project. This chapter explains what this information is and how it can be provided and checked.

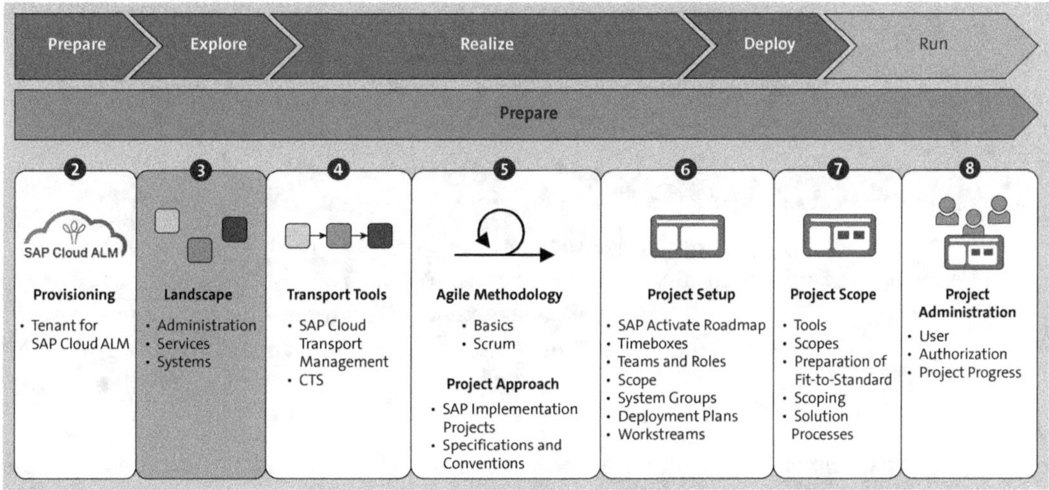

Once you have completed the steps in Chapter 2, you will be able to log in to SAP Cloud ALM and should already be familiar with the general functionality of the application. We will now show you how we connect the technical system landscape in our example scenario. We will use an SAP S/4HANA on-premise landscape.

Certain information must be available in SAP Cloud ALM to carry out the implementation project and for its subsequent operation. This includes, for example, the data for change management or automated test execution in the implementation area and the data required for monitoring in the operation area. To ensure this information is available, configurations are required to connect or activate the data exchange. You can document the implementation of these activities using the "Assign Landscape" roadmap task, which you can find within the configuration of the SAP Cloud ALM task source.

When managing a project landscape, a distinction is made between the systems and services groups (Section 3.2). Both groups are managed via the Landscape Management app in the **Administration** app group (see Figure 3.1). Access to this app should only be provided to administrators. You can find more information on authorizations in Chapter 8.

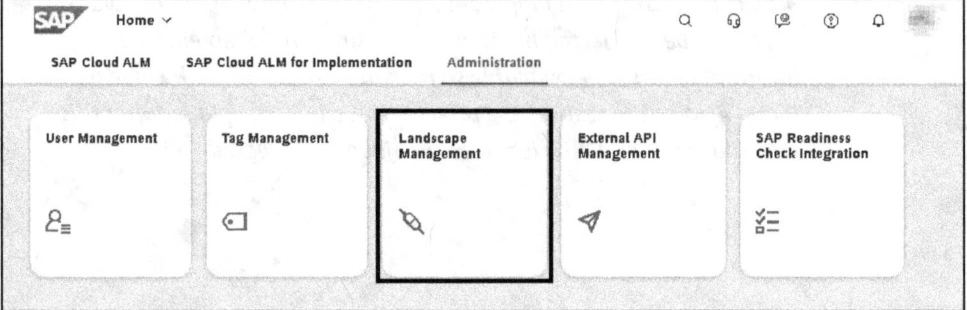

Figure 3.1 App for Managing the Landscape

In addition to systems and services, business services are also managed here. You can find more information on this group in Chapter 19, Section 19.1.

[+]

Expert Portals for SAP Cloud ALM

SAP offers *expert portals* for SAP Cloud ALM, where you can find detailed information on various areas of the SAP solutions. In the SAP Cloud ALM for operations expert portal, you can look in the Landscape Management section for up-to-date information on managing the landscape: *https://support.sap.com/en/alm/sap-cloud-alm/operations/expert-portal.html*.

This chapter begins by outlining a project landscape that forms the basis for the implementation. Further systems and services are presented for the ALM operation area, for which configuration examples are used. We also present the individual areas of the Landscape Management app.

3.1 Project Landscape Management

Your SAP solution is implemented in the project landscape. Various system and service landscapes can be used for this, which often consists of a development, quality/test, and productive environment.

Figure 3.2 shows the project landscape that we use for the sample configurations. It is also the basis for the presentation of the practical realization of the ALM implementation area.

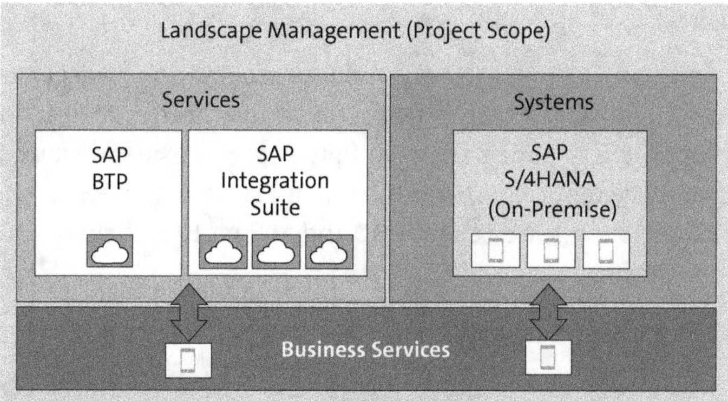

Figure 3.2 Landscape Management with Project Scope According to SAP Cloud ALM

In the following sections, the selected SAP solutions are discussed in more detail as reference examples.

[«]

SAP Solutions

There are SAP products, such as SAP S/4HANA Cloud, that can be configured for a specific business context. In addition to the associated applications and databases, interfaces are often also part of this solution. With regard to the management of these components, a distinction is made between a cloud-centric and an on-premise infrastructure.

3.1.1 Example: SAP Integration Suite

We will use SAP Integration Suite consisting of three tenants as an example of the integration scenarios for cloud services. Similar to the system landscape, the *service landscape* is defined by service roles such as test and production. The services in this landscape are automatically registered in SAP Cloud ALM as opposed to how it works for on-premise systems. In Section 3.2.1, you will find an illustration of this as an example.

As a platform, SAP Integration Suite enables the integration of on-premise and cloud-based applications and processes. The requirements for such integrations are omnipresent during the implementation project and also during operation. Because SAP Integration Suite bridges the on-premise and cloud worlds, it is the perfect example to use for our project landscape.

Certain configuration steps must be carried out to enable changes to be managed in SAP Integration Suite using SAP Cloud ALM. You can find this information in Chapter 4, Section 4.1. You will find sample integration scenarios in Part VI of this book.

3.1.2 Example: SAP S/4HANA

As mentioned, we use on-premise SAP S/4HANA for the implementation project carried out in this book. Figure 3.3 shows the transport routes of the three-system landscape. *Transport routes* define where a change, i.e., a configuration, is to be transferred. Here, DS4 has the development system role, QS4 is the quality system, and PS4 is the production system. Each system has at least a client 000 and a client 100. In client 100, the SAP solution is configured, tested, and used productively.

[»]

Client

According to SAP, a *client* is a "separate unit within an SAP system in terms of commercial law, organization, and data, with separate master records and an independent set of tables" (source: SAP Help Portal, *http://s-prs.co/v589701*).

Transport management must be configured to distribute the configurations. Figure 3.3 shows the transport routes for transportable changes between the systems.

As you can see, there is a consolidation path from **DS4** for a target group to **QS4** for SAP objects with the standard **SAP** transport layer and for customer objects based on the **ZDS4** transport layer. In **QS4**, the client **150** is supplied in addition to client **100**. **PS4** is linked to a delivery route from **QS4**. In this example, the transport routes from QS4 to PS4 are defined on a client-specific basis.

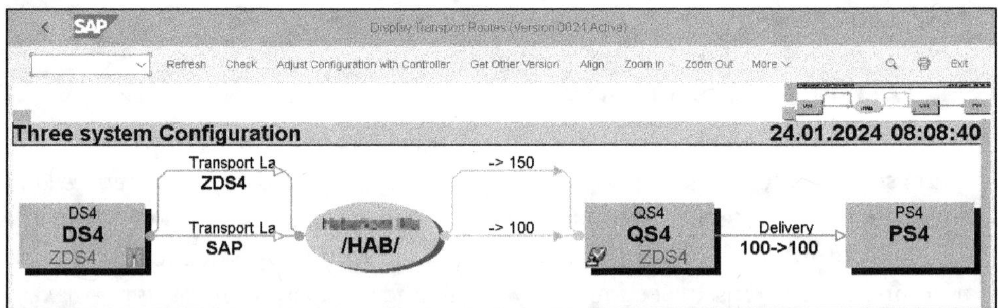

Figure 3.3 Delivery Routes of the Changes for the Three-System Landscape

[»]

Consolidation Routes, Transport Layers, and Delivery Routes

To make changes transportable, you must set up a *consolidation route* for each affected transport layer. The starting point (source) of these consolidation routes is the development system, the transport destination is the quality assurance system (in a two-system landscape, the production system).

All development projects that are carried out in an SAP system and transported on the same transport routes are combined in one *transport layer*. The objects delivered by SAP belong to the SAP transport layer.

Once you have imported your developments into the quality assurance system, you also want to transfer them to your production system, possibly even to several SAP systems (e.g., additional training systems). To do this, you need to create *delivery routes*.

This type of client-specific configuration is common and is also supported by SAP Cloud ALM through integration with Change and Transport System (CTS). Target groups, like the one named **/HAB/** in the example, are also supported when integrating CTS with SAP Cloud ALM.

Target Groups

Target groups bundle several client-dependent transport targets under one symbolic name. Target groups can be specified as transport targets when defining consolidation and delivery routes.

Certain configuration steps must be carried out to enable the management of changes in SAP S/4HANA on-premise using SAP Cloud ALM. This information can be found in Chapter 4, Section 4.2.

3.2 Managed Landscape Groups

The management of the landscape information, which is necessary for implementation and operations, is divided into three groups:

- Systems
- Services
- Business services

Systems and services are SAP solutions that you manage in SAP Cloud ALM for specific ALM functions, such as change management, automatic test execution, or monitoring. *Business services* refer to the ability of these SAP solutions to monitor a logically separated scope. Business services are not explained in this chapter and are therefore not shown in Figure 3.4, in which the project landscape to be managed is shown graphically. Instead, we will focus here on just the systems and services landscape groups and the information that is available or can be managed for them.

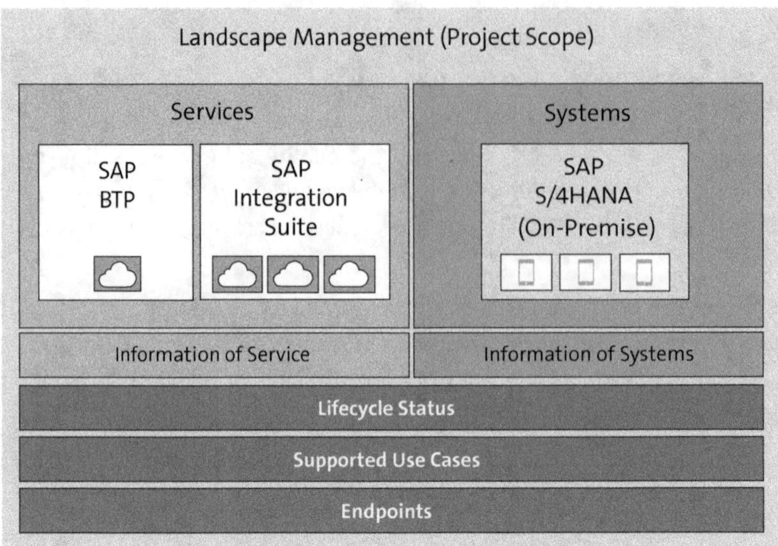

Figure 3.4 Services and Systems and Their Manageable Information

3.2.1 Services

The group of managed services consists of cloud-specific SAP solutions. In our project landscape, this is SAP Integration Suite. To manage the services in your landscape, first open the Landscape Management app. Now select the service that you want to manage.

To preselect the relevant service, click on the **Select a Scope** button in the top menu on the right. In the subsequent window of the same name (see Figure 3.5), select the dedicated **Service Type** based on the **Managed Object Type** service, such as **SAP Integration Suite (Cloud Integration)**.

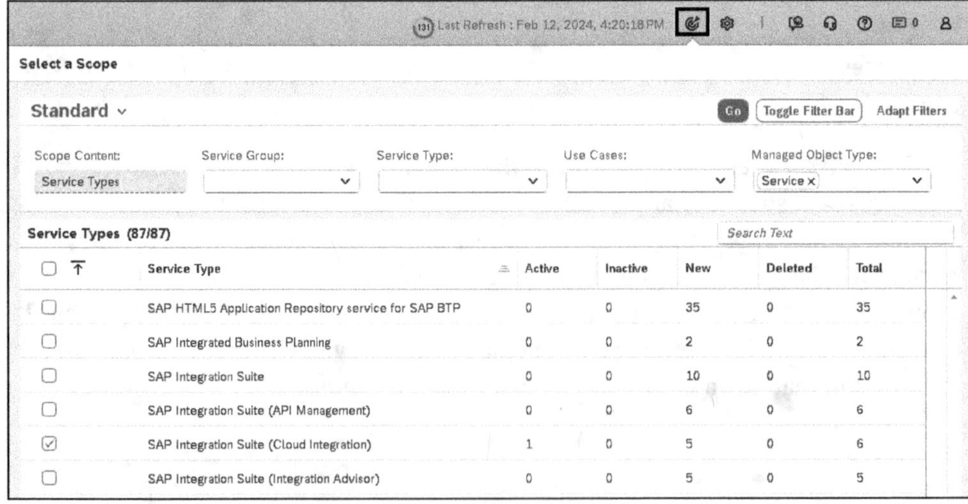

Figure 3.5 Preselection of Services Based on the Service Type

Depending on the preselection you have made, the available services are displayed in a graphical overview on the **Home** menu. Here you can see the total number of services and their current status. In Figure 3.6, you can see that five services of the preselected service type **SAP Integration Suite (Cloud Integration)** are available. One service has the status **Active**, one service is **Inactive**, and the other four services have the status **New**.

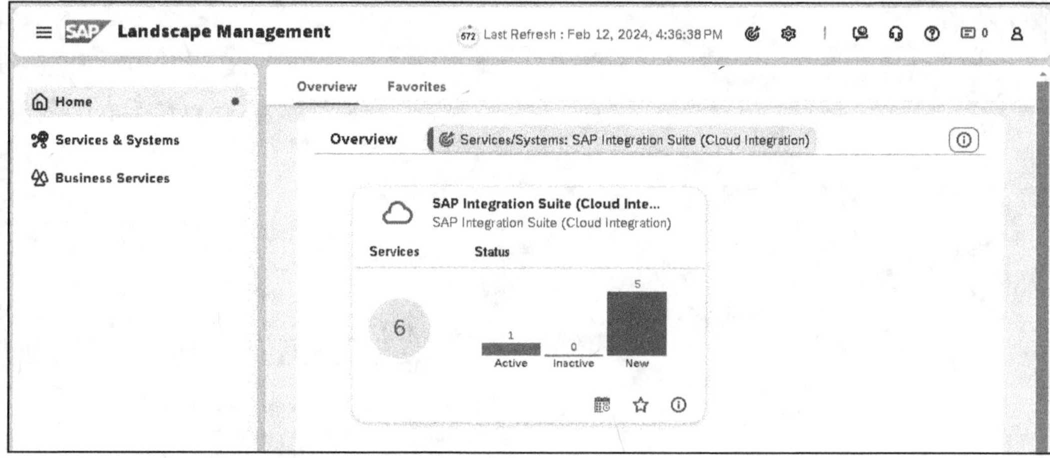

Figure 3.6 Preselected Services on the Home Menu

You can mark the service type as a favorite using the star symbol. It could therefore be a good idea to first preselect all available service types to have them visible in the overview and then mark the service types required for the project as favorites. You will then be able to find the service types marked as favorites on the **Favorites** tab in the **Home** menu. The **Info** button provides you with detailed explanations of the displayed status values. To access the detailed view of the preselected services, navigate to the **Services & Systems** menu (see Figure 3.7).

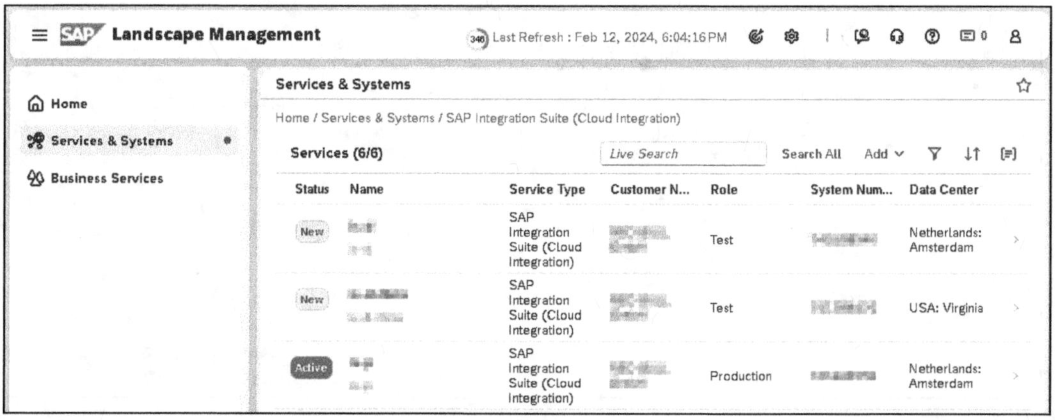

Figure 3.7 Overview of the Managed Services

You have already seen which services are supported by SAP Cloud ALM by preselecting the scope. Once you have selected all service types, the number of available services can become confusing. You can use the **Filter** option in the **Services & Systems** menu to limit the display of services again (see Figure 3.8).

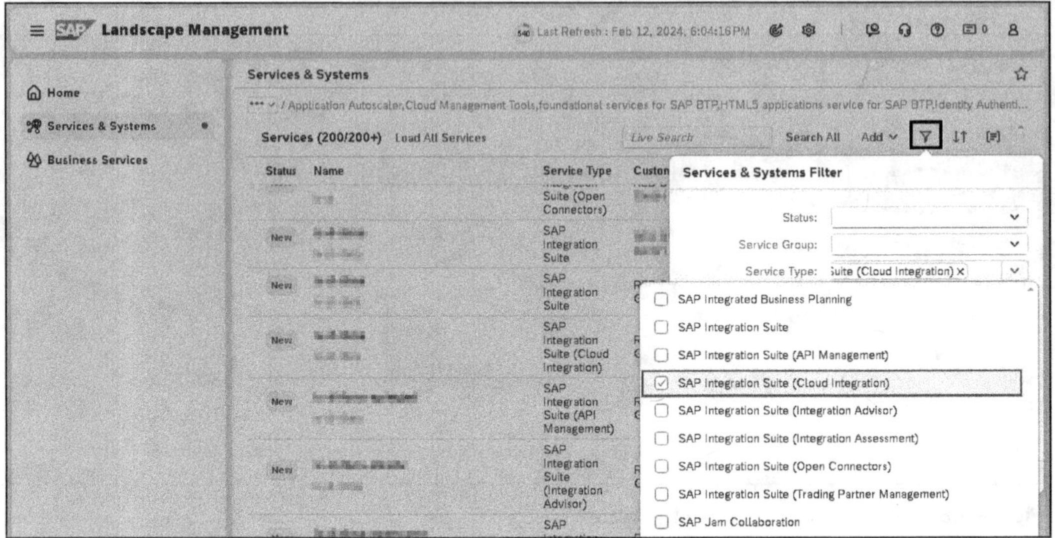

Figure 3.8 Filter the Supported Service Types

The services are imported into SAP Cloud ALM via the *System Landscape Information service* using the relationships to the SAP customer number in SAP Cloud ALM. This happens once a day and depends on which services are subscribed. You can view the status of the import for the services via the gear icon in the top-right menu bar. Figure 3.9 shows this option.

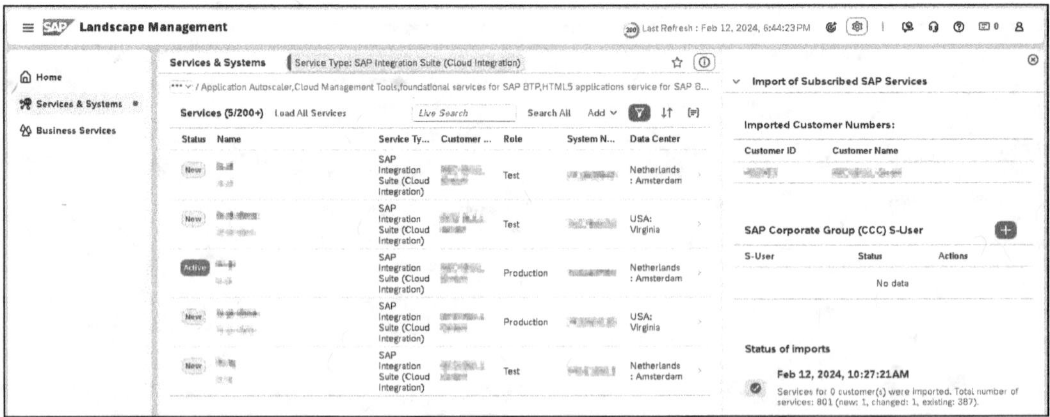

Figure 3.9 Show Import Status of Subscribed Services

In addition to the automatic import, you also have the option of adding a service manually. To do this, click on the **Add** button and choose **New Service**. You will then be asked to fill in the mandatory parameters: **Name**, **System Number**, **Service Type**, **Role**, and **Root URL**. The service is differentiated on the basis of the service type. The *service type* is the SAP product on which the available service is based. These mandatory parameters are described in Table 3.1.

Parameter Name	Description
Name	Name for the new service
System Number	Enter the system number that is available in SAP for Me. It is a unique identifier or alias for the tenant and is based on the tenant's name.
Service Type	Choose from the dropdown list—e.g., **Managed Gateway for Spend & Network**
Role	Choose from the dropdown list—e.g., **Development**, **Test**, or **Production**
Root URL	Enter the URL that needs to be used to access the API

Table 3.1 Mandatory Parameter to Create a New Service

Save your settings, and the new service will appear in your list. Further information and which services need to be created manually can be found at *https://help.sap.com/docs/cloud-alm/setup-administration/services*.

To integrate services from multiple SAP customer numbers, skip to Section 3.5.

In the header area of the service overview (see Figure 3.8), you will see the total number of all registered services in brackets, depending on the preselected service types. You can use the **Live Search** button to search for a service name. The service overview provides you with the following service information:

- Status (lifecycle)
- Name (of the service)
- Service type (SAP product)
- Customer name
- Role (tenant)
- System number
- Data center

To obtain further information on a service, you must select it from the overview. Then you will find a wide range of service information on the **Details** tab, grouped according to **General Details**, **Product Details**, and **Additional Details** and, if available, **URLs**. We have illustrated this service information using SAP Integration Suite (Cloud Integration) as an example (see Figure 3.10).

Use the **Edit a Service** button to adjust the description or role of the imported service. The **Endpoints** tab shows if an authenticated connection for certain use cases of the service have already been set up for SAP Cloud ALM (see Section 3.3.3). The use cases themselves are also displayed in detail in the **Supported Use Cases** tab. Here you will find the use cases' configuration status and runtime status (see Section 3.3.2). The **Tags** tab provides you the option to enter additional data—for example, a responsible administrator. This tag information is applicable as service filter. If the service is in use by a business service, you can find that information in the **Where-Used List** tab. The **Action Logs** button shows the change history for the selected service.

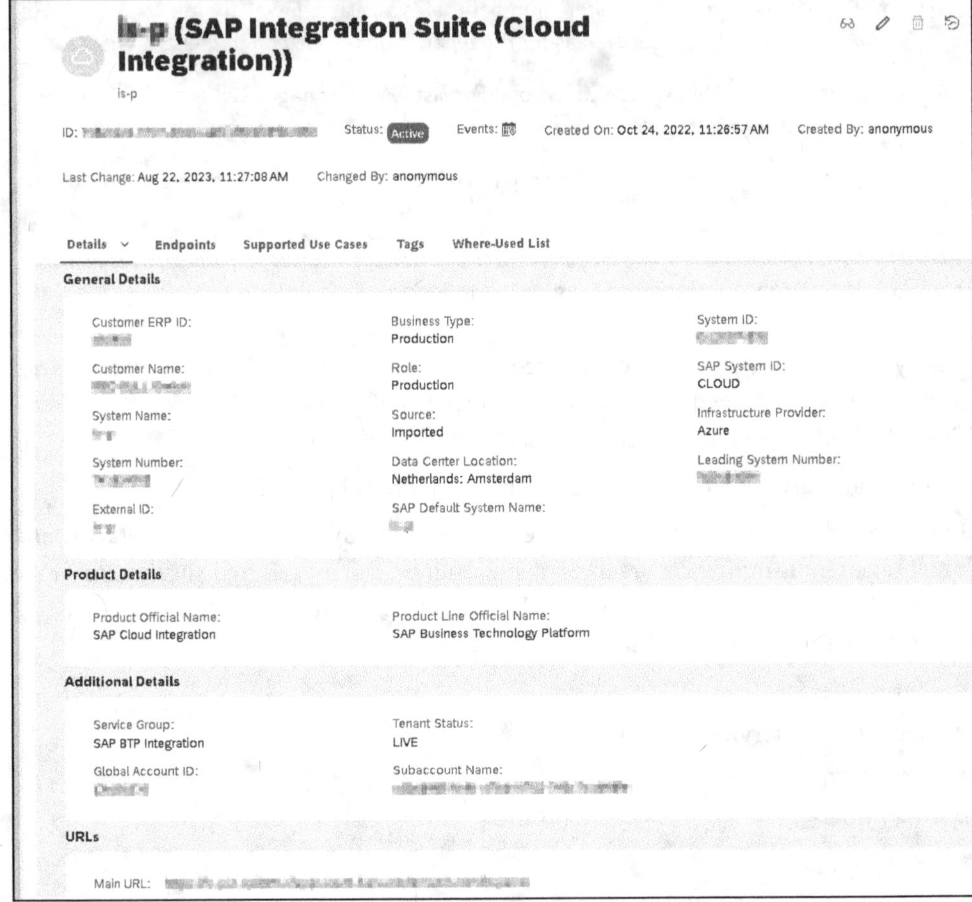

Figure 3.10 Detail Information View of the Selected Service

3.2.2 Systems

The group of managed systems consists of on-premise ABAP systems. To display existing and manageable systems, open the Landscape Management app and then select the **Services & Systems** menu on the left-hand side. Ensure that you have previously selected a service type based on the **Managed Object Type** using the preselection in the **Select a Scope** maintenance window (see Figure 3.5).

The way to register ABAP systems in SAP Cloud ALM is to use Transaction /SDF/ALM_SETUP (Set Up Integration with SAP Cloud ALM). Information on this configuration can be found in Chapter 4, Section 4.2. Other system types, such as the cloud connector, SAP Focused Run, and SAP Process Integration, must be added manually using the **Add** button (see Figure 3.11), as no other registration option is available for them.

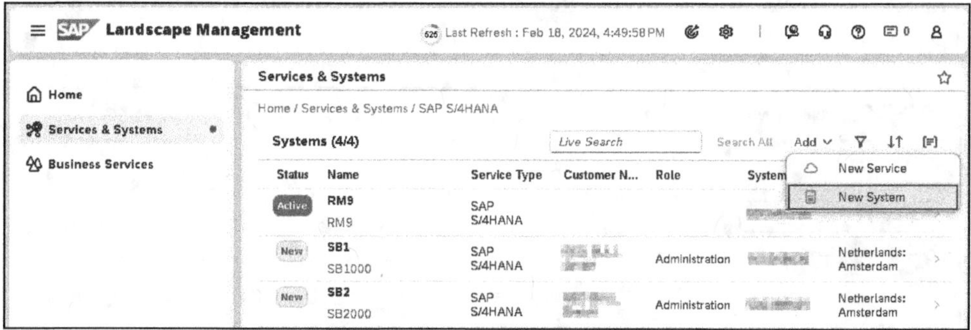

Figure 3.11 Possibility to Add a New System

To add a new system, some mandatory parameters are required, as shown in Table 3.2.

Parameter Name	Description
Product	Select from dropdown list—e.g., **SAP Focused Run**
System ID	System ID of the system (three-character string)
Role	Select from dropdown list—e.g., **Development**, **Test**, or **Production**
Virtual Host and Port	Virtual host and port maintained in the cloud connector connecting to this system in the form <host>:<port>
System Number	System number that is available in *SAP for Me*
Installation Number	Installation number that is available in *SAP for Me*. Alternatively, you can find the installation number in the ABAP system by clicking **Status** in the system menu.
Client	Default client of the system

Table 3.2 Mandatory Parameters to Add a New System

After being registered in the respective ABAP system or added via SAP Cloud ALM, the system information is also displayed in the **Services & Systems** menu. Further information on setting up a system can be found in SAP Help Portal at *https://help.sap.com/docs/cloud-alm/setup-administration/systems*.

Similar to what is available for the services, the following information is available for the systems in the overview in the **Service & Systems** menu:

- Status (lifecycle)
- Name (system ID)
- Service type (SAP product)
- Customer name
- Role
- System number
- Data center

The SAP S/4HANA systems of the example project landscape are shown as examples in Figure 3.12.

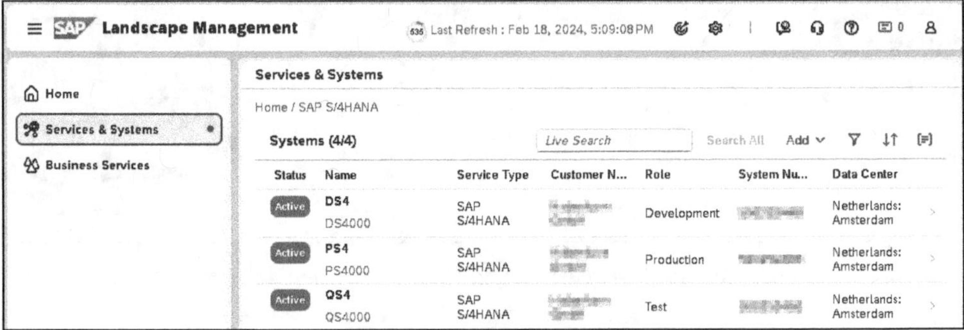

Figure 3.12 Overview of the Manageable SAP S/4HANA Systems (Project Scope)

If you select one of the systems, further system information is displayed on the **Details** tab, divided into **General Details**, **Product Details**, **Additional Details**, and **URLs** (see Figure 3.13).

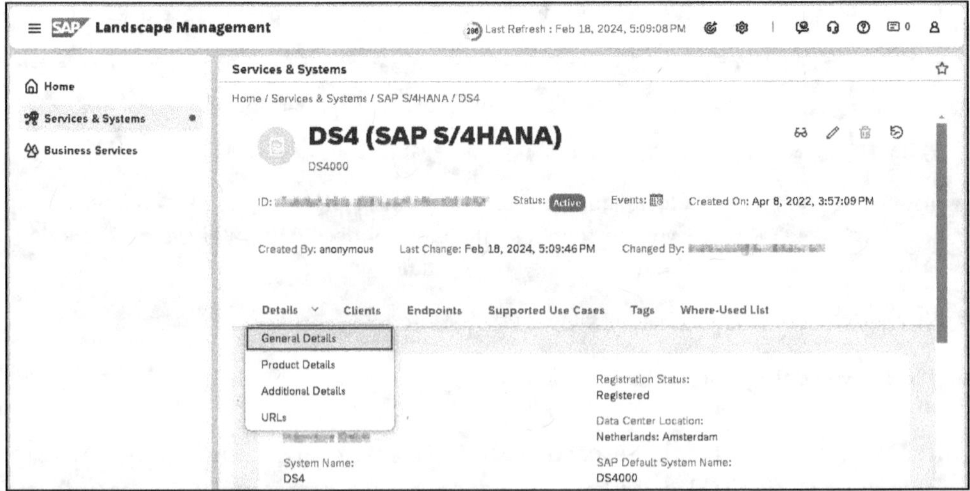

Figure 3.13 Detail Information View of the Selected System

On the **Clients** tab, you will find the manageable client entries for the selected system. Each client is displayed with its own **Lifecycle Status** (see Figure 3.14).

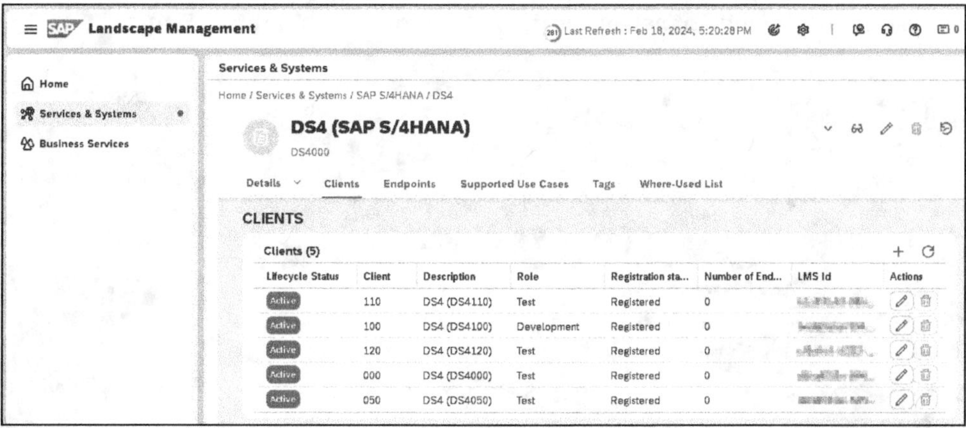

Figure 3.14 Client Information for the Selected System

It is also possible to adjust the description and role for each client. To do this, use the **Edit Client** button. The **Endpoints** tab contains the necessary information for other systems to establish an authenticated connection to the client.

On the **Supported Use Cases** tab, you can see from the configuration status and runtime status whether use cases are already active. Based on the **Runtime Status**, you can see whether data is being delivered for the respective use case. The data exchange for a use case between SAP Cloud ALM and an ABAP system is client-specific. You can see which client the use case receives the data for by clicking the **Display** button in the **Actions** column. (Further information on use cases is provided in Section 3.3.2.) With help of the **Tags** tab, you can maintain additional information regarding dedicated systems. For example, all relevant project scoped systems are tagged with **LM: Project Services and Systems** and can be use as filter criteria in certain searches (see Figure 3.15). The **Where-Used List** tab shows the business function where the system is in use.

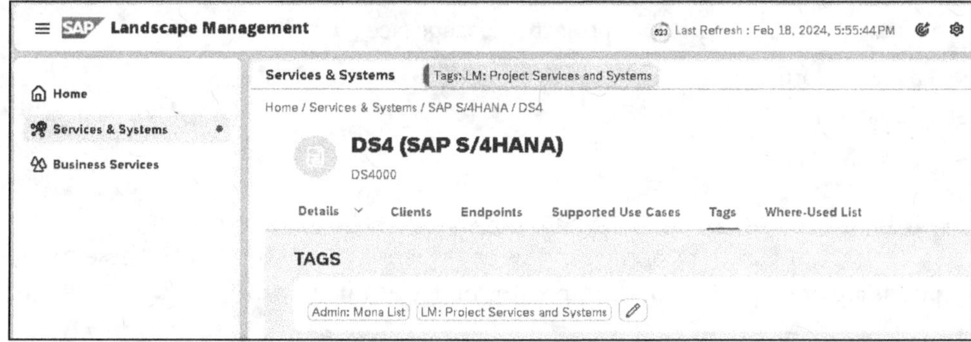

Figure 3.15 Use Tags as Filter Criteria

In addition to the information tabs presented here, the system detail view also provides the option of editing the selected system. To do this, select the **Edit System** button. In the **Edit System** window, you can adjust the **Description**, the **Role**, and—if necessary—the **Cloud Connector Location ID** and the **Logon URL** (see Figure 3.16).

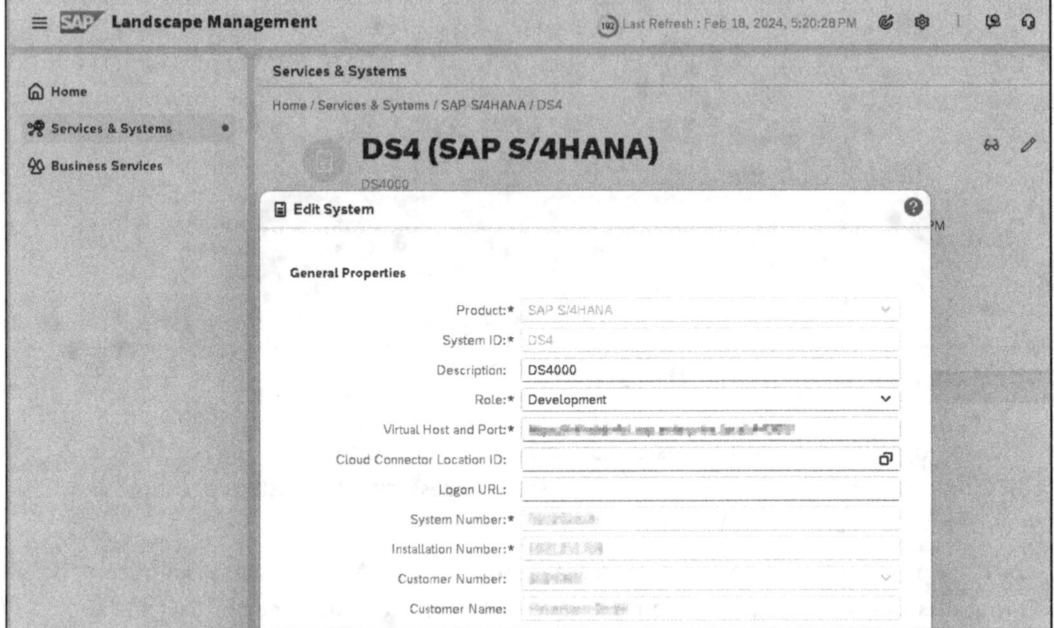

Figure 3.16 Edit Information of a System

In the top menu bar, you will also find the **Action Logs** button. Changes in system actions, such as the registration of a client, are displayed here in chronological order.

3.3 Manageable Information

Within the systems and services landscape groups are various information areas that are relevant for managing your project landscape (see Figure 3.4):

- Lifecycle status
- Use cases
- Endpoints

These three areas are relevant for both the services and the systems. They are used to display the configurations of the services or systems by SAP Cloud ALM. These configurations are prerequisites for certain SAP Cloud ALM functions.

However, there is also information that is only relevant for the administration of one of the two groups. These pieces of service or system information are described in Section 3.2.1 and Section 3.2.2 in more detail.

3.3.1 Lifecycle Status

The first information area, *lifecycle status*, indicates whether at least one use case is currently in utilization for a system or service (see Figure 3.17).

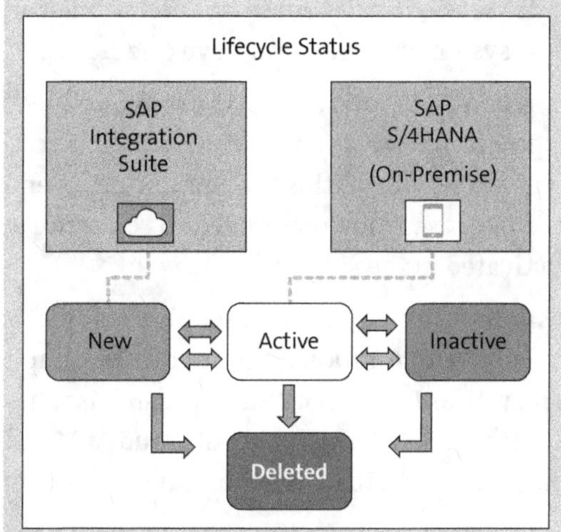

Figure 3.17 Lifecycle Status Flow Change of a Service or System

The lifecycle status is shown for each service or system on the **Home** menu in the Landscape Management app in the **Overview** and **Favorites** tabs (see Figure 3.18). For this example of the **SAP S/4HANA** service type, three systems have a lifecycle status of **Active,** and one has a lifecycle status of **New**.

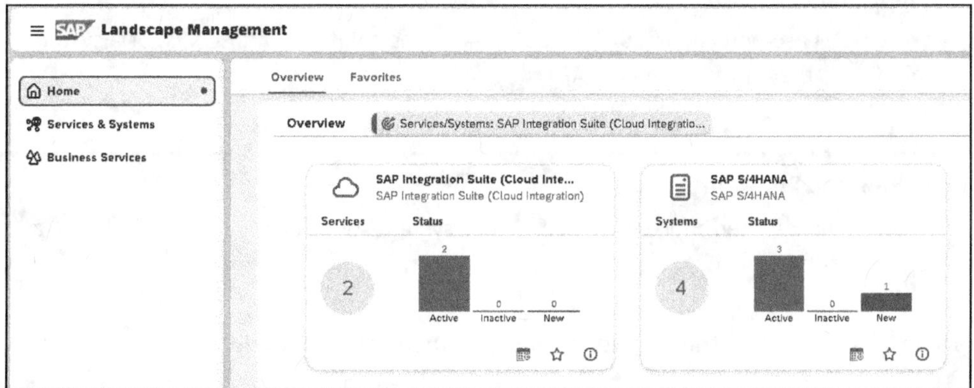

Figure 3.18 Lifecycle Status in the Home Menu

Likewise, the lifecycle status is shown in the **Services & Systems** menu for the selected service type and in the detail view of each service and system. There are four status values, as follows:

- **New**
- **Active**
- **Inactive**
- **Deleted**

These status values describe the status of the service or system in relation to the configuration status of the use case. A service or system can only ever have one status.

When a service or system is initially registered in SAP Cloud ALM, it is assigned a lifecycle status of **New**. The lifecycle status changes from **New** to **Active** as soon as a use case is activated. For services with a pull-based use case, you do this by configuring an endpoint (see Section 3.3.3). If it is a push-based use case, you do not need to configure an additional endpoint; the use case is only activated in the service.

The activation of use cases for systems is similar. Pull-based use cases also require the configuration of an endpoint. Such an endpoint configuration is required, for example, for the integration of SAP Solution Manager or SAP Focused Run. A push-based use case, on the other hand, is activated in the system itself without an SAP Cloud ALM endpoint; the use case receives the **Active** configuration status after configuration is complete (see Section 3.3.2).

The lifecycle status is an aggregated status that is determined by the configuration status of the use case. If an active use case is deactivated and no further use case is active for the service or for the system, the lifecycle status changes to Inactive. The **Deleted** status can also occur for services. This status occurs when a cloud service has been canceled and deleted in SAP BTP.

Pull-Based Procedure

In the pull-based procedure, SAP Cloud ALM requests data from a service or a system. This data is then transmitted to SAP Cloud ALM.

Push-Based Procedure

With the push-based procedure, a service or system sends data to SAP Cloud ALM at regular intervals.

Table 3.3 summarizes the individual status values for the lifecycle.

Status	Color Display	Description
New	Blue	Services or systems in **New** status have no configured use cases. Go to the system details and check the **Supported Use Cases** section to learn which use cases can be set up for the service or system.
Active	Green	Services or systems in **Active** status have at least one use case that is configured and active. *Active* can mean that data collection is executed or that active event rules exist. Go to the system details and check the **Supported Use Cases** section to see the status of the use cases.
Inactive	Yellow	Services or systems in **Inactive** status have no active use cases and at least one use case that is configured but not activated. *Inactive* can mean that data collection is currently paused or that event rules exist but are deactivated. Go to the system details and check the **Supported Use Cases** section to see the status of the use cases.
Deleted	Red	Services with the **Deleted** status have been unsubscribed by the customer. However, to avoid unintentional data loss, the configuration of the use cases has not been deleted automatically. Before the services can be removed from the landscape management, you have to remove the configuration for the use cases.

Table 3.3 Displayed Color and Description of Lifecycle Status Values

3.3.2 Use Cases

Certain use cases must be active for data exchange between SAP Cloud ALM and services or systems. The basis here is the configuration of a push- or pull-based use case.

An active or inactive use case is defined from the configuration status and the runtime status. The *configuration status* of a use case indicates whether it has been configured and whether data can be retrieved or provided. The *runtime status* indicates whether the data is being provided without errors. You can access the area for use cases via the Landscape Management app. Navigate to the **Services & Systems** menu on the left-hand side and select the desired entry to access the detailed view. Then open the **Supported Use Cases** tab. In Figure 3.19, you can see the **Runtime Status** with a value of **Success** for the activated use cases **Exception Monitoring** and **Health Monitoring** for the service **is-p (SAP Integration Suite [Cloud Integration])** are active.

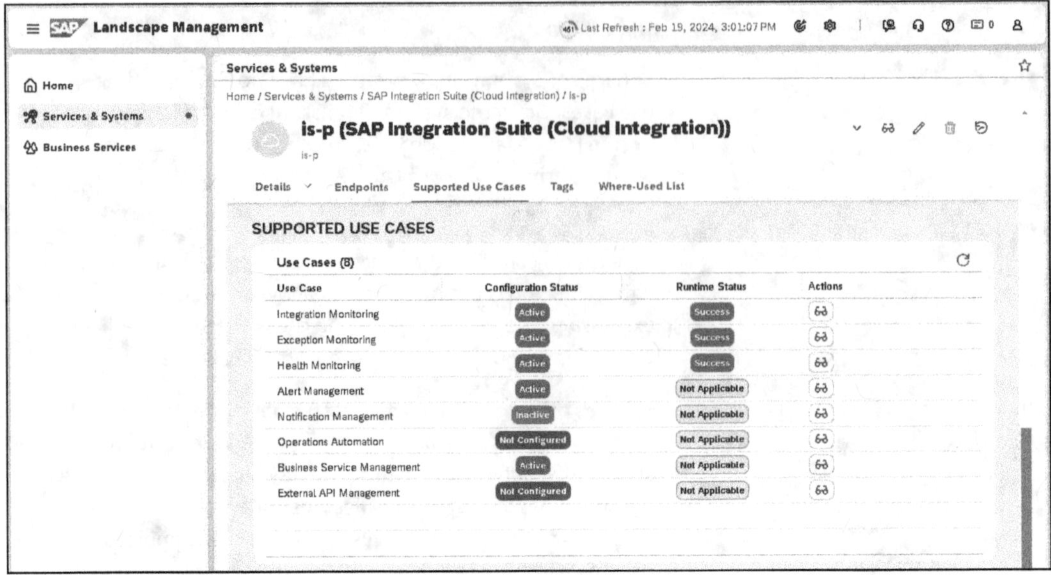

Figure 3.19 Configured Use Cases of a Certain Service.

For each use case, a configuration status and a runtime status as shown in Table 3.4 can be displayed.

Configuration Status	Condition
Not Configured	The use case was not configured, or the configuration was removed.
Active	The use case was configured and activated for this service.
Inactive	The use case was configured, but the data collection was deactivated.
Runtime Status	**Condition**
Only applies to use cases in which data is collected.	
Not Applicable (Initial)	The data collection is not active. This means the configuration status is **Not Configured** or **Inactive**). Note: Some use cases don't have actual data collection. For them, the runtime status will always be **Not Applicable**.
Success	The data collection was successful.
Error	An error occurred during the data collection, or the service has not sent data for an unusually long time.

Table 3.4 Configuration Status and Runtime Status of a Use Case

An overview of which SAP solutions are supposed by SAP Cloud ALM and their use cases is available in the SAP Help Portal at *https://help.sap.com/docs/cloud-alm/setup-administration/supported-solutions*. In the table displayed in Figure 3.20, note that you can filter the first column by **SAP Solution**. In the second column, you can filter by the **SAP Cloud ALM Area** (**Operation** or **Implementation**). The third column is used to narrow down choices on the basis of an **SAP Cloud ALM Capability** (in this example, **Test Management**) and usually shows an available use case. Whether this is a use case depends on whether data is exchanged between SAP Cloud ALM and a service or a system. For a better indication, refer to the fourth column: **Notes**.

In Figure 3.20, for example, under **SAP Cloud ALM Capability**, we filtered for **Test Management** without limiting it to a particular **SAP Solution** or an **SAP Cloud ALM Area**.

As you can see, the **Test Management** capability is supported for many SAP solutions. For SAP S/4HANA Cloud, there is also a use case that enables the exchange of test-related data: **Test Automation**.

Hide/Show Columns ▼ | Search entire table

SAP Solution	SAP Cloud ALM Area	SAP Cloud ALM Capability	Notes
Filter: [No Selection] ▼	Filter: [No Selection] ▼	Filter: Test Management ▼	Search column
Intelligent Enterprise – Implementation	Implementation	Test Management	
SAP Ariba	Implementation	Test Management	
SAP Fieldglass	Implementation	Test Management	
SAP Integrated Business Planning for Supply Chain (SAP IBP)	Implementation	Test Management	
SAP Intelligent Robotic Process Automation (iRPA)	Implementation	Test Management	
SAP Marketing Cloud	Implementation	Test Management	
SAP S/4HANA (on-premise)	Implementation	Test Management	
SAP S/4HANA Cloud, private edition	Implementation	Test Management	
SAP S/4HANA Cloud, public edition	Implementation	Test Management	More under Integrating the Test Automation Tool for SAP S/4HANA Cloud.

Figure 3.20 Search Use Case by SAP Cloud ALM Capability

3.3.3 Endpoints

The third group of manageable information for services and systems is endpoints. They are configured depending on a pull-based use case for services and systems and enable a technical connection between SAP Cloud ALM and the service or system. This involves communicating via SAP Cloud ALM using the pull mechanism in a service or system in order to retrieve or schedule data from there, which is initiated by the configured use case.

Existing endpoints can be found in the detailed view of a selected service or system. After opening the Landscape Management app, navigate to the **Services & Systems** menu on the left-hand side. Then select the desired system or service entry and open the **Endpoints** tab. In Figure 3.21, you can see an example of this view in SAP Integration Suite.

In this example, you can see that three use cases are configured for the endpoint named **is-p_01**. Regarding the connectivity, the **Root URL** and the **Authentication** are maintained. You can check whether the existing endpoints can be reached by SAP Cloud ALM using the **Ping Connection** button in the **Actions** column. Existing endpoints can also be adjusted using the **Edit** button in the **Actions** column. You can also define new endpoints using the **Add** button. A configuration example for creating an endpoint can be found in Chapter 19, Section 19.3.2. In the example in this section, the endpoint creation for the configuration of integration and exception monitoring is assumed.

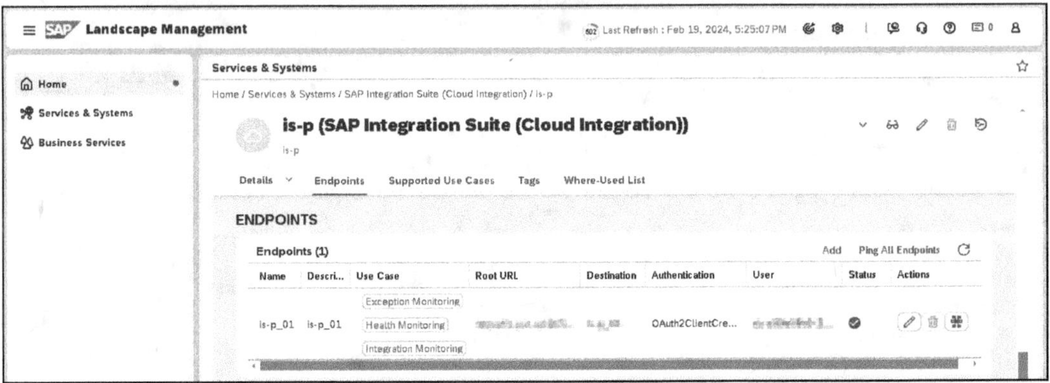

Figure 3.21 Configured Endpoint for SAP Integration Suite

3.4 Service Key Enabling and Maintenance

For the communication of push-based use cases, you need to establish a service key or binding credentials. The credentials of this communication are realized by an SAP BTP cockpit instance on the SAP Cloud ALM API service. Depending on the provisioning date of your SAP Cloud ALM tenant, the service key/binding is already available in the settings of the Landscape Management app. If your date is October 16, 2023 or later, the service key/binding is ready to use and established for operation and implementation use cases. For SAP Cloud ALM tenants whose provisioning was done before that date, you need to create or update the service key/binding as described at *https://help.sap.com/docs/cloud-alm/setup-administration/enabling-sap-cloud-alm-api.*

Ensure that your user has the Global Account Administrator role and is assigned to the SAP Cloud ALM subaccount. Open the SAP BTP cockpit and choose the subaccount of your SAP Cloud ALM. Check to ensure the availability of the entitlement for the SAP Cloud ALM API service. Next, navigate to the **Instances and Subscriptions** menu and click on the **Create** button and maintain the required basic information in step 1 (see Figure 3.22). Choose **SAP Cloud ALM API** for **Service**, **standard** for the **Plan** parameter, and **Other** for the **Runtime Environment**. Enter the name of your choice for the last parameter, **Instance Name**.

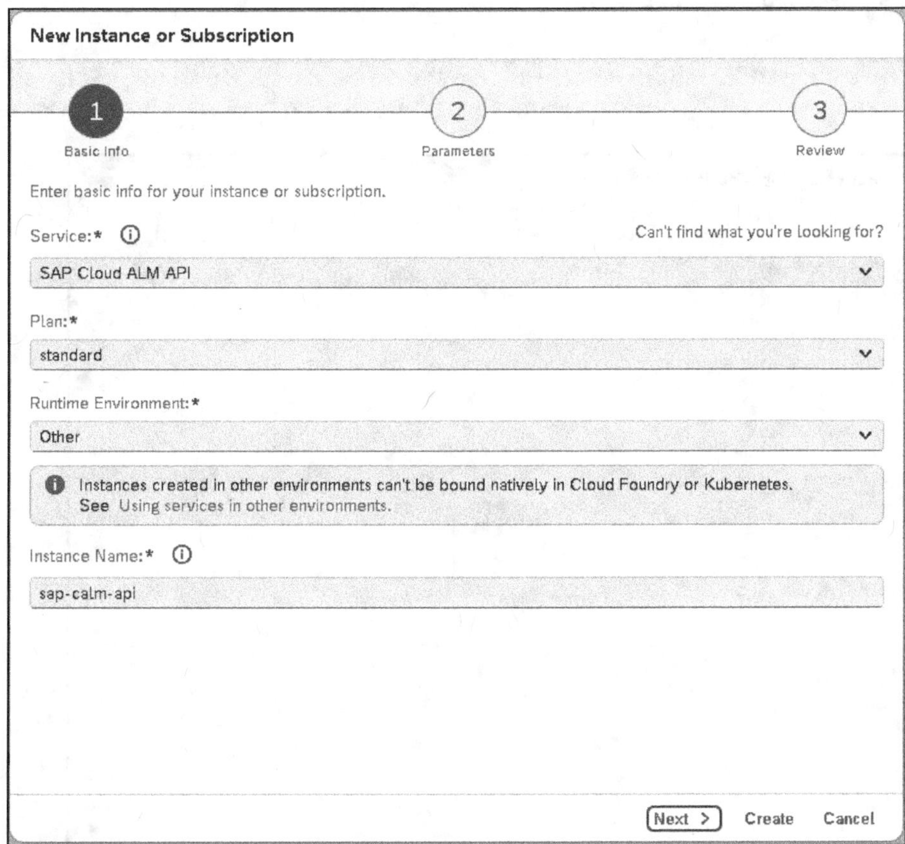

Figure 3.22 Basic Information of New Instance

Click on the **Next** button to continue with step 2. Copy the JSON code from Listing 3.1 in the text editor and replace the XSAPPNAME parameter with your instance name.

```
{
    "xs-security": {
        "xsappname": "<your-instance-name>",
        "authorities": [
            "$XSMASTERAPPNAME.imp-cdm-feature-display-ui",
            "$XSMASTERAPPNAME.imp-cdm-feature-manage-ui"
        ],
        "oauth2-configuration": {
            "credential-types": [
                "binding-secret"
```

```
            ]
        }
    }
}
```

Listing 3.1 Required JSON Code

By clicking on the **Next** button, you will get an overview of your entered information in step 3 (see Figure 3.23). To finalize the instance setup, click on the **Create** button.

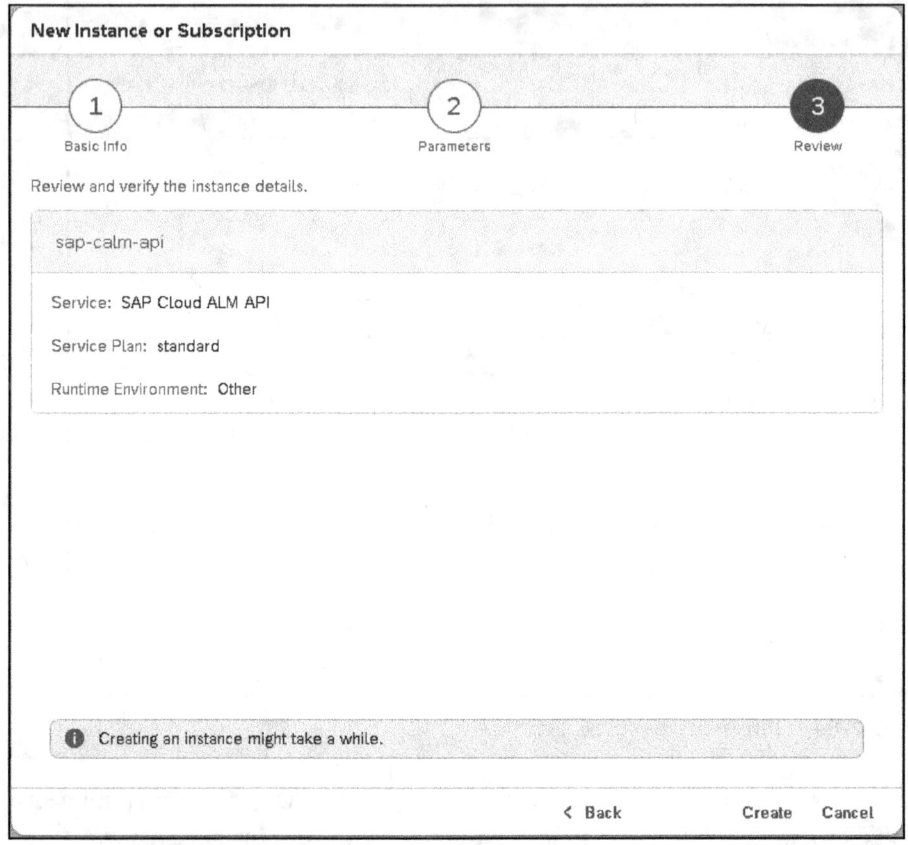

Figure 3.23 Review of Information for New Instance

The new instance is generated, and a new entry appears in the overview of instances with the **New** status. Click on the **Actions** button and choose **Create Service Binding** (see Figure 3.24). Download this service binding for an upload into SAP Cloud ALM.

After creation or updating the instance, you can upload the service binding to a dedicated section in the Landscape Management app. Click on the **Configuration** button (see Figure 3.25). Enter a **Name** for the service key and click **Browse...** to select the **File Name** of your service key/binding.

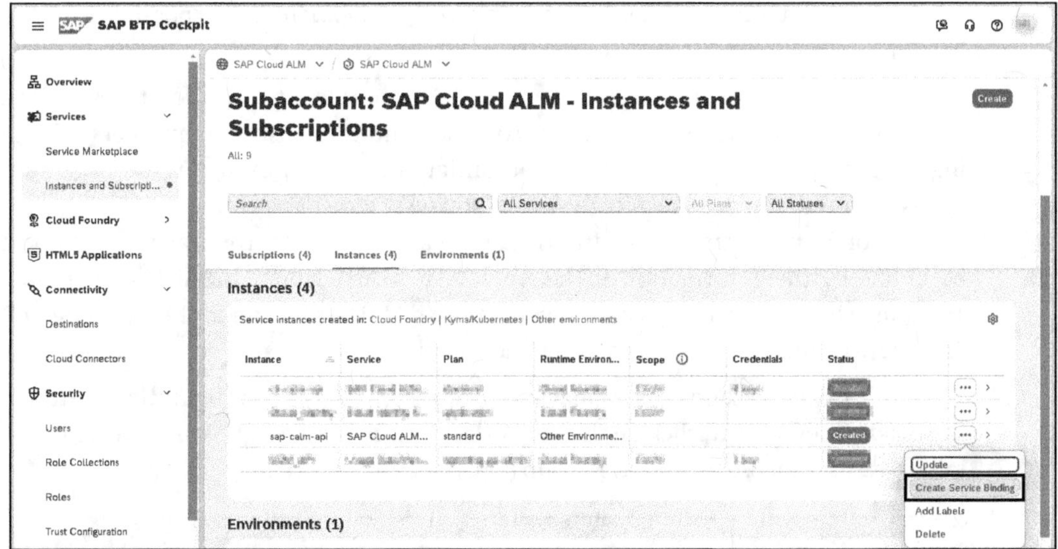

Figure 3.24 Run Action Create Service Binding for New Instance

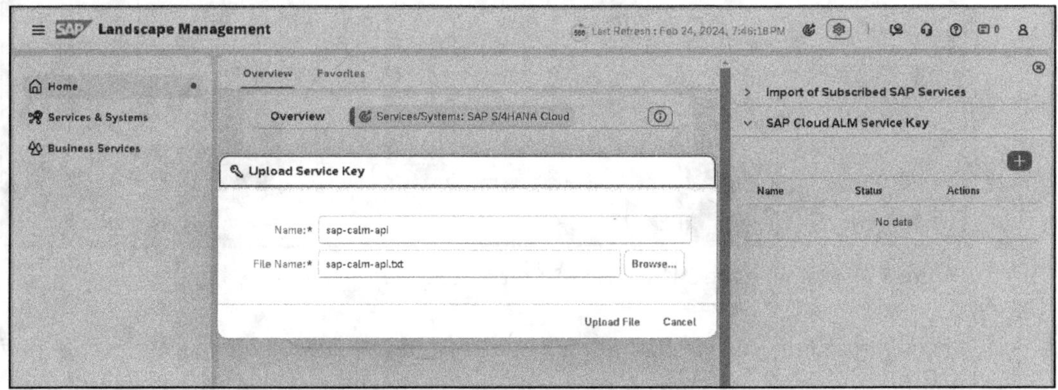

Figure 3.25 Upload Service Key into Landscape Management Configuration

After uploading the key, you will run a test to check whether the connection is working. Every SAP Cloud ALM administrator now has access to the service key. There is no need to store the key in a file system or require additional access to SAP BTP cockpit. You can find further information about managing your SAP Cloud ALM service key at *https://help.sap.com/docs/cloud-alm/setup-administration/service-key*.

3.5 Scenario with Multiple SAP Customer Numbers

In principle, the registered services and systems associated with a specific SAP customer number are managed in SAP Cloud ALM. Corporate groups with subsidiaries may have several SAP customer numbers and will need to manage the associated services and systems centrally. Such centralized management is also possible with the help of SAP Cloud ALM.

Services are, as already described in Section 3.2.1, imported into SAP Cloud ALM via the System Landscape Information service. To import multiple customer numbers belonging to the corporate group, an additional configuration in the Landscape Management app is required. The prerequisite for this is an *S-user* who has a connection to all these SAP customer numbers and is authorized to read service data from the SAP's support backbone. Further information on the prerequisites can be found in SAP Note 3070306. The S-user is stored with a password using the **Add S-User** button in the **Maintain SAP Corporate Group (CCC) S-User** maintenance window.

To do this, open the Landscape Management app and press the **Configuration** button in the header menu. Next, click on the **Add S-User** button on the right-hand side to maintain the S-user for SAP Corporate Group (CCC) (see Figure 3.26).

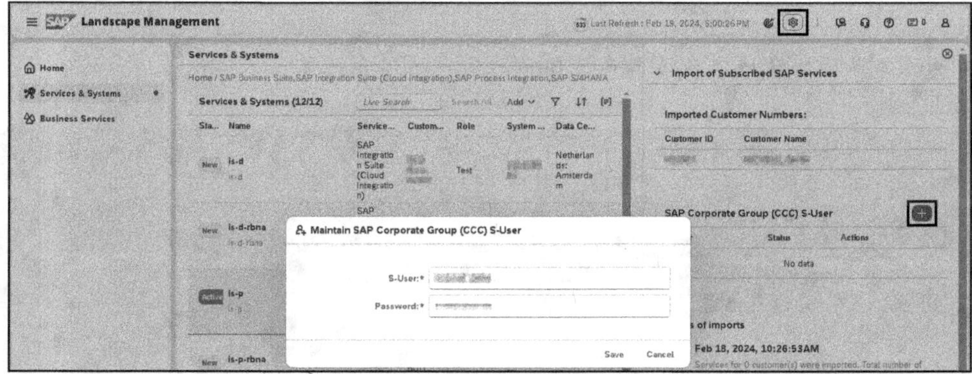

Figure 3.26 Maintain the S-User for SAP Corporate Group (CCC)

A note regarding this configuration can be found in the online document *Setup and Administration for SAP Cloud ALM* in SAP Help Portal: *https://help.sap.com/docs/cloud-alm/setup-administration/services.*

3.6 Summary

In this chapter, we have explained what information is required for your project landscape in order to use it with the help of SAP Cloud ALM during the implementation project and also for operations.

You have learned in detail how to manage the project landscape of our example scenario SAP S/4HANA on-premise and SAP Integration Suite using the **Service & Systems** landscape groups. In the next chapter, we will return to these services and systems and explain how you can make them available to the integrated transport tools of SAP Cloud ALM.

Chapter 4
Transport Tools of SAP Cloud ALM

The setup and seamless integration of the tools used to perform transport management are crucial for the central control of changes to different landscapes. Transport tools are the essential functions for establishing controlled change enablement.

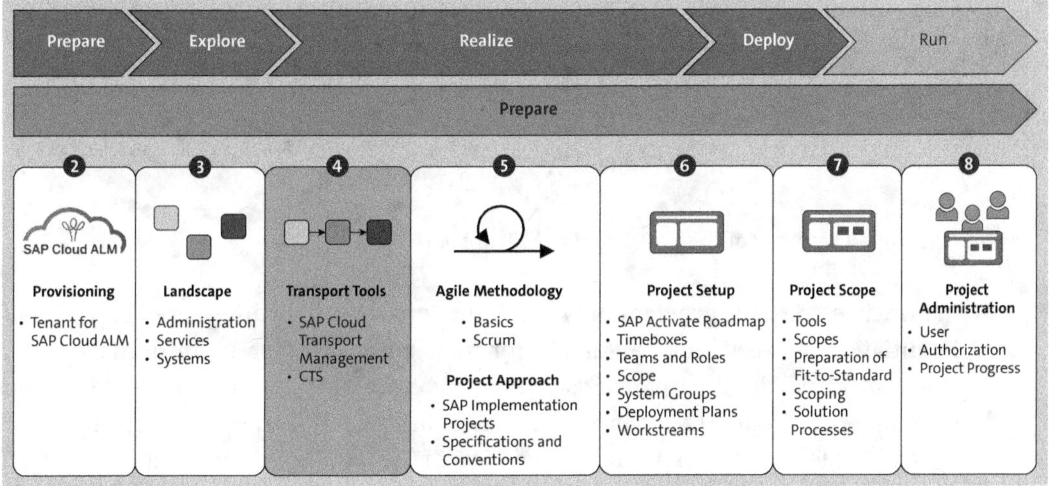

In Chapter 3, the example project landscape was examined from the perspective of categorization by services and systems. You learned what is required to be able to use certain ALM capabilities in SAP Cloud ALM. One of these ALM capabilities is the control and documentation of all changes (change enablement) to the associated systems and services as part of the project. This chapter will give you an overview of the landscapes that can be managed by SAP Cloud ALM and illustrate the transport tools used and how they are set up. Here, we will use our example scenario to illustrate how to set up transport management for Cloud Integration and for SAP S/4HANA. The measures described ensure that transport management can be carried out via features once it has been set up.

SAP Cloud ALM itself has no native integration in terms of transport control for SAP solutions. The actual functions for controlling cloud-native transports or ABAP system native transports are provided by the SAP Cloud Transport Management's adaptation transport organizer (ATO) and Change and Transport System (CTS) tools. The tools are used to define transport landscapes and provide the basic functions for transport control and logging. Figure 4.1 shows a simplified representation of the integration of these tools.

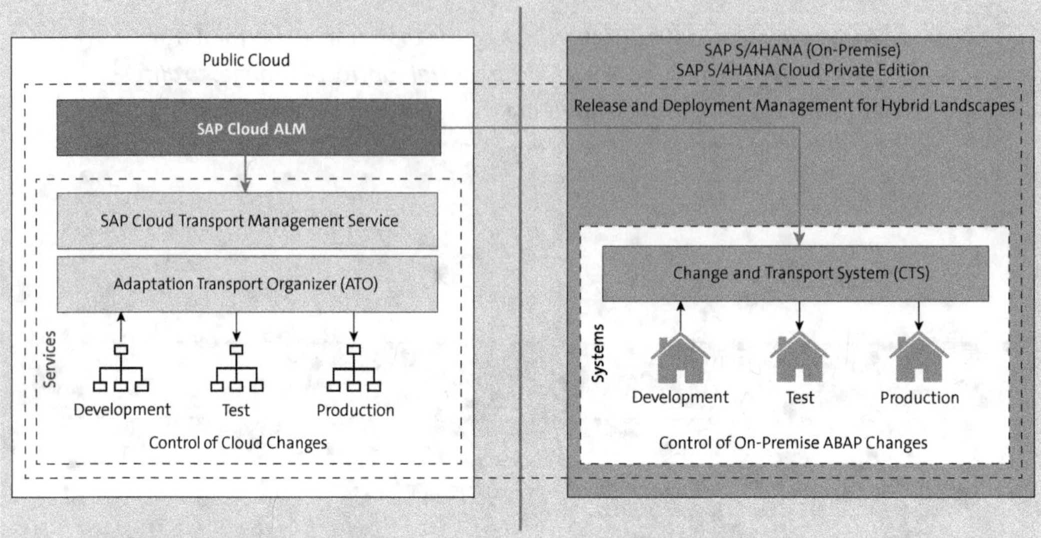

Figure 4.1 Overview of the SAP Cloud ALM Transport Tools

In principle, the tools provide equivalent functions. The main difference lies in the differentiation between their suitability for managing classic on-premise ABAP-native landscapes and cloud-native landscapes. If different SAP solutions are used in your company, SAP Cloud ALM enables you to establish a central cross-landscape transport control. Table 4.1 provides a detailed overview of the tools and their suitability for common SAP applications.

Transport Management Tool	Area of Application
Change and Transport System (CTS)	SAP S/4HANA Cloud Private Edition
	SAP NetWeaver Application Server for ABAP (on premise)
SAP Cloud Transport Management	SAP Integration Suite
	SAP BTP, Cloud Foundry environment
	SAP BTP, Neo environment
	SAP Analytics Cloud network package content
Adaptation transport organizer (ATO)	SAP S/4HANA Cloud Public Edition

Table 4.1 Transport Management Tools and Their Areas of Application

Depending on transport deployment tool you're using, you'll have access to different transport controls in SAP Cloud ALM, as shown in Table 4.2.

Transport Control in SAP Cloud ALM	SAP Cloud Transport Management	CTS	ATO
Create transport	No	Yes	No
Assign/unassign transport	Yes*	Yes	Yes*
Create/export transport of copy	No	Yes	No
Release transport	No	Yes	No
Import transport or transport of copy	Yes	Yea	No
* Released transport only			

Table 4.2 Supported Transport Control

As you can see in the table, the highest integration regarding transport control by SAP Cloud ALM is available with the CTS transport deployment tool. You can find the latest information regarding integration of the transport tools for SAP Cloud ALM at *https://help.sap.com/docs/cloud-alm/applicationhelp/integration-of-deployment-tools*.

4.1 SAP Cloud Transport Management

SAP Cloud Transport Management is the recommended option for the delivery of software (applications and application content) on SAP BTP. As an essential component of SAP Cloud ALM, SAP Cloud Transport Management is activated together with SAP Cloud ALM on the same subaccount as part of provisioning. Alternatively, the use of SAP Cloud Transport Management is also possible independently of SAP Cloud ALM but is associated with additional usage costs when using the standard service plan, which do not apply when used as part of SAP Cloud ALM. A free service plan is also offered as a free tier service, the use of which is associated with restrictions in terms of support, maximum storage capacity, and storage time of the transport data.

[«]

Current and Future Setup of SAP Cloud Transport Management

Concerning the current setup of SAP Cloud Transport Management with SAP Cloud ALM, if you are already using SAP Cloud Transport Management independently of SAP Cloud ALM on another account, the existing service cannot be subsequently linked to SAP Cloud ALM. Furthermore, no full migration of existing configurations is supported, so in this situation the service obtained as part of SAP Cloud ALM must be set up as described in the following sections. Concerning the future setup of SAP Cloud Transport Management Service with SAP Cloud ALM, SAP plans to introduce another recommended integration approach in the second half of 2024. As a baseline, this setup will include the following:

- Integration of an already enabled SAP Cloud Transport Management instance
- Connection of several SAP Cloud Transport Management instances with the same SAP Cloud ALM tenant

For the technical configuration, you will require the following:

- Setup for specific SAP solutions based on SAP Cloud Transport Management configuration guides
- The service key for your SAP Cloud Transport Management instance
- The destination of your SAP Cloud ALM subaccount to connect the SAP Cloud Transport Management instance (HTTP connection with OAuth 2.0)

In principle, the delivery of changes within SAP BTP is also possible independently of SAP Cloud Transport Management. The common alternatives are manual export and import or using the extended Change and Transport System (CTS+). But these procedures are associated with significant limitations. Although manual export and import do not require any configuration, they also offer no transparency regarding the changes made. In contrast, CTS+ must be set up on an on-premise ABAP system (often SAP Solution Manager) that handles both the delivery of content and logging. In this case, delivery is restricted to multitarget application archives (MTAR archives). This means that not all SAP BTP artifacts are supported.

In contrast to these options, SAP Cloud Transport Management enables the control and logging of transports between subaccounts of cloud environments as part of a cloud service. Various subaccounts are combined into transport landscapes to ensure the orderly delivery of changes. The transport landscapes are not limited to a global account, meaning that delivery to subaccounts that are assigned to a different global account is also possible. Changes from a source account can be recorded directly from an application. The upload of archives from a local file system is also available, provided this option has been activated as part of the landscape definition. The service supports the following object types:

- Multitarget application archive (MTA)
- Delivery units
- Application contents
- SAP BTP, ABAP environment

[+]

Supported Content

A detailed overview of the content supported in connection with various cloud environments can be found at *https://help.sap.com/docs/cloud-transport-management/sap-cloud-transport-management/supported-content-types.*

4.1.1 Prerequisites

The measures and configuration steps that must be carried out by SAP Cloud ALM to ensure transport management using SAP Cloud Transport Management are explained in the following section. Use the SAP Help Portal at *https://help.sap.com/docs/cloud-alm/setup-administration/transport-management*. Under the **SAP Cloud Transport Management Service** menu, you will find up-to-date information on the prerequisites and setup.

Subscribing to SAP Cloud Transport Management

As we have described in detail, SAP Cloud Transport Management is subscribed to automatically as part of the provisioning process. If the service has not yet been subscribed on your tenant, the following steps are required to set it up in the SAP BTP cockpit:

1. Navigate to your global account and select your SAP Cloud ALM subaccount.
2. Navigate to **Entitlements** and click on the **Edit** button.
3. Use the search bar to search for "Cloud Transport Management." If no entry is found, use the **Add Service Plans** button.
4. Select the **standard** plan and confirm your selection.
5. SAP Cloud Transport Management is now listed under **Entitlements**. Save your previous changes.
6. Navigate to **Instances and Subscriptions** and click on **Create**.
7. Select SAP Cloud Transport Management with the **Standard Service Plan** and confirm the dialog.

You have now subscribed to SAP Cloud Transport Management.

Authorizations to Use SAP Cloud Transport Management

SAP Cloud Transport Management provides several authorization roles to enable configuration and administration, export and import, or simply to view the current status. SAP Cloud ALM only requires the assignment of the following role sets in terms of configuration:

- `TMS_LandscapeOperator_RC`
- `TMS_Viewer_RC`

These role sets or the roles included are limited to the rights to set up the landscapes. The actual use of transport management within SAP Cloud Transport Management is not required, as these functions are controlled passively by SAP Cloud ALM.

[»]

Authorization for Direct Use of SAP Cloud Transport Management

If you want to verify the functions of SAP Cloud Transport Management independently of SAP Cloud ALM or make them available as part of any emergency process, you can optionally authorize additional activities within the service. The required roles are assigned to the application name *alm-ts*.

Set Up SAP Cloud ALM API

The SAP Cloud ALM API is used to establish a connection between managed systems or tenants with an SAP Cloud ALM application (in this case, SAP Cloud Transport Management). In Chapter 3, Section 3.4, you learned how to perform the required service binding with the JSON file that contains the data you can access in the Landscape Management app. Click on the **Configuration** button and navigate into the **SAP Cloud ALM Service Key** section. Click the **Display Service Key** action and make note of the following parameters:

- `endpoints.api` (URL for addressing the SAP Cloud ALM API)
- `uaa.clientid` (login data for the service)
- `uaa.clientsecret` (login data for the service)
- `uaa.url` (URL for generating the OAuth token)

You have now established the necessary prerequisites to have the transport control of cloud environments managed by SAP Cloud Transport Management. In the next section, you will set up the integration of SAP Cloud Transport Management in SAP Cloud ALM.

4.1.2 Setting up Transport Management for Cloud Integration

The steps required to connect a cloud environment to SAP Cloud Transport Management are illustrated in this section using the example of Cloud Integration, part of SAP Integration Suite. Figure 4.3 illustrates the key component that are required as part of the connection.

In simple terms, connectivity must first be established between the connected tenants and SAP Cloud Transport Management. This is followed by the creation of transport nodes, each representing a subaccount, as part of the landscape configuration. Transport routes are used to define the order in which the transports are delivered in the landscape. Use the Cloud Transport Management app in the **SAP Cloud ALM for Implementation** app group (see Figure 4.2) to carry out the configuration steps indicated. The following sections explain the setup of a three-tier transport landscape in detail.

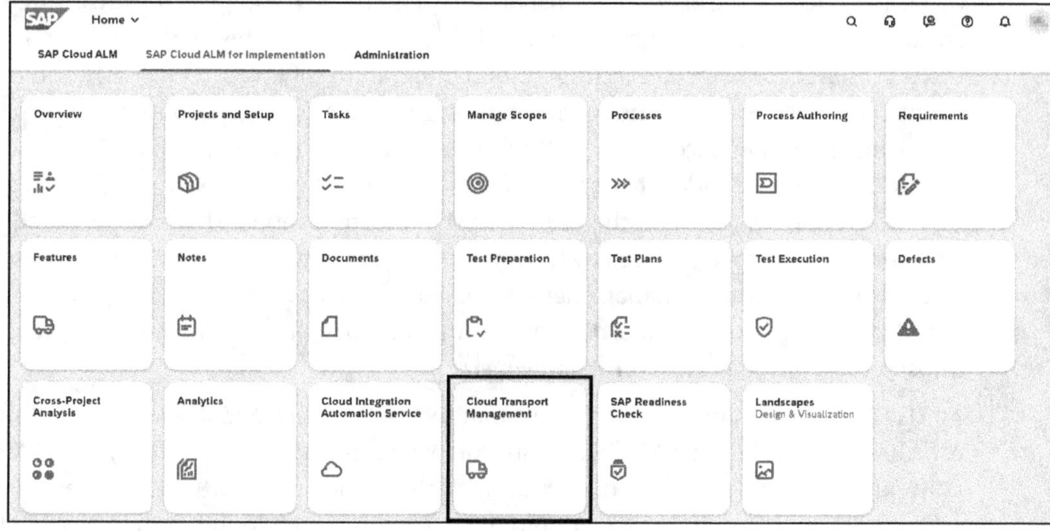

Figure 4.2 App to Set Up Cloud Transport Management

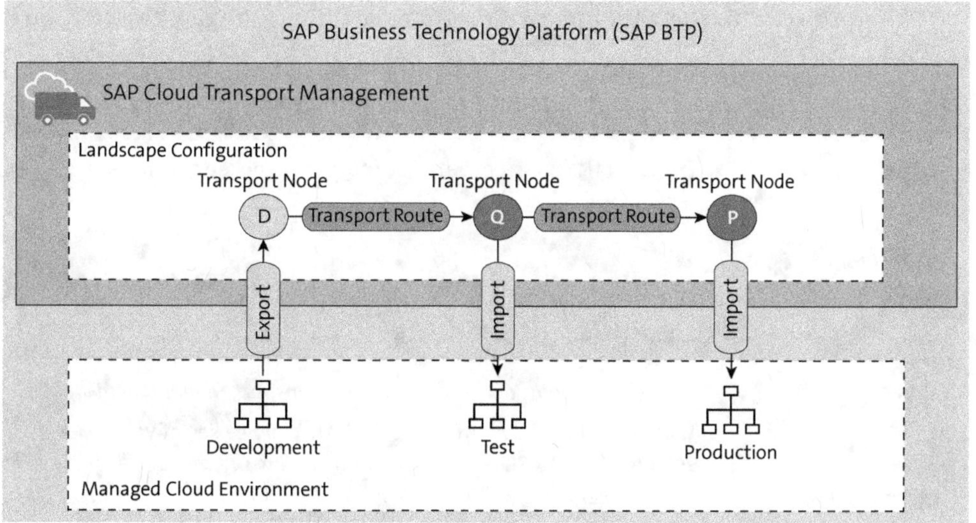

Figure 4.3 Integrating a Landscape into SAP Cloud Transport Management

Different Configuration Steps for Each Target Environment

The following sections clarify the general procedure for landscape connections. Please note that there may be marginal differences when setting up connectivity for the respective target environments.

Setting Up Transport Nodes for the Export of Source Tenant

This step is required to enable the direct data exchange of an application with the SAP Cloud Transport Management service and is performed on the tenant's subaccount that is intended for the export of transports. Depending on the application area, different services may be responsible for packaging the change and the subsequent export. In the present context of connecting a Cloud Integration subaccount, the SAP Content Agent service is required. This service makes it possible to export content from the Cloud Integration subaccount and thus make it available to SAP Cloud Transport Management service for further processing. Another prerequisite is setting up the process integration runtime and then selecting the transport mode. The transport mode influences which process is used by the integrated transport functions and must be set to SAP Cloud Transport Management. Detailed instructions on these steps can be found at *https://help.sap.com/docs/cloud-integration/sap-cloud-integration/content-transport-using-cloud-transport-management*.

An HTTP connection must be defined so that the SAP Content Agent service setup can establish a connection to SAP Cloud Transport Management. For this purpose, the SAP Content Agent service uses a connection with the name `TransportManagementService`. All the information required to set up the connection can be found in the SAP Cloud ALM service key. Please note that the suffix `/imp-cdm-transport-management-api/v1` must be added to the URL. For this example, we use the authentication option `Oauth2-JWTBearer`. You can find detailed information on maintaining the export connection in Table 4.3.

Parameter	Notes
Name	`TransportManagementService` (note that the letters are case-sensitive).
Type	`http`
Description	You are free to choose the name of the connection (e.g., "CTMS export connection").
URL	*https://<endpoint api>/api/imp-cdm-transport-management-api/v1* Take the value `endpoints.api` from the service key of your SAP Cloud ALM API instance and add the suffix */imp-cdm-transport-management-api/v1*.
Proxy Type	`Internet`
Authentication	`Oauth2JWTBearer` The choice of authentication method is up to you.

Table 4.3 Maintenance Instructions for Creating the HTTP Connection for Export

Parameter	Notes
Client ID	Take the value `uaa.clientid` from the service key of your SAP Cloud ALM API instance.
Client Secret	Take the value `uaa:clientsecret` from the service key of your SAP Cloud ALM API instance.
Token Service URL	Take the value `uaa.url` from the service key of your SAP Cloud ALM API instance.
Additional Properties	▪ Property: `SourceSystemId` ▪ Value: `<Transport Node ID>` Create a new entry and note that it is case-sensitive. The value describes the transport node of the source account (the transport node is created later as part of the landscape configuration in SAP Cloud Transport Management).

Table 4.3 Maintenance Instructions for Creating the HTTP Connection for Export (Cont.)

[!]

Check the Connection

Please note that the connection check fails with an HTTP 404 error code. The address of the endpoint is only completed during the actual call as part of an export. In this respect, the check result can be ignored. To check the connectivity, carry out a sample export instead (this requires the landscape configuration in the later step).

[«]

Transport Migration

If you have already used the SAP Cloud Transport Management service on another subaccount before implementing SAP Cloud ALM, only the SAP Cloud Transport Management connection needs to be adjusted to point to the SAP Cloud ALM API instance. This ensures that all future exports are directed to SAP Cloud Transport Management of SAP Cloud ALM. Transports already exported to the previously used separate service instance of SAP Cloud Transport Management remain unaffected by this change.

Setting Up the Import Connection

As with exporting, a connection is required for importing transports. You maintain this connection in the SAP BTP cockpit of the SAP Cloud ALM subaccount. To create the connection, you need the connection data and login information for the subaccounts on which the exported content is to be imported later. Detailed information on maintaining the connection is listed in Table 4.4.

Parameter	Notes
Name	You are free to choose the name (e.g., "CPI_PRD_DEPLOY").
Type	`http`
Description	You are free to choose the name of the connection (e.g., "CTMS import connection").
URL	Select one of the following options to contact the SAP Cloud Deployment service: ■ **Use of organization names and space** Sample: *https://deploy-service.cfapps.<default-domain>/slprot/<myorg>/<myspace>/slp* Please note that any spaces and special characters within your organization and space name must be encoded (e.g. spaces must be replaced by *%20*). ■ **Use of the GUID of your Space** Pattern: *https://deploy-service.cfapps.<default-domain>/slprot/<my-space-guid>/slp* The Cloud Foundry Command Line Interface (CLI) is required to determine the GUID. You can determine the required GUID using the `cf space <space name>-guid` command.
Proxy Type	**Internet**
Authentication	**Basic Authentication** The choice of authentication method is up to you.
User	Specify a platform user for the import. The specified user requires at least the **SpaceDeveloper** authorization in the target tenant.
Password	Enter the password of the platform user.

Table 4.4 Maintenance Instructions for Creating the Import Connection in the SAP Cloud ALM Subaccount

Landscape Configuration

The landscape configuration enables the arrangement of subaccounts into a transport landscape within SAP Cloud Transport Management. A transport node is created for each subaccount involved in the transport process. You can use transport routes to specify the delivery sequence. *Transport routes* are a directional connection between two transport nodes, which means that a transport is automatically added to the import queue of the target node as a result of the import on the source node.

To start the landscape configuration, open the Cloud Transport Management app and click on the **Landscape Wizard** menu (see Figure 4.4).

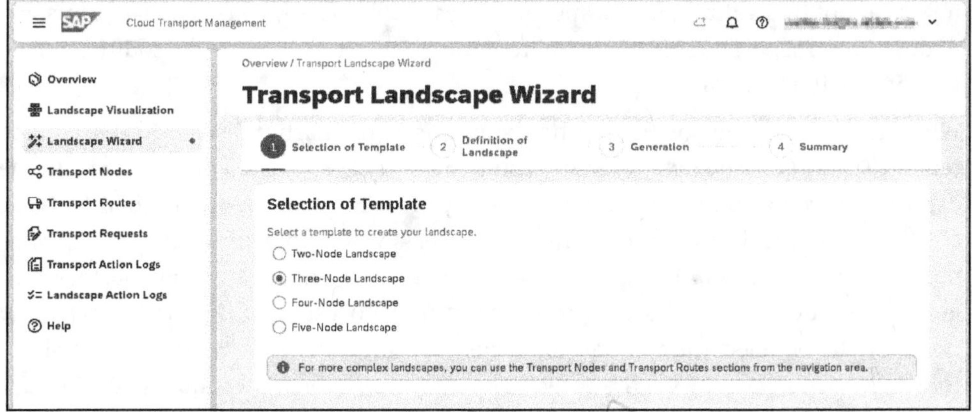

Figure 4.4 Transport Landscape Wizard

The **Transport Landscape Wizard** (see Figure 4.5) offers a simplified landscape definition option for common landscape constellations such as a three-stage transport landscape.

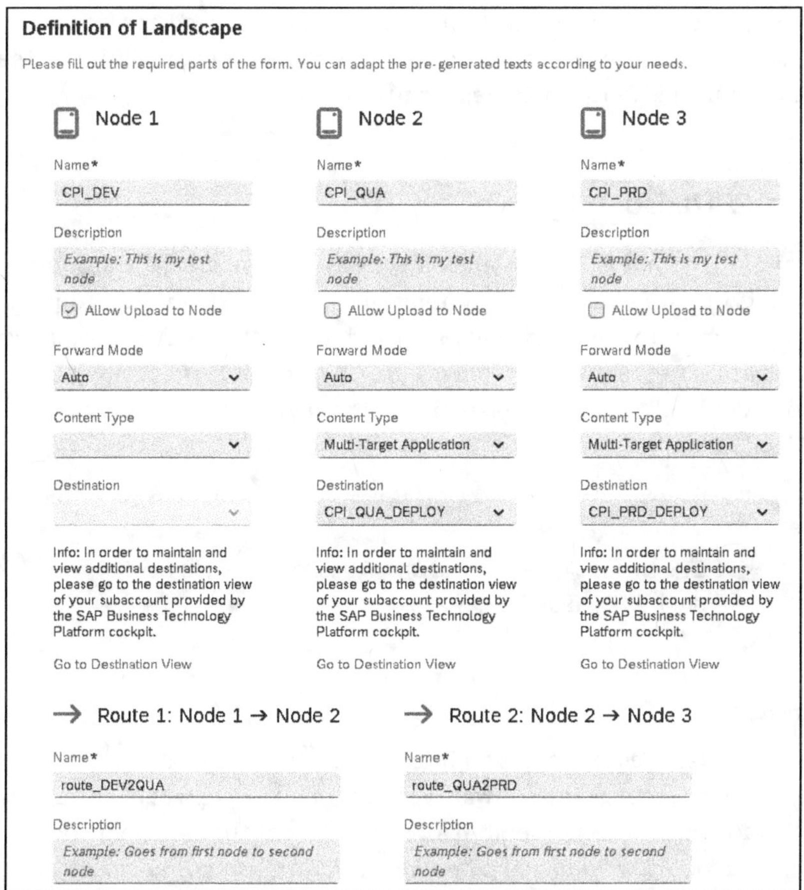

Figure 4.5 Configuration of a Three-Node Landscape

If you cannot find suitable landscape templates in the wizard, you can create transport nodes and transport routes manually to define new landscapes or extend the definition of existing landscapes.

When creating a transport node, the naming is particularly relevant for the source tenant: The name used here must be stored in the `SourceSystemId` property within the export connection that was previously set up. The checkbox **Allow Upload to Node** enables the manual creation of transport requests using uploads. In connection with the usage of SAP Cloud ALM, leave the **Forward Mode** at the default value **Auto**, which guarantees automatic forwarding to the next transport node during the import of a transport. If you select the **Manual** value, there is no automatic forwarding, so you have to trigger the forwarding via the **Forward** button within the import queue. The specification of the **Content Type** is not relevant for nodes that represent a source tenant; it is required to set up the import for a target tenants. At this point, assign the previously created HTTP connections for the import to the corresponding nodes.

Using the landscape wizard, the corresponding transport routes are automatically defined for the specified transport nodes. The default options in the landscape wizard always result in sequential delivery of the transport nodes. If you need to set up more complex landscapes, you can create additional landscape elements manually under the **Transport Nodes** menu and **Transport Routes** menu options.

4.2 Change and Transport System

Change and Transport System (CTS) combines functions that enable the transport of changes to SAP software for classic ABAP systems. In this context, you can combine systems and their clients into transport landscapes and thus create a binding sequence for the targeted distribution of software changes to SAP systems. In the context of the consideration of SAP Cloud ALM, our explanations here are limited to the integration of CTS in SAP Cloud ALM.

[»]

Basic Knowledge of CTS

For a detailed look at CTS, the book *SAP Change and Transport Management* by Armin Kösegi and Rainer Nerding (SAP PRESS 2013) is a good option.

4.2.1 Prerequisites

Now we will describe all the requirements for integrating CTS into SAP Cloud ALM. This includes both the general settings and software versions on the system to be connected as well as the characteristics in SAP Cloud ALM. Please look at the SAP Help Portal at *https://help.sap.com/docs/cloud-alm/setup-administration/transport-management*. In the menu item **SAP S/4HANA Cloud, Private Edition and On-Premise Systems**, you will find up-to-date information on the requirements and setup.

Technical Requirements

To connect a system or system landscape to SAP Cloud ALM, the following technical requirements must be met:

- SAP Basis release SAP_BASIS 7.40 SP 20 or higher has been installed.
- ST-PI 7.40 SP 18 or higher has been installed (pay attention to the SAP Notes listed in Table 4.5 related to the dependency of the installed ST-PI support package version, which must also be observed).

ST-PI SP Version	SAP Notes
18 and 19	3196078 and 3322679
20 and 21	3240966 and 3322679
22 and higher	3310406 and 3322679

Table 4.5 SAP Notes, Depending on ST-PI-SP Version

- The profile parameter `icm/HTTPS/client_sni_enabled` must be set to `TRUE`. Further information can be found in SAP Note 510007.
- The profile parameter `ssl/client_ciphersuites` needs to be set up according to section 7 of SAP Note 510007.
- Successful import of the certificates `DigiCert Global Root CA`, `DigiCertGlobalRootG2`, and `DigiCertRSA4096RootG5` using Transaction STRUST for the SSL client (anonymous) and the SSL client (standard). Further information on setup is available at *https://support.sap.com/en/alm/sap-cloud-alm/operations/expert-portal/setup-managed-services/setup-abap/setup-strust.html*).

Additionally, consider that the last system at the transport track will always be treated as a production system, and transport management system (TMS) landscape configuration needs to be consistent. A client-specific transport route setup is recommended and can be establish by the TMS parameter **CTC**.

[«]

Client-Specific Transport Routes

For details on how to set up client-specific transport routes, refer to SAP Help Portal at *http://s-prs.co/v589702*.

To prepare the following setup, you must check whether the users have sufficient authorizations for the background job and for carrying out the setup. An overview is provided in Table 4.6.

User	Transaction PFCG Roles
Carrying out the setup	SAP_SDF_ALM_SETUP
Execution of the background job	SAP_SDF_ALM_METRIC_PUSH_FND SAP_BC_TRANSPORT_ADMINISTRATOR

Table 4.6 Overview of User Authorization for the SAP Cloud ALM Setup

The system and client for which these users must be provided depends on the respective use cases to be activated. For the transports-based use cases, you will find a template in Table 4.7. Both users with the respective authorization role must be provided for the relevant system and client.

Set up SAP Cloud ALM API

As a prerequisite, the SAP Cloud ALM service key needs to be available. In Chapter 3, Section 3.4, you will find the setup. Open the Landscape Management app and click the **Configuration** button to ensure the availability of the SAP Cloud ALM service key. Use the **Download Service Key** action to save the JSON file (see Figure 4.6).

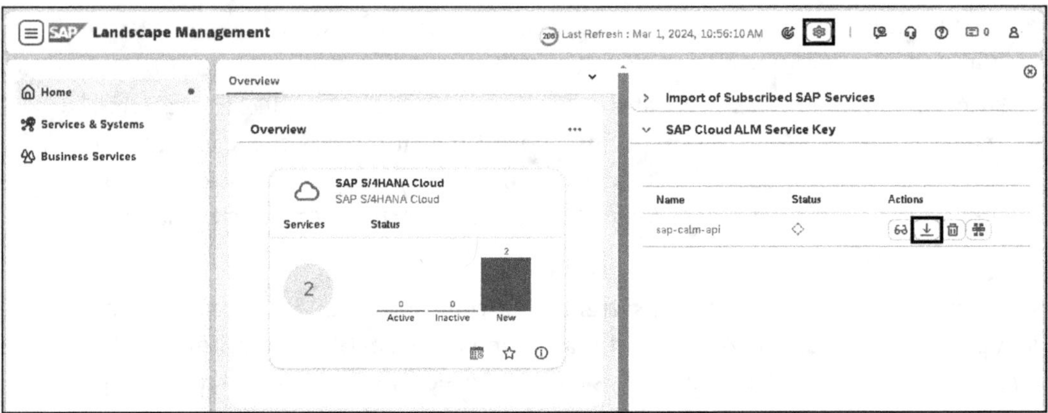

Figure 4.6 Download SAP Cloud ALM Service Key

4.2.2 Setting up Transport Management for SAP S/4HANA

If the prerequisites described above are met, you can start with the setup to integrate the transport management of the managed system.

There are currently four use cases available for the integration of transport management from SAP S/4HANA (on-premise) in SAP Cloud ALM, which must be activated in the respective system clients. To achieve full integration of transport management, you must activate the use cases listed in Table 4.7 to activate the respective use cases. For more information on the reference landscape used here, see Chapter 3, Section 3.1.2.

System and Client/ Use Case	DS4 000	DS4 100	QS4 000	QS4 100	QS4 140	PS4 000	PS4 100
Create and export transports and transport of copy		X					
Import transports and transport of copy			X			X	
Read transportation landscape	X						
Read transports	X						

Table 4.7 Activation Use Case per System and Client

The transport control-related use cases are as follows:

- **Read and export transports for each client**
  This use case is responsible for creating and releasing the transport request in the respective development client. In the example shown in Table 4.7, this is the development system DS4 with client 100. The transport request is created in SAP Cloud ALM using the feature shown as an example in Chapter 17, Section 17.3.1 and Chapter 17, Section 17.3.3.
- **Import transports**
  With the help of this use case, transport requests can be imported into the QS4 test system and the PS4 production system. The job `/SDF/CALM_CDM_IMPORT_TRANSPORTS` is scheduled in the target system for the import. The corresponding feature activity is described as an example in Chapter 17, Section 17.3.5 and Chapter 17, Section 17.3.7.
- **Read landscape**
  Here, information is transferred from the transport landscape to SAP Cloud ALM. The exemplary activation takes place in the DS4 system in client 000. As changes rarely occur here, the period for data transfer to SAP Cloud ALM can also be set higher.
- **Read transports**
  This use case is also activated in client 000 of the DS4 development system. It allows you to add existing transport requests to the feature regardless of their status.

Carrying Out the Registration of System and Client

The registration of the on-premise ABAP systems with clients in SAP Cloud ALM and the activation of the use cases for transport management are carried out using Transaction /SDF/ALM_SETUP (Set Up Integration with SAP Cloud ALM). As already mentioned in Table 4.7, this must be run in the corresponding system client.

For example, the registration for the DS4 development system for client 100 is described as “next.” To do this, call up the setup user (see Section 4.2.1) and then call up

Transaction /SDF/ALM_SETUP in the development client. If this setup has never been carried out before, the configuration view looks as shown in Figure 4.7.

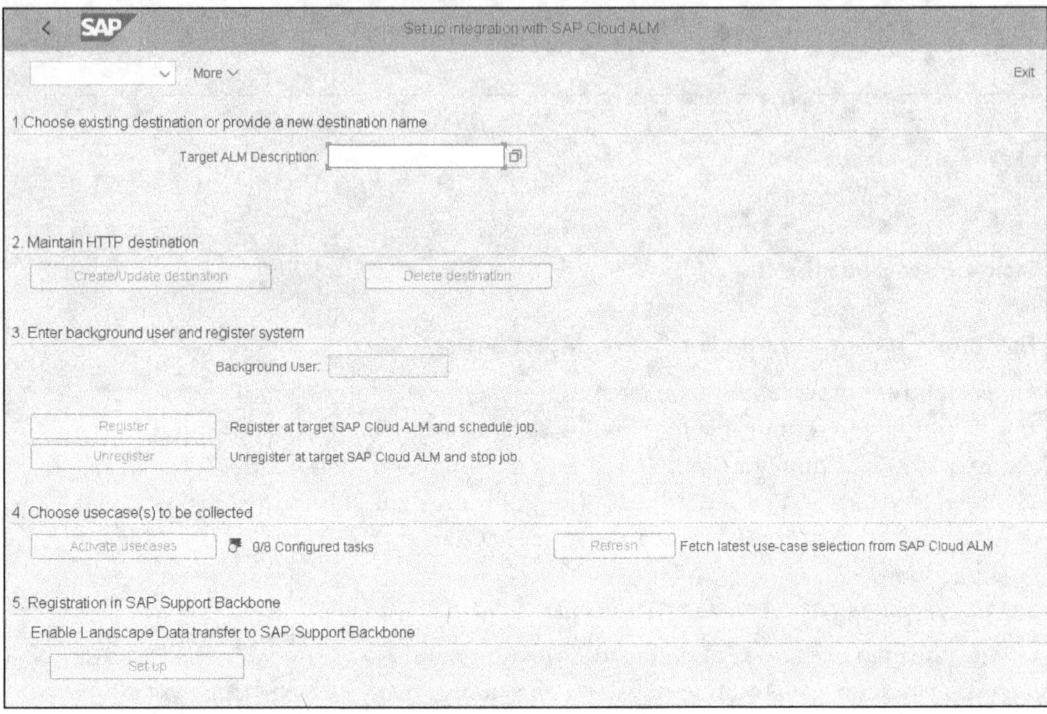

Figure 4.7 Report Setup Integration with SAP Cloud ALM

The setup is divided into four areas, which we will explain in the following. The connection to SAP Cloud ALM is configured in areas 1 and 2: Enter a name for the SAP Cloud ALM connection in the **Target ALM Description** field. After entering the name, the **Create/Update destination** button becomes active in area 2 and can be clicked (see Figure 4.8). The connection data is stored in the popup window **Modify HTTP destination for getting OAuth token**. To do this, click the button **Paste Service Keys and** copy the service key into the text field of the **Paste Service Keys JSON** popup window, described in Section 4.2.1.

Confirm the connection data with the **Next** button. The HTTP connection with the name ZSDF_<name ALM connection> is created for the target SAP Cloud ALM entered in area 1.

Enter the name of the background user in the third section of the configuration (see ❶ in Figure 4.9). You can find out which authorizations this background user requires in Section 4.2.1. The registration of the system is initiated with the **Register** button ❷. This schedules the background job CALM Auto Discovery CLOUD_ALM, which registers the system in the Landscape Management app.

In section 3, you also have the option of canceling the registration of the system and stopping the job. To do this, use the **Unregister** button.

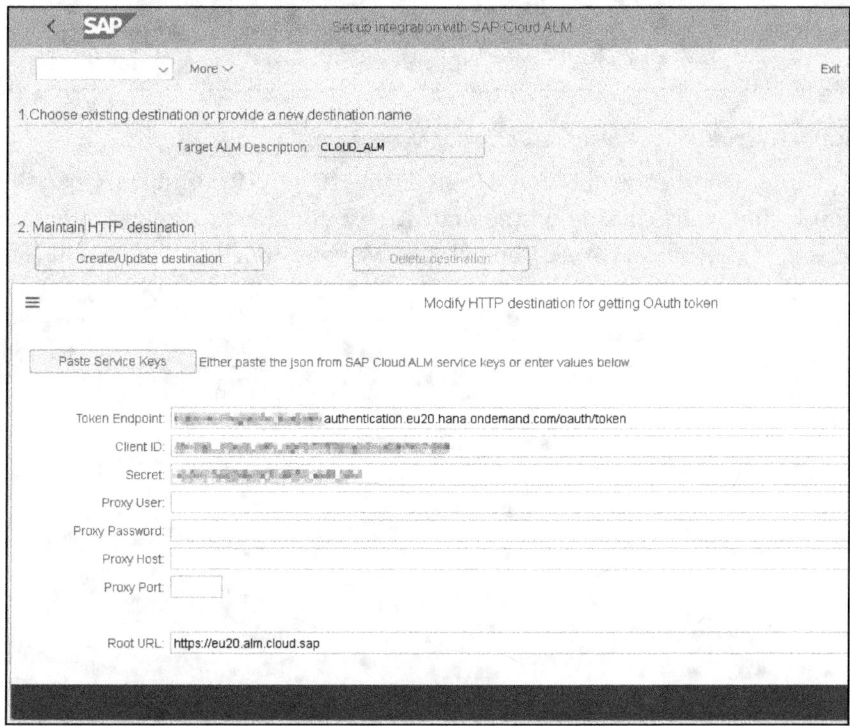

Figure 4.8 Maintain Authentication Data

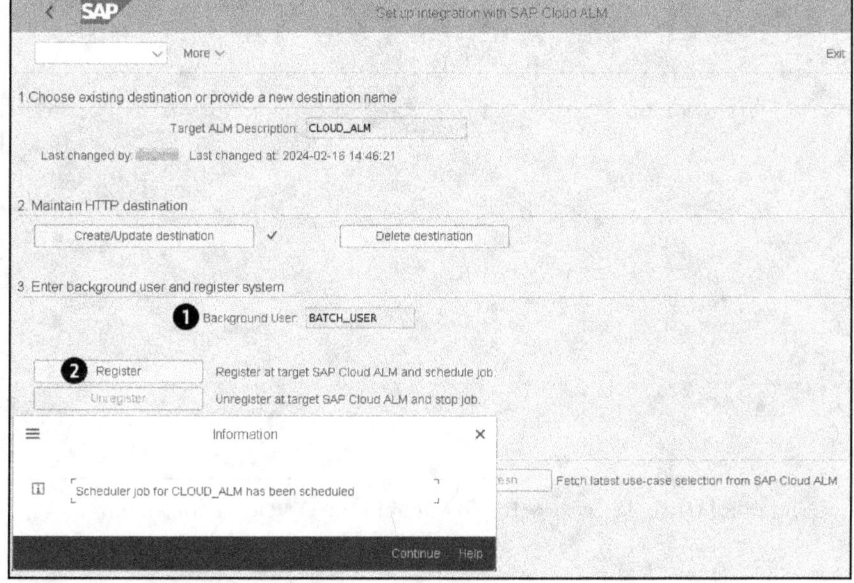

Figure 4.9 Maintain Background User and Perform Registration

[»]

Initial Registration of Systems and Clients

The initial registration (sections 1 to 3) of system DS4 and client 100 needs to be performed for each system and client that should be managed by SAP Cloud ALM.

Carrying Out Activation of the Use Cases

The use case for transport control is now selected and activated in section 4. Open the **Available tasks** window by clicking on the **Activate usecases** button (see ❶ in Figure 4.10) and select the checkbox **Feature Deployment: Manage Transports per Client** (see ❷ in Figure 4.10).

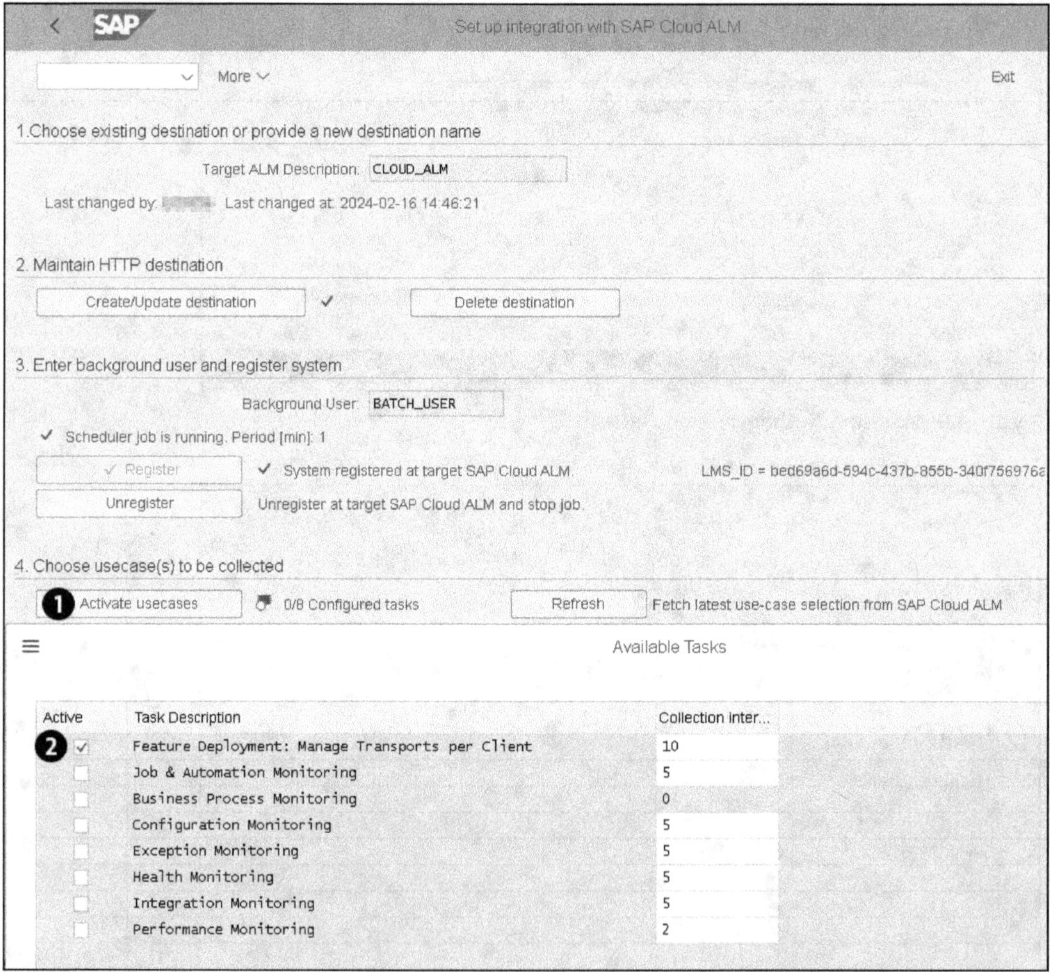

Figure 4.10 Activation of Use Case Feature Deployment in the Development System DS4 Client 100

In the **Collection interval** column, you define how often the data is transferred to SAP Cloud ALM or how often a job is scheduled for transport control. The time is specified in minutes and can be adjusted at any time. For the use case activated in the illustration, you see that 10 minutes is specified. The activated use case is used to create and release the transport request. If these transport controls are to be carried out in the managed system at shorter intervals, the collection interval must be reduced accordingly.

Click the **Next** button to confirm your selection, and the use case will then be activated. This completes the setup for the DS4 client 100 development system.

[«]

Tips and Troubleshooting

You can troubleshoot transport controls that aren't working properly at *https://help.sap.com/docs/cloud-alm/setup-administration/issues-and-solutions-on-premise?locale=en-US*. For managed systems, you can use the report /SDF/CALM_CDM_CONN_DIAGNOSTIC to check which use cases are active in the managed system and whether there are any problems with data transfer to SAP Cloud ALM. For more information, check out *https://help.sap.com/docs/cloud-alm/setup-administration/solutions-setup-check?locale=en-US*.

Repeat this setup for the development client DS4 000 as well as the clients for the test system and for the production system. Please note that certain use cases must be selected here. You can find out what they are by referring to Table 4.7.

4.3 Adaptation Transport Organizer

Transport tool integration using the adaptation transport organizer (ATO) allows you to perform transport management in SAP Cloud ALM for changes in SAP S/4HANA Cloud Public Edition. This integration enables exported changes from a development tenant to be include in the feature workflow of SAP Cloud ALM. The changes are imported into the subsequent tenants directly by the transport control within SAP S/4HANA Cloud. In addition to the change documentation, the linked feature can also be used to monitor the import status. To do this, use the Feature Traceability app, which can be opened via the Analytics app. The deployment status then provides you with a color-coded display of the tenants into which the change has already been imported. Another integration between the feature and the change in SAP S/4HANA Cloud is the controllability of the quality status. The change can only be imported into the production tenant once the quality status has been set to **Approved** (see Figure 4.11).

Setting the quality status to **Approved** is controlled by the feature as soon as it has been set to **Ready for Production** via the Feature app (see Figure 4.12).

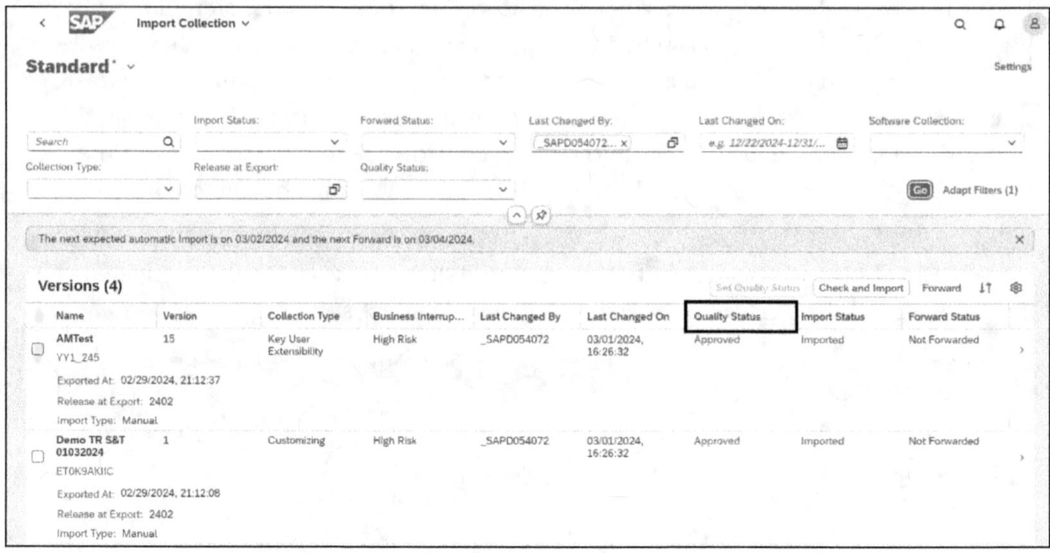

Figure 4.11 SAP S/4HANA Cloud Change with Quality Status Approved

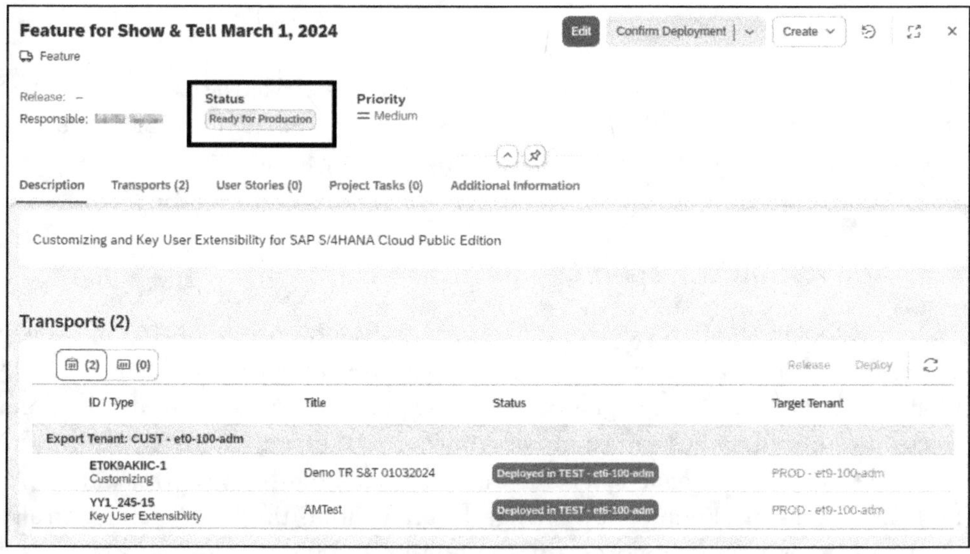

Figure 4.12 Feature with Status "Ready for Production"

If you want to use the integration shown here in your implementation project for SAP S/4HANA Cloud Public Edition, certain configuration steps are required. Step 1 is the creation of a communication system for your SAP Cloud ALM tenant within each SAP S/4HANA Cloud tenant. The required parameter values can be found in the JSON file created in Chapter 3, Section 3.4. In the second step, create the communication agreement based on the SAP_COM_0690 scenario and activate the required outgoing services that transmit the transport status information to SAP Cloud ALM:

- Deployment management export
- Deployment management import
- Deployment management retry
- Deployment management status change

For detailed and up-to-date information on this setup, follow the instructions on the SAP Help Portal page: *https://help.sap.com/docs/cloud-alm/setup-administration/adaptation-transport-organizer*

4.4 Summary

Setting up the transport tools for the underlying project landscape is an important prerequisite for the integrated requirement-to-deploy approach. Without this function, the requirements cannot be rolled out to the systems during the realize phase and the deploy phase in a process-supported manner. In this chapter, we explained in detail how to set up transport management for a system landscape and a service landscape using our example scenario. In addition to managing the project landscape and providing the transport tools, implementation using the agile methodology is of central importance. For this reason, in the next chapter we will go into detail about the agile methodology using Scrum and explain how it is supported within SAP Cloud ALM. We will also provide you with detailed recommendations regarding the specifications and conventions for the project setup based on our example scenario.

Chapter 5
Project Approach with Agile Methodology

5

In this chapter, we give you an introduction to the basics of agile project management and explain the extent to which SAP Cloud ALM supports the Scrum method. Furthermore, we present the most important project approaches in SAP implementation projects and based on them, explain our recommendations for the project setup.

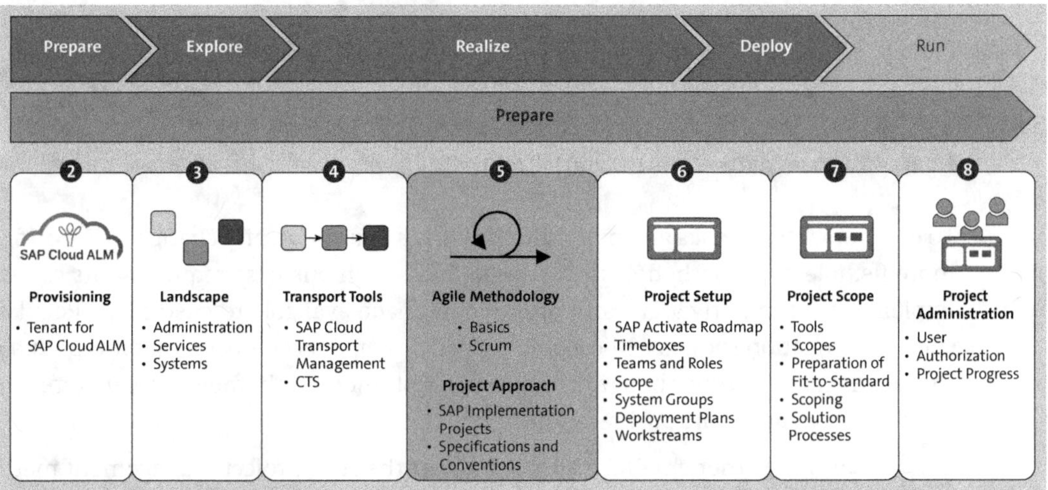

The last two chapters dealt with setting up the project landscape and connecting the systems to the SAP Cloud ALM transport management system (TMS). Following these technical topics, we will now turn our attention to the organizational methodology. It is important to familiarize yourself with the basics of agile methodology in advance so you can establish the project culture in the project team before starting the implementation. We will also look at the various characteristics of SAP implementation projects. Building on this, we use our example scenario to present our recommendations on requirements and conventions for the subsequent project setup.

IT projects can generally be implemented in three different ways: using a classic waterfall approach, an agile approach, or a mix of both methods (hybrid approach). The waterfall-based approach (see Figure 5.1) has long since faded into the background. The extensive, consecutive phases from analysis to implementation prevent business requirements from being tested and approved by stakeholders in a timely manner. The greatest danger is that conceptual errors are carried through to the implementation phase. In the test phase, the undesirable developments then become visible for the first time and have to be rectified through costly corrective measures in the form of change requests. As a result, the project budget is exceeded, and the milestones defined in the project plan can no longer be met. In some cases, the requirements are already outdated at this point and do no longer correspond to the intended business benefit.

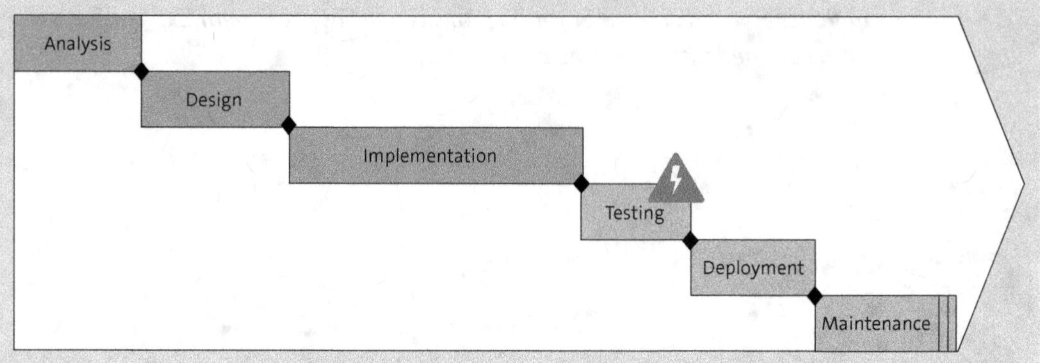

Figure 5.1 Phases of the Classic Waterfall Methodology

To counteract these risks and meet the challenges of a constantly changing IT world, a more flexible *agile* methodology is required, in which business requirements can be implemented promptly and transparently and made available to customers. Regular, short coordination rounds without overloaded design phases ensure that it is possible to react flexibly to scope changes and specific customer requirements to generate tangible added value for the business.

In this chapter, we therefore take a closer look at the agile project management methodology. In addition to a general introduction, we examine the functional possibilities of SAP Cloud ALM. We provide you with an outlook and recommendations on how theory and practice can be harmonized to ensure sophisticated application lifecycle management. We will take a closer look at the following points in particular:

- Basics of agile methodology with Scrum
- Roles and responsibilities in the project approach
- Support for project planning by defining time frames (timeboxing)
- Creation of requirements, tasks, and features
- Implementation of sprints and releases
- Guidelines and conventions for working methods
- Project orientation and choice of project approach in the SAP environment

5.1 Basics of Agile Methodology with Scrum

To establish a uniform understanding of the agile approach in the company, project management according to *Scrum* is a good choice. As SAP Cloud ALM also uses Scrum functions and terminology, we will provide you with the central principles of this methodology in relation to SAP implementation projects using SAP Cloud ALM.

As an agile project management method, Scrum comprises, as shown in Table 5.1, three roles (responsibilities), five events, and three artifacts, which are described in more detail in the rest of this chapter using a use case.

Roles	Events	Artifacts
▪ Product owner ▪ Scrum master ▪ Development team	▪ Sprint planning ▪ Daily Scrum ▪ Sprint review ▪ Sprint retrospective ▪ Product backlog refinement	▪ Product backlog ▪ Sprint backlog ▪ Product increment

Table 5.1 Roles, Events, and Artifacts in Scrum

5.1.1 Before the Start

Before a company decides to implement an agile project using the Scrum model, the entire team involved must be trained in advance. To this end, it is advisable to conduct basic Scrum training for all participants to sharpen their general understanding and clarify the specific terminology. It is also extremely important to jointly define the right Scrum setup for the upcoming project; i.e., to derive your own good practices from best practices. It does not always make sense to proceed strictly according to the textbook. Your own good practices include criteria for the general Scrum processes such as the sprint length, the duration of the various Scrum events or the frequency of the daily Scrum—a daily does not always have to be daily, for example. You should also define rules regarding organizational topics, such as the tools used, the storage locations of documentation, or the criteria for the definition of ready and the definition of done. With the *definition of ready*, you formalize criteria that must be fulfilled before a requirement can be handed over to development. The *definition of done*, in turn, defines criteria and activities that are important for the acceptance of implemented requirements. These criteria help to establish a standardized procedure and create a common understanding between requesters and developers.

Scrum projects are carried out cyclically in *sprints*. The implementation is worked on iteratively over a period of two to four weeks. The aim is to achieve a result, the *product increment*, at the end of each sprint.

5.1.2 Stakeholders and Requirements

In general, every project begins with the formulation of requirements raised from different areas of the company or from stakeholders that should bring business value to the company. The stakeholders thus represent the customers and initiators of the project. Organizationally, they are viewed as a role outside of the Scrum method. Requirements can also be described as a *vision* in a more generic context. These include, for example, the improvement of an existing solution, the introduction of new products, or the focus on new business areas (see ❶ in Figure 5.2).

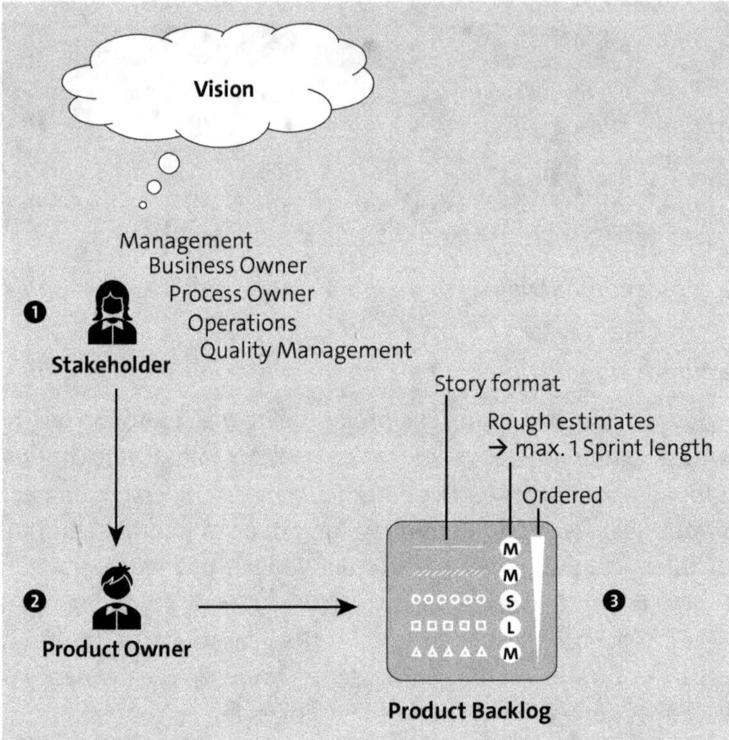

Figure 5.2 Scrum Method: Stakeholders, Product Owner, and Product Backlog

There are different groups of stakeholders:

- **Management**
  The management of a company sets the direction for the introduction of new products, systems, or applications that are needed to implement management strategies and achieve competitive advantages. For example, a cloud-first strategy can lead to an extensive transformation project.
- **Process owner**
  Process owners are responsible for certain business processes in the company and are responsible for their control, planning, and continuous optimization. This can include both organizational and functional improvements.

- **Business users**
  Users in the company who carry out various activities in business processes as part of their daily work should also have the opportunity to submit suggestions for improvements or new requirements for IT systems or applications. Usually, these requirements serve as input for projects, but they should be analyzed, evaluated, refined, and documented in advance with the help of a separate requirements engineering process.
- **Operation (monitoring)**
  New requirements can also be derived from the monitoring of business processes, applications, or systems that might contribute to the stabilization or improvement of performance.
- **Quality management**
  Quality management can also set requirements that contribute to the implementation of measures for planning, controlling, and optimizing processes.
- **Governance**
  Corporate governance refers to the legal and factual regulatory framework for the management and supervision of a company. To guarantee this and to be able to react to internal and external changes, new requirements may also have to be placed on IT systems or applications.

5.1.3 Product Owner and Product Backlog

The requirements of the stakeholders are collected by the *product owner*. The product owner is authorized to make decisions on their behalf regarding implementation to maximize the benefits of the product. The results are approved or rejected by the product owner at the end of each sprint (❷ in Figure 5.2).

The product owner enters all requirements into the *product backlog* and writes them in *story format*, specifying the role, function, and benefit. The product backlog thus represents a list of the desired requirements that have been defined in the form of *user stories*. A rough estimate using the specification of *story points* helps to determine the complexity in advance. Stories should be designed in such a way that they can be implemented within a sprint. The product owner prioritizes, refines, and updates the product backlog regularly. Criteria such as benefit, risk, costs, and dependencies play a key role here. The stories in the product backlog are sorted so that the top elements can be considered for implementation in the next sprint (❸ in Figure 5.2).

5.1.4 Sprint Planning and Sprint Backlog

During the sprint execution, the development team carries out all the necessary work to implement the requirements or stories. The team consists of a maximum of nine people and should be made up of all the necessary experts required for product development—i.e., developers, consultants, administrators, analysts (❹ in Figure 5.3).

It is the task of the *Scrum master* to coach and enable the development team to focus on creating the product increment. Disruptive obstacles are registered in an *impediments backlog*. The Scrum master initiates appropriate measures to remove obstacles. In addition, the Scrum master moderates all Scrum events and meetings and ensures that all participants follow the defined Scrum rules (❺ in Figure 5.3).

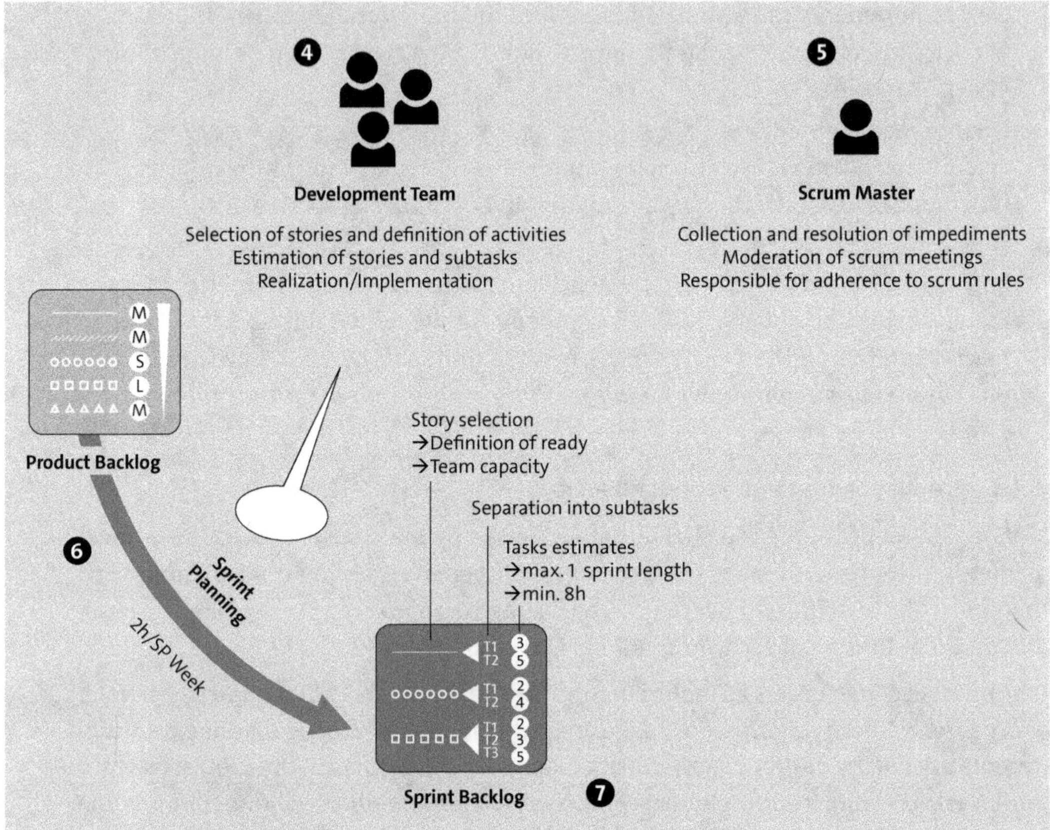

Figure 5.3 Scrum Method: Sprint Planning and Sprint Backlog

The Scrum cycle begins with the *sprint planning* (❻ in Figure 5.3). This is the first Scrum event and usually lasts two hours per sprint week (e.g., 4 hours for a sprint cycle of two weeks). In this meeting, the product owner presents their preselection of stories from the product backlog. For the development team to efficiently record and evaluate requirements and content, the stories should fulfill the *definition of ready* criteria—i.e., corresponding criteria defined in advance that characterize a story as ready for implementation.

The development team comments on the feasibility and reprioritizes the entries if necessary. The team also takes into account its own capacity in terms of workload and availability of team colleagues within the upcoming sprint. It is advisable to divide extensive stories into individual *subtasks*. This not only sharpens the understanding of which activities and efforts are necessary for realization; it also enables better tracking of activities and the degree of completion. The stories eligible for the sprint are summarized in the *sprint backlog* (❼ in Figure 5.3).

5.1.5 Sprint Execution and Daily Scrum

Immediately after sprint planning, the sprint is started, and the execution phase with a defined *sprint length* begins (❽ in Figure 5.4).

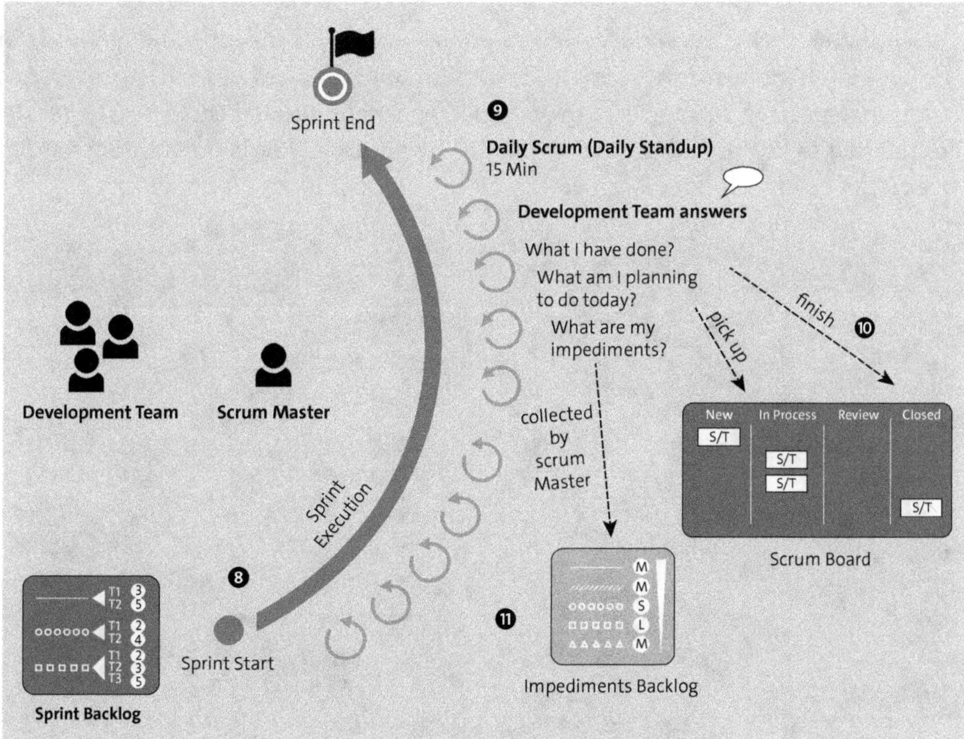

Figure 5.4 Scrum Method: Sprint Execution and Daily Scrum

In the form of short daily meetings lasting 15 minutes, the *daily Scrum* or *daily standup*, the completion status of the sprint backlog is discussed. The members of the development team answer the following questions one after the other:

1. What have I finished?
2. What do I want to do today?
3. What are my obstacles?

The development team is self-organized, and the members themselves decide which stories or subtasks should be processed next (❾ in Figure 5.4).

For simple control within the daily Scrum, it makes sense to use a graphical *Scrum board*. This allows the individual stories or subtasks to be discussed efficiently and moved back and forth between the status values (board columns). Furthermore, the board enables the Scrum master and the product owner to keep an eye on the execution status and assess whether the sprint goal can be achieved (❿ in Figure 5.4).

Obstacles are registered by the Scrum master in the impediments backlog. This keeps them transparent and allows the actions taken to eliminate them to be easily monitored (⓫ in Figure 5.4).

5.1.6 Sprint End: Sprint Review, Product Increment, and Sprint Retrospective

The sprint ends with the *sprint review meeting*. During this meeting, the members of the development team present the results in *show-and-tell sessions* to everyone involved, especially, of course, to the product owner and the stakeholders. The results are referred to as the *product increment*. The event usually lasts one hour per sprint week (⓬ in Figure 5.5).

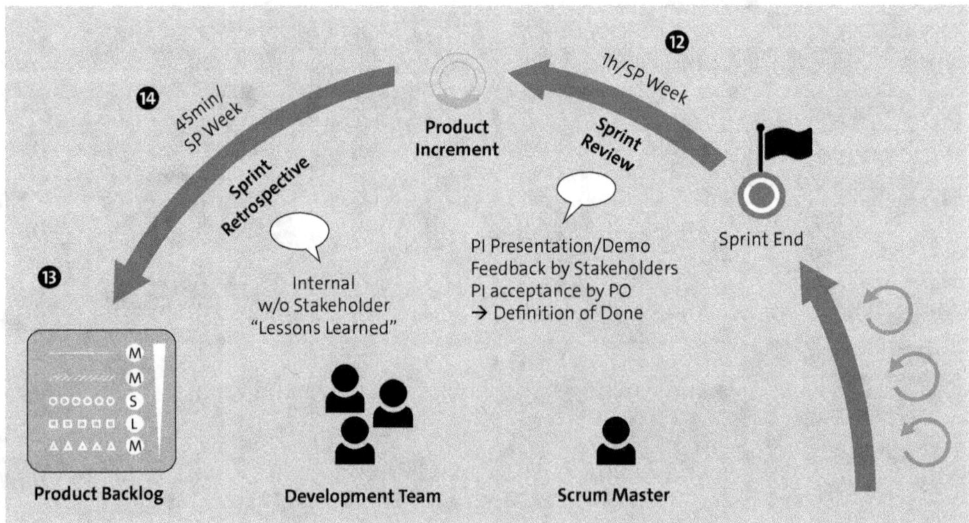

Figure 5.5 Scrum Method: Sprint Review, Product Increment, and Sprint Retrospective

The product owner is responsible for accepting or rejecting the created product increment according to the defined definition of done criteria (DoD for short). DoD is a kind of checklist that must be fulfilled for each story before a product is considered ready for release or finished. This can be, for example, documentation created, a test script, or successful testing on a quality assurance system. Unfinished or unaccepted stories are transferred back to the product backlog or passed directly to the next sprint (⓭ in Figure 5.5).

The final meeting of each sprint is for the *sprint retrospective*. In this internal meeting, the focus is on general collaboration between the Scrum members. It usually lasts 45 minutes per sprint week. The following questions can be dealt with in this meeting (⓮ in Figure 5.5):

- How do we work together?
- Are there conflicts or communication problems?
- What are the lessons learned?
- What can we improve in our Scrum process?
- Do we need to adapt the Scrum approach (e.g. number or duration of daily Scrums)?

5.1.7 Product Backlog Refinement

In an agile approach, new requirements can emerge at any time to further refine current product increments or set focus on new functions (⓯ in Figure 5.6).

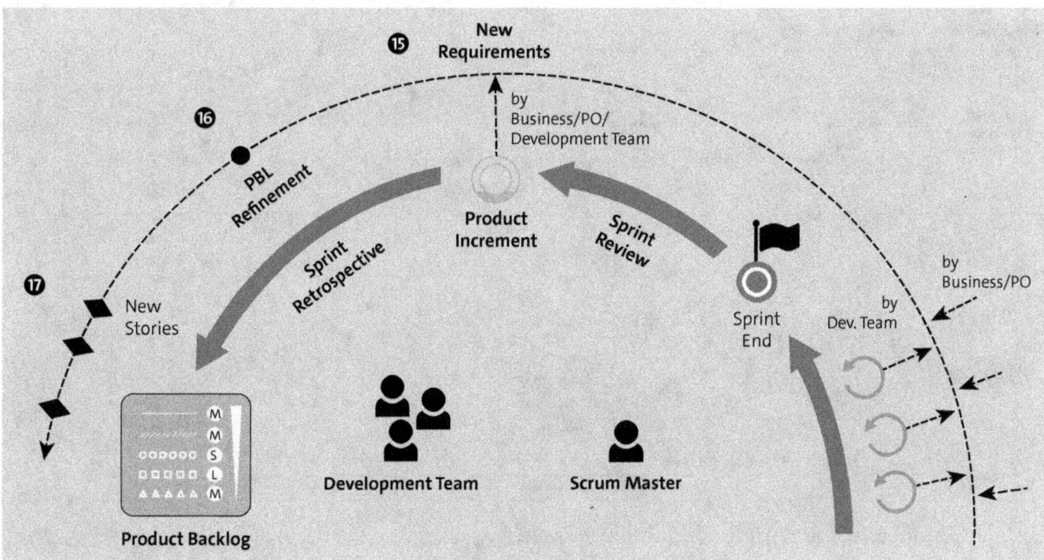

Figure 5.6 Scrum Method: Product Backlog Refinement

This is done in two ways, as follow:

- By the development team itself, during the implementation phase
- By the stakeholders involved and the product owner, usually during the show-and-tell sessions

It is important that new requirements are first entered into the product backlog as new stories. It is not permitted to provide these directly to the development team during the sprint execution; otherwise the entire sprint planning would become useless. The development team itself decides whether there is still free capacity to implement further stories from the product backlog. If necessary, it is advisable to schedule an additional meeting to discuss new requirements together and to clarify content and questions. This meeting is called the *product backlog refinement* (⓰ in Figure 5.6). Finally, new requirements are included in the product backlog as user stories and can be implemented in one of the next sprints (⓱ in Figure 5.6).

In Figure 5.7, you can see the entire Scrum cycle.

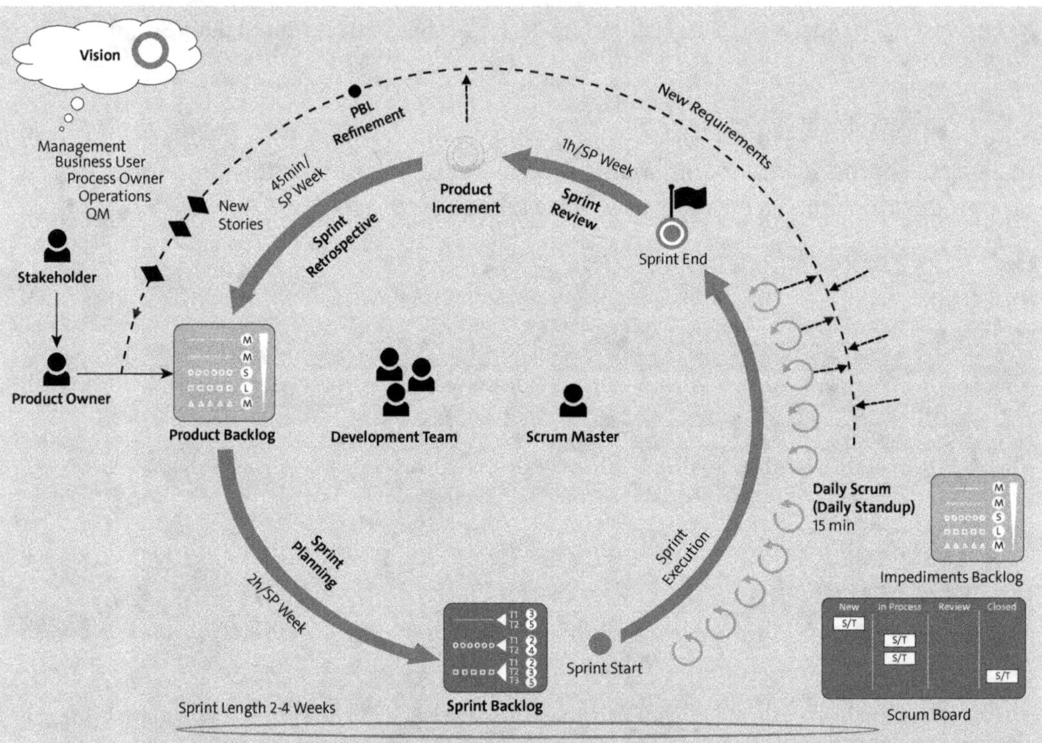

Figure 5.7 Scrum Method: Complete Cycle

5.2 Support for Scrum in SAP Cloud ALM

In the previous section, we looked at the central principles of the Scrum method. Building on this, we will now explain the functions provided within SAP Cloud ALM to support agile project management with reference to Scrum.

Table 5.2 compares the various Scrum elements with the corresponding functions in SAP Cloud ALM.

Scrum Elements	SAP Cloud ALM
Scrum roles: ■ Product owner ■ Scrum master ■ Development team	■ Scrum roles are not available as standard, but can be created manually ■ Roles can be activated for your own Scrum team ■ Users can be assigned to the roles of a team
Stakeholder requirement	■ Requirements can be created manually or from business processes or solution processes ■ Requirements can be created out of predefined templates
Story	■ User stories can be created manually or from existing requirements ■ User stories can be provided with story points
Subtask	■ User stories can be divided into subtasks.
Sprint	■ Sprints can be created as timebox elements ■ The start and end date must be entered manually
Sprint planning	■ Timebox elements (sprints) can be assigned to user stories ■ Sprint end is noted in the user story ■ User stories can be divided into subtasks
Daily Scrum	■ Task list with filtering or grouping according to the current sprint ■ Status switching of user stories directly in the task list ■ Status switching of user stories by using the included scrum board (cards view) ■ Burn-up diagram for monitoring the trends of stories ■ Progress analysis by means of user story traceability ■ Progress analysis of sprints using project progress reporting
Sprint review	■ Traceability analysis and analysis of the test execution of features, requirements, and user stories ■ Task list or scrum board (cards view) with filtering or grouping according to the performed sprint ■ If necessary, visualization of the event as a milestone in the project overview
Sprint retrospective	■ Creation of impediments as a project task ■ Tracking of impediments using a task list or cards view and filtering by project task type ■ If necessary, visualization of the event as a milestone in the project overview

Table 5.2 Scrum Functions in SAP Cloud ALM

Scrum Elements	SAP Cloud ALM
Product backlog refinement	■ No functional support available
Product backlog	■ Task list with filtering according to the prepare phase ■ Rough sorting according to priority ■ Display of story points and effort
Sprint backlog	■ Task list or scrum board (cards view) with filtering or grouping according to the current sprint ■ Progress analysis by means of user story traceability ■ Progress analysis of sprints using project progress reporting
Product increment	■ Creation of features ■ Assignment of transports to features ■ Assignment of features to user stories and requirements

Table 5.2 Scrum Functions in SAP Cloud ALM (Cont.)

5.2.1 Agile Approach in the SAP Activate Method

In the SAP Cloud ALM project settings, you can assign the SAP Activate roadmap for your project in the Projects and Setup app under the **General Information** tab. SAP Activate sometimes provides a set of various *roadmap tasks* that are assigned to the various phases of the method. The tasks contain valuable information on technical and organizational activities that play a role in the project plan. We recommend that you familiarize yourself with the roadmap tasks provided. Decide for yourself which of the tasks are of interest to you and could be used iteratively in the course of your project.

In Table 5.3, we have summarized the provided SAP Activate roadmap tasks for you with reference to the agile project management methodology in the various phases.

SAP Activate Phase	Roadmap Tasks with an Agile Context
Prepare	■ Prepare tools and methodology for project support: SAP Solution Manager, SAP Cloud ALM, and SAP Activate ■ Set stakeholder expectations for agile project ■ Define project standards
Explore	■ Review and complete requirements ■ Update and prioritize backlog ■ Estimate backlog items effort ■ Conduct scrum meetings ■ Conduct release and sprint planning meeting ■ Validate release and sprint plan against scope of work (SoW)

Table 5.3 SAP Activate Roadmap Tasks with Agile Context

SAP Activate Phase	Roadmap Tasks with an Agile Context
Realize	■ Confirm scope and requirements ■ Plan and execute agile sprints ■ Plan sprints and solution reviews ■ Conduct sprint planning meeting (iterative) ■ Manage sprints ■ Conduct sprint review meeting ■ Signoff sprint results ■ Conduct sprint retrospective ■ Perform Scrum-of-Scrums meeting (iterative) ■ Communicate sprint scope and plan (iterative) ■ Communicate sprint result (iterative)
Deploy	■ Prepare product backlog for next release/sprint ■ Update the release and sprint plan
Run	■ Initiate a new innovation cycle

Table 5.3 SAP Activate Roadmap Tasks with Agile Context (Cont.)

Within the SAP Activate method, the agile approach mainly takes place in the realize phase, as shown in Figure 5.8. To start the agile development of the solution, defined requirements are needed, which were entered during the fit-to-standard workshops in the explore phase. All stakeholder requirements can also be referred to as a *requirements backlog*. Once they have been approved or confirmed, the corresponding user stories are created by the responsible product owner.

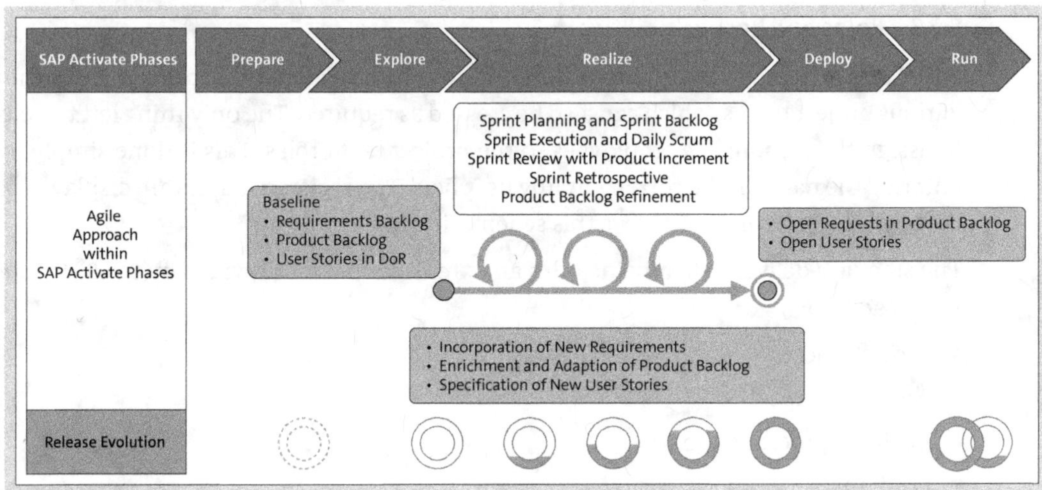

Figure 5.8 Agile Approach within the SAP Activate Phases

All user stories represent the product backlog. As soon as the first stories meet the definition of readiness criteria, you have created the *baseline*, or starting point, for the solution to be implemented. The realize phase is ready to begin. During the realization, new requirements can be identified and incorporated, which in turn can lead to an extension or adaptation of the product backlog.

Once the minimum requirements for the release have been implemented, the deploy phase begins, and with it the transfer to productive operation. At this point, there are usually still open requirements or stories. You will need to transfer them to the planning for the subsequent release and implement them there in a new realize phase.

5.2.2 Requirements, User Stories, and Features

New requirements are mainly created from the business context of a business process but can also be created without reference to it. These requirements can be of an organizational, technical, or functional nature. User stories are created from the requirements in an agile context. They can in turn be divided into subtasks and assigned to a feature (see Figure 5.9).

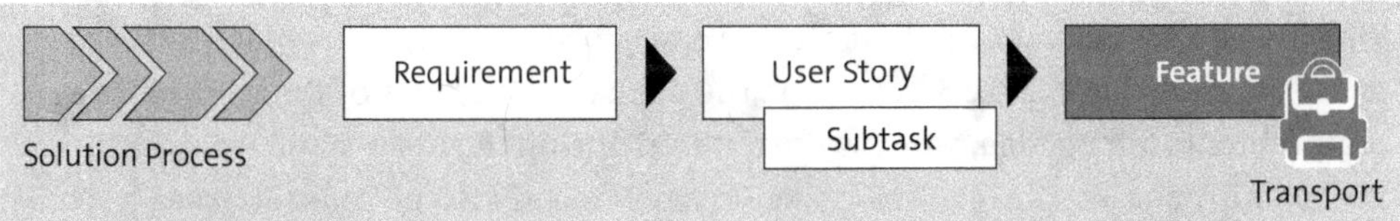

Figure 5.9 From the Requirement to the Feature

5.2.3 Roles and Responsibilities

SAP Cloud ALM offers the option of defining your own roles and activating them within various project teams, which can also be created as required. The only thing left to do is to assign the responsible employees to the project team roles. This is done simply by entering the name in the corresponding user field. The registered users are displayed as default values. Figure 5.10 shows this schematically.

The standard delivery already includes an extensive role set provided by the SAP Activate roadmap:

- Project lead
- Analytics expert
- Business process expert
- Configuration expert
- Data migration expert
- Education expert
- Enterprise architect

- Extensibility expert
- Integration implementation expert
- Organizational change management expert
- Output management and forms expert
- System administrator
- Technology expert
- Testing expert

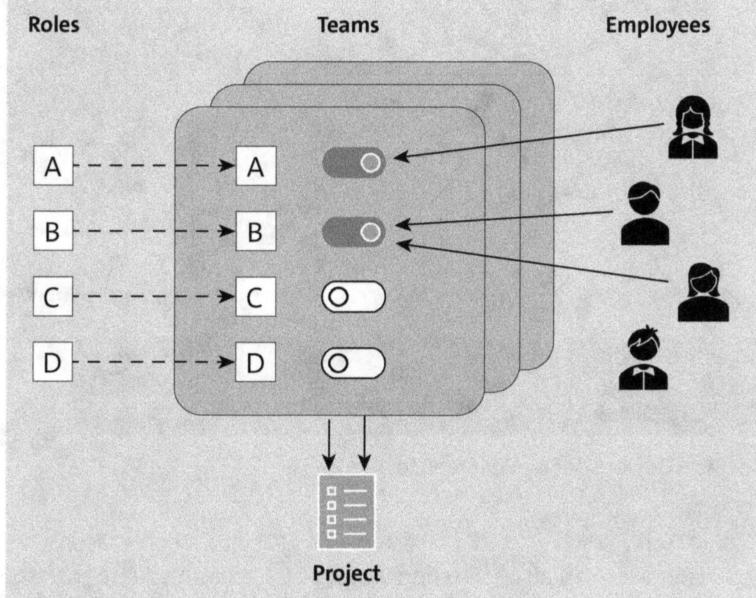

Figure 5.10 Team, Role, and Employee Assignments in SAP Cloud ALM

When a new project is created, these project roles are activated for the standard team "PMO team." It is the only team on a project containing the project lead role. Employees who are assigned to this role have special authorizations within the system. For example, they can create new projects, define timeboxes, create teams, and assign employees to them. The PMO team cannot be deleted and is marked with a special icon: .

If the project is to be carried out using the Scrum method, the roles for the product owner and scrum master must be added to the system. The colleagues in the development team can be mapped via an equivalent new role if you choose to use role names other than the ones supplied in advance.

5.2.4 Timeboxing

IT projects usually run within a time frame with a planned start and end. As already mentioned, projects can be managed using the three project methods (waterfall, agile, or hybrid). SAP Cloud ALM functionally supports the definition of phases, sprints, and milestones, which usually also play a significant role in SAP implementation projects. Within SAP Cloud ALM, these three elements are referred to as a *timebox*. A timebox can in turn be assigned to a task in the project (see Figure 5.11).

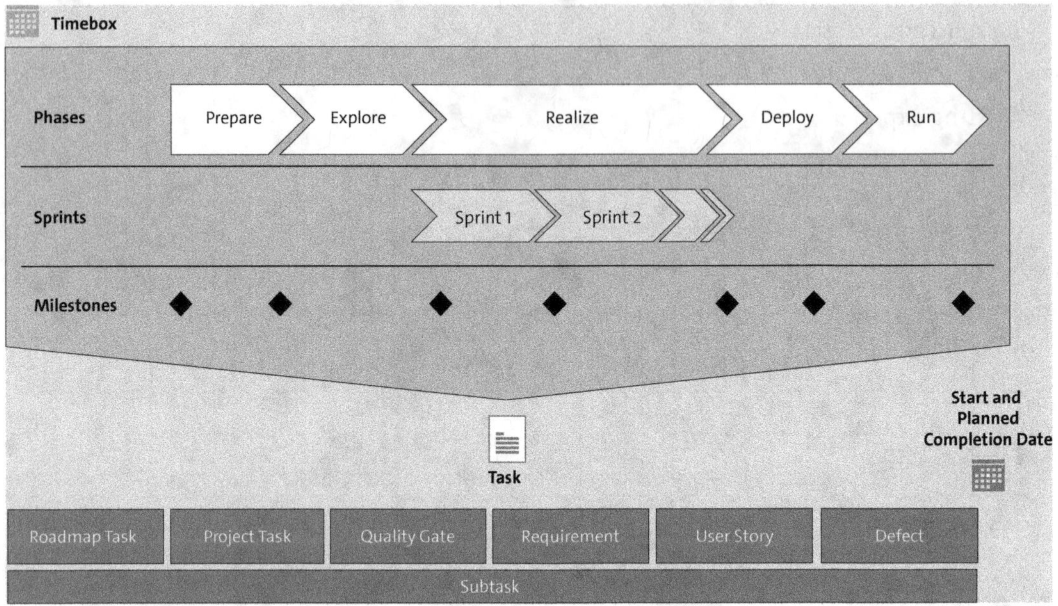

Figure 5.11 Timeboxing and Options for Assignment to a Task

If specified in the timebox, the start date and the planned end date are automatically referenced in the task. The tasks can also be assigned a type to accommodate the various project management methods.

Depending on the project requirements and associated working method, you can use all three timeboxing elements flexibly. In the pure waterfall approach, for example, the use of phases and milestones is essential, while the agile methodology requires the use of sprints. In the hybrid approach, all three elements are used.

The project timeboxes with the respective time frames can be maintained centrally within the project properties. Any changes made to the scheduling of phases, sprints, or milestones are automatically passed on to the referenced tasks.

The following timebox elements are already available in the standard settings of a newly created project without SAP Activate roadmap tasks:

- **Phases**
  - Prepare
  - Explore
  - Realize
  - Deploy
  - Run

- **Milestones**
  - Kickoff
  - Planned go-live
  - Actual go-live

In the project settings on the **Timeboxes** tab in the Projects and Setup app, you can deactivate all phases of a project in edit mode—except prepare—using a slider (see Figure 5.12). If required, you can also create your own phases here by specifying a name and a date range. The predefined milestones cannot be deactivated, but it doesn't matter as long as they are not specified by date and assigned to various tasks. In display mode, a counter behind the respective timebox element provides a quick overview of the number of all assigned tasks.

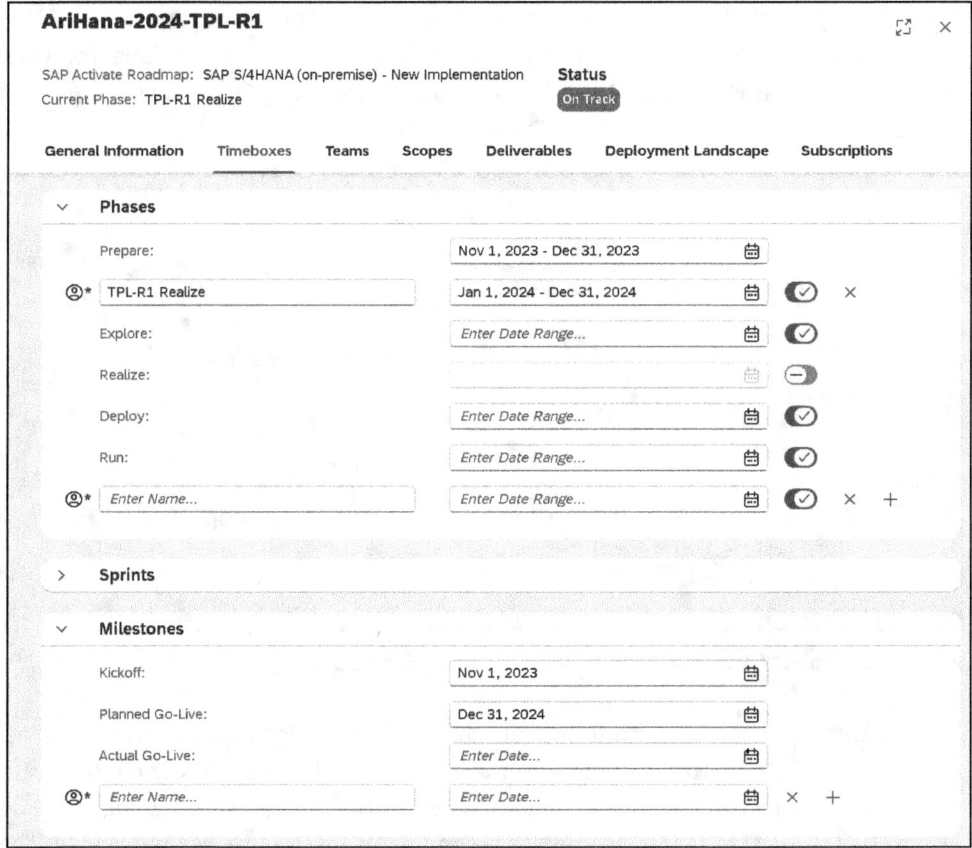

Figure 5.12 Project Settings: Edit Mode for Timeboxes

In SAP Cloud ALM, no sprints are preconfigured by default. You can freely define them using the **Create** button, specifying a sprint name and the date range. However, the date range must be entered manually for each sprint, taking into account the desired sprint length.

Roadmap Templates

SAP Cloud ALM contains various roadmap templates for the implementation of different SAP business products based on the SAP Activate method. Selecting a template in the general project settings provides product-specific tasks, but sometimes also a separate set of standard phases within the timeboxes. Further information can be found at *https://go.support.sap.com/roadmapviewer/*.

The defined phases and time specifications have a direct influence on the general project reporting and the tracking of project progress on the project overview dashboard.

The current project phase is displayed in the **Project** tile within the Overview app. If the phase exceeds the defined end, the days overdue are displayed as additional information. If various time settings have been made for the timeboxes, these appear in chronological order in the **Upcoming Events** tile (see Figure 5.13).

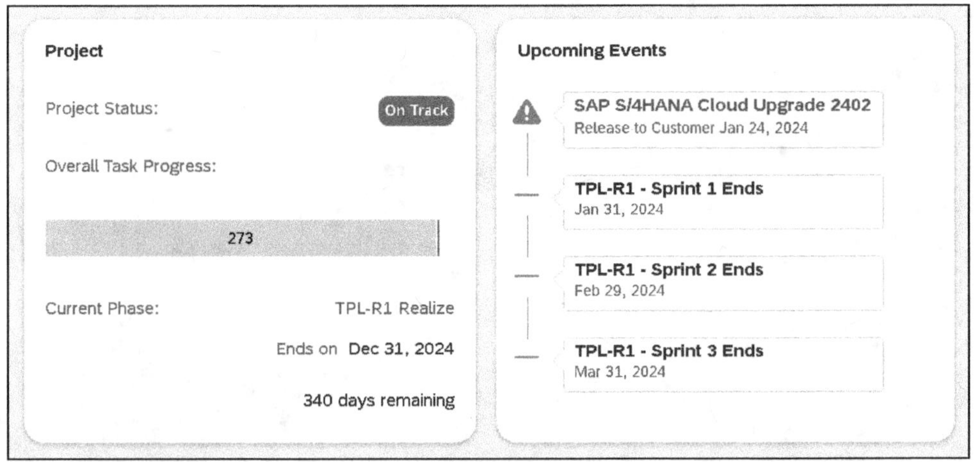

Figure 5.13 Project Overview: Project and Upcoming Events Tiles

The system assigns a sprint or milestone to the corresponding phase based on the specified completion date. This happens automatically when the task is assigned to a sprint or milestone. For a task without a timeboxing element, it chooses the fallback prepare phase. Therefore, this task represents an element in the requirements backlog. The product backlog reflects elements that have already been confirmed for implementation and are explicitly assigned to the realize phase.

The timebox element and the corresponding phase are listed as an attribute in the **Additional Information** section of a task (see Figure 5.14). Within the results list in the Tasks app, both values can be seen in the **Timebox** column attribute.

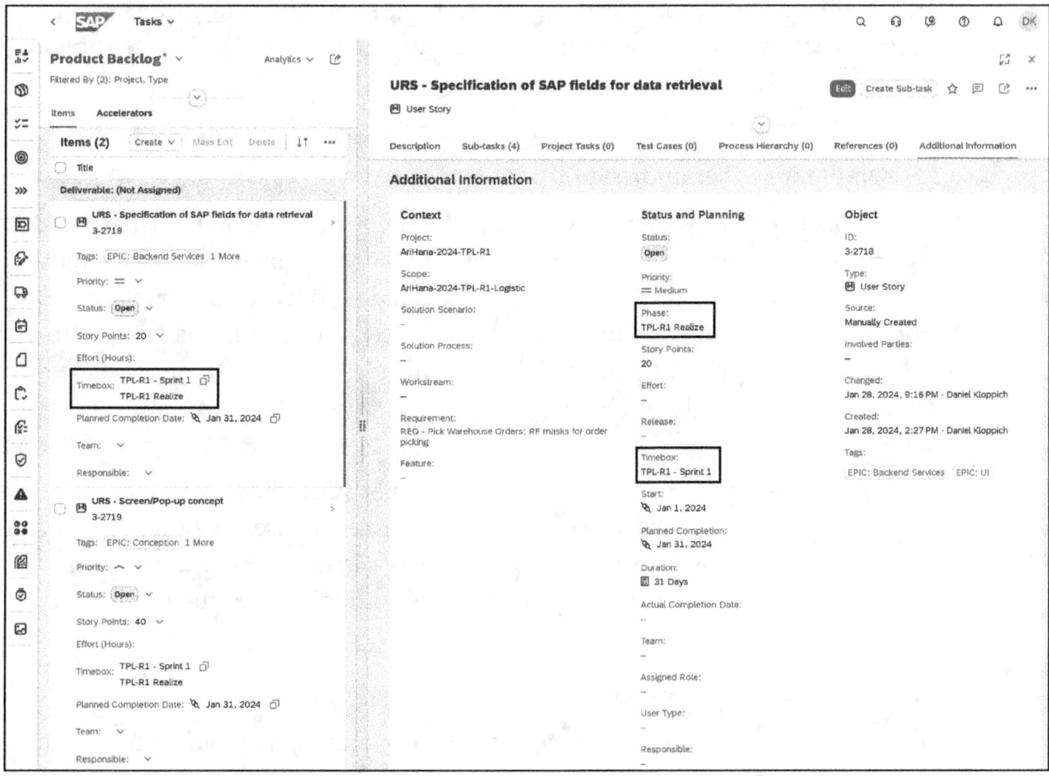

Figure 5.14 Task List and Task Form: Interaction of Timebox and Phase

Nevertheless, it is advisable to synchronize the sprint name with the name of the phases, especially when it comes to continuous development of a release approach. You can find more information on this in Section 5.4.

5.3 Project Approaches for SAP Implementation Projects

Which project approach do you choose—purely agile or hybrid? When choosing the right project approach for implementation projects in the SAP environment, the orientation of the future business solution, its further development, and the consideration of the associated landscapes play a central role. A purely agile approach supports the rapid delivery of functions, especially in the cloud environment. Within the SAP Activate methodology, the agile approach takes place in the realize phase. A hybrid approach combines the advantages of agile product development with those of the waterfall methodology, so sprint execution is based on sequential project phases. The hybrid approach is usually suitable for projects focusing on the delivery of major releases. The development phase is usually followed by a consolidated test phase, the delivery phase, and a maintenance phase. To a certain extent, all other phases of the

SAP Activate method can also be managed in an agile manner using sprints in a hybrid approach. There are various scenarios you can choose for your implementation project, which we will now present individually.

5.3.1 On-Premise Scenario/Hybrid Scenario

The *on-premise scenario* essentially deals with the introduction and further development of an on-premise product that is operated in the company's own data center, with a hosting partner or in a private cloud. If other cloud products from a public cloud are introduced in addition to the main product, this scenario can also be referred to as a *hybrid scenario*. A hybrid scenario therefore includes a mix of context-related on-premise and cloud solutions that map a common business context and implement end-to-end processes.

In an on-premise scenario, developments are usually bundled, delivered, and consolidated in the form of releases (minor/major). The project phases in an SAP Cloud ALM project can, for example, reflect the different release versions. A hybrid approach is particularly suitable here to ensure the release delivery after a successful test phase.

5.3.2 Cloud-Centric Scenario

The *cloud-centric scenario* is another approach. Here, only cloud products from a public cloud are introduced. In addition to your own individual adjustments to the solution, the specific delivery cycle of the cloud product also plays a key role. Features rolled out by the supplier are available immediately and can sometimes influence your own adaptions. In an ongoing implementation project, an ad hoc product update is all the more challenging. Here, the processes and functions under development of the initial product version must be manually compared with the changes provided by the subsequent version.

After the introduction or rollout of a cloud-centric solution, custom requirements are usually continuously developed and delivered iteratively in short cycles or sprints. This means that cloud-centric solutions are usually characterized by an agile project approach. Nevertheless, the product supplier's delivery cycle should be taken into account and coordinated with your own.

5.3.3 Single Deployment

With a *single deployment*, a specific customer solution (on-premise or cloud-centric) is implemented and rolled out. Additional functions, modules, requirements, etc. are implemented continuously after the initial deployment. In an on-premise landscape, the implementation of the new product release usually takes place in parallel with maintenance in a separate project landscape.

5.3.4 Template, Rollout, and Further Deployments

In a *template approach*, a global solution (on-premise or cloud-centric) is developed and then rolled out to various business units, such as branches, plants, or countries. In this way, the template provides processes and functions that are used in the individual business units (see Figure 5.15).

Local requirements are developed individually as localizations. They can be extensions to the template, for example, as results of necessary process variants, or they can be new developments. In principle, any changes to the template should be agreed on in advance with the unit responsible for the template. If possible, it makes sense to make all the changes available globally so that all company units can participate.

A template approach helps to establish standardized workflows, processes, and functions within all areas of the company. Similar requirements are channeled, orchestrated, and not developed redundantly. This allows development resources and expertise to be pooled more effectively, which ultimately also helps to reduce costs.

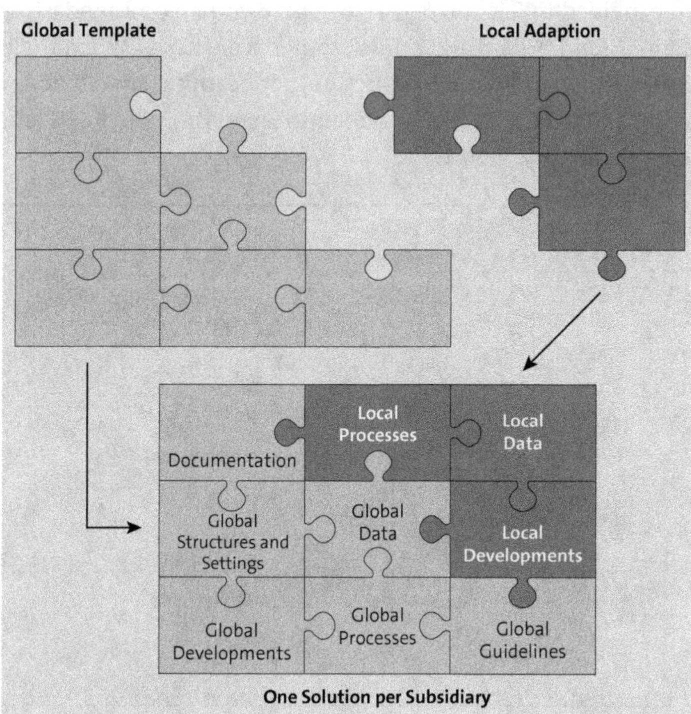

Figure 5.15 Template Approach

Additional mechanisms for protecting the template, also known as *template protection*, can be established to ensure the proper functionality of the provided components. Problems can arise when local adaptations are only transferred to the template at a later time, with continuous template enhancements realized through subsequent deployments to the separate business units. In this way, the template itself is generally not used productively because ongoing operation takes place in the various business units using template functions that have perhaps not been updated. In case of any issues, it is therefore important to clearly identify the source of the error—template or local adaptation. In any case, template-related bugs must also be corrected there and made available to all.

5.4 Specifications and Conventions for the Project Setup

SAP Cloud ALM is designed so that projects can be set up and managed flexibly according to your needs. Nevertheless, you should give some thought to the naming of your projects and their content before you start. A high degree of flexibility also entails a certain risk of quickly losing track later on in the course of the project or when managing it.

Based on the information provided in Section 5.3, we will give you recommendations for setting up projects in SAP Cloud ALM using our customer scenario as an example. You will get information on naming, timeboxing structure, and team setup so that you can make the appropriate settings in the Projects and Setup app within the **SAP Cloud ALM for Implementation** app group (see Figure 5.16).

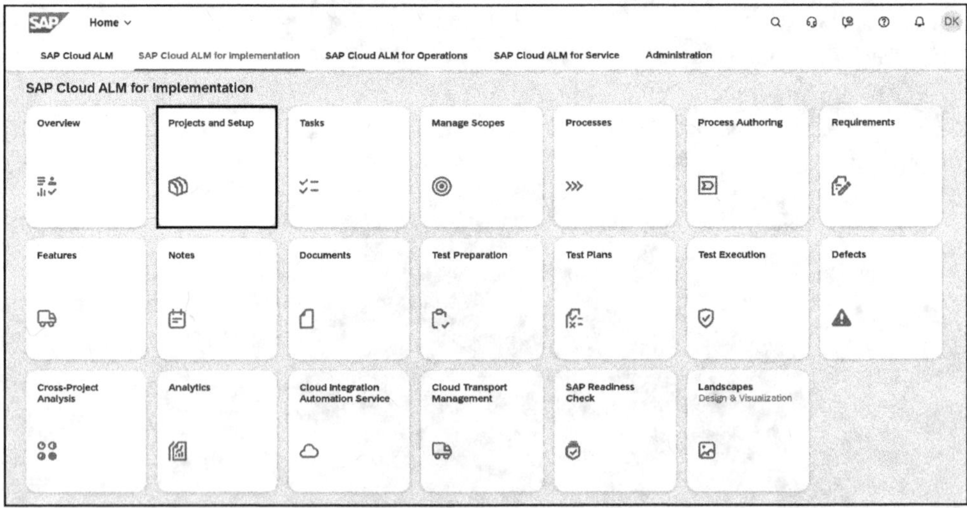

Figure 5.16 SAP Cloud ALM for Implementation: Projects and Setup

5.4.1 Project Setup for the Example Scenario

Before we start creating various projects in SAP Cloud ALM for our practical customer scenario (see Chapter 1, Section 1.4), it is essential to consider the interaction and business context of the solutions to be implemented.

The business areas of SAP S/4HANA and SAP Ariba use common process interfaces in combination. We will use SAP SuccessFactors to introduce various HR processes and functions that operate rather separately from the other two products.

It makes sense to combine products that are part of a common process context in a joint project. It allows end-to-end business processes and their common requirements to be managed, implemented, and tracked. For our example scenario, it looks like this:

- Project for SAP S/4HANA and SAP Ariba (hybrid scenario)
- Project for SAP SuccessFactors (cloud-centric scenario)

In the rest of this section, we will explain the conventions for the project setup using the hybrid scenario as an example.

Figure 5.17 illustrates the project structure to be created for the context "SAP S/4HANA and SAP Ariba" of the example scenario. We create a separate project in SAP Cloud ALM for each major template release so that the various template release statuses are documented and recorded. The two rollouts for the business units are managed in a joint ongoing project.

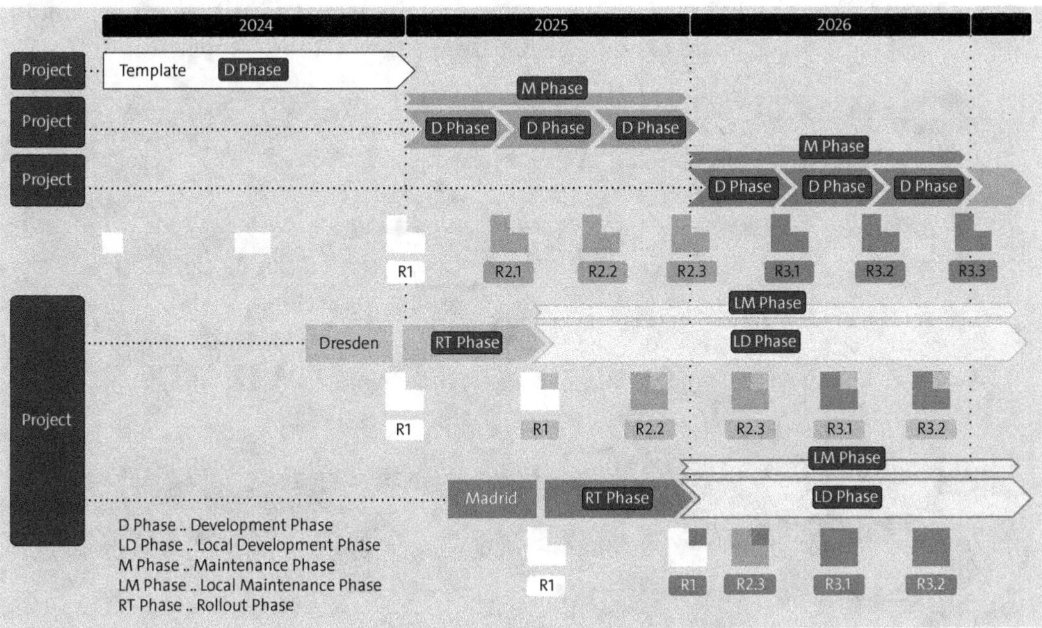

Figure 5.17 Projects in SAP Cloud ALM for SAP S/4HANA and SAP Ariba

The project for the initial template development (**R1**) follows the SAP Activate methodology with its phases, while the further template development (**R2.1** to **R2.3**) represents the continuous development of further functions without SAP Activate phases. With the help of Figure 5.18, we explain the interaction of the release developments and the first rollout in more detail:

❶ Release 1 is developed within the realize phase.

❷ A separate prepare and explore phase also begins at the Dresden plant in line with the SAP Activate method to record and evaluate company-specific requirements and planning the rollout.

❸ Template release 1 was completed at the end of the realize phase and handed over to the first plant for the rollout. The rollout of the template and the implementation of local requirements now take place in the plant-specific realize phase. From the template perspective, the deploy phase begins at this point.

❹ Once the "localized" release 1 has been completed, it is used productively in Dresden after the deploy phase. This marks the start of the run phase from a template and rollout perspective.

❺ Immediately after the completion of template release 1, open and new requirements are implemented in a new cycle in the form of releases 2.1 and 2.2. As the rollout is still in full swing at the time R2.1 is completed, the minor release R2.1 is not imported.

❻ Only after completion of R2.2 are both minor releases handed over to the plant in a bundle. The run phase of template release 1 ends with the deployment.

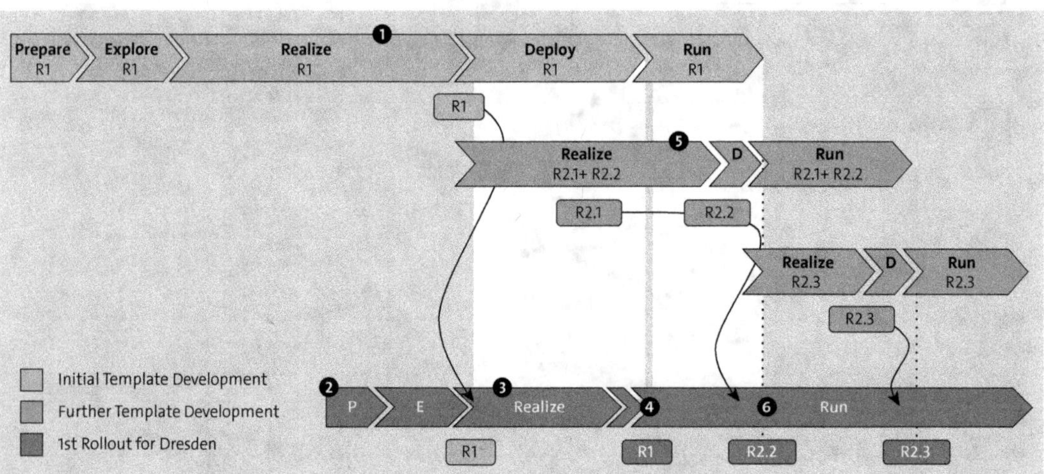

Figure 5.18 SAP Activate Phases of the Example Scenario

5.4.2 Hybrid Scenario: Initial Template Development

In this section, we use the example scenario to present our recommendations for the naming conventions of the project setup with regard to the creation of the first template release.

Naming Convention for Projects

For the initial template development, we recommend naming the project in SAP Cloud ALM according to the following scheme:

- <Name of the initiative>
- <Date of the planned completion date>
- <Template identifier>
- <Release identifier>

For our example scenario, the project name is therefore AriHana-2024-TPL-R1.

[«]

Name of the Initiative

Companies often assign specific names to their planned projects to differentiate them from solutions currently in use, to raise awareness, or to avoid confusion. In our example scenario, we have opted for an initiative name composed of the two products to be introduced (SAP S/4HANA and SAP Ariba): *AriHana*.

Naming Convention for Scope

To increase recognition within the various functional areas of SAP Cloud ALM and enable you to filter efficiently, it makes sense to prefix the project scope with the project name, followed by the corresponding line of business (LoB):

- <Project name>
- <Line of business>

For our example scenario, this results in the following name for the scope logistics: AriHana-2024-TPL-R1-Logistics.

Naming Convention for the Timebox Element Phases

For an initial release creation, it is advisable to use the predefined phases of the SAP Activate method. For the actual realize phase, we recommend creating a separate project phase in the timebox settings (see Section 5.2.4). For this phase, we assign a name identifier for the associated project. This makes it easier to differentiate between development components such as requirements, user stories, or features across projects in a shared task backlog. For better understanding, the original SAP Activate phase name "realize" can also be added:

- <Template identifier>
- <Release identifier>
- <Phase name>

The name in our example scenario is therefore TPL-R1 Realize.

[+]

SAP Activate Phases

A corresponding SAP Activate roadmap template can be added in the general information of a project (see Section 5.2.4). To use the associated tasks of the phases, you can activate the corresponding phases (prepare, explore, realize, deploy, and run) using the activation slider. The names of the phases themselves, however, cannot be changed.

Naming Convention for the Sprints Timebox Element

If several projects are implemented simultaneously in a company, the sprint characteristics (e.g. sprint length or weekdays of the various scrum meetings) may also differ. To allow the project participants to keep track of which projects and sprints they are active in, it makes sense to also include the release number before the sprint ID and sprint number. This can also prove helpful for establishing cross-project reporting:

- <Template identifier>
- <Release identifier>
- <Sprint identifier>
- <Sprint number>

In our example scenario, the following names for sprints are conceivable:

- TPL-R1 - Sprint 1
- TPL-R1 - Sprint N

[»]

Sprint Cycle in the Example Scenario

In our example, the solution is to be completed within one year using an agile project approach with the use of sprints. We have decided on a four weeks sprint length. This results in twelve sprints, which can already be planned in advance in the timebox settings by specifying the date ranges. This also allows various backlog items to be assigned to future sprints.

Summary: Project, Scope, Phases, and Sprints

Table 5.4 summarizes the proposed naming conventions for the initial template development in the example scenario with regard to the designations for project, scope, phases, and sprints.

<table>
<tr><th>Project and Scope</th><th>Phases</th><th>Sprints</th></tr>
<tr><td colspan="3">AriHana-2024-TPL-R1</td></tr>
<tr><td rowspan="9">AriHana-2024-TPL-R1-Logistics</td><td colspan="2">Prepare</td></tr>
<tr><td colspan="2">Explore</td></tr>
<tr><td rowspan="5">TPL-R1 Realize</td><td>TPL-R1 - Sprint 1</td></tr>
<tr><td>TPL-R1 - Sprint 2</td></tr>
<tr><td>TPL-R1 - Sprint 3</td></tr>
<tr><td>...</td></tr>
<tr><td>TPL-R1 - Sprint 12</td></tr>
<tr><td colspan="2">Deploy</td></tr>
<tr><td colspan="2">Run</td></tr>
</table>

Table 5.4 Naming Conventions for Initial Template Development

5.4.3 Hybrid Scenario: Further Template Development

The template release will be continuously developed. New requirements from the various divisions will be incorporated, developed, and delivered in the form of minor releases.

To preserve the previous release for audit purposes and ensure maintenance for older rollouts, we decided to manage the subsequent releases in a new project. In this way, processes, requirements, corrections, and even transports are documented for the individual project or release. The challenge with this approach is that the project information from Release 1 (such as scope and solution processes) cannot easily be transferred to a new project using standard tools. The third-party solution X-ProCo (Cross Project Cockpit) from NTT Data Business Solutions can be used for this.

[«]

NTT Data X-ProCo

NTT Data *Cross Project Cockpit (X-ProCo)* is an innovative add-on that enables the complete or selective export of project data from SAP Cloud ALM. It also provides comparison, archiving, and reporting functionality to easily process the project information.

This add-on makes setting up projects for upcoming rollouts or new releases a breeze. It allows realizing various project situations, such as solution implementation in different countries or branches, as well as in complex cloud or hybrid system landscapes.

In our example scenario, requirements are to be delivered to template consumers every four months. This results in a release cycle of three minor releases per year. The sprint length of four weeks remains unchanged. We created a separate realize phase with the associated sprints for each of these minor releases according to the same scheme as for the initial template development:

- <Template identifier>
- <Release identifier>

This looks as follows for our example scenario:

- TPL-R2.1
- TPL-R2.2
- TPL-R2.3

With the first rollout in a business unit, the template goes into maintenance. This means that issues that occur during or after the rollout of release 1 in ongoing productive operation must also be corrected there. However, as the template is already being further developed in a new release, we recommend creating a new phase in which all maintenance tasks can be referenced. This makes it easier to consolidate the provided corrections and deliver them to the dependent rollouts:

- <Template identifier>
- <Release identifier>
- <Maintenance identifier>

This looks as follows for our example scenario: TPL-R2 Maintenance.

No Sprints within the Maintenance Phase

In our scenario, we do not use sprints during the maintenance phase, as error corrections usually have to be delivered to the template consumers quickly. Using maintenance sprints as containers for the corrections would only make sense if they have a very small iteration and the errors are not critical. You can also correct template errors directly in the affected rollout and feed them back to the template before it is delivered to the other consumers from there.

The entire project structure is shown in Table 5.5.

SAP Activate Phases in Further Template Development

As subsequent releases represent continuous development of further template releases, the phases of the SAP Activate methodology no longer apply. If a future release describes larger development components, such as the introduction of new modules, processes, or product releases, the SAP Activate phases with the roadmap tasks should be considered again.

Project and Scope	Phases	Sprints
AriHana-2025-TPL-R2		
AriHana-2025-TPL-R2-Logistics	TPL-R2.1	TPL-R2.1 - Sprint 1
		TPL-R2.1 - Sprint 2
		TPL-R2.1 - Sprint 3
		TPL-R2.1 - Sprint 4
	TPL-R2.2	TPL-R2.2 - Sprint 1
		TPL-R2.2 - Sprint 2
		TPL-R2.2 - Sprint 3
		TPL-R2.2 - Sprint 4
	TPL-R2.3	TPL-R2.3 - Sprint 1
		TPL-R2.3 - Sprint 2
		TPL-R2.3 - Sprint 3
		TPL-R2.3 - Sprint 4
	TPL-R2 Maintenance	

Table 5.5 Naming Conventions for Template Development

5.4.4 Hybrid Scenario: Rollouts

Once the template has been completed, it should be rolled out in various company units, such as branches, plants, or countries. Local requirements usually have to be taken into account here. They might require changes or extensions to the template as well as the introduction of new processes and associated functions. To ensure that the rollout can still be managed and controlled efficiently using SAP Cloud ALM, you have the following options:

- Definition of a scope within the template project
- Definition of a rollout project with specific scope for each rollout or business unit
- Definition of a rollout project for each rollout or business unit

All of these options offer advantages and disadvantages, which you should consider in advance based on your own needs and requirements. Here, we continue our example scenario using the second approach. As already indicated in the further template development, the third-party solution X-ProCo from NTT Data Business Solutions is a good option for transferring the scope and solution processes from the template to the rollout project.

Naming Convention for Projects

For the project name of all upcoming rollouts, we recommend the following convention based on the template (see Section 5.4.2):

- <Name of the initiative>
- <Date of the planned last rollout>
- <Rollout identifier>
- <Template release identifier>

This looks as follows for our example scenario: AriHana-2026-RT-R1.

Naming Convention for Scopes

To retain the naming convention of the template and thus also promote clarity, we proceed in a similar way for the definition of the scope (see Section 5.4.2). We simply add the company unit identifier to these:

- <Project name>
- <Line of business>
- <Company unit>

This looks as follows for our example scenario:

- AriHana-2026-RT-R1-Logistics-Dresden
- AriHana-2026-RT-R1-Logistics-Madrid

Scopes

By using separate scopes for each business unit, the specific solution processes, requirements, user stories, test cases, features, etc. can also be defined, adequately differentiated and managed using a joint project. In this recommendation, this also includes the LoB level.

Naming Convention for the Phases Timebox Element

The rollout project is similar to the initial template development in terms how it uses separate phases. In our example, the rollouts take place one after the other and are represented by two separate realize phases for each rollout. After the rollout, the subsequent template releases are rolled out in plant-specific run phases. Further localizations also take place in these phases (see Figure 5.17). So the naming convention is:

- <Rollout identifier>
- <Template release identifier>
- <Company unit>
- <Phase>

This looks as follows for our example scenario:

- RT-R1-Dresden Realize
- RT-R1-Dresden Run
- RT-R1-Madrid Realize
- RT-R1-Madrid Run

To document the maintenance of the rollouts during ongoing productive operation, we use a separate maintenance phase for each rollout, analogous to the template, which runs parallel to the run phase (see Figure 5.17). In this phase, faults and error corrections to local versions are documented. Errors detected in template components must be communicated to those responsible for the template:

- <Rollout identifier>
- <Template release identifier>
- <Company unit>
- <Maintenance identifier>

This looks as follows for our example scenario:

- RT-R1-Dresden Maintenance
- RT-R1-Madrid Maintenance

SAP Activate Roadmap Tasks in the Rollout Project

You can, of course, also activate the phases of the SAP Activate method with the contained roadmap tasks for the rollout project. However, in the current case (one project for several rollouts), bear in mind that some tasks would actually be suitable for each rollout.

To assign the SAP Activate tasks to rollout-specific phases, the Tasks app offers an option for exporting and importing selected tasks to Excel. This allows you to assign the tasks to your own phases when editing the Excel file. Copies of the tasks are created for the referenced phases when you do the import.

Naming Convention for the Sprints Timebox Element

The various rollout projects can be carried out in parallel, with a time delay or sequentially. The latter is helpful to gather experience and knowledge from the first rollout and incorporate it into the next one. This means that the projects generally have different characteristics in terms of start date, duration, involved project members, and timing of the sprints.

With regard to the sprint designation, we use again the recommended scheme of the template:

- <Phase identifier>
- <Sprint identifier>
- <Sprint number>

This looks as follows for our example scenario:

- RT-R1-Dresden - Sprint 1
- RT-R1-Dresden - Sprint N
- RT-R1-Madrid - Sprint 1
- RT-R1-Madrid - Sprint N

Number of Sprints

The activities to be carried out for the rollout in a business unit are structured and implemented within the sprints. Once the rollout is completed, productive operation begins. Although this marks the end of the original project orientation, it does not usually mean the end of the project itself. Further local requirements are implemented in subsequent sprints as part of continuous development.

Summary: Project, Scope, Phases, and Sprints

Table 5.6 summarizes the proposed naming conventions for the rollouts of the example scenario with regard to the designations for project, scope, phases, and sprints.

Project and Scope	Phases	Sprints
AriHana-2026-RT-R1		
AriHana-2026-RT-R1-Logistics-Dresden	RT-R1-Dresden Realize	RT-R1-Dresden - Sprint 1
		...
		RT-R1-Dresden - Sprint 6
	RT-R1-Dresden Run	RT-R1-Dresden - Sprint 1
		RT-R1-Dresden – Sprint...
	RT-R1-Dresden Maintenance	
AriHana-2026-RT-R1-Logistics-Madrid	RT-R1-Madrid Realize	RT-R1-Madrid - Sprint 1
		...
		RT-R1-Madrid - Sprint 6
	RT-R1-Madrid Run	RT-R1-Madrid - Sprint 1
		RT-R1-Madrid – Sprint...
	RT-R1-Madrid Maintenance	

Table 5.6 Naming Conventions for Rollout Projects

5.4.5 Hybrid Scenario: Further Deployments

After the first template release has been made available, it is developed further in the form of individual minor releases (see Section 5.4.3). These can include new modules, processes, the refinement of existing functions, and the implementation of new functions that are requested by the business units, for example.

Usually, the template minor releases should only be delivered to a business unit after a successful rollout. You now have the following options for managing further deployments of the template with the associated changes (e.g. new scope and processes) in the rollouts:

- Continuation of the existing rollout project
- Creation of a new project for release 2

For the first option, you should change the project name. The new phases and sprints then refer to release 2. If you go for the second option, you must transfer the local changes to the solution processes to the new project in addition to creating the new project.

5.5 Summary

In this chapter, we have given you an introduction to the agile methodology with Scrum, which we will continue to refer to in this book when implementing the example scenario within the realize phase. We have clarified the basic concepts of Scrum and discussed the supporting functional possibilities of SAP Cloud ALM. Using the example scenario, you learned about our recommendations for specifications and conventions for the project setup. In the next chapter, we will put the conceptual suggestions into practice in the system using the example scenario. You will learn how to create the project and configure it accordingly.

Chapter 6
Project Setup

In this chapter, you will learn how to create projects in SAP Cloud ALM. In addition to selecting SAP Activate templates, you will be able to add your own content to existing timeboxes, project members, and project roles.

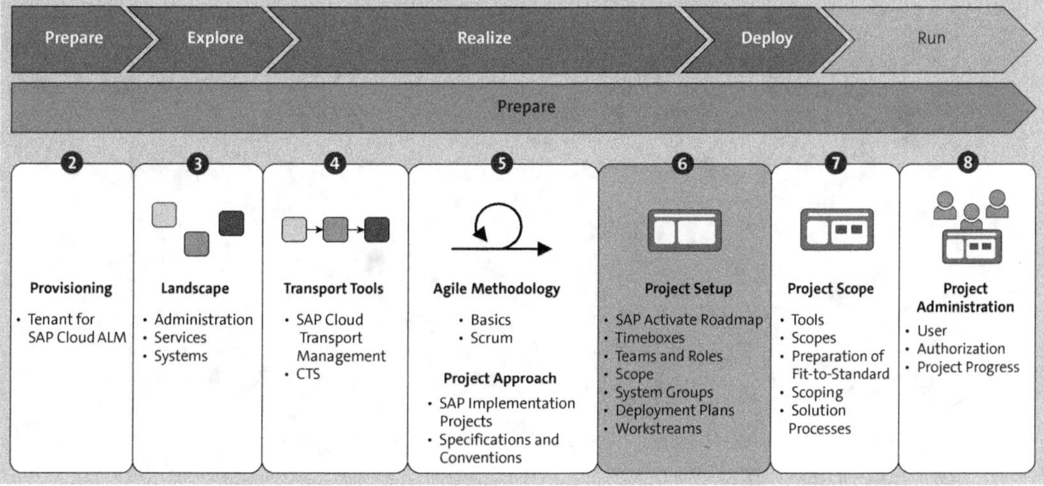

In the previous chapters, in addition to the introduction to SAP Cloud ALM and SAP Activate, you were also able to familiarize yourself with provisioning a tenant for SAP Cloud ALM. We also explained how you can set up the project landscape and the required transport tools, which will also play a role in this chapter. The last chapter focused on the agile project methodology as a central core element of the realize phase. We discussed the various project approaches and explained our recommendations for the specifications and conventions for implementing a hybrid scenario using our example scenario. We will return to these approaches in this chapter to illustrate the setup of a project on the system.

The project is a central entity in SAP Cloud ALM and a mandatory prerequisite for supporting the implementation of SAP products with SAP Cloud ALM. This chapter will deal in detail with the creation of a project and all the parameters to be defined for a project.

After successfully logging in to your SAP Cloud ALM tenant, navigate to the app group **SAP Cloud ALM for Implementation** and open the Projects and Setup app, as shown in Figure 6.1.

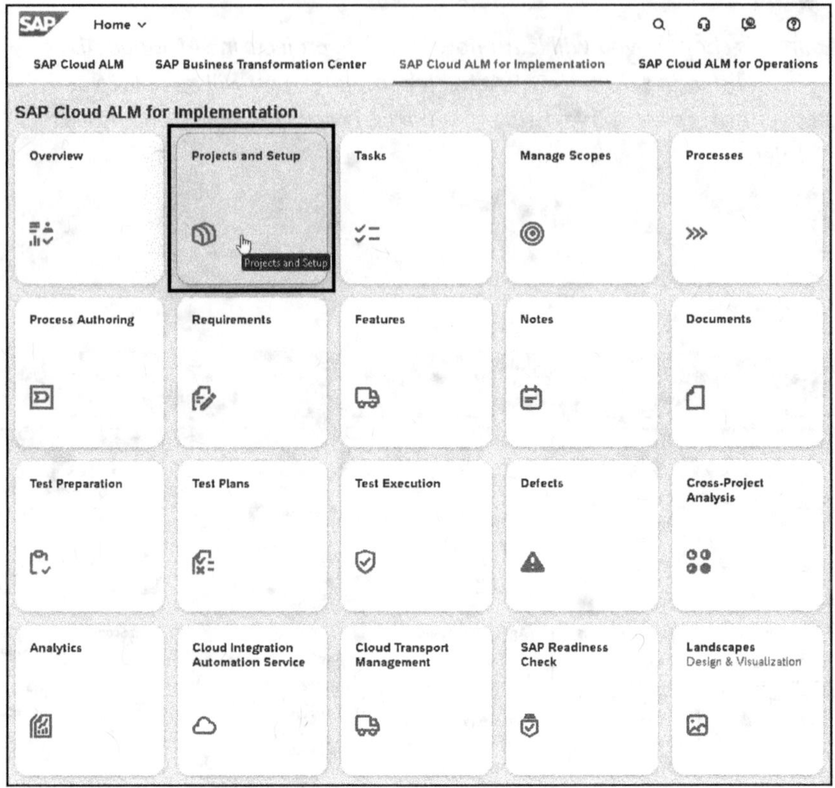

Figure 6.1 Projects and Setup App

6.1 General Information

Before we create a new project, let's first clarify the term "project" in the context of SAP Cloud ALM. A *project* is the container that contains all relevant information (i.e., entities and content). The following information is required in the first step:

- Name
- Dates (time periods)
- Team (project members)
- Scope
- System landscape

Once you have opened the Projects and Setup app, you will see all existing SAP Cloud ALM projects in the **Manage Projects** area on the left-hand side. To create a new project in your role as a project manager, select the project name as shown in Figure 6.2 select the **Create** button.

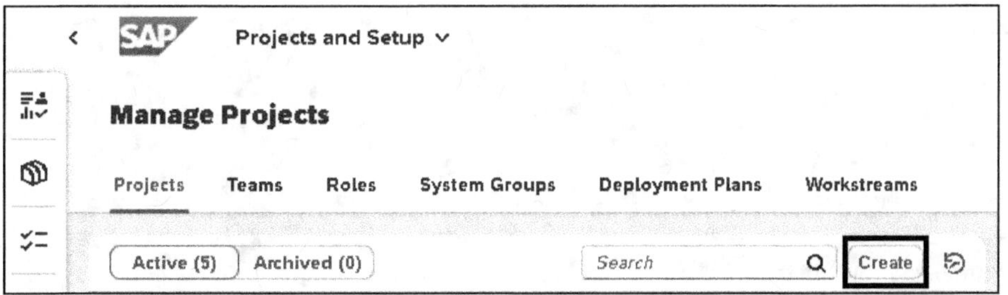

Figure 6.2 Create a New Project

If the button is not visible, you can use the menu with the three-dot symbol ••• (**Additional Options** menu) to access the **Create** button (see Figure 6.3). You can also see in the illustration how the **New Project** area and the **General Information** tab opens after clicking the **Create** button on the right.

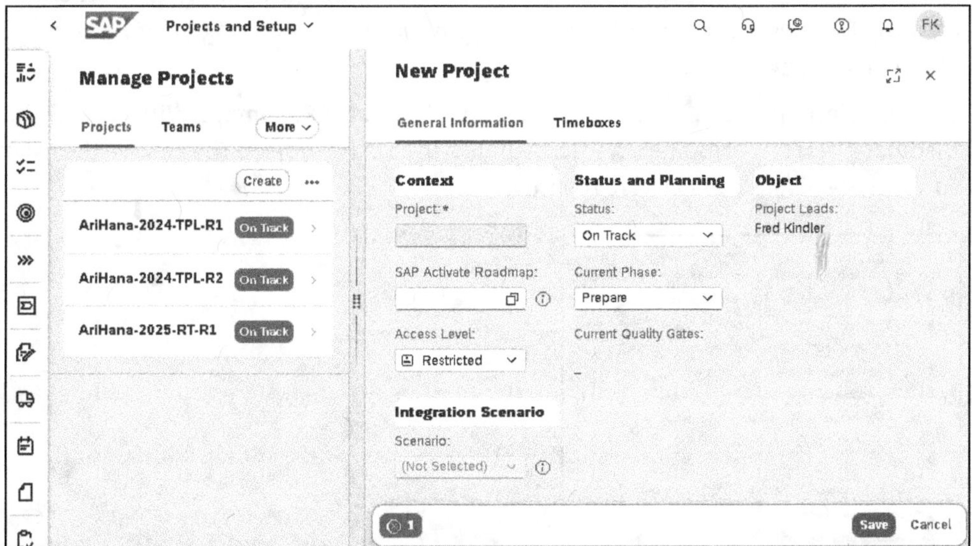

Figure 6.3 New Project: General Information

The first step is to assign a project name, which is also defined as a mandatory field. The project name used in this book has already been described in detail in Chapter 5, Section 5.4.2.

Show History Button

In case you want to see what was changed within your project: Press the **Show History** button and see all the information that was changed together with corresponding user and date/time information (see Figure 6.4).

Figure 6.4 Show History

6.1.1 SAP Activate Templates (Task Templates)

Since you learned about the SAP Activate methodology in Chapter 1, Section 1.3, you can use the **SAP Activate Roadmap** selection list to select the SAP Activate content you will use for your project. In this example, **SAP S/4HANA (on-premise) – New Implementation** is selected. The selection of SAP Activate templates is strongly recommended, because if you do not select a template, only basic tasks for setting up and using SAP Cloud ALM are available.

You can select between **Single Roadmaps** and **Combinable Roadmaps**. With combinable roadmaps, you can combine several roadmaps in a single project. You can see the list of currently available SAP Activate roadmaps in Figure 6.5.

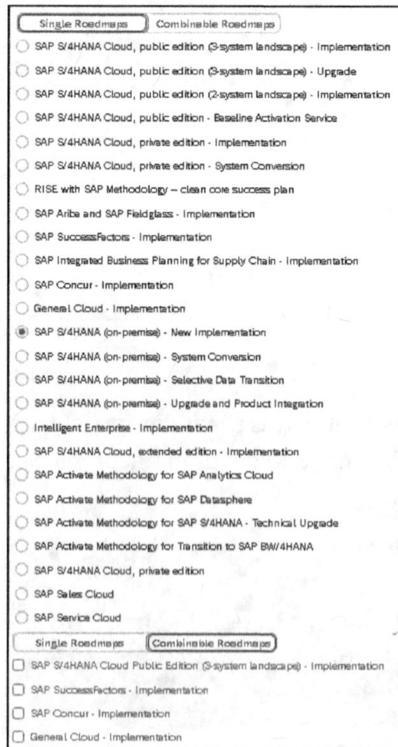

Figure 6.5 SAP Activate Roadmaps

The SAP Activate content provided is continuously expanded and is available in SAP Cloud ALM shortly afterwards. In your role as project manager, you should select the SAP Activate template for the SAP product that plays a leading role in your implementation project.

Below the list of roadmaps, you will see the **Access Level** selection list. The options are:

- **Public**
- **Restricted**
- **Private**

You have the option of restricting access to your project via the access level. This option has only been available, however, since December 2022. If you created projects before then, they are automatically defined as **Public**. Below this, in the section **Integration Scenario**, you will find the option to activate the integration of SAP Cloud ALM projects with multiple SAP Central Business Configuration projects (see Figure 6.3).

6.1.2 Status and Planning

In the right part of Figure 6.3, you can see the **Status and Planning** area in the project view. The SAP Cloud ALM project can have one of the following statuses:

- **On Track**
- **Needs Attention**
- **Critical**

By selecting the project status, you as the project manager can see in the project overview which projects you need to pay attention to. This is helpful if you are working on several projects at the same time. Below the **Status** section, you can see the **Current Phase**. You can change the phase by selecting a different phase in the selection list and then saving. The project manager is displayed to the right of **Status and Planning** in the **Object** area.

In this section, you have created an SAP Cloud ALM project and maintained general information. In the following, we will guide you through the most important settings and specifications.

6.2 Timeboxes

Every project needs a defined start and a defined end. All data (defined points in time or time periods in the SAP Cloud ALM project) is defined by timeboxes. These timeboxes can be assigned to tasks, phases, sprints, and milestones and are explained in this section.

6.2.1 Phases

Phases are defined by the SAP Activate methodology. The phases already delivered by SAP Cloud ALM are:

- Prepare
- Explore
- Realize
- Deploy
- Run

You can also add your own phases to the list using the plus symbol + (**Add Custom Phase**) to the right of the last phase. The new phase can also be deleted by clicking on the cross symbol × (**Remove Custom Phase**). The human symbol to the left of the phase name indicates that this phase was created individually and was not predefined by the SAP Activate roadmap or template. To define the period for the respective phase, select the calendar symbol on the right in the **Enter Date Range...** line. . With the first click, you define the start date, and with the second click, you define the end date of the phase. This results in the implementation period of the phase. You can also deactivate phases by clicking on the slider symbol to the right of the calendar symbol. This is useful if not all delivered phases are relevant for your project. In Figure 6.6 you can see the implementation of the template approach for our example scenario, AriHana 2024.

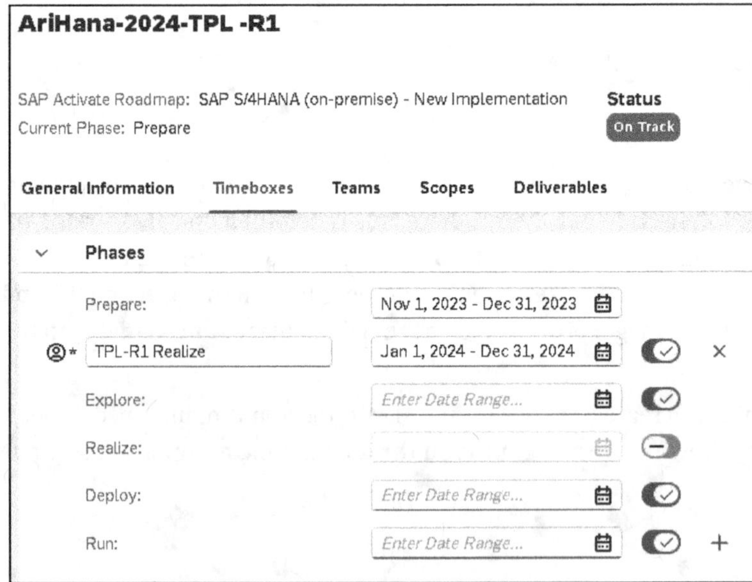

Figure 6.6 Timeboxes: Phases

6.2.2 Sprints and Milestones

Below the phases, you can define the sprints and store your sprint planning in SAP Cloud ALM. If your way of working is predetermined by stakeholders or similar outside decision-makers, you might not need a sprint plan and can just use the time periods of the phases. This is usually the case if the scope of the project is already defined at the start. If the scope of the project has not yet been finalized or if it is about the continuous improvement of an existing functionality, you may follow the agile project methodology approach. Experience shows that both approaches are mixed. This means that sprint planning is agreed for the implementation phase and the sprints do not exist at the hierarchy level of the phases. It is therefore not necessary to synchronize sprints with phases, but it is advisable—and recommended—to determine the appropriate phase based on the sprint period. If you don't link the sprint planning to the associated phases, you will need to use more effort to create the various filter settings. If you delete a sprint, the appropriate phase and its due date are used as a fallback solution. Figure 6.7 shows an example of sprint planning for the AriHana template project.

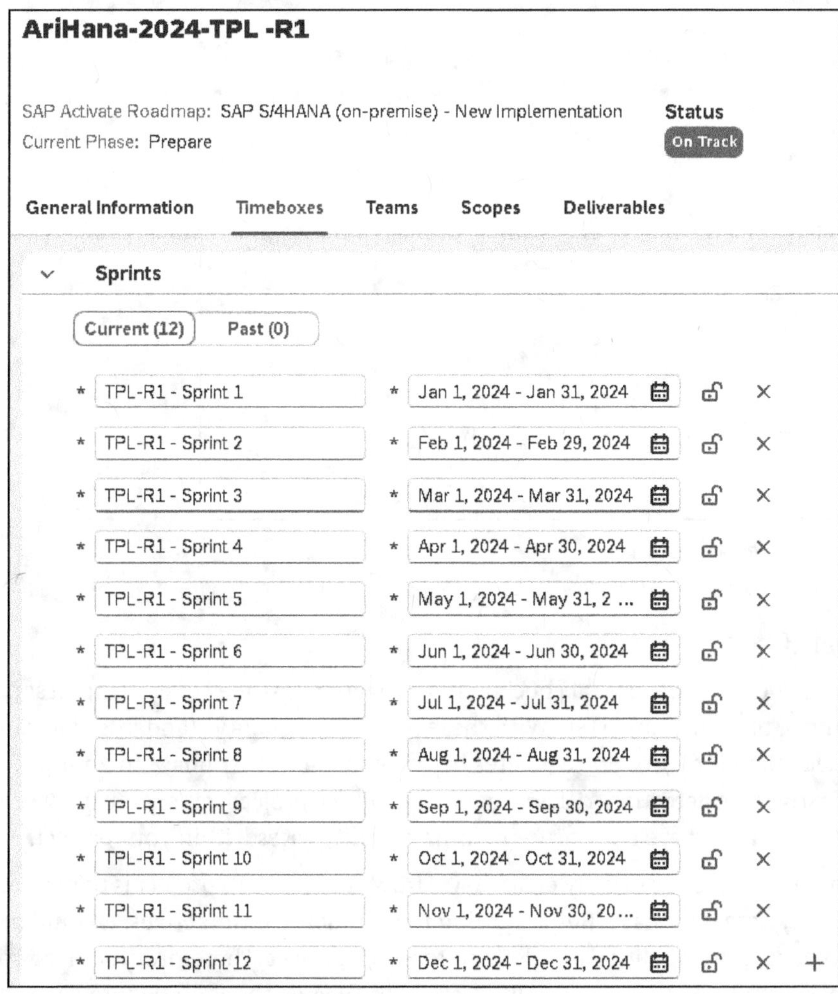

Figure 6.7 Timeboxes: Sprints

In contrast to phases and sprints, only the end date is set for milestones. Here you can add additional milestones by clicking on the + (**Add Milestone**) button to add additional milestones and the × (**Remove Milestone**) button to delete added milestones, as shown in Figure 6.8. The sorting here is chronological. If you assign a timebox (phase, sprint, or milestone) to a task, the end of the timebox is automatically defined as the due date of the task. This makes it easy for you as the project manager to adjust the time periods and due dates in the event of any project delays. If you change the due date of a task manually, it no longer updates automatically if a timebox changes.

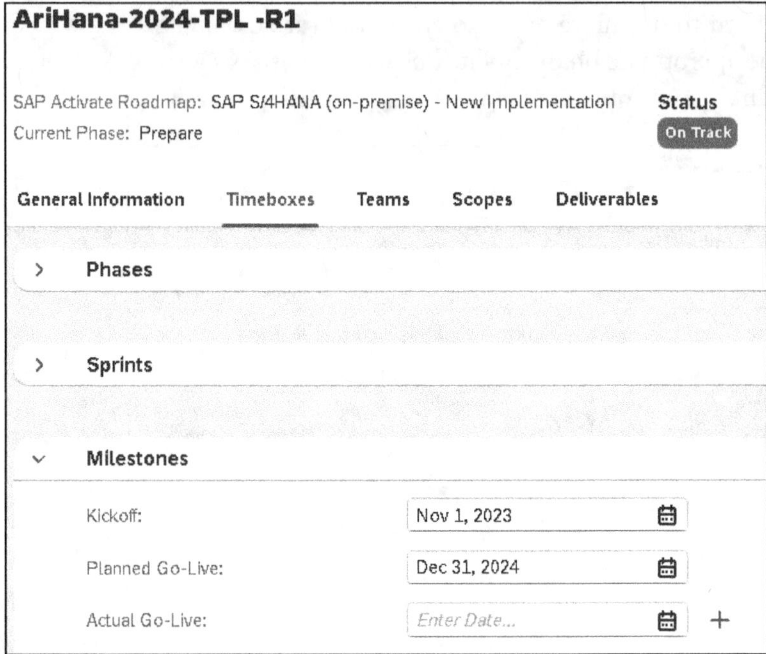

Figure 6.8 Timeboxes: Milestones

6.2.3 Quality Gates

Quality gates in projects are critical checkpoints or even milestones that are necessary to fulfill to move on to the next step. With these predefined quality standards, you can ensure to meet all mandatory prerequisites before you enter next phase in your project. Besides early issue detection, risk mitigation and quick project transparency as well as the governance aspect are reasons for planning quality gates within your project.

Our recommendation for our newly created AriHANA-2024-TPL-R1 project is to have at least one quality gate per phase. But according to your defined timeboxes, it could be necessary to have more quality gates defined for your project. Therefore, you need to go to the Tasks app and select the **Quality Gate** task **Type** as shown in Figure 6.9 and Figure 6.10.

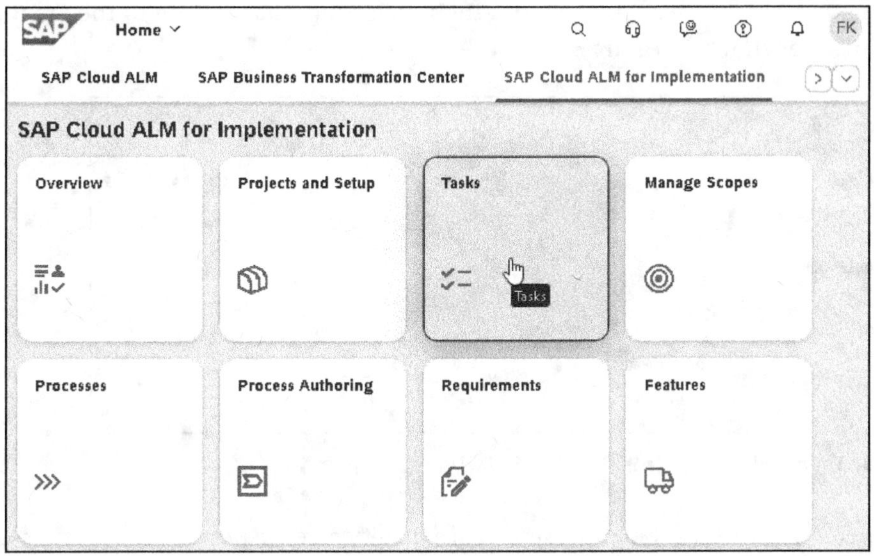

Figure 6.9 Tasks App

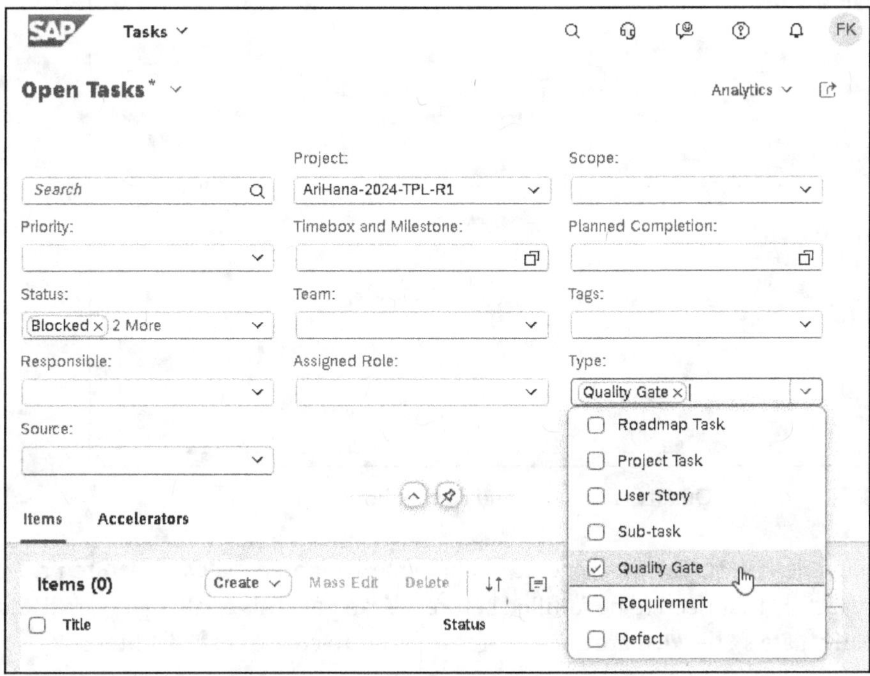

Figure 6.10 Select "Quality Gate" as Type

After that, you can create a new quality gate by selecting **Quality Gate** in the **Create** dropdown list as shown in Figure 6.11.

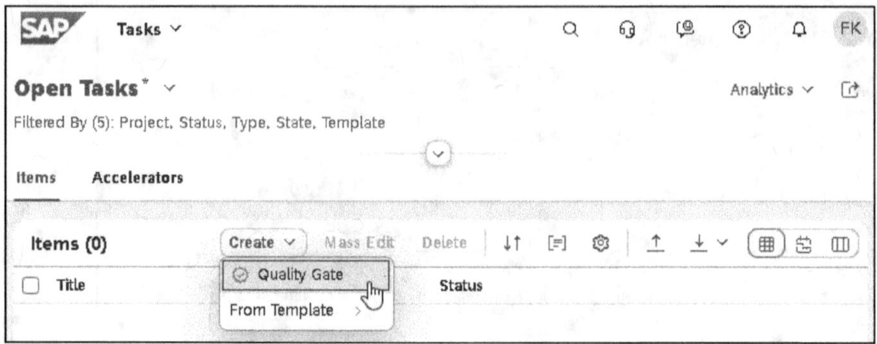

Figure 6.11 Create New Task from Quality Gate Type

In Figure 6.12, you can see the initial screen of a quality gate creation. You are required to give it a title, but a description is optional.

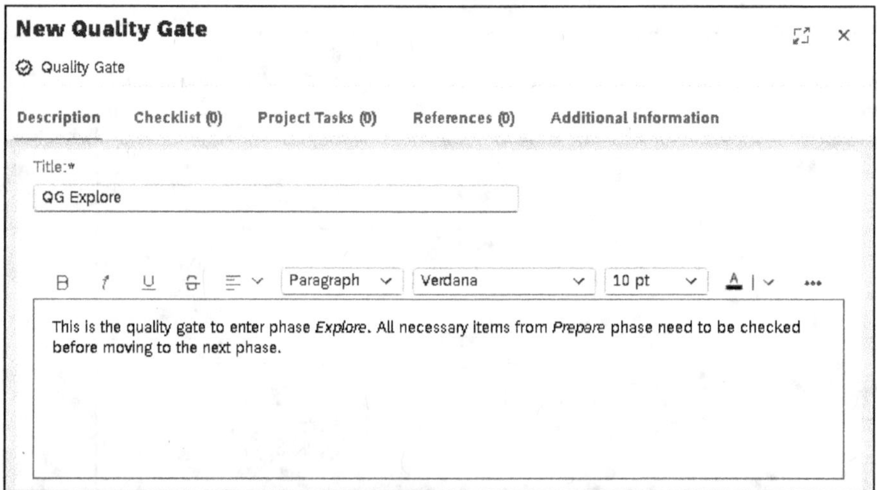

Figure 6.12 Create New Quality Gate: Title and Description

The measurable criteria for this specific quality gate are noted via checklist items. All necessary factors that need to be fulfilled before we can pass this quality gate are listed in the quality gate's **Checklist** area. There are two ways to create checklist items:

- Create them manually via the **Create** button in the **Checklist** area (as shown in Figure 6.13)
- Use the upload and download function (as shown in Figure 6.14)

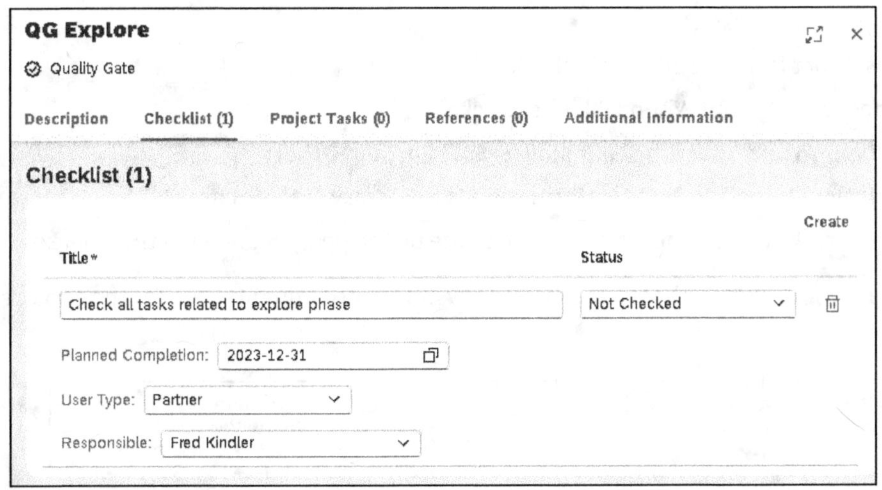

Figure 6.13 Create Checklist Item: Manual

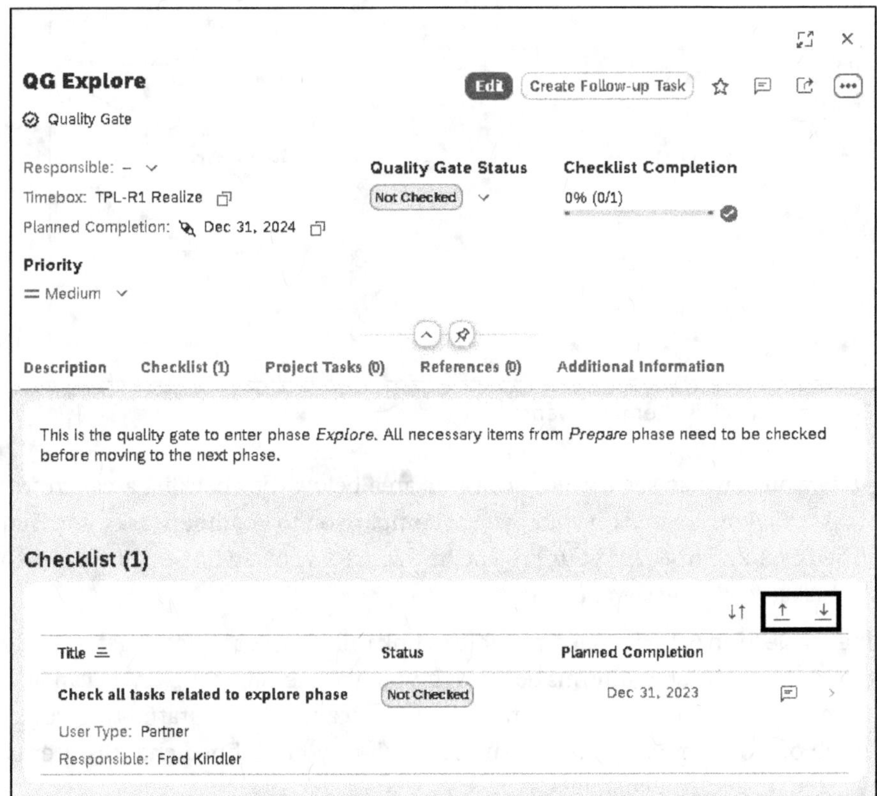

Figure 6.14 Create Checklist Item: Upload and Download Function

The following elements are part of a good checklist and the corresponding questions that can be answered there:

- **Title**: What is checked?
- **Status**: What is the result of this check?
- **Planned Completion**: When will it be checked?
- **User Type**: Who is checking? (**Employee**, **SAP**, **Partner**, **Other**)
- **Description**: What additional information is necessary for this check?
- **Comments**: What comments have been made by the people following this checklist?

In Figure 6.15, you can see how it looks in the system for our quality gate for the explore phase.

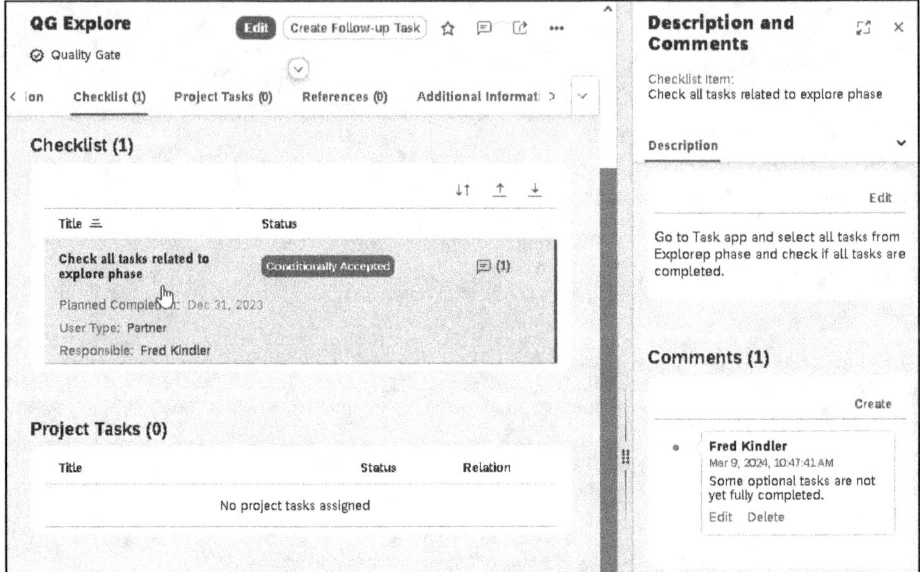

Figure 6.15 Create Checklist Item: Elements

In Figure 6.15, you can also see the **Project Tasks** area below the checklist area. *Project tasks* are tasks that you created on your own, in comparison to Roadmap tasks, and that you identified as mandatory for your project. In this area, you can link quality gates to your project tasks via the **Assign** button.

As explained in Section 6.2.2, it is also possible to link milestones to quality gates. To do this, switch to the **Additional Information** tab. If you like the Gantt view, you can also see quality gate milestones visualized that way. You can assign a status that corresponds to the outcome of your check result for each of your defined checklist items. The following statuses are available:

- **Not Checked**
- **Accepted**
- **Conditionally Accepted**
- **Not Accepted**
- **Not Relevant**

As all mandatory tasks for this quality gate are fulfilled in this case, the status is now **Conditionally Accepted**.

If follow-up activities are necessary, you can create them via the **Create Follow-up Task** button. With this button you can create a project task that is then automatically assigned to this quality gate. In our example, we created a new project task to check optional tasks related to the explore phase. Once this project task is completed (status **Done**), you can switch the quality gate status to **Accepted**.

Usually, you will have some kind of team for reviewing such imported quality gates when moving from one phase to each other. Participants on this review team can be part of the following user groups or their representatives:

- Relevant project stakeholder
- Subject-matter experts or functional experts
- Project leads and members

It is their responsibility to examine the completeness, accuracy, and compliance of deliverables for the checklist elements of this quality gate. Therefore, it is important to set an appropriate status for each checklist item based on result of the check. If you identify potential risks or deviations within the check of a quality gate, you should also create a project task as follow-up to document and track it. After your analysis of the quality gate checks, you can set the status based on the result as shown in Figure 6.16.

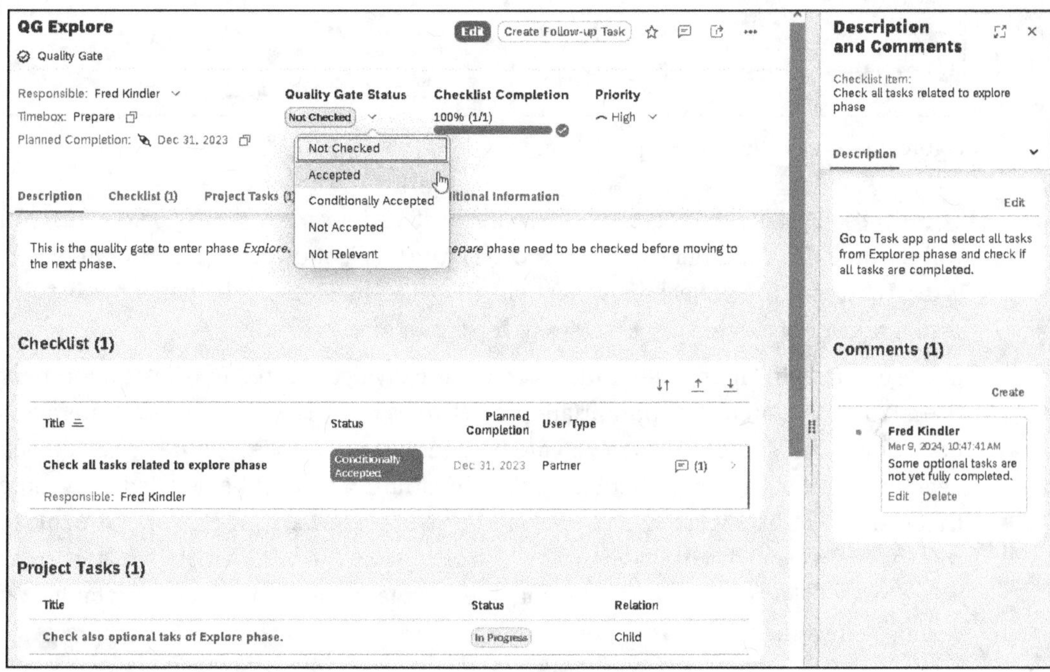

Figure 6.16 Create Checklist Item: Result Status

6.3 Teams

In the previous chapters, you have been introduced to the topic of roles and responsibilities. This section is about assigning project roles to the responsible persons. First, we will show you how to create your own teams and project roles.

6.3.1 Creating Teams

In the left-hand section of the Projects and Setup app, we have been working in the **Projects** tab. Now navigate to the second tab, **Teams**, and you will see that a team has already been created automatically when the project was created: the **PMO Team**. The PMO team is the only team in which the role of the project manager exists.

For this reason, the team cannot be deleted and is marked with a PMO symbol. You can create a new team via the **Create** button. Figure 6.17 shows the creation of a team for development.

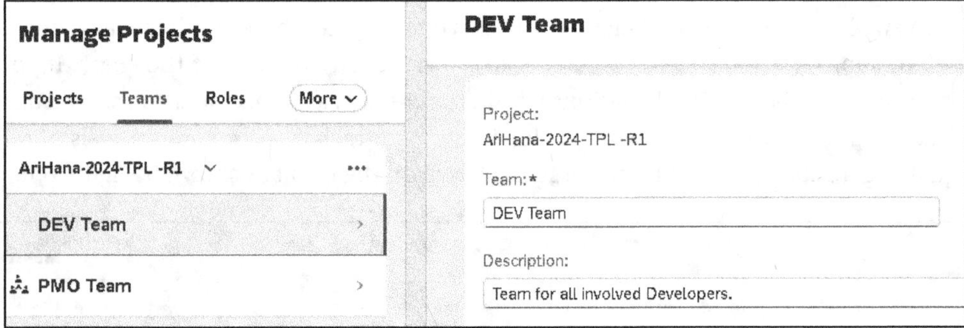

Figure 6.17 Create New Development Team

The team's name is a mandatory field, while the description is optional. Below this, you will see the role assignment areas shown in Figure 6.18 is shown. The slider symbol is used to define the project roles that exist in this team. Please note that you must also assign the newly created project roles to the teams after creation.

In the example in Figure 6.18, some roles are already deactivated for the development team. You can assign the appropriate project roles directly to your project members here, or you can do it later when you are back in the detailed view of the project. When deactivating the roles, both here in the administration and in the detailed view, note that no more positions can be assigned after deactivation. Existing user assignments to this role are removed. If user assignments already exist, they will remain in place. In the upper area of Figure 6.18, you can see that the **By Roles** choice is selected by default. You can also select the **By Users** assignment.

Figure 6.18 Role Assignment for the Development Team

After switching to **By Users**, you will see the **Create** button on the right-hand side of the window. Click on it and then select the user on the left-hand side. On the right-hand side, you can select the corresponding project roles for this user via the selection list, as shown in Figure 6.19.

The users defined here can see their team members and edit them themselves. They are also responsible for the task assignments according to their role as defined in the role properties of the tasks. There is no direct link between the roles listed here and the authorization roles. If a user is assigned to a team on a project for which they do not have authorization, the system automatically sends a request to the administration team. The team then decides whether to approve the user. Administrators also have the option of granting users individual authorization for projects and tasks. This gives these users access to projects and tasks, but they are not listed here.

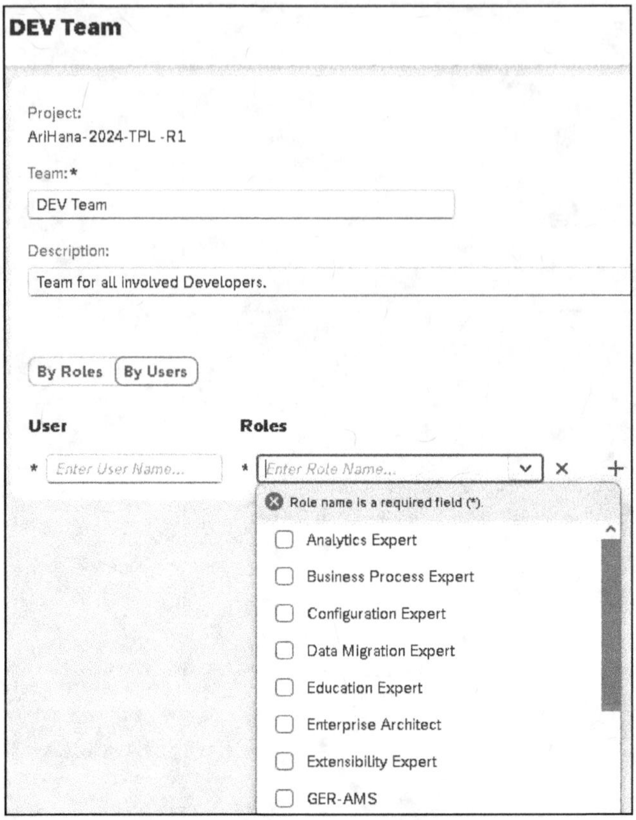

Figure 6.19 Role Assignment According to Users

6.3.2 Create Your Own Roles

Now navigate to the third tab, **Roles**, in the **Manage Projects** section. There you will see project roles that have already been created automatically when you created the SAP Cloud ALM project (see Figure 6.20):

- **Analytics Expert**
- **Business Process Expert**
- **Configuration Expert**
- **Data Migration Expert**
- **Education Expert**
- **Enterprise Architect**
- **Extensibility Expert**
- **Integration Implementation Expert**
- **Organizational Change Management Expert**
- **Output Management and Forms Expert**

- **Project Lead**
- **System Administrator**
- **Technology Expert**
- **Testing Expert**

To use the agile project approach you learned in Chapter 5 and also to meet your customer-specific requirements, you can add your own project roles in SAP Cloud ALM. To do this, use the **Create** button and see the details to be maintained on the right-hand side, as shown in Figure 6.20. Because of the ⓐ symbol, you can see that this is a customer-specific role.

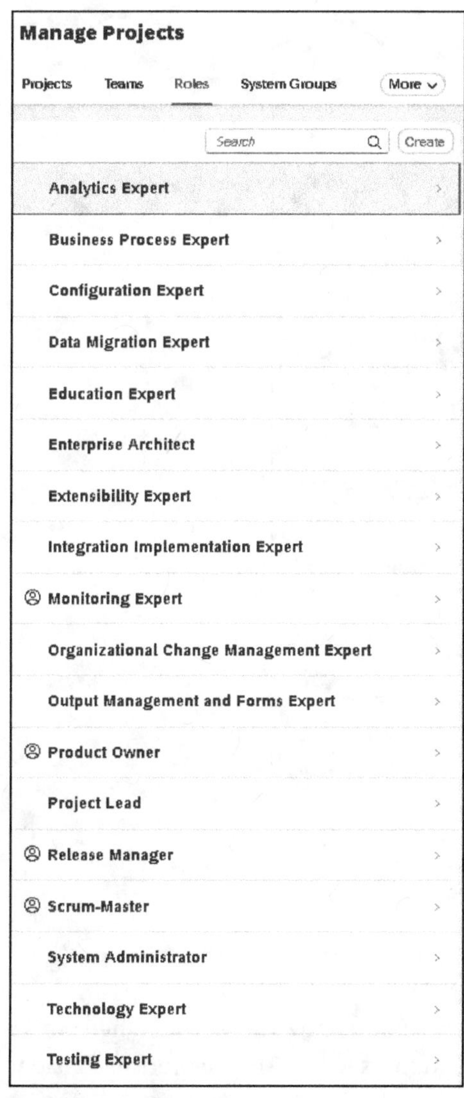

Figure 6.20 Create New Project Role

We have added the following roles for the AriHana project in our practical example:

- **Monitoring Expert**
- **Product Owner**
- **Release Manager**
- **Scrum Master**

To give you an idea of how the assignment of project members to the roles in the project can look, take a look at Figure 6.21 to see the assignments for our AriHana project.

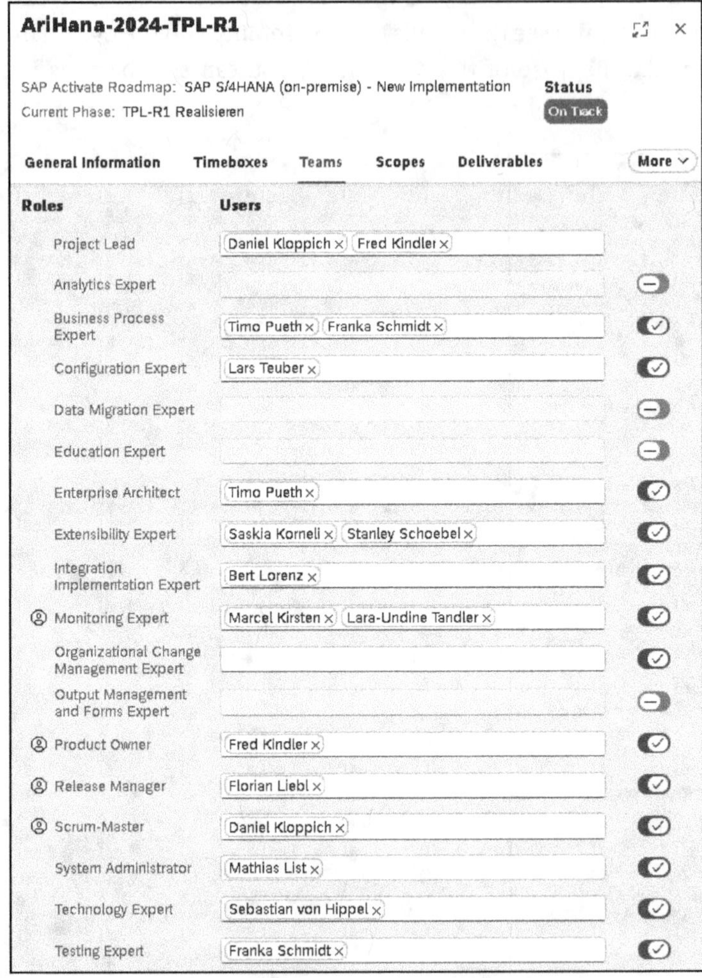

Figure 6.21 Role Assignment in the AriHana Project

6.4 Scopes

The previous section dealt with the project members and how they are assigned project roles. The configuration of scopes is explained in this section. The theoretical consideration of project scopes is discussed in Chapter 7. As explained in Chapter 5, Section 5.4, our AriHana template project starts with the Logistics module, as shown in Figure 6.22.

Figure 6.22 AriHana Template Project Release 1

A better example is the AriHana rollout project shown in Figure 6.23.

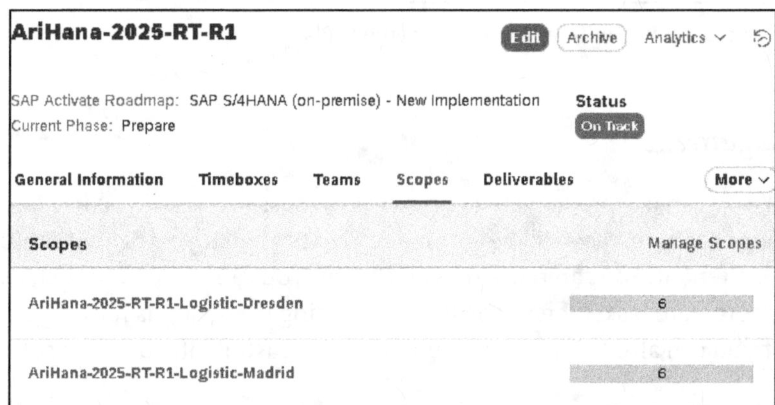

Figure 6.23 Scope of the AriHana Rollout Project

Here you can see two scopes, one for the location in Dresden and the second for the location in Madrid. If you select the **Manage Scopes** link, you will immediately jump to the Manage Scopes app, where you can create or change scopes as described in Chapter 7.

6.5 Deployment Landscape and Deployment Plan

In the last step of this project setup, we will show you how to integrate your managed systems into the project context. To do this, you need to have defined the system groups and the deployment plan. This is described in detail in Chapter 17.

In Figure 6.24, you can see that the system groups have also been assigned by selecting the AriHana deployment plan. Further information on deployment plans can also be found in Chapter 17.

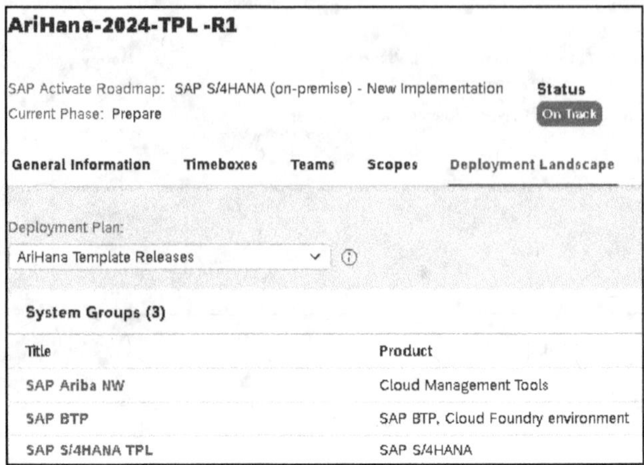

Figure 6.24 Assign Deployment Landscape and Deployment Plan

6.6 Tag Management

In SAP Cloud ALM, you can use predelivered tags as well as custom tags. If you use tags in your SAP Activate roadmap, they will be automatically assigned to template tasks in your SAP Cloud ALM tenant. You can reuse these tasks, but you cannot rename, delete, or unassign (from template tasks) them. The benefit of using these tags is for filtering and grouping. But you can also think about using them for custom attributes in solution processes.

Tags Possibilities

Custom tags are currently available for tasks, sub-tasks, user stories, requirements, solution processes, and test cases.

From our perspective, it is essential to think about tags from the very beginning of your project. You will see many benefits during your project if you give clear guidance to project team members. Two roles are available for tag management:

- Tag administrator: Create, update, merge, and centrally delete tags; view tag usage; and organize tags into groups
- Tag viewer: Can view tags, tag groups, and tag usage

You should also think about naming conventions and when tags will be used for better handling of all entities. It's a good practice, for example, to use tags for a risk assessment.

In our example, we have a roadmap task called Setup Project. Because it has a high risk level, we can create a tag to reflect that by clicking in the empty **Tags**: line and selecting the **click here** link, as shown in Figure 6.25. In the resulting popup window, we can enter the new tag label "risk_high," as shown in Figure 6.26.

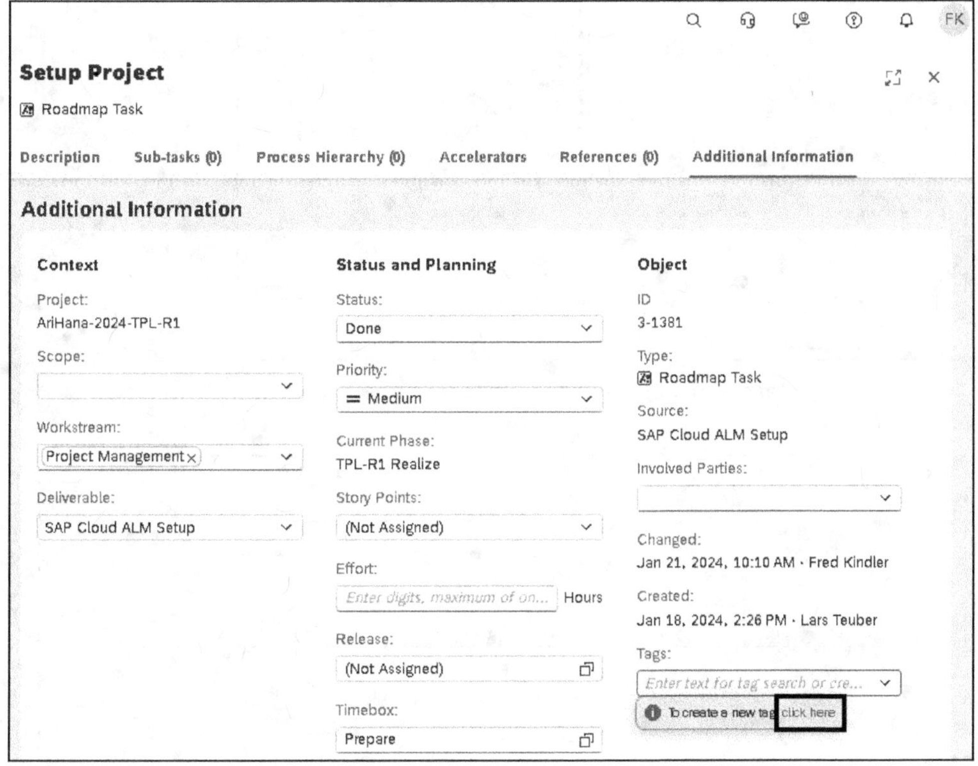

Figure 6.25 Create New Tag

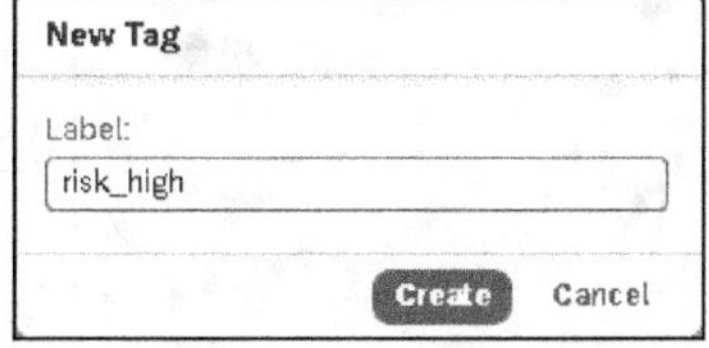

Figure 6.26 Create New Tag Label

If you need an overview about all tags and tag groups, you can use the Tag Management app in the **Administration** area, as shown in Figure 6.27.

In the Tag Management app, you can see all tags as well as their creator and the source of all tags. Our previously created tags were created by the end user, as shown in Figure 6.28.

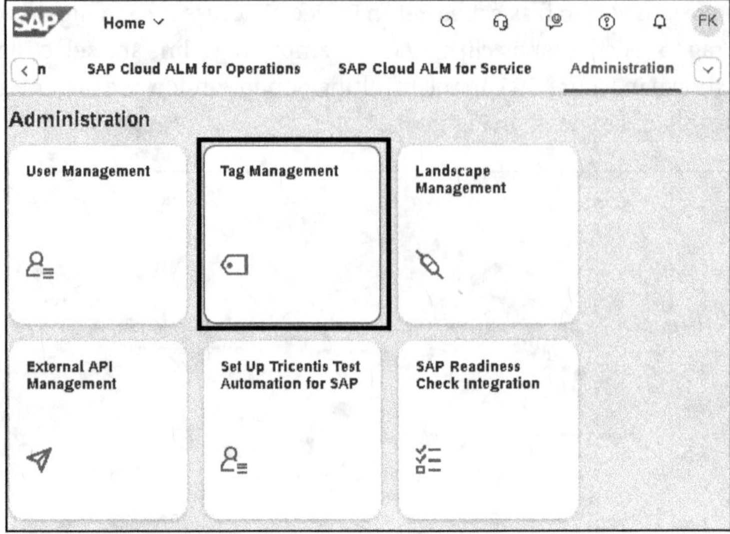

Figure 6.27 Tag Management in Administration Area

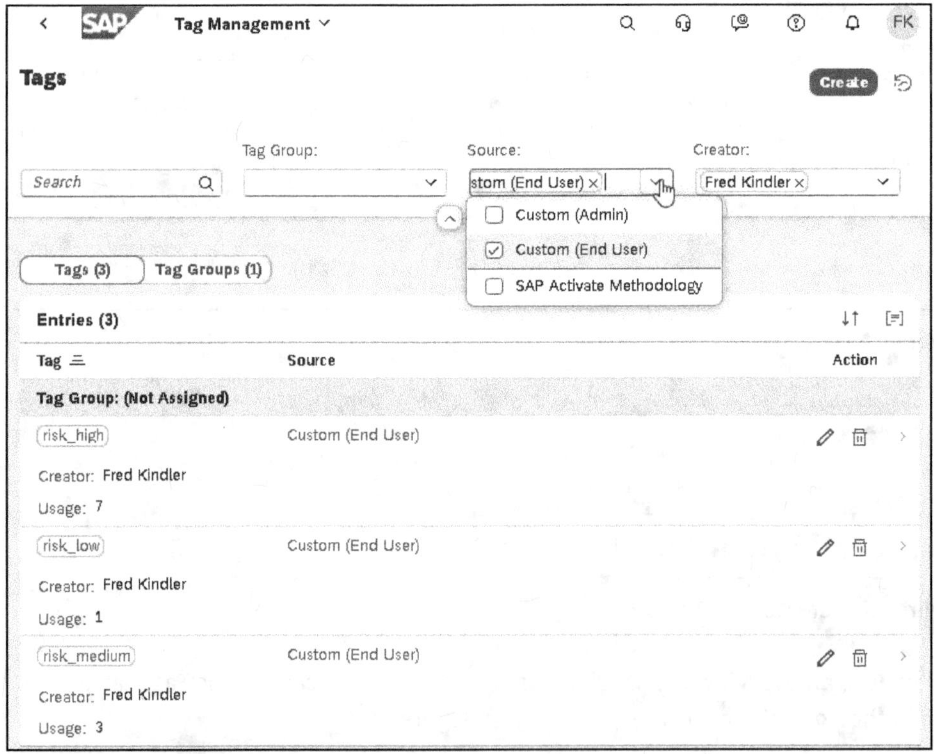

Figure 6.28 Tag Management App: Tags

If you want to create tags that offer multiple tag values for a specific aspect, you should use tag groups. Tag groups are optional, but if a tag belongs to a specific **Tag Group**, the group's title will be displayed before the individual tag label, e.g., **Risk: Low**, **Risk: Medium**, and **Risk: High**, as shown in Figure 6.29.

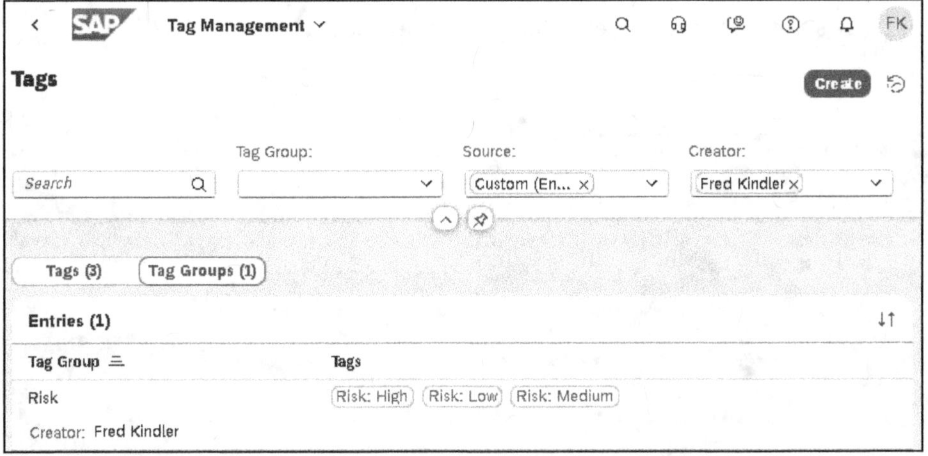

Figure 6.29 Tag Management App: Tag Groups

Merge Custom Tags

You can also merge custom tags. If you merge two tags, the assignments of both tags are combined, and the duplicated entry is deleted. Predelivered tags cannot be merged.

Tags can be used for searching, and not only for dedicated entities. You can use this kind of search for many types of entities cross-project. You can use the search field in header or many other applications to launch a search based on tags. You simply navigate to the **Home** screen and click on the magnifier glass button . After that, you can choose which entities you want to search for—in this case, **Tags**, as shown in Figure 6.30.

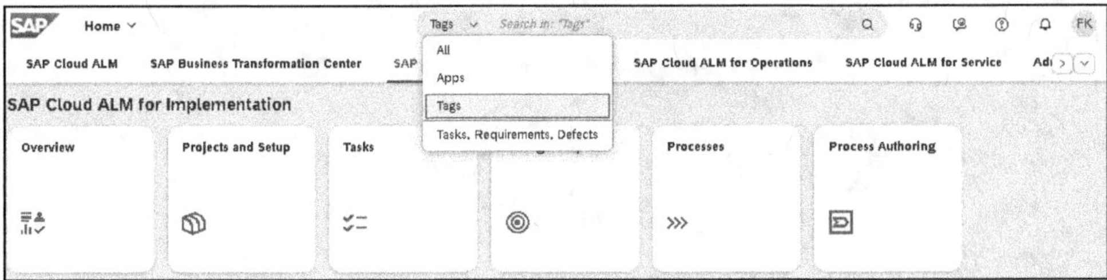

Figure 6.30 Search for Tags

Finally, you will find all related entities that are assigned to the corresponding tag. If you follow our recommendation for naming conventions and abbreviations, you can easily identify which type the results belong to. You can see an example in Figure 6.31.

The result list can be filtered via the filter button [filter icon]. Here, we have searched for "risk" as the tag and afterward can easily filter for different risk level: low, medium, and high. In the filter area you can also filter for object types (e.g., requirements, features, roadmaps tasks, solution processes, etc.) and projects as shown in Figure 6.32.

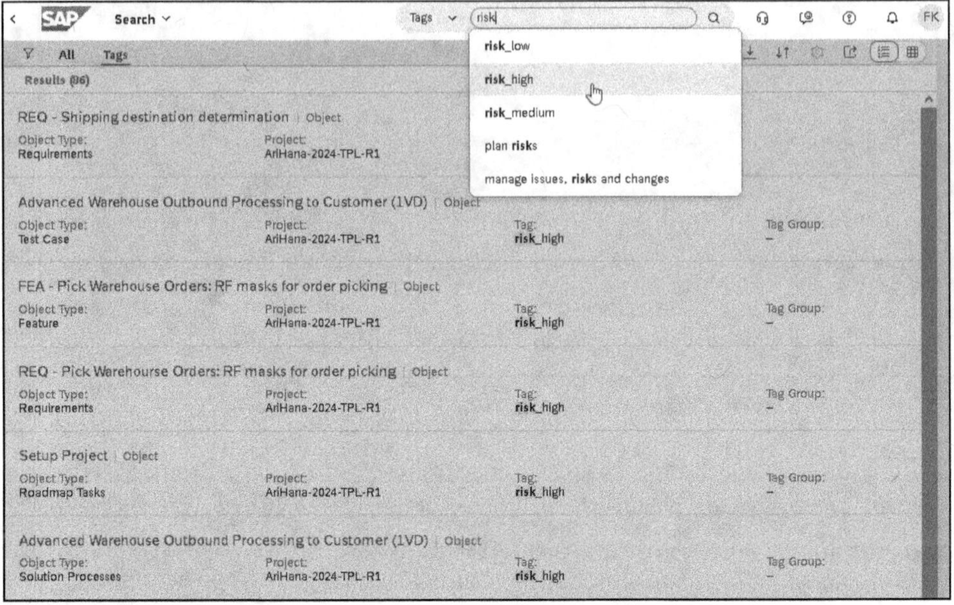

Figure 6.31 Search for "risk_high" Tags

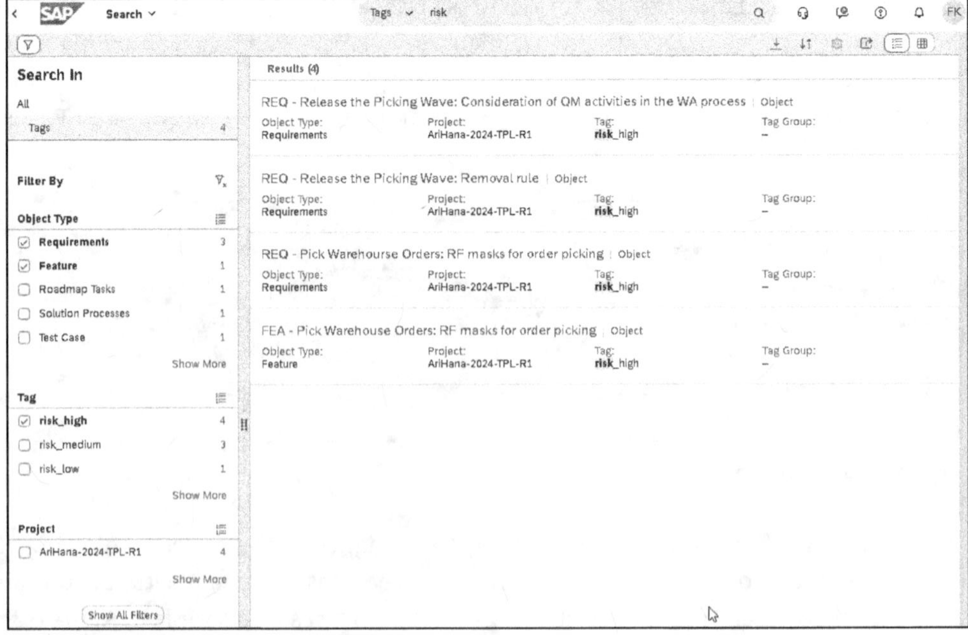

Figure 6.32 Search for Tags: Filter

You can also directly jump into a search result if you click on the description of the relevant entity. Another easy option is the share button . You can share this search via email, but a more useful choice is the **Save as Tile** option. After selecting **Save as Tile** you see a popup as shown in Figure 6.33. After saving your search as tile, you can easily jump to your search once you've started SAP Cloud ALM.

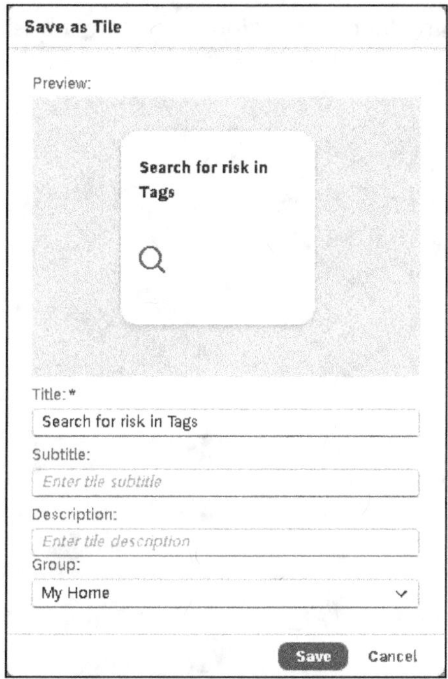

Figure 6.33 Search for Tags: Save as Tile

> **Tags for Result Lists**
>
> By default, tags are not displayed in lists. But you can activate it via the **Personalization Settings** button . After selecting tags and saving, tags will be displayed in the results list. You will also be able to search for just part of a tag label, e.g., "risk" instead of "risk_high." If there are more than five tags assigned to a solution process, the first five occurring alphabetically are shown in full, and any further tags can be viewed by selecting the **More** link, where *n* shows the exact number of additional tags.

6.7 Summary

The project is the central bracket for all project activities within the requirement-to-deploy approach. This is where all processes and their requirements are documented, from creation and implementation through to the rollout of the developed functions. In this chapter, we have used our example scenario to show you how to create a project in SAP Cloud ALM for the hybrid scenario with the associated entities such as time-boxes, teams, scopes, the deployment landscape, and the deployment plan. The project is now technically prepared. In the following chapter, we will look at the content of the solution to be implemented. We will present tools that can help determine the appropriate project scope from legacy SAP systems. After defining the scope or activating the required solution scenarios, the associated SAP Best Practice processes are provided in SAP Cloud ALM. The focus of this chapter is on the documentation of the target business processes (solution processes).

Chapter 7
Define Project Scope

In this chapter, we show you how to define the project scope. We also explain how to prepare a fit-to-standard workshop. In this context, we look in particular at the significance of the selectable scopes and how they relate to the solution processes. Finally, we present two tools that will help you determine the project scope.

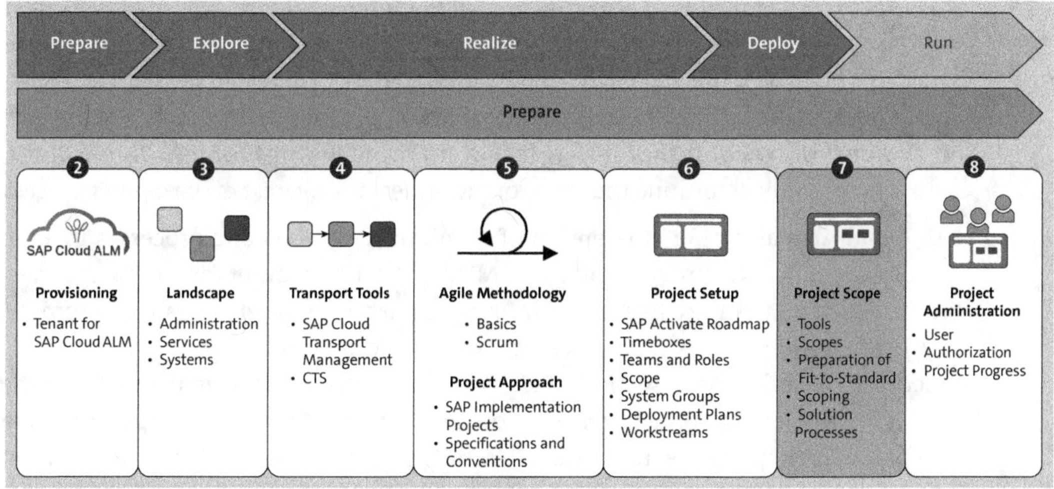

The previous chapter dealt with project creation in SAP Cloud ALM. For the example scenario, we created a project for the hybrid scenario in the system, considering our recommended guidelines and conventions from Chapter 5, Section 5.4. We went into detail about project properties such as timeboxes, teams, scopes, deployment landscape, and the deployment plan. In the further progress of the prepare phase, we will now focus on the first content-related aspect of the solution to be implemented. We will assign relevant solution scenarios to the project in our example scenario and add the core processes for the required business context. After providing these SAP Best Practices processes (also known as *solution processes*) in SAP Cloud ALM, we will explain the preparatory measures for discussing the selected business processes in the downstream process analysis workshops (fit-to-standard and fit-gap) during the explore phase (evaluation).

At the end of this chapter, we will introduce you to two SAP tools that will help you to determine the correct scope of SAP legacy systems for your project: SAP Readiness Check and SAP Business Transformation Center.

The documentation of business processes helps customers create transparency and makes processes repeatable and traceable. It creates competitive advantages and is fundamental for all types of projects, regardless of the industry or business area in which they operate. Processes are the single source of truth and enable everyone involved to gain a better understanding of the company's workflows. This chapter describes how SAP Cloud ALM can be used to document and manage business processes. We will first discuss how the project lead (i.e., the project manager) defines the scope of the project, which is necessary for further work with the processes, and will then explain how to prepare a fit-to-standard workshop. In the course of the project work, you must define tasks and solution processes. We will show you how to do it and what needs to be taken into account. Finally, the filter functions relating to the business process group and the country or region are shown within the solution processes. Among other things, this serves to simplify the search within the processes. The knowledge conveyed in this chapter serves to provide you a better understanding of the project and process structure that will be detailed in the following chapters. The challenge here is to correctly determine the scope of the project for the rest of the overall process.

SAP Cloud ALM also offers the option of monitoring projects and processes in a targeted manner. We will provide you with SAP Best Practice content for rapid implementation, which you can use to start your fit-gap or fit-to-standard workshops promptly and create requirements and notes. It is advisable to proceed in several steps, as described in the following sections. Before you can start with the actual work in your project and achieve the project goal later on, the first step is to define the project scope, is of particular importance in SAP Cloud ALM. The definition of the scope within a specific project is also known as the scoping process, project scoping, or just *scoping*. After scoping, the scope then consists of a selected set of solution scenarios that represent *solution processes*. The scope is also defined by a collection of solution processes that are processed together in the selected project. SAP Cloud ALM supports the scoping process by allowing you to define scopes that act as containers for solution processes to be managed together.

In the course of project scoping, the project process scope is then finalized by adding solution processes to the original scopes. These processes are grouped into a scope, and then solution scenarios with a specific content version are assigned to the scopes. The processes are then selected from these solution scenarios. This is the actual scoping process. Processes are usually assigned to the same scope if they use the same application system landscape, require a coordinated upgrade cycle, or serve the same functional purpose, such as finance or human resources processes. These activities are explained in more detail in the following sections and illustrated using examples.

7.1 Create Scope

Access to the applications mentioned is always via the SAP Fiori launchpad. There you can record, create, display, and manage processes in the **SAP Cloud ALM for Implementation** area. The Projects and Setup, Processes, and Process Authoring apps are essential for creating scopes and recording processes as well as process creation. They therefore form the starting point for process management and the subsequent steps. The content that needs to be selected for this is provided by *SAP Enterprise Architecture Reference Library* (SAP EA Reference Library).

Several steps are necessary to define the scope of the project and thus form an important basis for all subsequent processes. The project management team should therefore agree with the business process experts on the scope of the project and which solution processes need to be included. To define this scope, we will show the necessary steps in detail below using the system. First, select the **SAP Cloud ALM for Implementation** area in the SAP Fiori launchpad (see Figure 7.1). The most important apps are available for selection in this area.

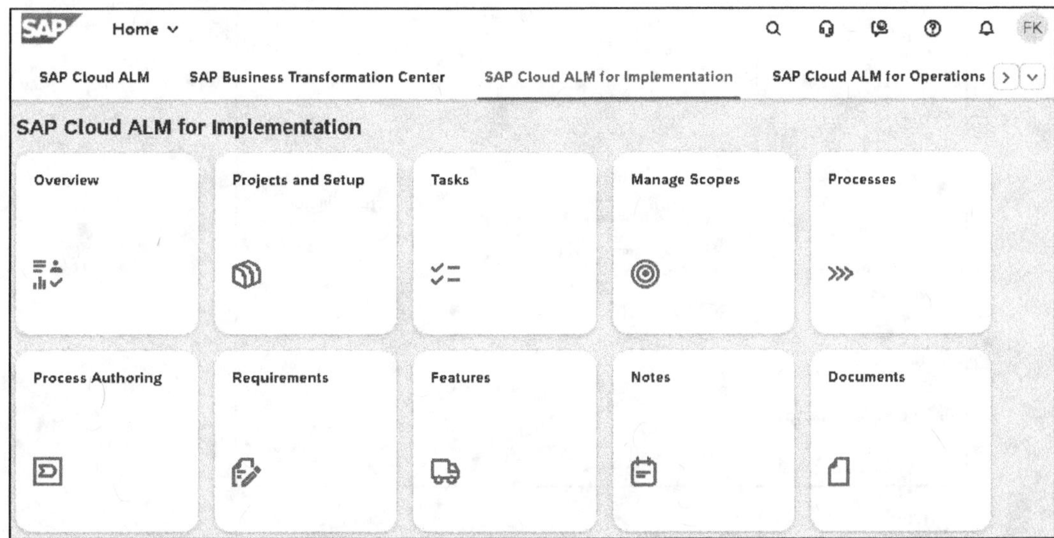

Figure 7.1 Launchpad for SAP Cloud ALM for Implementation

First, select the Projects and Setup app. In this app you will see all the necessary information about the projects and can edit the scopes. As you can see in Figure 7.2, we have selected our example project AriHana-2024-TPL-R1. In each project, the tabs **General Information**, **Timeboxes**, **Teams**, **Scopes**, **Deliverables**, **Subscriptions**, and **Deployment Landscape** are displayed.

As described in the previous chapters, a *roadmap template* must first be selected from the SAP Activate methodology. This converts the roadmap instructions into actionable and comprehensible tasks. As shown in Figure 7.2, the **SAP S/4HANA (on-premise) - New Implementation** template was selected for this project. Now you can define the *timeboxes*—i.e., the time periods resulting from a combination of phases and milestones. Optionally, you can also use sprints that follow the agile methodology. On the next tab, the teams are added and the people are assigned to the respective roles in the project. In addition, further phases, roles, tags, workspaces, and deliverables are available for the project. Optionally, you can use the Excel upload to quickly load content into the project. Defining the scope is one of the last steps in the project setup, which you perform on the **Scopes** tab, as shown in Figure 7.3. Click on the **Manage Scopes** button to start defining the scopes.

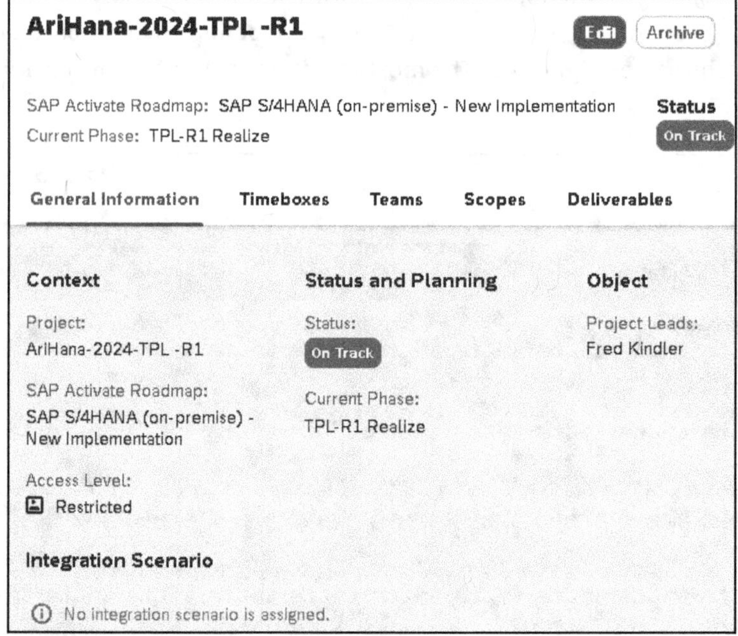

Figure 7.2 Projects and Setup

The implementation of these activities can be documented using the **Create Scopes and Define Business Processes Scope** roadmap task. You can find this roadmap task in the SAP Cloud ALM setup task source. There are no defined scopes at the start of a project. You can create them via the **Create** dropdown field. There are two options: You can create the scope directly (**Create • Scope**) or upload it from a file (**Create • Scope from File**; see Figure 7.4).

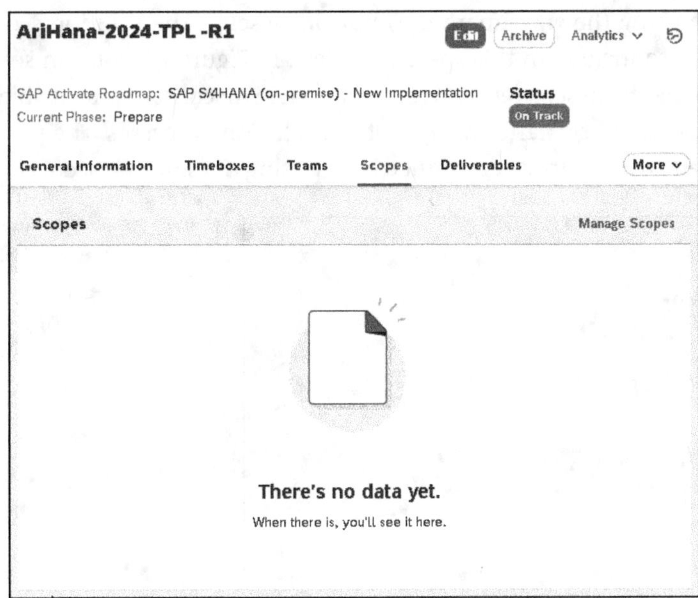

Figure 7.3 Define Scopes

Figure 7.4 Create Scope

First select **Create • Scope**. This takes you to the overview for scope creation (see Figure 7.5), where you can maintain general information about the scope. After that, you can add the scopes in the **Solution Scenarios** area.

You define the scope by using the checkmark button to select the solution scenario that you would like to include in this specific project. In Figure 7.6 you can see that the **Custom** scenario has been selected as an example. The SAP content is available via the solution scenarios. These scenarios consist of the solution processes. The processes that can be selected are therefore determined by selecting the solution scenario.

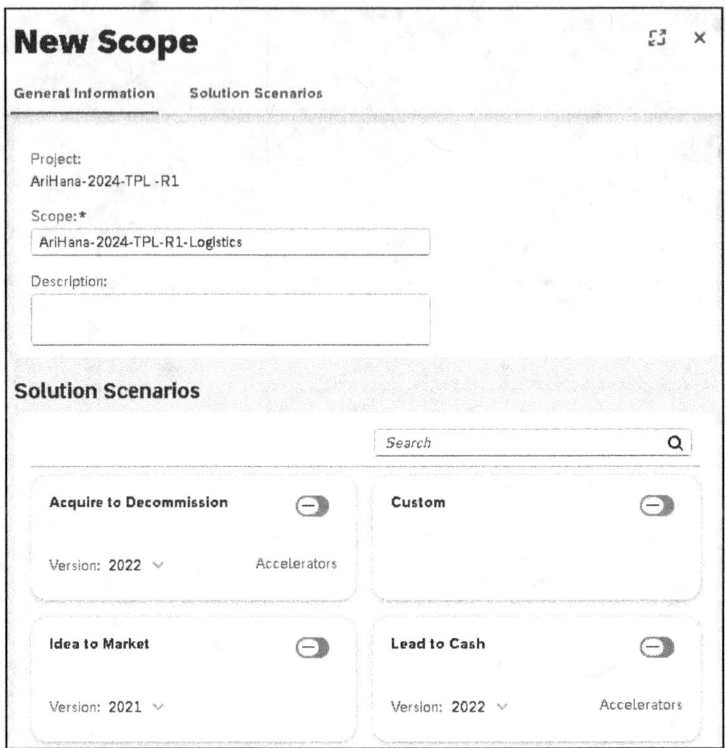

Figure 7.5 Create Scope (2)

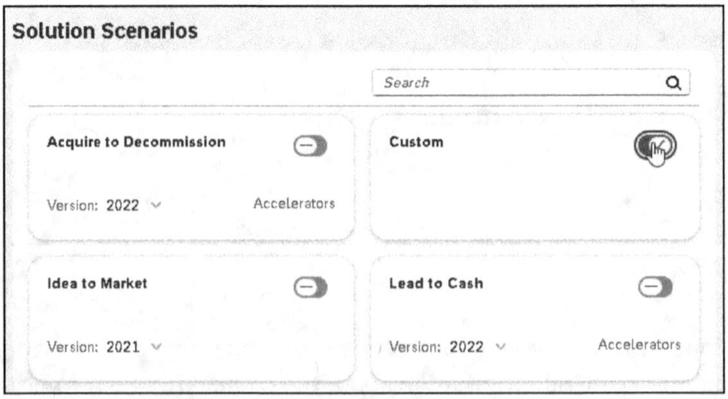

Figure 7.6 Select Custom Solution Scenario

For our example project, we have also activated the SAP Best Practices for SAP S/4HANA Cloud Private Edition scenario in version 2023 in order to have the currently available processes from the SAP Best Practices content for the implementation ready for our AriHana template project (see Figure 7.7).

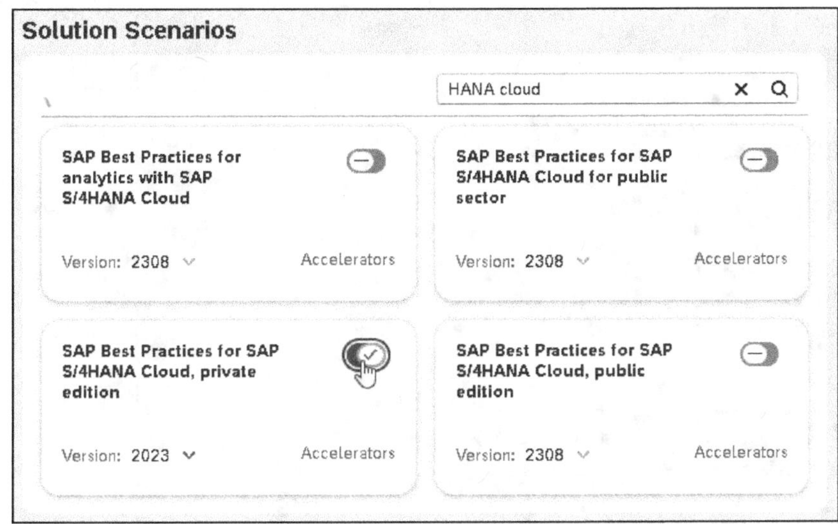

Figure 7.7 Select SAP S/4HANA Solution Scenario

In Figure 7.8, you can see the message that the system displays as soon as a scenario has been selected and you want to save it. However, you are not forced to include corresponding processes from a scenario that is no longer required in the scope. Processes that are not or no longer required can therefore also be set as "not relevant."

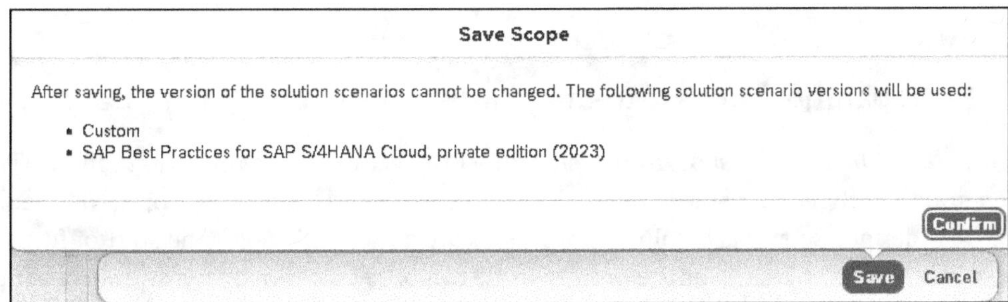

Figure 7.8 Information Message

[!]

Information on One-Time Activation

When activating the scenarios, make sure that solution scenarios can no longer be deselected after they have been selected and saved.

Now we will enter further information for the scope of the AriHana-2024-TPL-R1 project. In the mandatory **Scope** field, we enter the name of the scope: "AriHana-2024-TPL-R1-Logistics". In the **Description** field, we add the following description: **S/4HANA + ARIBA Template Release 1 LoB Logistics** (see Figure 7.9) and click **Save**.

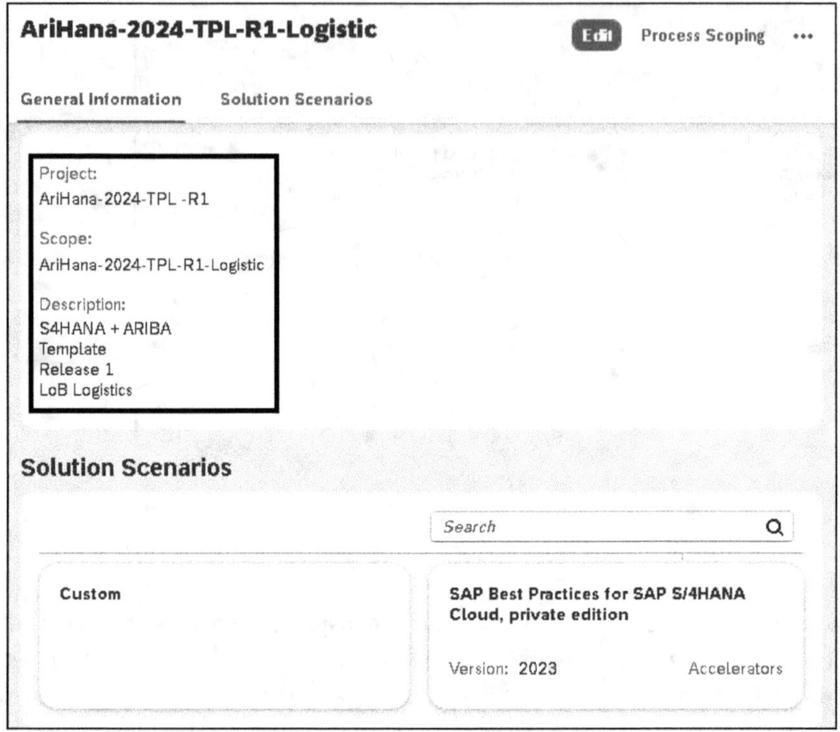

Figure 7.9 Fill In General Information of the Scope

7.2 Preparation for a Fit-to-Standard Workshop

In a *fit-to-standard workshop*, the possible functions are explored, the future solution is examined with regard to the company's requirements, and the project scope is defined. Clearly defined scopes and solution processes are a prerequisite for the start of fit-to-standard workshops. The entire project scope can only be defined after the successful completion of the fit-to-standard workshops at the end of the exploration phase. The project scope consists then of the determined process scope and the requirements from the workshops. By selecting the SAP Activate template or the SAP Activate roadmap in the project setup, several task templates are generated in your project. These tasks also include the planning and implementation of a fit-to-standard workshop. Once the project scope has been defined, the scoping process is complete. In Figure 7.10 you can see the status of the project after the scoping process has just been completed.

Manage Scopes

Project:

Search

AriHana-2024-TPL -R1

Scopes (1)

Create

AriHana-2024-TPL-R1-Logistic

Figure 7.10 Manage Scopes

To start working on the processes, the following requirements must be met, as already explained:

- An SAP Cloud ALM project was created.
- The initial scope of the project was defined as the SAP Best Practices for SAP S/4HANA Cloud Private Edition solution scenario.

You have two options for accessing the process editing area in SAP Cloud ALM. You can view and access all available functions via the project overview, as shown in the example in Figure 7.11.

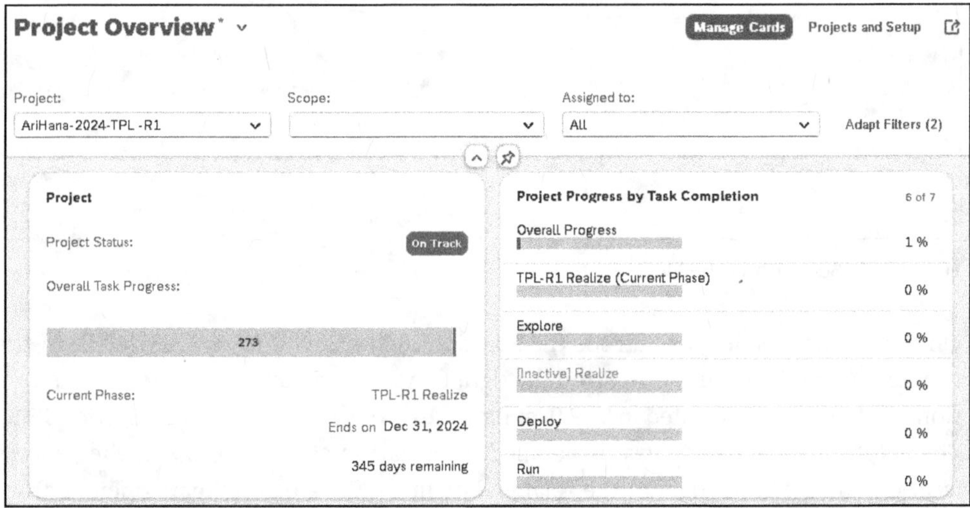

Figure 7.11 Project Overview

To do this, select the button in the selection panel of the sidebar. In this view, you will now find various tiles for the respective applications in the project. Use the filter settings to specify which project and possibly also which scope you want to view. Click on the corresponding tile to go directly to the respective application. You can also click directly on the **Solution Processes per Scope** or **Solution Process Statuses** tile to go directly to the solution process scope area of your project. This overview also shows a comprehensive status of the current number of processes and tasks as well as their status. The second way to access the processes is to select the Processes app from the SAP Fiori launchpad. If you are not in the launchpad, you can access the processes at any time via the panel on the left-hand side.

Once you have completed the initial setup of the project in SAP Cloud ALM and defined the solution scenarios, it's time to start working with the processes.

In SAP Cloud ALM, you have several options for working with a solution process. Let's look at what is available for a separate solution process.

In our project, we will use the **SVF Advanced Warehouse Outbound Processing to Customer (1VD)** process as an example, as shown in Figure 7.12.

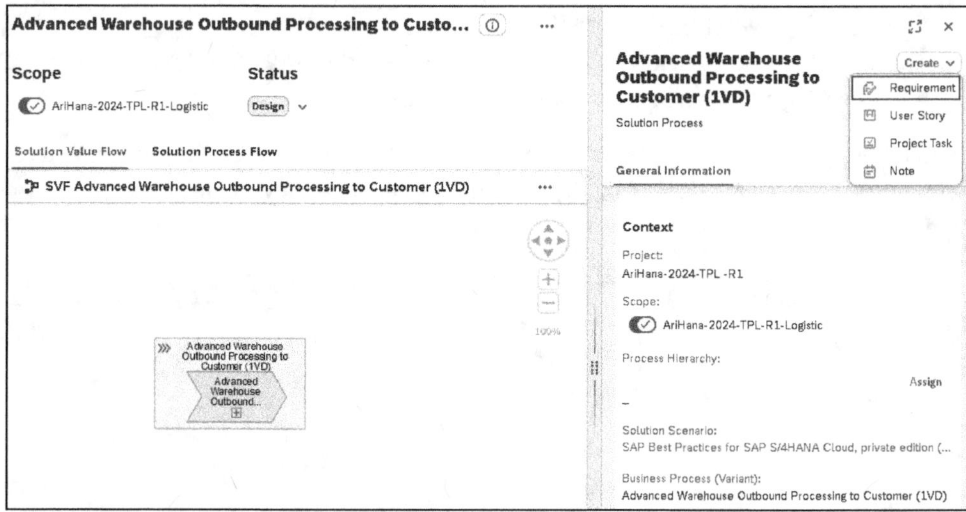

Figure 7.12 Solution Value Flow

On the left-hand side, you can see the **Solution Value Flow** and various details of the process on the right-hand side. From the right pane, you can also jump to setup instructions and test scripts located in SAP Best Practices Explorer (see Chapter 1, Section 1.3.3). The solution value flow gives us detailed information about the role of the process (scope element) in the overall site-related data flows. Clicking on **Create** opens a drop-down menu in the detail area on the right-hand side, which you can use to assign the following elements:

- **Requirement**: Expectations of the company that must be met
- **User Story**: Defining requirements from the user's perspective
- **Project Task**: Units for teamwork and clarifications
- **Note**: Optional additional information

Such a detailed structure of the solution process helps the project team—and in particular, the person responsible—to implement the scope item thoroughly and to track all associated tasks and process flows. Another important basis in the preparation for the fit-to-standard workshop is a general overview of the processes and their status. To obtain an overview of the current status of the processes, SAP Cloud ALM provides an analytical view that can also be helpful for a fit-to-standard workshop. To use this, select the Analytics app within the Process Overview app in the top-right area (see Figure 7.13).

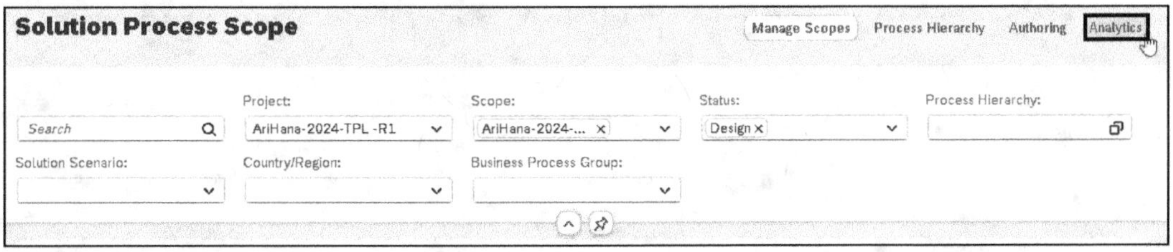

7

Solution Process Scope
Manage Scopes | Process Hierarchy | Authoring | Analytics
Search | Project: AriHana-2024-TPL-R1 | Scope: AriHana-2024-... | Status: Design | Process Hierarchy:
Solution Scenario: | Country/Region: | Business Process Group:

Figure 7.13 Solution Process Scope

You can also access this view from the SAP Fiori launchpad. To do this, select the Analytics app and then click on the **Solution Process Traceability** area, as shown in Figure 7.14.

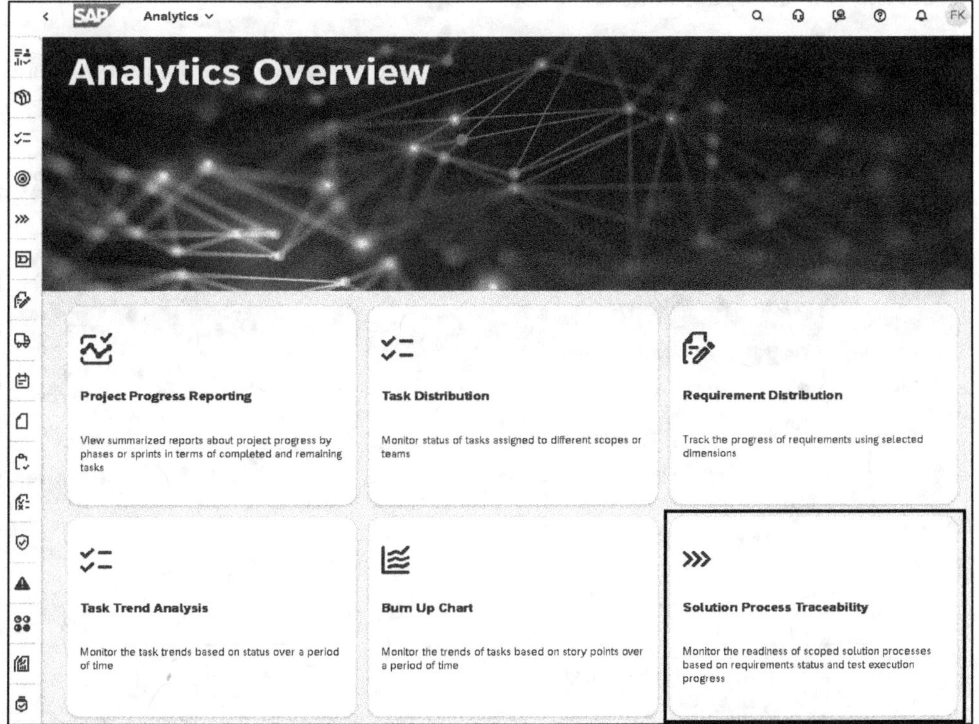

Figure 7.14 Solution Process Analytics App

This takes you to the traceability overview of the solution processes, in which you can also see the current status in connection with the existing requirements and test cases. As you can see in Figure 7.15, you can view the associated requirements by status for each individual process.

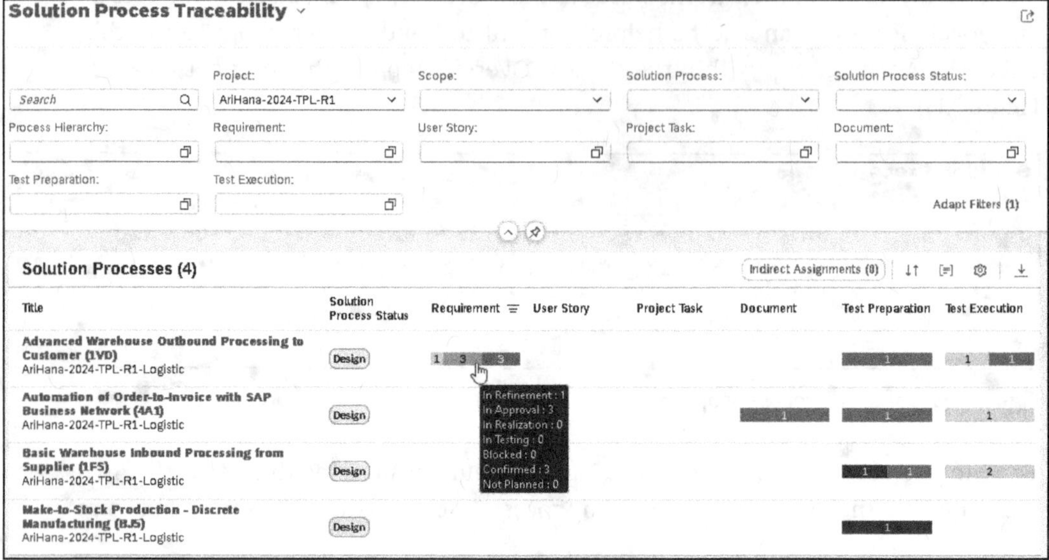

Figure 7.15 Traceability of Solution Processes

If you click on the bar shown, you will receive a detailed overview of the requirements (see Figure 7.16), as well as the test cases that have already been created (see Figure 7.17).

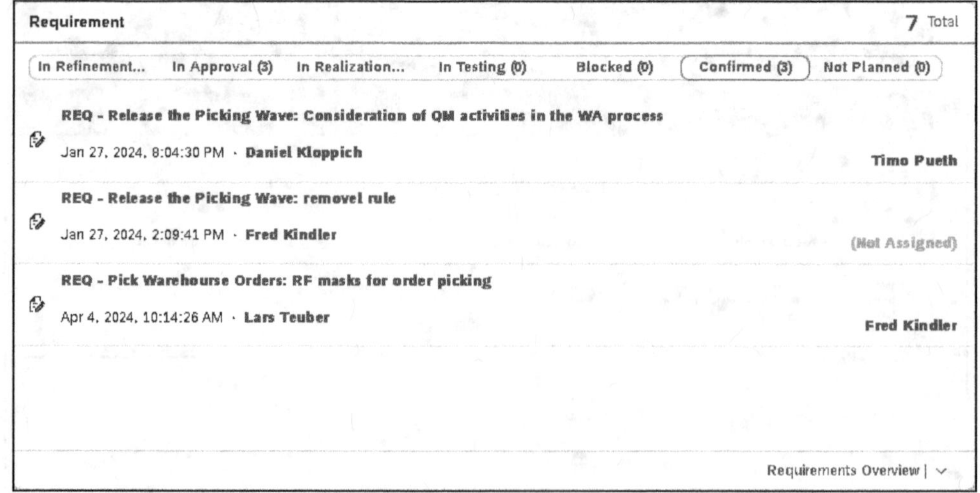

Figure 7.16 Traceability of the Requirements

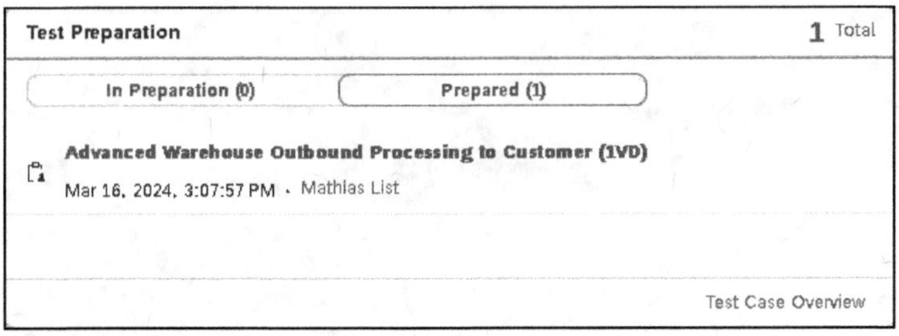

Test Preparation 1 Total
In Preparation (0) Prepared (1)
Advanced Warehouse Outbound Processing to Customer (1VD)
Mar 16, 2024, 3:07:57 PM · Mathias List
Test Case Overview

Figure 7.17 Traceability of Test Cases

7.3 Define Tasks

In the previous sections, we explained how to set up a project and therefore also how to define the timebox. When you set up the timebox (the time frame for the project), you can choose between phases, sprints, and milestones. After defining the template and timeboxes, you can assign timeboxes to specific tasks via the Tasks app by clicking on the (**Select Timebox**) button. It is located next to the currently set timebox, to assign any timebox to a specific task (see Figure 7.18 and Figure 7.19).

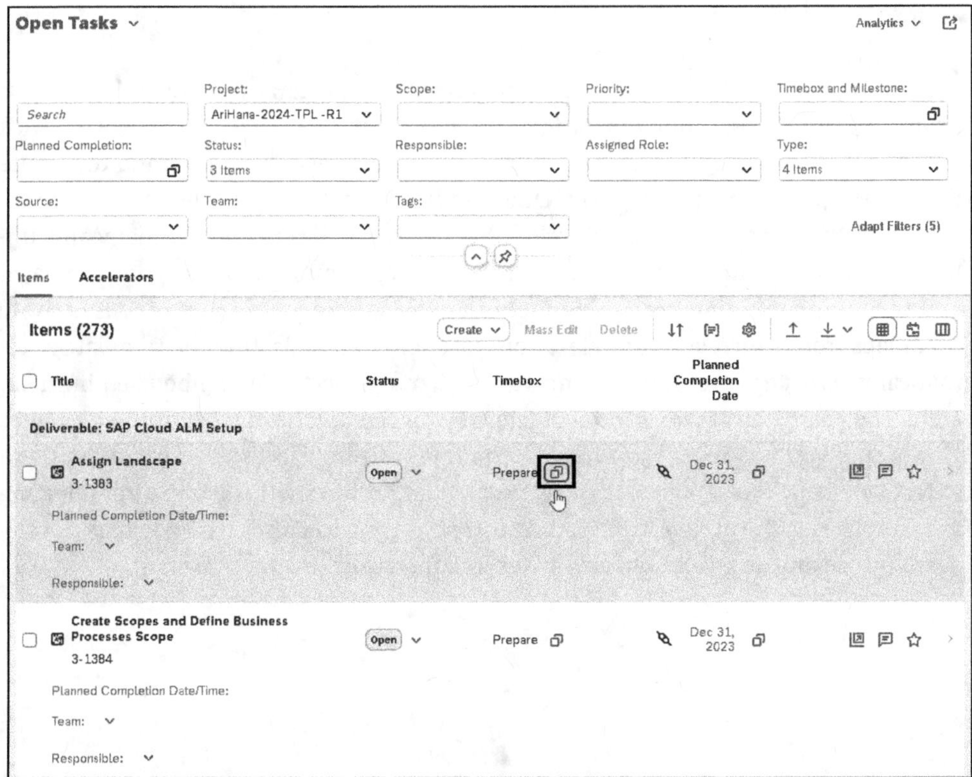

Figure 7.18 Timeboxes

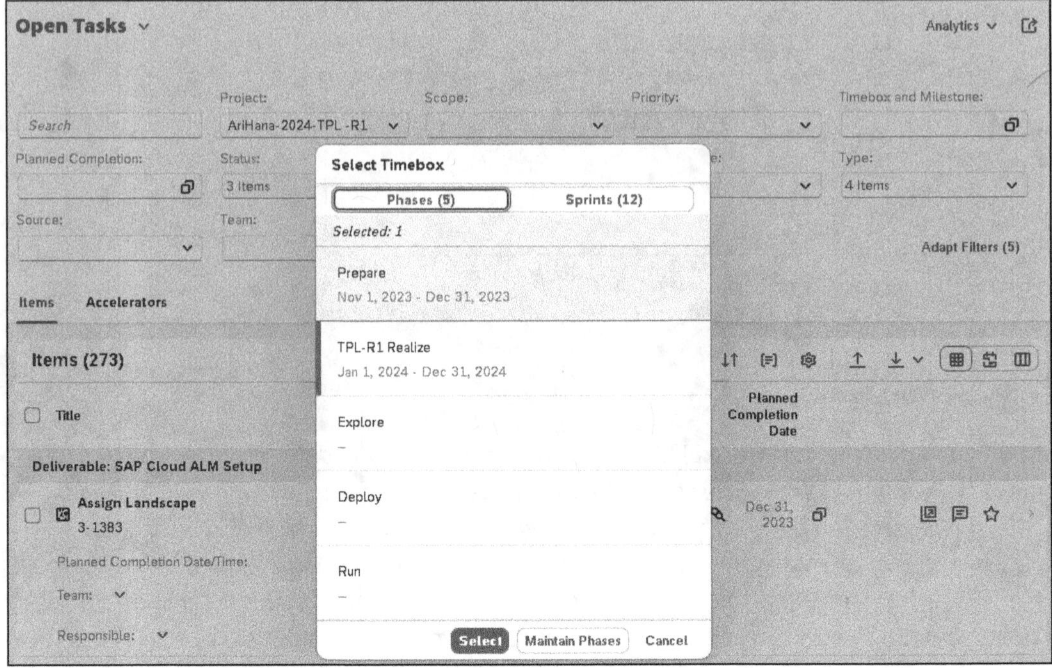

Figure 7.19 Select Timebox

This automatically sets the end date of the timebox as the new due date of the task. For example, if you assign the preparation phase to a task, the duration of this task ends with the selected phase. However, you can also create tasks manually and distribute them to the predefined project roles. Other task types that you can select and define are project tasks, user stories, and subtasks. These tasks can be created manually and, unlike the template, can also be deleted. All of these task types can be assigned to phases and sprints. As explained earlier, SAP Cloud ALM contains a number of SAP Activate roadmaps from which you can select your task template. Selecting the template automatically activates several task templates for your project that can be used immediately. The generated tasks cannot be deleted, and the description cannot be changed. However, if you no longer want to use certain tasks from this list, you can mark them as **Not Relevant**. You can display the generated task list in the Tasks app after selecting the template, as shown in Figure 7.20. At this point, you can also mark a task as not relevant by clicking on the drop-down arrow and then selecting **Not Relevant**.

Figure 7.20 Adjust Task Status

7.4 Define Solution Process

As described in Section 7.2, you define the scope of the project in the system as part of the fit-to-standard workshop during process scoping. To define the project scope, add individual solution processes to it. You have the option of changing the scoping again and setting the status of the process. The following status values can be selected in the Processes app throughout the project:

- **Design** (before the fit-to-standard process is completed)
- **Realization** (if the process is still in configuration)
- **Production** (the process is already being used by end users)
- **Maintenance** (adjustments are required, but the process is still in use)
- **Obsolete** (the process should no longer be used)

When you have finished making your selection, you can complete the process by clicking on the **End Scoping** button and, if necessary, continue by clicking on the **Edit Scope** button.

For our example project, the data shown in Figure 7.21 were selected in the scope AriHana-2024-TPL-R1-Logistics, and the **Design** status was defined.

Solution Processes (4) Edit Scope

Title	Scope	Status	Process Hierarchy	Business Process (Variant)	Country/Region
Scope: AriHana-2024-TPL-R1-Logistic					
Advanced Warehouse Outbound Processing to Customer (1VD) SAP Best Practices for SAP S/4HANA Cloud, private edition (2023)	AriHana-2024-TPL-R1-Logistic	Design		Advanced Warehouse Outbound Processing to Customer (1VD) Supply Chain	Australia · 24 More
Automation of Order-to-Invoice with SAP Business Network (4A1) SAP Best Practices for SAP S/4HANA Cloud, private edition (2023)	AriHana-2024-TPL-R1-Logistic	Design		Automation of Order-to-Invoice with SAP Business Network (4A1) Sales	Australia · 38 More
Basic Warehouse Inbound Processing from Supplier (1FS) SAP Best Practices for SAP S/4HANA Cloud, private edition (2023)	AriHana-2024-TPL-R1-Logistic	Design		Basic Warehouse Inbound Processing from Supplier (1FS) Supply Chain	Australia · 25 More
Make-to-Stock Production - Discrete Manufacturing (BJ5) SAP Best Practices for SAP S/4HANA Cloud, private edition (2023)	AriHana-2024-TPL-R1-Logistic	Design		Make-to-Stock Production - Discrete Manufacturing (BJ5) Manufacturing	Australia · 41 More

Figure 7.21 Select Solution Processes

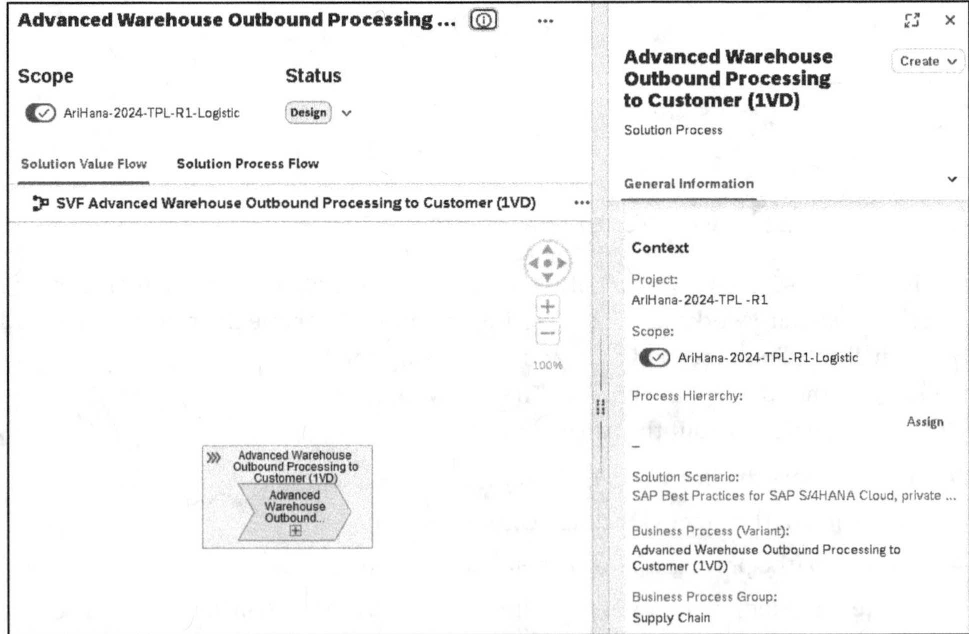

Figure 7.22 Solution Value Flow of a Process

Before selecting a solution process, you can call up the detailed view to see the process displayed graphically and the process flow descrWibed. In Figure 7.22, the use of example process **Advanced Warehouse Outbound Processing to Customer (1VD)** is shown. First, the **Solution Value Flow** is shown, which we will discuss in Chapter 9, Section 9.3. On the tab next to it, you can see the detailed **Solution Process Flow** in the form of a diagram. We will go into the solution process flow in more detail in Chapter 9, Section 9.4.

If you enlarge the window on the right-hand side, you can also view general information and a description of the process (see Figure 7.23).

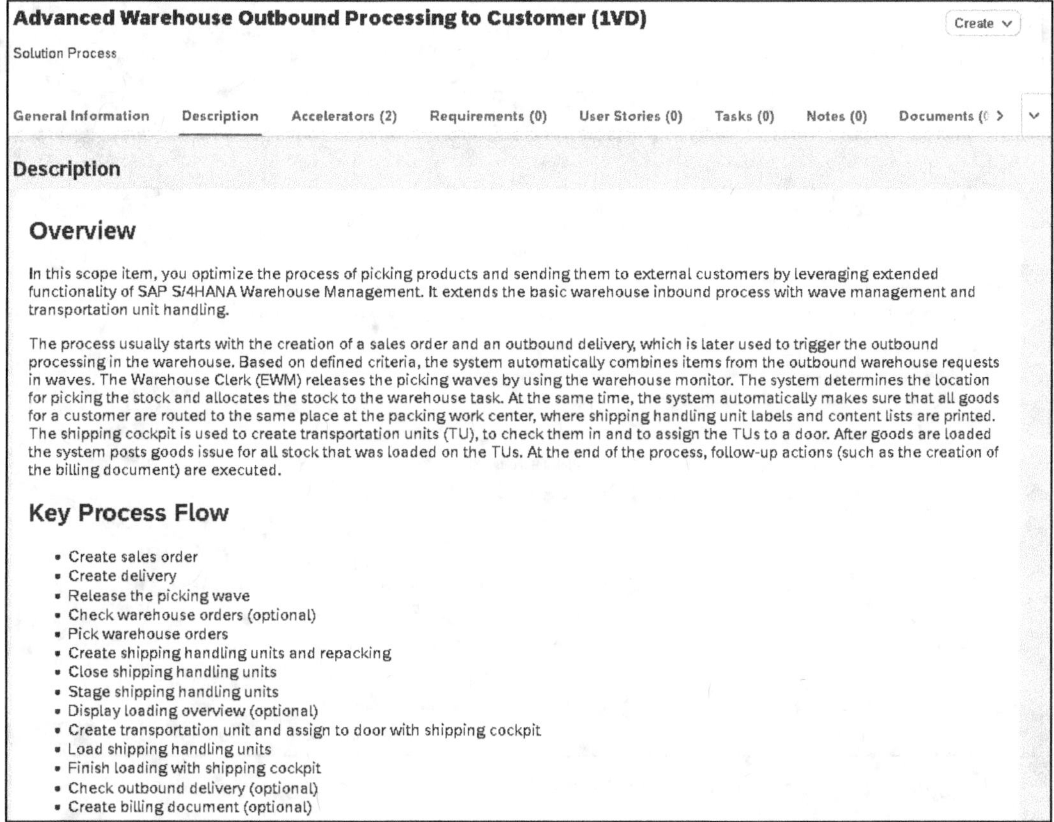

Figure 7.23 Process Description

On the other tabs, you can display the elements created in each case (**Accelerators**, **Requirements**, **User Stories**, **Tasks**, **Notes**, and **Documents**) for the solution process flow. In Chapter 9, you will find further information on and differences between these tabs.

Solution processes can be grouped into business process groups for a better overview. For example, you can have all processes from the **Human Resources** area aggregated into the **Human Resources** group, and all financial processes into the **Finance** group. In this way, you can search for specific processes that are only used in **Asset Management**, **Finance**, or **Human Resources**, for example. This function enables a targeted selection of processes within a business area.

The **Solution Processes** view offers the option of filtering processes according to the business process group (see Figure 7.24). To do this, simply click the **Business Process Group** field [v] button in the **Business Process Group** field and select the business areas you want to filter by from the dropdown menu.

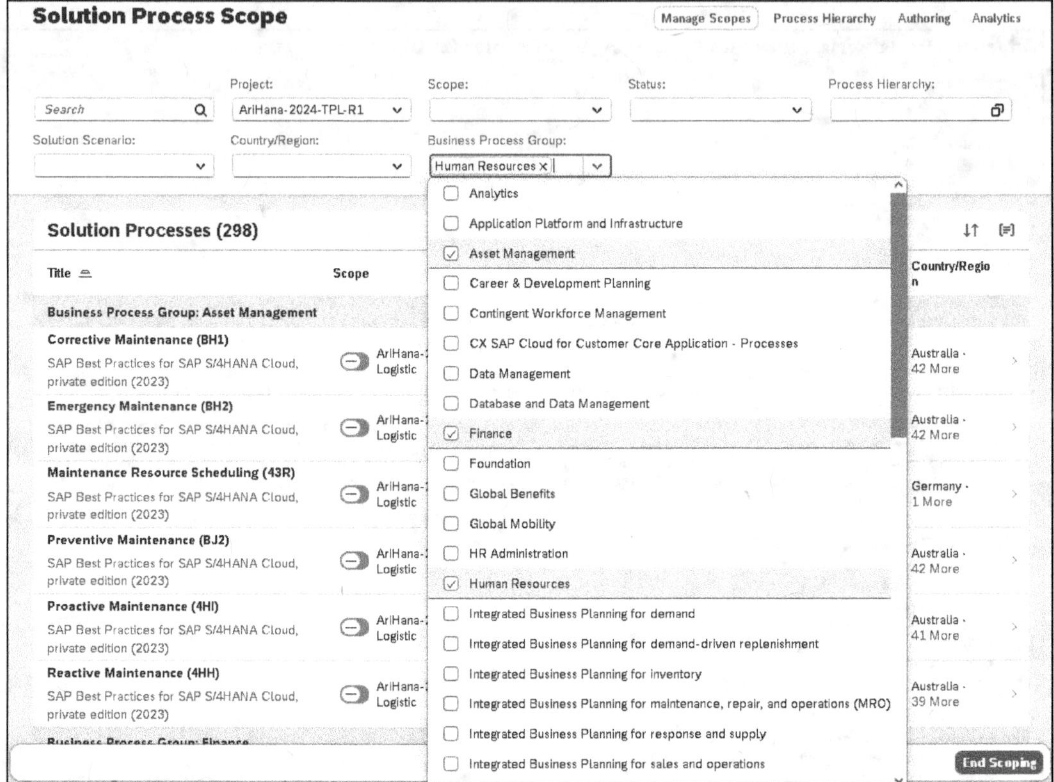

Figure 7.24 Region and Business Process Group

For our example, we have selected three areas. The result of this filtering is shown in Figure 7.25.

Within the solution process scope window, you can filter not only by the business process group, but also by the countries or regions for which the solution process has been created. Figure 7.26 shows an example of the filter setting for **Austria**, **Germany**, and **China**.

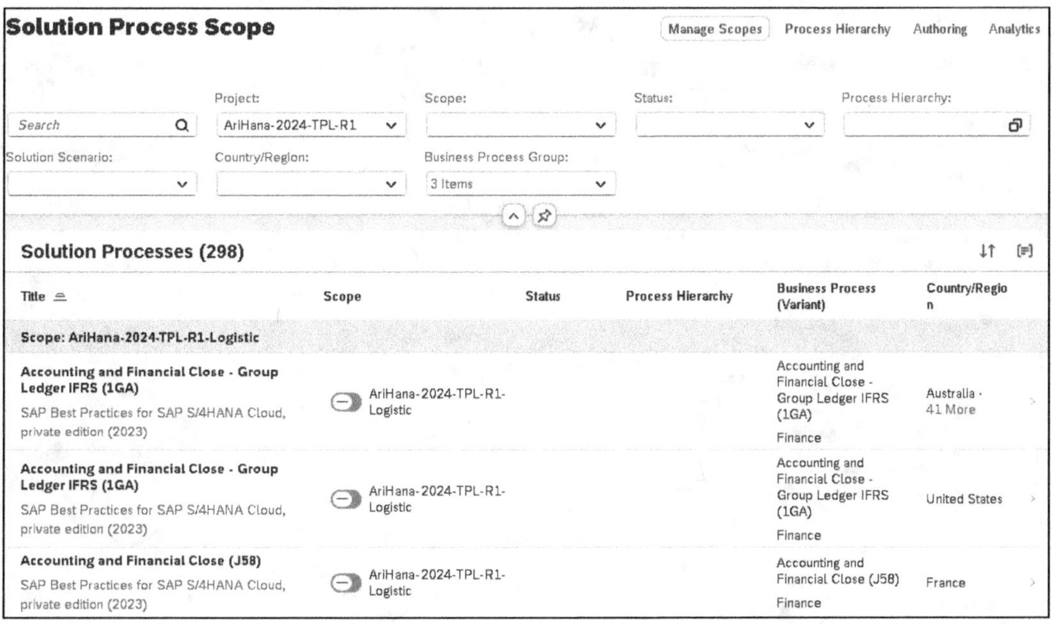

Figure 7.25 Filtering by Business Process Groups

Solution Processes (418)

Title	Scope	Status	Process Hierarchy	Business Process (Variant)	Country/Region
Scope: AriHana-2024-TPL-R1-Logistic					
Accounting and Financial Close - Group Ledger IFRS (1GA) SAP Best Practices for SAP S/4HANA Cloud, private edition (2023)	AriHana-2024-TPL-R1-Logistic			Accounting and Financial Close - Group Ledger IFRS (1GA) Finance	Australia · 41 More
Accounting and Financial Close (J58) SAP Best Practices for SAP S/4HANA Cloud, private edition (2023)	AriHana-2024-TPL-R1-Logistic			Accounting and Financial Close (J58) Finance	Germany
Accounting and Financial Close (J58) SAP Best Practices for SAP S/4HANA Cloud, private edition (2023)	AriHana-2024-TPL-R1-Logistic			Accounting and Financial Close (J58) Finance	China
Accounting and Financial Close (J58) SAP Best Practices for SAP S/4HANA Cloud, private edition (2023)	AriHana-2024-TPL-R1-Logistic			Accounting and Financial Close (J58) Finance	Austria
Accounting for Incoming Sales Orders (2FD) SAP Best Practices for SAP S/4HANA Cloud, private edition (2023)	AriHana-2024-TPL-R1-Logistic			Accounting for Incoming Sales Orders (2FD) Finance	Australia · 41 More

Figure 7.26 Filtering by Country and Region

The result of this filtering is shown in Figure 7.27. If all countries or regions are available for a solution process, the term **Generic** is displayed in the **Country/Region** column. In this way, you can also display special processes that explicitly apply to Germany, for example, and can only be used in this country for legal or tax reasons.

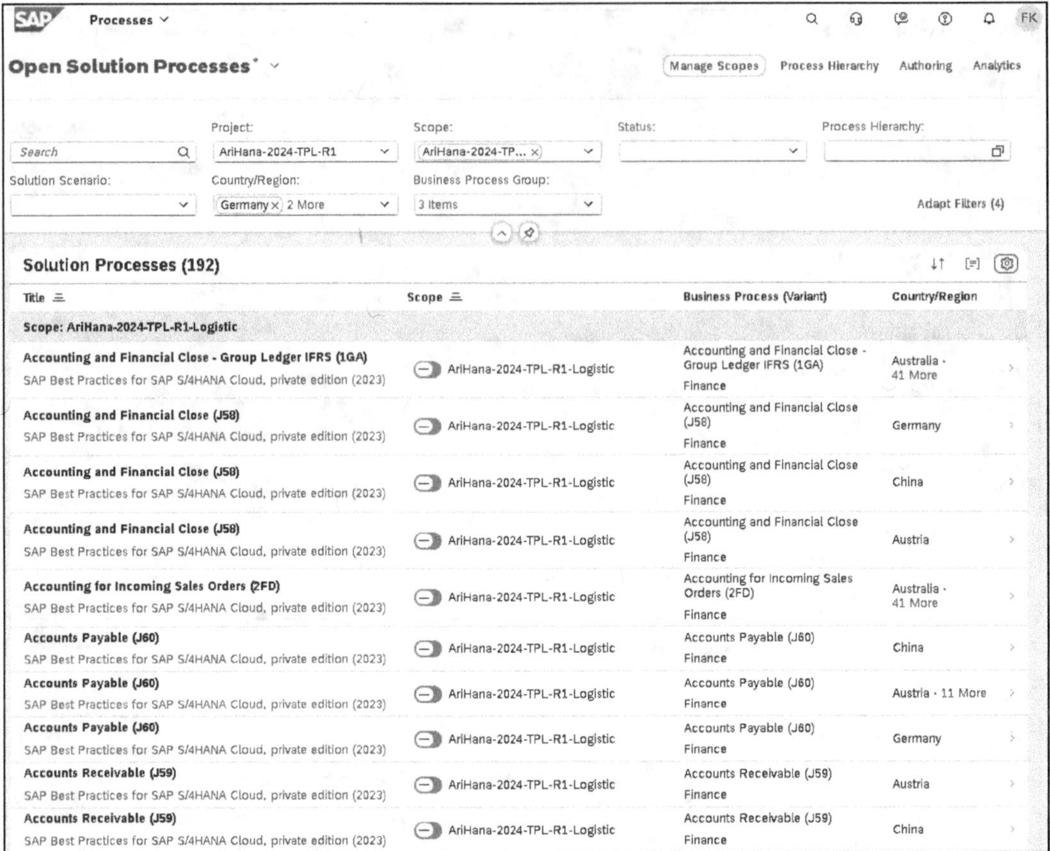

Figure 7.27 Filtering by Specific Countries

7.5 Tools for Scope Definition of the Project

If you are planning to migrate an existing SAP solution, you should consider two tools provided in SAP Cloud ALM: SAP Readiness Check and SAP Business Transformation Center. With these tools, you can analyze various data for specific scenarios, such as the scope of functions used, function usage, conversion compatibility, or deviations from the standard. These tools can in turn help you identify the right project scope for SAP Cloud ALM without having to take unused legacy data into account.

7.5.1 SAP Readiness Check

With *SAP Readiness Check*, SAP provides a suite of tools with which you can plan your path to digital transformation by comparing the information from your existing SAP solutions with the target solutions. We will explain the features in this section, as these resources will help you determine the right scope for your project in SAP Cloud ALM. The following scenarios are supported:

- SAP S/4HANA conversion: SAP Readiness Check for SAP S/4HANA provides a comprehensive overview of conversion compatibility and the preparatory steps required to convert an SAP ERP system 6.0 to SAP S/4HANA.
- SAP S/4HANA upgrade: SAP Readiness Check for SAP S/4HANA upgrades provides an overview of the compatibility of the existing system with the upgraded applications provided by the target release and the required preparation steps.
- SAP BW/4HANA conversion: SAP Readiness Check for SAP BW/4HANA provides an overview of the conversion compatibility and the necessary preparation steps for converting an SAP BW 7.x system to SAP BW/4HANA.
- SAP ERP usage and data profiling: SAP Readiness Check for SAP ERP usage and data profiling provides valuable data on implemented functions, component usage, and organizational structure, helping customers gain an overview of the structure and usage of their SAP ERP 6.0 system from a functional perspective.
- SAP Customer Experience solutions: SAP Readiness Check for SAP Customer Experience solutions shows how an existing SAP CRM 7.0 system is used, how it deviates from the standard, and what needs to be considered when planning the transition to the latest SAP Customer Experience solutions in a modern, scalable, and supportable target architecture.
- SAP SuccessFactors solutions: SAP Readiness Check for SAP SuccessFactors solutions shows how an existing SAP ERP Human Capital Management 6.0 system is used and how it deviates from the standard. These insights can then be used to scope and plan the transition to SAP SuccessFactors solutions.
- SAP Cloud ALM: SAP Readiness Check for SAP Cloud ALM provides insight into the application lifecycle management capabilities used in your SAP Solution Manager system. The scenario also provides availability details and information on equivalent capabilities in SAP Cloud ALM or other SAP products to consider when planning the transition from SAP Solution Manager to SAP Cloud ALM.

Within SAP Cloud ALM, you can use the SAP Readiness Check app from the **SAP Cloud ALM for Implementation** area to perform your analyses, as shown in Figure 7.28. Please note that an initial setup for SAP Cloud ALM is required to use SAP Readiness Check. You can find details on this in SAP Note 3344480.

After starting the app, you have the option of creating new analyses, opening your current analyses, and obtaining further information on the supported scenarios (see Figure 7.29).

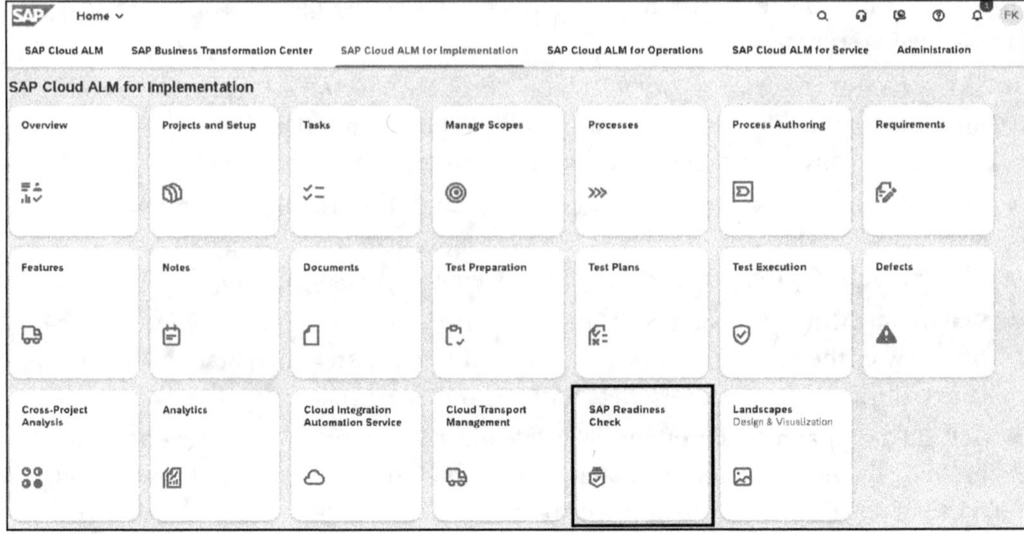

Figure 7.28 SAP Cloud ALM for Implementation App Group

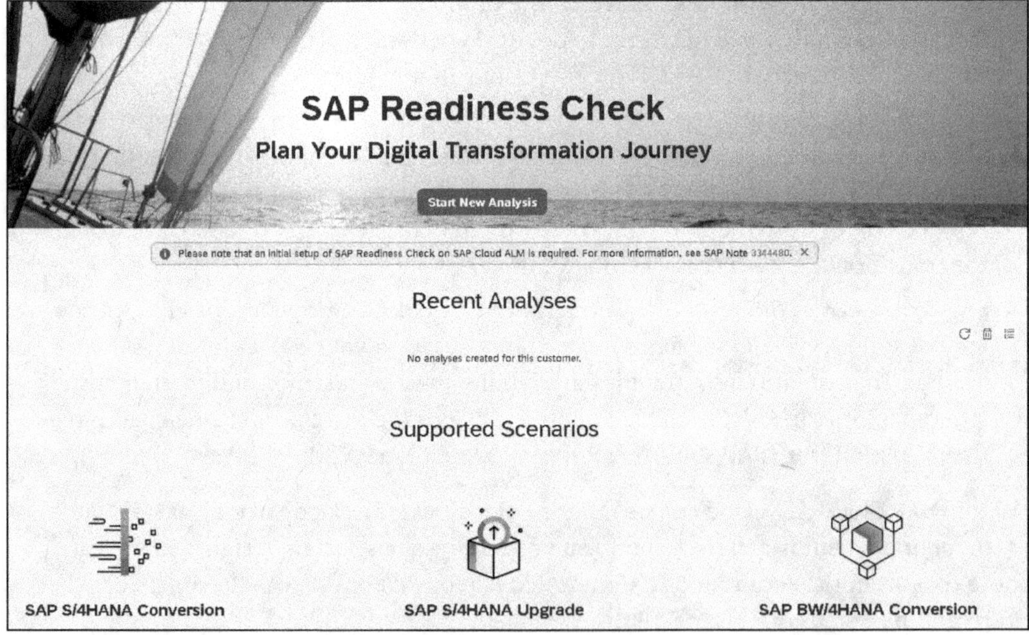

Figure 7.29 SAP Readiness Check

To start an analysis, you must first upload the results as a ZIP archive to SAP Cloud ALM using the **Start New Analysis** button. SAP recommends generating the results from the production system using the SAP Readiness Check analysis framework to obtain an exact image of the functions used. After the upload, the recorded data record is processed and is available for you to perform analysis.

If you are using SAP Solution Manager as an ALM tool, you can use SAP Readiness Check to analyze the functional scope used and thus prepare the transition to SAP Cloud ALM. Although this scenario is not actually relevant for determining the scope in the context of our example project, it nevertheless describes an important function of SAP Cloud ALM, which we will discuss in more detail.

The system analysis provides you with an overview of the scenarios and functions used within SAP Solution Manager. It also shows whether the equivalent functions are available in SAP Cloud ALM or whether they are planned for implementation in the future. This gives you a good overview of how you are currently using your SAP Solution Manager, which helps you to plan the transition of the individual ALM functions to SAP Cloud ALM. The transition can take place either completely or for each individual functionality.

SAP Readiness Check can help you answer the following questions, for example:

- Which functions are used on the current SAP Solution Manager system?
- Is there an equivalent function in SAP Cloud ALM, and what is the planning/delivery status of this function?
- Is there any further information on this function?

Based on the analysis of the source system, the result is presented in three groupings (see Figure 7.30):

Figure 7.30 SAP Readiness Check for SAP Cloud ALM

- **Usage in SAP Solution Manager**
  This category provides an overview of the scenarios and capabilities used in the analyzed source system. The functions are categorized according to their relevance for a migration. You can access the detailed analysis (see Figure 7.31) by clicking on the group title.

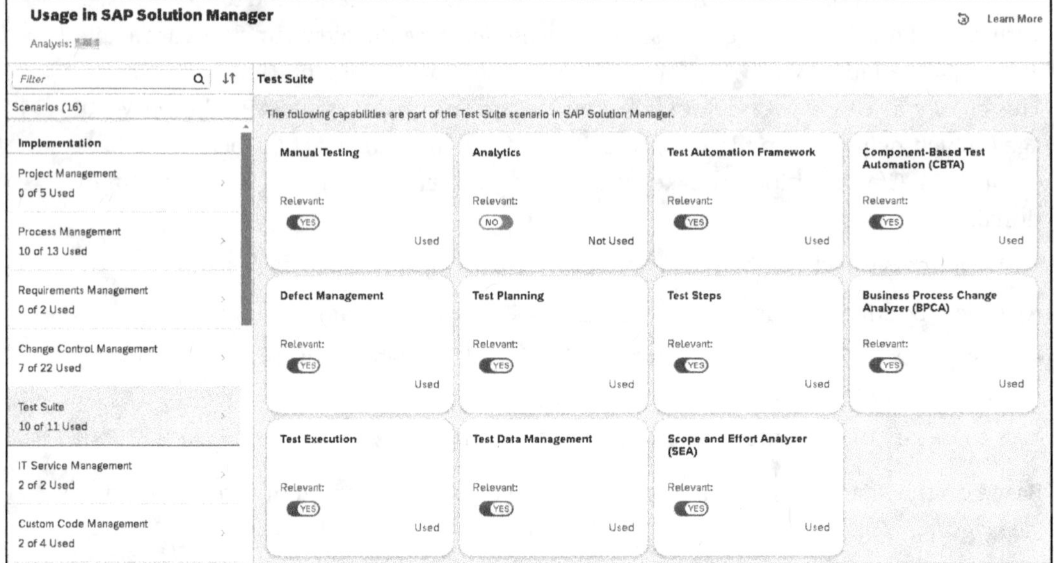

Figure 7.31 Analysis: Usage in SAP Solution Manager

- **Availability of Relevant Capabilities**
  This category shows all functions in the source system that were previously classified as relevant in the **Usage in SAP Solution Manager** category. Their availability status in SAP Cloud ALM is also shown here. The detailed analysis (see Figure 7.32) can be accessed by clicking on the group title.
- **SAP Cloud ALM Capabilities**
  In this category, you will find a list of the functions that are currently available in SAP Cloud ALM. Both comparable functions in SAP Solution Manager and functions that are only available in SAP Cloud ALM are displayed. You can call up the detailed analysis (see Figure 7.33) by clicking on the group title.

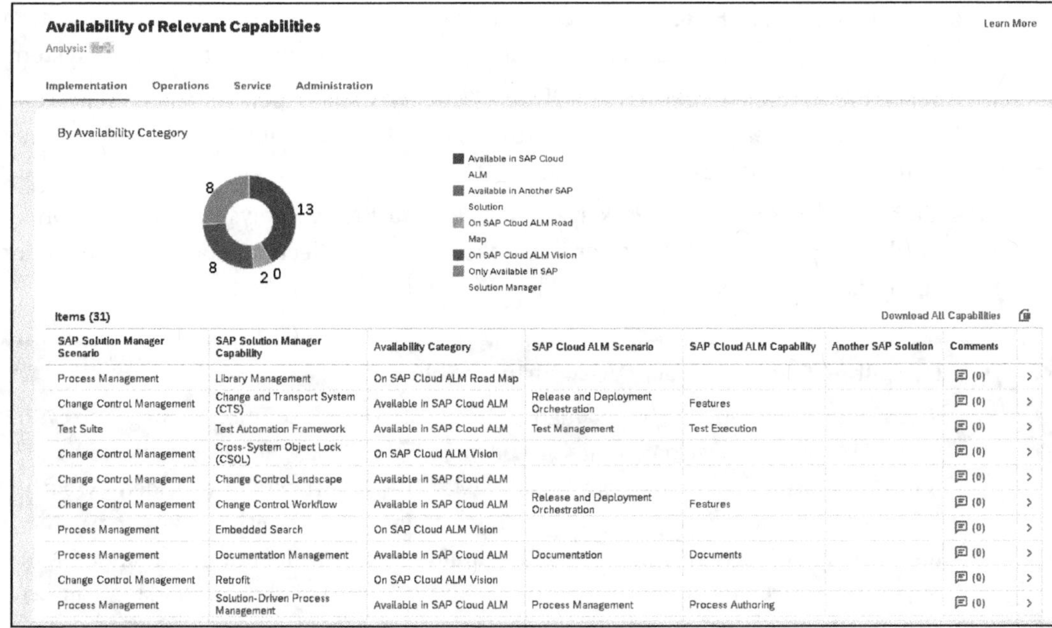

SAP Solution Manager Scenario	SAP Solution Manager Capability	Availability Category	SAP Cloud ALM Scenario	SAP Cloud ALM Capability	Another SAP Solution	Comments
Process Management	Library Management	On SAP Cloud ALM Road Map				(0)
Change Control Management	Change and Transport System (CTS)	Available in SAP Cloud ALM	Release and Deployment Orchestration	Features		(0)
Test Suite	Test Automation Framework	Available in SAP Cloud ALM	Test Management	Test Execution		(0)
Change Control Management	Cross-System Object Lock (CSOL)	On SAP Cloud ALM Vision				(0)
Change Control Management	Change Control Landscape	Available in SAP Cloud ALM				(0)
Change Control Management	Change Control Workflow	Available in SAP Cloud ALM	Release and Deployment Orchestration	Features		(0)
Process Management	Embedded Search	On SAP Cloud ALM Vision				(0)
Process Management	Documentation Management	Available in SAP Cloud ALM	Documentation	Documents		(0)
Change Control Management	Retrofit	On SAP Cloud ALM Vision				(0)
Process Management	Solution-Driven Process Management	Available in SAP Cloud ALM	Process Management	Process Authoring		(0)

Figure 7.32 Analysis: Availability of Relevant Capabilities

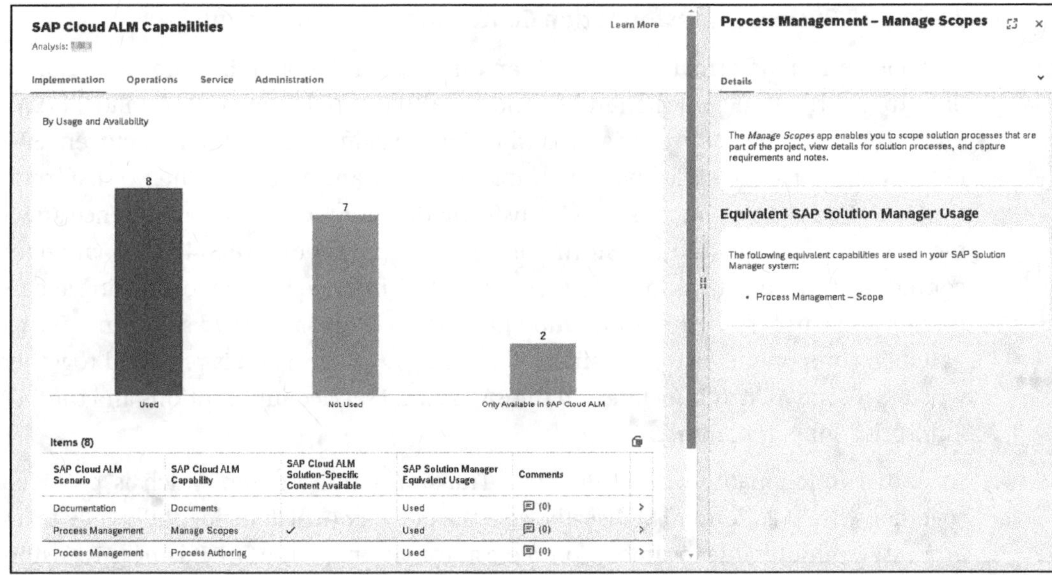

SAP Cloud ALM Scenario	SAP Cloud ALM Capability	SAP Cloud ALM Solution-Specific Content Available	SAP Solution Manager Equivalent Usage	Comments
Documentation	Documents		Used	(0)
Process Management	Manage Scopes	✓	Used	(0)
Process Management	Process Authoring		Used	(0)

Figure 7.33 Analysis: SAP Cloud ALM Capabilities

To start the analysis using the SAP Readiness Check, follow these steps:

1. Follow the instructions in SAP Note 3236443 for installing the check framework. To obtain a complete overview of the functions used, we recommend running the program on the production system.

2. Execute report /SDF/ RC_ALM_COLLECT_DATA via Transaction SA38 or Transaction SE38 to initialize the data collector. For large amounts of data in the source system, it is advisable to schedule the report in the background.
3. Once the analysis is complete, the results can be downloaded. These are grouped into various XML files and stored as a ZIP file.
4. Start the SAP Readiness Check app and select **Start New Analysis**. Upload the previously downloaded ZIP file; after a few seconds, the collected data will be ready for analysis.

Consulting Services of NTT DATA Business Solutions

If you need even more support when switching from SAP Solution Manager to SAP Cloud ALM, you are welcome to get in touch with us. The experienced consultants at NTT DATA Business Solutions are familiar with the ALM functions provided by both solutions. We will analyze the results of SAP Readiness Check with you and derive the necessary activities in the form of a roadmap for your individual transformation journey. We will explain the differences and explain how your way of working can be implemented from SAP Solution Manager in SAP Cloud ALM.

7.5.2 SAP Business Transformation Center and Digital Blueprints

SAP Business Transformation Center is an offering for the planning and execution of data transformation projects (end-to-end) with an initial focus on SAP ERP to SAP S/4HANA. It is aimed at organizations who want to gain insight into their current SAP ERP system, consider a selective data transformation, and accelerate their transformation to SAP S/4HANA. SAP Business Transformation Center provides a reference architecture for migration and transformation in the context of SAP S/4HANA, provides content and best practices for this, and, above all, is intended to support scaling at customers and partners via *transformation factories*. It is an SaaS cloud solution offering based on components of SAP Business Technology Platform and is provided together with SAP Cloud ALM. It is included in the SAP's cloud service subscriptions and the SAP Enterprise Support contract.

An initial functionality of SAP Business Transformation Center, which is provided together with SAP Cloud ALM directly with the delivery (which incidentally only takes around 20 minutes due to almost complete automation!), is the SAP Business Transformation Center digital blueprint (see Figure 7.34). SAP is primarily addressing enterprise architects, solution architects, and people in sales roles who are evaluating the topic of SAP S/4HANA transformation or migration. The SAP Business Transformation Center digital blueprint is primarily intended to be helpful in the scoping phase, i.e., the definition and delimitation of the scope of the data to be transferred to SAP S/4HANA as part of a selective data migration, and to prepare further steps towards extract, transform, load (ETL) tasks.

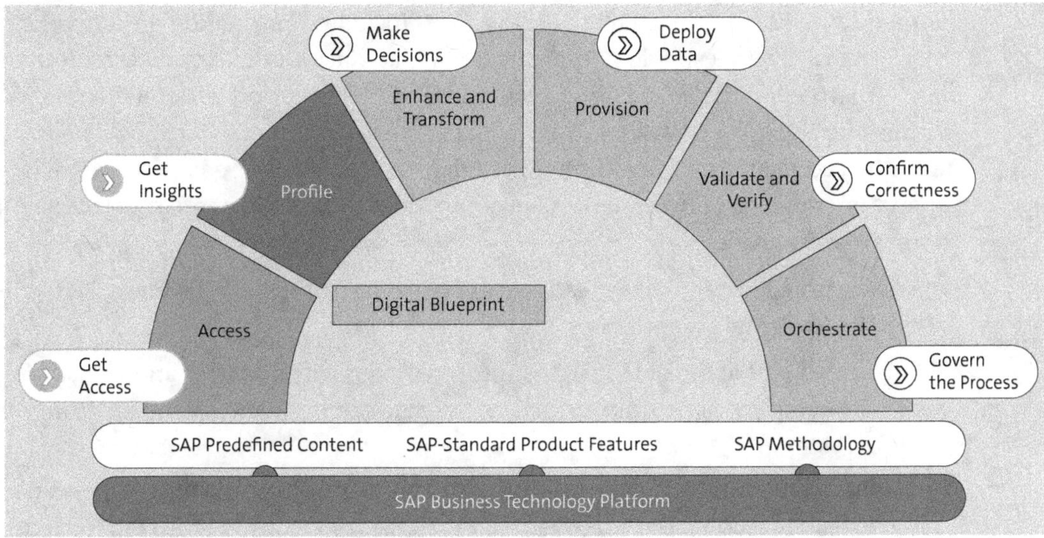

Figure 7.34 SAP Business Transformation Center at a Glance

[«]

Selective Data Transition

If you would like to find out more about selective data migration as a migration path to SAP S/4HANA, take a look at the **Selective Data Transition Engagement** landing page at *http://s-prs.de/v946405*.

SAP Business Transformation Center Digital Blueprint

In practice, we have mostly encountered traditional tools and methods in workshops and projects (e.g., Microsoft Excel-based evaluations and complex customer-specific selection and filter reports). These are usually laboriously developed in ABAP and laboriously extracted from the source system. Often the manual and step-by-step analysis of customizing master data and transaction data takes place before such a workshop can even be started—and then usually without versioning or reliable validation of the analysis results and always with the question: How can these results actually be kept reliable, consistent, comprehensible, and documented for the further project procedure? With special functions that SAP is planning along the roadmap for future features, SAP Business Transformation Center aims to provide answers to precisely this question.

SAP Business Transformation Center promises to be a *guided transformation platform*, based on SAP BTP. This means that SAP, as the manufacturer, provides its customers and partners knowledge about the application (in addition to the existing documentation and the various existing tools in the SAP product portfolio), transformation scenarios (patterns) and the data model (and here in particular, current changes or special features) as well as the guided methodology. This includes information on data consistency, potentially undesirable side effects, incompatibilities, etc., to provide customers and SAP partners with standardized, secure, and efficient support as part of a transformation journey to SAP S/4HANA. SAP Business Transformation Center should help significantly accelerate the transformation and at the same time reduce the necessary support interactions between customers and SAP in order to increase satisfaction on all sides of the project.

Let's take a brief look at the key features and objectives of the SAP Business Transformation Center digital blueprints:

- Data-driven and actionable insights into your company's master and transaction data to be migrated and transformed, as the data can be analyzed directly from the source system
- Support for data scoping workshops with specialist departments or application experts led by architects at the customer or consulting partner. With SAP Business Transformation Center digital blueprints, you can describe a visual story about the situation and support the discussions with drilldowns—based on figures, data, and facts from the system.
- The interactive and guiding application for defining and recording the scope of data for a transformation project allows you to select the right organizational units and transformation objects, including a time frame.

The functionality of the solution includes the following aspects:

- Selection of company codes, transformation objects, and transaction data to be migrated
- Data-based information and recommendations, e.g., to exclude company codes without up-to-date data
- Insights into the applications used and the underlying number of data records
- Documentation of the data scope (blueprint), including in spreadsheet format for further use in existing migration tools

To use SAP Business Transformation Center digital blueprints, you must meet the following requirements:

- SAP ERP as source system for data transfer
- SAP Readiness Check (usage and data profiling) as a basis (included in every standard installation of SAP ERP)
- Ready for immediate use for customers with an existing SAP Cloud ALM tenant

In Table 7.1, you can see an overview of the functional scope of SAP Business Transformation Center.

Function	Use
Upload the analysis file and creation of the digital blueprint	Prepare your data for planning a transformation project. Upload the analysis file with the data exported from the SAP ERP source system and save it. Create and manage the digital blueprint that contains the technical and business details of your planned transformation project.
Planning the scope of your data transformation project	Plan your transformation project: ■ Refine the company code scope in the digital blueprint. ■ Refine the transformation object scope in the digital blueprint. ■ Show technical and business details of the digital blueprint.

Table 7.1 Functional Scope of SAP Business Transformation Center

Practical Application Example for the Digital Blueprint

For this book, we have selected a practical example to illustrate some of the functions of digital blueprints. We start with an overview of the five applications currently available in SAP Business Transformation Center on the SAP Cloud ALM dashboard (see Figure 7.35). From this overview, you can perform the individual steps along the apps from left to right, from uploading the file generated in the source system from the readiness check, to creating a digital blueprint (several blueprints can refer to one source file!) and selecting the company codes and transformation objects that are to be included in the scope of the transformation. You can jump back and forth and restart actions without losing any data. Observe how the changed selection of company codes and transformation objects affects your digital blueprint in the overview.

After you have uploaded a file via **Manage Analysis Files**, you can create your first digital blueprint based on SAP Business Transformation Center in the Manage Digital Blueprints app—simply assign a descriptive name for your digital blueprint, create an optional description, and select the corresponding file uploaded in the previous step from the selection list.

Clicking on one of the company codes or a transformation object will then take you to a detailed view where you can find further useful information on dependencies, the object itself, the data volume, and the current scoping status—which you can, of course, also adjust. Perhaps you would like to know who made a particular decision to include or exclude an object and when? A glance at any comments with the creator and timestamp provides an immediate answer (see Figure 7.36).

Home
SAP Cloud ALM
SAP Business Transformation Center
SAP Cloud ALM for Implementation
SAP Business Transformation Center
Manage Analysis Files
Manage Digital Blueprints
Select Company Codes
Select Transformation Objects
Digital Blueprint Overview
Manage System Scans
ABAP Custom Code to SAP BTP
Select Scanned Tables

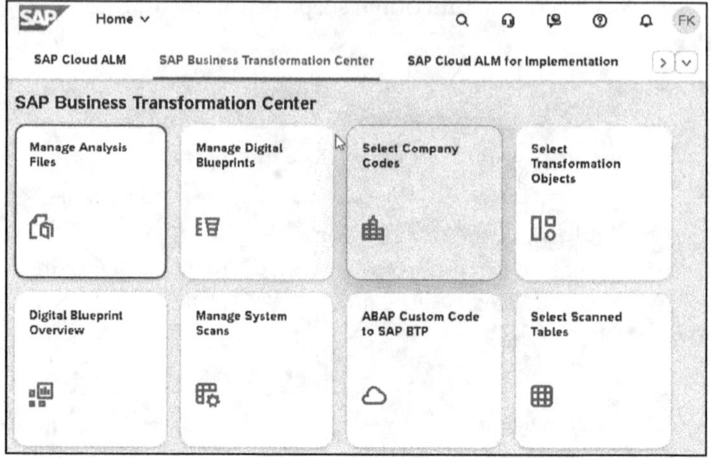

Figure 7.35 SAP Business Transformation Center

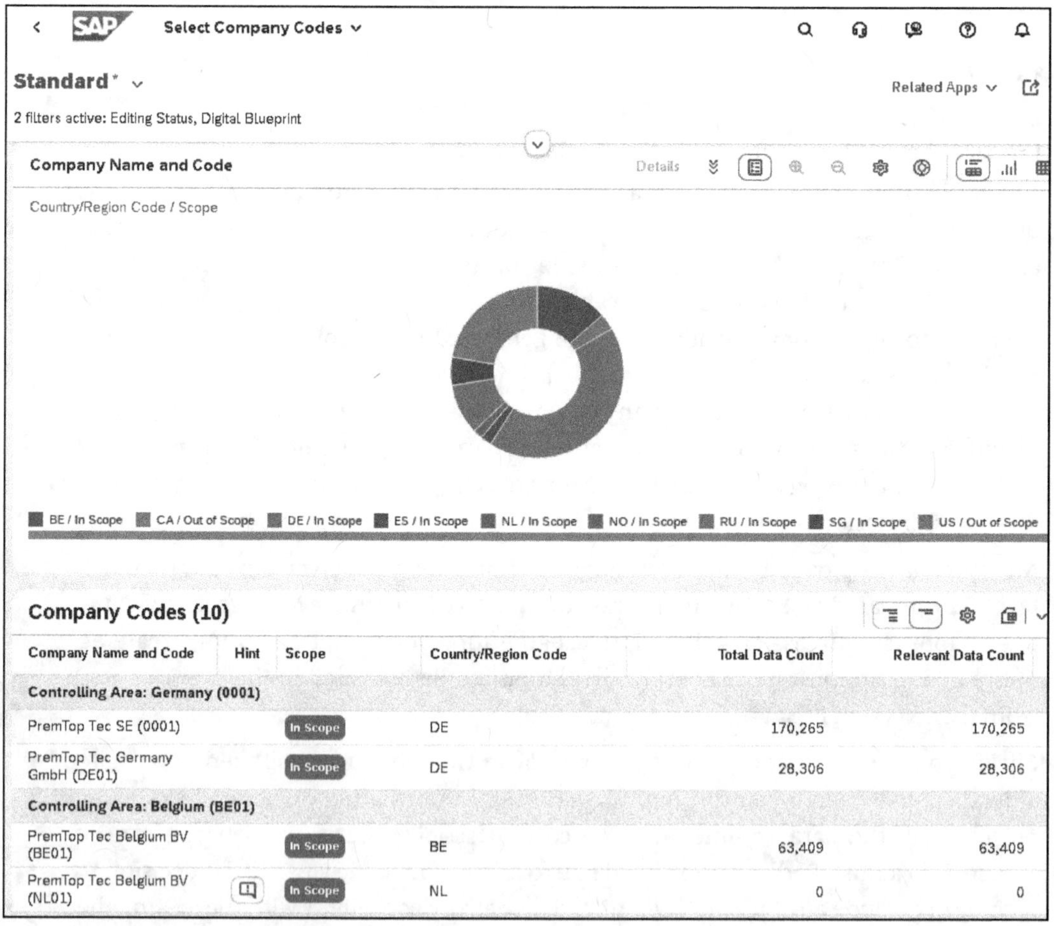

Select Company Codes
Standard*
Related Apps
2 filters active: Editing Status, Digital Blueprint
Company Name and Code
Details
Country/Region Code / Scope
BE / In Scope
CA / Out of Scope
DE / In Scope
ES / In Scope
NL / In Scope
NO / In Scope
RU / In Scope
SG / In Scope
US / Out of Scope
Company Codes (10)
Company Name and Code
Hint
Scope
Country/Region Code
Total Data Count
Relevant Data Count
Controlling Area: Germany (0001)
PremTop Tec SE (0001)
In Scope
DE
170,265
170,265
PremTop Tec Germany GmbH (DE01)
In Scope
DE
28,306
28,306
Controlling Area: Belgium (BE01)
PremTop Tec Belgium BV (BE01)
In Scope
BE
63,409
63,409
PremTop Tec Belgium BV (NL01)
In Scope
NL
0
0

Figure 7.36 Insights into the Organizational Structure and Data Volume

Of particular interest here are the notes that you may have already discovered in the **Hint** column. Here, SAP Business Transformation Center recommends certain actions based on sets of rules or points out special dependencies and constellations that you should pay attention to (but of course you don't have to if you know what you are doing!).

If you now select **SAP Application Component** in the **Select Transformation Objects** area, this overview will help you to put together your project team for the planned transformation, as you can see at a glance which modules are currently in use and get an overview of your company's application profile (see Figure 7.37).

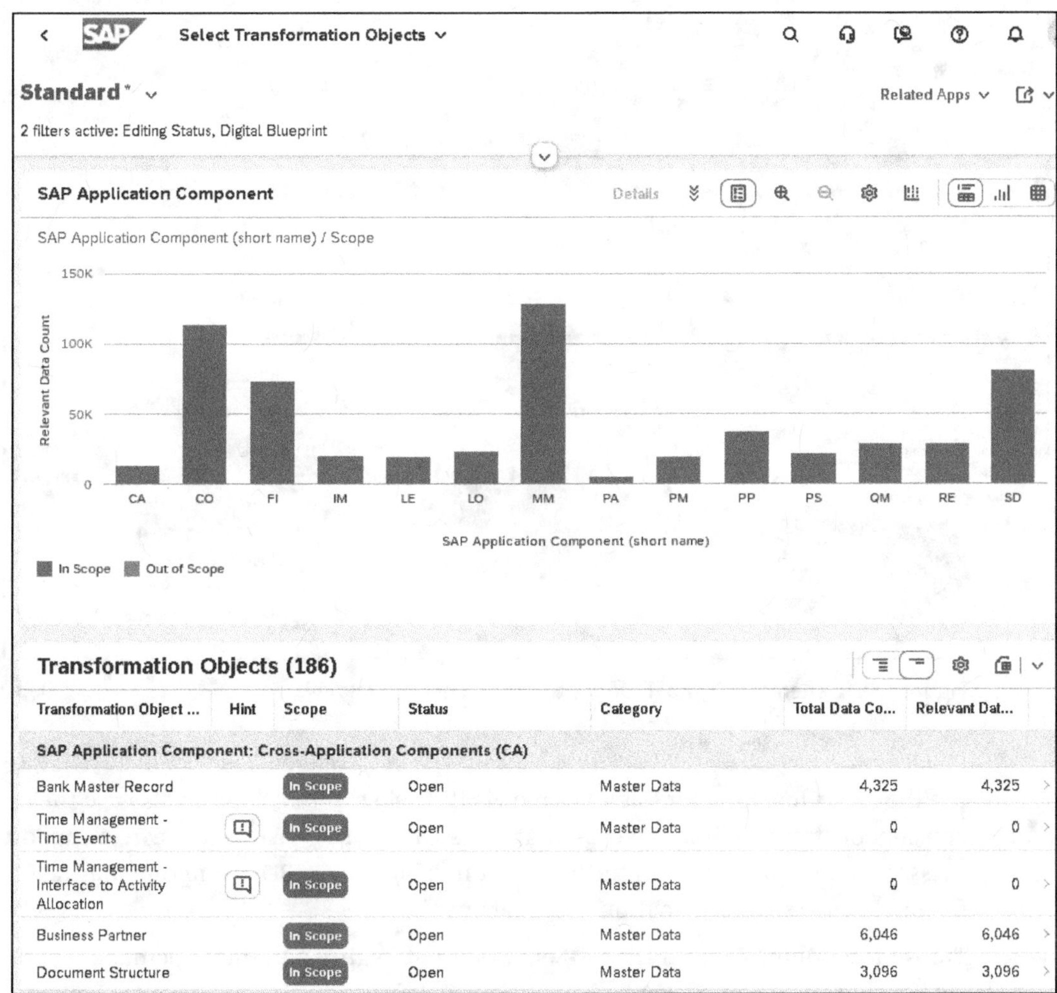

Transformation Object ...	Hint	Scope	Status	Category	Total Data Co...	Relevant Dat...
SAP Application Component: Cross-Application Components (CA)						
Bank Master Record		In Scope	Open	Master Data	4,325	4,325
Time Management - Time Events		In Scope	Open	Master Data	0	0
Time Management - Interface to Activity Allocation		In Scope	Open	Master Data	0	0
Business Partner		In Scope	Open	Master Data	6,046	6,046
Document Structure		In Scope	Open	Master Data	3,096	3,096

Figure 7.37 SAP Application Components Used in the System at a Glance

In the last step, once you have selected and delimited the organizational units (especially company codes) and transformation objects (master data and transaction data with delimitation to individual financial years), you can now display the summary of your generated digital blueprint. The result looks like Figure 7.38.

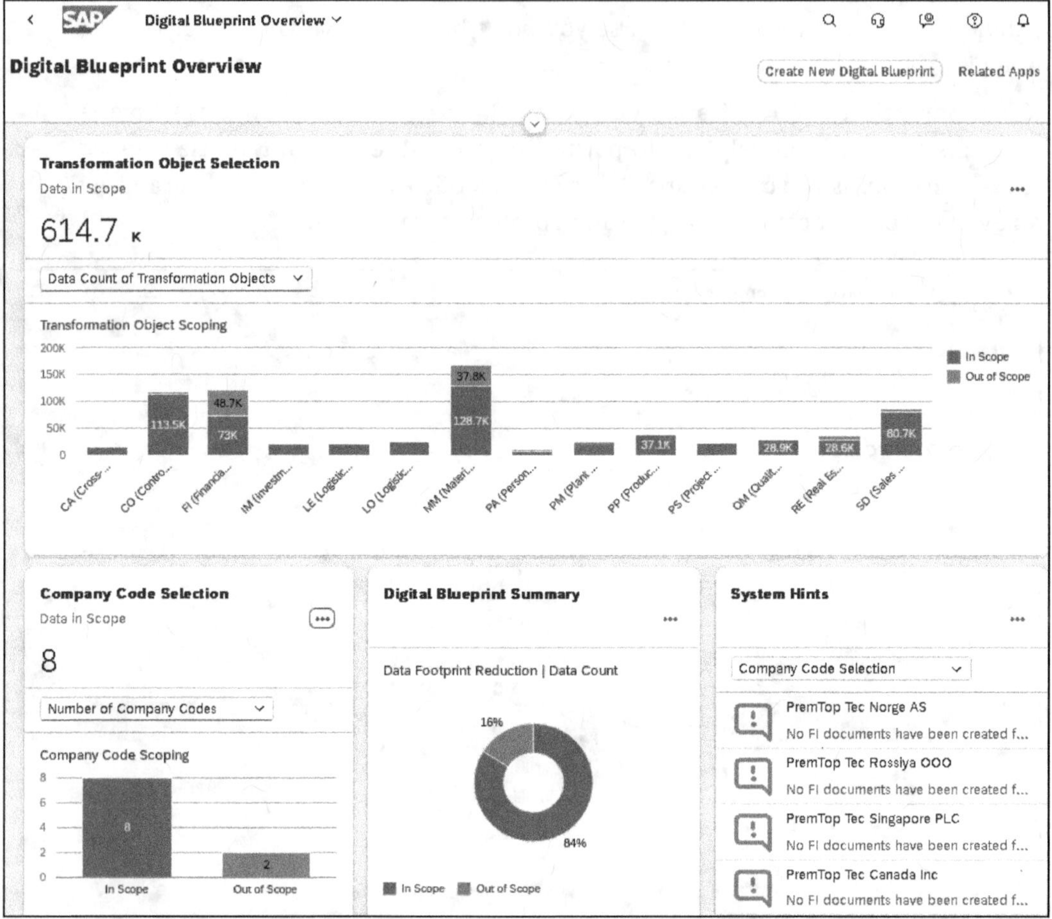

Figure 7.38 Generated Digital Blueprint in the General Overview

Finally, two practical tips: You can get context-specific, concrete instructions on how to use the individual functions at any time via the help symbol. You can also use the in-app support function and the chat at any time. It is also possible to create a support message to SAP directly from the application using the CA-BDT component available for SAP Business Transformation Center at any time.

Transformation projects in the SAP environment begin and remain demanding and challenging projects for everyone involved. SAP Business Transformation Center is designed to help you make these projects more standardized, secure, and efficient. This section has given you an initial impression of how you can use SAP Business Transformation Center to prepare, support, and implement your transformation projects in the

SAP environment using the SAP Activate methodology and on the basis of SAP technology out of the box. The SAP Business Transformation Center digital blueprint is currently available for this purpose. In the future, even more comprehensive tools for scoping, mapping, monitoring, and—last but not least—transformation are to be added. These will be provided on an ongoing basis in accordance with the SAP cloud deployment model. You can also try out a demo version of the SAP Business Transformation Center digital blueprint on the SAP Cloud ALM website.

More details on planned new functions and their availability can also be found regularly in SAP Roadmap Explorer and the product roadmap for SAP Cloud ALM (see *https://roadmaps.sap.com/* under the search term "SAP Business Transformation Center") as well as in SAP Support Portal at *https://support.sap.com*.

Due to the cloud-based architecture and the associated release strategy, the functional scope of SAP Business Transformation Center is constantly being expanded through regular, sometimes smaller and sometimes larger updates (on average, there is something new to discover every 14 days)—so it remains exciting!

[+]

Recommendation

We recommend that you regularly check SAP Note 3328319 for up-to-date information on SAP Business Transformation Center before using SAP Business Transformation Center in your project.

7.6 Summary

In this chapter, you learned from our example scenario how to assign a scope for the logistics area to the project created for the initial template development. The relevant business processes were then activated for this project scope (process scoping), which are later considered in the subsequent process analysis workshops (fit-to-standard and fit-gap) within the explore phase (see Chapter 11). Finally, in this chapter we presented two tools for determining the correct project scope from legacy SAP systems. However, these tools do not play a significant role in our example scenario, as a greenfield approach is used to switch from an outdated non-SAP ERP system in the logistics area to SAP S/4HANA.

After this chapter, you will have completed the preparatory measures of the prepare phase with regard to providing the content-related business context. In the next, final chapter of the prepare phase, we will take a final look at the administration of the project. Here, we will focus on the onboarding of future project participants, the associated authorization management, and the project reporting options offered by SAP Cloud ALM.

Chapter 8
User Management and Project Reporting

This chapter deals with central project administration regarding how to manage involved users as well as how to track the various phases of your projects. The user management aspect includes user creation, assignment of required authorization roles, and granting access to SAP Cloud ALM. We explain in detail which project reporting capabilities are offered for which use case.

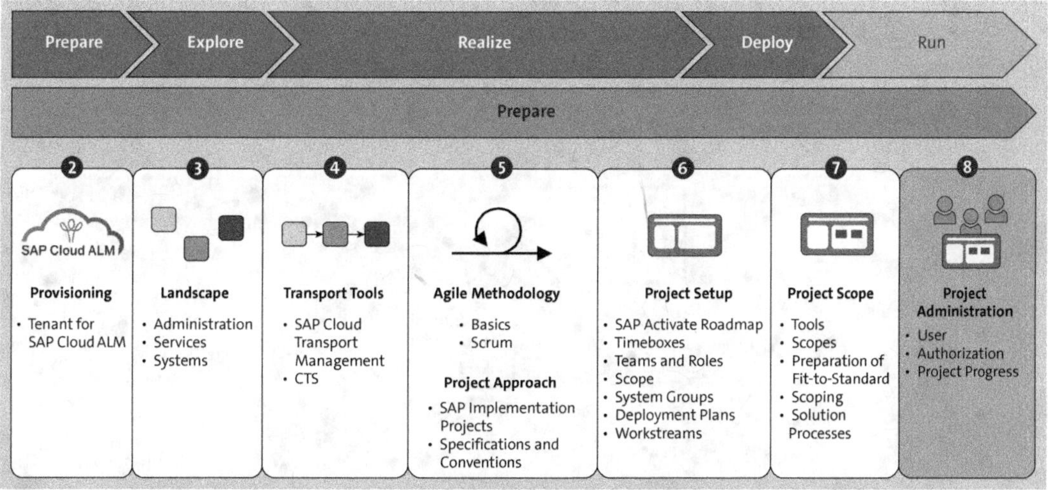

The last two chapters have mainly dealt with setting up a project, defining the scope, and selecting the solution processes that play a role for the solution that should be implemented. We have shown in detail the important steps in the system by using our example scenario from the introductory chapter. In general, these chapters conclude the preparatory measures of the prepare phase with regard to the provision of the content-related business context. To ensure that all project participants can actually work on this project in the next phases of SAP Activate with SAP Cloud ALM, this chapter explains the onboarding of the involved project participants. We will show in detail how users are created and get equipped with the right authorization roles. Furthermore, we explain the preparation and assignment of individual role collections. Within the second part of this chapter, we examine the options for retrieving the current status of project progress and project reporting of SAP Cloud ALM.

8.1 Project Onboarding

Project onboarding is a crucial process that allows you to set up and successfully establish a new project within SAP Cloud ALM. Various steps are carried out as part of this process, including setting up users and roles, configuring systems, and assigning authorizations (see Figure 8.1).

In this chapter, we explain in detail the whole procedure of adding users and granting access to your SAP Cloud ALM projects:

- Create users in Identity Authentication
- Add registered users to SAP Cloud ALM
- Add users to project roles and teams
- Equip users with authorization roles for working within SAP Cloud ALM
- Define project access permissions

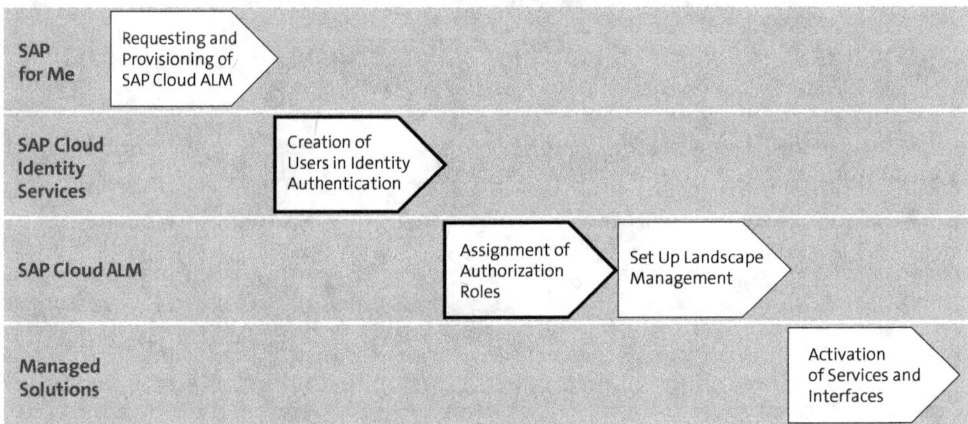

Figure 8.1 Project Onboarding Process: User Onboarding

8.1.1 Identity Authentication Service

The onboarding of every user who will work with SAP Cloud ALM takes place via the Identity Authentication service. This service is provided by SAP specifically for cloud applications. To be able to act as an administrator within the administration console of SAP Cloud Identity Services and, among other things, create or maintain new users or user groups, you must have been designated as an administrator for your SAP Cloud ALM tenant. When you were assigned to the authentication tenant, you were automatically assigned the necessary administrative permissions that you require as an administrator of the respective identity authentication tenant. This also gives you the option of assigning administrative rights to additional colleagues.

Before your users can logon to SAP Cloud ALM, they must be registered in the Identity Authentication service by using the administration console. As soon as you are registered as an administrator within the Identity Authentication service, you can provide other colleagues access to SAP Cloud ALM. Therefore, click on the User Management app in the **Users & Authorizations** app group (see Figure 8.2) in the administration console of SAP Cloud Identity Services.

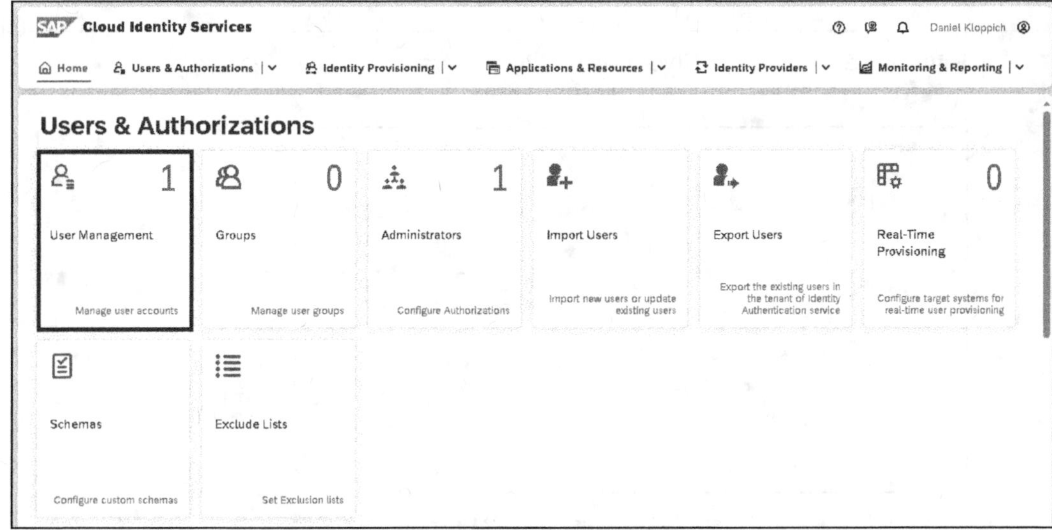

Figure 8.2 SAP Cloud Identity Services: User Management

Administration Console of SAP Cloud Identity Services

The URL has the following pattern:

```
https://<tenant ID>.accounts.ondemand.com/admin
```

The first administrator created for the Identity Authentication tenant received an activation email containing a direct link to the administration console.

After clicking on the app, you will find the **Add User** button (see Figure 8.3), which allows you to register all the employees who should be able to logon later to SAP Cloud ALM. To do this, you must fill in the popup window that appears with all the relevant information such as user name, email address, and user type (see Figure 8.4). The newly created user is identified via the defined email address. This is important for the subsequent creation of users and role assignments within SAP Cloud ALM (refer to Section 8.1.3).

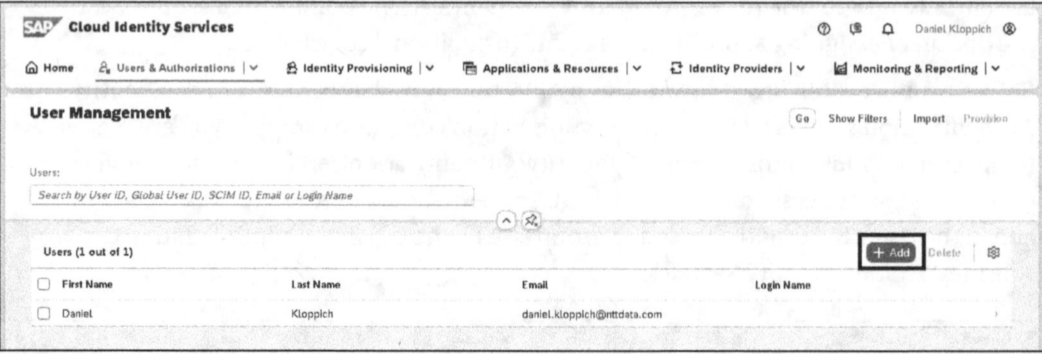

Figure 8.3 SAP Cloud Identity Services: Adding Users

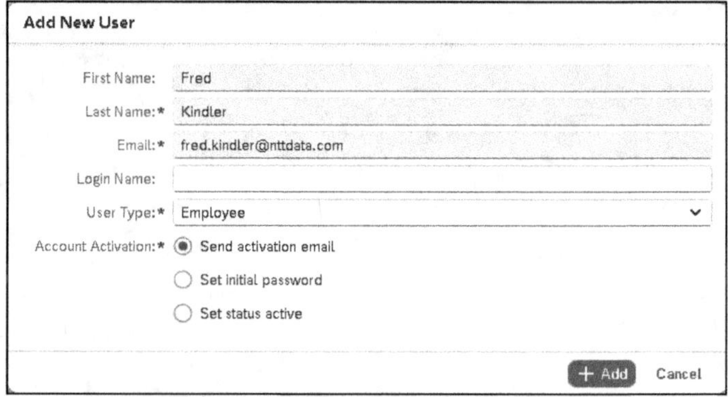

Figure 8.4 SAP Cloud Identity Services: Specify User Credentials

Once you have entered all the information, confirm the user creation by clicking on the **Add** button. This starts the user's onboarding process by sending an activation email to the specified email address (see Figure 8.5 and Figure 8.6).

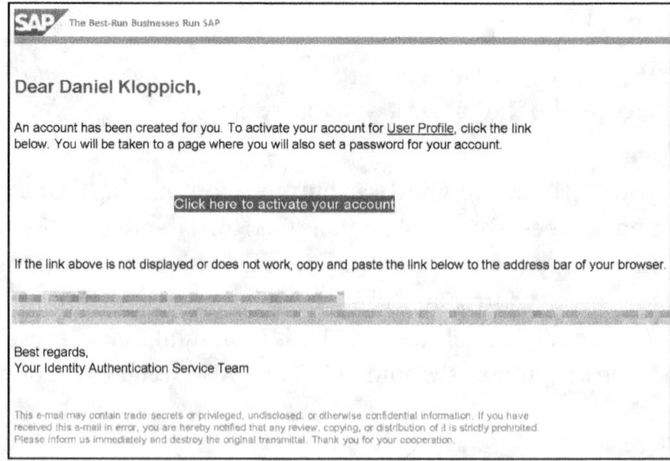

Figure 8.5 SAP Cloud Identity Services: Identity Authentication Activation Email

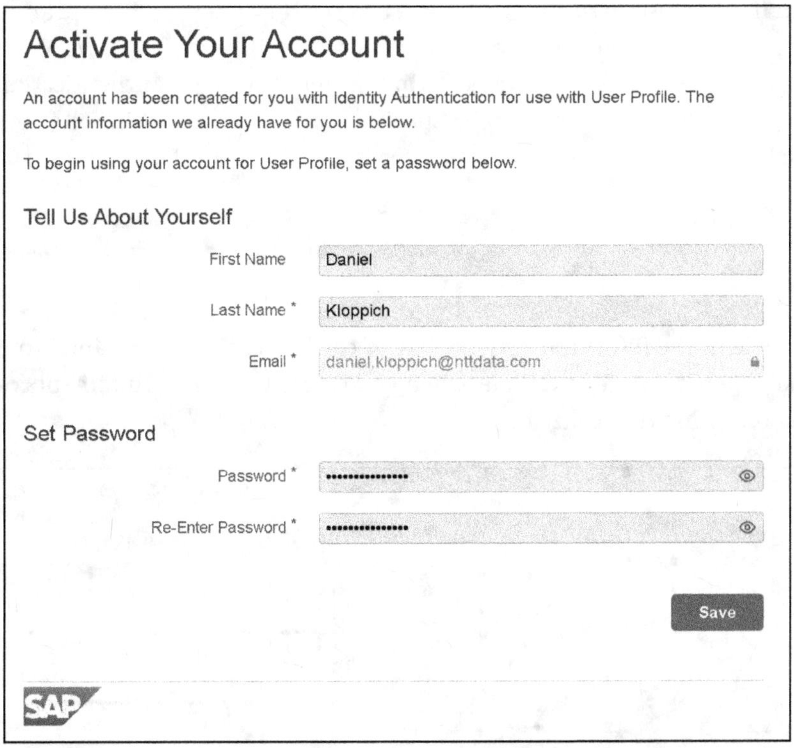

Activate Your Account
An account has been created for you with Identity Authentication for use with User Profile. The account information we already have for you is below.
To begin using your account for User Profile, set a password below.
Tell Us About Yourself
First Name
Daniel
Last Name *
Kloppich
Email *
daniel.kloppich@nttdata.com
Set Password
Password *
Re-Enter Password *
Save
SAP

Figure 8.6 SAP Cloud Identity Services: Identity Authentication Account Activation

8.1.2 Add Users to SAP Cloud ALM

The next step is to add the user to SAP Cloud ALM. Open the User Management app within the **Administration** app group of SAP Cloud ALM (see Figure 8.7). To add the newly registered user and maintain the corresponding roles, you have to navigate first from the **Overview** area to the **Users** area by pressing the button on the right side of the screen.

SAP
Home
DK
SAP Cloud ALM for Implementation
SAP Cloud ALM for Operations
SAP Cloud ALM for Service
Administration
Administration
User Management
Landscape Management
External API Management
SAP Readiness Check Integration

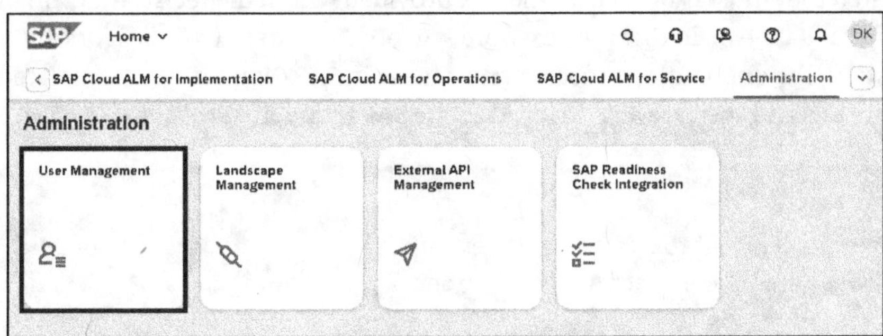

Figure 8.7 SAP Cloud ALM: User Management App

After clicking on the **Add User** button (❶ of Figure 8.8) the **New User** area is appearing on the right-hand side. Now you have the possibility to enter the email address of the colleague who has been previously registered within the Identity Cloud Authentication Service (❷ of Figure 8.8). A user type can be specified optionally. For this, following types are available:

- Employee
- SAP
- Partner
- Other
- (Not assigned)

The type is available in the filter settings, so you can easily identify consultants from SAP or an official SAP partner, for example, when you want to assign requirements or user stories later on within the realize phase.

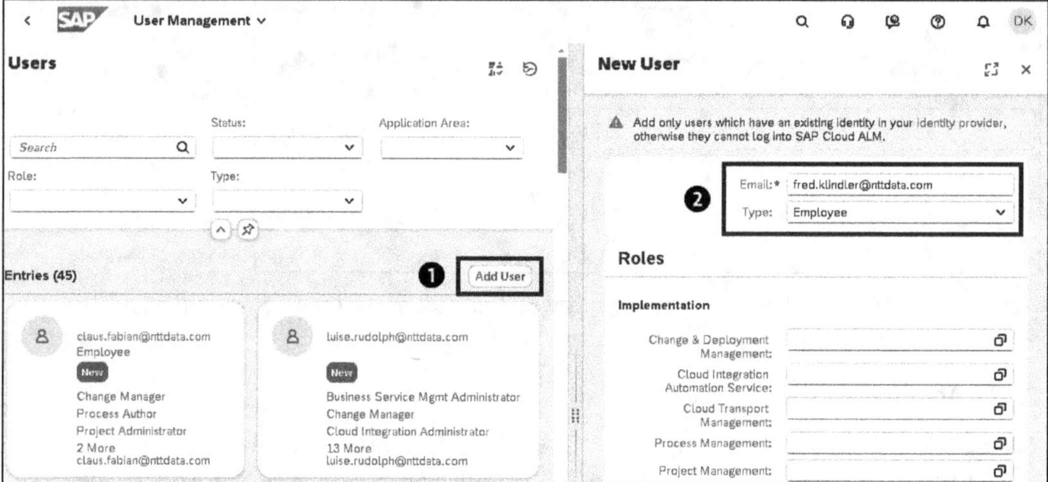

Figure 8.8 SAP Cloud ALM: Add New User

After saving, the user will now receive the email **Welcome to SAP Cloud ALM** (see Figure 8.9). The invited user must now click on the link provided within the registration email and set their user data—including the password—to productive status (see Figure 8.10). With these steps, the onboarding process of new users in SAP Cloud ALM is completed for the time being, and they are able to log in to the SAP Cloud ALM tenant.

[+]

Welcome Email

The welcome email is sent to the affected user with introductory information for SAP Cloud ALM when a user administrator performs one of the following tasks:

- Adding a user
- Approving a request for a new or previously deactivated user
- Reactivating a user

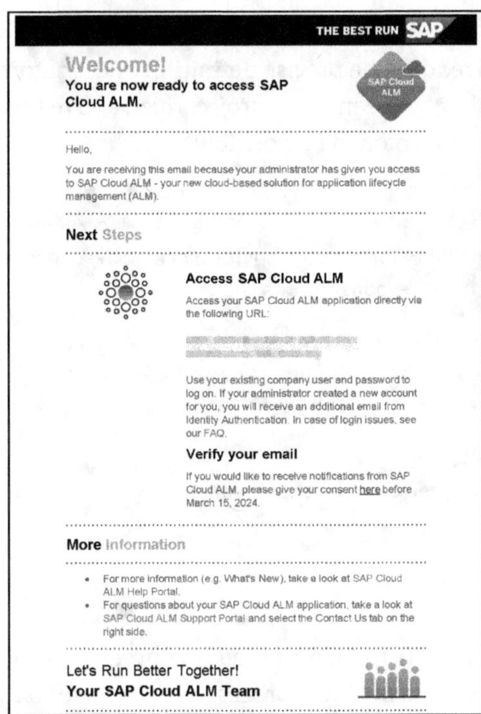

Figure 8.9 SAP Cloud ALM: Welcome Email

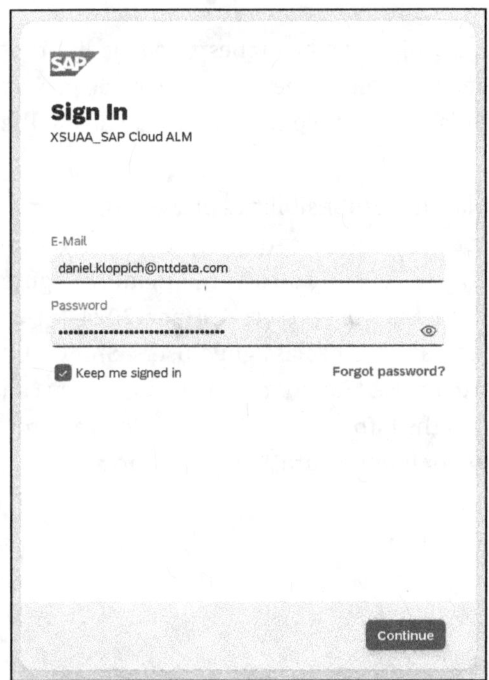

Figure 8.10 SAP Cloud ALM: Account Activation

Users, Project Roles, and Teams

In SAP Cloud ALM, projects are created that require the precise definition of the actors working in them. SAP Cloud ALM distinguishes between *users*, *project roles*, and *teams*. Table 8.1 shows you a brief definition of the three named responsibilities.

Responsibility	Description
User	Carries the executive responsibility for certain activities; can be described as an individual project member (responsible or assignee).
Project role	Provides a direct connection to the project created in SAP Cloud ALM. Several users can be assigned to specific predefined or customer-specific project roles. Project roles can be found under the term "assigned role" in SAP Cloud ALM.
Team	A defined set/group of roles or persons. The exact team organization depends on the project.

Table 8.1 Responsibilities: User, Project Role, and Team

In the Projects and Setup app, you can assign users to teams based on numerous project roles, such as project lead, analytics expert, and business process expert. As soon as a user is assigned, a request is triggered in the User Management app to assign the user to the respective role collection. The assignment of dedicated authorization roles is explained in more detail in Section 8.1.3.

A team consists of several project roles, and a project can be orchestrated by different predefined teams. When using the predefined PMO team, the existence of the project manager role is ensured at the same time. It is not possible to accidentally delete the PMO team delivered as standard.

Using the data shown in Table 8.1, in particular the responsibilities of the project manager, it is possible to assign individual tasks to a specific person, project role, or team and to manage projects of different scales in an efficiently structured manner. Figure 8.11 briefly shows the default **PMO Team** of our example project **AriHana-2024-TPL-R1** and how it is composed of various project roles as well as assigned users. SAP Cloud ALM offers various pre-delivered roles that can be used for your own purposes. You can simply check out the role description by using the **Info** icon ⓘ. You will find the Projects and Setup app within the **SAP Cloud ALM for Implementation** app group.

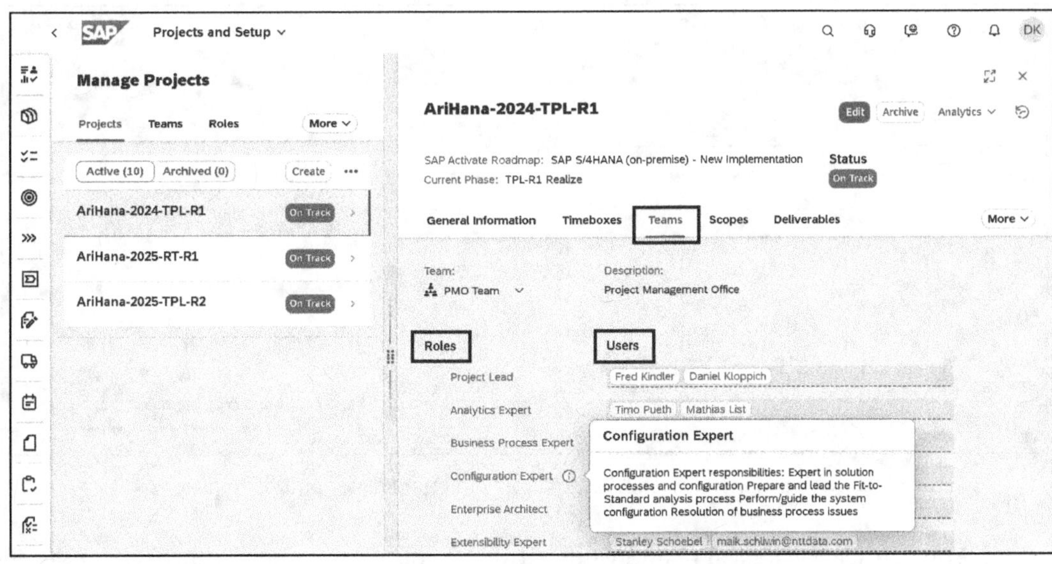

Figure 8.11 Manage Projects: Team Composed of Project Roles and Assigned Users

Authorization Roles

In addition to the project roles, there are also *authorization roles* in SAP Cloud ALM. These authorization roles and their assignment to the users determine the functionalities that the respective user is able to perform in the system. This can be assigned to the project management function. Figure 8.12 shows the assignment of specific authorization roles for a selected user. Authorization roles are assigned by using the User Management app of the **Administration** app group (see Figure 8.7 and Figure 8.8). A distinction is made here between the following authorization roles with corresponding characteristics of activities:

- **Project Administrator**
  Manage projects, deployment plans, tasks, scopes, requirements, features, notes, test cases, defects, and tags
- **Project Manager**
  Manage projects, deployment plans, tasks, scopes, requirements, features, notes, test cases, and defects
- **Project Member**
  View projects and deployment plans. Manage tasks, scopes, requirements, features (with restrictions), notes, test cases, and defects
- **Project Viewer**
  Only view permission of the abovementioned elements, no administrative tasks

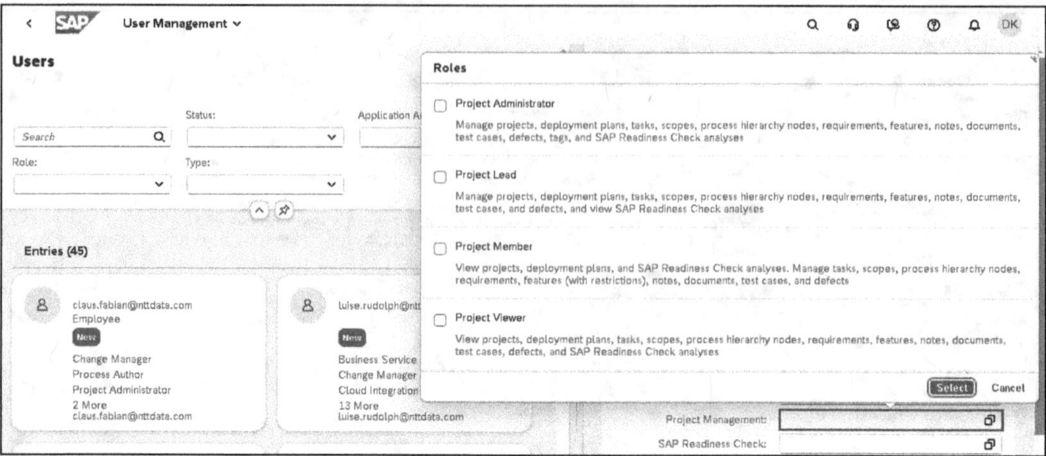

Figure 8.12 Assignment of Authorization Roles in User Management App

You can see the logical relationship between project roles and authorization roles in Table 8.2. Note that the authorization roles are user-specific, but not project-specific.

Project Role	Authorization Role
Project manager	Project Manager
All roles except project manager	Project Member
Customized role	Project Member

Table 8.2 Project Roles and Authorization Roles

Compared to the way authorizations are assigned within SAP Solution Manager 7.2, in SAP Cloud ALM the authorizations of individual persons are managed by assigning them to their respective project roles using predefined role sets. To date, user administrators have not been able to intervene at authorization object level to adjust the delivered authorizations for specific customers or projects. The project roles delivered in SAP Cloud ALM are based on the SAP Activate methodology. One way of defining customer-specific role collections is described in Section 8.1.4.

You can assign roles to users directly via user management in SAP Cloud ALM or as corresponding role collections in the SAP BTP cockpit. Role collections in SAP Cloud ALM are delivered predefined and ready to use. Each application role in SAP Cloud ALM represents a role collection in your SAP BTP subaccount, which in turn corresponds 1:1 to the respective role template. This means that you don't have to configure any role collections yourself. Instead, you can simply assign the relevant role collections to the users based on their tasks and requirements. This assignment not only automatically assigns authorizations according to administrative, edit, or display access rights, but also assigns the available apps in the SAP Cloud ALM launchpad.

An overview of all predefined SAP Cloud ALM authorization roles and associated role collections can be found in Table 8.3. This should also serve as a supplement to the authorization matrix mentioned in Chapter 2, Section 2.3.3.

Scenario	Capability	Role	
		Role Collection and Template	**Description**
Cross topics	Cross application	■ Global Administrator: Cross Global Administrator ■ x_calm_GlobalAdministrator ■ FlexOperator	■ Create and manage all SAP Cloud ALM objects, except for personal or sensitive data ■ This role provides the complete set of authorizations. We recommend assigning suitable roles to your users.
		■ Global Auditor: Cross Global Auditor ■ x_calm_GlobalAuditor	■ View all SAP Cloud ALM objects, except personal or sensitive data ■ This role is to be used for general audit purposes
	Landscape management	■ Landscape Management Security Administrator: Landscape Management Security Administrator ■ x_landscape_SecurityAdministrator	■ Manage certificates
	User and access management	■ User Administrator: User & Access Management Administrator ■ x_uam_UserAdministrator ■ FlexOperator	■ Create and manage users ■ Assign and approve authorizations
		■ User Viewer: User & Access Management Viewer ■ x_uam_UserViewer	■ View users and authorization assignments
	Tag management	■ Tag Administrator: Tag Management Administrator ■ x_tag_TagAdministrator	■ Create, update, merge and centrally delete tags ■ View tag usage ■ Organize tags into groups
		■ Tag Viewer: Tag Management Viewer ■ x_tag_TagViewer	■ View tags, tag groups and tag usage

Table 8.3 Overview of Predefined Roles

Scenario	Capability	Role	
Service	Service collaboration	■ Services Administrator: Service Collaboration Administrator ■ `sdc_sc_ServicesAdministrator`	■ Execute service-specific administration tasks ■ Configure and execute services ■ View service execution results
		■ Services Expert: Service Collaboration Expert ■ `sdc_sc_ServicesExpert`	■ Configure and execute services ■ View own and shared service execution results
		■ Services Viewer: Service Collaboration Viewer ■ `sdc_sc_ServicesViewer`	■ Browse through the service explorer ■ View the shared service execution results
SAP Business Transformation Center	Scoping	■ Digital Blueprint Administrator: Digital Blueprint Administrator ■ `btc_bdts_DigitalBlueprintAdministrator`	■ Create and edit digital blueprints and all associated tasks ■ Company code scoping ■ Transformation object scoping
		■ Digital Blueprint Viewer: Digital Blueprint Viewer ■ `btc_bdts_DigitalBlueprintViewer`	■ Consume digital blueprint content and all associated tasks ■ View company code scoping and transformation object scoping
Implementation	Cloud integration automation service	■ Cloud Integration Administrator: CIASIntegrationAdmin ■ `Destination_Administrator` ■ `Destination_Configuration_Administrator` ■ `ExecutionSummaryRole` ■ `InboxRole` ■ `PlanningRole` ■ `WorkflowAdmin` ■ `WorkflowContextAdmin` ■ `WorkflowDeveloper` ■ `WorkflowInitiator` ■ `WorkflowParticipant`	■ Access My Inbox ■ Work on assigned tasks ■ Plan, monitor, manage, and review integration scenario execution

Table 8.3 Overview of Predefined Roles (Cont.)

Scenario	Capability	Role	
		■ Cloud Integration Expert: CIASIntegrationExpert ■ Destination_Administrator ■ Destination_Configuration_Administrator ■ InboxRole ■ WorkflowAdmin ■ WorkflowContextAdmin ■ WorkflowDeveloper ■ WorkflowParticipant	■ Access My Inbox ■ Work on assigned tasks
		■ Cloud Integration Viewer: CIASIntegrationMonitor ■ Destination_Configuration_Viewer ■ Destination_Viewer ■ ReadOnlyRole ■ WorkflowContextViewer ■ WorkflowParticipant ■ WorkflowViewer	■ View scenario execution monitoring ■ Work on assigned tasks
	Cloud transport management	■ Transport Management Operator (TMS_LandscapeOperator_RC) ■ LandscapeOperator	■ Create, edit, and delete transport nodes and transport routes
		■ Transport Management Viewer (TMS_Viewer_RC) ■ Viewer	■ View transport nodes and transport routes
	Change and deployment management	■ Change Manager: Change & Deployment Management Change Manager ■ imp_cdm_ChangeManager	■ Approve features for production deployment
		■ Deployment Manager: Change & Deployment Management Deployment Manager ■ imp_cdm_DeploymentManager	■ Deploy transports into test and production systems

Table 8.3 Overview of Predefined Roles (Cont.)

Scenario	Capability	Role	
	Process management	▪ Process Author: Process Management Process Author ▪ imp_pm_ProcessAuthor	▪ Create, edit, and delete custom solution processes
		▪ Process Hierarchy Author	▪ Create, edit, and delete process hierarchy nodes
	Project management	▪ Project Administrator: Project & Task Management Administrator ▪ imp_pjm_ProjectAdministrator	▪ Manage projects, deployment plans, tasks, scopes, process hierarchy nodes, requirements, features, notes, documents, test cases, defects, tags, SAP Readiness Check analyses ▪ The permissions within implementation projects of users with this role additionally depend on the access level of the project and the team assignment.
		▪ Project Lead: Project & Task Management Project Lead ▪ imp_pjm_ProjectLead	▪ Manage projects, deployment plans, tasks, scopes, process hierarchy nodes, requirements, features, notes, documents, test cases, defects ▪ View SAP Readiness Check analyses ▪ The permissions within implementation projects of users with this role additionally depend on the access level of the project and the team assignment
		▪ Project Member: Project & Task Management Project Member ▪ imp_pjm_ProjectMember	▪ View projects, deployment plans, and SAP Readiness Check analyses ▪ Manage tasks, scopes, process hierarchy nodes, requirements, features (with restrictions), notes, documents, test cases, defects ▪ The permissions within implementation projects of users with this role additionally depend on the access level of the project and the team assignment

Table 8.3 Overview of Predefined Roles (Cont.)

Scenario	Capability	Role	
		▪ Project Viewer: Project & Task Management Project Viewer ▪ `imp_pjm_ProjectViewer`	▪ View projects, deployment plans, tasks, scopes, process hierarchy nodes, requirements, features, notes, documents, test cases, defects, SAP Readiness Check analyses ▪ The permissions within implementation projects of users with this role additionally depend on the access level of the project and the team assignment
	SAP Readiness Check	▪ Readiness Check Analysis Administrator: Readiness Check Analysis Administrator ▪ `imp_rc_AnalysisAdministrator`	▪ Create, edit, and delete SAP Readiness Check analyses
		▪ Readiness Check Analysis Viewer: Readiness Check Analysis Viewer ▪ `imp_rc_AnalysisViewer`	▪ View all SAP Readiness Check analyses
Operation	Business service management	▪ Business Service Management Administrator: Business Service Management Administrator ▪ `ops_bsm_BusinessServiceManagementAdministrator`	▪ Create and manage business services ▪ Plan events
		▪ Business Service Management Viewer: Business Service Management Viewer ▪ `ops_bsm_BusinessServiceManagementViewer`	▪ View business services, their configuration, and the event calendar

Table 8.3 Overview of Predefined Roles (Cont.)

Scenario	Capability	Role	
	Configuration and security analysis	■ Configuration Monitoring Administrator: Configuration Monitoring Administrator ■ `ops_csa_ConfigurationMonitoringAdministrator`	■ Administrate managed objects, configuration settings ■ Display application data
		■ Configuration Monitoring Analyst: Configuration Monitoring Analyst ■ `ops_csa_ConfigurationMonitoringAnalyst`	■ Display application configuration data
	Health monitoring	■ Health Monitoring Administrator: Health Monitoring Administrator ■ `ops_hm_HealthMonitoringAdministrator`	■ Create, edit, delete, and manage configuration settings, including the display of monitoring data
		■ Health Monitoring Viewer: Health Monitoring Viewer ■ `ops_hm_HealthMonitoringViewer`	■ View health overview and metrics of cloud services
	Integration and exception monitoring	■ Integration Architect: Integration Monitoring Integration Architect ■ `ops_im_IntegrationArchitect`	■ Configure integration monitoring
		■ Integration Owner: Integration Monitoring Integration Owner ■ `ops_im_IntegrationOwner`	■ View the messages in integration monitoring
		■ Integration Owner Sensitive: Integration Monitoring Integration Owner Sensitive ■ `ops_im_IntegrationOwnerSensitive`	■ View the messages in integration monitoring, including data defined as sensitive

Table 8.3 Overview of Predefined Roles (Cont.)

Scenario	Capability	Role	
	Job and automation monitoring	■ Job Monitoring Administrator: Job Monitoring Administrator ■ `ops_jm_JobMonitoringAdministrator`	■ Create, edit, delete, and manage configuration settings, including the display of monitoring data
		■ Job Monitoring Consumer: Job Monitoring Consumer ■ `ops_jm_JobMonitoring-Consumer`	■ View the job monitoring application for the maintained global configurations
	Business process monitoring	■ Process Executor: Business Process Monitoring Process Executor ■ `ops_bm_ProcessExecutor`	■ Consume KPIs and the respective process content (including sensitive data) ■ Process alerts ■ Additionally, access to particular business processes and data criteria can be restricted for users with this role by setting up access groups
		■ Process Manager: Business Process Monitoring Process Manager ■ `ops_bm_ProcessManager`	■ Consume KPIs and the respective process content (including sensitive data) ■ Process alerts ■ Maintain specific configurations ■ Additionally, access to particular business processes and data criteria can be restricted for users with this role by setting up access groups
		■ Process Monitoring Administrator: Business Process Monitoring Administrator ■ `ops_bm_ProcessMonitoringAdministrator`	■ Maintain global and specific configurations ■ Consume KPIs and the respective process content (including sensitive data) ■ Process alerts

Table 8.3 Overview of Predefined Roles (Cont.)

Scenario	Capability	Role	
		■ Process Monitoring Consumer: Business Process Monitoring Consumer ■ ops_bm_ProcessMonitoringConsumer	■ Consume KPIs and the respective process content (including sensitive data) ■ View alerts ■ Additionally, access to particular business processes and data criteria can be restricted for users with this role by setting up access groups
		■ Process Monitoring Viewer: Business Process Monitoring Viewer ■ ops_bm_ProcessMonitoringViewer	■ Consume KPIs and non-sensitive process content, without access to alert information ■ Additionally, access to particular business processes and data criteria can be restricted for users with this role by setting up access groups
	Real user monitoring	■ Real User Analyst: Real User Monitoring Analyst ■ ops_rum_RealUserAnalyst	■ View utilization and performance of requests
		■ Real User Analyst Administrator: Real User Monitoring Analyst Administrator ■ ops_rum_RealUserAnalystAdministrator	■ View utilization and performance of requests, including sensitive data like the user ID ■ Configure the Real User Monitoring app
		■ Real User Analyst Sensitive: Real User Monitoring Analyst Sensitive ■ ops_rum_RealUserAnalystSensitive	■ View utilization and performance of requests, including sensitive data like the user ID
	Synthetic user monitoring	■ Scenario Administrator: Synthetic User Monitoring Administrator ■ ops_sum_SyntheticUserMonitoringAdministrator	■ Create, edit, delete, and manage configurations ■ Manage global application settings ■ View performance and availability of scenario executions

Table 8.3 Overview of Predefined Roles (Cont.)

Scenario	Capability	Role	
		■ Scenario Expert: Synthetic User Monitoring Scenario Expert ■ ops_sum_SyntheticUserMonitoringExpert	■ Create, edit, delete, and manage configuration settings ■ View performance and availability of scenario executions
		■ Scenario Viewer: Synthetic User Monitoring Scenario Viewer ■ ops_sum_SyntheticUserMonitoringViewer	■ View performance and availability of scenario executions
Custom	Display all	■ ViewerAll	■ Read-only access to content of SAP Cloud ALM

Table 8.3 Overview of Predefined Roles (Cont.)

You can find the SAP-provided role overview in the SAP Help Portal at *https://help.sap.com/docs/cloud-alm/setup-administration/role-collections*.

8.1.3 Create and Authorize Users

It is important to carefully manage user creation and permissions to ensure that the right users have the necessary access rights to perform their tasks effectively, while ensuring the security and protection of sensitive information.

The SAP BTP cockpit offers the authorization administration team ready-made role collections for this purpose. These enable the assignment of content-defined and process-related authorizations to the respective project users or user groups. These role collections are differentiated according to their risk level in various (sub)scenarios. To give an example, for the User & Access Management administrator role, two role sets are provided in SAP Cloud ALM for direct use: the administrative role collection for the person responsible for user and access management, and the display authorizations for all other users. These users are authorized to perform display activities but cannot perform any administrative activities and therefore receive the User & Access Management Viewer role set.

Each role collection has an assignment to the respective persons, the actual assignment of the role, a template if necessary, and an alphanumeric application identification. In the editing environment, it is also possible to add delivered roles exclusively to your own, newly created role collections. This means that your own specific definitions of authorizations are retained during updates; for example, during an upgrade.

8.1.4 Create Your Own Role Collections

Using the example of a comprehensive display workstation, we will show you how to create a customer- or project-specific role collection. We will show you an example of how you can make all relevant SAP Cloud ALM subscenarios available to your specific users or user groups in restrictive display mode by assigning a role collection.

In the SAP BTP cockpit, a new role collection is created in the subaccount using the **Create** button, as shown in Figure 8.13.

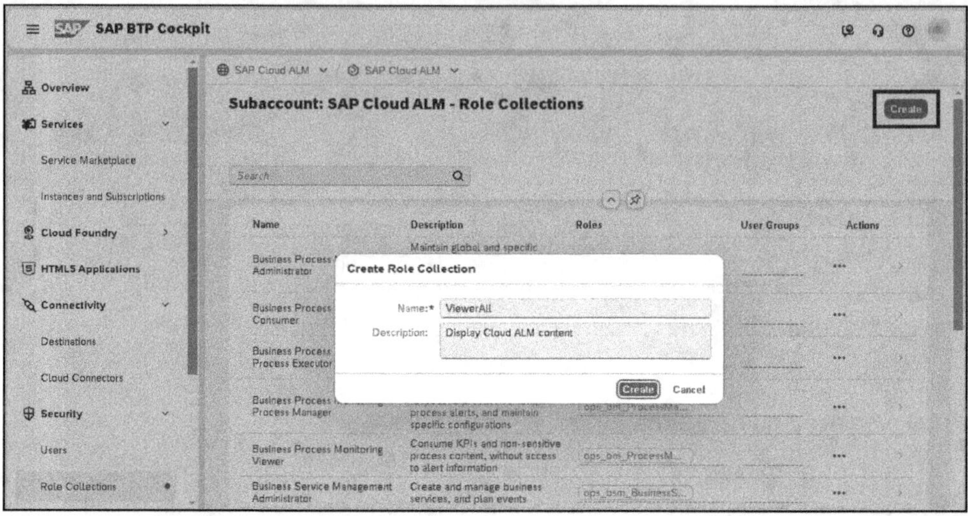

Figure 8.13 Creating Your Own Role Collection

In the popup window, assign a **Name** for the role collection, in this case "ViewerAll". Enter a corresponding **Description** and confirm the creation via the **Create** button.

From the existing role collections, which SAP Cloud ALM offers by default for all relevant scenarios in the corresponding characteristics (display, edit, administrative, etc.), you can select exactly those that should be taken into account for the **ViewerAll** display workstation.

Figure 8.14 shows an example of this selection filtered by attribute **Role Template** and value **Viewer**. You can select the standard role templates to be taken into account according to your requirements for the role collection to be created.

After selecting all the desired elements, click on the **Add** button. Your own, newly defined role collection is now available in the relevant sub-account of the SAP BTP cockpit, as shown in Figure 8.15. There are further editing and mapping options as well as the option to make assignments to users or user groups directly in the SAP BTP cockpit. The assignment of the previously selected role templates is also displayed.

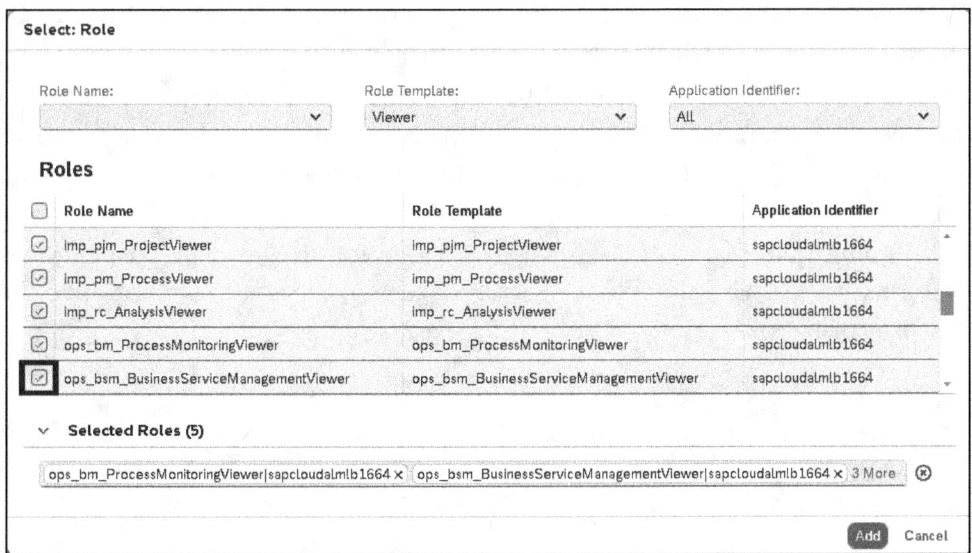

Figure 8.14 Selection from Available Role Templates for Your Own Role Collection

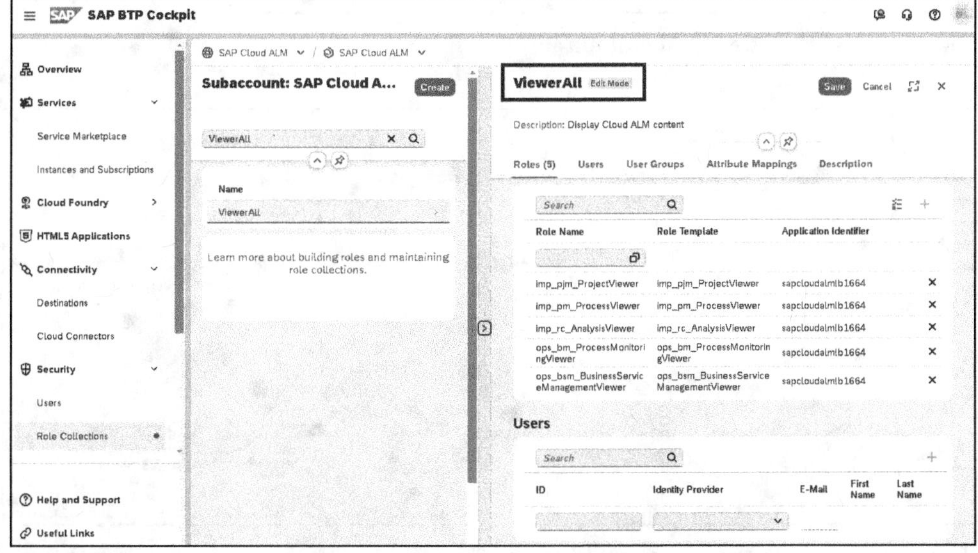

Figure 8.15 Resulting Role Collection for "ViewerAll"

8.1.5 Manage Users and Assign Roles

There are two ways to assign standard and custom role collections to users for SAP Cloud ALM:

- Assignment via the SAP BTP cockpit
- Assignment in SAP Cloud ALM via the User Management app

We recommend the second approach, as the User Management app of SAP Cloud ALM provides an advanced role request and assignment feature.

First, access SAP Cloud ALM by using the link included in the **Welcome to SAP Cloud ALM** email and your login credentials. Use the password that you set during the account activation process in the Identity Authentication service. It is not possible to logon with other login information, such as that of an S-user.

As an administrator, you have been assigned to the role collection of the User Administrator so that you have access to the User Management app of the **Administration** app group (see Figure 8.16).

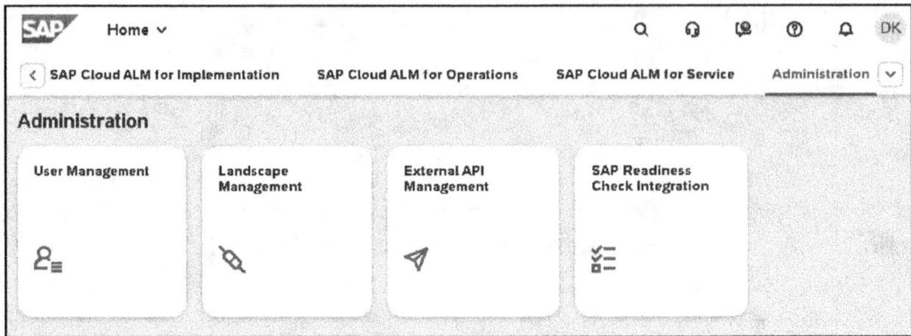

Figure 8.16 Administration App Group

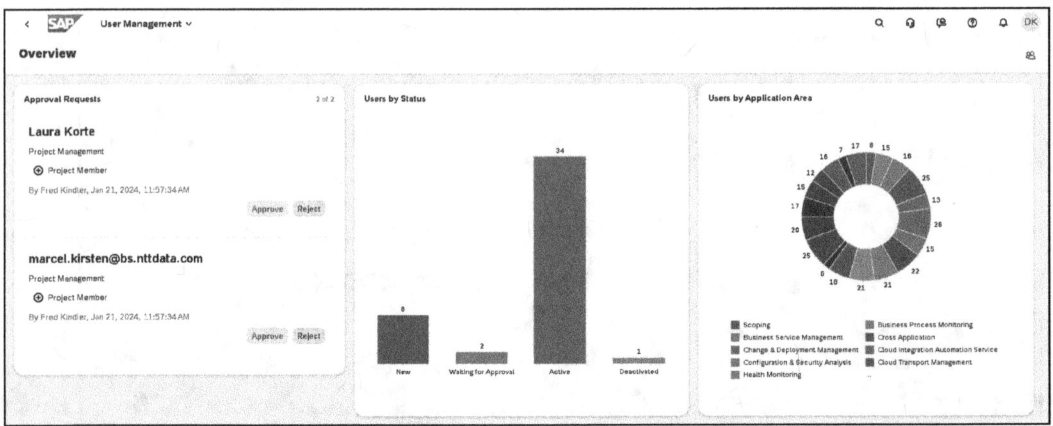

Figure 8.17 User Management App: Overview Page

Click on the **Users** icon located on the top right corner of the screen to access the overview of all named users. There you can use the filter bar to search for specific users. The following filters can be used for this purpose:

- **Search:** User name, email address
- **Status: Active, Deactivated, Expiring Soon, New,** or **Waiting for Approval**
- **Application Area:** Task or scenario-related (e.g., business process monitoring, modeling, and tag management)

- **Role** (e.g., Change Manager, Deployment Manager, or Project Member)
- **Type: Employee, SAP, Partner, Other**, or (not assigned)

The definition of the different types can be found in Table 8.4.

Type	Definition of
Employee	Company employees
SAP	SAP consultant
Partner	Consultant of an official SAP partner
Other	Partner from a company that is neither an official SAP partner nor SAP itself
(Not Assigned)	Default value

Table 8.4 User Types

Click on a user to open a detailed window on the right-hand side of the screen, which you can also expand to full screen. In this detailed view, you will find all relevant administrative information (see Figure 8.18). Here you can edit, delete, or deactivate the user using the corresponding buttons. The change history for the respective user is also visible.

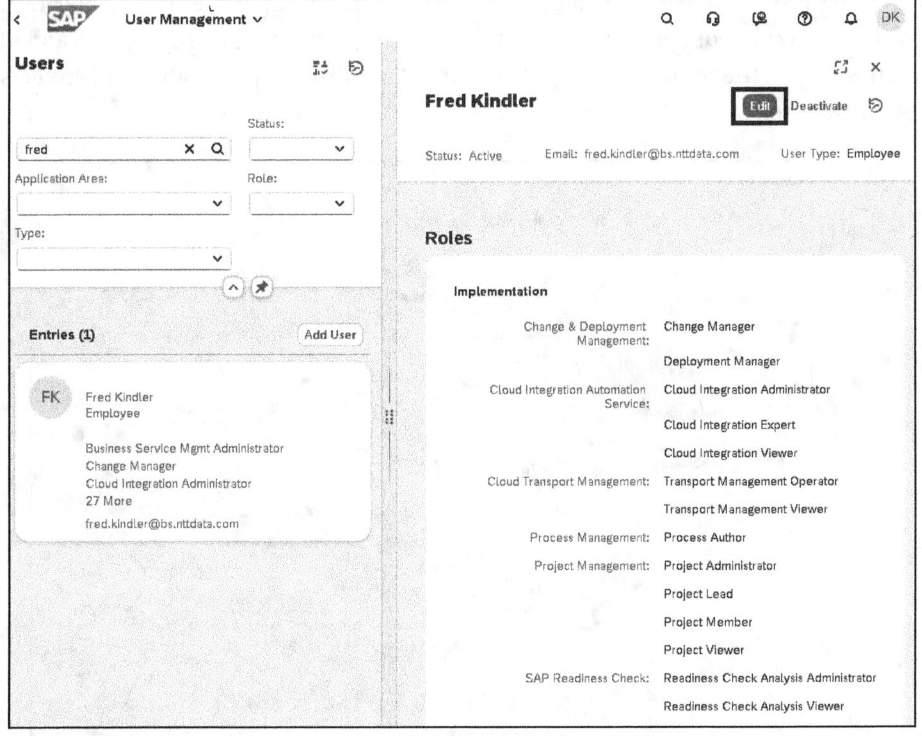

Figure 8.18 Manage Existing Users

For maintaining the authorization roles, you have switch first to edit mode via the **Edit** button. The standard roles can be assigned to the application areas provided by the **Roles** area. Your own prepared role collections can be assigned in the **Custom Roles** field. At this point, you can, for example, add the role collection **ViewerAll** previously created in the SAP BTP cockpit to the selected user (see Figure 8.19). Complete your selection by clicking **Select** and then **Save**.

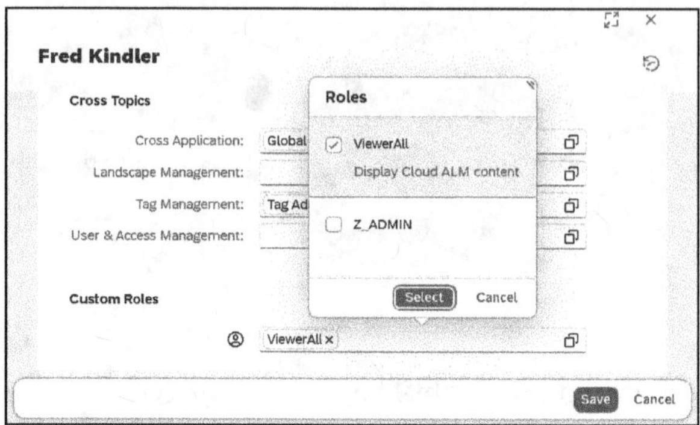

Figure 8.19 Assign a Custom Role to a User

8.1.6 Approve User Requests

Approval requests are created by application owners in the corresponding application stating that a user needs specific roles. In the following example, an employee is added to a team's role on the project **AriHana-2024-TPL-R1** within the **Teams** tab of the Projects and Setup app (see Figure 8.20).

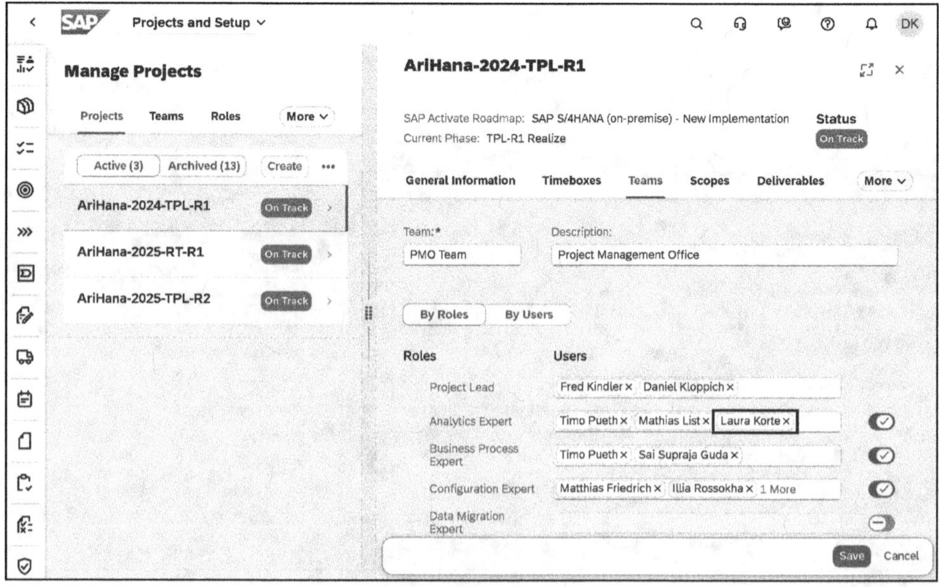

Figure 8.20 Projects and Setup: Add Team Members

If an approval request is created, all users with authorization to approve the request are notified via the SAP Cloud ALM launchpad notification 🔔 located on the top right corner of the screen. To process the approval request, you can click the corresponding notification. Furthermore, all requests are displayed within the **Overview** screen of the User Management app. As a user administrator, you process these requests by approving or rejecting them (see Figure 8.21).

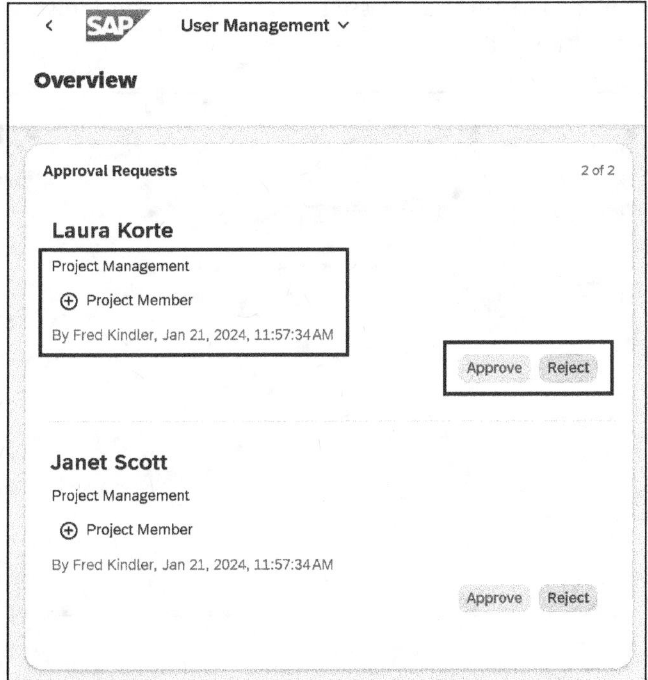

Figure 8.21 User Management: Card Approval Requests

You can navigate from the **Approval Requests** card to a list of all users with pending approval requests by clicking on the title area. Select any approval request to display the complete details of the affected user.

[+]

Launchpad Notification and Multiple User Requests

If multiple open approval requests for the same user and application area exist, the user administrator can only process them in the detail view of the user. This prevents the processing of an approval request without knowing the other relevant requests.

8.1.7 Define Project Access Permissions

Project user authorizations are defined by the project access levels, the roles users are assigned to in the user management, and the roles users are assigned to in project management.

In the Projects and Setup app, the corresponding access level can be maintained for a project, as shown in Figure 8.22:

- **Public**: This project is publicly accessible by all users with project management authorizations.
- **Restricted**: This project is restricted and only project members can make changes within the project (display for non-project members is allowed)
- **Private**: This project is private and can only be accessed by project members (display for non-project members is not allowed).

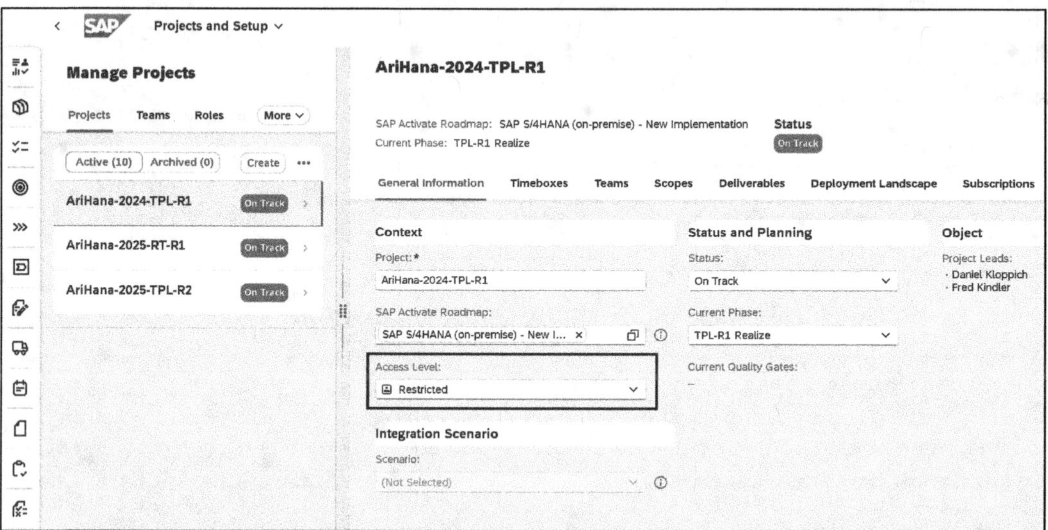

Figure 8.22 Project Access Permissions

[»]

Default Access Level

When a project is created, the access level is set to **Restricted** by default. This can be changed afterwards.

Within each access level, the combination of role assignments in the User Management app and the team assignment in a project defines the project permissions for a user.

When you set the access level for a project to **Public**, user authorizations are only affected by the roles that are assigned in the User Management app. This is not the case for the access levels **Restricted** and **Private**. Here, user authorizations are both affected by the roles that are assigned in the User Management app and the team assignments made in the Projects and Setup app.

[«]

Authorization Changes

Any authorization changes to projects or assignments to projects or teams only apply to the individual users after they have logged out of SAP Cloud ALM.

8.2 Project Progress and Project Reporting

With the Overview, Tasks, Analytics, and Cross-Project Analysis apps, SAP Cloud ALM provides advanced capabilities for tracking the progress of a project as well as for project reporting in multiple ways (see Figure 8.23). These apps can be found within the **SAP Cloud ALM for Implementation** app group. In the subsequent sections, we describe these apps in more detail. You can use them to call up and track the current status of your project.

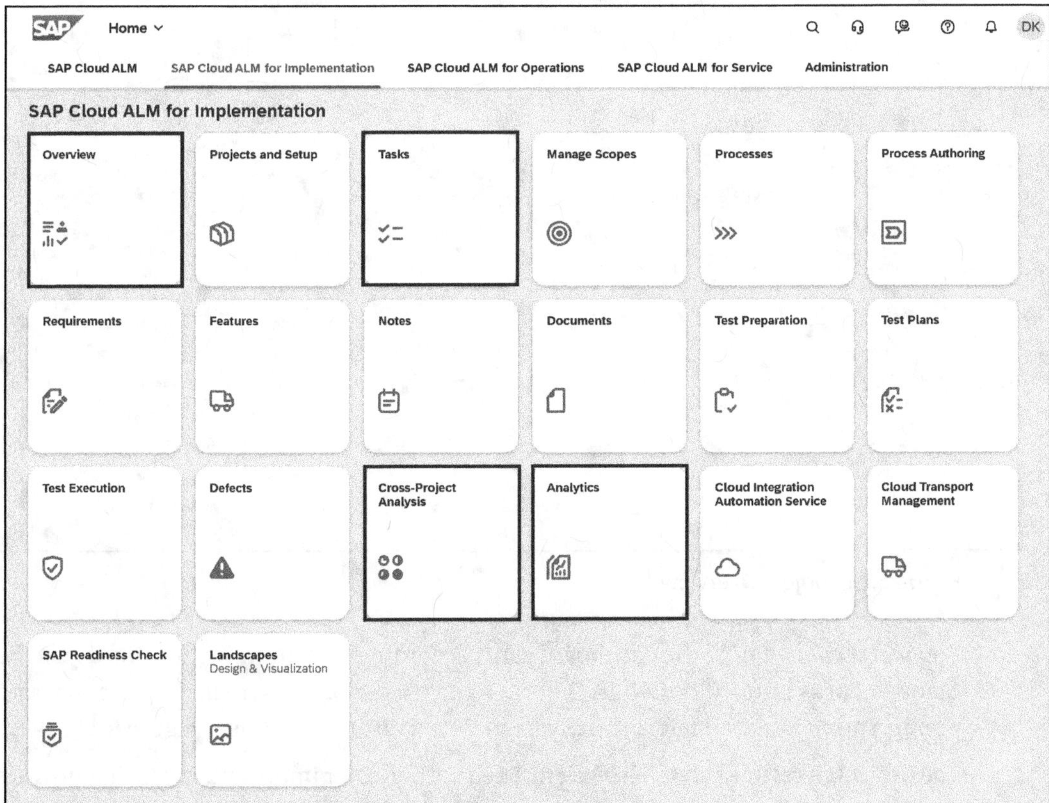

Figure 8.23 Apps for Tracking the Project Progress and Project Reporting

8.2.1 Project Overview

The Overview app represents a project dashboard orchestrated by various cards providing a broad range of information regarding the selected project (see Figure 8.24). You can use the filters on the header to limit or specify the presented results within the cards (you can filter for a specific scope or timebox). The project manager is able to use this dashboard to display the overall status using the filter **Assigned to** and value **All**, while the project members typically use value **Me** or **My Roles** to visualize information regarding elements such as tasks, requirements, test cases, etc., relevant for them.

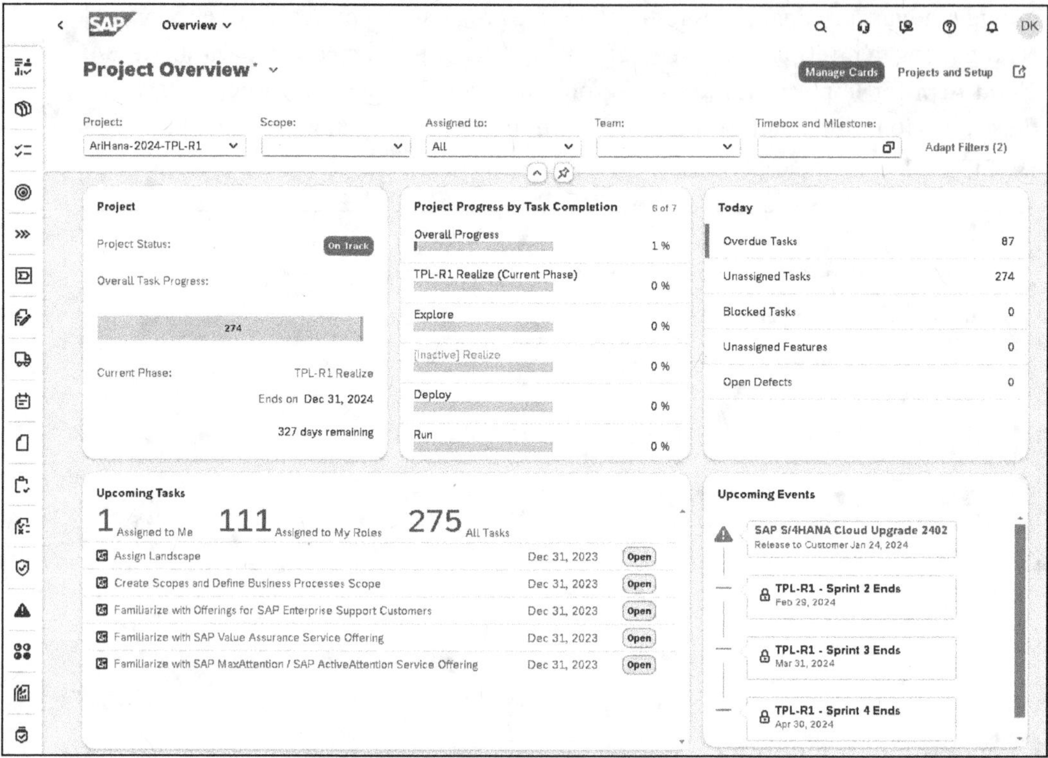

Figure 8.24 Project Overview

You can show and hide the dashboard cards according to your individual needs via the **Manage Cards** button located in the top-right corner of the screen. In the appearing popup, simply use the sliders to activate or deactivate the cards (see Figure 8.25).

The project overview can so also be enriched with further important cards that provide you information on **Requirement Distribution, Features, Documents, Test Case Preparation, Test Execution, Failed Manual Test Runs** as well as **Defects** (see Figure 8.26). It is possible to filter most of this information according to various criteria. For example, the requirements can be displayed regarding to their approval status or priority.

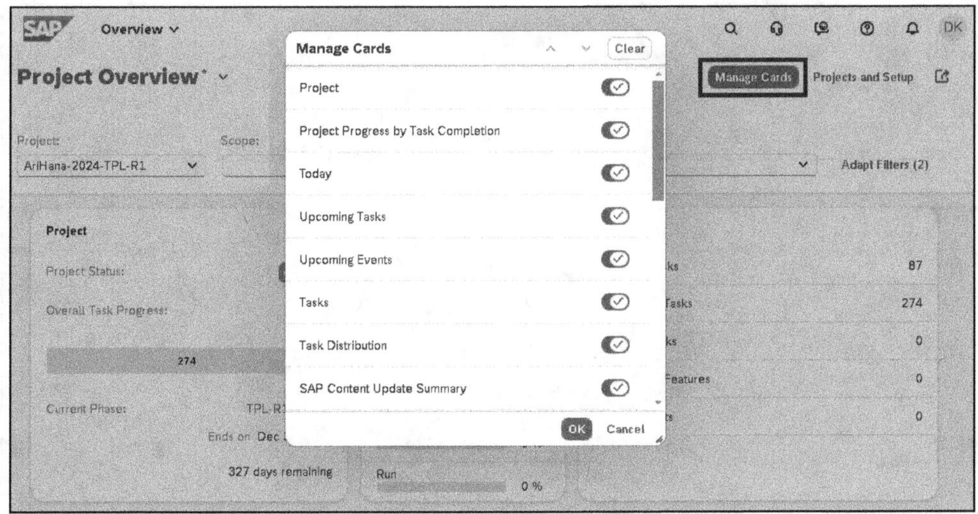

Figure 8.25 Project Overview: Manage Displayed Cards

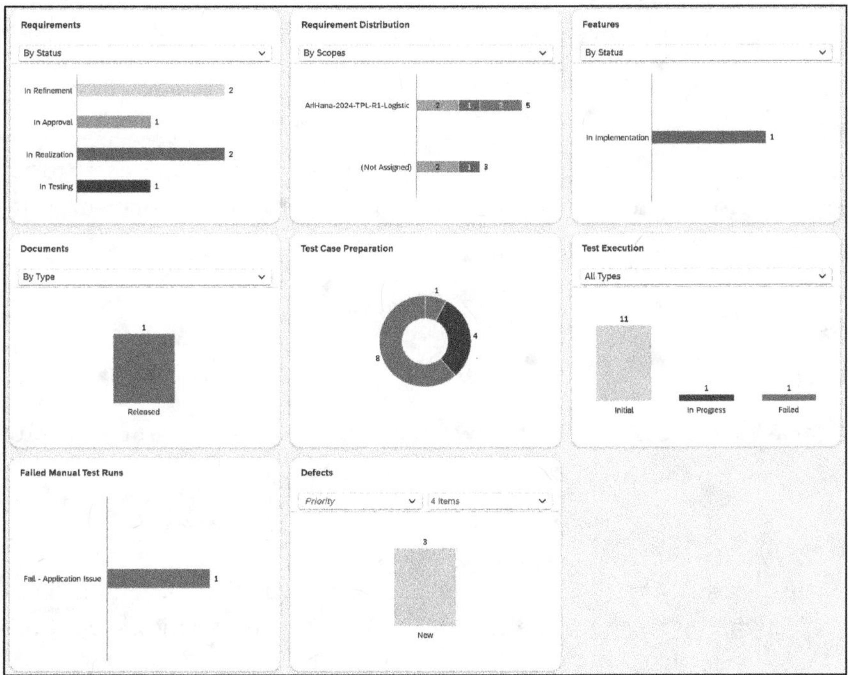

Figure 8.26 Project Overview: Further Cards

Some cards compose the information of a specific topic. You can use them to drill down to display the details and obtain further information. For example, the **Unassigned Tasks** entry of the **Today** card navigates you to the Tasks app. Here, all open tasks are displayed, taking into account the filter for the statuses **Open** and **In Progress** as well as the filter for the respective person responsible **(Not Assigned)** (see Figure 8.27).

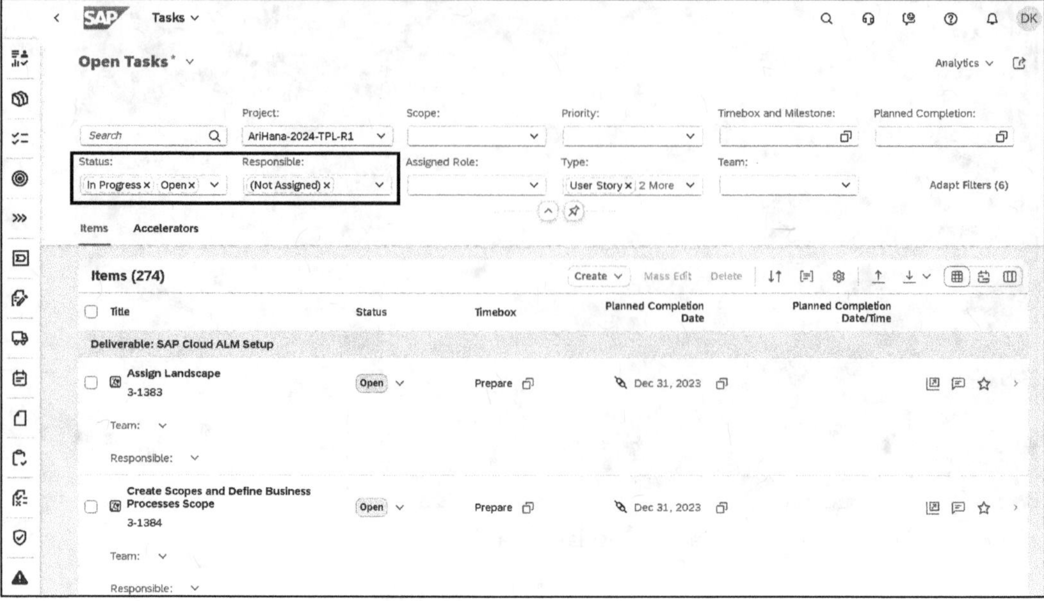

Figure 8.27 Project Overview: Navigation to "Unassigned Tasks"

8.2.2 Tasks

The Tasks app is usually used to list all task types such as requirements, user stories, project tasks, etc., according to specified filter criteria. The app provides following three data visualization functions:

- **Table View** (default)
- **Gantt Chart View**
- **Cards View**

Next to the **Table View**, the **Gantt Chart View** and the **Cards View** support typical project management activities.

Gantt Chart View

With the help of the **Gantt Chart View** (see Figure 8.28), the project manager can visualize the project activities on the timeline. Dependencies between activities, such as predecessors or successors, can be easily traced.

Cards View

The **Cards View** (see Figure 8.29) enables simple status changes of tasks (e.g., requirements, user stories, and subtasks) via drag-and-drop on a board. This can be used in agile meetings such as the daily Scrum to easily update the sprint backlog.

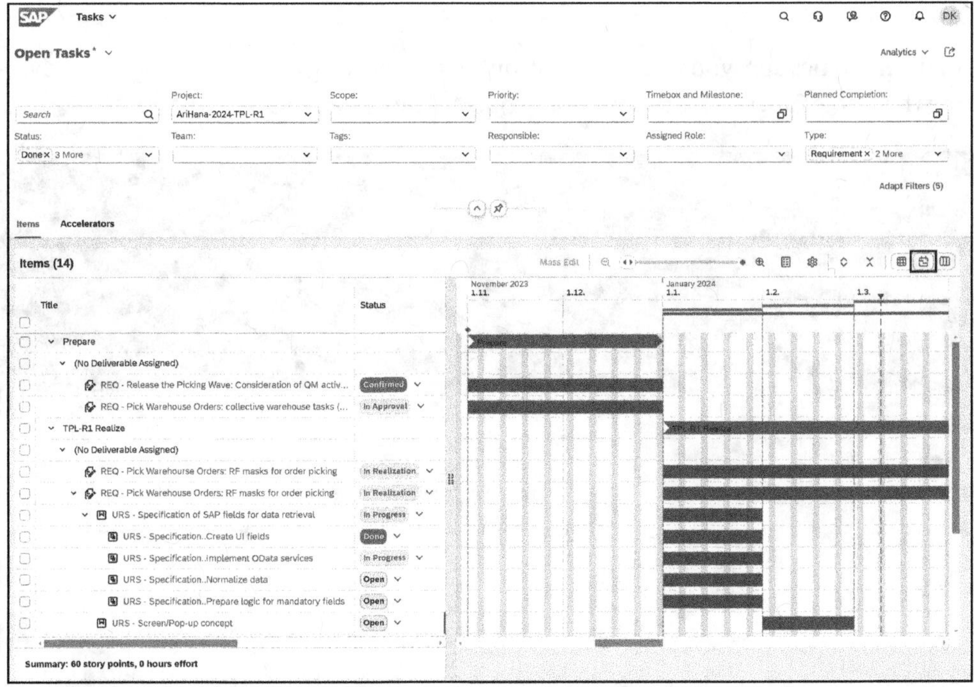

Figure 8.28 Tasks: Gantt Chart View

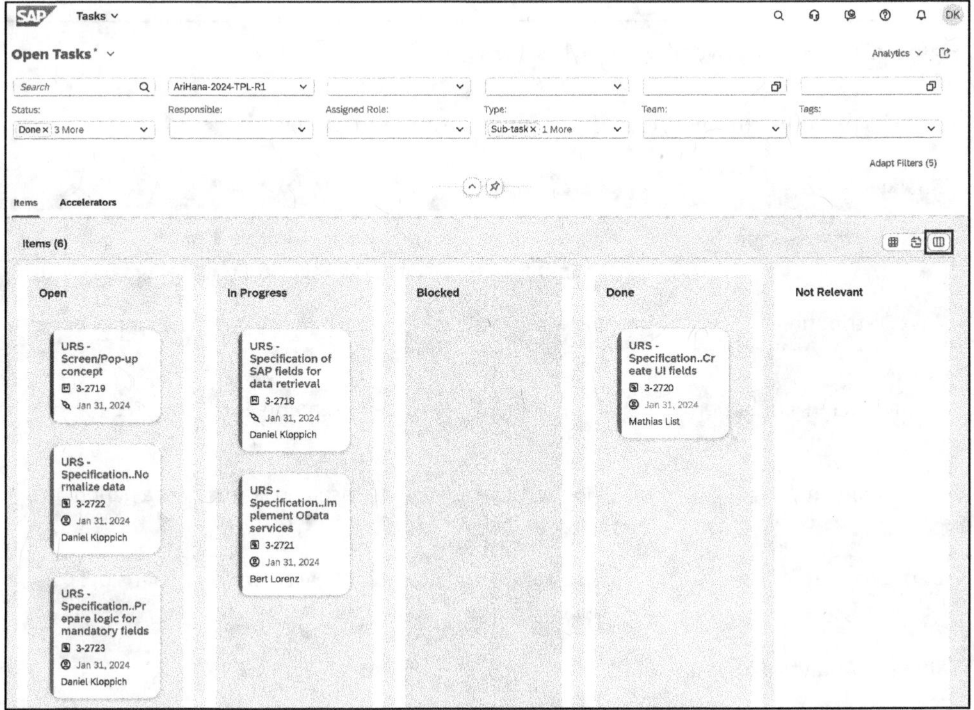

Figure 8.29 Tasks: Cards View

8.2.3 Analytics

In the Analytics app, you will find many different reporting and analysis options covering the whole functional scope of SAP Cloud ALM (see Figure 8.30).

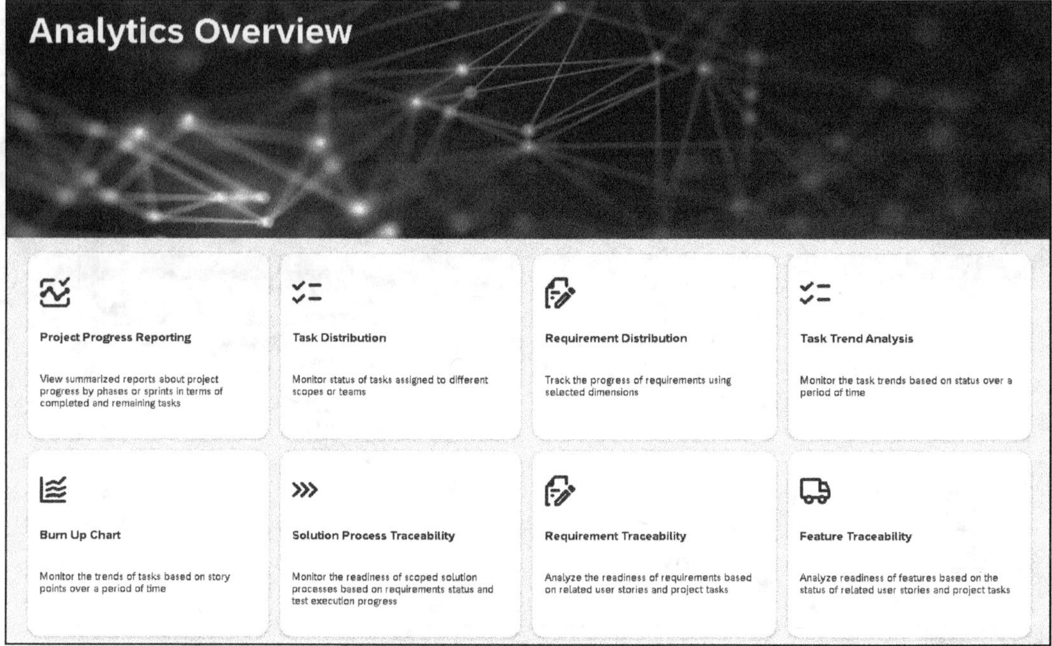

Figure 8.30 Analytics Overview: Anaylics Apps

Table 8.5 provides an overview of the available analytics apps.

Analytics App	Description
Project Progress Reporting	View summarized reports about project progress by phases or sprints in terms of completed and remaining tasks
Task Distribution	Monitor status of tasks assigned to different scopes or teams
Requirement Distribution	Track the progress of requirements using selected dimensions
Task Trend Analysis	Monitor the task trends based on status over a period of time
Burn Up Chart	Monitor the trends of tasks based on story points over a period of time

Table 8.5 Analytics Apps

Analytics App	Description
Solution Process Traceability	Monitor the readiness of scoped solution processes based on requirements status and test execution progress
Requirement Traceability	Analyze the readiness of requirements based on related user stories and project tasks
Feature Traceability	Analyze readiness of features based on the status of related user stories and project tasks
User Story Traceability	Monitor the readiness of user stories based on the progress of related items such as subtasks and test executions
Test Execution Analysis	Analyze the progress of test execution over a period of time
Defects Reporting	Analyze overall trends for defects

Table 8.5 Analytics Apps (Cont.)

We will go into more detail on some of these in this book. For example, the Requirement Distribution and Requirement Traceability apps are described in more detail in Chapter 10.

For the general time planning overview, there is the Project Progress Reporting analytics app. Here, you get full insights of all tasks among your project phases. The table view displays following task information:

- **Planned Tasks**
- **Completed Tasks**
- **Overdue Tasks**
- **Remaining Tasks**
- **Planned Total Remaining Tasks**
- **Actual Total Remaining Tasks**

If the tasks are broken down by phase, you can quickly determine which tasks from the previous phase, for example, are still open and therefore overdue in relation to the planned completion date (see Figure 8.31). A breakdown by sprints, deliverables, and milestones is also possible.

Within all display variants, you can specify the information by using the filter options, for example, to display everything relating to your project, a specific scope, or task type (see Figure 8.32).

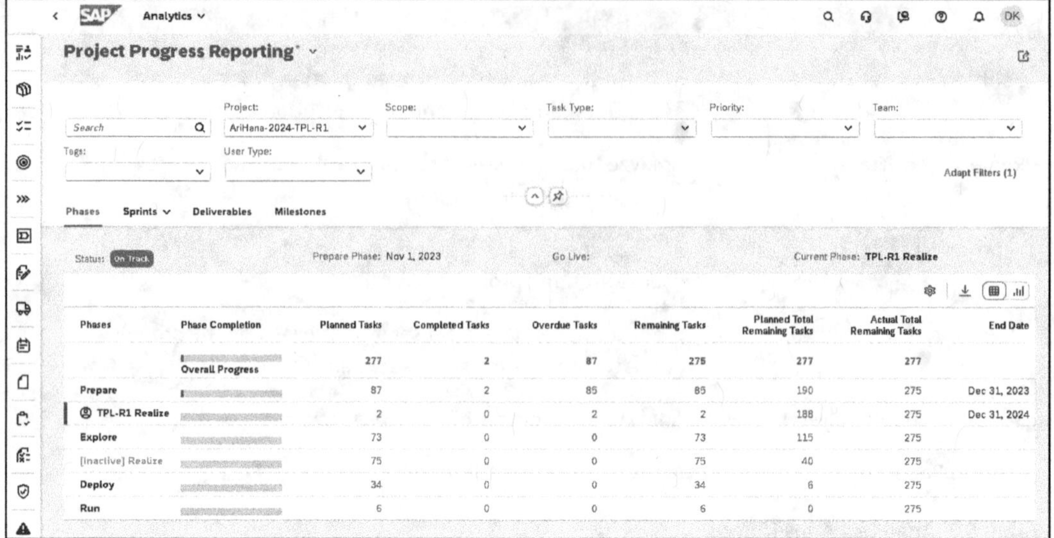

Figure 8.31 Project Progress Reporting: Table View by Phases with Filter on "Project"

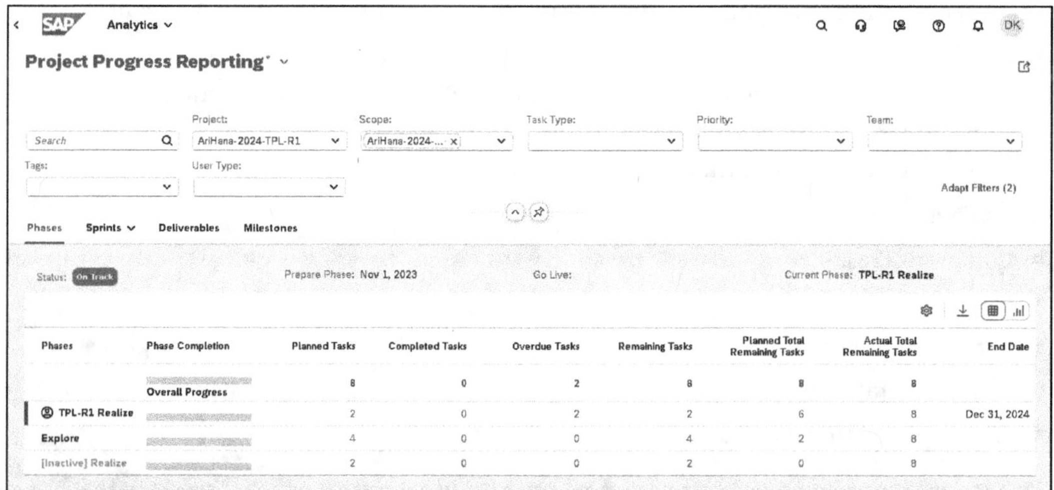

Figure 8.32 Project Progress Reporting: Table View by Phases with Filter on "Project" and "Scope"

Figure 8.32 shows two remaining tasks for the phase **TPL-R1 Realize**. Clicking on the number opens an overview in which these items are listed in more detail (❶ of Figure 8.33). You can drill down further by pressing the **Tasks Overview** button (❷ of Figure 8.33). Thus, the entries are displayed within the Tasks app. Here, you are able to investigate the circumstances, finding out who is responsible, or adjust the status or the planned completion date.

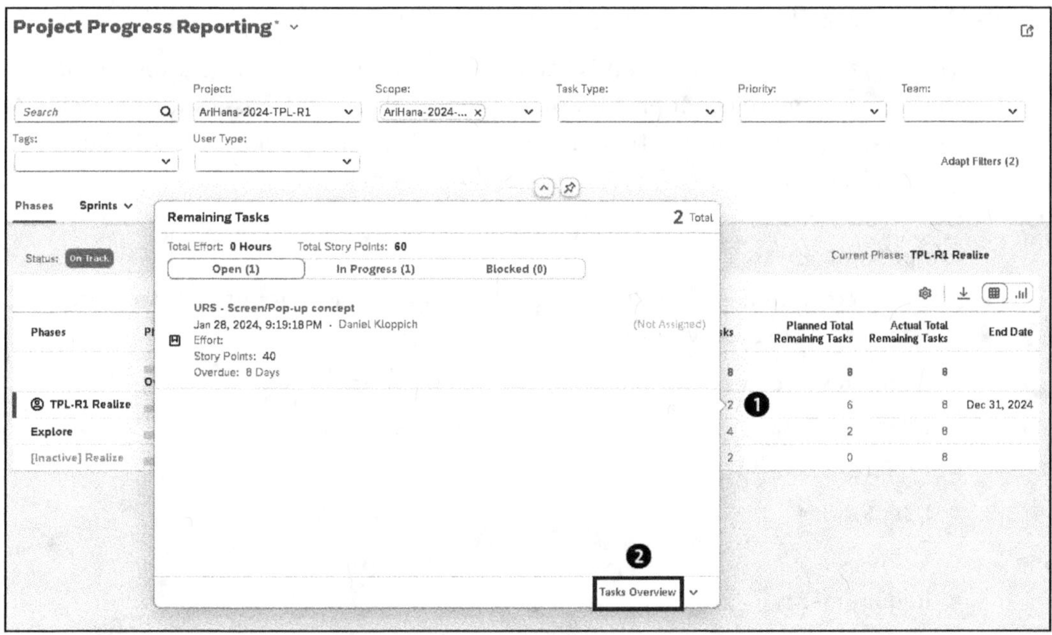

Figure 8.33 Project Progress Reporting: Details on Tasks

To display the Project Progress Reporting app as a table using the **Table View** button, it is also possible to analyze the data in a diagram using the **Chart View** button (see Figure 8.34). This view represents a **Burn Down Chart** useful for visually tracking the extent to which the actual remaining tasks correspond to the planned remaining tasks.

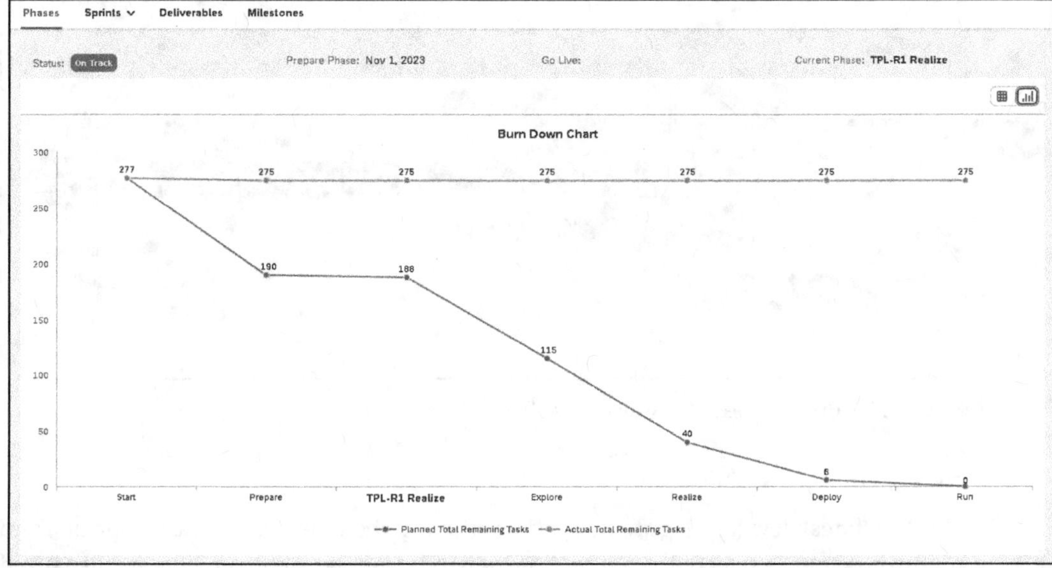

Figure 8.34 Chart View by Phases (Burn Down Chart)

8.2.4 Cross-Project Analysis

The Cross-Project Analysis app is mainly used by project managers responsible for more than one project. It enables them to compare characteristics for multiple SAP Cloud ALM projects at one time. The Cross-Project Analysis app displays the project information in two views, **Projects Readiness** and **Timelines**.

Projects Readiness

The **Projects Readiness** view (see Figure 8.35) displays the amount and status values of the main SAP Cloud ALM entities:

- **Solution Processes**
- **Roadmap Tasks**
- **Project Tasks**
- **User Stories**
- **Subtasks**
- **Requirements**
- **Features**
- **Test Preparations**
- **Test Plans**
- **Test Executions**
- **Defects**

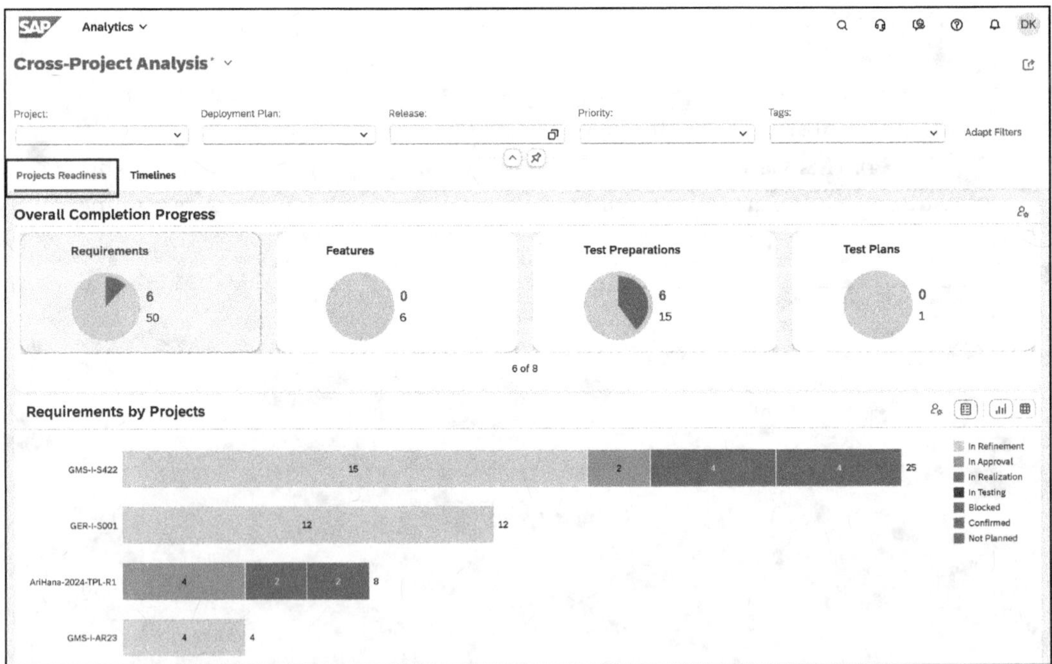

Figure 8.35 Cross-Project Analysis: Projects Readiness

Timelines

The **Timelines** view (see Figure 8.36) shows the timebox elements phases, sprints, and milestones for the selected projects. Time dependencies between projects are visualized and can be easily linked. In our example, the template realization and the corresponding rollouts can be planned. In addition, the delivery of upgrades of your considered SAP solutions is displayed. This makes it possible to overlay upgrades and subsequent necessary activities within your current running project, such as testing or the synchronization of updated SAP Best Practices content.

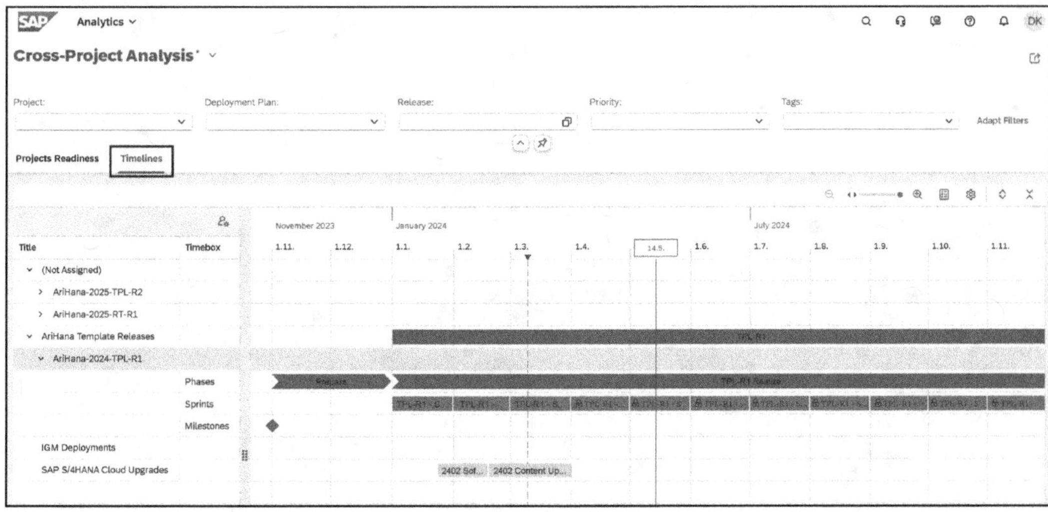

Figure 8.36 Cross-Project Analysis: Timelines

8.3 Summary

In this chapter, we explained the onboarding procedure, including the subsequent management of users as well as the assignment of dedicated authorization roles within SAP Cloud ALM. Furthermore, we examined the difference between project and authorization roles as well as the interaction between project, team, project roles, and users. You learned how to prepare your own role collection within the SAP BTP cockpit. Finally, we provided details on how to track the progress of your projects and how the various project reporting functions can be used.

With this chapter, we end the prepare phase of SAP Activate in our book and move on to the explore phase. Your involved project members should now be able to logon to SAP Cloud ALM and participate in the project. Bear in mind that, in addition to providing access, project onboarding must also include measures on how to use the tool. This may include role-based trainings and the provision of related materials. The next phase, explore, is characterized by the initial requirements analysis of the solution to be implemented. In the first step, we use our example scenario in the context of process management to describe how you can adapt the solution processes provided by SAP to your individual business needs to design your target processes.

PART II

Explore

Chapter 9
Process and Document Management

Within SAP Cloud ALM, solution processes are of particular importance because they are the starting point for both requirements creation and the fit-to-standard workshops. In the previous chapters, you learned how to select and use solution processes. But what does process management mean, and how is it used? In this chapter, we will answer these two questions, among others, and show you how to use solution processes.

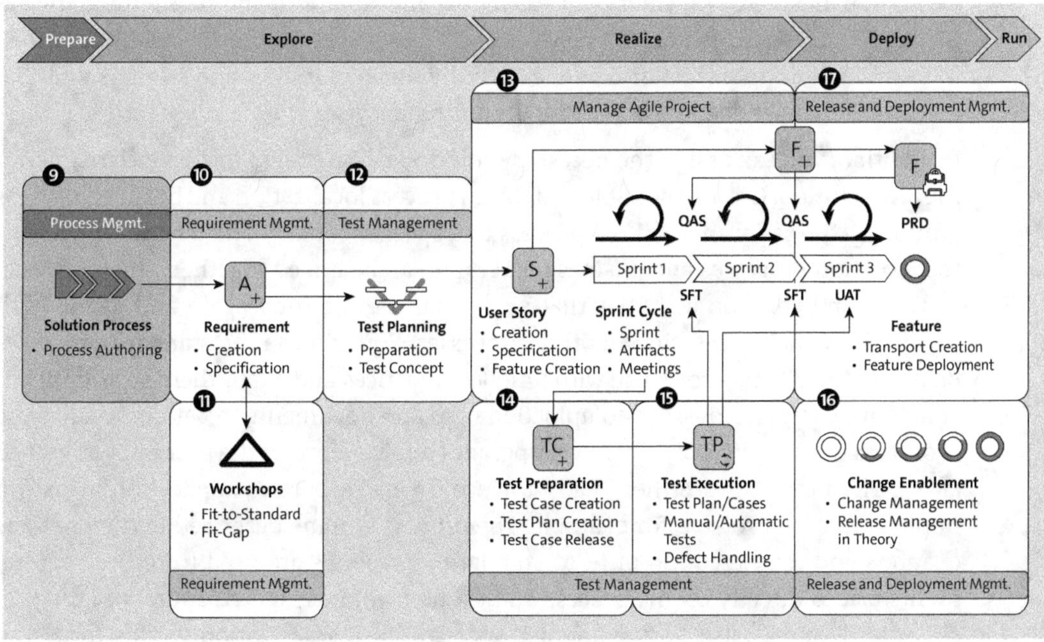

In the previous chapters, we presented the most important preparatory measures for the start of your project with SAP Cloud ALM. In the SAP Activate methodology, the associated activities are part of the prepare phase. In Chapter 7, we provided the project scope to be implemented in our example scenario as well as the relevant SAP Best Practices processes. We revisit those solution processes in this chapter and explain the functions provided in SAP Cloud ALM as part of process management. In this and the following three chapters, we will discuss the most important points within the explore phase, which in their entirety deal with the initial recording of requirements for the implementation project. The documented requirements and the scoped (SAP Best Practices) solution processes form the basis for the subsequent realize phase.

With SAP Cloud ALM, the individual project members can not only select and use predefined solution processes as templates, but also design their own solution processes, adapt them to their needs at any time, and then implement them. The documentation of the solution processes should ideally support the organizational and business objectives of the project. The solution processes provided by SAP are made available in graphical form with solution value flows and solution process flows. This not only enables a better representation of the process flow, but also facilitates the creation of requirements, user stories, and comments. The specific functions and processes are explained in more detail in the following sections of this chapter.

But the application goes far beyond pure process representation. In particular, you have the option to create your own business and solution processes and also to visualize them using BPMN (Business Process Model and Notation) and thus extend the existing SAP practices. This functionality is known as *process authoring* and is described in Section 9.5 in more detail. You create individual solution value streams and diagrams and manage the content across the entire lifecycle.

In addition, you can assign the necessary solution components to create clarity about the SAP solution. It is also possible to define a process localization and business process group, e.g. the assignment of the process to a specific country and business area. Further solution processes and business processes can be linked together. The individual business and solution activities that you create can be used again and again, both within the same process and in other processes. You can use the standard processes delivered by SAP in accordance with SAP Best Practices and adapt them to your needs at any time. It is also possible to upload images and documents to better visualize the processes used. You can also manage upgrades and versions of the processes as you go along. The processes are then ready for use in the respective application areas—in implementation projects, for example. The process scoping described in the previous chapters and the execution of fit-to-standard workshops are used to record requirements placed directly on the process and to manage the respective status of the processes. The aim is also to document the entire business solution in this form and always keep it up to date. The result is a constantly evolving *solution documentation* that represents a single source of truth for all further processes and therefore serves as the starting point for requirements, tasks, and tests. The knowledge imparted in this chapter is crucial for the further use of the solution processes and the associated creation of requirements.

9.1 Processes in SAP Cloud ALM

When defining processes, a distinction is made between solution processes and business processes. A *solution process* represents a concrete implementation of a business process. The solution process therefore provides a detailed description of how the business process was designed or will be designed in the solution. But it is also possible to have two solution processes that represent the same business process. This can be seen in the general process of *sourcing* and can be implemented, for example, in the central sourcing solution process within SAP S/4HANA Cloud or via SAP Business Network in the "Sourcing in SAP Business Network" solution process. This means that there can be different *process variants* with different activities. The solution process, in turn, has a relationship to the solution component, which is then the anchor point for describing what the business application and the products that support the solution process are.

In the context of SAP Cloud ALM a process also consists of a series of steps that together represent a closed business function—for example, the acceptance of a sales order or the creation of a goods receipt posting with a clearly defined start and end. Typically, these processes are used in fit-to-standard workshops. SAP Cloud ALM offers a broad range of SAP standard content. The included processes are delivered with predefined solution process flows in BPMN format. The graphical editor of the Process Authoring app can also be used to create new, user-defined solution processes to reflect existing business process variants. The user-defined solution processes can also be assigned to specific countries or regions in which they are valid. Furthermore, it is possible to create the corresponding requirements, user stories, and project tasks for single activities or the whole process directly in the app.

Both business processes and solution processes can occur in different variants. The *business process variant* (e.g., the sourcing process) consists of various functions, which are described in the model as business activities and which are also a specific business process from the company's perspective. The *solution process variant* can also be understood as the implementation of this process from the IT perspective. Among other things, it consists of the necessary solution components, SAP Fiori apps, data elements, and physical systems that are necessary to implement this process. The specific products and solutions are defined or formulated, and then those same products and solutions implement and execute the processes. The business process with its business activities forms the functional scope of the process.

SAP Cloud ALM uses *SAP Enterprise Architecture Methodology*. This framework is a methodology and toolset that defines how to create and use an enterprise architecture. It provides principles and practices for creating and using the architecture description of a system. Within this methodology, a distinction is also made between business processes and solution processes. The solution process represents the core concept of all activities in SAP Cloud ALM.

9.2 Documents

The purpose of the Documents app is to simplify the creation, management, tracking, and reuse of existing documentation within SAP Cloud ALM to be the central source of truth for your solutions. With this central app for documentation purposes, you can develop a holistic approach for all relevant documentation and its corresponding and necessary relations to different entities, such as solution processes or process hierarchies. Furthermore, it is essential to connect documents to requirements, user stories, project tasks, test cases, and features.

You will find the Documents app in the **SAP Cloud ALM for Implementation** app group, as shown in Figure 9.1.

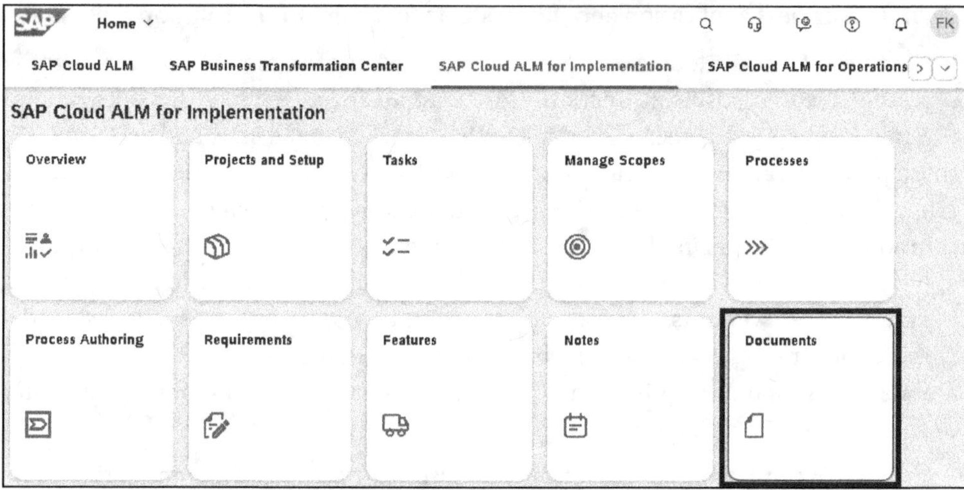

Figure 9.1 Documents App

After starting the app, you will see a list of all documents in one central place. You have the ability to filter and group documents by owner, status, change date, priority, process hierarchies, etc. and adapt views based on your individual needs. The filtered documents can also be exported to an Excel spreadsheet via the corresponding button.

In terms of our example scenario, we want to create a technical design document for an interface between SAP S/4HANA and SAP Business Network via SAP Integration Suite, managed gateway for spend management and SAP Business Network. To create a dedicated document, select the **Create** button as shown in Figure 9.2.

[»]

Project Relations in the Documents App

Make sure you have preselected the correct project, as the document will be created in connection with this project. It is currently not possible to move or remove the project link.

Documents
Standard*
Search
Editing Status:
All
Project:
AriHana-2024-TPL-R1
Scope:
Type:
Status:
Priority:
Owner:
Solution Process:
Process Hierarchy:
Adapt Filters (2)
Documents
Create
Upload
Delete
Document
Status
No data found. Try adjusting the search or filter parameters.

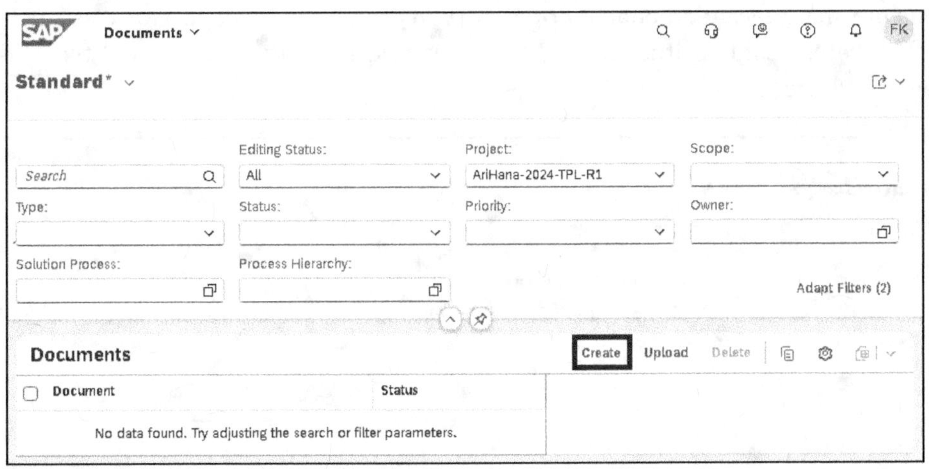

Figure 9.2 Documents App: Create Document

Now define a title for the document and start to specify the dedicated **Content** within the text area, as shown in Figure 9.3.

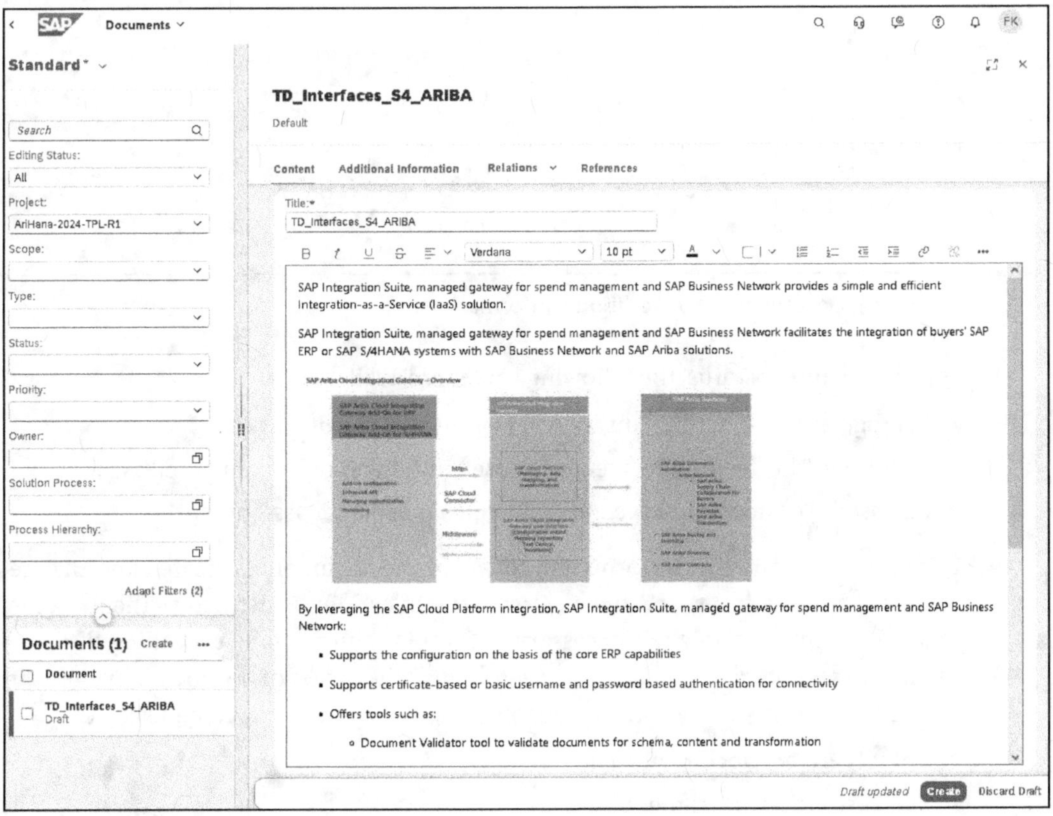

Documents
Standard*
Search
Editing Status:
All
Project:
AriHana-2024-TPL-R1
Scope:
Type:
Status:
Priority:
Owner:
Solution Process:
Process Hierarchy:
Adapt Filters (2)
Documents (1)
Create
Document
TD_Interfaces_S4_ARIBA
Draft
TD_Interfaces_S4_ARIBA
Default
Content
Additional Information
Relations
References
Title:*
TD_Interfaces_S4_ARIBA
Verdana
10 pt
SAP Integration Suite, managed gateway for spend management and SAP Business Network provides a simple and efficient Integration-as-a-Service (IaaS) solution.
SAP Integration Suite, managed gateway for spend management and SAP Business Network facilitates the integration of buyers' SAP ERP or SAP S/4HANA systems with SAP Business Network and SAP Ariba solutions.
By leveraging the SAP Cloud Platform integration, SAP Integration Suite, managed gateway for spend management and SAP Business Network:
Supports the configuration on the basis of the core ERP capabilities
Supports certificate-based or basic username and password based authentication for connectivity
Offers tools such as:
Document Validator tool to validate documents for schema, content and transformation
Draft updated
Create
Discard Draft

Figure 9.3 Documents App: Specify Content

In the next tab page, **Additional Information**, you can find more details for your document (in which context it is created, as well as priority, status, owner, and type), as shown in Figure 9.4.

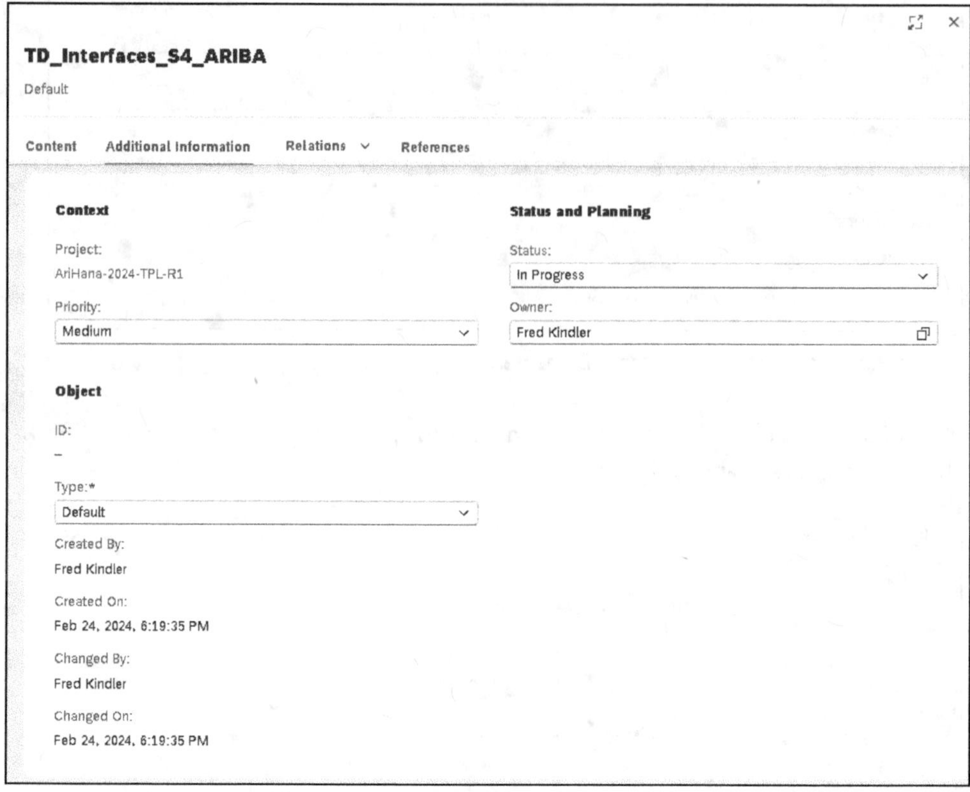

Figure 9.4 Documents App: Additional Information

For the document status, the following values are available:

- **In Progress** (the default): You are still working on this document.
- **In Review**: The document is ready for use but not yet published or released.
- **Released**: The document is completed and available for assignment.

By default, date and time information for when the document is created and changed are saved automatically, and the creator of the document is defined as the owner. As this field is optional, it is not necessary to fill this column while uploading the document via a spreadsheet. You can, however, change the owner of the document. Another field is mandatory: the **Type** field. The following standard types can be selected:

- **Default Project Documentation**
- **Business Process Document**
- **Solution Design Document**

- **Functional Specification**
- **Technical Design Document**
- **Configuration Guide**
- **Interface Specification**
- **Test Document**
- **End-User Documentation**
- **Training Material**
- **Fact Sheet**

Though you have to define a type when creating a document, you can change it at any time thereafter.

9

In the next tab page, **Relations**, you can connect documents directly to solution processes and process hierarchies. Process hierarchies will be explained in more detail in Section 9.6. In our example, this document is related to the **Automation of Order-to-Invoice with SAP Business Network (4A1)** solution process, as shown in Figure 9.5. You need to select the **Assign** button and search in the resulting popup for your solution process, but note that only scoped solution processes can be selected. This relation becomes even more important when you want to track completeness of your documentation within your project. Furthermore, SAP is planning the capability to see related documents in the Solution Process Traceability app. You can also delete existing relations via the **Unassign** button.

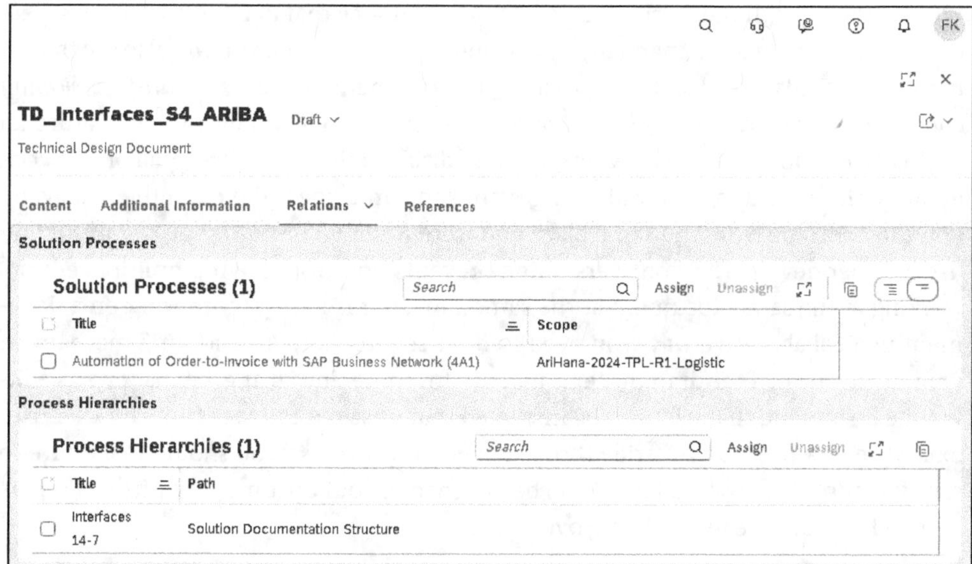

Figure 9.5 Documents App: Relations

In the next tab page, **References**, you can link external document sources (e.g., .docx from MS Word, .pptx from MS Powerpoint, or PDFs), as shown in Figure 9.6.

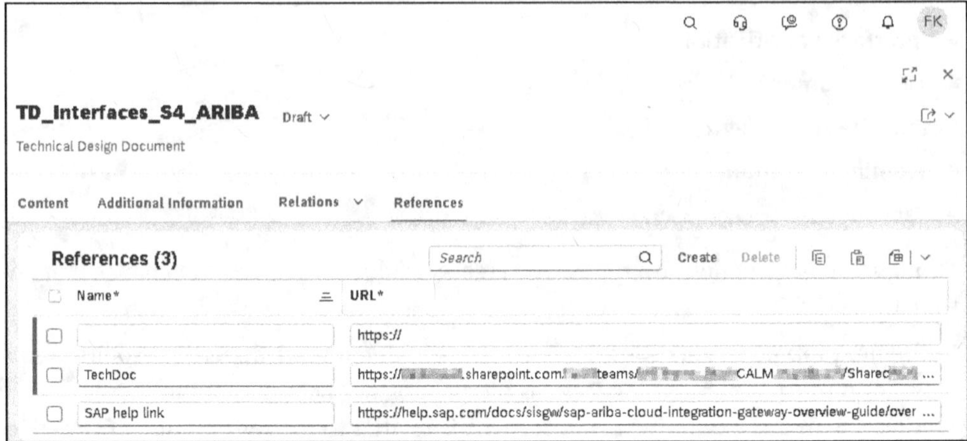

Figure 9.6 Documents App: References

For the creation of a reference, you need to select the **Create** button and enter a **Name** as well as the corresponding **URL**. SAP's general idea is that documentation for cloud solutions should be created directly in SAP Cloud ALM. Document management integration into the SAP Document Management service on SAP BTP and MS Sharepoint is planned. Therefore, stay tuned for what they will release at a later date.

While working with the Documents app, you will notice that everything will be saved, even if you do not save it manually by clicking the **Save** button. You will also see that all mandatory fields are filled. This is because the Documents app uses the *draft versioning concept* of SAP Fiori Elements. This prevents data loss and will eventually feature an exclusive lock for document owners. If your editing activity is interrupted or you accidentally close the app, unsaved changes are kept, and when you open the app again, you can continue from the same place you were when you left. This also works if you use another device. The draft versioning prevents you from having multiple editors working in the same document at the same time. If it is necessary to work on a document in a collaborative way, you need to do it sequentially. This locking mechanism expires, however, if the draft owner is inactive for some time. If another user starts editing the document, that user will be informed that unsaved changes from another user exist. The new user can override those changes, however, so you should make sure to save your document so that an ID will be assigned to your document and it can then be released without the fear of being overwritten.

[+]

Notes vs. Documents

The Notes app has been present since the beginning with SAP Cloud ALM. After the Documents app was released, SAP announced that the functionality of the Notes app will increasingly be migrated to the Documents app. After a certain time, the Notes app will be retired.

The upload function is another convenient way to upload documents in your SAP Cloud ALM tenant. You need to click the **Upload** button after you open the Documents app. In the resulting popup windows, you can use the documents' metadata and, after adding all of your documents, you need to select the created spreadsheet and then you will be able to upload this list of documents.

[!]

Recommendations

Some recommendations for uploading a list of documents via a spreadsheet are as follows:

- Only Microsoft Excel files with .xls or .xlsx extension are supported.
- Maximum allowed file size for a workbook is 5 MB.
- Maximum allowed number of rows is 1.000.
- Spreadsheets should only be stored in secure locations with restricted access.
- Column titles must not be changed.
- Title is a mandatory field, with maximum 255 characters.
- Check everything carefully, especially that URLs link to trusted targets, before you start the upload.

9.3 Solution Value Flow

In SAP Cloud ALM, we distinguish between two flowcharts, the solution value flow and the solution process flow. The *solution value flow* diagram usually describes the business scope of a user-defined solution process. It helps you understand what functionality is being delivered without getting lost in too many details. The *solution process flow* represents the functional scope provided by the solution variant and is built from business processes, as well as from reusable business activities. The value flow is intended as a visual description of the solution process and can be expanded at any time. The *process flow diagram* shows each individual process step in detail and the responsibilities of the project members. This helps you understand each employee activity and the business requirements needed for a desired configuration or development. In Figure 9.7, you can see the solution value flow of the process **Advanced Warehouse Outbound Processing to Customer (1VD)** process.

In the diagram of the solution process, you can create a requirement or a note to enter change requests directly in the process. To do this, click on the desired element and then select **Create Requirement**. Figure 9.8 shows that you can also download the test cases supplied by SAP as accelerators in Word or Excel as a file. *Accelerators* are links to various other resources provided by SAP, such as test cases or instructions. In addition, requirements, user stories, project tasks, and comments can be created here via the dropdown **Create** menu.

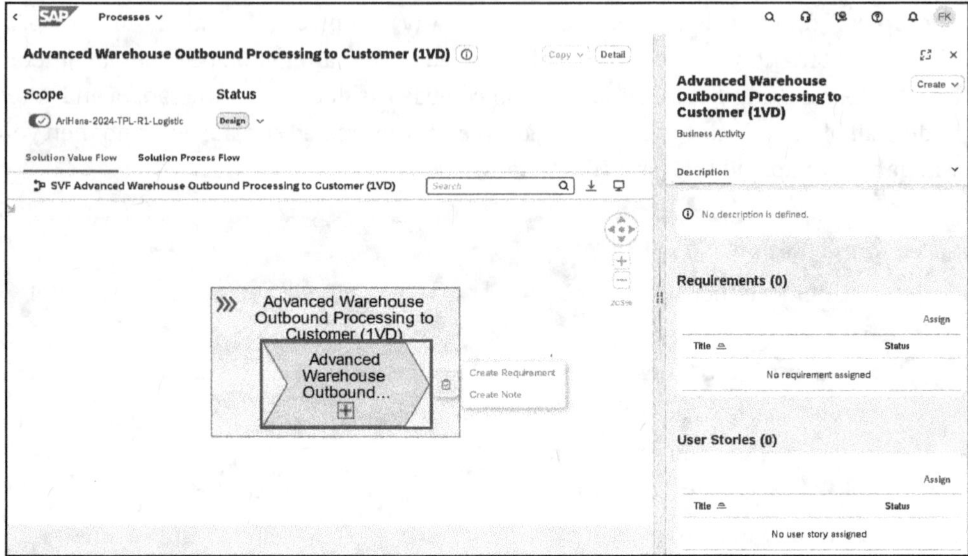

Figure 9.7 Solution Value Flow of a Selected Solution Process

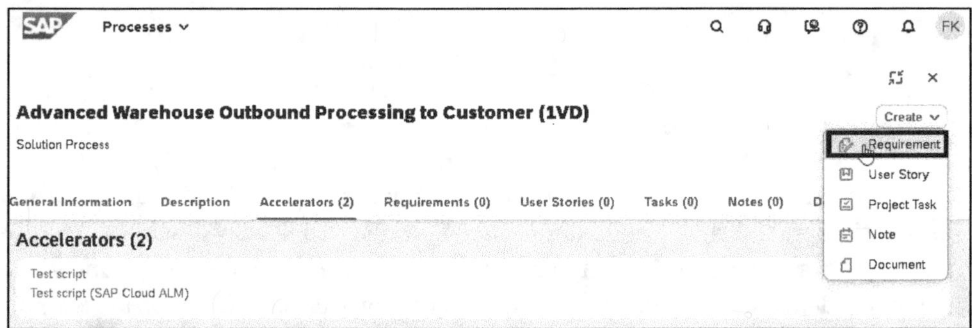

Figure 9.8 Create a Requirement out of the Solution Process

9.4 Solution Process Flow

The solution process flow represents specific, exemplary process flows of solution activities in SAP Cloud ALM. Solution process flows can be described by one or more solution process flow diagrams. The process flow represents all instances of the process, mainly for SAP implementations. The solution process flow diagram represents the flow of the solution process and is mainly mapped in the form of BPMN diagrams. With the process authoring function, which you can access both via the Process Authoring app and via the Processes app, you can use additional BPMN modeling functions to describe or adapt the desired process execution flow. All symbols used in modeling are later also relevant for other SAP Cloud ALM functions such as requirements management or testing, as they can be used to create requirements and notes as well as for test execution. As you can see in Figure 9.9, you can also create a requirement or a note in the solution process flow diagram. To do this, click on the desired element, and the **Create** icon will appear.

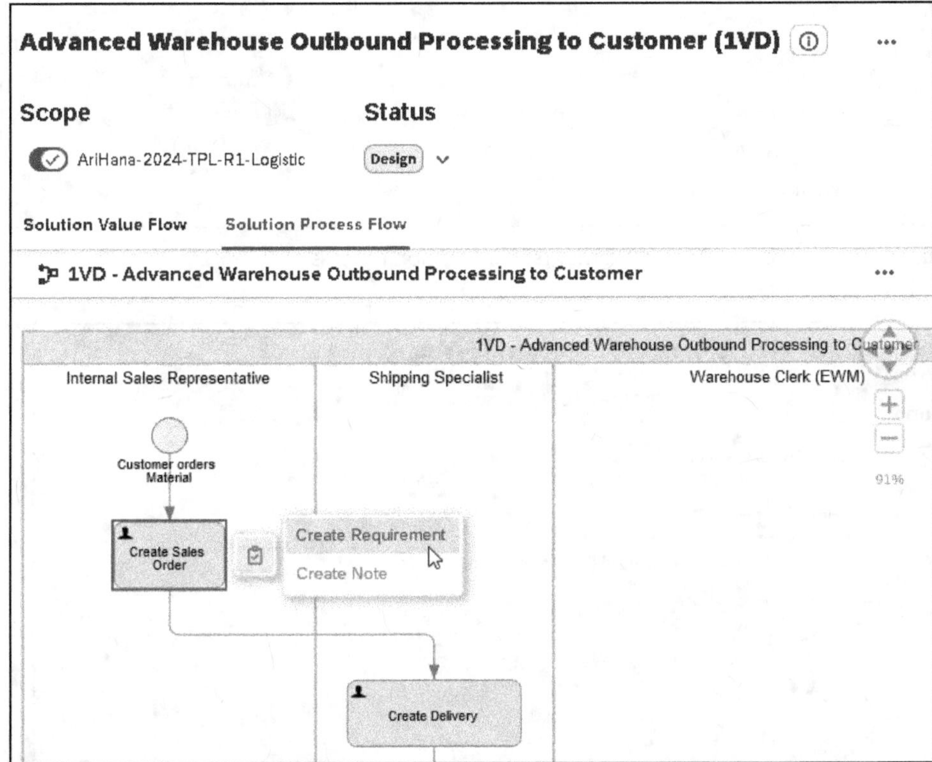

Figure 9.9 Create a Request via the Solution Process Flow

In Figure 9.10, you can see the result if several requirements have already been assigned. All you have to do is scroll down to the requirements on the right-hand side of the screen or click directly on **Requirements**.

In SAP Cloud ALM, it is possible to download diagrams (see Figure 9.11). To do this, click on the ⤓ button and select the option **PDF**, **SVG**, or **BPMN**. The solution process flow is now available as a PDF, SVG, or BPMN file.

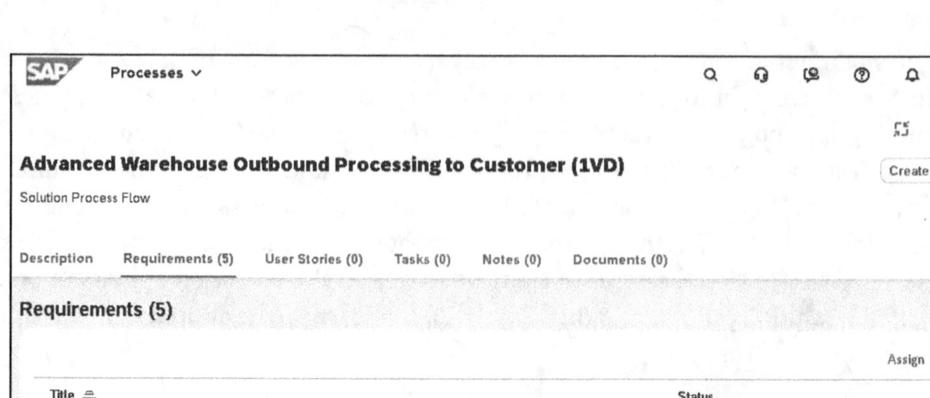

Figure 9.10 Overview of All Requirements of a Solution Process

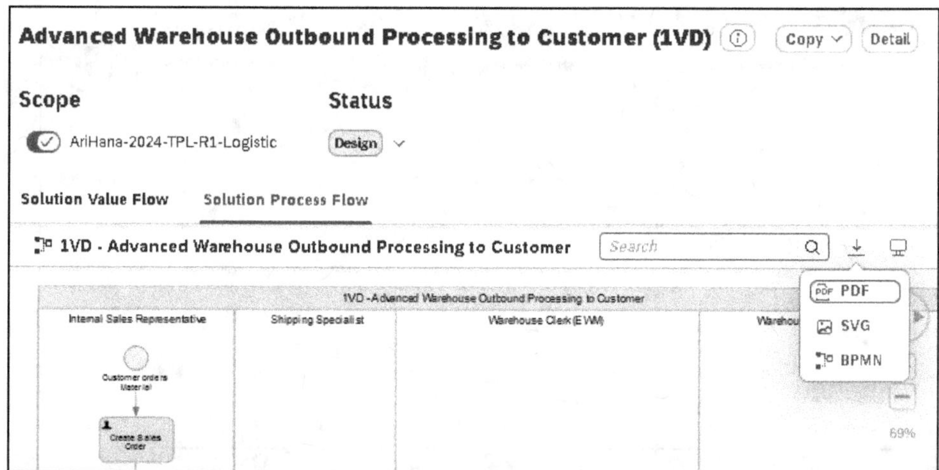

Figure 9.11 Download Diagram of the Solution Process Flow

You can also copy the diagram and transfer it to a new solution process. To do this, select the diagram of the solution process flow, click the **Copy** button, and select **Solution Process Flow Diagram** next to the process name in the upper area, as shown in Figure 9.12. In this example, the original **Automation of Order-to-Invoice with Ariba Network (4A1)** process has been copied to create a new version for the Dresden site.

You can now change the predefined process flow and create your own process from the template. In Figure 9.13, you can see how the process is copied. The process initially has the status **Draft** until it is published after saving by clicking the **Publish** button (see

Figure 9.14). You then have the option of creating a new draft using the **Create Draft** button.

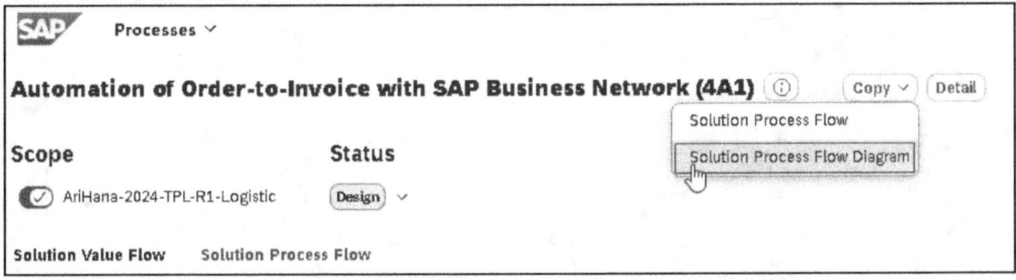

Figure 9.12 Copy of the Solution Process Flow Diagram

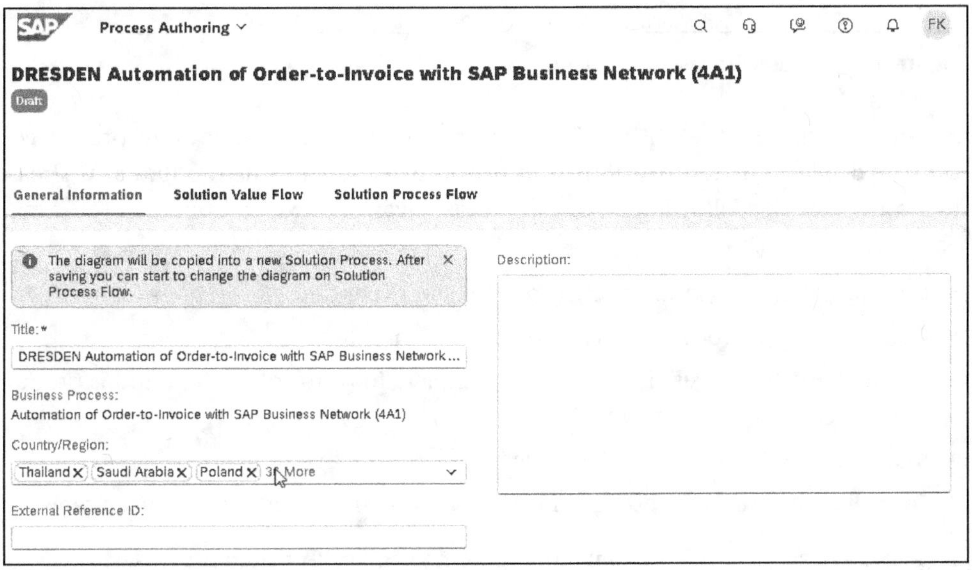

Process Authoring

DRESDEN Automation of Order-to-Invoice with SAP Business Network (4A1)

Draft

General Information | Solution Value Flow | Solution Process Flow

The diagram will be copied into a new Solution Process. After saving you can start to change the diagram on Solution Process Flow.

Title: *

DRESDEN Automation of Order-to-Invoice with SAP Business Network...

Business Process:

Automation of Order-to-Invoice with SAP Business Network (4A1)

Country/Region:

Thailand, Saudi Arabia, Poland, 3 More

External Reference ID:

Description:

Figure 9.13 General Information: Copied Diagram

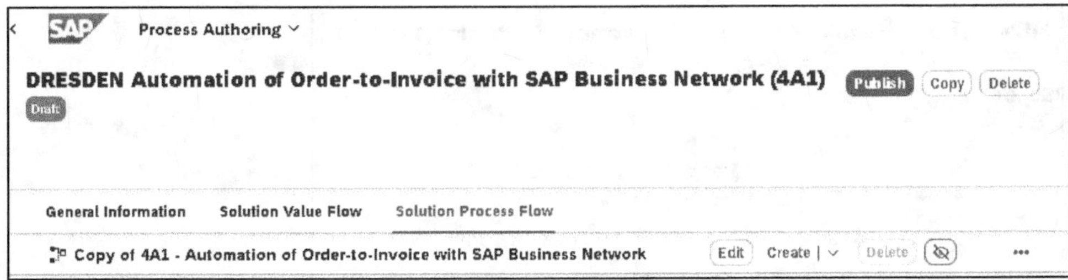

Figure 9.14 Publish Copied Solution Process

In the copied new version of the process, you now have an overview of the other existing versions (see Figure 9.15) and can customize the diagram and thus the process via authoring. This option is explained in Section 9.5.

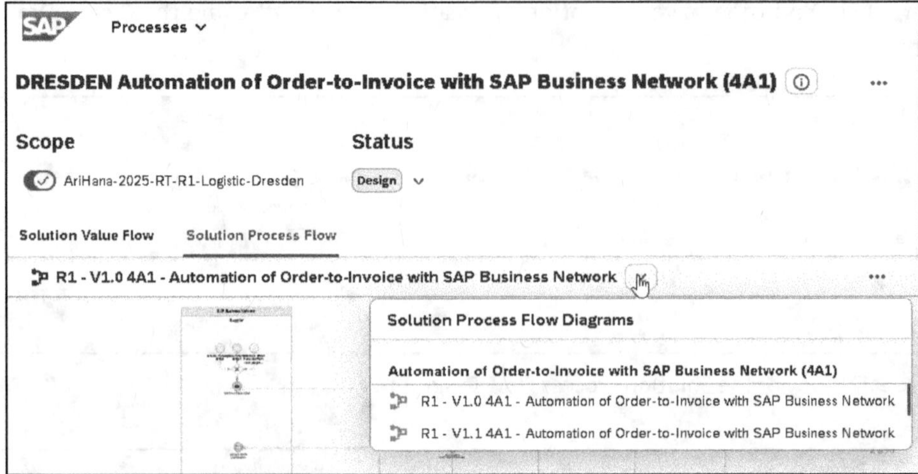

Figure 9.15 Solution Process Flow Diagrams

As you progress with the editing of your processes during the course of the project, you can also display the maturity of the solution process by selecting the status. As shown in Figure 9.16, the following status values are available, which you already learned about in Section 9.3:

- **Design**: The process is still being modified and expanded.
- **Realization**: The process is already being implemented.
- **Production**: The process is considered complete in terms of content and is productively used.
- **Maintenance**: The process can be revised after going live.
- **Obsolete**: The process is no longer used in this version or has already been replaced.

The status selection applies equally to predefined and self-created processes.

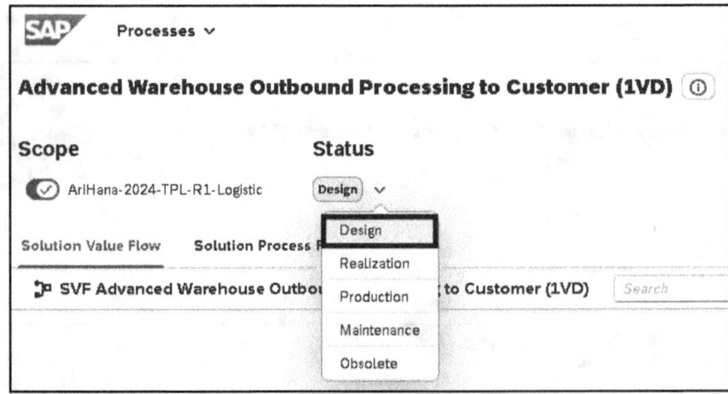

Figure 9.16 Status Change of a Solution Process

Another advantage of a process copy is that it can be subsequently transferred to another project. This means that a project copy is not necessary to transfer the processes already created to another project and continue processing them there.

9.5 Process Authoring

Process authoring allows you to create and maintain your own user-defined solution processes. This allows you to model your customer processes yourself or extend the content of SAP Best Practices. This gives your company full flexibility in the design of the processes and makes it independent of the predefined SAP content.

Process authoring follows a simple best-practice process for versioning processes. The solution processes are initially saved in a draft version and later in an active version. During editing, the **Draft** status is used, in which the process is not active and cannot be used in projects. If you assign the **Published** status, the processes can be used but the process version is not editable. A special feature of this status scheme is that the values of **Draft** and **Published** can also exist in parallel. This is because an already published (activated) version of a process can be inserted into a project while the draft version is still being worked on and then represents the new variant of the process as it progresses. To do this, the draft version must be activated, which automatically updates the existing active version. Solution processes can therefore exist in three possible states. As SAP Cloud ALM focuses more on the SAP solution, editing is also more about solution processes than business processes, but these can be linked to the business process variants.

Figure 9.17 shows the relationship between the different parts of the process content.

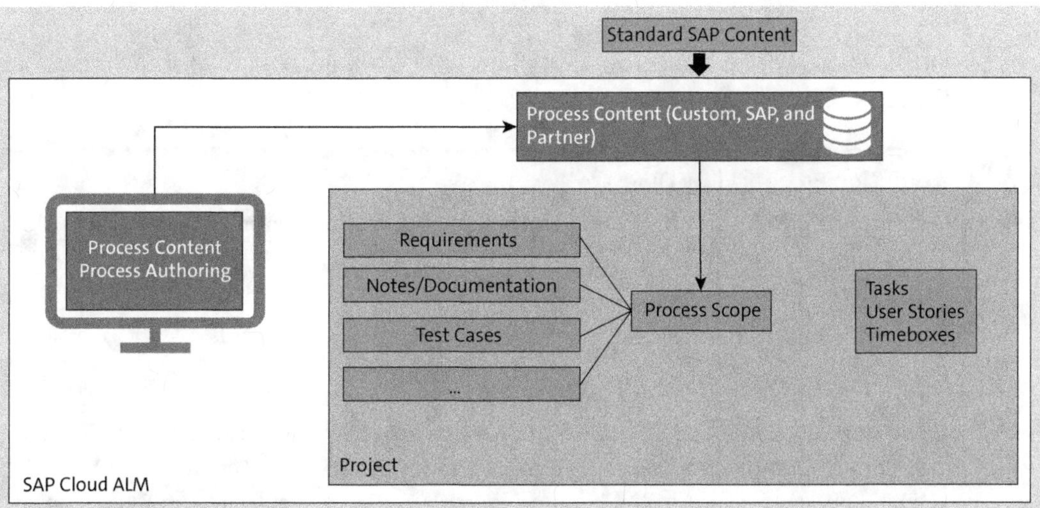

Figure 9.17 Overview of the Business Process Contents

As you can see, the delivered SAP standard content can be used to enrich your own process content. In addition, the processes from authoring can also be part of it, as you can see on the left-hand side of the image. The usable content therefore results from both predefined and self-created processes. This overall content is then used for process scoping, in which parts of the content are selected for a specific project to proceed with the creation of requirements, notes, and test cases (see Chapter 7). Process authoring is a way of providing the content to be used for the subsequent creation of requirements and test cases. The processes created in this way are part of the overall process scope.

With the Process Authoring app, you can create your own processes freely and without a template. You can see in Figure 9.18 a list of processes that have already been created and are displayed in two different status values: **Draft** and **Published**. Click on the **Create** button to create a new user-defined solution process or model a solution process flow diagram or a solution value flow diagram.

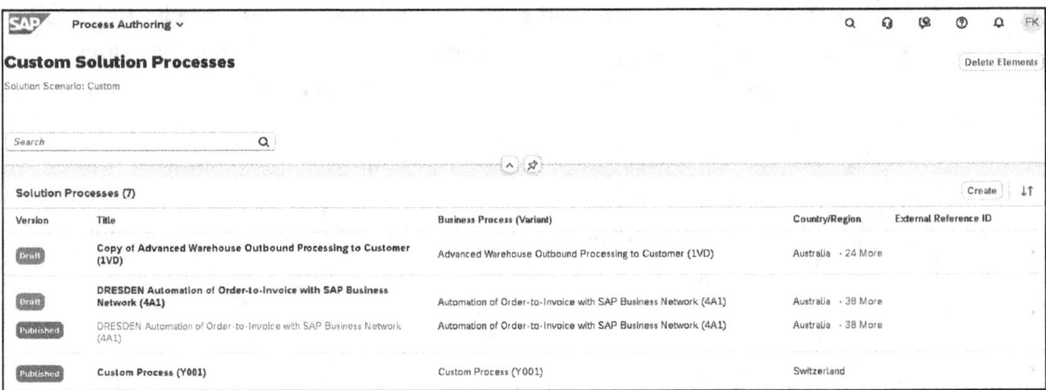

Figure 9.18 User-Defined Solution Processes

At this point, we would like to create a new solution process as an example; to do this, click on **Create**.

Solution Process Flow Diagram

Right beside the **Create** button, you have more options for creating a new solution process flow diagram via the ⌄ button:

- Create
- Create from BPMN File
- Create from SVG File

You can upload multiple BPMN files and add them to the solution process flow. The different solution process flow diagrams can be renamed or deleted at a later point in time. You could also use the Custom Process API, which is available on SAP Business Accelerator Hub, to import BPMN files via API. These different possibilities are shown in Figure 9.19.

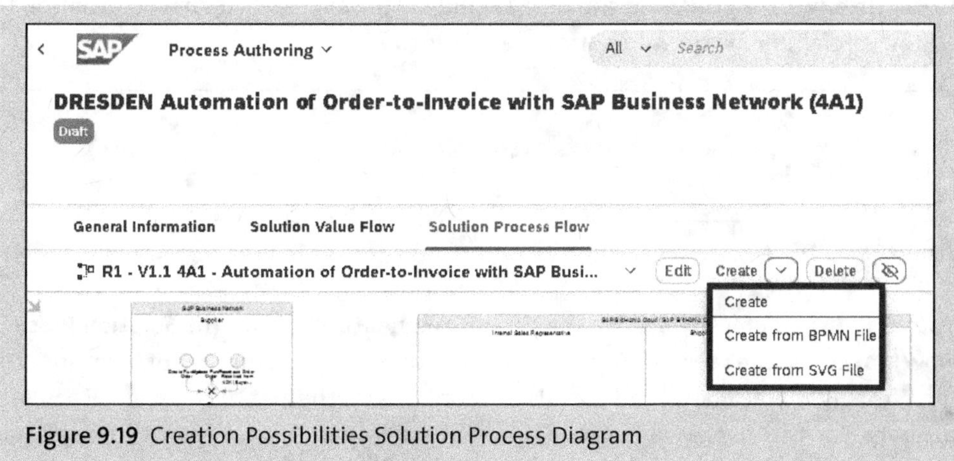

Figure 9.19 Creation Possibilities Solution Process Diagram

First, enter a **Title** for your new solution process. Then select the associated **Business Process** via the dropdown field or create a new business process variant, which you can then reference using the **Create Assignment** button (see Figure 9.20 and Figure 9.21).

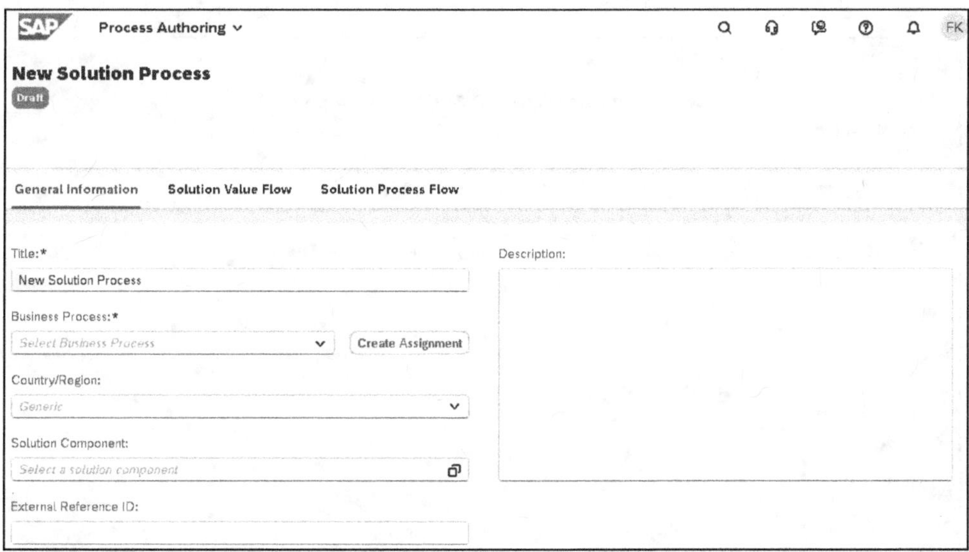

Figure 9.20 Create New Solution Process

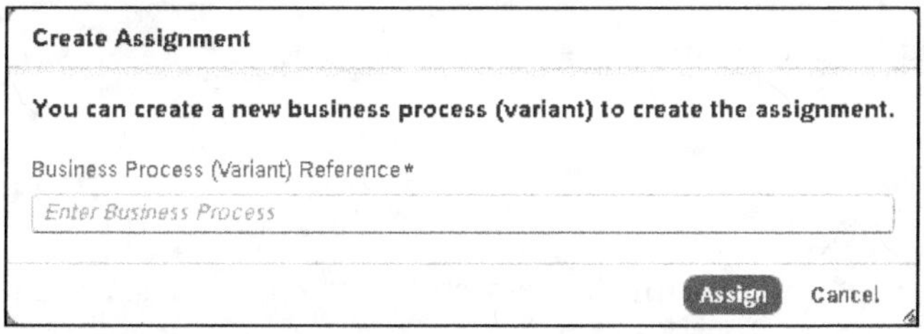

Create Assignment

You can create a new business process (variant) to create the assignment.

Business Process (Variant) Reference*

Enter Business Process

Assign Cancel

Figure 9.21 Create New Business Process Variant

If you already have a URL that could be helpful for this process, you have the option to add an accelerator via the **Add** button (see Figure 9.22).

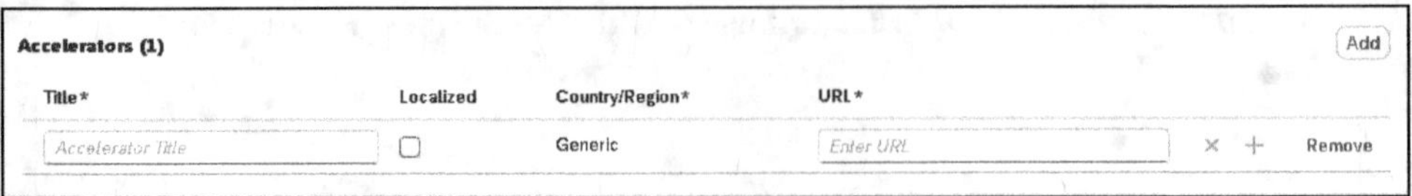

Figure 9.22 . Add Accelerator

You can now graphically display the flow of the new process on the **Solution Process Flow** tab. As you can see in Figure 9.23, you have only the most important/relevant artifacts from BPMN available, and not all of the options for the graphical design of a BPMN diagram. You can display gateways, data elements, communication paths, and systems and also map additional stakeholder groups using additional lanes. Each element in your diagram can be named exactly, and the different variants of a process can also be displayed. If you want to edit individual elements later, perhaps to add additional process steps, it is possible without restriction in the draft version ,and you can download the finished diagram in PDF, SVG, or BPMN format.

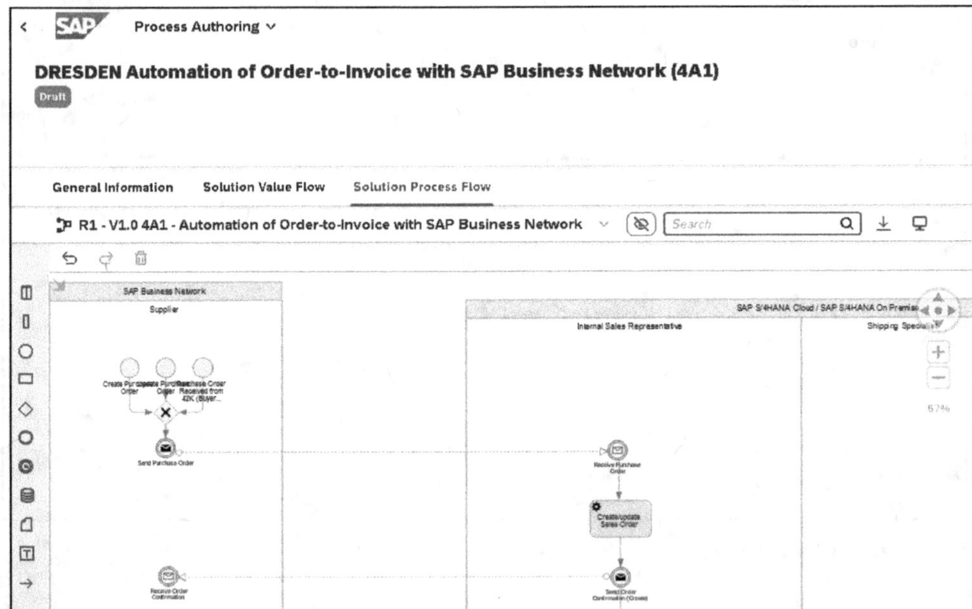

Figure 9.23 Solution Process Flow in Draft Status

9.6 Process Hierarchy

SAP Cloud ALM also offers the option of creating a hierarchy for the processes. You can create a *process hierarchy* that reflects your organization and structures your business into smaller pieces that can be handled easier. The benefit for you is that you can improve your productivity. If you organize your processes with this hierarchical structure, you can improve traceability, make navigation easier, and improve collaboration between your project team members. The process hierarchy is available cross-project (i.e., you can align a common structure with other SAP Cloud ALM projects).

As you can see in Figure 9.24, higher-level processes are mapped in a kind of hierarchy within the organization with associated subordinate processes. The main process—for example, logistics—is at the top level, while one or more subprocesses, such as the warehouse, are at the level below.

You access the process hierarchy via the Processes app. In the app, you can click on the **Process Hierarchy** button at the top, as shown in Figure 9.25.

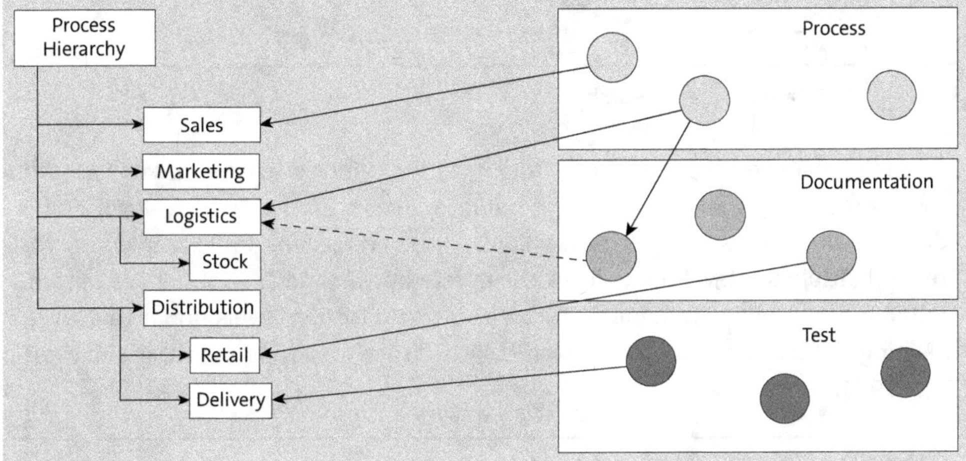

Figure 9.24 Process Hierarchy

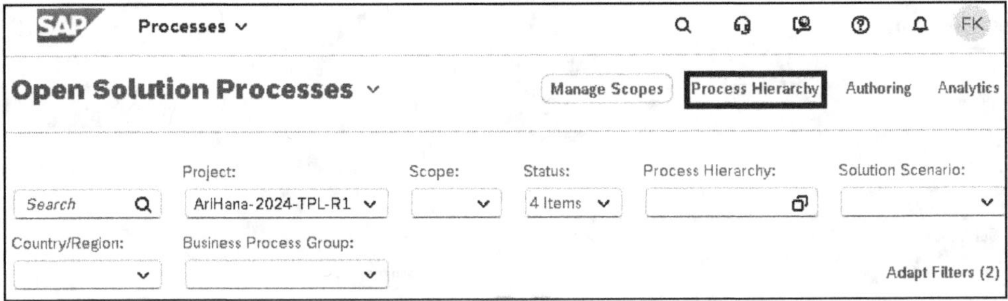

Figure 9.25 Scope of the Solution Process

In the **Process Hierarchy** view, you can maintain a hierarchy that reflects your business process organization. To do this, use the **Create** button.

You have the following options (see Figure 9.26):

- Create a process hierarchy (**Create** button)
- Edit the title and description of a process hierarchy node (**Edit** button)
- Delete a process hierarchy node (**Delete** button)

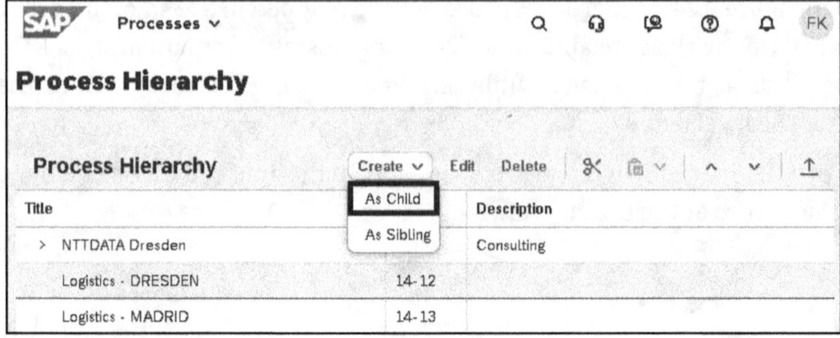

Figure 9.26 Create Process Hierarchy

The individual levels of the process hierarchy are referred to as *nodes* in SAP Cloud ALM. Three sublevels can be created under a level or node. Two top-level nodes—**Logistics - DRESDEN** and **Logistics - MADRID**—were created for our example project. To create multiple top-level nodes, click the **Create** button and then select the option **As Sibling** in the dropdown menu. You can also create further levels under this level by selecting **As Child**. You can see an example with three top levels and three sublevels in Figure 9.27.

Processes

FK

Process Hierarchy

Process Hierarchy — Create | Edit | Delete

Title	ID	Description
NTTDATA Dresden	14-2	Consulting
ALM	14-3	
S/4HANA	14-4	
Logistics - DRESDEN	14-12	
Goods receipt	14-14	Processing of all incoming goods
Purchase Order	14-19	Delivery and material receipt
Create inbound delivery	14-20	
Scan delivery documents	14-21	
Goods issue	14-18	Processing of all outgoing goods
Logistics - MADRID	14-13	

Figure 9.27 Overview of the Process Hierarchy

In this example, the **Logistics - DRESDEN** level has been enhanced by the **Goods receipt** level with the **Purchase Order** and the lowest levels of **Create inbound delivery** and **Scan delivery documents. Goods issue** is on the same hierarchy level as **Goods receipt**. If you want to delete individual nodes, you must proceed in order from the lowest to the highest level.

[!]

Maximum Values Process Hierarchy

The maximum depth for hierarchy levels is 10. The maximum length for a title is 255 characters. The maximum number of rows in a workbook to be uploaded is 10,000. The maximum file size is 5 MB. The maximum length for a description is 2,500 characters.

Once you have created your process hierarchy, you can filter and display your processes according to the respective process hierarchy. For example, if you select the **Process Hierarchy** filter in the solution process scope (see Figure 9.28), you will then see the processes of this selection (see Figure 9.29).

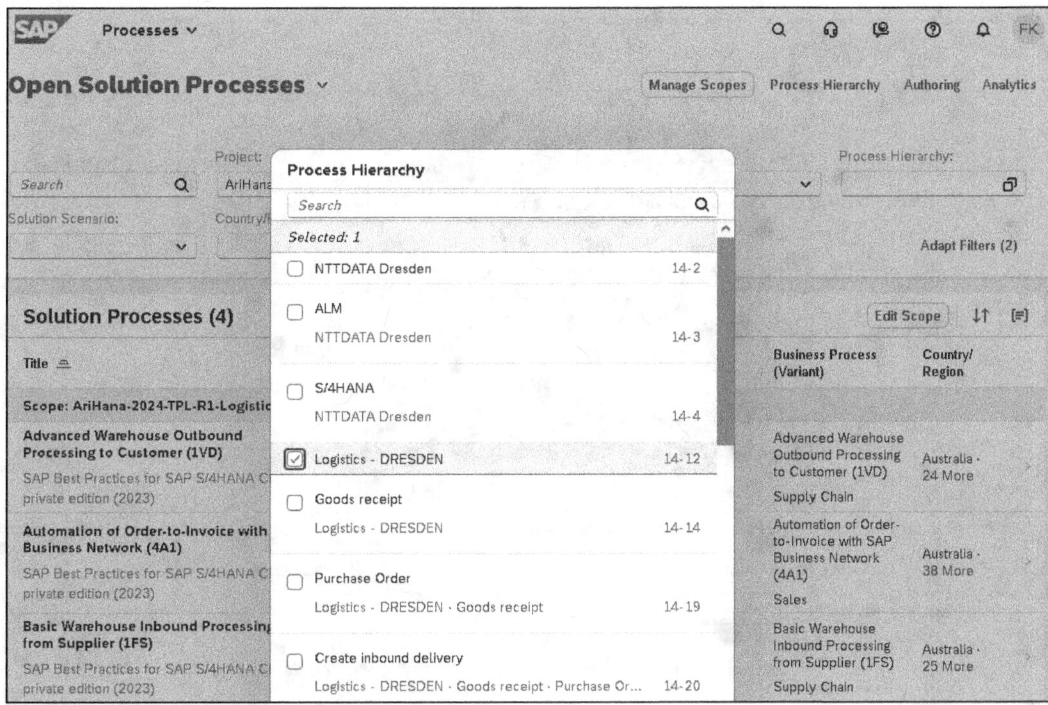

Figure 9.28 Hierarchy Selection

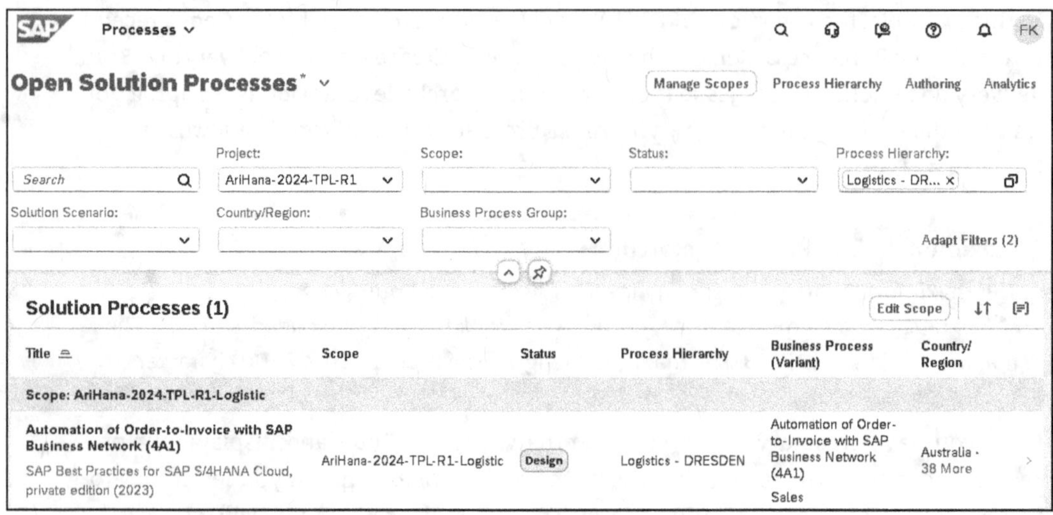

Figure 9.29 Using the Process Hierarchy as a Filter

[+]

Upload Process Hierarchy via Worksheet

If you click the **Upload** button, a popup will be displayed where you can also download a worksheet template (see Figure 9.30). When you are finished with your process hierarchy, you can upload it again. There are also good examples for process hierarchies available for download, e.g., line of business, based on the business capability model.

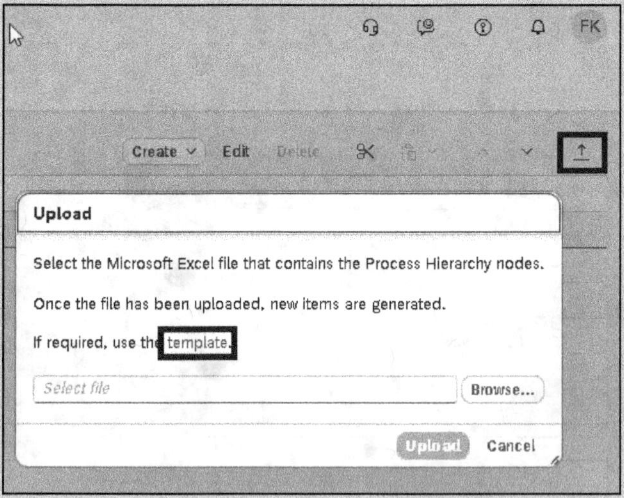

Figure 9.30 Upload Process Hierarchy

When using process hierarchies, it is recommended that you update and maintain process hierarchies on a regular basis to ensure that they will improve your work with processes and entities that come along. It is planned that the process hierarchy and process flows will be synchronized with SAP Signavio in the future, but the Processes API for linkage between SAP Cloud ALM and SAP Signavio is currently not available.

You might have noticed in Figure 9.5 back in Section 9.2 that a process hierarchy node was already linked in the created document. In addition to the linkage of solution processes to the process hierarchy, it is also possible to assign documents. To assign a solution process to a process hierarchy node, open your solution process and click the **Assign** button in the **General Information** tab right beside **Process Hierarchy**, as shown in Figure 9.31. In the resulting popup, select the process hierarchy node that you want to assign and click the **Assign** button.

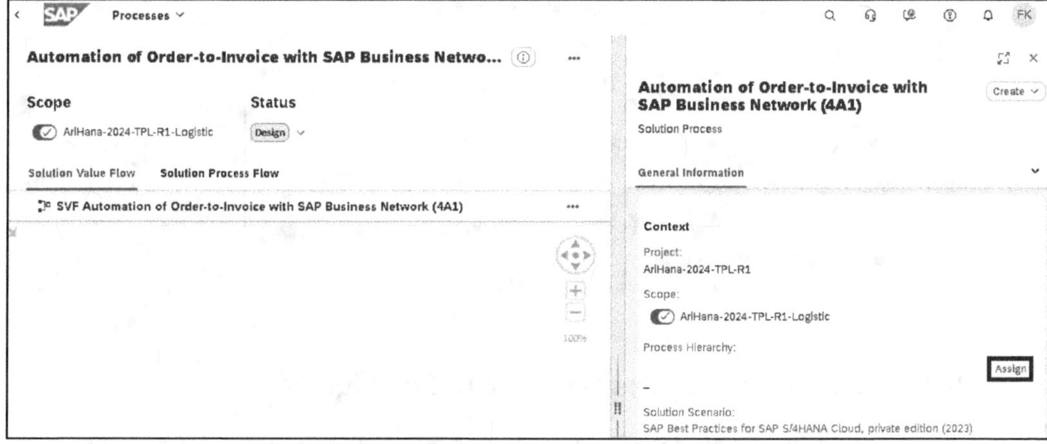

Figure 9.31 Assign Process Hierarchy to Solution Process

The assignment of process hierarchy can be used in solution processes, documents, requirements, and defects.

[+]

Process Hierarchy Author

In February 2024, a new role became available: Process Hierarchy Author. This role can create, edit, and delete process hierarchy nodes. This role was previously integrated into the Project Administrator and Project Lead roles, and for newly created projects it will now not be automatically included. The separation of the project management and process hierarchy roles allows you to manage your implementation projects in a more granular way. If an existing project lead or project administrator continues to need the authorization, no action is required. But if an existing project lead or project administrator doesn't need this role, you can remove them in the User Management app, as shown in Figure 9.32.

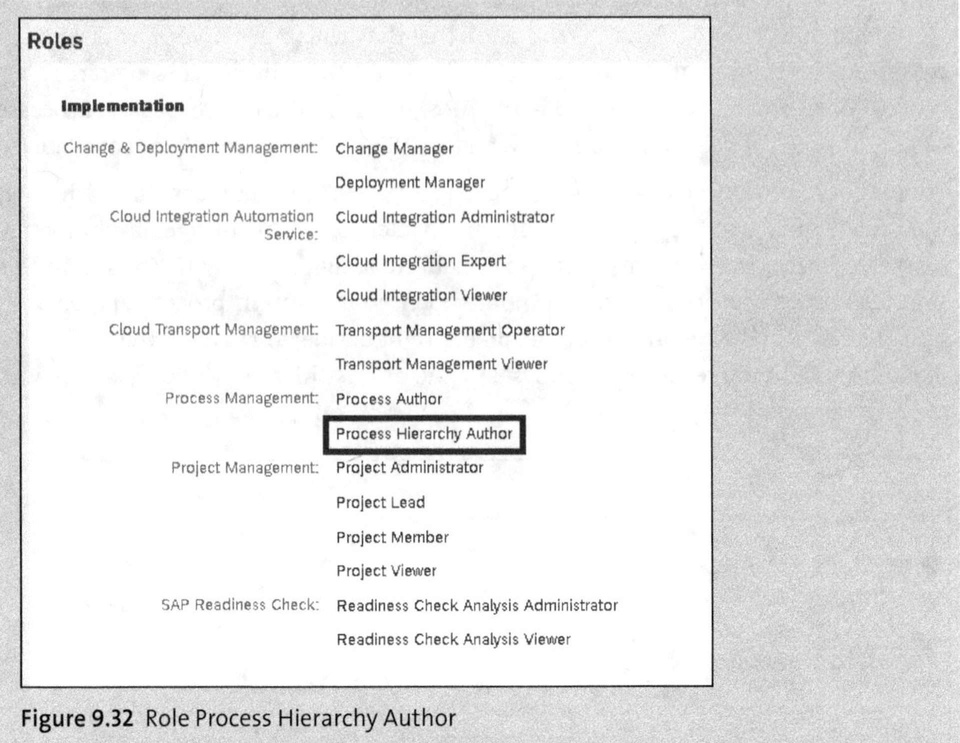

Figure 9.32 Role Process Hierarchy Author

9.7 Outlook on the Integration with SAP Signavio

SAP Signavio Process Manager is an SAP product used to model and manage business processes. The web-based product uses the industry standard Business Process Model and Notation 2.0 (BPMN 2.0) for modeling. The modeled processes can be expanded to include additional parameters such as costs, duration, frequency, and resources. This enables you to create process simulations and thus identify the risks of changes in the process. Transformations can thus be supported and optimized. For SAP Signavio customers, SAP Signavio Process Manager is the preferred source for company-wide process documentation.

Integrating this product into SAP Cloud ALM would be a great advantage, as processes that have already been modeled in SAP Signavio Process Manager could be synchronized to SAP Cloud ALM to derive requirements for implementation. Conversely, transferring the solution processes and process hierarchies from SAP Cloud ALM to SAP Signavio would also be a desirable scenario. For example, solution processes recorded in SAP Cloud ALM could be synchronized to SAP Signavio and enriched with the aforementioned parameters. A possible integration is currently on the SAP Cloud ALM roadmap as a product vision. It remains to be seen whether it will ultimately be implemented.

9.8 Summary

In this chapter, you learned about the process and document management functions of SAP Cloud ALM. Using our example scenario, we introduced you to options for visualizing and editing solution processes. With process authoring, we showed you how you can derive a specific process variant for the Dresden plant from a template process. These plant-specific adaptations are also associated with corresponding requirements, which are dealt with as part of requirements management. Requirements management is described in detail in the next chapter. Project-specific requirements will be included in a logistics process of our example scenario.

Chapter 10
Requirements Management

Requirements management deals with the creation, formulation, and specification of technical requirements. It represents the needs of the stakeholders and thus forms the basis for the topics of implementation and test management.

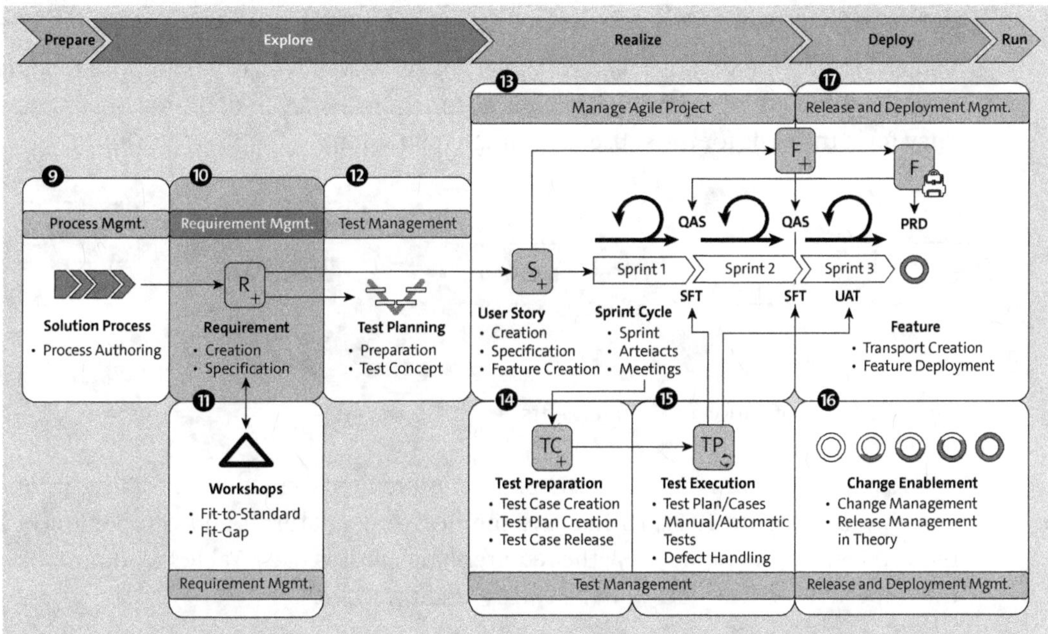

Based on the logistics process for the template release of our example scenario from the last chapter, we will explain in this chapter how you, as a specialized department, can incorporate project-specific requirements into the solution process during the explore phase (evaluation) as part of requirements management. These requirements are further specified in the SAP Activate methodology through fit-to-standard or fit-gap workshops, used for test planning and later transferred to the realize phase, where they are implemented in the form of user stories. Requirements also serve as a test criterion for implementation acceptance.

In this chapter, we explain requirements management in more detail. First, in Section 10.1 we give you a general overview of requirements, including the workflow and basic functions. Then, in Section 10.2 we delve into the creation of requirements in the system. Both sections will be explained from both the conceptual "Where?" and the technical "How?" standpoint. In Section 10.3, we use an example to specify a requirement and show best practices for the description. The steps required to transfer the requirements to the implementation are described in Section 10.4, and then we will look at the options for enriching requirements during implementation in Section 10.5.

10.1 Overview of Requirements

A *requirement* is an expectation that must be fulfilled or to obtain something specific. It is a functional gap or need identified by the customer during an implementation project. It can also reflect (for example) the specific need to improve an existing solution or to focus on new business areas.

Usually, requirements are collected as part of fit-to-standard workshops (see Figure 10.1) and are of an organizational, technical, functional, or even nonfunctional nature. They form the basis for the subsequent implementation.

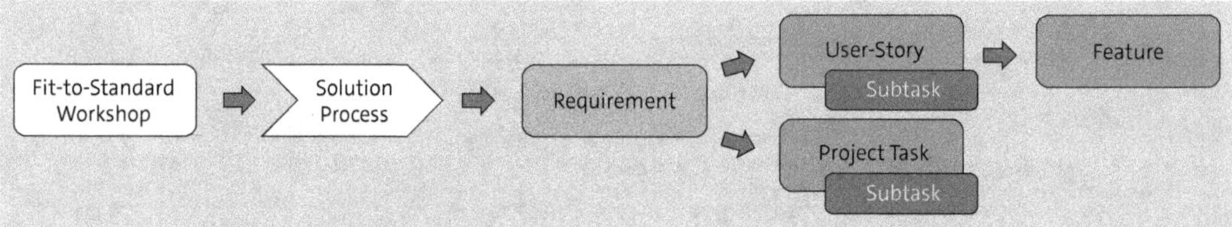

Figure 10.1 Classification of Requirements

The requirements workflow consists of the requirements' creation, specification, and approval (see Figure 10.2). To indicate the degree of maturity and completion of a requirement in SAP Cloud ALM, the requirement status is used. Table 10.1 summarizes the respective status values and the corresponding activity.

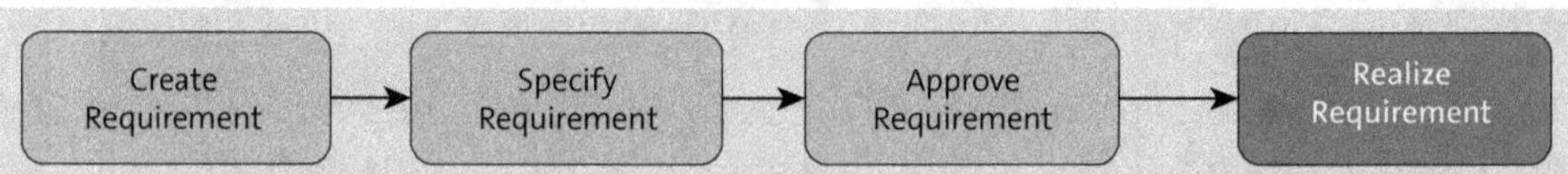

Figure 10.2 Requirements Workflow

Status Value	Activity
In Refinement	The requirement is still in draft form.
In Approval	The request is fully described and ready for approval.

Table 10.1 Status Values and Activities for Requests

Status Value	Activity
In Realization	The requirement is currently being implemented.
In Testing	The request is ready for testing or is currently being tested.
Blocked	The request will currently not be processed further.
Confirmed	The requirement has been fully implemented and completed.
Not Planned	The requirement is not relevant to the current scope of the project.

Table 10.1 Status Values and Activities for Requests (Cont.)

In Figure 10.3, you can see the available status values in the search filter, which you can use to filter the requirement by status.

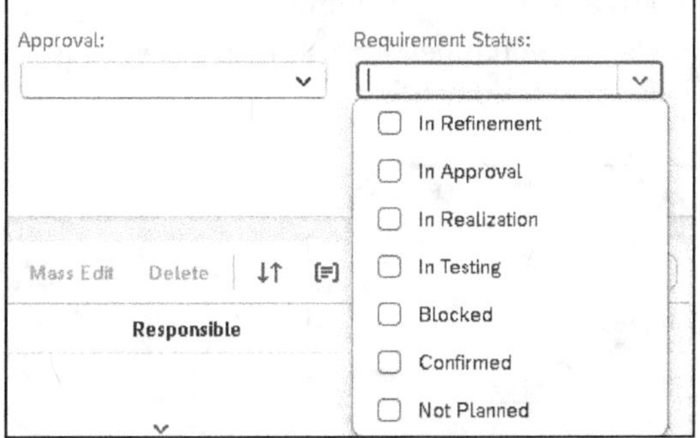

Figure 10.3 Status Values of a Request

The different statuses of the requirement (**Requirement Status** dropdown) do not follow a prescribed sequence. Each requirement can return to a different status at any time if necessary.

Requirements may need to be approved before they can be implemented. Therefore, the **Approval** attribute (see Figure 10.4) is in place, which can have various values. You can find a more detailed explanation of approvals and approval workflows in Section 10.4.

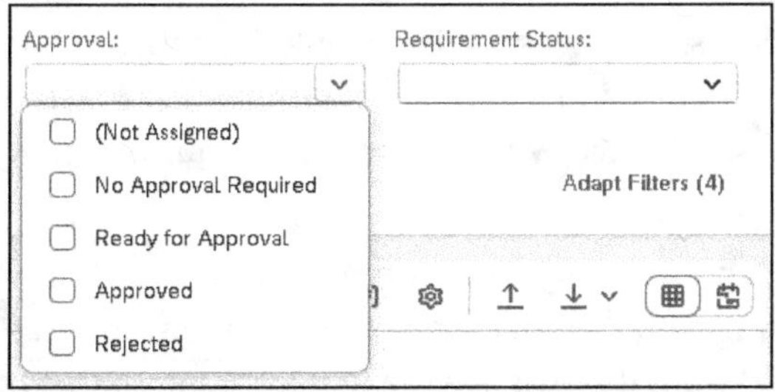

Figure 10.4 Status Values for Determining the Approval of a Requirement

Before we explain the creation of a requirement in more detail using a specific example (see Section 10.2), we will first look at a few general functions relating to requirements. In Figure 10.5, the following functions are shown:

- **Mark as Favorite** ☆
- **Show Comments**
- **Copy**
- **Set to Inactive** ◇
- **Set to Obsolete**
- **History**

You can use the favorites function to find content more quickly. Copying allows you to reuse content for other requirements or to minimize the redundant writing effort for similar requirements. You can use the comment function to enter comments or suggestions for improvement without having to adapt the requirement text directly. The history is a changelog in which you can track all changes for the respective requirement.

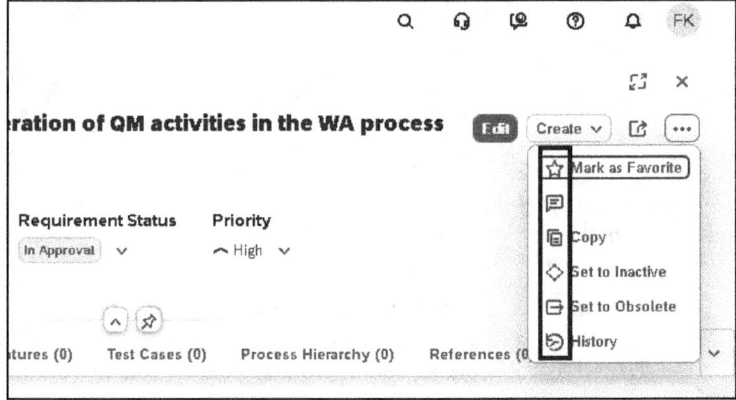

Figure 10.5 General Functions for Requirements

If a particular requirement or task is no longer relevant, you can set it to **Inactive**. This marks it as inactive in the system and prefixes its name with the notation **[Inactive]** (see Figure 10.6). However, its original status is retained, and you can still switch to edit mode and work on inactivated requirements. Maintenance is possible for both individual requirements and via mass changes. The decision to mark a requirement as inactive can reversed at any time by setting it back to Active, which returns the requirement to its original status.

If a particular requirement is no longer needed in your system, you can mark it as **Obsolete**. In contrast to the **Inactive** setting, you cannot switch into edit mode and edit the requirement. You do, however, have the ability to delete the requirement, as shown in Figure 10.7.

To delete a requirement, click on the ••• button and then on **Delete**. The prerequisite for this is that the requirement must already be marked as obsolete. Note: Once deleted, a requirement cannot be restored.

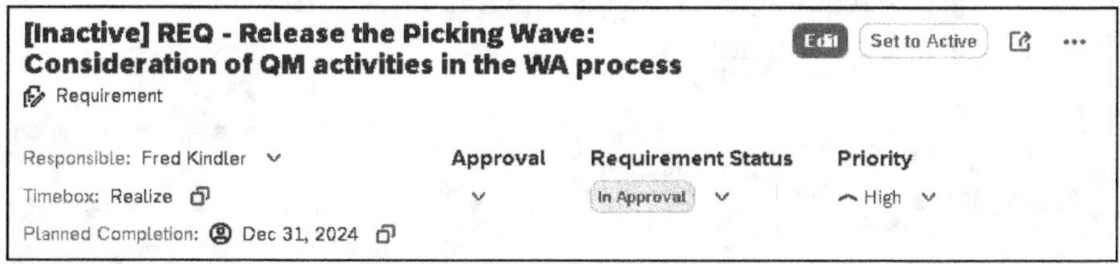

Figure 10.6 Requirement Marked as Inactive

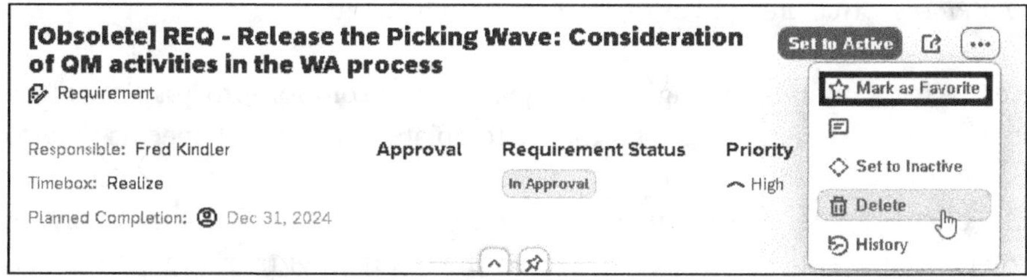

Figure 10.7 Requirement Marked as Obsolete

Delete

To prevent accidental deletion, you can only delete requirements if they have previously been given the status **Obsolete**. Requirements whose **Approval** attribute has been set to **Approved** can no longer be deleted, as **Approved** is a final status.

10.2 Creation of Requirements

In this section, we describe the creation of requirements. We address both the conceptual "Where?" and the technical "How?"

Potential requirements and gaps are identified in fit-to-standard workshops. The process of these workshops and the documentation of the results are described in Chapter 11.

As mentioned in Chapter 5, Section 5.2.2, new requirements are mainly created from the business context of a business process but can also be created without this reference. Technically speaking, there are four ways to create a requirement:

- Attachment via the solution process context
- Creation via the Requirements app (as a standalone requirement)
- Import via an Excel document
- Import via a programming interface (API)

In the solution process context, there are various points at which requirements can be created or assigned. In Figure 10.8, the darker fields represent the structural elements where this is possible. In the solution activity, however, the level of detail is usually too deep to be able to assign requirements in a meaningful way, and this would usually also involve a great deal of maintenance work.

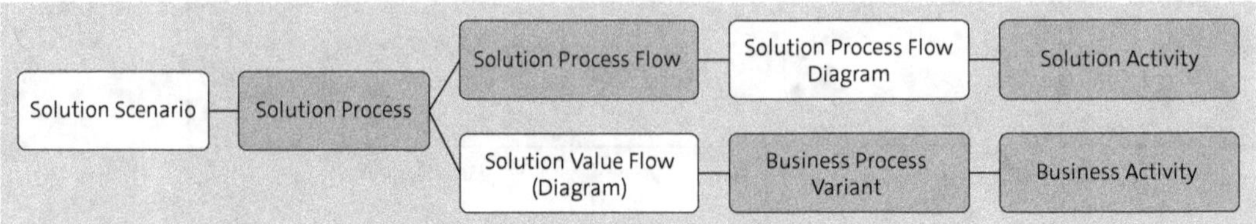

Figure 10.8 Process Structure Options for the Requirement Creation

We therefore recommend creating requirements in the solution process flow of the template if the requirement is relevant for the template. Only rollout-specific requirements should be created at rollout-specific points such as solution activities. The advantage of creating requirements in the solution process flow is that not only the requirements of the solution process, but also the requirements of the subordinate solution activities, are collected here. In Figure 10.9, you can see an example of this for the **Advanced Warehouse Outbound Processing to Customer (1VD)** solution process.

Advanced Warehouse Outbound Processing to Customer (1VD)

Create

Solution Process Flow

Requirements (5)

Requirements (5)

Assign

Title	Status	
REQ - Pick Warehouse Orders: collective warehouse tasks (from buffer to packing table)	In Approval	X
REQ - Pick Warehouse Orders: RF masks for order picking	In Approval	X
REQ - Release the Picking Wave: Consideration of QM activities in the WA process	Confirmed	X
REQ - Release the Picking Wave: removel rule	Confirmed	X
REQ - Shipping destination determination	In Approval	X

Release the Picking Wave

Create

Solution Activity

Requirements (2)

Requirements (2)

Assign

Title	Status	
REQ - Release the Picking Wave: Consideration of QM activities in the WA process	Confirmed	X
REQ - Release the Picking Wave: removel rule	Confirmed	X

User Stories (0)

Assign

Figure 10.9 Requirements Diagram of the Solution Process Flow and Solution Activity

It is a good idea to create a requirement as a standalone requirement—i.e., independently and without reference to a process—if it has been identified by the specialist colleagues and is to be recorded without having determined its specific solution process. To ensure that the requirement is not forgotten, you can record it first and assign it to a process later.

We explain the possibility of creating the system via the solution process context in Section 10.2.1 and the standalone creation in Section 10.2.2. Creating a requirement via the solution process context is the preferred method, as the solution process is automatically assigned in the requirement.

10.2.1 Creation via the Solution Process

You can create requirements either directly in the solution process or on its subordinate elements such as solution process flow, solution activity, or business process. To do this, open the Processes app (see Figure 10.10). In the solution process scope that opens, you can use the filter options to find the desired solution process and open it by left-clicking (see Figure 10.11).

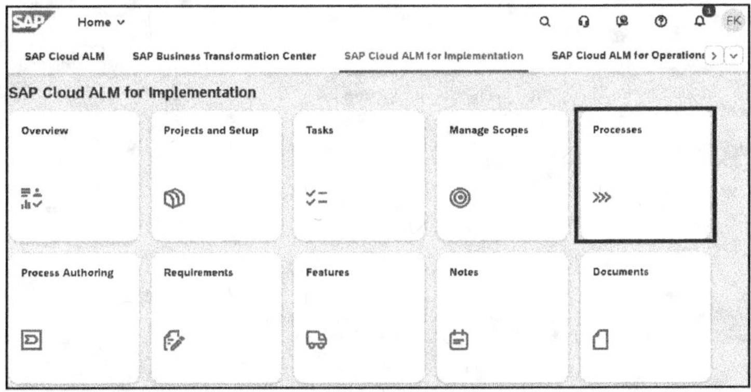

Figure 10.10 Overview of the Processes App

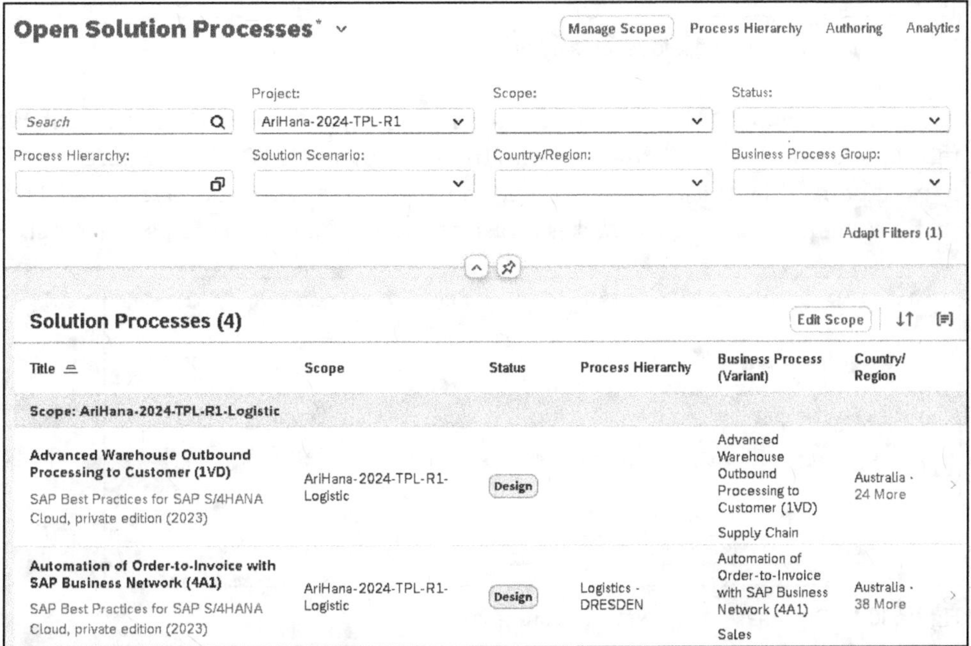

Figure 10.11 Selection of a Solution Process

Navigate from the solution value flow to the solution process flow by clicking on the corresponding tab. As shown in Figure 10.12, you will find the description and the linked elements on the right-hand side. All requirements for the solution process flow and its solution activities are listed in the **Requirements** area. Click on the **Create** button to open a selection list. With the selection of Requirement you will be taken to the creation screen for the requirement, which is also shown in Figure 10.12.

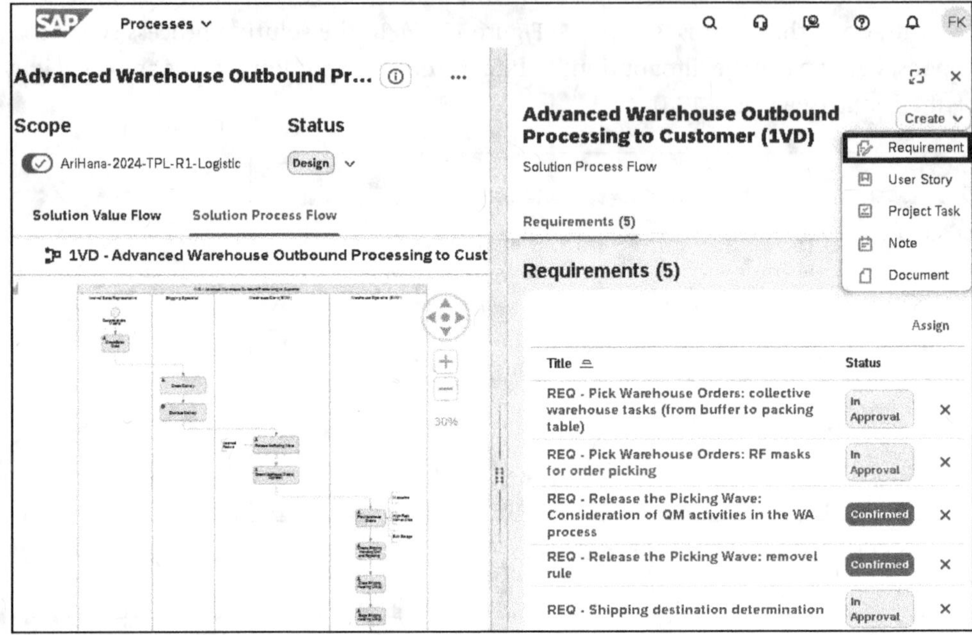

Figure 10.12 Creation of a Request via a Solution Process Flow

As shown in Figure 10.13, it is necessary to first define a title for the requirement. We recommend that you prefix the title with the abbreviation "REQ" (for "requirements") in front of the title. This naming convention makes it easier for you to recognize later that it is a requirement. It also makes it easier to search for requirements in general.

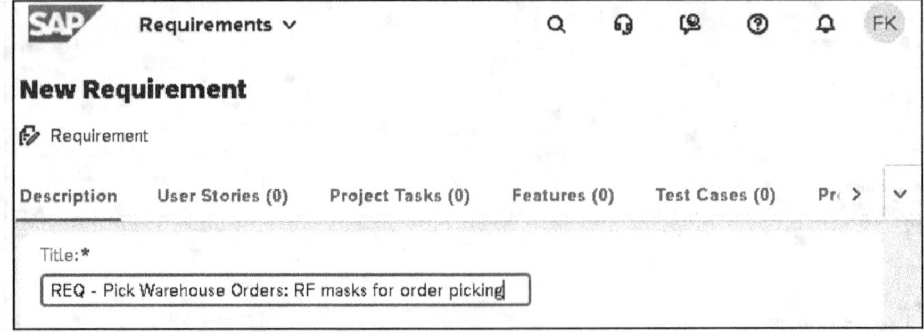

Figure 10.13 Designation of a New Requirement

A concrete example of a request title would be "REQ - Pick Warehouse Orders: RF masks for order picking," as the requirement was created out of the pick warehouse orders process step.

A draft is initially created for each requirement that arises during the discussions from the fit-to-standard workshops. The detailed requirements specification is described in Section 10.3. Details on fit-to-standard workshops can be found in Chapter 11.

10.2.2 Creation as a Standalone Requirement

In addition to the requirement creation via the solution process context, you can also create a standalone requirement. To do this, open the Requirements app to display a list of all open requests, as shown in Figure 10.14.

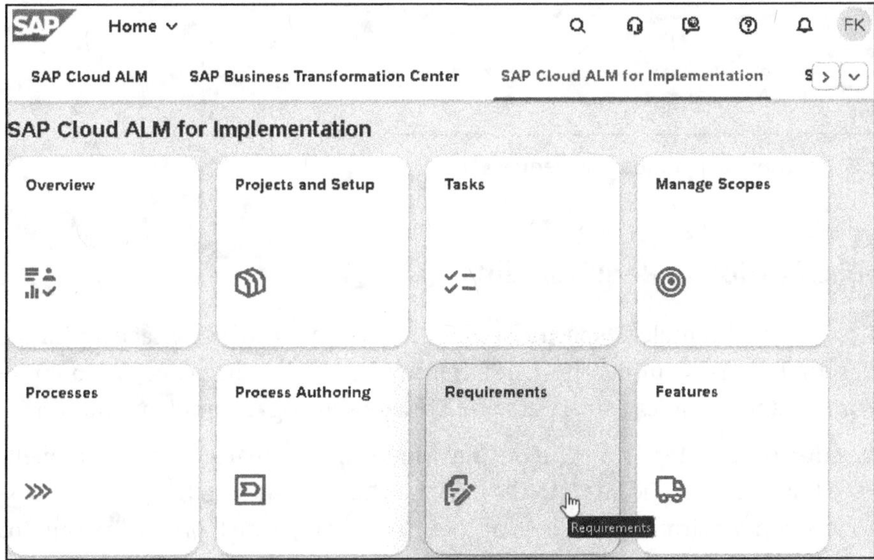

Figure 10.14 Overview of the Requirements App

You can reach the requirement creation screen by selecting the **Create** dropdown and choosing the Requirement option (see Figure 10.15).

As mentioned in Section 10.2.1, we recommend prefixing the request name with the abbreviation "REQ." After naming the requirement, save it. Further formulation and classification are explained in the following sections.

Referencing Requirement and the Solution Process

Requirements can also be added to the solution process at a later point in time via the **Assign** button.

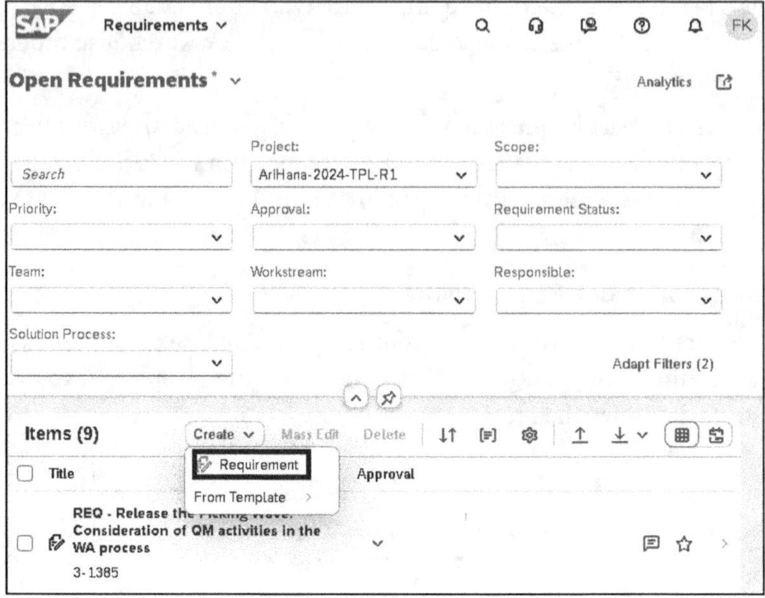

Figure 10.15 Creation of a Standalone Request

10.3 Specification of Requirements

This section uses an example to specify a requirement and shows best practices for the textual description. As requirements can differ in complexity and scope, a more precise description and classification are essential before starting the implementation.

The requirements collected from the business are used to create your *requirements backlog* (see Chapter 13, Section 13.2). Note that this does not refer to the product backlog from the Scrum environment (see Chapter 5, Section 5.2), which primarily contains the user stories for implementation.

10.3.1 Best Practices for Requirements Specification

The definition of a *priority* supports the initial classification and grouping of the requirement for the requirement backlog. The values **Low**, **Medium**, **High**, and **Very High** are available to you (see Figure 10.16).

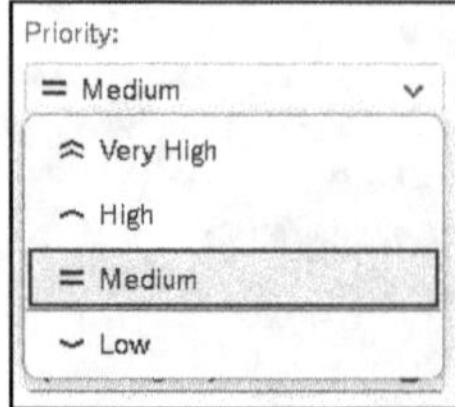

Figure 10.16 Define Requirement Priority

To make it easier for users to decide which priority to choose and to create a uniform understanding, we recommend classification according to the *MoSCoW method*. As shown in Table 10.2, requirements are systematically categorized into must, should, could, and would.

Priority	MoSCoW Classification
Very High	Must have (absolutely necessary)
High	Should have (should be fulfilled)
Medium	Could have (can be fulfilled)
Low	Would have/would/won't (not necessary, but would be nice)

Table 10.2 Classification of Priority According to the MoSCoW Method

In addition to the priority, we recommend providing at least three short pieces of information (see Figure 10.17) to document the requirement:

- A description of the requirement
- Comments
- A description of the estimated effort

This information serves as an important basis for decision-making during implementation planning. As there are no separate fields for this information, you can include it in the free text.

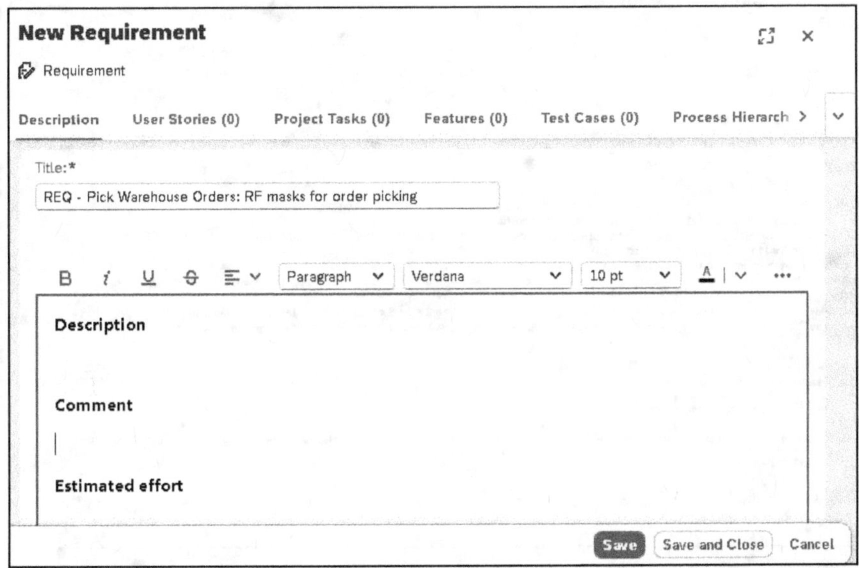

Figure 10.17 Recommended Text Modules for the Requirements Description

The description includes the specific details that are required. Do not hesitate to use graphics, tables, or lists if it helps to improve understanding.

Further details, such as targets or implementation proposals, are provided in the form of comments. You can also state the consequences of not implementing the requirement.

The indication of a roughly estimated effort also helps with the subsequent structuring and planning of the implementation. We recommend either a classification according to T-shirt sizes (S, M, L, XL) or Fibonacci numbers (1, 2, 3, 5, 8, 13, 21, etc.). T-shirt sizes are known to most people and are therefore a familiar unit of measurement. In contrast, Fibonacci numbers help to assign each requirement a unique position in the backlog so that several equivalent requirements do not compete for scheduling.

10.3.2 Text Example for Requirements

In relation to the previously created requirement **REQ – Pick Warehouse Orders: RF masks for order picking**, Figure 10.18 shows the three text modules: **Description**, **Comment**, and **Estimated effort**.

The **Description** section contains the detailed explanation for displaying additional information that is not contained in the SAP EMW standard picking screens. The expected objectives and the expected disadvantages in the event of non-implementation are described under **Comment**. The **Estimated effort** for programming is indicated with the T-shirt size M.

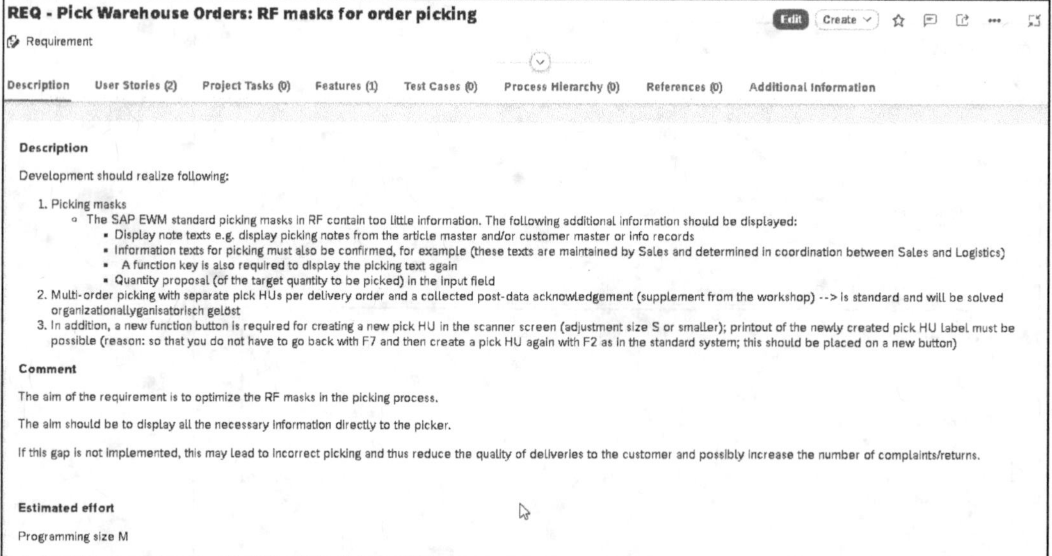

Figure 10.18 Example of the Textual Description of a Requirement

10.3.3 Further Information on Requirements

In addition to the textual descriptions, we recommend filling in the **Assigned Role**, **Responsible** (person), and **Workstream** attributes for the requirement definition on the **Additional Information** tab (see Figure 10.19).

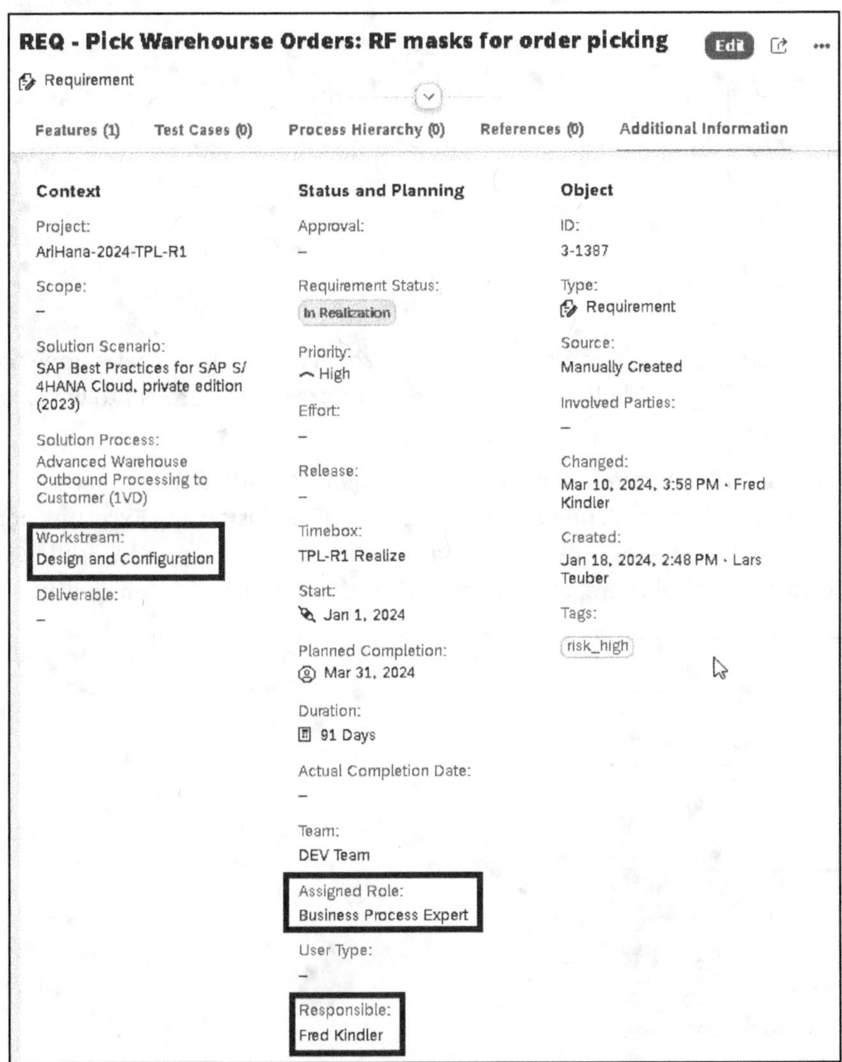

Figure 10.19 Requirement Definition: Additional Information

A requirement should be assigned to a role so that it is clear which project role is responsible for the requirement. This could be the business process expert, for example. The assigned role can change as the requirement progresses further. A detailed list and explanation of roles can be found in Chapter 5, Section 5.2.3.

The **Responsible** field (see Figure 10.20) refers to the person who is responsible for the further processing and technical specification of a requirement. This can be the person who initially created the requirement or another person. This information also helps you to quickly find your own tasks to be processed via the search. The proposed selection options for the responsible person are influenced by the selection of the assigned role.

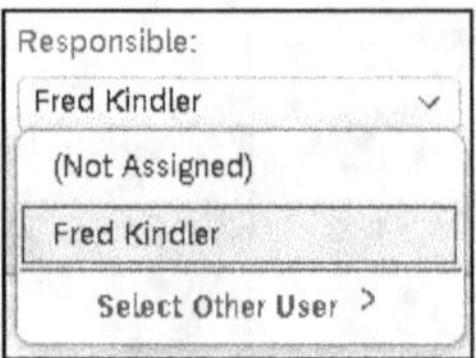

Figure 10.20 Responsible Person for the Requirement

The **Workstream** attribute represents a grouping of services or results that belong to the same area of activity. This allows you to specify the area in the project to which a requirement belongs thematically, for example.

The entries for the **Workstream** (see Figure 10.21) come from SAP Activate methodology. However, they can also be defined on a customer-specific basis using Excel upload. As requirements can affect not only one, but also several work areas and workflows, multiple selection is possible here (e.g., **Design and Configuration** and **Extensibility**).

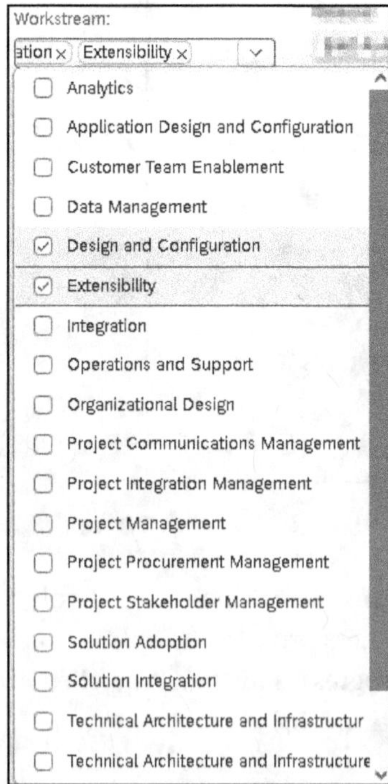

Figure 10.21 Working Area of the Request

The attributes **Scope**, **Timebox**, **Start** date, and **Planned Completion** date (see Figure 10.22) are inherited via the solution process if there are entries there. Otherwise, you can enter or adjust them here.

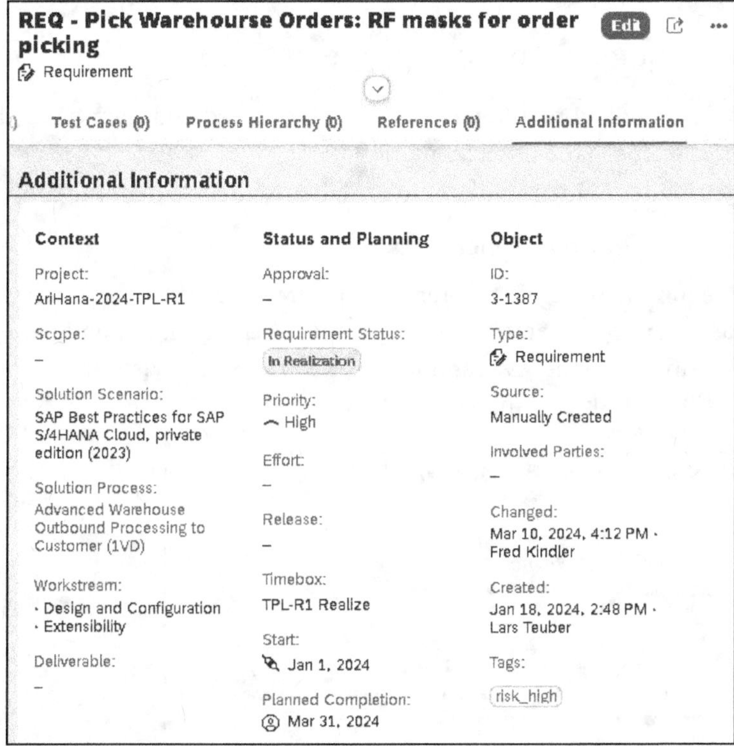

Figure 10.22 Additional Information: Requirement

In addition, as shown in Figure 10.23 within the **References** area, you can use the **Add** button to add links to the request—for example, to your company SharePoint. By assigning a meaningful name and a URL, it is possible to link to further information.

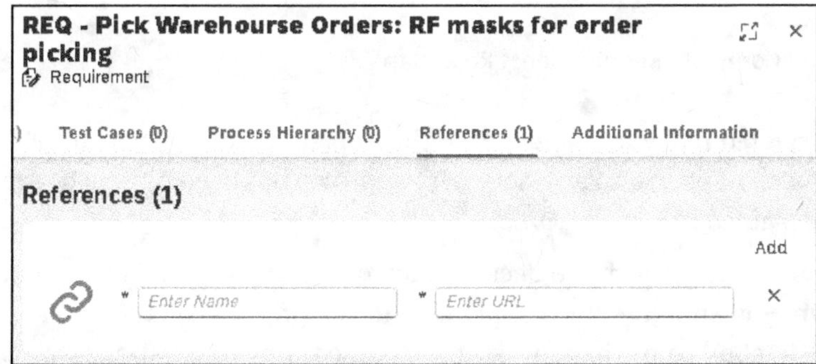

Figure 10.23 Create References for Requirements

Using tags also helps to further specify requirements (as described in Chapter 6, Section 6.6). You should define a convention in advance to ensure a uniform approach. An example would be **Risk: Low**, **Risk: Medium**, and **Risk: High**.

10.4 Handover of Requirements for Implementation

In this section, the fully described requirement is finally reviewed and then handed over to the colleagues or teams in the implementation.

10.4.1 Preparation for Handover to Implementation

You should check the requirements again beforehand so that they can be transferred to the implementation and an implementation can be planned. A change is possible without changing the requirements status. As a reminder, you can use the **Review and Complete Requirements** roadmap task, shown in Figure 10.24.

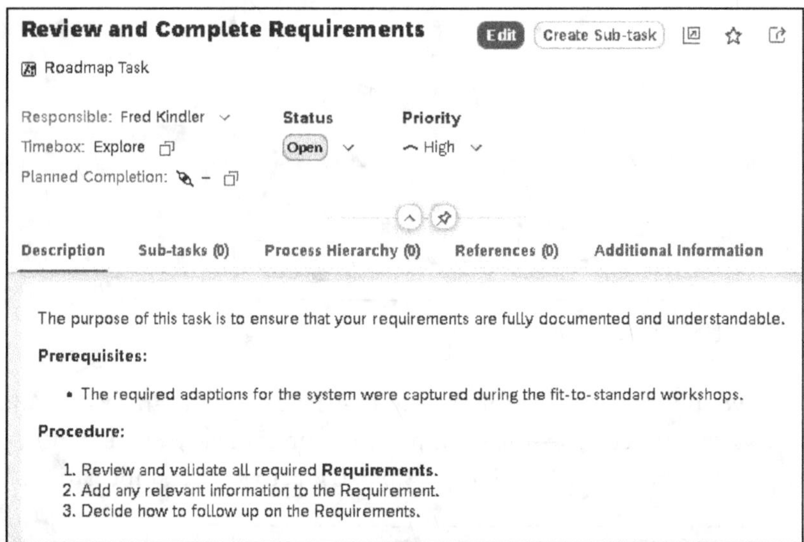

Figure 10.24 Review and Complete Requirements Roadmap Task

A request can be considered ready for handover to the implementation when all relevant attributes have been filled in and any necessary approval has been obtained. The approval process is explained in Section 10.4.2.

The product owner can use the priority to decide which user stories should be created first. If already known, you should enter the planned development team (**Team**) and specify the element **Timebox** to further specify the planned implementation. This is shown in Figure 10.25. Alternatively, the product owner can provide further information about these points.

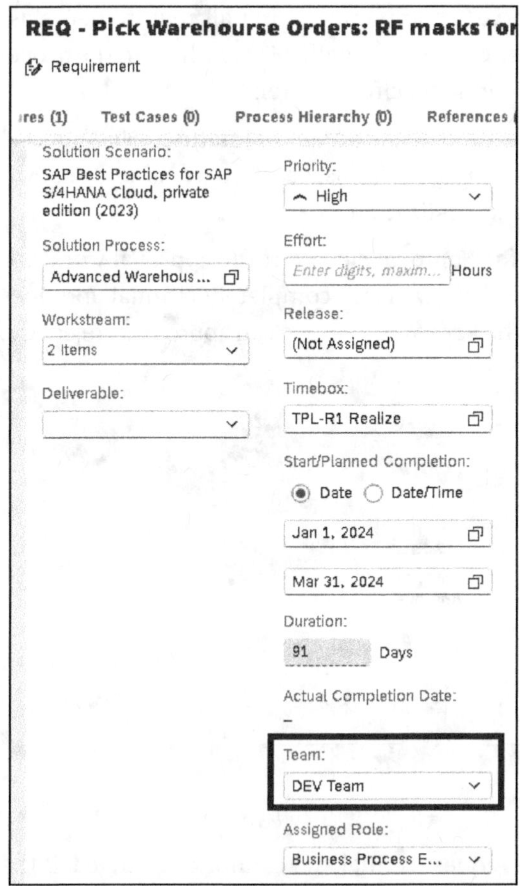

Figure 10.25 Preparation of the Requirement for Handover to the Implementation

You can use the F4 help for the **Timebox** field to specify details for the timebox element, e.g., a specific sprint.

10.4.2 Requirement Approval

In addition to the requirement status, you can also define an approval status for your requirements. Approval takes place via the **Approval** attribute (see Figure 10.26).

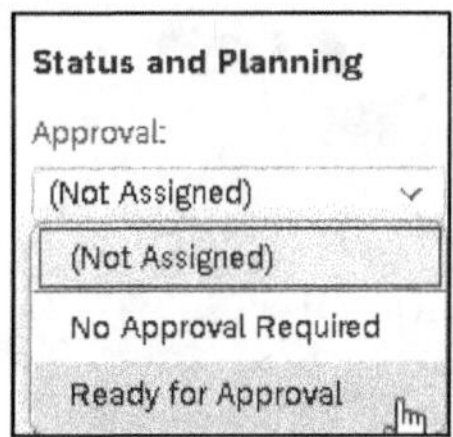

Figure 10.26 Requirement Approval

With regard to requirements approval, we recommend having all requirements assigned to the approved template to ensure consistent quality. Depending on the procedure, approval may not be required for rollout-specific requirements.

If you do not require approval for your requests, set the approval status to **No Approval Required**. It is also possible to renounce the general usage of approval and instead switch the requirement status directly to **In Realization**.

By setting the approval status to **Ready for Approval**, you start the approval workflow (see Figure 10.27). At the moment, this workflow is still completely manual and does not lead to an automatic status change. The approving person can approve or reject the request.

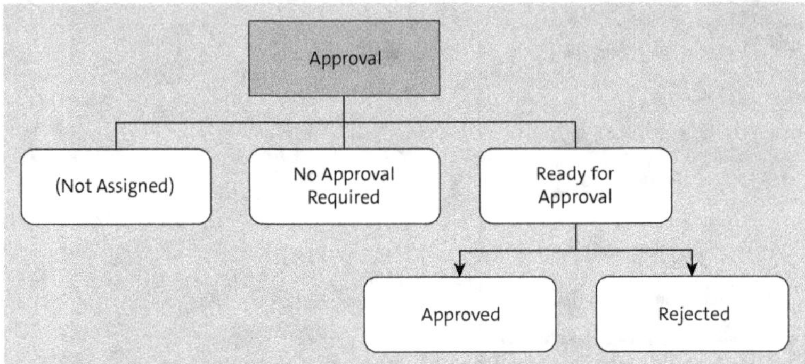

Figure 10.27 Approval Workflow

The final decision to accept or reject the request is final and cannot be changed. If the request does not meet the criteria for a positive approval, but should still be given the option of resubmission, you as the approver should add comments to the request instead of setting it directly to **Rejected**.

Approval Status and Request Status

The approval statuses **Approved** and **Rejected** are final status values that cannot be changed later.The approval workflow does not automatically change the request status, as implementation can also begin at a later point in time. In exceptional cases, implementation may also begin before final approval.

10.4.3 Possible Follow-Up Elements

Based on the formulated and, if applicable, approved requirements, implementation can be carried out using one of the following elements (see Figure 10.28):

- Project task
- User story
- Feature

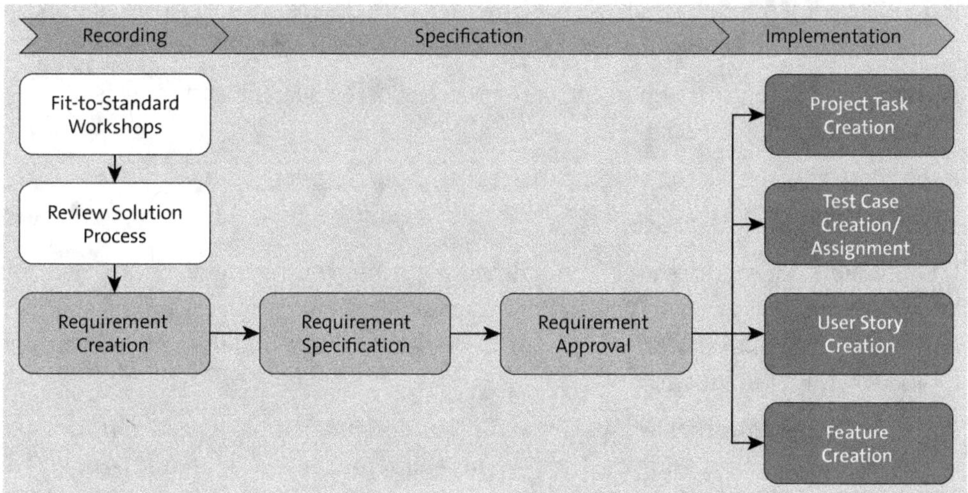

Figure 10.28 Classification of the Requirement and Possible Follow-up Elements

User stories can be created either directly after the user has approved the requirements or, following the Scrum approach, by the product owner as soon as implementation begins. Starting from the requirement, it is possible to create **Project Task**, **User Story**, and **Feature** choices in the dropdown **Create** menu, shown in Figure 10.29.

The test case must also be taken into account. This is usually only written during implementation, but it can still be assigned to a requirement at an early stage.

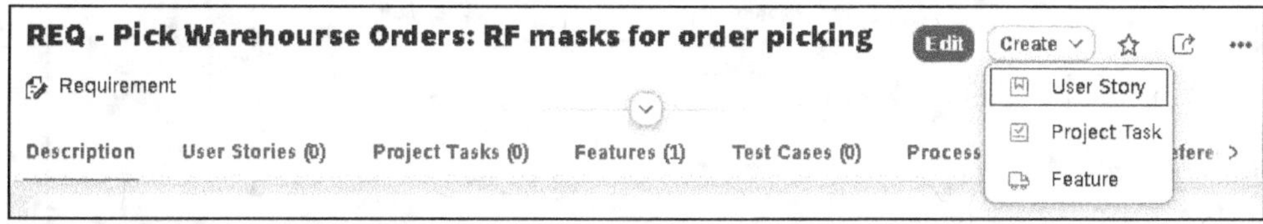

Figure 10.29 Subsequent Elements for Implementing a Requirement

Two general approaches are conceivable with regard to planning the next steps:

- Requirement with test cases
- Requirement with user stories, features, and test cases

In the first variant, the technical elements (user story, feature) are not created in SAP Cloud ALM. The implementation is based on the respective requirement. If a requirement is too complex and implementation comes to a standstill as a result, you should consider making an exception. To make it easier for the development teams to deal with the requirement, it should be broken down into manageable pieces with the help of user stories.

In the second variant, each requirement is assigned to one or more user stories, and each of them has its own test cases and features. The user story test cases check the individual configurations that are implemented with the features. The overall package of the requirement is thus tested for functionality by the sum of the test cases.

To have complete documentation, we recommend the second option. It is also described in the following chapters.

10.4.4 Outlook on User Stories

It is advisable not to create user stories directly as a follow-up element, but to leave this task to the product owner. It is his or her responsibility to maintain the product backlog and to present the user stories in detail so that they can be understood and comprehended by the Scrum team.

The product owner can view all approved requirements that have been confirmed as ready for development using the list of open requirements (see Figure 10.30) or via the traceability of requirements (see Section 10.5).

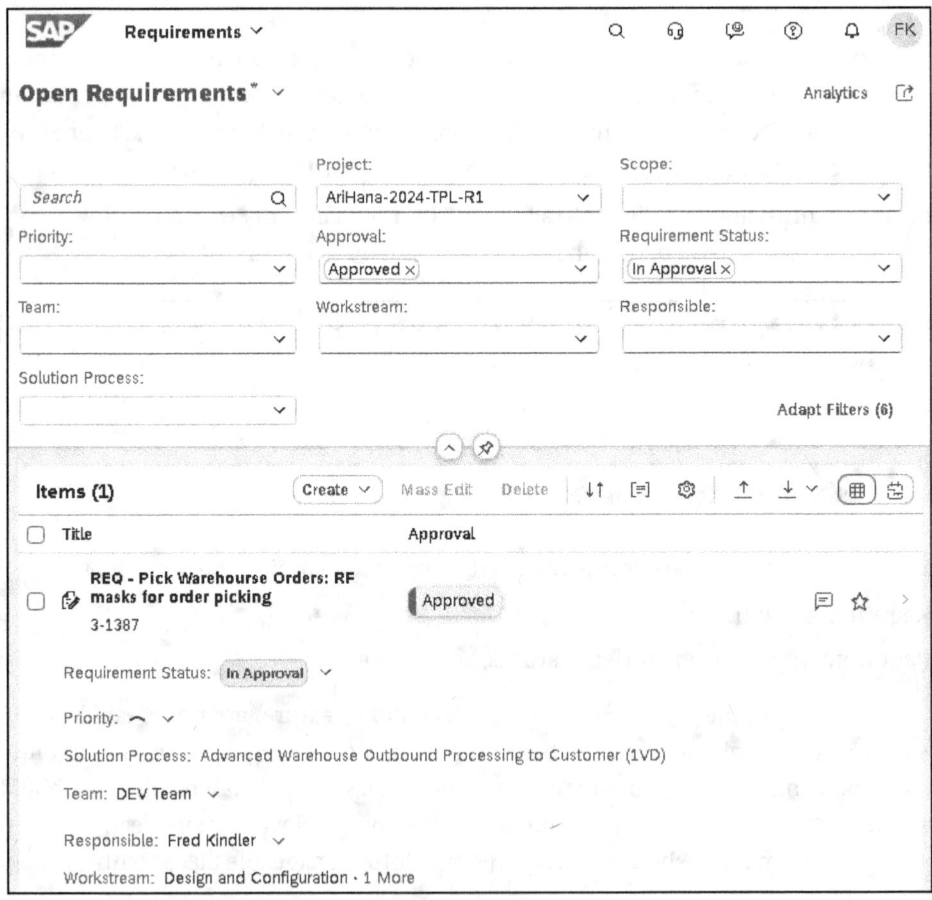

Figure 10.30 Requirements Ready for Implementation

From this requirements backlog, he or she builds a product backlog by creating user stories based on the requirements. A detailed description of the topic of user stories can be found in Chapter 13, Section 13.3.

10.5 Enrichment of Requirements during Implementation

Requirements and user stories can be enriched with one or more features for implementation. You can find more detailed explanations in Chapter 17, Section 17.2.

You should also assign test cases to requirements. Test cases are usually only created during implementation and are later assigned by responsible persons from test management. The requirement serves as a collection point for the test cases of the implemented stories or features, as the final acceptance is normally carried out via a user acceptance test (UAT), which is performed at requirement level. A detailed description can be found in Chapter 12 and Chapter 14.

SAP Cloud ALM offers you the Analytics app for this purpose in the area of analysis and evaluation options. You can find them in the **SAP Cloud ALM for Implementation** section. With regard to requirements, the Requirement Distribution app and Requirement Traceability app are available (see Figure 10.31 and Figure 10.32).

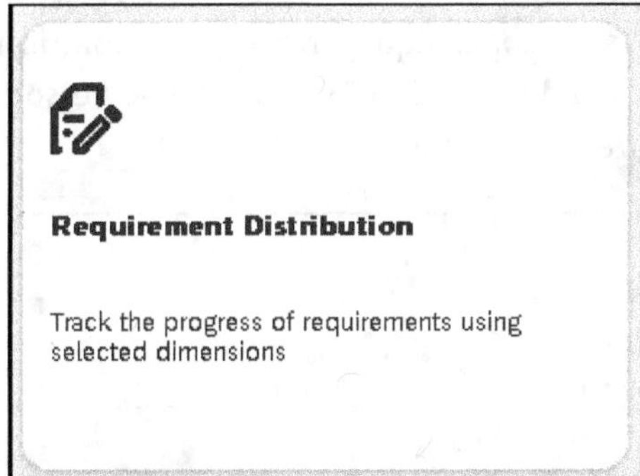

Figure 10.31 Requirement Distribution in the Analytics App

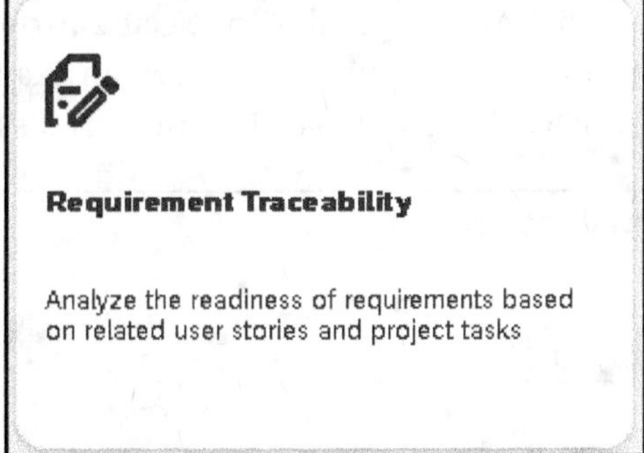

Figure 10.32 Requirements Traceability in the Analytics app

The **Requirement Traceability** overview helps you to analyze, for example, when it is the right time to change the status of a requirement, as the elements belonging to the requirement, such as user stories or test executions, are displayed collectively (see Figure 10.33). Navigating to the respective elements allows a detailed view. Among other things, it is possible to track the implementation progress on the basis of the feature progress by using the filters available at the top.

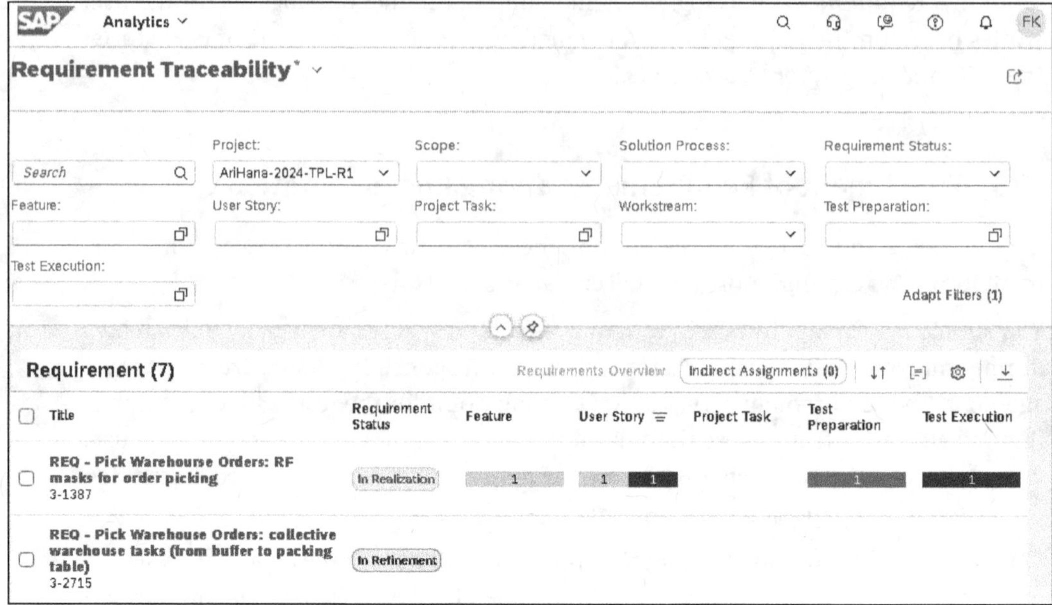

Figure 10.33 Overview of the Traceability of Requirements

With the app for analyzing the **Requirement Distribution**, you can display and evaluate requirements according to selected attributes (e.g., scope, requirement status, solution process, etc.). Figure 10.34 shows an example of a distribution **By Responsible Person**, which is further divided into the various requirement statuses.

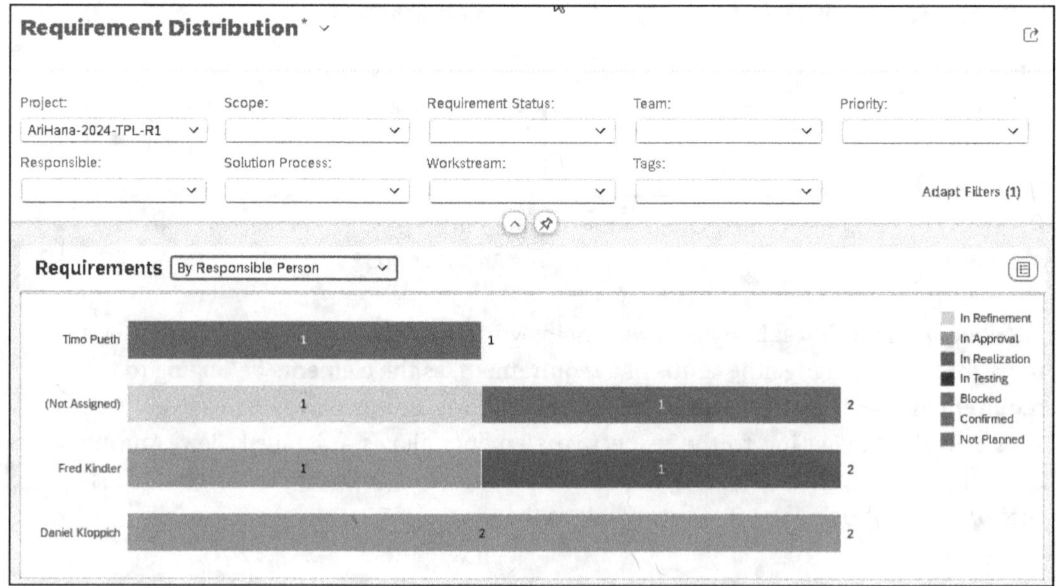

Figure 10.34 Overview of the Distribution of Requirements According to Responsible Person

Regarding the deploy phase (import), it should be noted that you must check the requirements for complete implementation after the go-live of the release. If a requirement has been implemented accordingly and completely fulfilled, we recommend setting the requirement status to **Confirmed**.

10.6 Summary

In this chapter, we have gone into detail about the possibilities of SAP Cloud ALM within requirements management. We have created and specified project-specific requirements for the template release for a logistics process in our example scenario, and we have outlined our recommendations for the description of requirements. In the next chapter, we explain how you can use fit-to-standard or fit-gap workshops to analyze and further specify your individual requirements. Chapter 12 deals with how the captured requirements can be used for test planning. Chapter 13 explains the handover to the realize phase based on the creation of user stories.

Chapter 11
Fit-to-Standard and Fit-Gap Workshops

In this chapter, we look at the differences between fit-to-standard and fit-gap workshops and present both approaches. The workshops are an essential part of the implementation project and serve to validate the processes and their requirements.

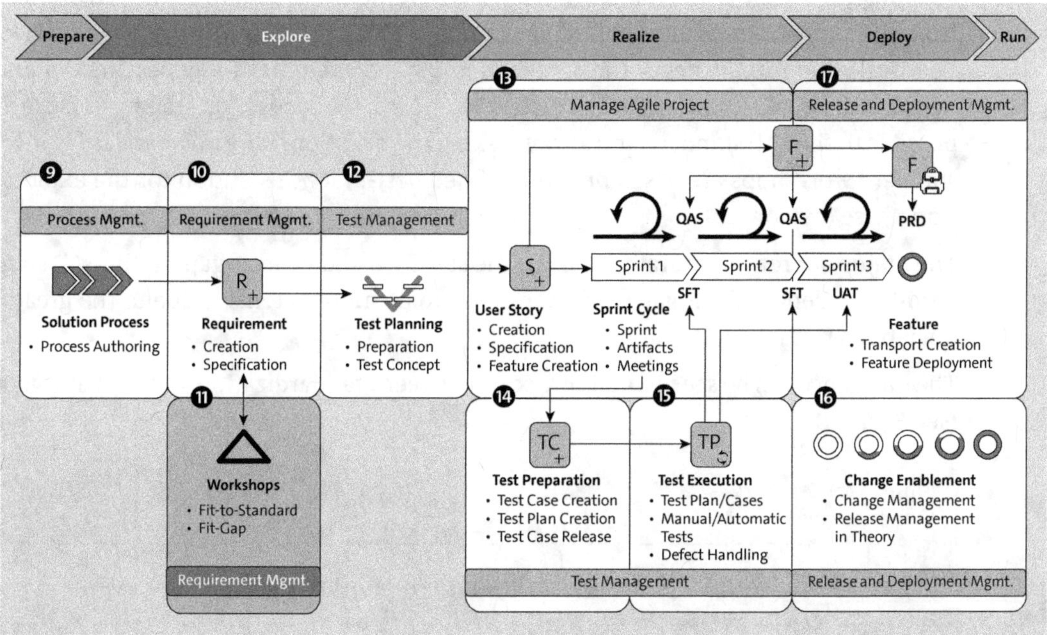

We recommend that you take a look at the explore phase in SAP Activate. To do this, use SAP Cloud ALM with the available source, SAP Activate methodology. There you will be provided with a variety of roadmap tasks, depending on the SAP solution to be implemented, for preparing and conducting the fit-to-standard and fit-gap workshops. Use the roadmap tasks in your project to document progress and responsibilities. Onboarding the project team, defining the scope, solution scenarios and solution processes for the next iteration (timebox) from the prepare phase are important prerequisites for the subsequent fit-to-standard workshops in the explore phase.

The aim of these workshops is to determine the scope of the project. This includes the scope of processes and the collected requirements and document/notes. With this knowledge, you can estimate the effort required for implementation (configuration adjustments, developments, integrations, etc.). In this chapter, we go into detail about how you can methodically specify adequate requirements as part of requirements management. Use the helpful information from Chapter 9 and Chapter 10 as a basis for the documentation of these workshops. Let's first look at the differences between fit-to-standard and fit-gap workshops.

11.1 Differences Between Fit-to-Standard and Fit-Gap Workshops

With SAP Activate, the focus is shifting to fit-to-standard workshops, which have replaced the traditional blueprint workshops as part of the old ASAP methodology. One advantage of the fit-to-standard workshops is that the SAP standard is focused on right from the start and individual adjustments to the standard SAP solution are minimized. The workshop format helps you to get to know the customer's business model and define additional requirements. With fit-to-standard, the focus is on staying as close as possible to the SAP standard and always asking the question: Why not standard? Fit-to-standard workshops are based on a predefined best-practice solution for the exploration phase (explore).

The *fit-gap* approach generally weakens the standard and enables gaps in the SAP standard to be identified. Fit-gap workshops tend to be more extensive due to the greater deviations from the SAP standard.

The difference can be seen as a sliding scale between standardization and allowing gaps (see Figure 11.1).

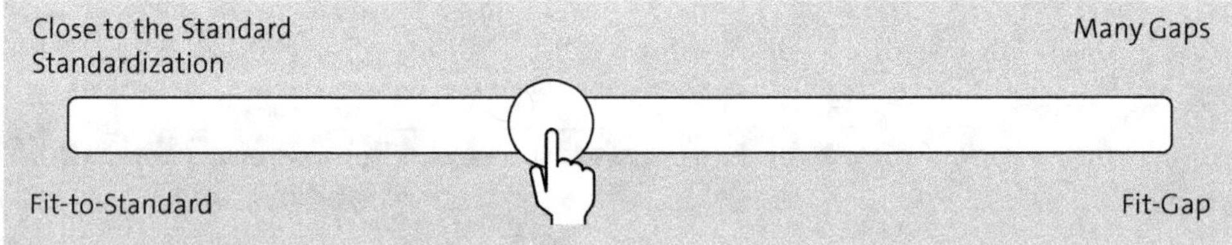

Figure 11.1 Sliding Scale of Standardization

There are compromises that need to be made, of course, such as when deciding where to keep your own practices and where it makes sense to follow the standard delivered in the SAP software.

Fit-to-standard attaches great importance to close adherence to the standard. Fit-gap was previously mainly used by SAP for on-premise software, but all SAP Activate roadmaps are now standardized to fit-to-standard.

Blueprinting from the ASAP days has gradually evolved into fit-gap and finally into fit-to-standard (see Figure 11.2). The term *fit-gap* is still used, but in recent years, there has been a greater shift toward fit-to-standard.

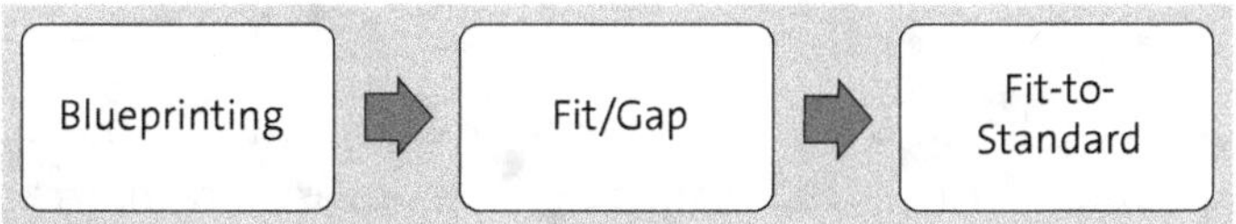

Figure 11.2 Evolution from Blueprinting to Fit-to-Standard

Which approach you choose depends on both the SAP S/4HANA product (see Table 11.1) to be implemented as well as the chosen project method. Our world of project implementation is currently moving toward simplifying and developing agile approaches. The fit-gap method is gradually becoming too cumbersome and often adds additional scope that deviates from the SAP standard.

Product	Workshop	ALM Tool Used	Check Against
SAP S/4HANA Cloud Public Edition	Fit-to-standard	SAP Cloud ALM	SAP Best Practices
SAP S/4HANA Cloud Private Edition	Fit-to-standard	SAP Cloud ALM	Enterprise management layer for SAP S/4HANA Cloud, private edition/SAP Best Practices and SAP Model Company content
SAP S/4HANA (on premise)	Fit-to-standard or fit-gap	SAP Cloud ALM or SAP Solution Manager plus Focused Build	Enterprise management layer for SAP S/4HANA/SAP Best Practices and SAP Model Company content

Table 11.1 Workshop Approaches and ALM Tools

When introducing SAP S/4HANA Cloud Public Edition, fit-to-standard workshops are conducted with SAP Cloud ALM. The test is carried out against SAP best practices.

For SAP S/4HANA Cloud Private Edition, fit-to-standard workshops are conducted with SAP Cloud ALM. The test is carried out against the enterprise management layer for SAP S/4HANA Cloud Private Edition/SAP Best Practices and SAP Model Company content.

In SAP S/4HANA on-premise projects with extensive gaps and development requirements, you can opt for Focused Build for SAP Solution Manager. This add-on supplements SAP Solution Manager with additional standardized functions. Focused Build refers to a fit-gap rather than a fit-to-standard, although SAP Best Practices can still be used.

Fit-gap in the context of SAP Cloud ALM is typically used for rollouts with a template approach, where local requirements are checked against a template. For more information, see Chapter 5, Section 5.3.4 and the following sections, as well as Chapter 7, Section 7.4.

11.2 Fit-to-Standard Workshops

In this section, we explain the procedure for fit-to-standard workshops (see Figure 11.3).

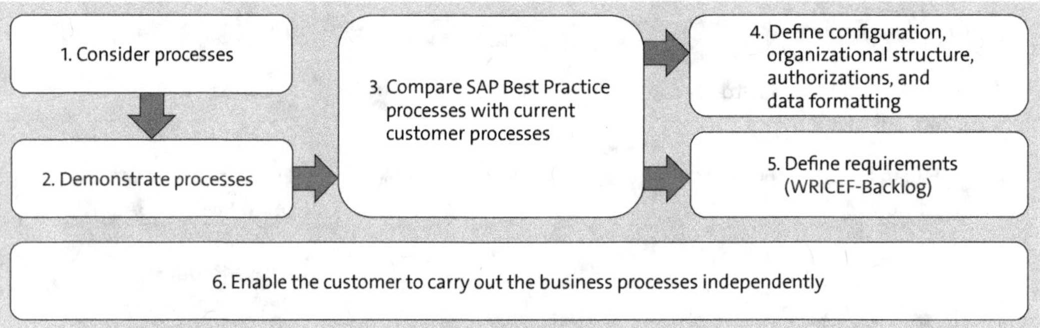

Figure 11.3 Implementation of a Fit-to-standard Workshop

Fit-to-standard workshops are divided into six steps:

1. View processes (review)
2. Demonstrate processes (walkthrough)
3. Align SAP Best Practices processes with current customer processes
4. Define configuration, organizational structure, authorizations, and data formatting
5. Define requirements (workflow, report, interface, conversion, enhancement, forms—in short, WRICEF backlog)
6. Enable the customer to carry out the business processes independently

In summary, the processes are explained, demonstrated, discussed, and trained in the workshops on the basis of the solution process diagrams. This ensures that the customer processes match the standard processes delivered by SAP. Gaps, requirements, and notes are collected in the workshops. The individual steps of the workshops are explained in more detail later in this chapter.

Remote Fit-To-Standard Workshops with Mural Support

Fit-to-standard workshops are often planned and carried out remotely, especially for geographically dispersed teams. To prepare and conduct the workshops, you can use prepared templates in SAP Activate with the Mural software.

Mural is a web platform that allows teams to work freely with virtual Post-it notes, among other things. The Post-it notes can be freely moved, arranged, or clustered. The templates for Mural prepared in SAP Activate guide you through a fit-to-standard workshop and allow distributed teams to work together on the workshop result.

11.2.1 Definitions and Basics

Fit refers to all process requirements that can be covered by a configuration of the standard SAP solution. A *gap* means a customized development. This subdivision is shown in Figure 11.4 once again. The SAP- and ABAP-specific abbreviations RICEFW and WRICEF stand for:

- Reports
- Interfaces (interfaces)
- Conversions (data format conversion)
- Enhancements (extensions or modifications, i.e., changes to the standard software)
- Forms
- Workflows (automation of work processes)

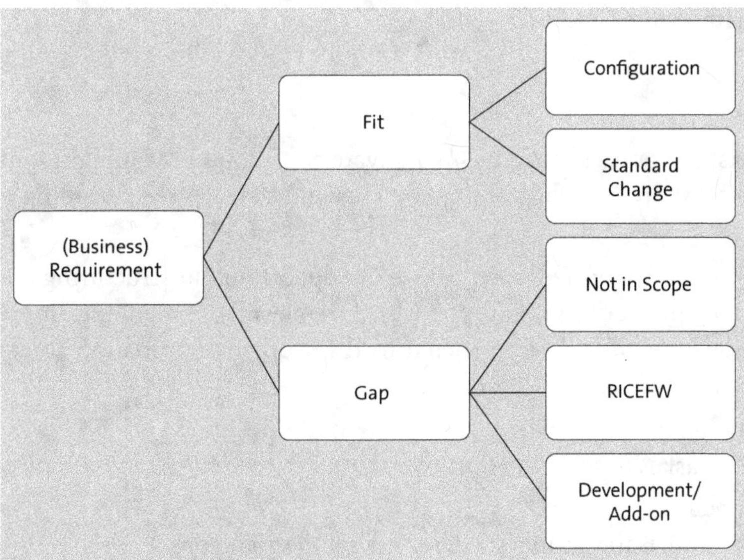

Figure 11.4 RICEFW Localized as a Gap

The term *WRICEF* sounds outdated and obsolete in the age of RISE with SAP. Attempts have been made to create a translation of the development object types from the various historical eras for orientation purposes. For example, the original meaning of WRICEF can be extended to include some of the cloud-native modern services offered as part of SAP BTP:

- Reports include SAP S/4HANA embedded analytics queries, CDS views, SAP Analytics Cloud, SAP HANA, and SAP Datasphere.
- Interfaces for integrating OData, APIs, and other SAP Integration Suite functions.
- Conversions to SAP S/4HANA migration cockpit and selective data transition.
- Enhancements to open up a whole new world of development options, including in-app extensibility, parallel extensibility, low-code/no-code, and even SAP Intelligent Robotic Process Automation (SAP Intelligent RPA).
- Forms and workflows for the integration of SAP Build Process Automation.

Some benefit of these innovations are as follows:

- Key users can implement identified gaps using in-app extensibility and thus convert gaps into fits.
- By using low-code/no-code platforms, SAP Build Process Automation workflows, and SAP Intelligent RPA, the implementation of identified gaps can be checked.

To help the ecosystem change and reflect modern technology in implementations, the broader term *extensibility* has been added to the workgroup name in SAP Activate. This term replaces *developments* and *custom code extensions* and thus better expresses the new course, according to Kamugisha, Adnette (2019): SAP Activate Methodology Workstreams Names update, *http://s-prs.de/v946407*.

11.2.2 Project Tasks

Fit-to-standard workshops are described in SAP Activate. SAP Cloud ALM supports you in conducting the workshop by means of project tasks and a structured recording of requirements.

Each project scope contains scope-specific tasks that support you in structuring and organizing the project. These tasks provide you with a framework for planning, implementation, and follow-up and can be restricted to the scope determination source using a filter, shown in Figure 11.5:

- **Schedule Fit to Standard**
  The purpose of this task is to schedule the fit-to-standard workshops.
- **Conduct Fit to Standard**
  The purpose of this task is to conduct the fit-to-standard workshops.
- **Review and Complete Requirements**
  The purpose of this task is to ensure that your requirements are fully documented and understandable.

- **Approve Requirements by Generating Implementation Tasks**
  The purpose of this task is to approve requirements by generating implementation tasks. In doing so, you hand over a requirement to the realization.
- **Confirm Scope and Requirements**
  The purpose of this task is to obtain confirmation that the scope has been implemented and the requirements are provided.
- **Define Test Cases**
  The purpose of this task is to define the required test cases that provide valid test coverage for your current scope. You should create test cases (manual or automated) based on the current scope (business processes and solution processes) and the approved requirements.

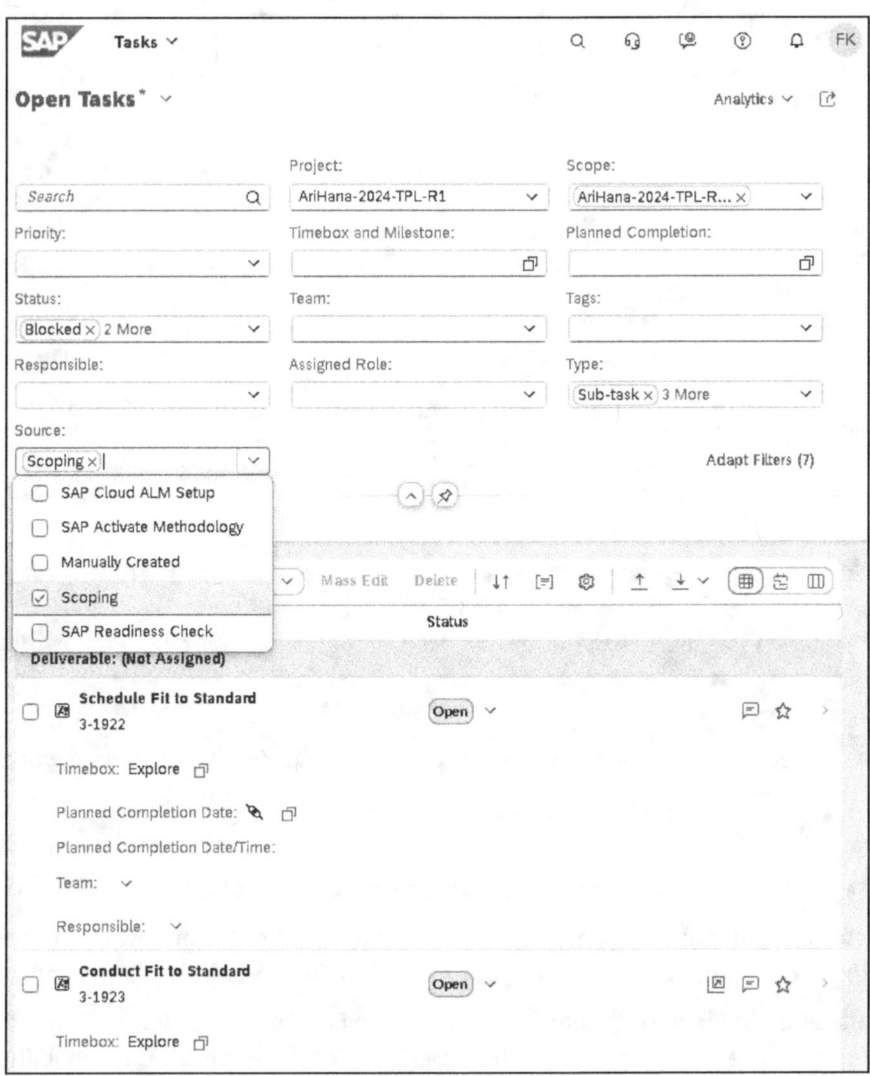

Figure 11.5 Tasks from the Scope Determination

Before workshops are planned and carried out, roadmap tasks from SAP Activate are available to you, which are preselected at project header level. Details on the selection options can be found in Chapter 6, Section 6.1.1.

[+]

Task Descriptions and Accelerators

Each task has a descriptive text, as shown in Figure 11.7. A planned completion date must be added manually. Linked accelerators and additional links also support the understanding of the task.

Other tasks relevant to the fit-to-standard workshops can be restricted to deliverables or deliverable results using the filter (see Figure 11.6).

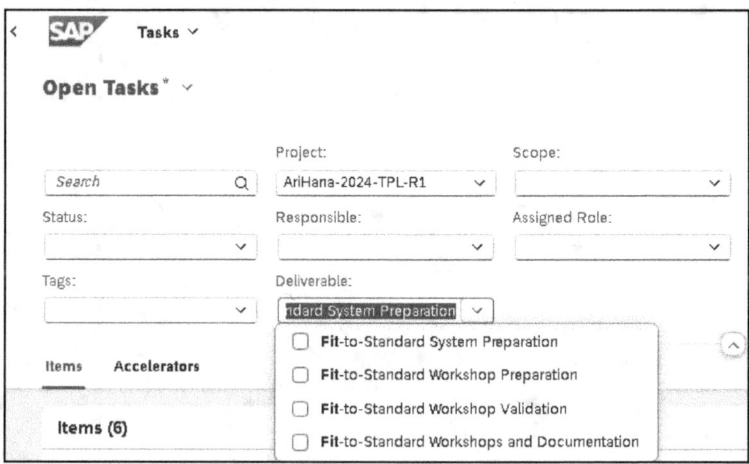

Figure 11.6 Relevant Tasks for Fit-to-Standard

Predecessor and Successor Relationships Already Stored

For the roadmap tasks, it is helpful that predecessor and successor relationships to other roadmap tasks are already stored (see Chapter 7, Section 7.2).

The following sections describe the six steps for conducting the fit-to-standard workshops in more detail.

11.2.3 Steps 1 to 3: View, Demonstrate, and Compare Processes

The first steps include the independent work and preparation of the project team with the help of the documents and the need to familiarize themselves with SAP Activate.

The roadmap task in Figure 11.7, **Conduct Project Team Enablement on Fit-to-Standard** for the project team enablement deliverable describes useful prerequisites for enabling the project teams prior to the workshops.

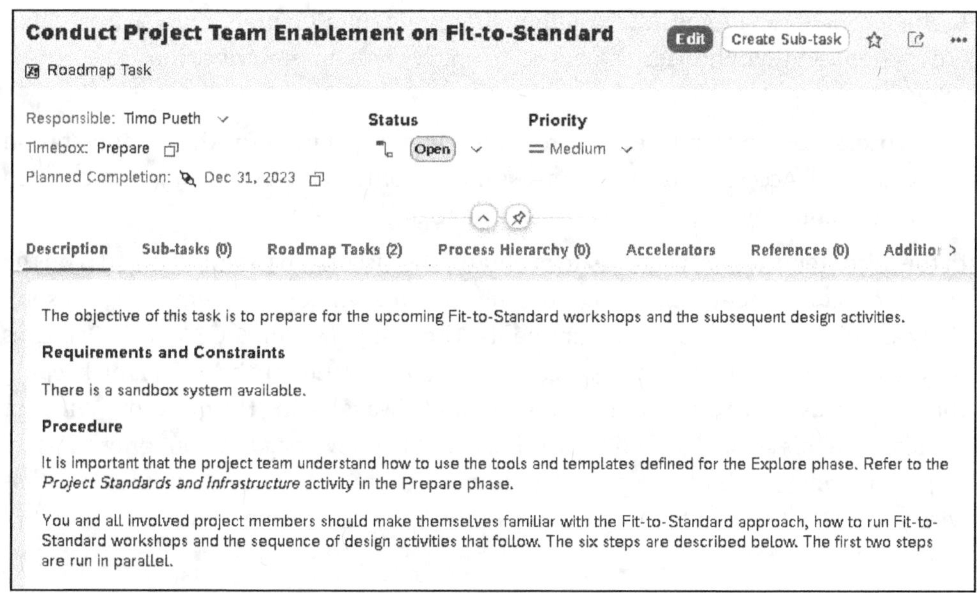

Figure 11.7 Empowerment of the Project Team Before the Workshop

At the beginning of the fit-to-standard workshops, a consultant describes the solution process (review) as a first step. The first step is to take a theoretical look at best practices to explain the terminology and establish a common understanding of the process. The extensive contents of the solution processes in SAP Cloud ALM are used for visualization and explanation. In addition to SAP processes, customer-specific processes or process content from SAP partners can also be used (see Figure 11.8).

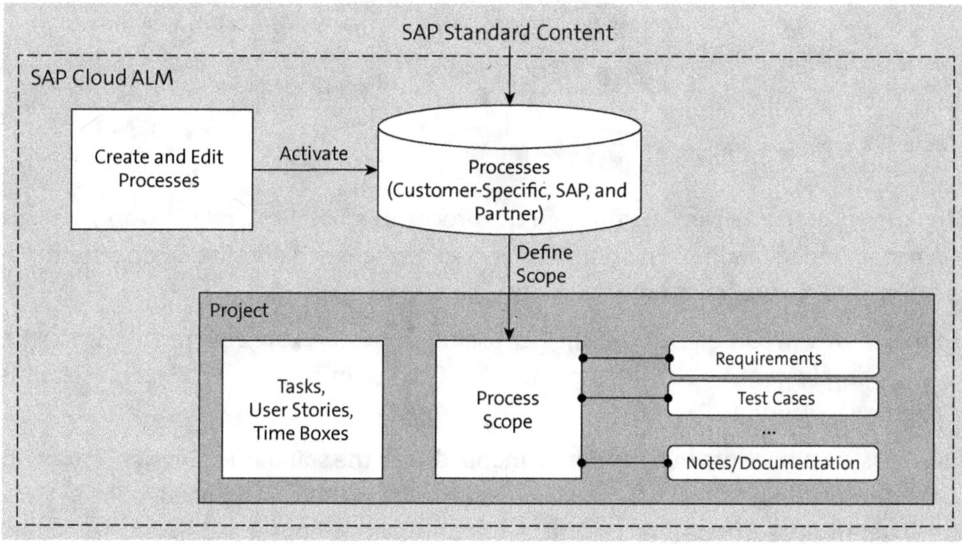

Figure 11.8 Determine Process Scope

In the second step, the consultant demonstrates the solution process using a prepared SAP system and emphasizes the areas in which configuration decisions need to be made.

The process diagrams are used as a reference and orientation for the demonstration (walkthrough). Access to the on-premise sandbox or the cloud starter system can also be supplemented by guided simulations, online training, etc.

In the third step, the consulting team facilitates the discussion to better understand the customer requirements and to support the customer in how the demonstrated solution can map these customer requirements. The aim is to compare SAP Best Practices processes with current customer processes. With ft-to-standard, the focus is on keeping the solution as close to standard as possible and always asking the question: Why not standard? In this step, no concrete solution should be developed yet, but only a discussion about whether the process fits. Initially, only note down the requirements, otherwise you will not complete the workshops within the time frame.

11.2.4 Step 4: Identify Requirements

In the fourth step, the workshop participants identify the requirements. Results should be documented in requirements and in important content in notes or documents. A brief comparison between task and note/document is shown in Table 11.2.

Notes/Documents	Requirements
▪ Notes/documents are time-independent and part of the process documentation. ▪ Notes/documents are used for additional knowledge or insights about the solution process that is important to capture.	▪ Requirements are implementable and (after approval) are converted into tasks to facilitate fulfillment. ▪ A requirement describes in detail what the solution process must fulfill to meet customer requirements.

Table 11.2 Notes vs. Requirements

The common thread running through the processes plays a key role in workshop participants' understanding. The requirements can be created and linked directly to the processes to ensure traceability later on.

In this fourth step, the results of the discussion from step 3 are concretized as to what is to be implemented. The documentation of requirements is described in detail in Chapter 10, Section 10.2.

Business requirements that cannot be mapped with the standard scenarios or individual extensive change requests can be noted in bullet points and mapped to the SAP system later in an additional, separate fit-gap analysis workshop, depending on feasibility.

11.2.5 Step 5: Define Configurations

In the fifth step of the workshop, the specific configuration, organizational structure, authorizations, data preparation, and other master data are defined and described in detail. The review and validation of the collected requirements is also part of this step. Good consulting follows the principle of staying as close as possible to the SAP standard and encouraging a cost-benefit analysis. The customer is responsible for providing the content of the requirements.

11.2.6 Step 6: Empower the Customer

The aim of the sixth and final step in the workshop is to enable the customer to carry out the business processes independently in a starter SAP system. By enabling the customer to deal with all the business processes in the scope, important prerequisites for the transition to the realize phase have been created.

The following results from the fit-to-standard workshops are now available:

- Solution process scope for the upcoming implementation.
- Requirements such as configurations, extensions, and integrations to be delivered in the implementation.

Together, these results form the project scope.

11.3 Fit-Gap Workshops and Rollouts

In addition to fit-to-standard workshops, fit-gap workshops are also used in some projects. There are various possible uses for fit-gap workshops:

- Introductory workshops for SAP S/4HANA on premise
- Rollouts with a template approach

According to SAP Activate, fit-to-standard workshops are the workshop format you should prefer when introducing SAP S/4HANA or setting up a template. We will therefore only briefly discuss the use of fit-gap workshops for rollouts with a template approach below.

The fit-gap method is used to identify areas in which the template or a business process fits (or does not fit) the requirements for the respective country and company. Simplified, this is shown in Figure 11.9. Typical gaps are, for example, legal requirements in other countries or the need to adapt processes due to different sizes of departments or groups of people than an app serves.

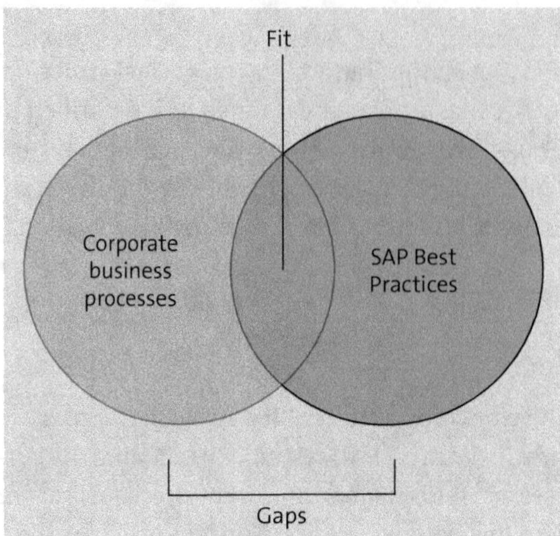

Figure 11.9 Fits and Gaps

The aim of a fit-gap analysis is to identify the best practices for an organization. This also includes everyday work processes, which are examined for their efficiency. They are compared with the processes from a standard, whereupon changes are made to the systems or the template after weighing the possible options for implementation.

In the case of rollouts with a template approach, the requirements of the countries to be delivered are iteratively compared with the processes of the template and documented in fit-gap workshops (see Figure 11.10).

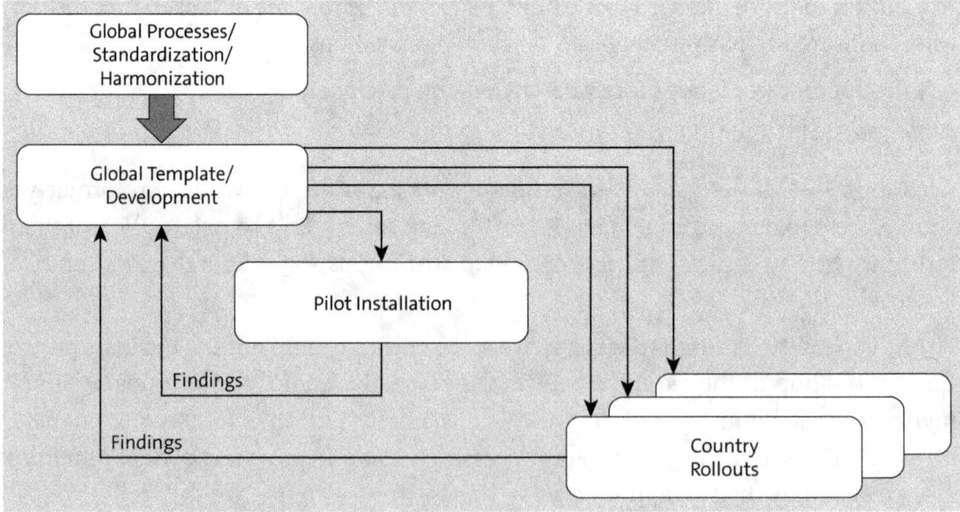

Figure 11.10 Pilot Installation and Rollouts by Country

In preparation for a fit-gap workshop, it is helpful to prepare an up-to-date template system for initial training or to keep the existing template system up to date. Sample data of the location should be created or imported in this system to be able to concentrate on understanding the processes.

From the perspective of organizational change management, it is a success criterion for a template approach if global *process owners* present the template processes or support them significantly. A presentation by process owners and not by IT supports the identification of future users with the template processes. Involving local consultants at an early stage can help you to prepare the content of the SAP standard and to translate the business processes into the local language and for the market.

11.4 Summary

In this chapter, we have gone into detail about the differences between fit-to-standard and fit-gap workshops. We presented the advantages of fit-to-standard workshops for the agile project methodology. The steps were explained to you so you could define methodically adequate requirements and thus to be able to better estimate the effort required for implementation in the realize phase. In the next chapter, we will turn our attention to test management, taking a closer look at test planning for the recorded requirements.

Chapter 12
Test Management

In this chapter, test management in SAP Cloud ALM is explained and the support options for test planning are outlined. You will also gain insight into the methodology and technical integration of test automation in SAP Cloud ALM.

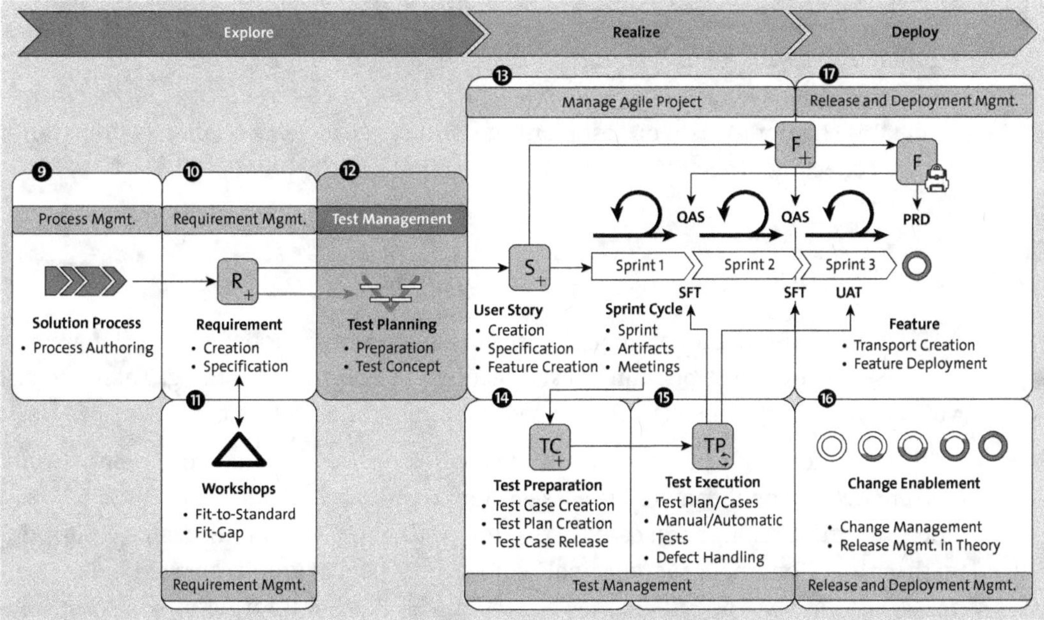

After gaining an initial insight into the process and requirements management of SAP Cloud ALM in Chapter 9 and Chapter 10, in this chapter you will learn about the functions of test management functions and how SAP Cloud ALM supports you in planning tests, preparing tests, and executing manual and automated tests.

Test management in the SAP methodology includes all tool-supported activities of a test process, taking into account the company-specific test strategy. The test management functions of SAP Cloud ALM therefore range from test planning to the creation, supported execution, and logging of test cases to the elimination of errors that occur and the monitoring and evaluation of test phases. The typical test management process is shown in Figure 12.1. The individual activities of a test management process are not carried out strictly one after the other but overlap in their execution times.

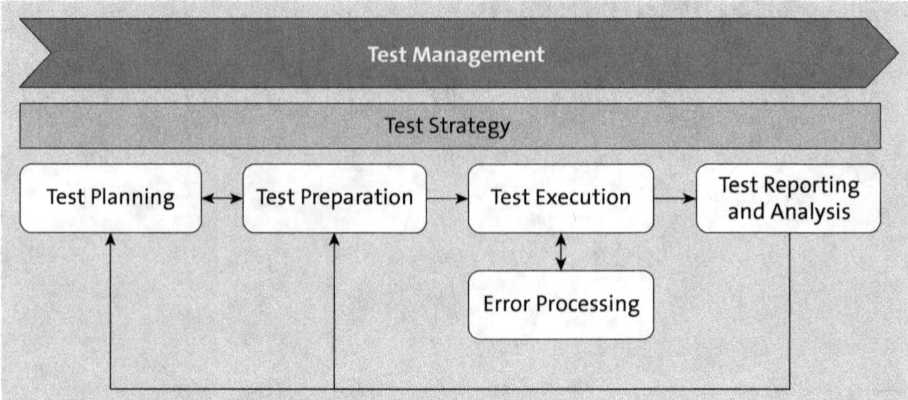

Figure 12.1 Test Management Process

While this chapter focuses on test planning, you can read more about test preparation and test execution in SAP Cloud ALM in Chapter 14 and Chapter 15.

12.1 Test Management in SAP Cloud ALM

SAP Cloud ALM provides a range of test management functions by different apps for test case creation, test planning, test execution, and error processing. There are also various dashboard apps that can be used to monitor and analyze the testing progress.

With the functions in SAP Cloud ALM, test cases can be executed both manually and automatically. For automated test execution, there is a *Test Automation API* to create and execute automated test cases in the SAP testing tool. A system-specific description of these functions using test automation for SAP S/4HANA Cloud can be found in Chapter 14, Section 14.2.4 and in Chapter 15, Section 15.2. Further information on connecting a test automation tool can be found in Section 13.3.

During a project, different types of testing are carried out depending on the project progress. The following types of testing should be considered:

- Functional tests
- Integration tests
- Acceptance tests
- End-to-end tests

A *functional test* is used to check individual business requirements or functions. An *integration test*, on the other hand, ensures that services or modules of an application interact smoothly. *Acceptance tests* (also known as *user tests*) check whether a system has been configured in accordance with the defined requirements and focus on the future user's perspective. *End-to-end tests* are the most complex of all testing types and focus on the proper functioning of user workflows within a complete business process, possibly even across company boundaries. All testing types can, but do not necessarily have to, be carried out within an implementation project. The selection and scheduling of these testing types defines one of the essential tasks in test management.

Figure 12.2 shows an example of how these testing types can be used within a project. In our example project, a functional test of the implemented user stories is carried out after each sprint. In the acceptance test that takes place later, the test cases from the functional test are then reused to check the implemented requirements.

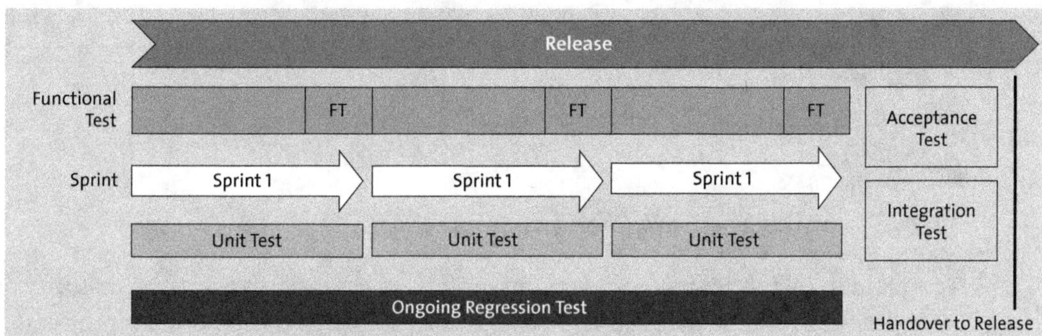

Figure 12.2 Testing Types within an Implementation Project

SAP Cloud ALM offers a direct link between the test cases and requirements and user stories created in your project so that you can also visualize the type of testing to be carried out. The solution process does not allow you to link to the test case directly, but all assigned requirements have their own test cases and are part of the test scope. Test plans can be used for structuring several test cases for a certain test type. As shown in Figure 12.3, you can define the type of testing to be carried out for each test case and adapt it according to the progress of the project. For example, you could link your test case to a user story for a functional test or with a requirement for an acceptance test. Solution processes are validated by integration tests. For this testing type, the test plan bundles all test cases delivered by dedicated requirements. You can find more information on linking design and build processes with test cases in Chapter 14, Section 14.3.

To document errors that occur during testing, SAP Cloud ALM provides the defect process, but it cannot currently (as of April 2024) be linked to a correction transport. As a workaround, the feature process described in detail in Chapter 17 is being used in test management to perform a correction transport. A detailed insight into the defect process in SAP Cloud ALM can be found in Chapter 15, Section 15.3.

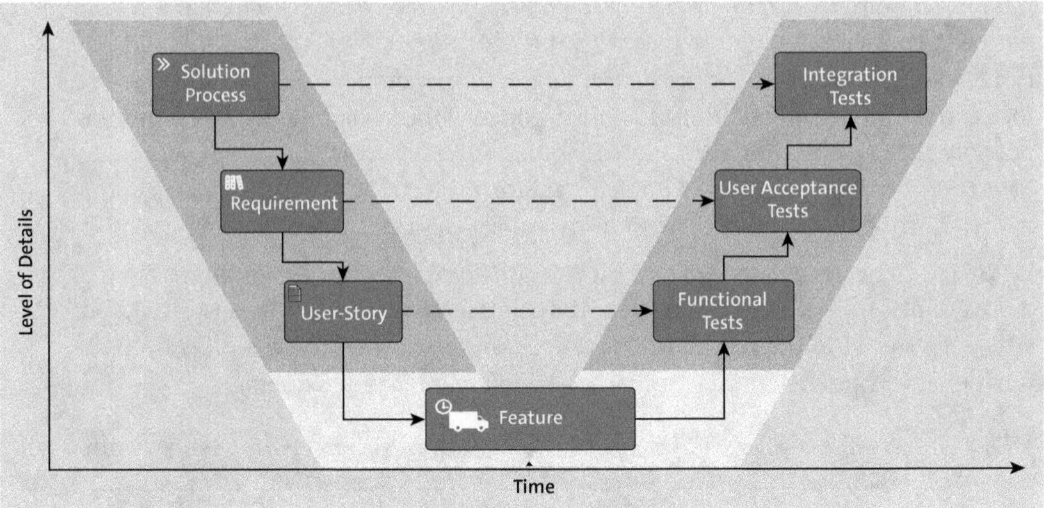

Figure 12.3 Processes and Their Assignment to Test Types

12.2 Planning the Test Preparation

Test planning activities required for your project are essential to check the maturity of your project within the defined test cycle. It does not matter whether your project is a purely technical conversion, a combined conversion project with the provision of new functions, or a new implementation project. In addition to defining a test strategy, the test planning includes decisions that significantly influence the preparation and execution of tests. You should pay particular attention to company-specific guidelines as well as design and build processes.

Due to the conceptual nature of test planning, this activity only takes place to a limited extent in SAP Cloud ALM itself. Nevertheless, SAP offers a wide range of tools to support you in test planning and creating a test strategy, which we present to you in the following sections.

12.2.1 SAP Activate and Test Management

SAP Cloud ALM also provides template tasks for test management using the SAP Activate methodology. To do this, open the Tasks app in the **SAP Cloud ALM for Implementation** app group. In the window that opens, filter for the source **SAP Activate Methodology** and type **Roadmap Task**. Next, filter by **Deliverable** (**Test Execution**, **Test Planning**, and **Test Preparation**), as shown in the example in Figure 12.4. The SAP Activate phases, which can be defined using the **Timebox** and **Milestone** fields, are another useful way of limiting the scope.

Assign the roadmap tasks to the corresponding scope and maintain a responsible or team and define a start and planned completion date.

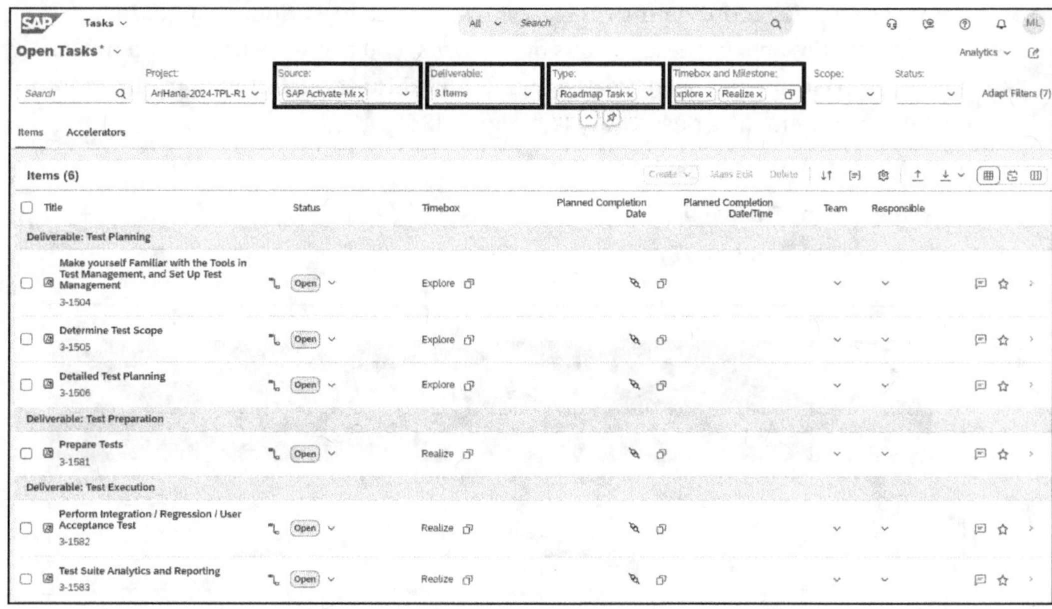

Figure 12.4 Roadmap Task Related to Test Management

SAP Cloud ALM also allows you to upload SAP Best Practices test cases and use them for test case creation. You can find a detailed explanation of this in Chapter 14, Section 14.2.2 and Section 14.2.3.

12.2.2 Determining Test Scope

An important part of test planning is to determine the scope of the testing. You should do this as early as possible and then initiate the creation of the necessary test cases and test plans. You should also determine whether your test phase should only use manual tests or whether the use of automated tests is also sensible and applicable. If you decide to use automated tests, you must take care of the technical connection for automated test execution in good time. This also includes the availability of suitable test scripts for execution.

You should also define whether and to what extent the SAP Best Practices content should be used to create test cases. Include the test planning options in SAP Cloud ALM when defining your test strategy (see Chapter 14, Section 14.4). To minimize the effort involved in creating test cases, you can reuse the test documents from tests that have already been carried out.

12.2.3 Dashboards for Test Preparation

In addition to initiating the creation of test cases, it is the test manager's task to monitor the test cases and thus ensure that the test phase starts on time. SAP Cloud ALM also provides tools for coordinating test case creation with its various dashboard functions in the Analytics app. In these various dashboards, you can see whether and how many test cases have already been created and finalized for the selected items like solution process, requirement, or user story (see Figure 12.5):

- **Solution Process Traceability**
- **Requirement Traceability**
- **User Story Traceability**

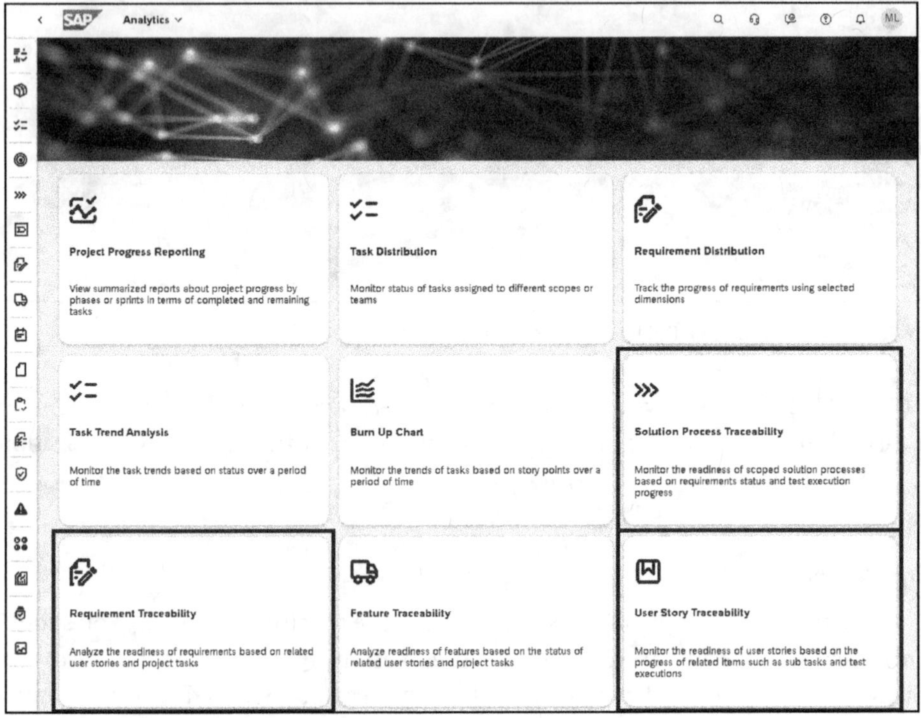

Figure 12.5 Dashboards for Test Case Preparation

You can get a deeper insight into the use of this analysis functionality for test management in SAP Cloud ALM in Chapter 15, Section 15.3.3. Please note that test cases are only displayed for requirements or user stories if they have been linked to the respective solution process. For more information, see Chapter 14, Section 14.3.

12.2.4 Tagging Function in Test Management

SAP Cloud ALM provides tag functionality for solution process and design and build processes. Tags can be created directly at dedicated design and build processes by the Tags field and reused in every process. We recommend that you set up your test management required tags within an administrator app named Tag Management. In this app, you can group your additional tags. Once a tag has been created, it can be used within the filter options of the various applications. You will find plenty of information about tag management in Chapter 6, Section 6.6.

The use of tags also offers several advantages in test management. For example, you can assign a tag to all requirements, user stories, features, etc., that are to be realized in a certain sprint: sprint 3, for example. You should then also assign this tag to each associated test case. This allows you to display all processes that are to be implemented in this sprint in the Tasks app. In the Test Execution app, your testers can filter for all test cases that are relevant for the current sprint. In addition to the time classification, also consider the use of tags in the business process context (e.g., purchasing).

12.3 Test Automation in SAP Cloud ALM

In addition to the creation and execution of manual tests, SAP Cloud ALM offers integration for automated testing. SAP Cloud ALM distinguishes between two types of test automation provider:

- Test automation with SAP S/4HANA Cloud Public Edition
- Test automation with SAP Test Automation by Tricentis or other third-party provider

Communication between SAP Cloud ALM, the test automation provider, and your system landscape (systems under test is often abbreviated SUT in this context) takes place via a provided interface, the Test Automation API (see Figure 12.6).

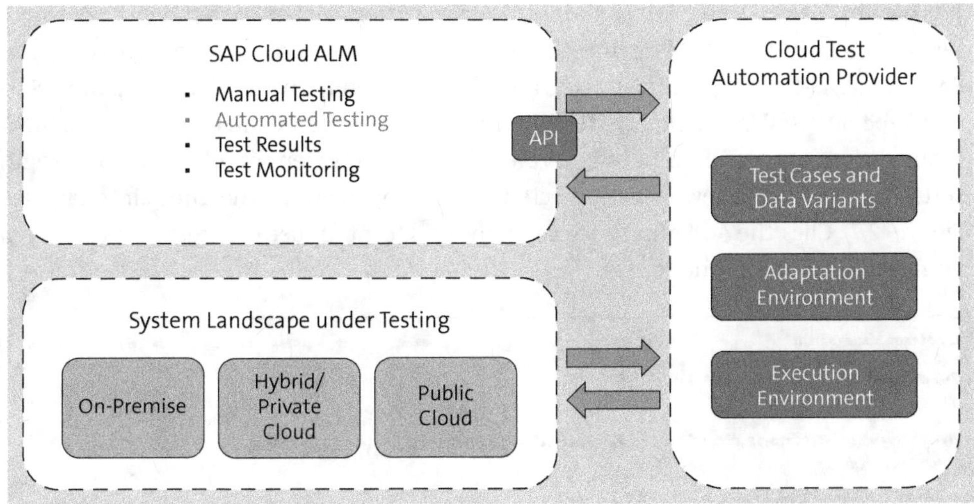

Figure 12.6 Communication Between SAP Cloud ALM, Provider, and SUT

Automated test cases are created as a shell via SAP Cloud ALM and physically created in the test automation provider. While you initiate the execution of your test scripts via

SAP Cloud ALM, the actual execution takes place in your automation provider. The test data for your test case is also stored at the provider. All status feedback data is sent back to SAP Cloud ALM and can be viewed there both in the Test Execution app and in the respective test dashboards.

12.3.1 Test Automation Tool for SAP S/4HANA Cloud

SAP S/4HANA Cloud Public Edition has integrated test automation functions that are available directly in each customer's quality system. After linking SAP S/4HANA Cloud with SAP Cloud ALM, predefined and directly usable test scripts are also provided for most standard business processes. SAP S/4HANA Cloud also offers directly usable post-upgrade test cases in the event of system upgrades. The test automation tool for SAP S/4HANA is a fully cloud-native solution with a no-code approach that requires no installation and uses standard content provided based on best practice content. The tool also has a recording function that allows you to create your own scripts. The test scripts run in the background so that you can continue your business processes undisturbed.

The following prerequisites must be met to connect the test automation tool in SAP S/4HANA Cloud to SAP Cloud ALM:

- SAP Cloud ALM tenant is provided.
- A dedicated SAP S/4HANA Cloud tenant is registered in SAP Cloud ALM tenant.
- An implementation project is created in your SAP Cloud ALM tenant.

It is also necessary to set up an endpoint for the interface between the SAP S/4HANA tenant and SAP Cloud ALM tenant. To do this, open the Landscape Management app in SAP Cloud ALM in the Administration app group. Now filter by **Service Group** or by **Service Type** SAP S/4HANA Cloud. After filtering you see several entries of SAP S/4HANA Cloud depending on different tenant roles. Chose the **Test** tenant role to continue with setup. In the detail view of the selected system, navigate to the **Endpoints** tab (see Figure 12.7). Click the **Add** button to create the endpoint for use case test automation of your SAP S/4HANA Cloud tenant.

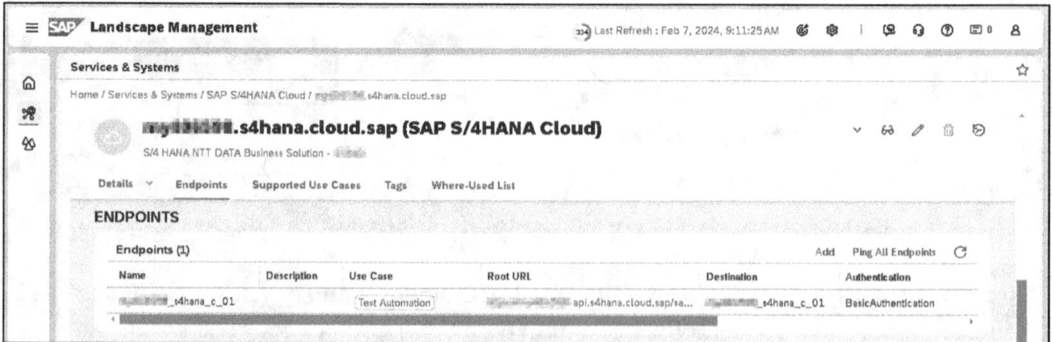

Figure 12.7 Configure Endpoint for Use Case Test Automation

Table 12.1 provides you with an overview of the mandatory information to be maintained for the basic authentication type.

Mandatory Parameter	Description	Example
Endpoint Name	Name of the maintained endpoint	SAPS4_TAUT
Use Case	Customer-specific name of the use case	test management
Root URL	URL of the SAP S/4HANA Cloud system	*https://xxxxx.ondemand.com/sap/bc/http/sap/...*
User	User who exists in your SAP S/4HANA Cloud system and is sufficiently authorized	AUTH_CALM
Password	Password created and generated in the SAP S/4HANA Cloud system	******

Table 12.1 Mandatory Settings for the Test Automation API

A communication arrangement is required in your SAP S/4HANA Cloud system. To do this, log in to your SAP S/4HANA Cloud tenant and open the Communication Arrangements app in the **Communication Management Administration** app group. Now create the communication scenario SAP_COM_0620. Enter the system ID for the communication system, configured in the Communication Systems app, and the user name, configured in the Maintain Communication Users app.

[+]

Authorizations to Create a Communication User

If the Communication Agreements app is not displayed, you do not have the authorization to create a communication user. Open the Maintain Business Users app and add the SAP_BR_CONF_EXPERT_BUS_NET_INT role to your user.

You can find the latest information regarding setup of integration with SAP S/4HANA Cloud Public Edition at SAP Help Portal: *https://help.sap.com/docs/cloud-alm/setup-administration/test-automation-tool-for-s4hana-cloud.*

12.3.2 SAP Test Automation by Tricentis with SAP Cloud ALM

The test automation interface shown in Figure 1.6 is not only available for the SAP S/4HANA Cloud Public Edition integrated test automation tool, but also gives third-party providers the opportunity to offer an integration for test automation in conjunction with SAP Cloud ALM. Tricentis is one of the first vendors to offer such an integration. SAP Test Automation by Tricentis provides a platform for continuous testing that allows you to design and execute automated end-to-end software tests.

Before you start with the automated setup of SAP Test Automation by Tricentis, you should familiarize yourself with the usage rights. The following is the status as of April 2024:

- A term license for SAP Test Automation by Tricentis is granted with SAP Enterprise Support, cloud editions, with SAP Enterprise Support, and with SAP Product Support for Large Enterprises. This term license is granted until July 31, 2026.
- SAP Test Automation by Tricentis can be used to test and validate business processes.
- SAP Test Automation by Tricentis may not be used to automate productive business processes.

The following technical key figures must also be observed:

- 5 named users
- 500 test runs per calendar month
- 2.5 GB memory
- 5 execution agents managed by the customer
- Retention period of 12 months for test results

The current usage rights can be found at *https://support.sap.com/en/alm/usage-rights.html* in the SAP Test Automation by Tricentis menu.

For the automatic setup of SAP Test Automation by Tricentis to function correctly, the following prerequisites must be met by the user:

- The user has the SAP Cloud ALM Global Administrator role.
- The user has an S-user for his email address which has the Edit Cloud Data role.

In addition, prerequisites regarding the identity provider used must be observed:

- Use of the standard SAP identity provider.
- The hierarchy of the SAP Cloud ALM application in the SAP identity provider has not been changed.

Incorrect Setup or Missing Prerequisites

In the event of an incorrect setup or a missing prerequisite, create a customer message on the SAP component SV-CLM-IMP-TM.

The automatic setup of SAP Test Automation by Tricentis in SAP Cloud ALM is carried out in four steps:

1. Setup of the SAP Test Automation by Tricentis application
2. Provisioning of the SAP Test Automation by Tricentis tenant
3. Updating the SAP Test Automation by Tricentis application
4. Configuration of the endpoint for Test Automation

Open the **Administration** app group in SAP Cloud ALM and open the Set up Tricentis Test Automation for SAP app (see Figure 12.8).

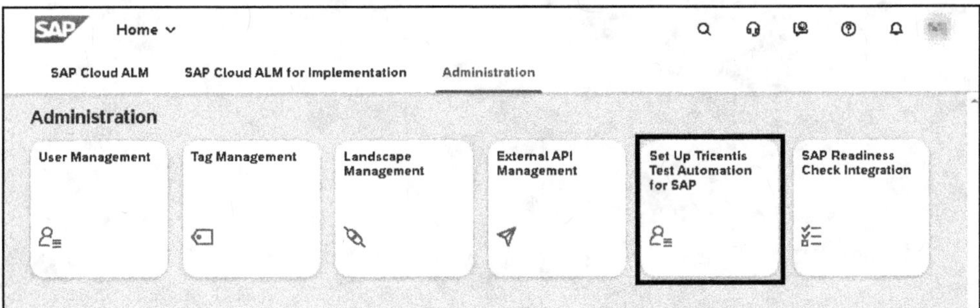

Figure 12.8 App for Starting the Automatic Setup

To begin, you must agree to the terms of use and receive the SAP Test Automation by Tricentis order document. The previously mentioned four-step automatic setup then starts.

After completing the setup, call up the application SAP Test Automation by Tricentis via the same app in SAP Cloud ALM. To do this, click on the link in the final setup screen (see Figure 12.9).

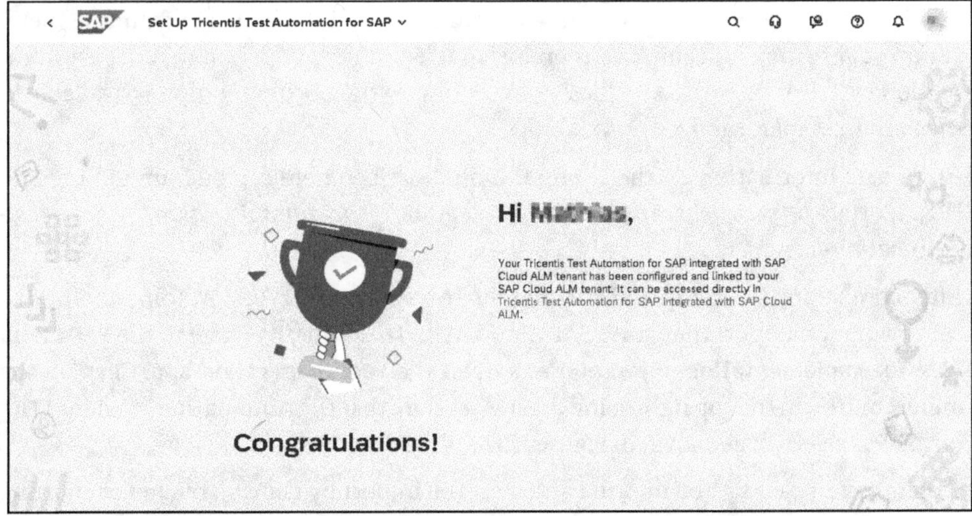

Figure 12.9 Final Screen of Setup with Link to SAP Test Automation by Tricentis

The start page of the SAP Test Automation by Tricentis application opens as shown in Figure 12.10.

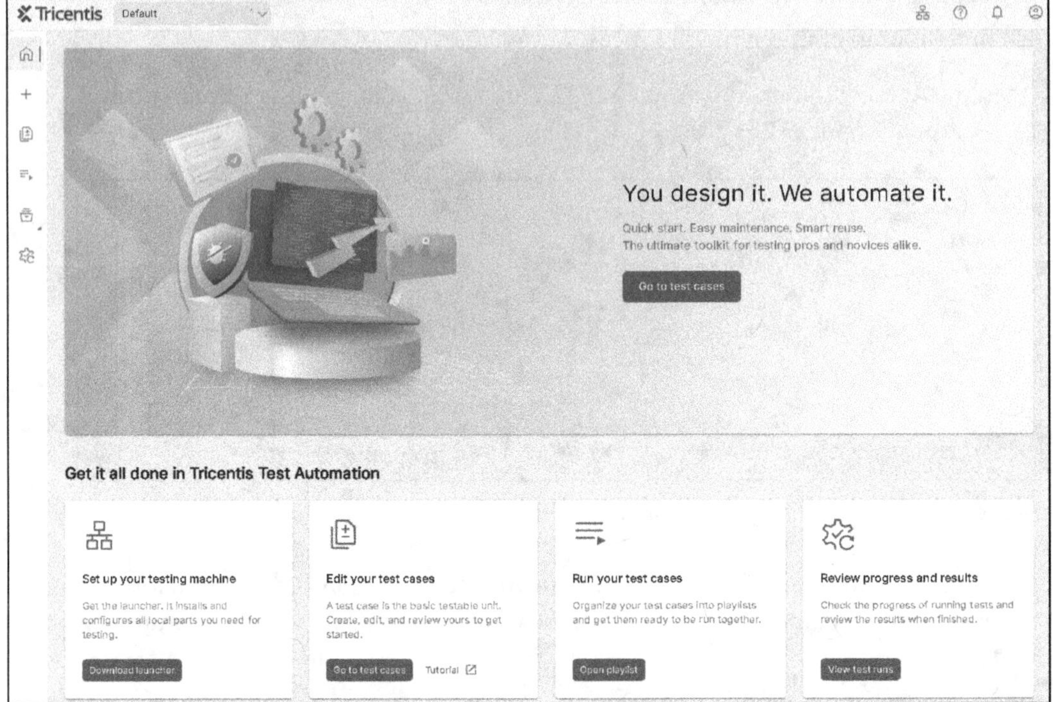

Figure 12.10 Start Page of SAP Test Automation by Tricentis

To create the test cases in SAP Test Automation by Tricentis, you still need the launcher, which is installed on your computer. You can download the launcher from the menu **Set up your testing machine** and then install it. For an introduction to SAP Test Automation by Tricentis with SAP Cloud ALM, a jump to the documentation is available in the **Resources** menu above.

Up-to-date information on the configuration described here can be found in the SAP Help Portal: *https://help.sap.com/docs/cloud-alm/setup-administration/tricentis-test-automation-for-sap.*

After completing the setup and creating the test cases in SAP Test Automation by Tricentis, you can import them into SAP Cloud ALM. To do this, navigate to the **SAP Cloud ALM for Implementation** app group and open the **Test Preparation** app. Click on the **Import** button in the top right-hand area. Make sure that the **Automation Provider TTA_for_SAP** is selected, as shown in Figure 12.11.

Select the test case to be imported and start the import by clicking the **Import** button. In the next step, select the scope and solution process for which the test case is to be imported. Imported test cases are displayed in the **Imported** area, as shown in Figure 12.12.

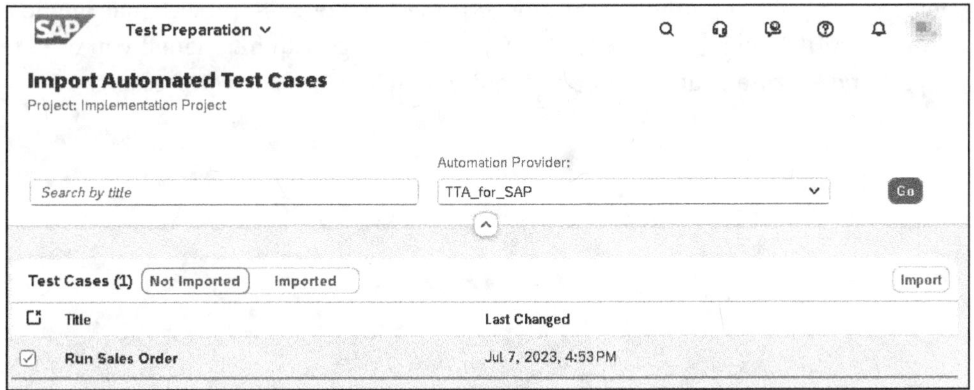

Figure 12.11 Import of Automated Test Cases

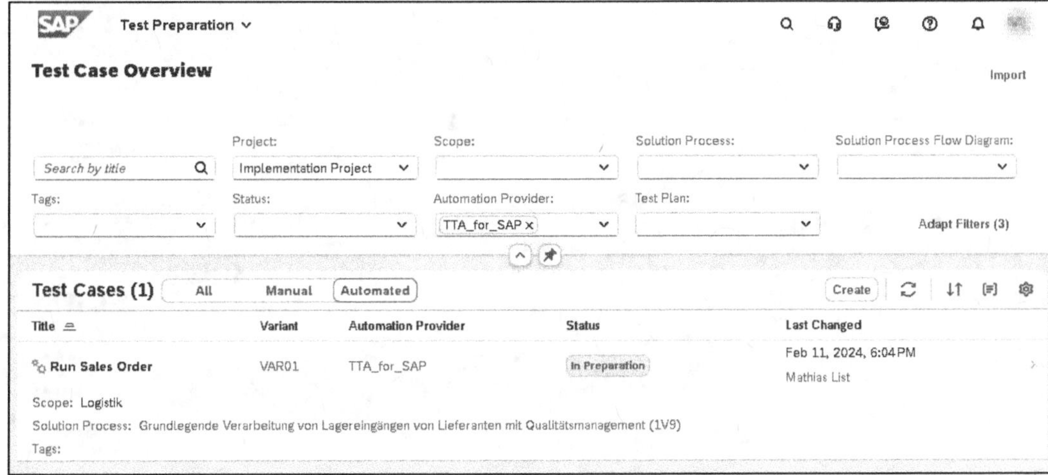

Figure 12.12 Imported Test Case of SAP Test Automation by Tricentis

You can find out how to assign these test cases to requirements or user stories or bundle them into test plans to organize your testing in Chapter 14.

12.4 Summary

Based on the fundamentals of the explore phase (evaluate), including all defined solution processes (see Chapter 9), the specified requirements, and associated user stories (see Chapter 10), you can now start thinking about the proper concept of test management. In this chapter, we provided you with the theoretical principles that play a role in *test preparation* and *test execution* in the realize phase. We also explained the technical means for executing automated tests. With this chapter, we have completed the most important activities of the explore phase for the implementation project of our example scenario. In the following chapter, we will move on to the realize phase and explain the processing of requirements and user stories in the agile project management approach. At this point, you can apply on the knowledge acquired here if you need to prepare and execute your tests during requirement and user story processing.

PART III

Realize

Chapter 13
Agile Implementation of User Stories

This chapter represents the beginning of the realize phase according to the SAP Activate methodology. We demonstrate agile implementation using user stories and take a closer look at the creation, specification, sprint allocation, and completion of user stories.

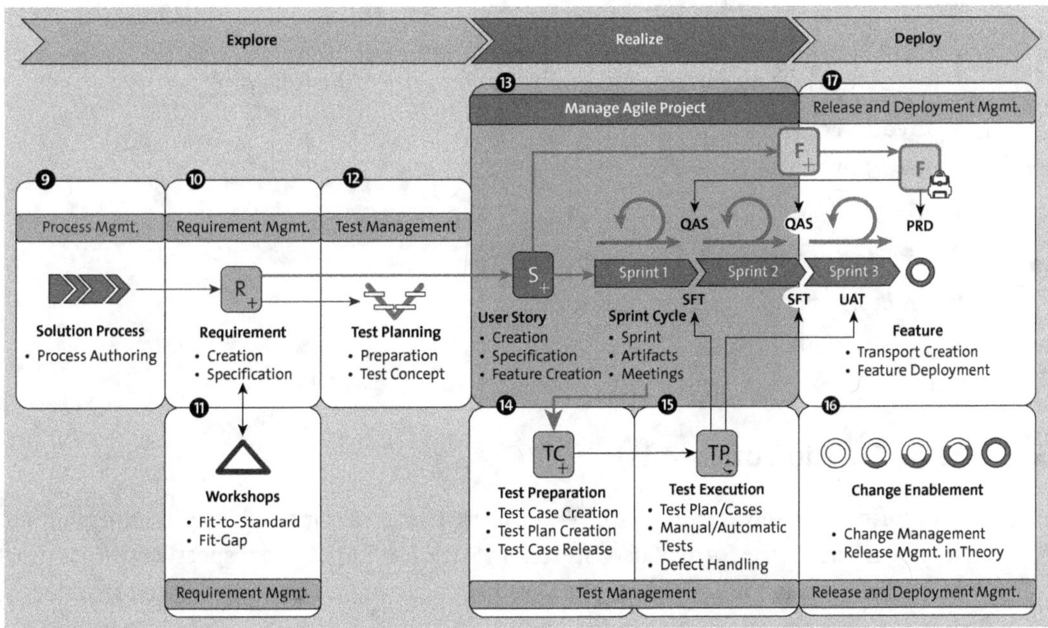

Within the explore phase, we looked in Chapter 9 and Chapter 10 at the business context of the target solution (in our example, the logistics line of business). The solution processes and the registered initial requirements play a key role here. The requirements represent functions within the solution processes that are to be implemented by the development team as part of the realize phase. The product owner creates the user stories from these requirements during the explore phase, taking into account the definition of readiness criteria. The set of finished solution processes, requirements, and user stories therefore forms the basis of the solution to be implemented. Based on our example scenario, this chapter takes a detailed look at the Scrum project methodology with the associated meetings, artifacts, and roles. The focus is on the agile implementation of the user stories. Please consider the working method of the agile project methodology as described in Chapter 5.

There are already a large number of ALM tools on the market that aim to manage agile project methodology with software support. With SAP Cloud ALM, SAP is providing its own tool that differs from the current solutions in the following ways:

- Ability to import SAP Best Practices processes for the solution to be implemented
- Consolidated processing of requirements from analysis to productive usage (requirements to deploy approach)
- Many possible integration options thanks to a large number of APIs
- End-to-end integration right down to the transport layer

This chapter builds on the basic agile principles of Scrum with its artifacts, roles, and events and presents the functional possibilities of SAP Cloud ALM in a practical way. The structure of this chapter corresponds fundamentally to the recommendation on how a Scrum setting can be implemented. We address the following points:

- Creation of sprints
- Creation of user stories
- Assignment of user stories to sprints
- Division of user stories into subtasks
- Processing user stories within sprints
- Completion of sprints

13.1 Creation of Sprints

The sprint is a central event within the Scrum methodology and is used to implement requirements in the form of user stories. In SAP Cloud ALM, the respective sprints must be defined in advance by the responsible product owner. They are planned seamlessly, starting with the sprint planning meeting and ending with the sprint review and the subsequent sprint retrospective (see Chapter 5, Section 5.1).

To ensure that the sprint characteristics fit adequately into the release cycle, the sprint setup should be coordinated with the responsible release management team. The important aspects that should be clarified in advance include the length of the sprints, the number of sprints per release, and the underlying implementation landscape. Detailed information on this can be found in Chapter 16, Section 16.4.1.

As a first step, sprints must be created in SAP Cloud ALM as timebox elements. They are created within the Projects and Setup app out of the **SAP Cloud ALM for Implementation** app group (see Figure 13.1).

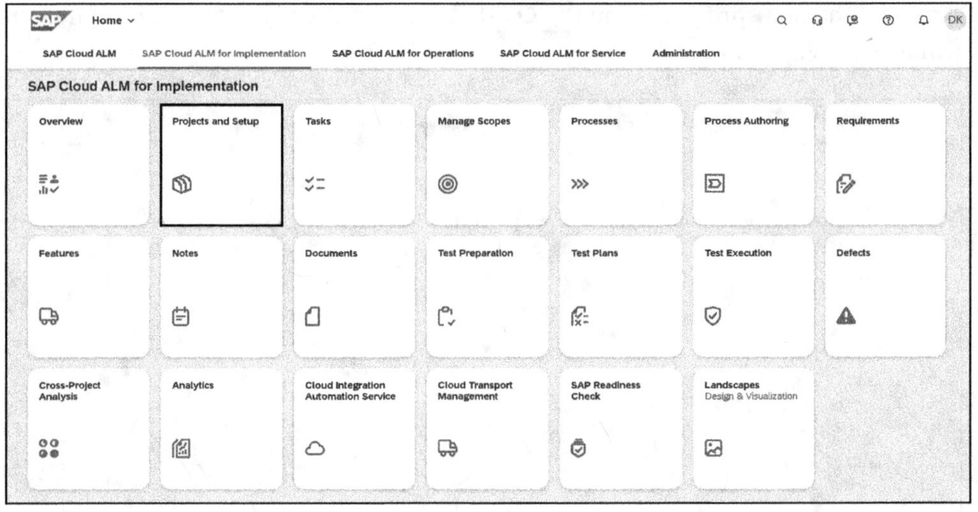

Figure 13.1 SAP Cloud ALM for Implementation App Group

First, select the desired project from the project list (see ❶ in Figure 13.2). For our example scenario, we first want to define the sprint structure for the release 1 template project and select therefore **AriHana-2023-TPL-R1** (see ❷ in Figure 13.2). Then switch to the **Timeboxes** tab. In the **Sprints** section, you will be informed that you can add new sprints by clicking the **Edit** button (see ❸ to ❺ in Figure 13.2).

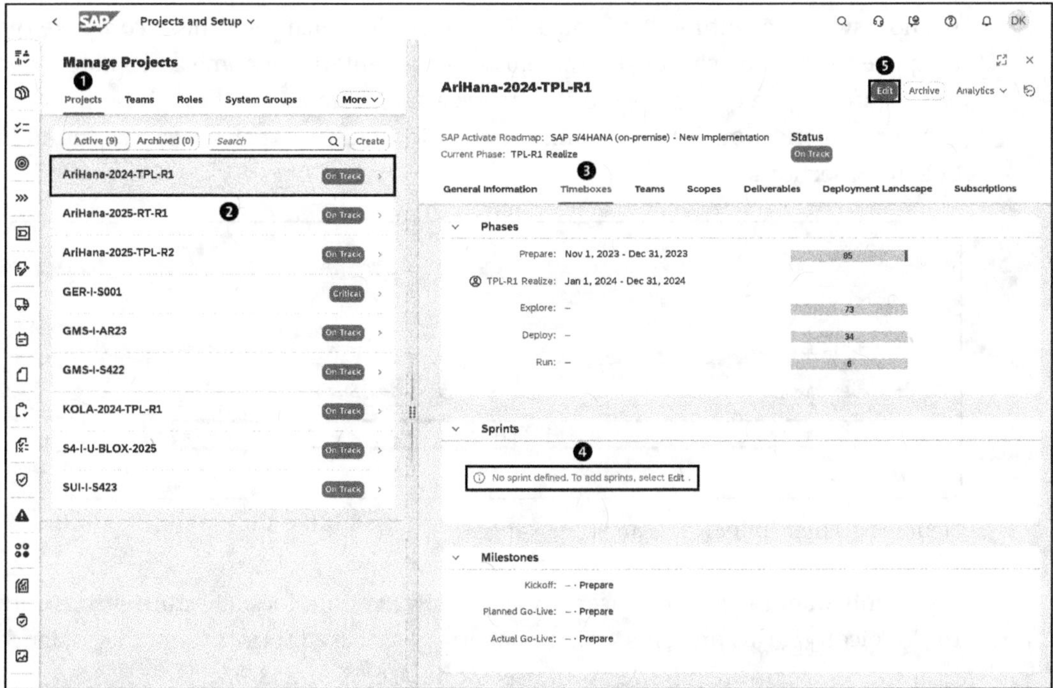

Figure 13.2 Timeboxes and List of Sprints Defined in Projects and Setup App

To create the first sprint, click on the **Create** button in the **Sprints** section of the next window (see Figure 13.3).

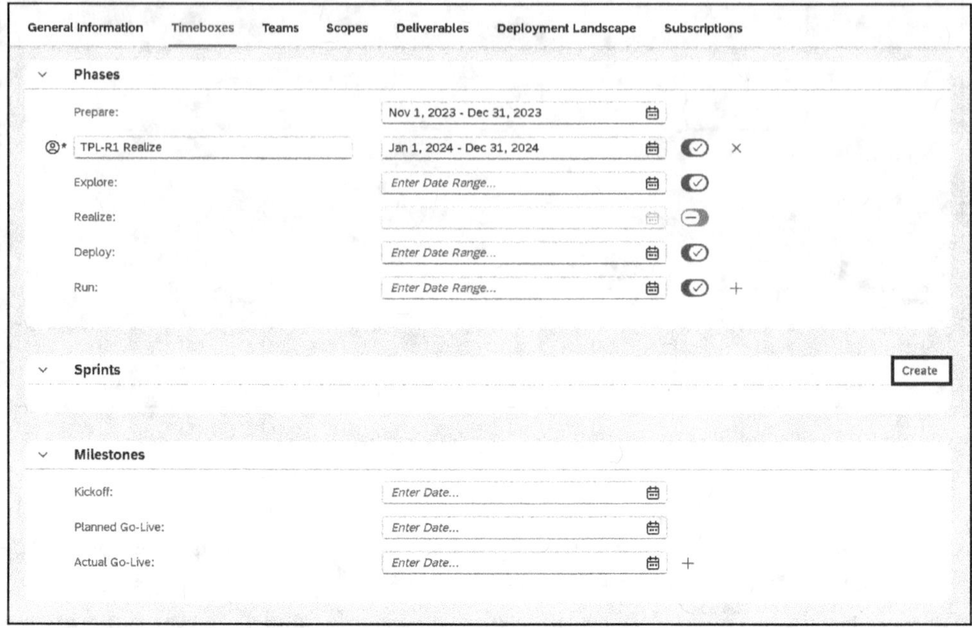

Figure 13.3 Create New Sprints

We now switch to edit mode to add all the sprints in which we want to complete the template release 1 by clicking on the plus icon and entering a name and a date range (see Figure 13.4). In our setup, the sprints are arranged monthly throughout the year. Our recommendation for the naming convention `TPL-R1 - Sprint *` can be found in Chapter 5, Section 5.4.2.

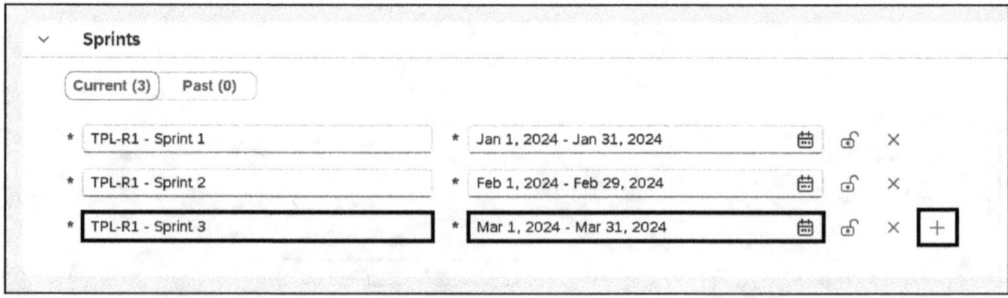

Figure 13.4 Enter Sprint Name and Date Range

If you only want to plan your user stories in active sprints, lock all future sprints with the lock icon 🔒. This prevents assignment errors, as locked sprints cannot be assigned to any tasks, such as requirements or user stories (see Figure 13.5).

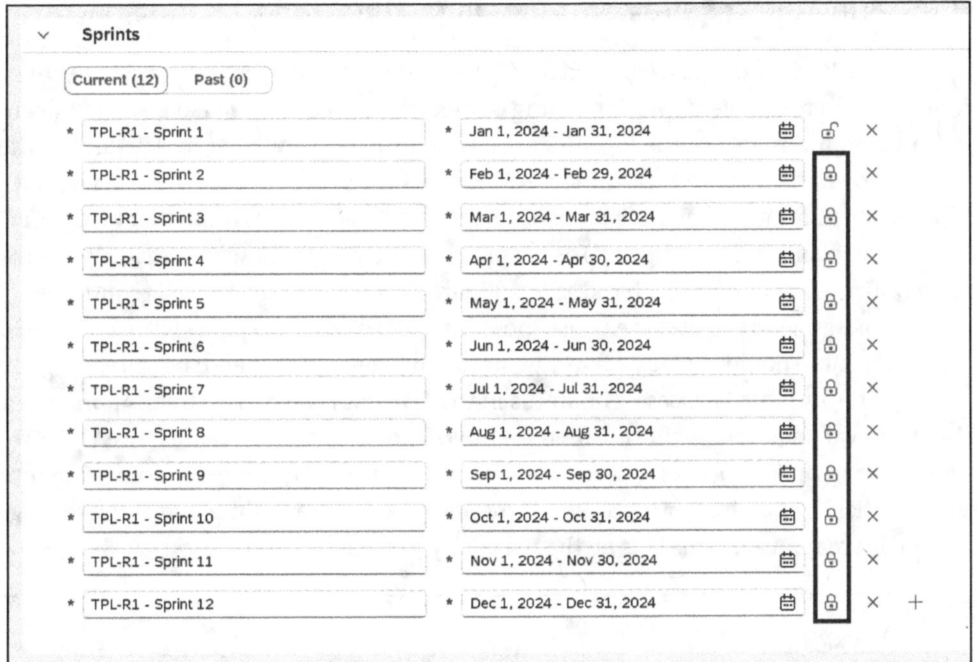

Figure 13.5 Timeboxes: Planned Sprints

In display mode, the sprints are grouped into current and past elements according to their defined date range (Figure 13.6 ❶). If the sprint range is within a defined phase, this is displayed directly after the respective sprint (Figure 13.6 ❷ and ❸).

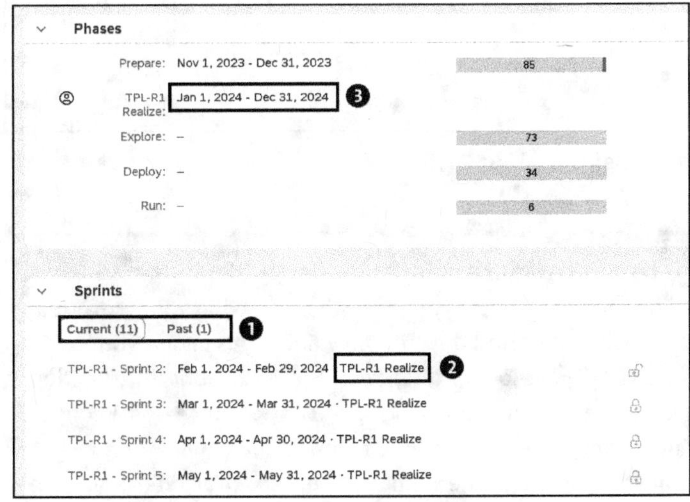

Figure 13.6 Timeboxes: Current vs. Past Sprints

13.2 From the Requirements Backlog to the Product Backlog

In Chapter 10, Section 10.4, we explained how to specify requirements and confirm or release them for implementation. Using our example scenario, we have already documented requirements in the logistics area, which we will now refer to. The task of the product owner is now to turn these requirements into corresponding user stories for their implementation and thus build up the product backlog. To implement this effectively, we recommend using the Tasks app to define a specific search that represents the confirmed requirements backlog (see Figure 13.7). We assume that all requirements are still assigned to the prepare phase (possibly also the explore phase). The confirmed requirements should already have the status **In Progress** if they have already gone through the approval process. The status in the **Approval** column is either **Approved** or **No Approval Required.** We recommend that you proceed as shown in Figure 13.7. Save your search for later use, as new requirements have to be continuously converted into user stories. In our search for a requirements backlog, the highlighted filter settings under ❶ are relevant. This lists all the necessary requirements, as shown in ❷.

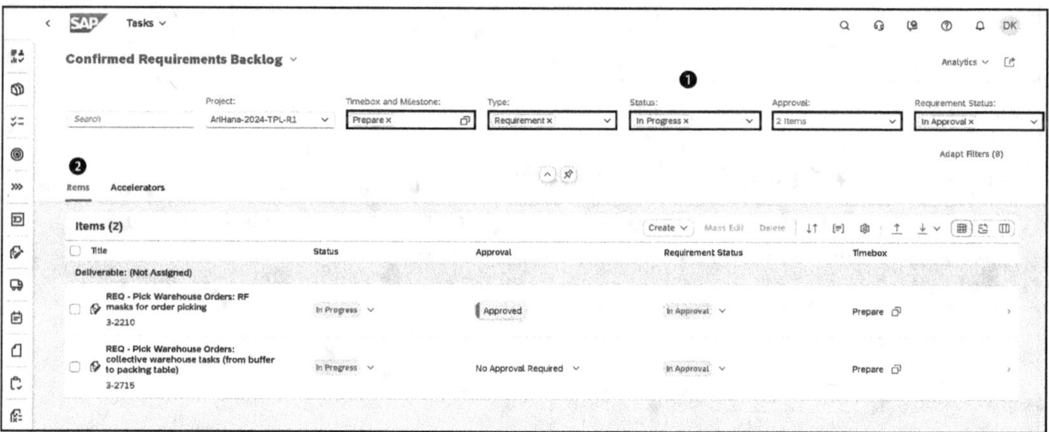

Figure 13.7 Settings of a Requirements Backlog

[»]

Save Search

In some search areas in SAP Cloud ALM, you have the option of saving your individual search settings with the displayed filters and the entered values. This allows you to call up these searches quickly and easily without entering the data again. But this function is not supported by all search areas. You will find it in the Tasks, Processes, Requirements, Documents, Test Plans and Defects apps. You can find the **Save** option by the downward-pointing arrow symbol next to the name in the header or filter area (see Figure 13.8 ❶). After clicking on the arrow, all the searches you have saved will appear in a list. The **Save As** button is also available there to save the currently defined search by using an individual name (see Figure 13.8 ❷).

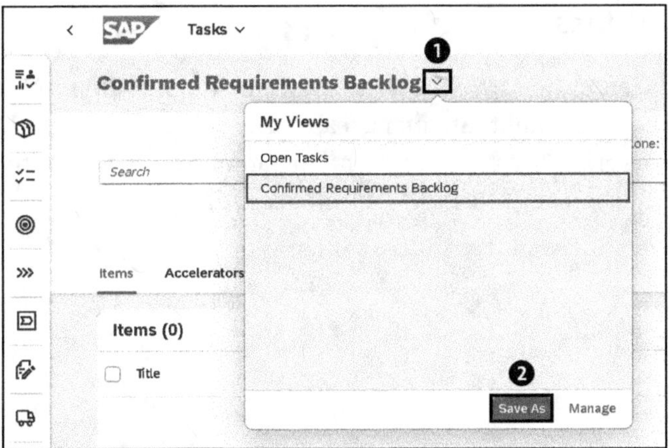

Figure 13.8 Save Defined Search for Quick Access

After the product owner has created one or more user stories out of a confirmed requirement, we recommend setting the related requirement to the realize phase. To do this, click on the search for the timebox element within the requirement and assign the corresponding realize phase **TPL-R1 Realize** from the selection of **Phases** (see Figure 13.9). This removes the requirement from the requirement backlog. Furthermore, all requirements in the realize phase indicate a reference to at least one associated user story. The creation of user stories is described in Section 13.3.1.

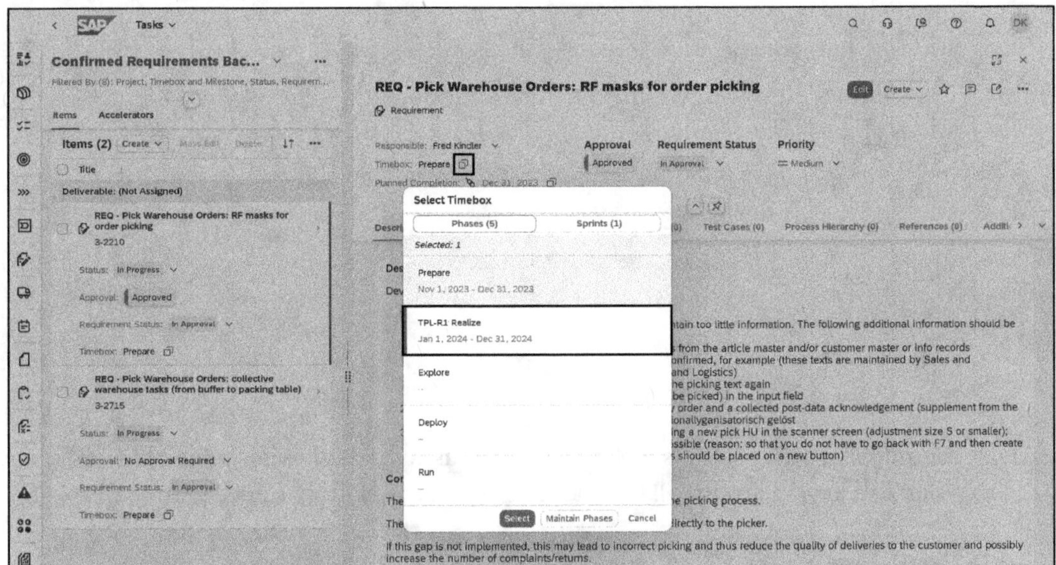

Figure 13.9 Assigning Requirements to the Realize Phase

13.3 Overview of User Stories

The *user story* is the central element of agile product development. SAP Cloud ALM offers the option of dividing a requirement into manageable user stories and subtasks that can be completed within a sprint. The user story contains, among other things, the following information, which plays a key role during sprint execution:

- Title
- Description
- Subtasks
- Project tasks
- Test cases
- Process hierarchy
- References
- Additional information
  - Status
  - Priority
  - Current phase
  - Story points
  - Effort
  - Release
  - Timebox (sprint)
  - Start date/planned completion date/duration
  - Team
  - Assigned role
  - User type
  - Responsible
  - Requirement
  - Feature
  - Tags

The status of a user story is reflected by various status values describing the workflow. A detailed explanation of all the status values of a user story can be found in Table 13.1.

Status	Description
Open	This represents the initial status that appears after the user story has been created. The user story is part of the product backlog until a sprint is added. The user story can be added to the sprint backlog but is not yet processed.

Table 13.1 Description of the Status Values of a User Story

Status	Description
In progress	This status is set by the developer as soon as the user story is being processed during the sprint execution. The user story should be assigned to a sprint.
Locked	This is a temporary status that indicates that processing of the user story is currently blocked by a circumstance.
Finished	This status indicates that the user story has been processed by the developer and that the DoD criteria have been met.
Not relevant	It can happen that user stories no longer play a role in the implementation of requirements. Therefore, this status is selected to exclude the user story from the sprint or product backlog. The user story can also be reactivated if required.

Table 13.1 Description of the Status Values of a User Story (Cont.)

13

It is possible to reach the various status values regardless of the currently set value. The most common workflow is presented in Figure 13.10.

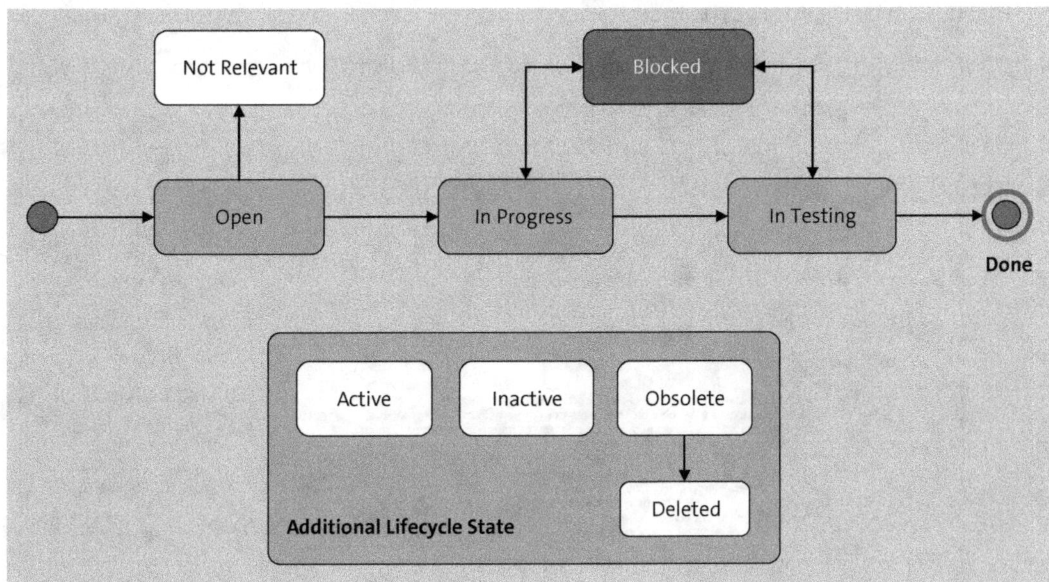

Figure 13.10 Common Workflow of a User Story

13.3.1 Create a User Story from a Requirement

In SAP Cloud ALM, user stories can be created flexibly and in different ways depending on your requirements. In general, the following options for creating a user story play a key role:

- **Creation from a requirement**
  In the *requirement to deploy* approach, you create the user stories directly from the solution process requirements to document the realization of the requirements.
- **Standalone creation**
  User stories can also be created individually without reference to the business process or a requirement. This can play a role in the following situations:
  - User stories that arise during implementation (e.g., measures that were not considered during planning)
  - User stories that arise during the sprint review meeting when the product increment is presented
  - User stories that arise from overarching topics

In the following, we will use our example scenario to show you how to create a user story from a requirement.

As described in Section 13.2, the product owner uses the requirements backlog to create a corresponding user story for a confirmed requirement. In our example, this concerns the requirement **REQ - Pick Warehouse Orders: RF masks for order picking**. The product owner now selects the entry **User Story** from the selection list presented by the **Create** button (see Figure 13.11).

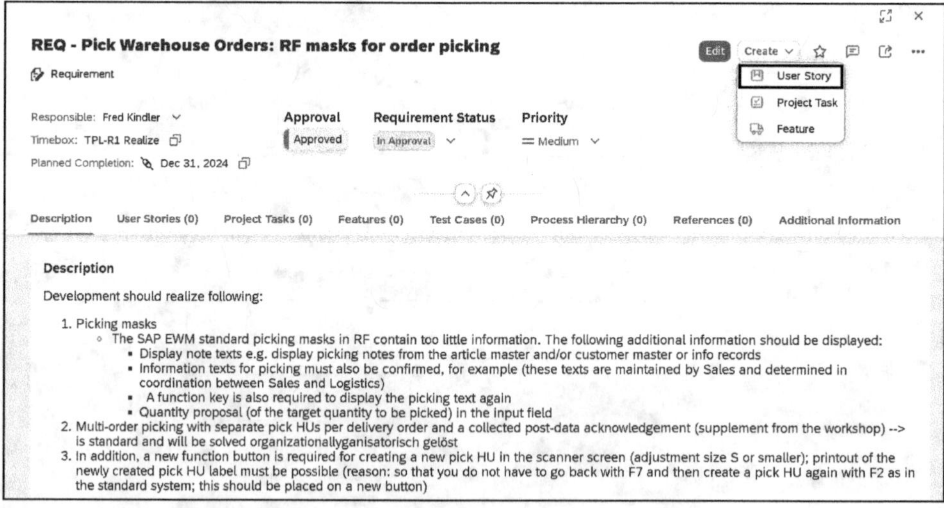

Figure 13.11 Creating a User Story out of a Requirement

The user story creation form is displayed. The **Title** field is mandatory and, similar to the requirement, we recommend adding a prefix to the title. We opt for the prefix **URS**. This allows the elements to be quickly distinguished from each other in generic searches. The prefix is followed by the technical action of the story. In our example, we name the user story as follows: **URS - Specification of SAP fields for data retrieval** (see Figure 13.12 ❶).

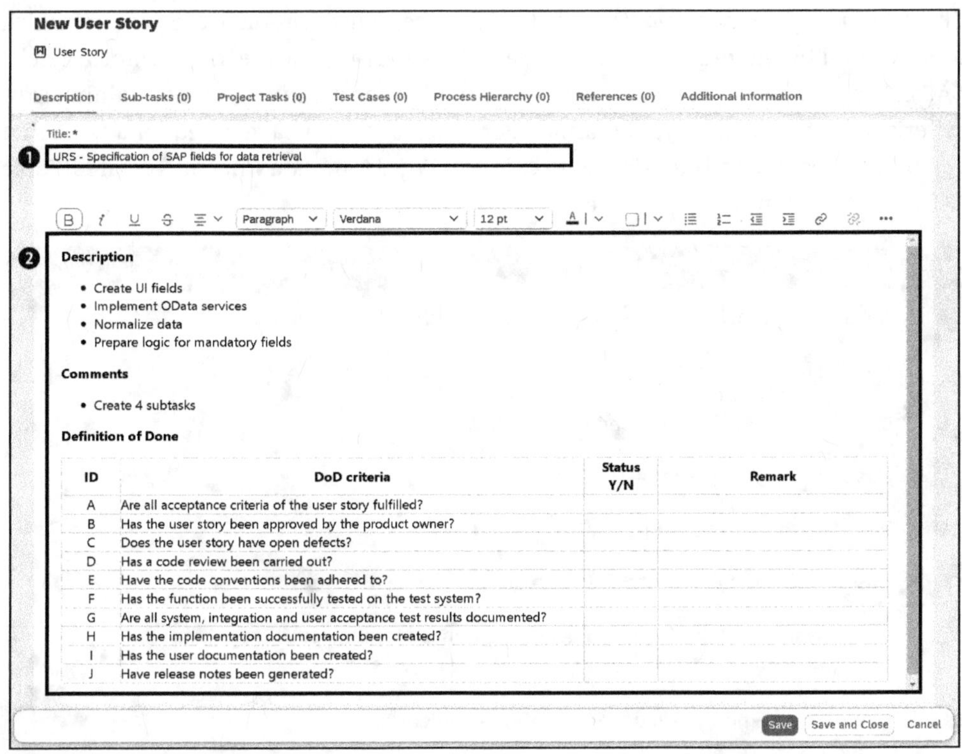

Figure 13.12 Create User Story

The content of the user story is described in detail within the **Description** section. If you want to proceed strictly according to Scrum, the description should be entered in user story format. The user story is formulated in a sentence that follows this format:

As a [user, role], I want [functionality], so that [value, benefit].

In other words, a user story answers the question: *Who* wants *what* and *why*? It is usually the task of the product owner to formulate this sentence when creating the user stories from the requirements. It is not uncommon that additional input from stakeholders and developers is added.

But in practice, the user story is rarely described consistently in user story format. It often happens that user stories are created by the development team and enriched with technical information about the planned work activities. As the person responsible for the company, you must find the best way for your project setting and describe it in the definition of ready criteria. In any case, we recommend establishing a standard so that the structure of the user stories is looking similar. In our example, we structure the user stories as follows (Figure 13.12 ❷):

- Description
- Comment
- Definition of done

The *definition of done criteria (DoD)* reflect the acceptance criteria of a user story, which are reviewed by the product owner during the sprint review meeting. Unfortunately, SAP Cloud ALM does not currently offer a function for defining and maintaining DoD criteria in a system-supported manner. For this reason, we have opted for a table that can be inserted into the description text from a Word template when it is created. Table 13.2 provides examples of possible DoD criteria.

ID	DoD Criteria	Status Y/N	Remark
A	Are all acceptance criteria of the user story fulfilled?		
B	Has the user story been approved by the product owner?		
C	Does the user story have open defects?		
D	Has a code review been carried out?		
E	Have the code conventions been adhered to?		
F	Has the function been successfully tested on the test system?		
G	Are all system, integration, and user acceptance test results documented?		
H	Has the implementation documentation been created?		
I	Has the user documentation been created?		
J	Have release notes been generated?		

Table 13.2 Example of Definition of Done Criteria

If you want to create a user story as a standalone element, open the Tasks app of the **SAP Cloud ALM for Implementation** app group. In the header area of the app, you must first set the **Type** filter to **User Story** (see Figure 13.13 ❶).

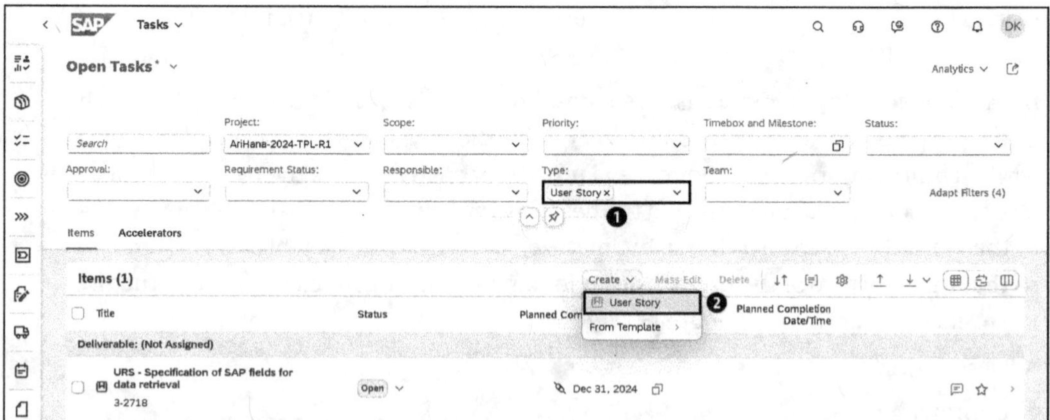

Figure 13.13 Standalone Creation of a User Story

You can then use the **Create** button to create a new user story without any reference to a requirement or solution process (see Figure 13.13 ❷).

Using the standalone creation, you also have the option of transferring predefined information from a template (see Figure 13.13, the **From Template** option). For the user story, this includes the title, description text, subtasks, scope, work area, priority, story points, effort, timebox, team, assigned role, and tags.

You can create own user story templates within the **Additional Information** tab using the **Use this item as a template** checkbox (see Figure 13.14).

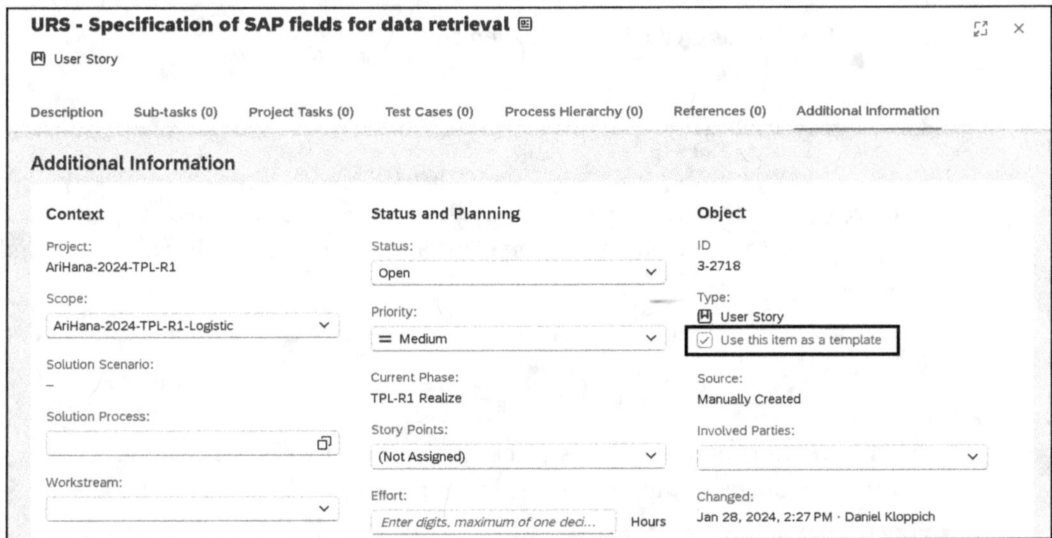

Figure 13.14 Create User Story Template

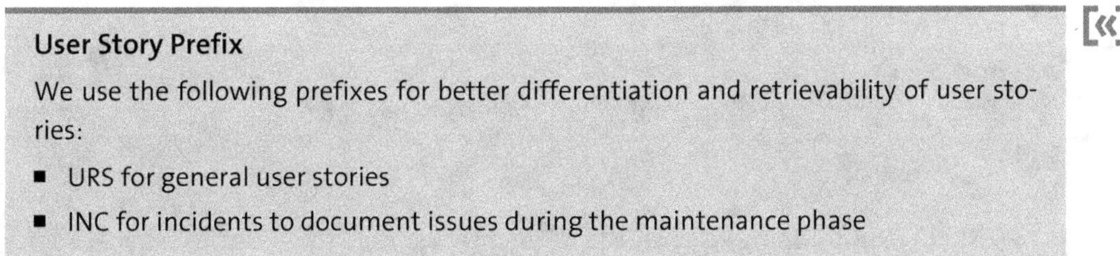

User Story Prefix

We use the following prefixes for better differentiation and retrievability of user stories:

- URS for general user stories
- INC for incidents to document issues during the maintenance phase

Once all user stories have been created out of a requirement, they are displayed in the corresponding user story list. Here you also have the option of assigning other user stories (e.g., manually created ones) or removing the assignment again via the icon ☒ (see Figure 13.15).

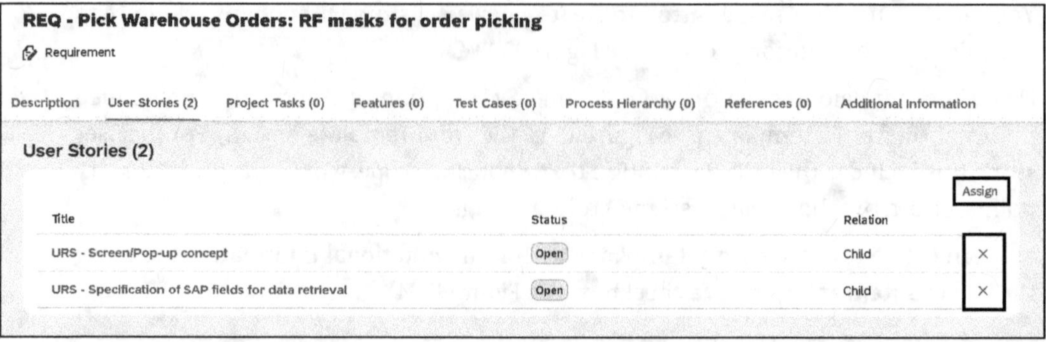

Figure 13.15 User Stories Created for a Requirement

13.3.2 Specify User Story

Once a user story has been created, further information must be added so that the definition of ready criteria are met. Some information is already transferred out of the original requirement to the created user story. This includes the scope, the workspace, and the tags.

The following points should be specified as a minimum in the **Additional Information** section (see Figure 13.16):

- **Priority**
  Prioritization of the user stories to be implemented within the product backlog according to urgency, importance, benefit, for example:
  Very High | High | Medium | Low
- **Story Points**
  Unit of measurement for assessing the complexity of a task, for example:
  1 | 2 | 3 | 5 | 8 | 13 | 20 | 40 | 100
- **Effort**
  Effort in hours required to implement the user story including the associated subtasks
- **Tags**
  Possibility of grouping user stories (e.g., in epics that represent extensive topics, functions, or work areas)
- **Timebox**
  Preselection of user stories by the product owner, which are to be prioritized for implementation according to the product owner's evaluation

You can also reference the following properties on the tabs of a user story:

- **Project Tasks**
- **Test Cases**

- Process Hierarchy
- References

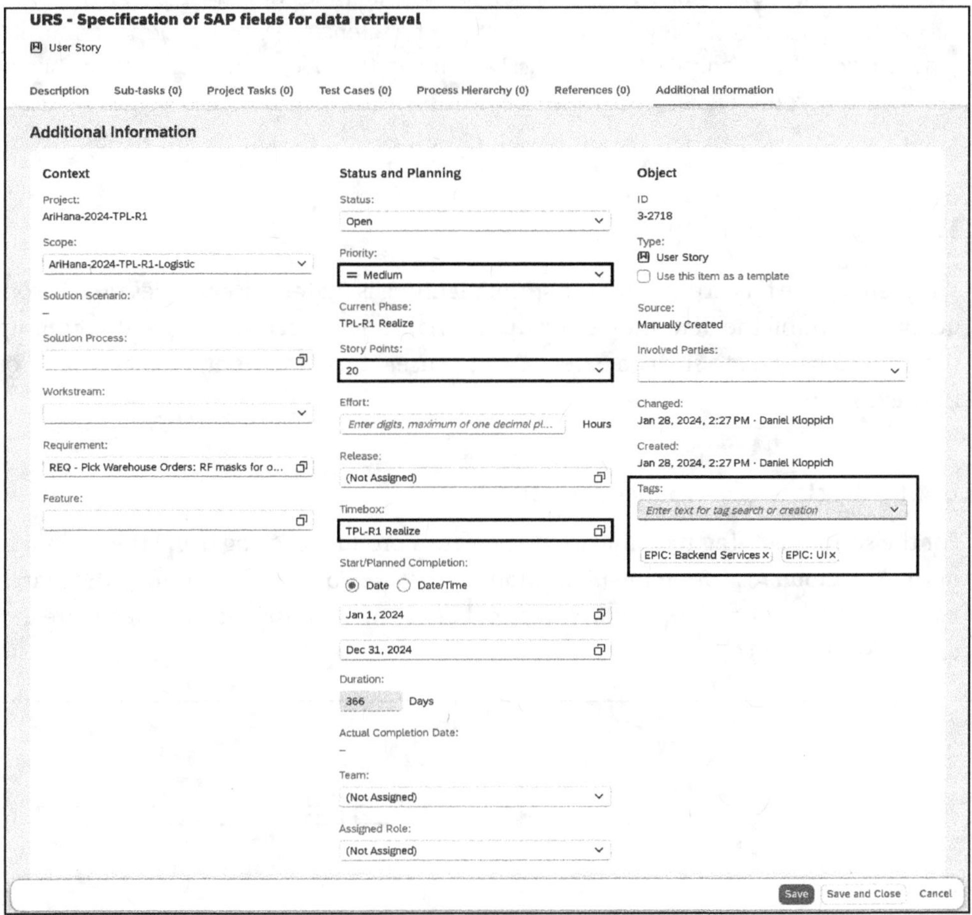

Figure 13.16 Specify User Story

If the product owner has maintained all the relevant information and this user story now meets the defined definition of ready criteria, it is ready for the next sprint planning meeting.

[«]

> **Project Tasks**
>
> In SAP Cloud ALM, it is not possible to link user stories with each other to make dependencies visible. You can use a project task for this purpose. This can serve as a bracket to link several user stories. In this way, a *Scrum of Scrums* (overarching team clarification and coordination with representatives from all teams) can track interdependent user stories in a separate backlog. The main advantage over using tags is that you can specify the status and log information.

[»]

References

SAP Cloud ALM does not offer the option of attaching files directly. It only supports links to other sources. This gives you the option of referencing further information such as functional or technical specifications, implementation documentation, instructions, or manuals from your document management system, such as SharePoint.

13.4 Sprint Planning

The prerequisite for carrying out the sprint planning is that the user stories in the product backlog fulfill the definition of ready criteria. The product owner has also built up the product backlog to such an extent that the user story stock is at least sufficient for the next sprint.

13.4.1 Structure of a Product Backlog

For the sprint planning meeting, we first create a product backlog using the Tasks app of the **SAP Cloud ALM for Implementation** app group. To do this, we define the search parameters so that only open user stories that are not yet assigned to a team are displayed (see Figure 13.17 ❶).

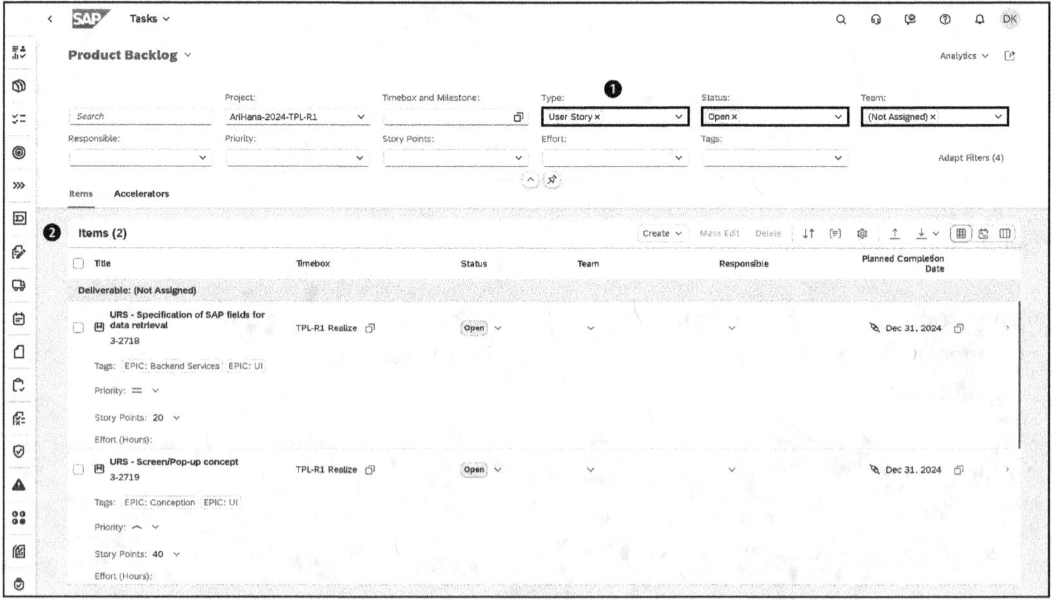

Figure 13.17 Structure of a Product Backlog

We assume that we will only assign responsibilities during the sprint planning meeting. All user stories are now listed accordingly in the results list (Figure 13.17 ❷). If a sprint assignment is already visible in the column **Timebox**, the user story has already been preplanned by the product owner. We recommend saving the search for upcoming planning meetings.

13.4.2 Assigning the User Story Selection to the Sprint

The high-priority user stories of the product backlog are now discussed together with the development team, taking into account the story points and the available capacities of the team colleagues. You can mark the selected user stories directly using the checkbox, as shown in Figure 13.18 ❶. You can use the **Mass Edit** function to assign the selection to the sprint in a consolidated manner (Figure 13.18 ❷ and Figure 13.19).

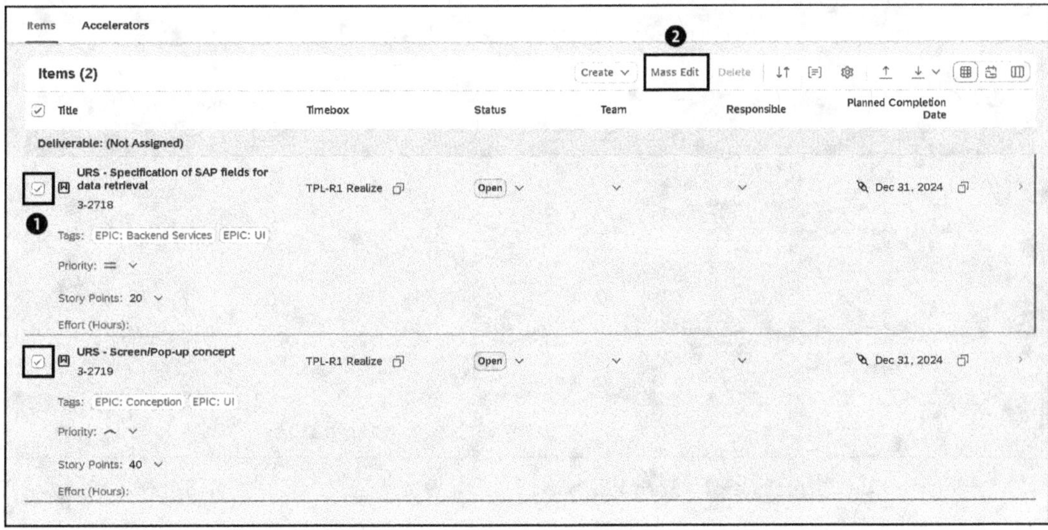

Figure 13.18 Select User Stories for the Next Sprint

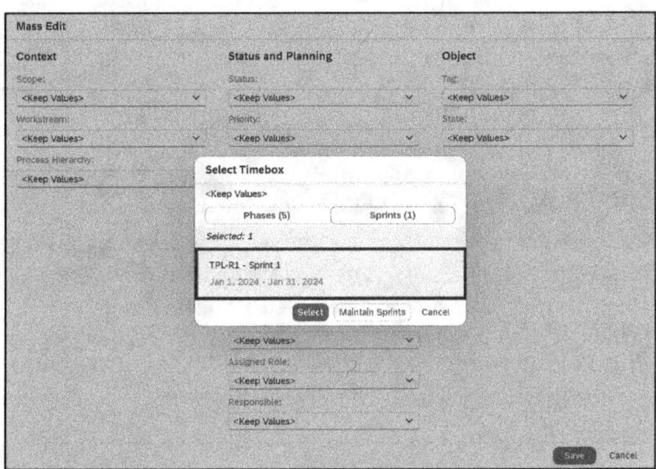

Figure 13.19 Assigning User Stories to the Sprint

13.4.3 Structure of a Sprint Backlog

The user stories were assigned to the sprint in advance. It now makes sense to create a *sprint backlog* in addition to the product backlog. This helps to better manage and track the worklist of a sprint. To do this, we also use the Tasks app of the **SAP Cloud ALM for Implementation** app group, only with different filter settings. We define the search so that it shows us all user stories of a specific sprint (see Figure 13.20 ❶). In the **Timebox** column ❷, the corresponding user stories are displayed.

Figure 13.20 Structure of a Sprint Backlog

13.4.4 Review of the User Story Selection

In the next step, the two parties (product owner and development team) take a closer look at the preselection of user stories to discuss the feasibility and the necessary steps for implementation. This involves looking at each user story in detail. Various aspects can be considered when determining the sprint goals (see Table 13.3).

Aspect of the User Story	Relevant Questions and Considerations
Content	Is the requirement clearly and comprehensibly defined, and do the developers know what is requested and what needs to be done? Further comments can be added to the user story if necessary.

Table 13.3 Define Sprint Goal

Aspect of the User Story	Relevant Questions and Considerations
Subtasks	For larger user stories, it is advisable to divide them into separate subtasks. This makes it easier to track the progress within the dailies than with a large user story that is in **In Progress** status for the entire sprint.
Story points and effort	The product owner may have roughly estimated the user story in advance to make a possible preselection of user stories to be implemented for the next sprint. Do the values previously estimated by the product owner need to be changed? By means of adequate estimates, better statements about the whole feasibility of all selected user stories can be made at the end of the planning phase.
Responsibilities	The responsibilities, such as the development team and the corresponding developers, are assigned directly to the user story. In an agile approach, it is also usual that the developers organize themselves and assign themselves to the user stories during the sprint execution according to demand or free capacity.
Capacity	Once all user stories have been considered, a final review of the selection takes place, taking into account the entered story points, the effort involved, the availability of the developers, and the activities to be carried out. If necessary, user stories are also put back into the product backlog or additional ones are added. The aim is to optimally fill the sprint backlog with user stories, taking feasibility into account.

Table 13.3 Define Sprint Goal (Cont.)

The sprint planning meeting is considered complete when no more user stories can be included in the sprint and all participants agree on the user stories assigned to the sprint. This defines the sprint goal.

Content of a User Story

We recommend that you clarify any questions regarding the content of the requirement in advance in the product backlog refinement meeting (see Chapter 5, Section 5.1.7). This usually takes a lot of time. It is therefore normally not possible to clarify requirements during the sprint planning meeting. It may also be necessary to ask the requesting stakeholders about the content in advance.

Allocation of Responsibilities

The responsible development team and the responsible developers can be assigned directly out of the sprint backlog. To do this, simply select the relevant team in the

Team column attribute of a user story using the arrow icon (see Figure 13.21). However, it is also possible to maintain the responsibilities on the **Additional Information** tab of the open user story.

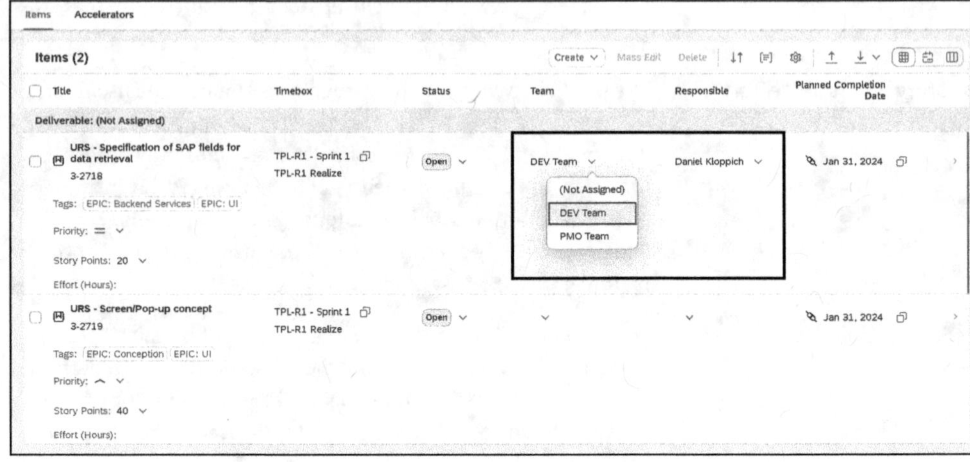

Figure 13.21 Sprint Planning: Assigning Responsibilities

Division of the User Story into Subtasks

As already mentioned, we recommend that you break down larger user stories into smaller subtasks. This allows you to better discuss the activities to be carried out in the daily Scrum meetings and track the degree of completion of the user story. The developer can also discuss the requirement in detail during the sprint planning meeting and describe the activities required for implementation. If you want to create a subtask for a user story, select the corresponding user story in the sprint backlog and click on the **Create Sub-task** button in the header area of the user story (see Figure 13.22).

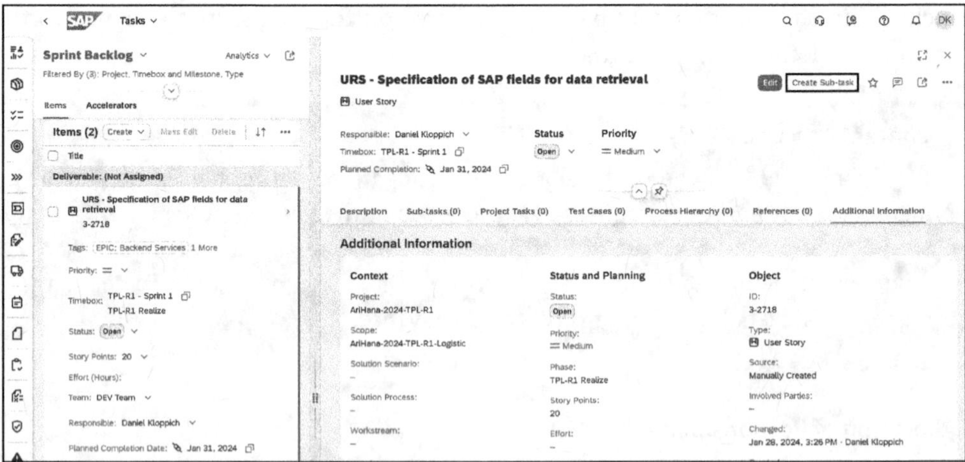

Figure 13.22 User Story: Create Subtask

The second option for creating a subtask is via the **Create** button of the **Sub-tasks** tab of the respective user story (see Figure 13.23). This button becomes available after switching into edit mode. Here you can create several subtasks directly, which are then displayed in list form. The sprint end date, which is taken from the user story, is automatically entered in the field of the **Planned Completion** column. You can simply fill in the fields for the **Team** and **Responsible** using the dropdown field of the related columns.

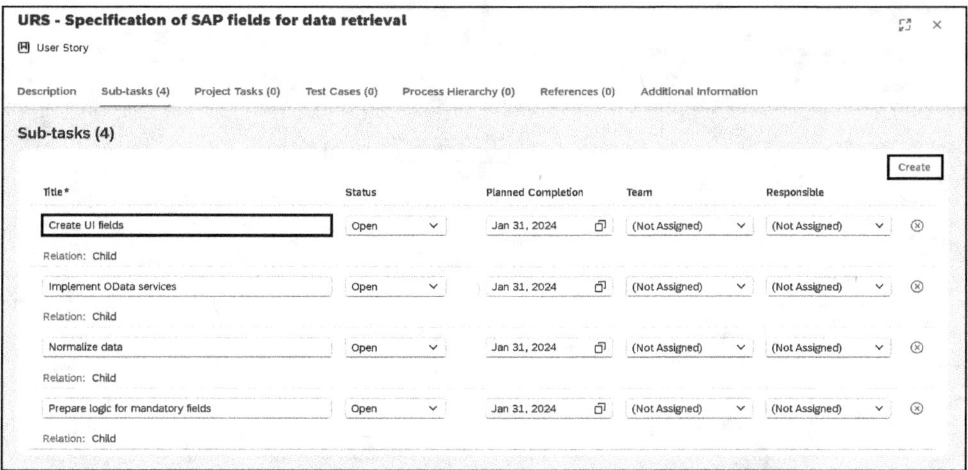

Figure 13.23 User Story: Create Subtask via the List

In the **Title** field, enter the name of the subtask as usual. We recommend that you use the first part of the title of the user story as a prefix, as shown in Figure 13.24. This increases the recognition effect with regard to the belonging of user stories and derived subtasks.

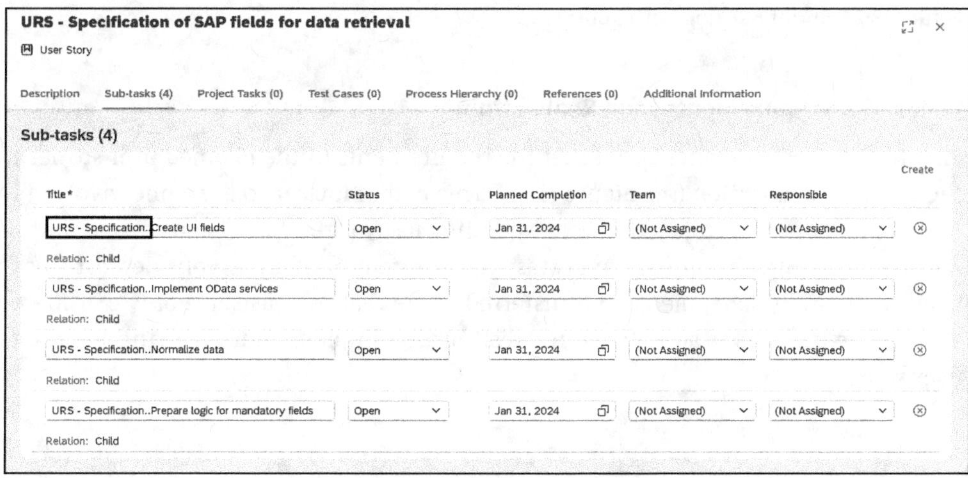

Figure 13.24 User Story: Specify Prefix for Subtasks

If you adhere to this naming convention, the user stories can be displayed bundled with their subtasks in the sprint backlog. To do this, add the **Sub-task** type to the filter settings of the sprint backlog (see Figure 13.25 ❶). In the settings for the results list, specify a sorting based on the title ❷. The degree of completion of the user story can now be better tracked based on the associated subtasks in the daily Scrum meeting ❸, as they are displayed in blocks based on the title.

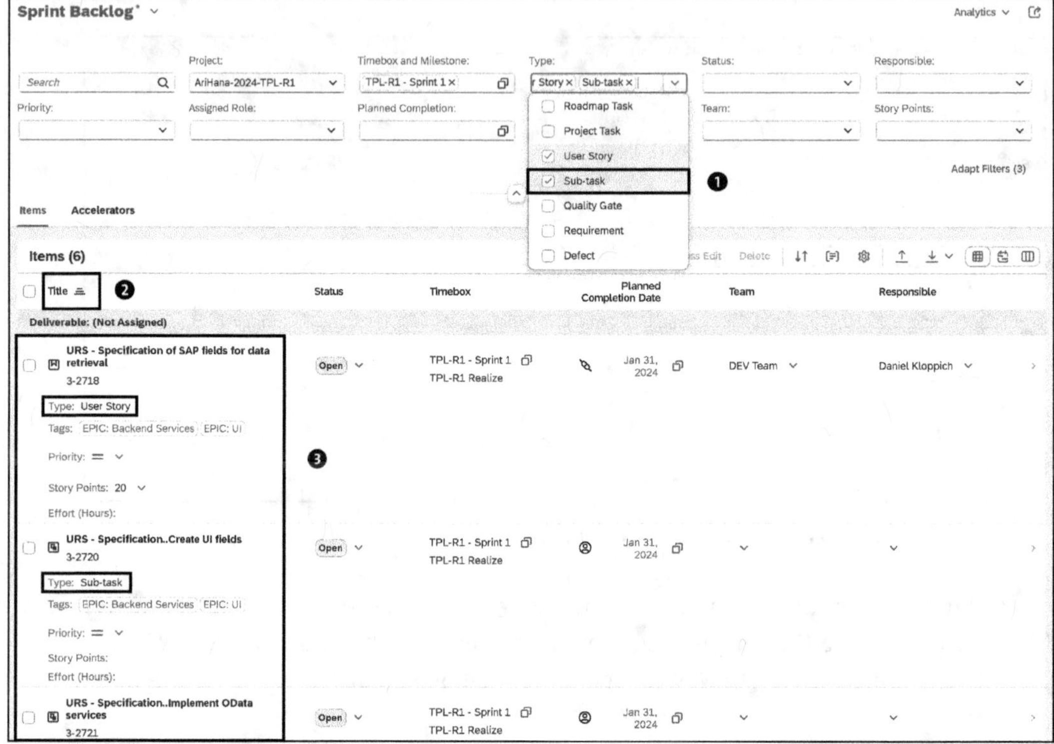

Figure 13.25 Sprint Backlog with Subtasks

13.4.5 Set Requirements to In Realization

Finally, the product owner should set the requirements of the planned user stories to the status **In Realization** (see Figure 13.26). This makes it clear to everyone involved in the implementation project that corresponding user stories have been created for this requirement. Furthermore, it indicates that the requirement is considered to have been successfully scheduled and transferred to the realize phase. It is only considered again when it comes to acceptance and testing after implementation by the associated user stories.

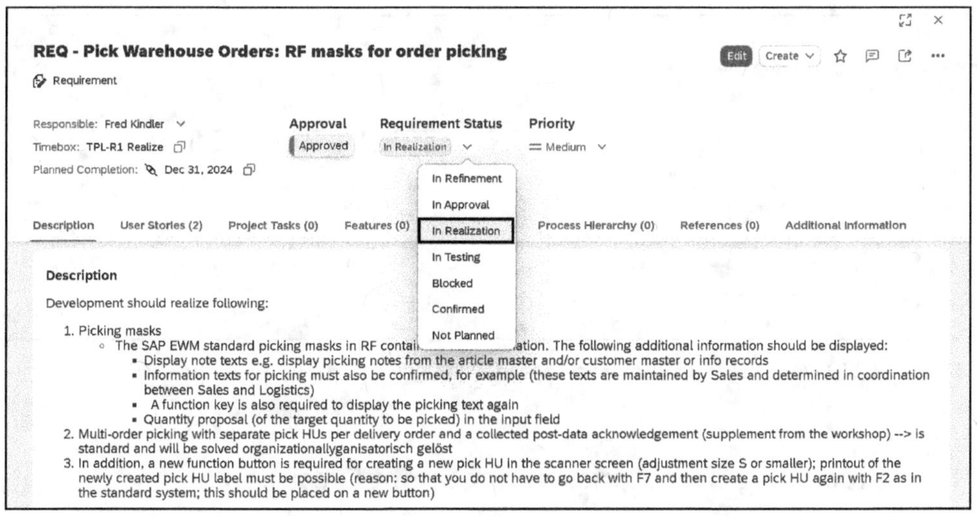

Figure 13.26 Sprint Planning: Set Associated Requirement to "In Realization"

13.4.6 Creating Features

In an agile context, it makes sense to create features as empty containers for implementing the requirements. *Features* are vehicles required for making the necessary changes in the form of transports in your system landscape. You should consider the following two options when creating a feature:

- The task of feature creation is assigned to the responsible product owner.
- The feature creation task is performed by the responsible release management team, which retains technical sovereignty over all change functions such as control, management, planning, and administration of the releases.

The advantage of both variants is that the developers do not have to worry about creating the features themselves during the sprint execution. They find a feature directly at the corresponding requirement from which they are able to create their needed transports for the implementation activities. Naming conventions or other documentation paradigms are therefore retained.

Ideally, a feature should be created during or shortly after the sprint planning meeting. This reduces the organizational effort directly in the sprint. Decide for yourself whether you want to provide the developers with the appropriate transports.

To easily list all requirements that are relevant for the creation of features, you can use the Analytics app available in the SAP Cloud ALM for Implementation app group. By selecting **Requirement Traceability**, you can display all requirements in status **In Realization** without a feature assignment (see Section 13.4.5) (see Figure 13.27). The creation of a feature is described in detail in Chapter 17, Section 17.2.2.

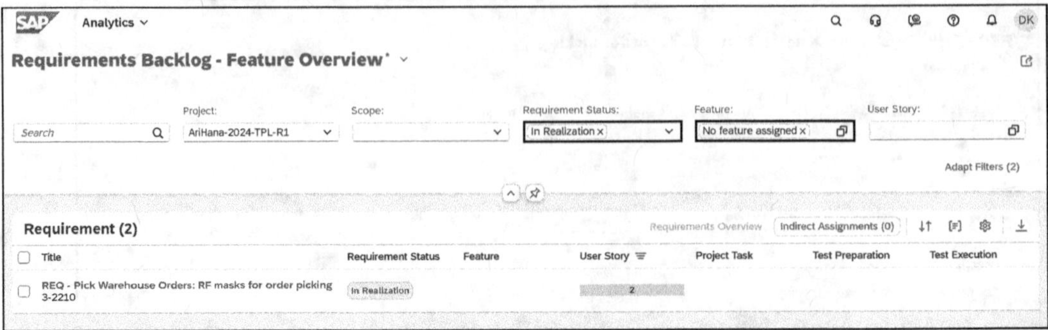

Figure 13.27 Overview of all Requirements to be Realized without a Feature

13.5 Processing User Stories during Sprint Execution

Following the sprint planning meeting, sprint execution begins. The developers implement their user stories and reflect on progress in daily meetings, the daily Scrums.

13.5.1 Processing the User Story by the Developers in the Sprint

To edit the user story, the developers open the current sprint backlog, independently select one of the user stories or subtasks to be implemented, and set the status to **In Progress** (see Figure 13.28 ❶ and ❷).

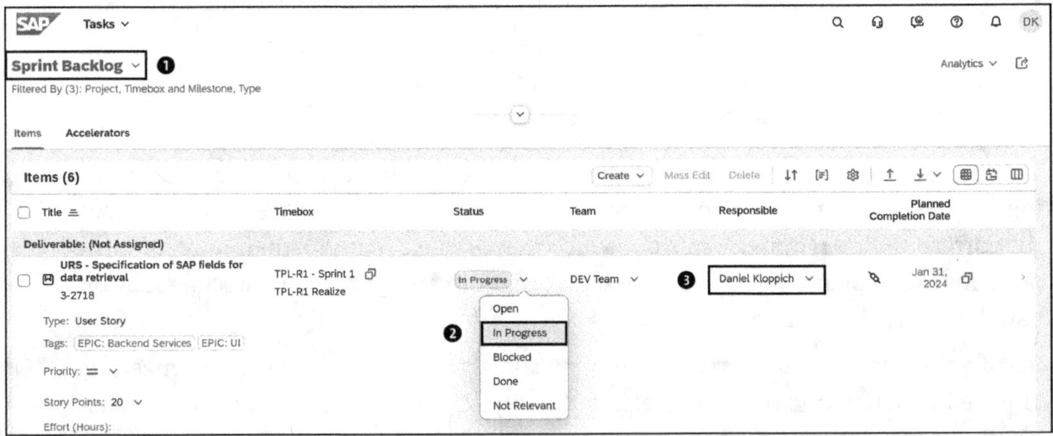

Figure 13.28 Sprint Execution: Developers Implement User Stories

If not already done during the sprint planning meeting (see Section 13.4.4), they also assign themselves as the person responsible via the selection list (see Figure 13.28 ❸). If a user story contains one or more subtasks, it is considered to be in progress as soon as the first subtask has been started and until the last subtask has been completed.

To enable developers to view and manage their specific worklist quickly and easily, they have the option to adjust the filter criteria of the sprint backlog so that only their own user stories and subtasks are displayed. To do this, you must enter following filter criteria (see Figure 13.29):

- **Type: User Story** and **Sub-task**
- **Responsible: Me**

If required, the **Timebox** (current sprint) and the **Status** (**Open** and **In Progress**) can also be used as filter criteria.

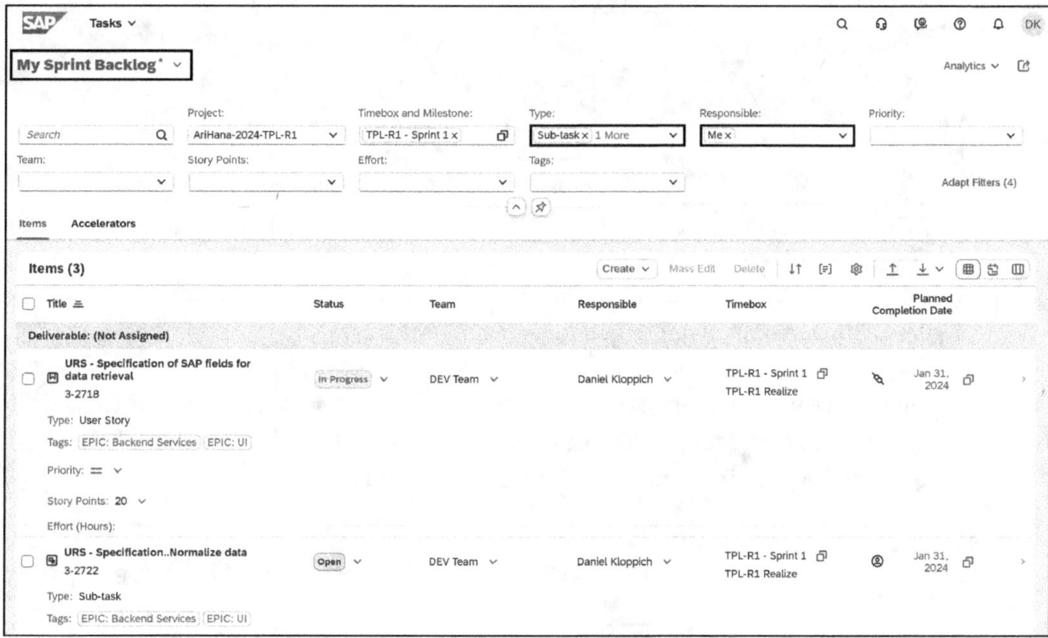

Figure 13.29 My Sprint Backlog: Worklist of a Developer

13.5.2 Creating Transports for Implementation

In general, developers require various transports on the corresponding development system for recording their implementation activities. The transports are necessary to save the changes of SAP systems and to transport or transfer them along the system line from the development system to the corresponding subsequent systems, such as the test and production systems. In this section, we present the necessary steps from the developers' perspective.

Navigate to the **Additional Information** tab of your user story (see Figure 13.30 ❶) and open the associated requirement via the link (see Figure 13.30 ❷).

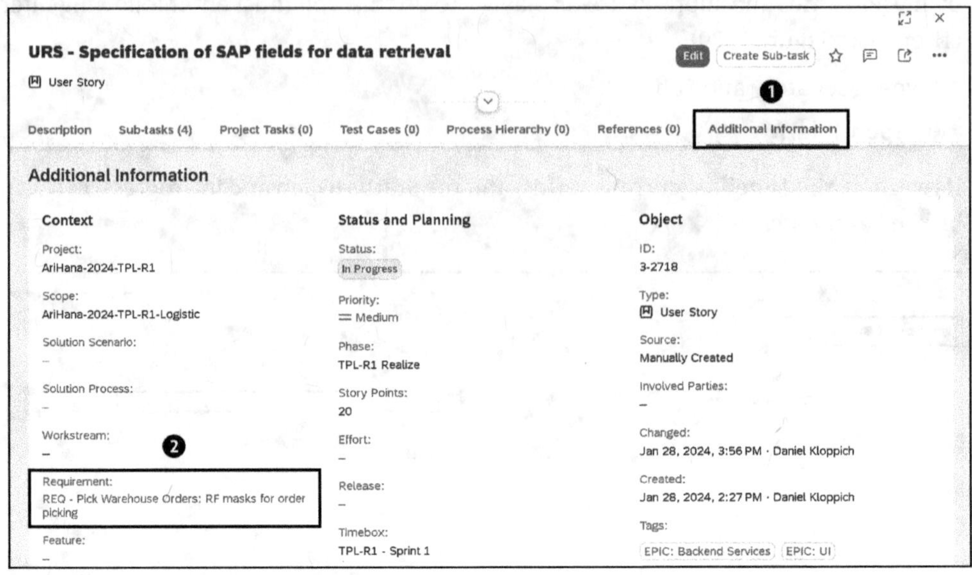

Figure 13.30 Open Associated Requirement from User Story

In the requirement that opens, switch to the **Features** tab ❶ and use the link to navigate to the corresponding feature ❷ (see Figure 13.31). At least one feature should already be assigned to the requirement, as described in Section 13.4.6.

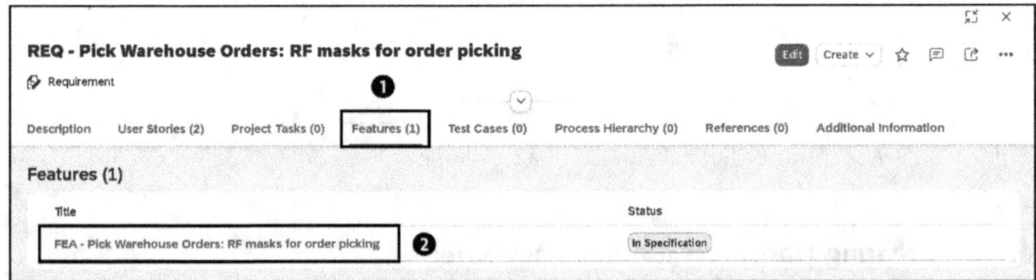

Figure 13.31 Open Associated Feature from Requirement

Now click on **Create • Transport** from the dropdown list, as shown in Figure 13.32.

In Figure 13.33, you can see the popup window for creating a transport. You can find detailed information on this in Chapter 17, Section 17.3.1.

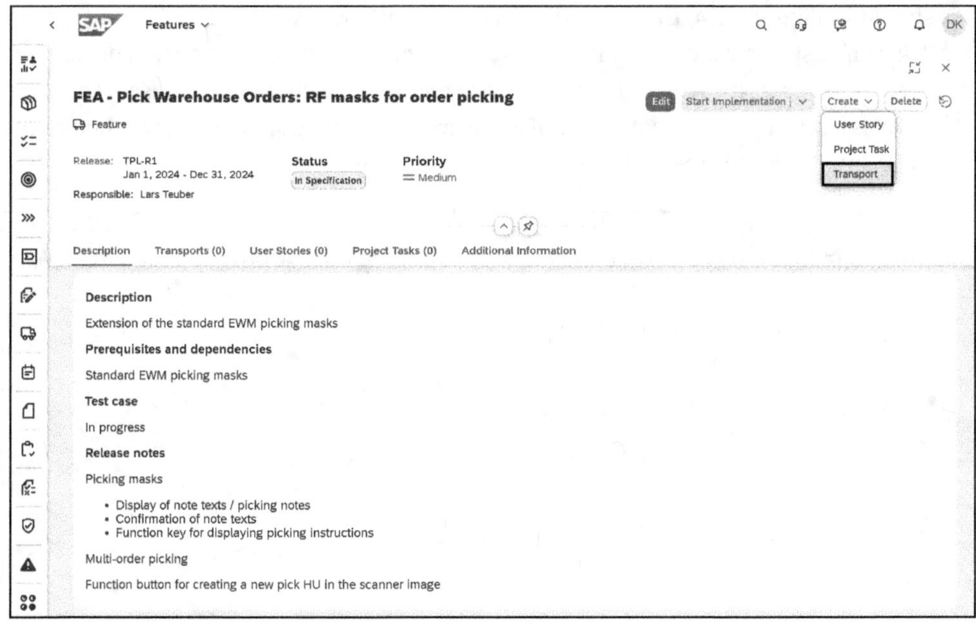

Figure 13.32 Implement User Story: Create Transport

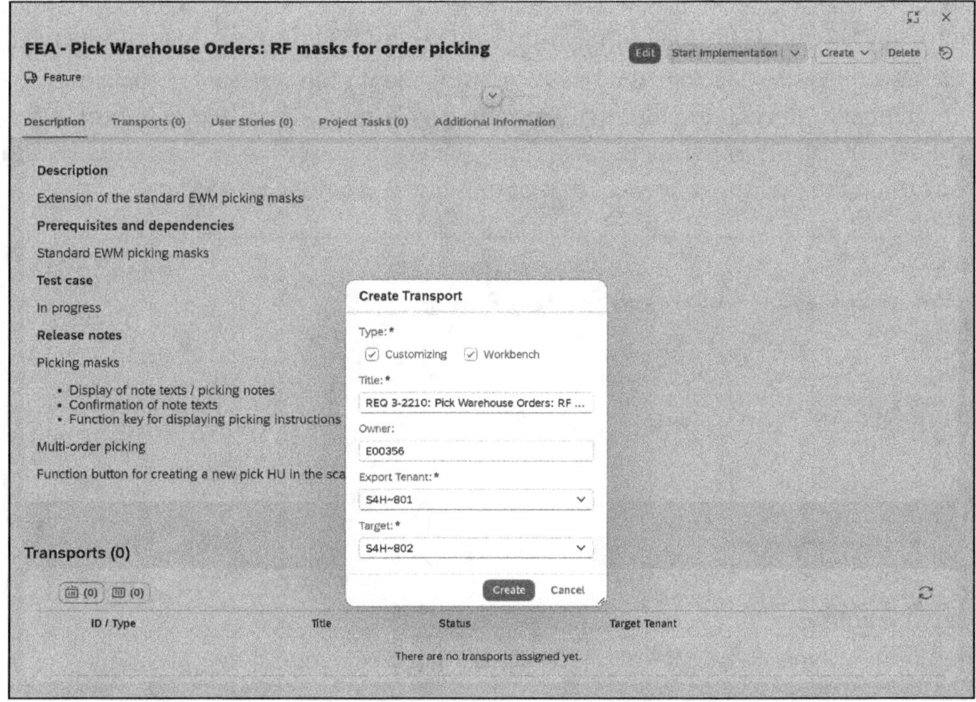

Figure 13.33 Implement User Story: Create Transport Popup

As shown in Figure 13.34, the created transports are now visible in the feature's transport list. The status **Modifiable** indicates to the developers that the transports have been created successfully, and they are able to record their changes to these transports. As soon as transports are created out of a feature, the feature should be switched to the status **In Implementation** via the **Start Implementation** action.

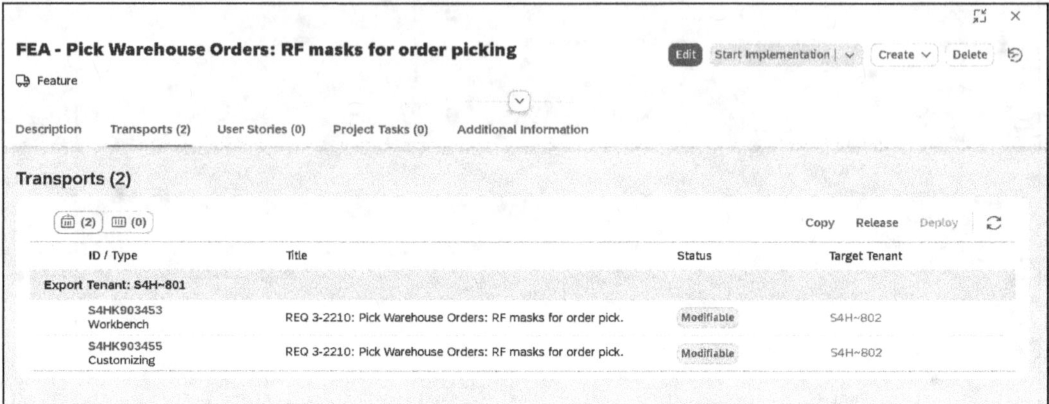

Figure 13.34 Implement User Story: Feature with Transport List

It makes sense for developers to link their user stories to the feature on the **User Stories** tab using the **Assign** button (see Figure 13.35). From a release management perspective, the advantage here is that the technical implementation (feature) is visible in the implementation documentation (user story). This aspect is particularly important for the delivery of technical functions. For the delivery of purely business context-related functions, the existing reference to the requirement is sufficient.

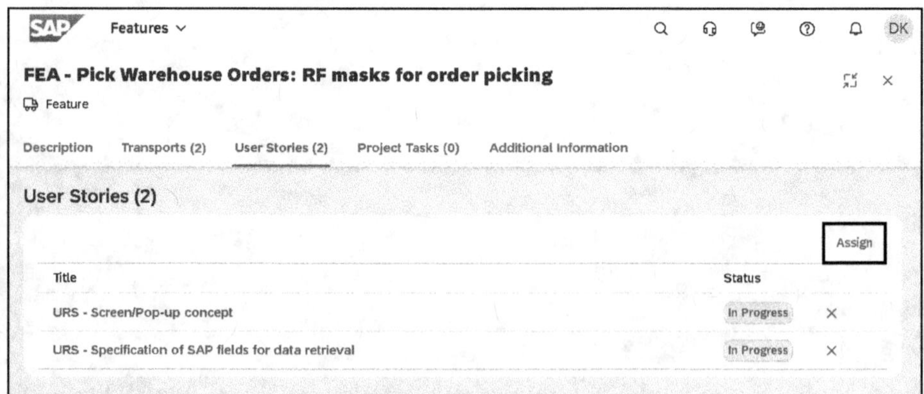

Figure 13.35 Implement User Story: Assign User Story to Feature

You can use the user story traceability function (in the Analytics app from the **SAP Cloud ALM for Implementation** app group) to easily display the user stories and the associated development components (features) (see Section 13.6).

13.5.3 Assigning Test Cases to a User Story

Tests are essential in software development, regardless of whether it is agile or traditional. In Chapter 14, we go into more detail regarding the creation of test cases and their link to the user stories. At this point, we would like to look at the linkage of user stories and test cases from the perspective of agile working methods. There are two basic approaches how test cases can be linked to a user story. The first is that during the implementation of the user story, corresponding test cases are created and assigned by the developer or by a responsible test management team, or that a separate user story deals with the topic of test script creation. The second option describes that existing test cases are merely linked to the user story. These test cases do already exist at the time of user story processing and are only adapted if necessary.

If you look at the granularity of test cases, a test case can be linked 1:1 to a user story. This is particularly the case if individual technical functions are to be tested independently of each other. The entirety of all test cases represents the requirement at a functional level. However, if a test case is assigned to a requirement, it covers the entire functional scope of a requirement from an end-to-end perspective.

Figure 13.36 shows how you can link a created test case to a user story via the **Test Cases** tab and the **Assign** button.

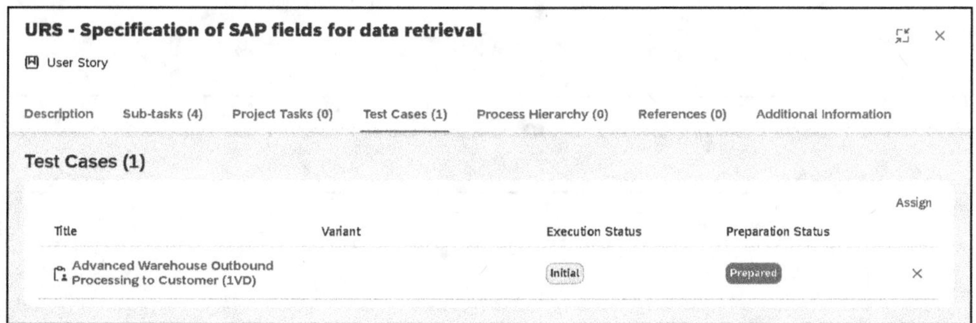

Figure 13.36 Assigned Test Case of a User Story

13.6 Daily Scrum: Daily Meetings

Daily Scrums are short daily meetings that extend over the entire sprint execution period (see Chapter 5, Section 5.1.5). In these meetings, the team members provide information on the progress of their work. To enable the *Scrum master* to moderate the meeting effectively, it is advisable to use the sprint backlog (see Section 13.4.3). Figure 13.37 shows an example of the sprint backlog with a completed subtask.

If you want to quickly display the completion rate of an user story's subtask, you can add the **Completion** column attribute to the list via the list settings ⚙ (see Figure 13.38).

Within the user story itself, a completion overview indicator for all corresponding subtasks is presented on the header, as shown in Figure 13.39.

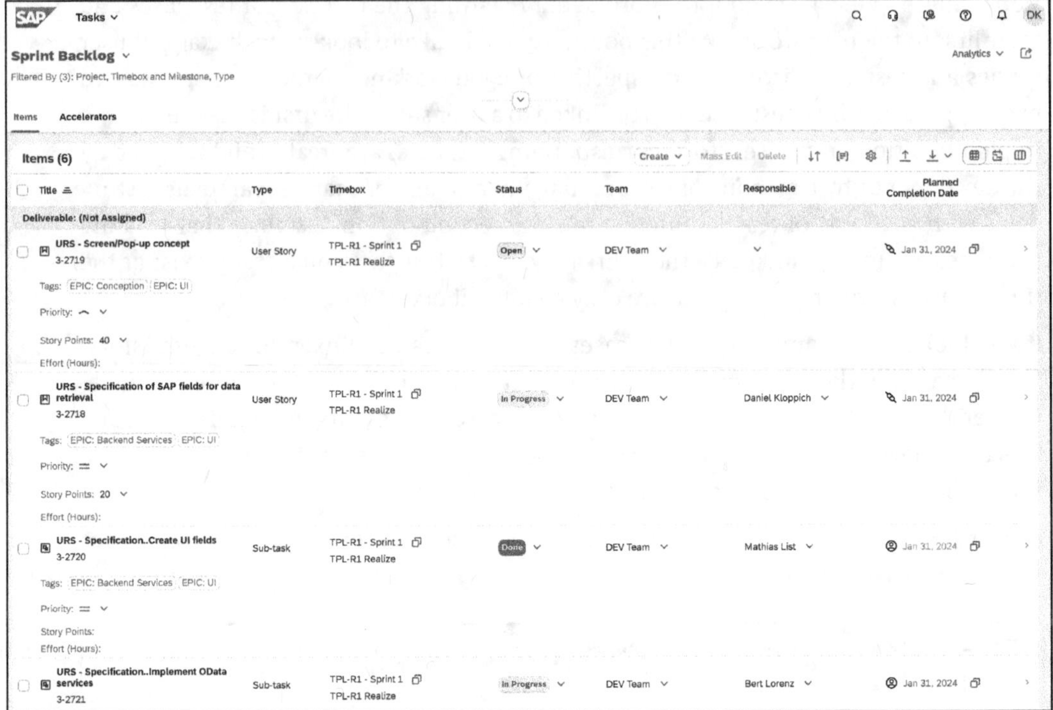

Figure 13.37 Sprint Execution: Daily Scrum

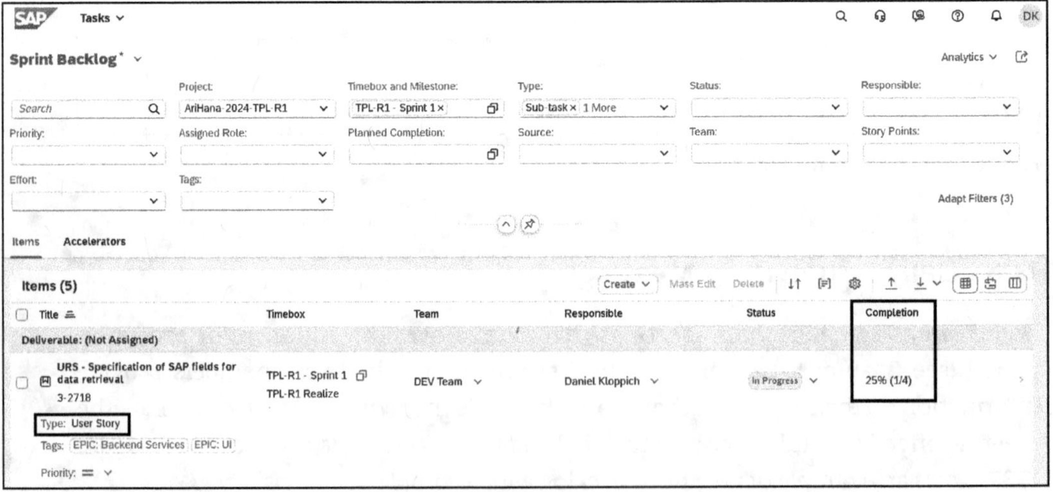

Figure 13.38 User Story: Completion Attribute

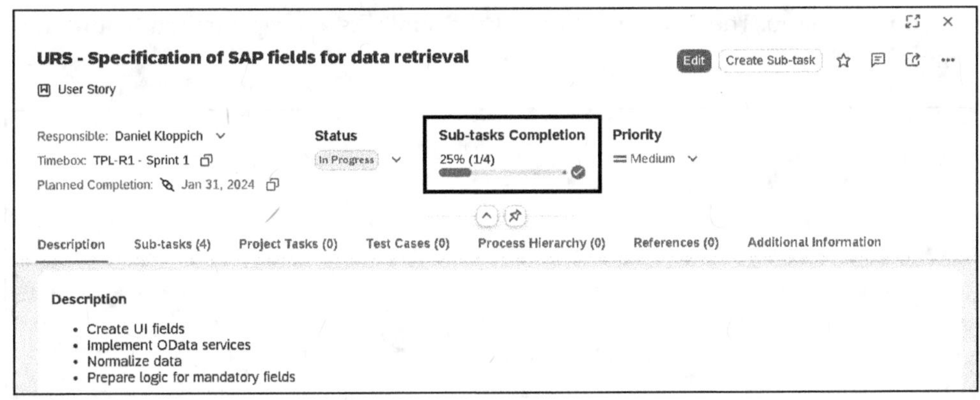

Figure 13.39 User Story: Subtask Completion Overview

Another option is provided by the **User Story Traceability** function, which can be accessed via the Analytics app of the **SAP Cloud ALM for Implementation** app group (see Figure 13.40).

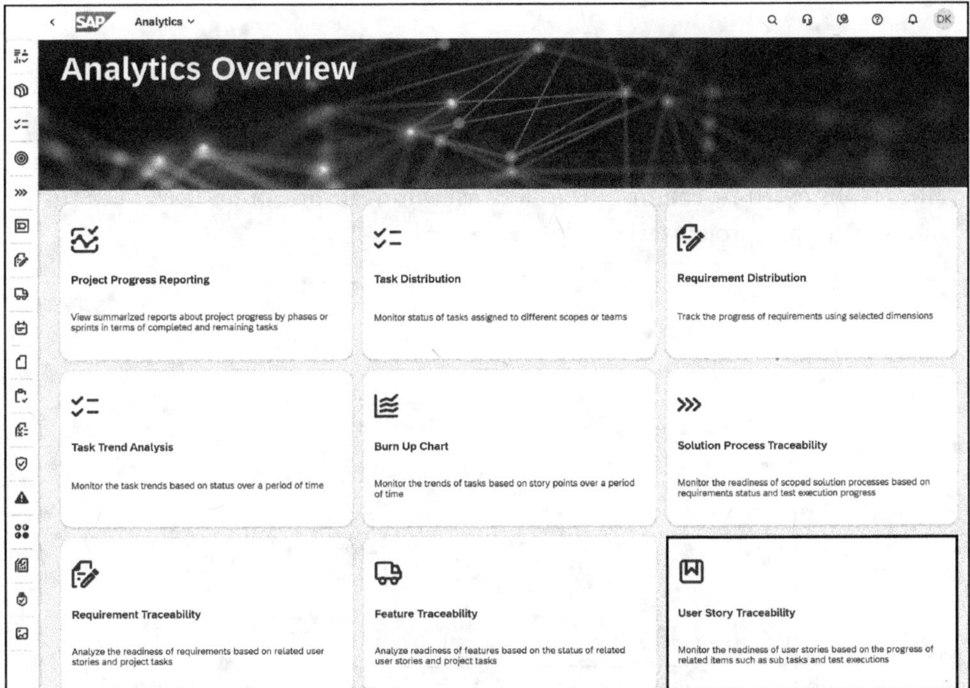

Figure 13.40 Analytics: User Story Traceability

All user stories are consolidated here with the relevant information such as requirements, features, subtasks, etc. By clicking on one of the entities, the Scrum master can jump directly to it. Figure 13.41 shows an example of the user story traceability for our example scenario. The different states of the subtasks are represented by different colors.

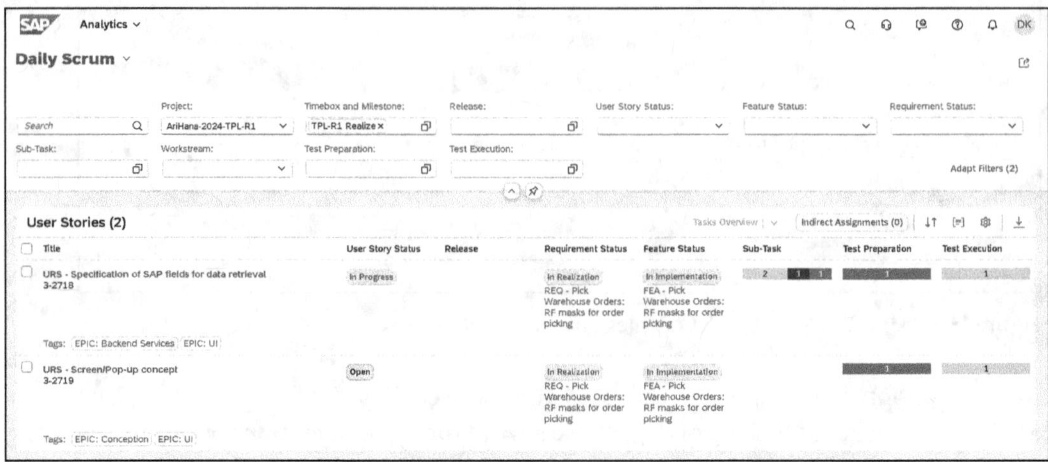

Figure 13.41 Daily Scrum: User Story Traceability

A Scrum board is used to visualize the sprint backlog and enables the graphical editing of user stories and subtasks in an easy way. SAP Cloud ALM does offer a Scrum board feature via the cards view within the Tasks app (see Figure 13.42). You can quickly drag and drop your user stories and subtasks during the daily Scrum to the desired status values represented by columns. The cards view is available for all **Task** types such as requirements and project tasks.

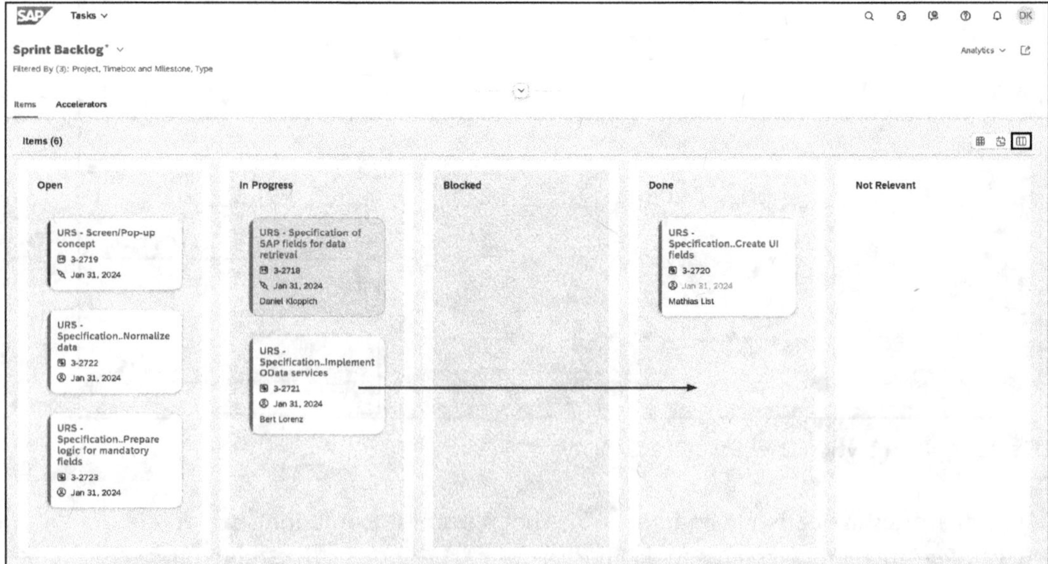

Figure 13.42 Scrum Board Offered by Cards View

External solutions can be developed and integrated via the provided APIs. Figure 13.43 shows an example of a Scrum board developed by Alexandre Sabatier, which he published on his GitHub repository (see *https://github.com/sabatale/prettier-kanban4-calm*).

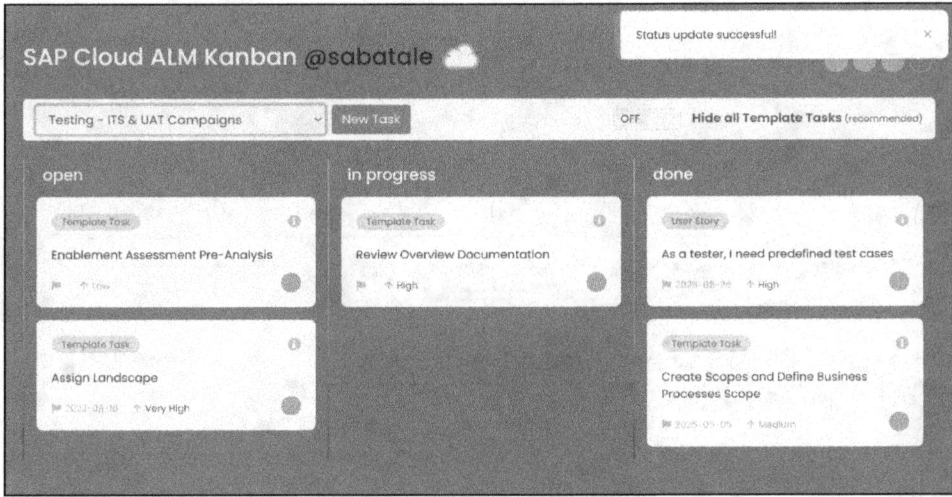

Figure 13.43 Scrum Board by Alexandre Sabatier

13.7 Sprint Review Meeting

The sprint execution ends with the *sprint review meeting*. Here, the developers present their work results or the product increment of the respective user stories to the product owner and the stakeholders one after the other. At the start of the meeting, many user stories are probably already marked with the status **Done**. The developers are required to finalize the user stories when the definition of done (DoD) criteria are fulfilled.

However, there is also the possibility that some user stories have not yet been completed. The reasons for this should then be worked through in retrospect to avoid a repetition. Dealing with such user stories can look like this:

- The status **Open** reflects that the user story has not yet been started. It can be assigned to the next sprint or transferred back to the product backlog.
- The user story has not yet been completed and still has the status **In Progress**. Normally, such a user story should be assigned to the new sprint. However, it can also be returned to the product backlog if necessary.
- The processing of a user story was blocked by a circumstance and the status was set to **Blocked**. The user story must be discussed and evaluated by the product owner.
- A user story that has been marked as **Not Relevant** must be evaluated by the product owner. If necessary, the user story can then be deleted.

Figure 13.44 shows an example of the sprint backlog at the time of the sprint review meeting. All user stories and their subtasks have been completed. The features are ready for consolidated testing on the test system. The test preparation for the upcoming execution has been completed.

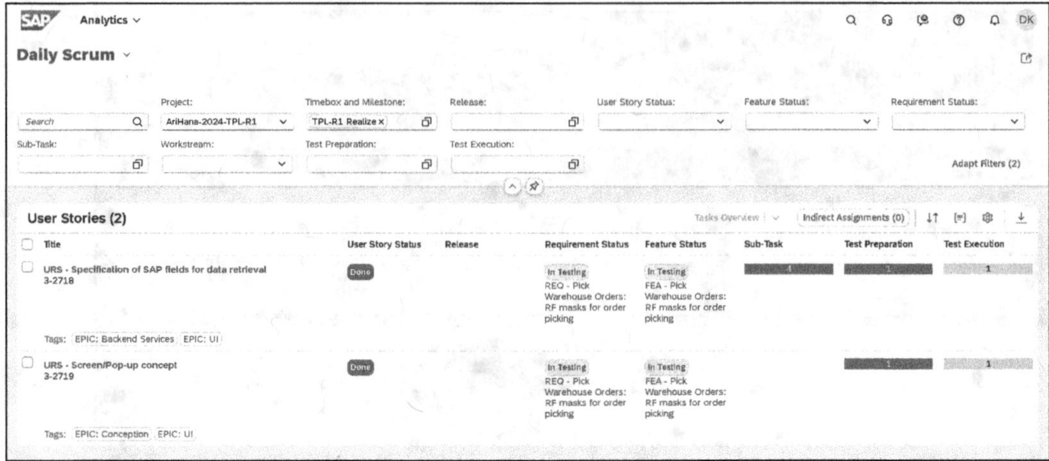

Figure 13.44 Sprint Review: User Story Traceability

In the sprint review meeting, the DoD criteria are checked for each user story. In our example, the product owner opens the individual user stories and examines the table with the DoD criteria (see Figure 13.45).

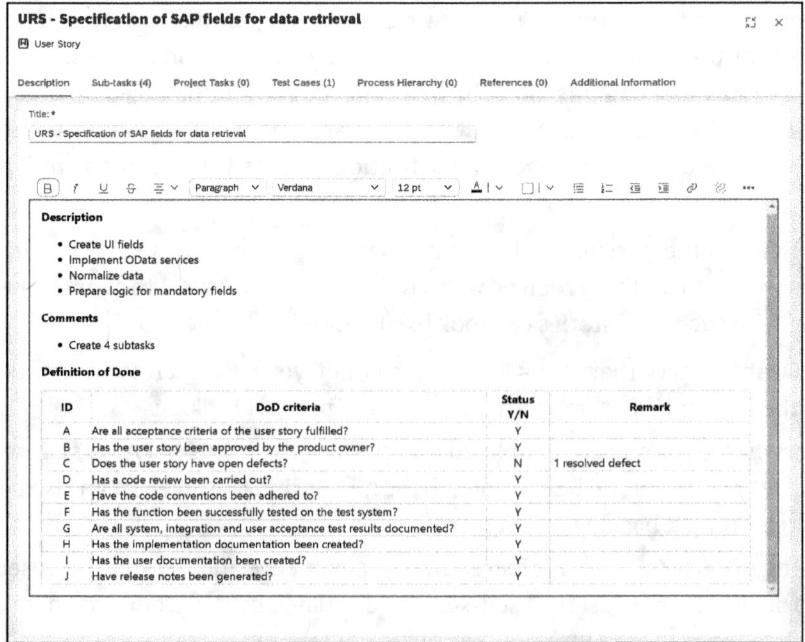

ID	DoD criteria	Status Y/N	Remark
A	Are all acceptance criteria of the user story fulfilled?	Y	
B	Has the user story been approved by the product owner?	Y	
C	Does the user story have open defects?	N	1 resolved defect
D	Has a code review been carried out?	Y	
E	Have the code conventions been adhered to?	Y	
F	Has the function been successfully tested on the test system?	Y	
G	Are all system, integration and user acceptance test results documented?	Y	
H	Has the implementation documentation been created?	Y	
I	Has the user documentation been created?	Y	
J	Have release notes been generated?	Y	

Figure 13.45 Sprint Review: Review of the DoD Criteria of a User Story

Ideally, all subtasks of a user story are completed, as shown in Figure 13.46.

URS - Specification of SAP fields for data retrieval

User Story

Edit | Create Sub-task

Description | Sub-tasks (4) | Project Tasks (0) | Test Cases (1) | Process Hierarchy (0) | References (0) | Additional Information

Sub-tasks (4)

Title	Status	Planned Completion	Team	Responsible
URS - Specification..Create UI fields Relation: Child	Done	Jan 31, 2024	DEV Team	Mathias List
URS - Specification..Implement OData services Relation: Child	Done	Jan 31, 2024	DEV Team	Bert Lorenz
URS - Specification..Normalize data Relation: Child	Done	Jan 31, 2024	DEV Team	Daniel Kloppich
URS - Specification..Prepare logic for mandatory fields Relation: Child	Done	Jan 31, 2024	DEV Team	Daniel Kloppich

Figure 13.46 Sprint Review: Review of the Subtasks of a User Story

Depending on the content of the user story, the creation of test cases can also be part of the user story in addition to the pure implementation of functions. If this is the case and has been defined as a DoD criteria, the test cases should also be stored in the user story on the corresponding tab. In the sprint review, the product owner checks the provided test cases. The test cases should have the status **Prepared** (see Figure 13.47). This status indicates that the test cases are available for further test planning and execution by the test management. A requirement can only be handed over to the test once all functions have been implemented by the associated user stories and the corresponding test cases have been prepared.

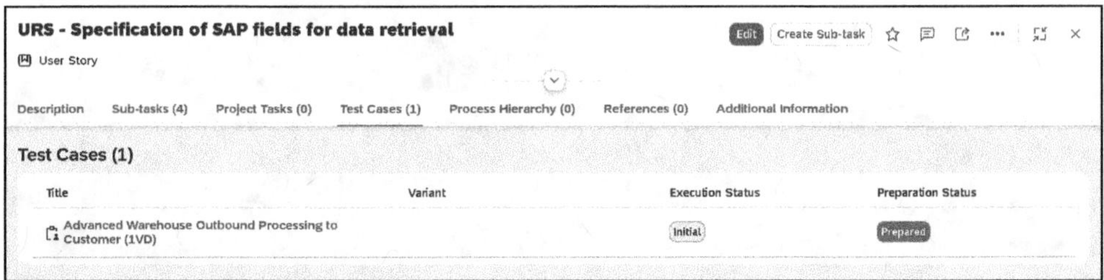

Figure 13.47 Sprint Review: Review of the Test Cases

13.8 Sprint Retrospective

The sprint retrospective is the last event of a sprint. It is used exclusively by the development team itself to assess and improve their working methods and collaboration. SAP Cloud ALM offers the option of documenting and tracking the derived activities via tasks of the **Project Task** type.

13.9 Summary

The processing of user stories in the realize phase is a central element of the agile project methodology. In this chapter, we used our example scenario to show you how to create corresponding user stories based on the requirements of the explore phase and implement them with Scrum. To do this, we took up the logistics solution process of Chapter 9 and the associated requirements from Chapter 10. You also gained an insight into how you can organize and manage Scrum meetings with SAP Cloud ALM. The collaboration between the various Scrum role, is such as product owner, developer, and Scrum master, was explained. Functions are implemented while the development team is working on user stories. This also includes the creation of test cases, which will be discussed in more detail in the next chapter. The developed functions are rolled out to the various systems (e.g., test and production systems) with the help of features and the associated transports. The deployment of features is dealt with in Chapter 17.

Chapter 14
Test Preparation

In this chapter, you will learn how to create test cases and test plans for different test purposes such as test phases or a solution walkthrough performed in SAP Cloud ALM. As the person responsible for creating the tests, you will learn to prepare the test data and assign it to dedicated SAP Cloud ALM processes.

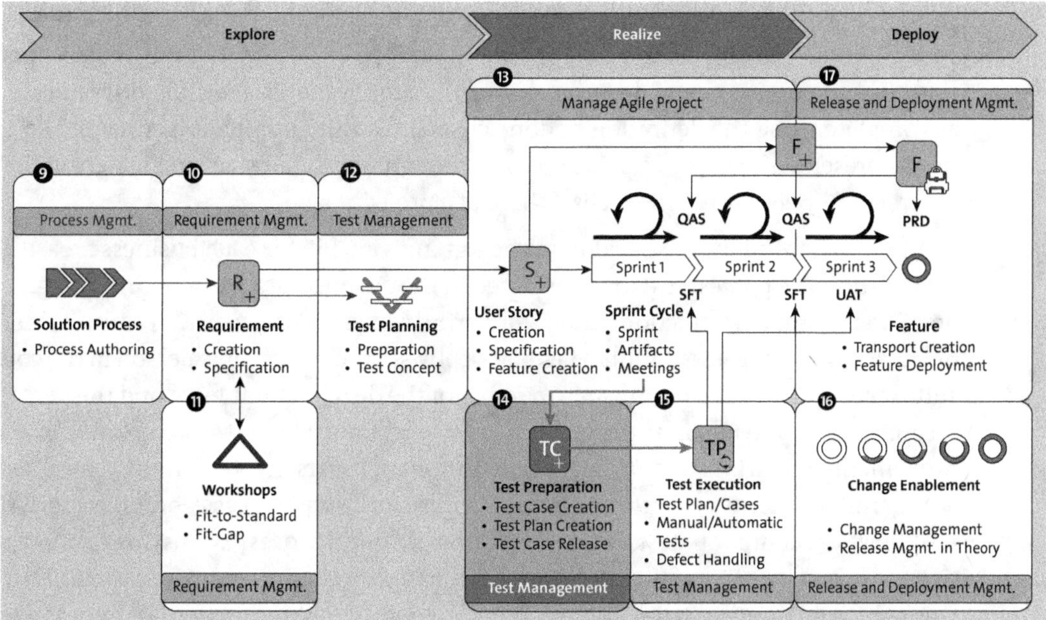

The continuous recording of new requirements and the associated creation and processing of user stories is an integral part of the realize phase. In the last chapter, you became familiar with the processes of the agile working method. While the development team works on the user stories, the requested functions are implemented technically. To ensure that they can be adequately tested by the testers in the future, you must also record the test cases in addition to developing them. In this chapter, we use our example scenario to show you the necessary steps for creating and releasing test cases as part of test management. We will show you how to link the existing requirements and user stories from the logistics area with the associated test cases. The link can be an important acceptance criterion for the definition of readiness criteria of the user story. As a test manager, you are responsible for test planning so that the upcoming test phase can be carried out successfully. We will show you how you can do this with SAP Cloud ALM.

Once a sprint or implementation project has been completed, the SAP solution needs to be validated with the recorded requirements during solution walkthroughs with the recorded requirements and to check its quality in various test cycles. While Chapter 12 provided an initial overview of the test management functions in SAP Cloud ALM and strategic test planning, this chapter focuses on preparing a test phase by test plans with test cases or supporting a solution walkthrough by using test cases.

14.1 Solution Walkthrough and Test Cycles

As explained in Chapter 10 and Chapter 13, the requirements are broken down into small task packages (user stories) to be implemented by the project team in several sprints during the implementation phase of a project. To ensure that the implementation corresponds to the defined user stories and requirements, it should be confirmed as complete using the definition of done criteria. To this end, the implemented solution is presented as part of a solution walkthrough or tested by several testers within several test cycles during the realize phase of a project.

Solution walkthroughs are usually carried out individually for each business area, e.g., for logistics. The processes to be considered in the business process are run through in detail to present new functions and their performance and to ensure an early transfer of knowledge. This presentation takes place between the responsible person in the consulting company or the responsible member of the development team and the person responsible for the business process who is also responsible for the acceptance of the implemented solution. If necessary, other project members involved are also included. For an initial solution walkthrough, you can use existing test cases provided by SAP Best Practices content to test the preconfigured solution processes. Missing configurations are documented and gaps are noticed as new requirements. A solution walkthrough can also result in the discovery of a previously missing process. In this case, an additional process must be included in the scope of the solution, and the necessary requirements and user stories must be defined.

Test cycles are iterating test phases with different test scopes and certain timeboxes. To schedule a test cycle with several testers, the following activities must be completed:

- The user stories were realized.
- The unit test was carried out.
- The requirement is ready for testing.

If the prerequisites for a test cycle are met, you have to investigate the test scope. You can see this test scope by looking at an overview of the requirements in the **Testing** status. To do this, use the Requirements app or Tasks app in the **SAP Cloud ALM for Implementation** app group and use the filter criteria for dedicated display in both cases. As shown in Figure 14.1, filter the choices within your project according to the business area (**Scope** field) that you want to view. In our case, this is the scope **AriHana-2024-TPL-R1-Logistics.** You can also filter by **Solution Process** if you want to set the test focus on specific processes.

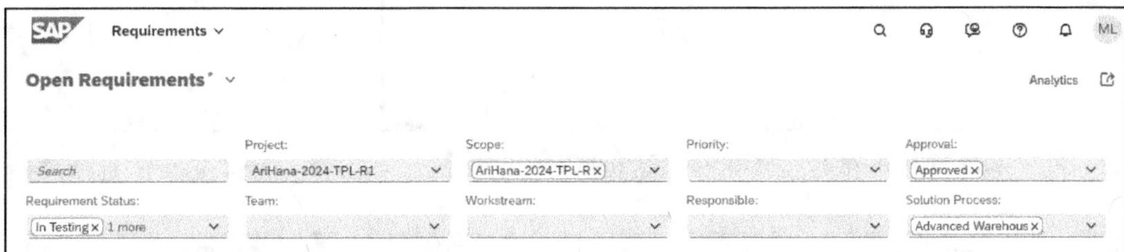

Figure 14.1 Defined Test Scope by Requirements

Any discrepancies, errors, and proposed changes identified during the test phase must be documented.

If, for example, a failed function is detected during the test, you document it in SAP Cloud ALM as a *defect*. You can find detailed information about defects in Chapter 15, Section 15.3.

14.2 Test Case Creation

In Chapter 12, you learned basic information about strategic test planning, which we will discuss in more detail below. An essential part of test preparation—in addition to test planning—is *test case creation*. The creation of test cases is a crucial prerequisite for test execution. In the following, we will show you how to create test cases in SAP Cloud ALM.

To do this, navigate to the **SAP Cloud ALM for Implementation** app group and open the Test Preparation app (see Figure 14.2).

The app displays all test cases already created for the selected project. As shown in Figure 14.3, you can narrow down the list of displayed test cases using additional filter settings.

You can also create new test cases in this app. SAP Cloud ALM offers two options here: You can create both manual as well as automated test cases. To do this, click on the **Create** button or import an automated test script using the **Import** button.

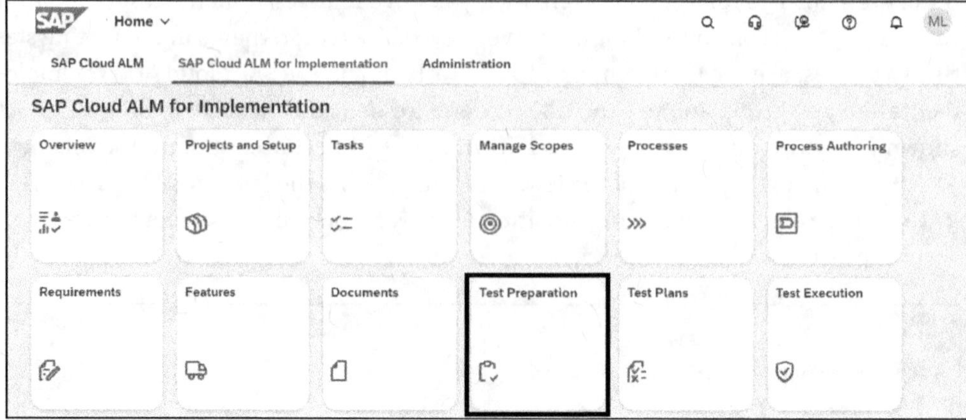

Figure 14.2 Launchpad with Test Preparation App

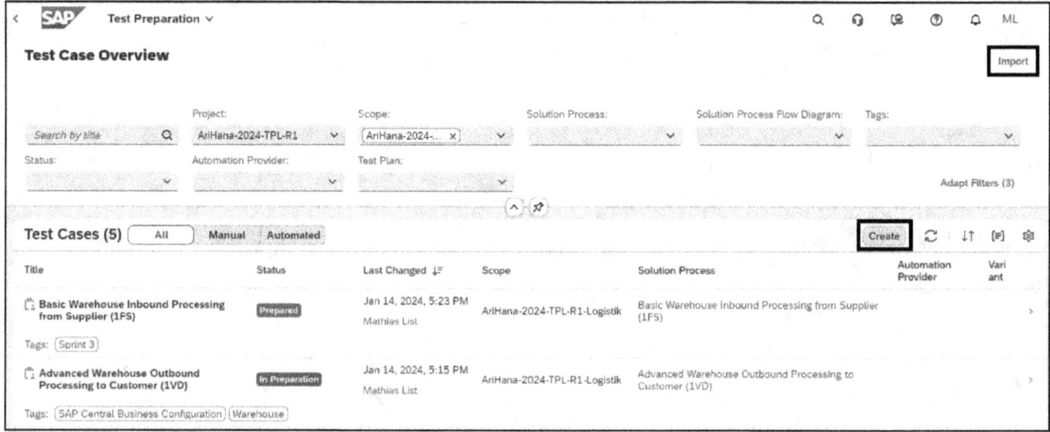

Figure 14.3 Test Case Overview in the Test Preparation App

14.2.1 Create and Edit Manual Test Cases

In SAP Cloud ALM, a test case is always assigned to a project and a scope. If you create a manual test case, you must always enter information into the **Scope** and **Project** fields in addition to the test case **Type** and **Title** fields. You can also assign your test case to a solution process and a solution process flow diagram. In this case, only the processes that are in the previously selected scope can be selected. If you select a process, the process diagram also becomes a mandatory entry.

As you can see in Figure 14.4, we have created a manual test case as an example. The scope is **AriHana-2024-TPL-R1-Logistics**, which is located in the **AriHana-2024-TPL-R1** project we created. To test a part of a process in the form of a functional test, we add the solution process **1VD - Advanced Warehouse Outbound Processing to Customer.** To enable a direct assignment between test case and process, in this example we named the test case the same as the process to be tested. Click on the **Save** button to create the test case.

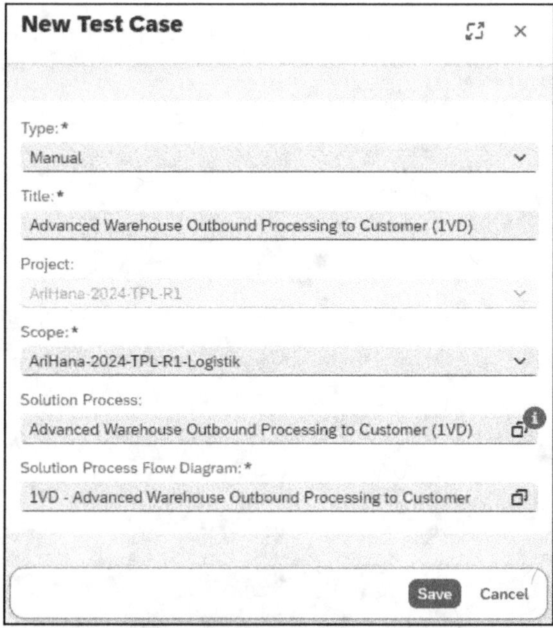

Figure 14.4 Create a Manual Test Case

14

[+]

Naming Test Cases

To make it easier to find and distinguish test cases in the various SAP Cloud ALM display options, it makes sense to include the test case type in the title. For example, add "SFT" in front of the actual test case name to visualize that the test case is a functional test (single functional test).

If a test case is assigned to a process during creation, this process serves as a template for test case creation. The activities/process steps of the selected process are transferred to the test case as test sections. These activities are displayed in the first structure view that opens after saving. Using the slider in the **Scope** column, you can, as shown in Figure 14.5, decide which of the activities you want to include as sections in the test case. It is also possible to adjust this structure later.

You can also add or remove additional test case sections using the **Add** and **Delete** buttons. You can edit existing sections using the pencil icon. When you add a new section, assign a **Title** and an **Activity Number** (see Figure 14.6), which determines the order of the test case sections. In this window, you also have the option of adding an application to the section in the form of a URL in the **Application** area. Confirm your entry by clicking **OK**.

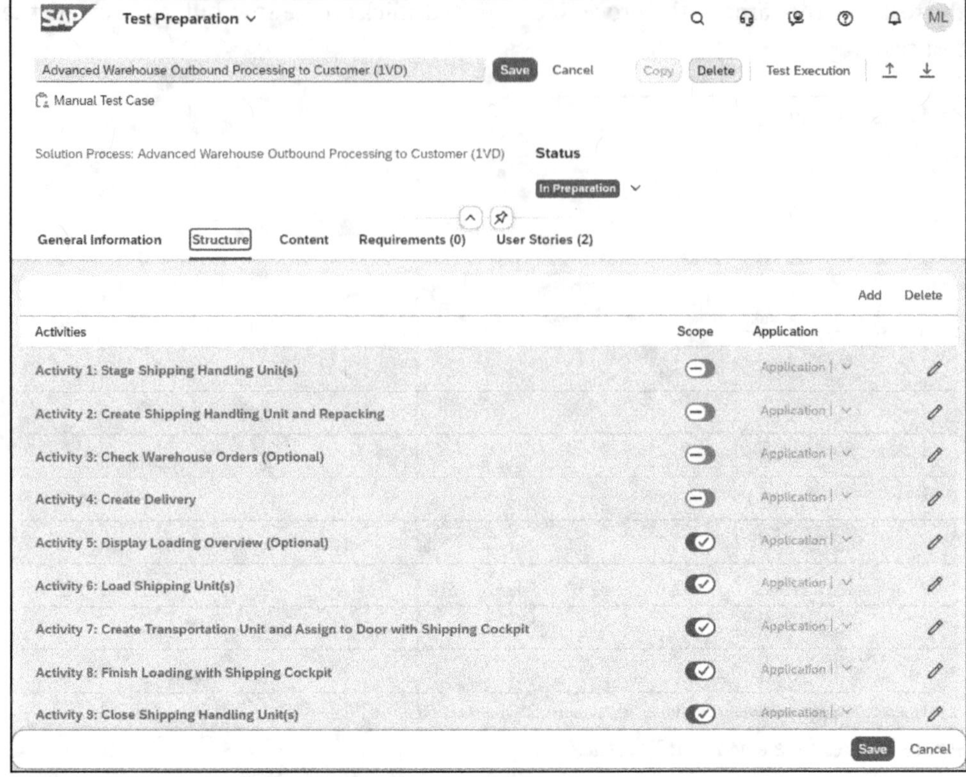

Figure 14.5 Selection of Process Activities

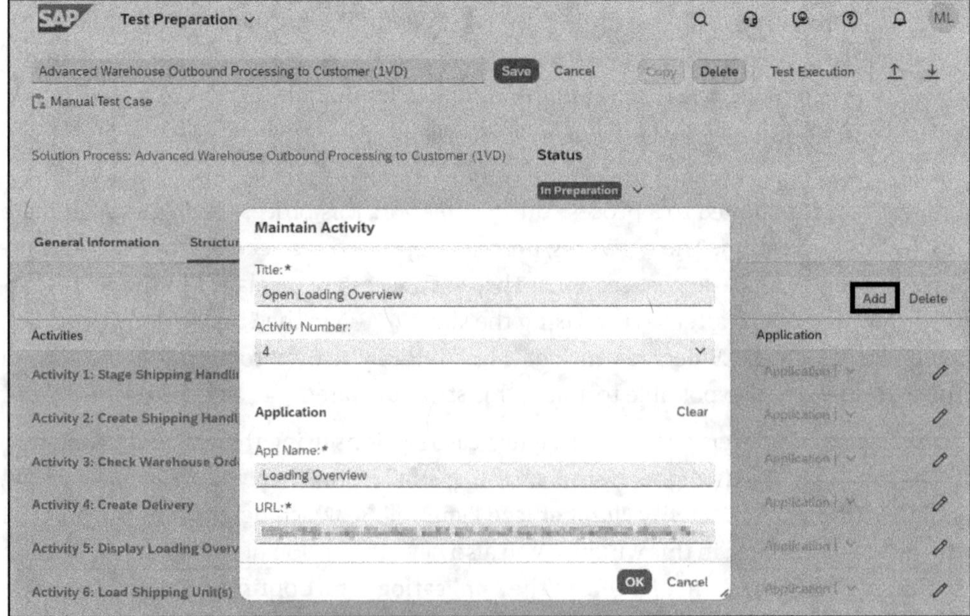

Figure 14.6 Add a Test Section

After saving, the actual content of the test case—which is based on the previously selected and/or created structure—is displayed on the **Content** tab. At this point, you fill your test case with the information required for execution. As shown in Figure 14.7 you can use the pencil icon to edit existing test steps and the **Add** button to add more test steps under the selected section. Use the **Delete** button to activate the delete mode in order to remove test steps. The **Application** button is used to jump to the maintained browser app. In addition to the free text search, the app offers you the option of navigating to the individual test steps of your test case via the dropdown **Navigation** button.

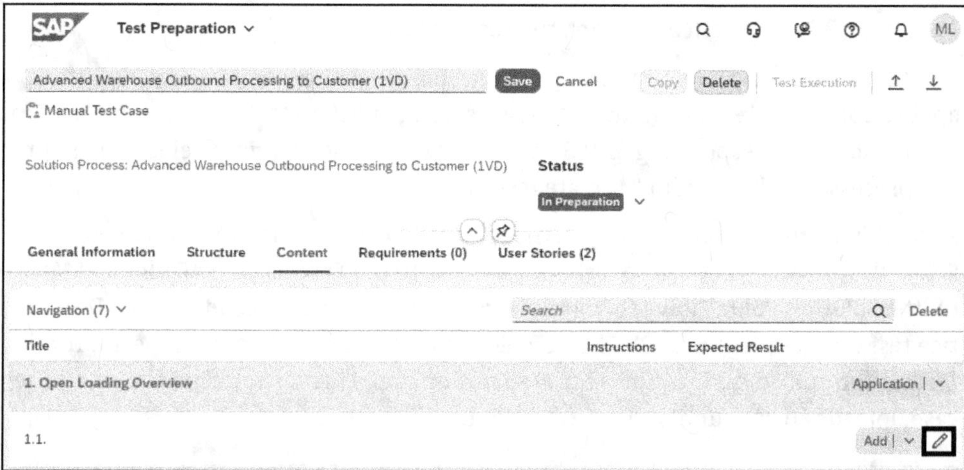

Figure 14.7 Edit a Test Case

For the content of your test case, a title and an instruction are expected for each test step, as shown in Figure 14.8.

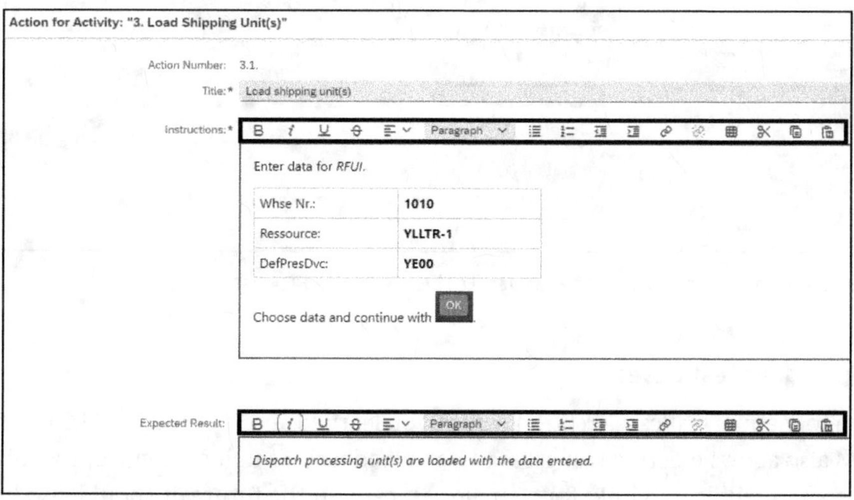

Figure 14.8 Formatting Options in the Test Step

You also have the option to maintain the expected result of the test case. Regarding the instruction and the expected result, SAP Cloud ALM offers a lot of freedom in the visual text design. For example, it is possible to format the entered text and insert bulleted lists, text inserts, and tables in the description texts. You can also integrate images into the available free text fields.

You can view general information about the test case on the **General Information** tab. In the **Object** area, you can also add tags to help you categorize the test cases. You can select existing tags via the dropdown menu or add new tags by entering them.

14.2.2 SAP Best Practice Content for Test Cases

SAP Cloud ALM enables you to integrate SAP Best Practices content directly into your application lifecycle management tool. This also applies to test case content. You can download the corresponding SAP Best Practices test cases for your selected best-practice processes and use them to create test cases.

To do this, open the Processes app in the **SAP Cloud ALM for Implementation** app group. Select the process you want to test and navigate to the **Test script (SAP Cloud ALM)** link on the **Solution Value Flow** tab via the **Accelerator** tab to download the SAP Best Practice test case (see Figure 14.9). The downloaded Excel file is already in the appropriate format for uploading into the **Test Preparation** app. Further information on the test case upload can be found in the next section.

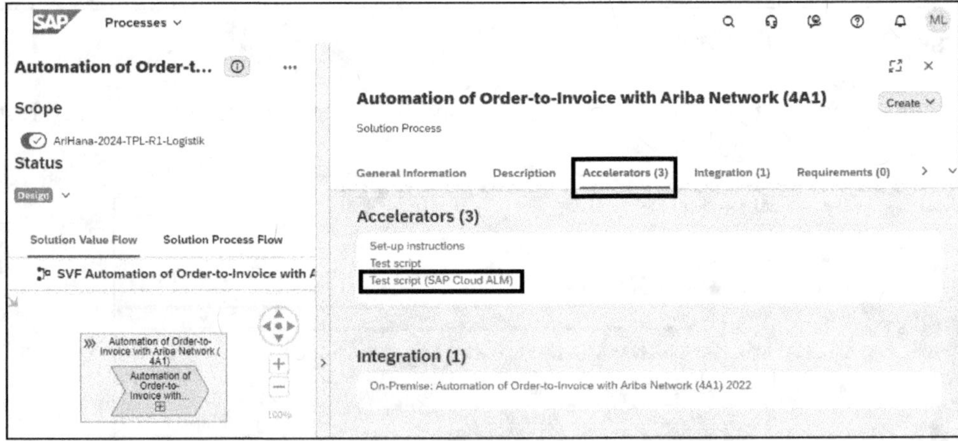

Figure 14.9 Download of an SAP Best Practice Test Case in the Process

14.2.3 Uploading Test Cases

In addition to manual test case creation, which you learned about in Section 14.2.1, SAP Cloud ALM also allows existing test cases to be uploaded. This upload is currently only possible as an Excel file and only for a single test case. In the future, it should also be possible to upload test cases in bulk uploads.

In Figure 14.10, you can see the mentioned test case. At the top right of the screen, note the buttons for both uploading and downloading the test case.

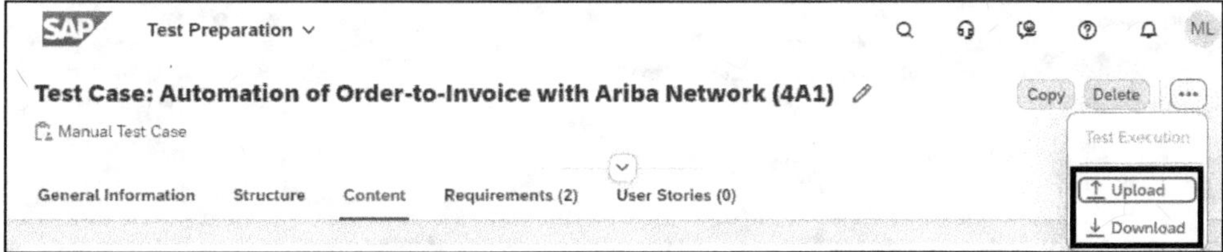

Figure 14.10 Upload and Download Test Case

If you are using an SAP Best Practices file, you must add certain fields (columns A to K) to the Excel file to match your test case (see Figure 14.11). To do this, download the test case you created. Copy the values from the downloaded file and paste them into the SAP Best Practices test case using Ctrl+C and Ctrl+V.

A	B	C	D	E	F	G	H	I	J	K
Test Case GUID	Test Case Name*	[Scope GUID]	[Scope Name]	[Solution Process GUID]	[Solution Process Name]	[Solution Process Flow GUID]	[Solution Process Flow Name]	[Solution Process Flow Diagram GUID]	[Solution Process Flow Diagram Name]	Test Case Status

Figure 14.11 Excel File of the SAP Best Practices Test Case

After you have clicked the **Upload** button shown in Figure 14.10 to start the upload process, click the **Browse** button and select the Excel file you want to upload (see Figure 14.12). Then click on the **Upload** button.

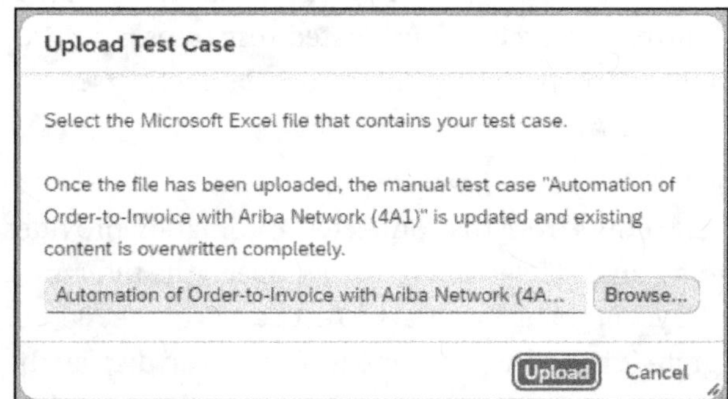

Figure 14.12 Upload a Test Case

Error during Test Case Upload

If the file cannot be uploaded in its current state, an error message will be displayed. This is usually due to incorrect formatting of the file or missing values in it. Read the error message carefully and solve the problem described in it. Then start the upload again.

After the successful upload, the test steps of the imported file are displayed on the **Content** tab (see Figure 14.13). Adjust these as required, as described earlier.

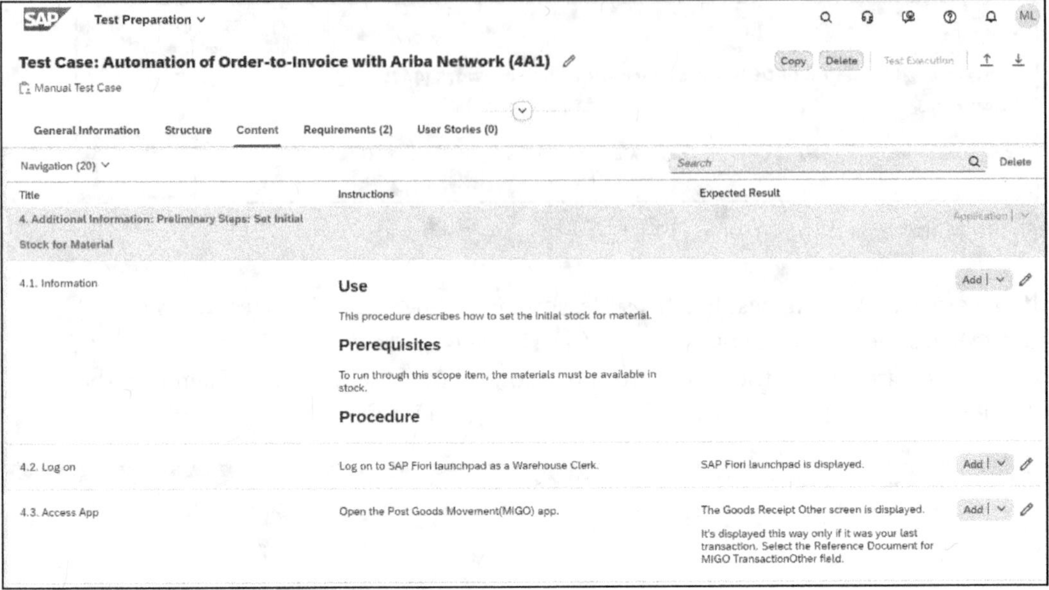

Figure 14.13 Test Case Contents After Upload

14.2.4 Create and Import Automated Test Cases

As mentioned in Chapter 12, Section 12.3, SAP Cloud ALM offers not only the ability to create manual test cases but also the option of directly integrating your test automation tools and thus create, synchronize, and execute automated test cases. As third-party integration is not yet available at the time of printing this book, this section focuses on test automation with SAP S/4HANA Cloud.

If, as described in Chapter 12, Section 12.3.1, all requirements for the integration of the tool are met, you can synchronize test scripts from the connected automation provider in the Test Preparation app, depending on the selection of your scope in the filter area. To do this, click the refresh icon (**Synchronize Automated Test Cases**) to the right of the **Create** button. If you have integrated several automation tools, you can display the respective test cases using the **Automation Provider** filter.

> **Synchronization of Test Cases**
>
> Please note that synchronization can always be carried out for the scope selected in the filter menu and the solution processes it contains, as well as for only one automation provider at a time.

The SAP S/4HANA Cloud test automation offers directly usable best-practice test cases of the selected solution processes as well as post-upgrade test cases that you can run for testing after a system upgrade.

You can use the **Import** button on the right-hand side of the screen to display all automated test cases from your test automation provider and transfer them to SAP Cloud ALM. This also includes customer-specific test cases. You can also create your own automated test cases using the **Create** button (see Figure 14.14).

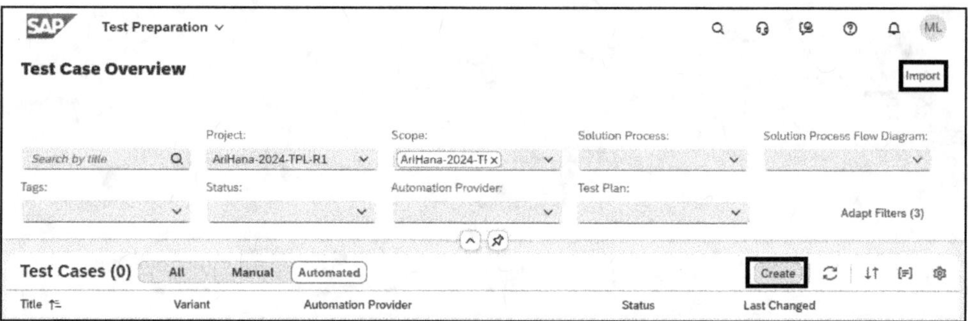

Figure 14.14 Importing and Creation of Automated Test Cases

The test case is created as a shell in SAP Cloud ALM. In the **New test case** window, in addition to the fields **Title**, **Project**, and **Scope**, you must now also fill in the **Automation Provider** and **Country/Region** fields and select the **Automated** option in the **Type** field (see Figure 14.15). Save your entries by clicking **Save**.

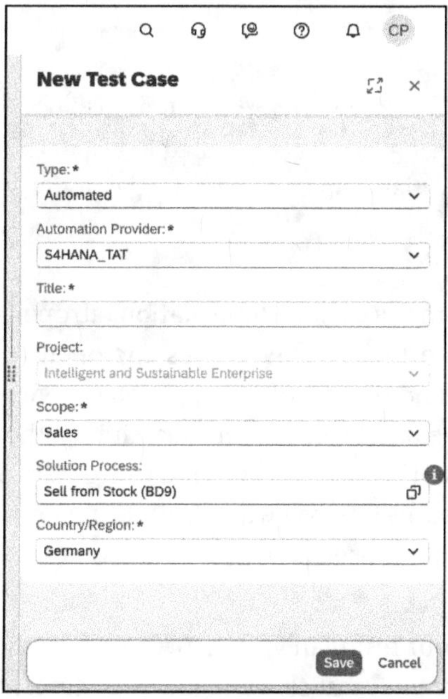

Figure 14.15 Automated Test Case

If an automated test case has been synchronized, imported, or created, you can open it in the Test Preparation app. The central information of the test case is now displayed there (see Figure 14.16). Click on the linked test case name or the linked test case variant to jump to the test automation provider.

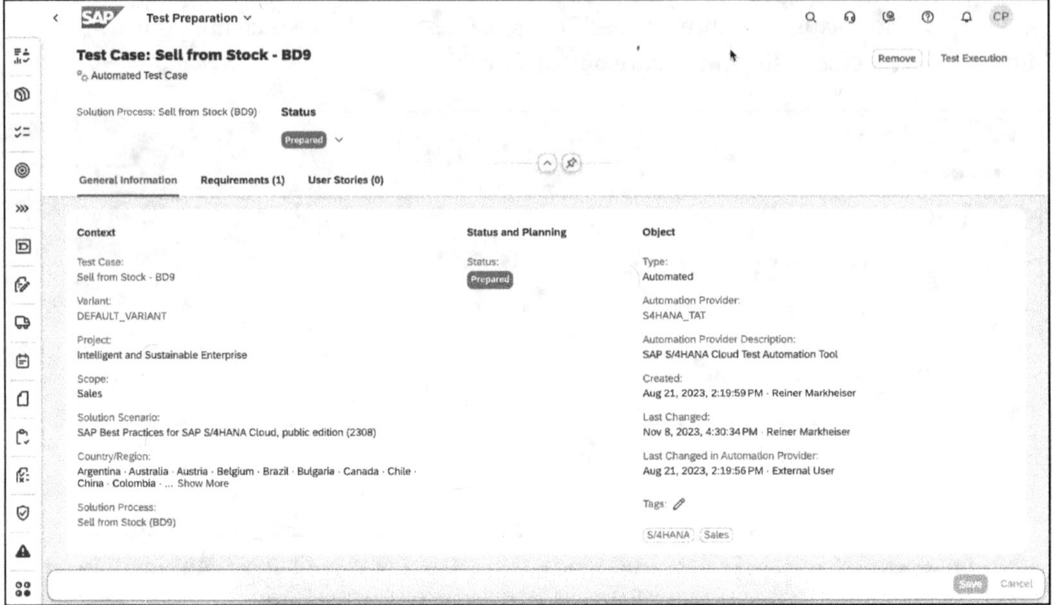

Figure 14.16 Automated Test Case in SAP Cloud ALM

Test Case Variants and Test Data

An automated test case always has at least one variant. The variants are used to determine which test data the automated test case is using. Test data is stored in test data containers within the test automation provider. A variant can use several test data containers.

Clicking on the test case name in the **Context** area in the **General Information** tab opens the dialog window shown in Figure 14.17 when the Test Your Processes app opens in your test automation provider. In the top right-hand section of the app, you can view the status of the last execution as well as the time and user. You can search for other test cases and jump to them via the left-hand area. The test case selected in SAP Cloud ALM is referred to here as a *test plan*. On the **Processes** tab, you will see all the test processes that are automatically run when this test plan is executed. A *test process* defines the test activities to be executed.

You can assign and view test data containers to your test plan via the **TDCs** tab. On the **Variants** tab, you can see all the variants created for your test case and jump to them. As shown in Figure 14.18, the individual test actions and their test data are displayed here.

If an automated test case is created, a variant with the name DEFAULT_VARIANT is automatically created. It is possible to create further variants.

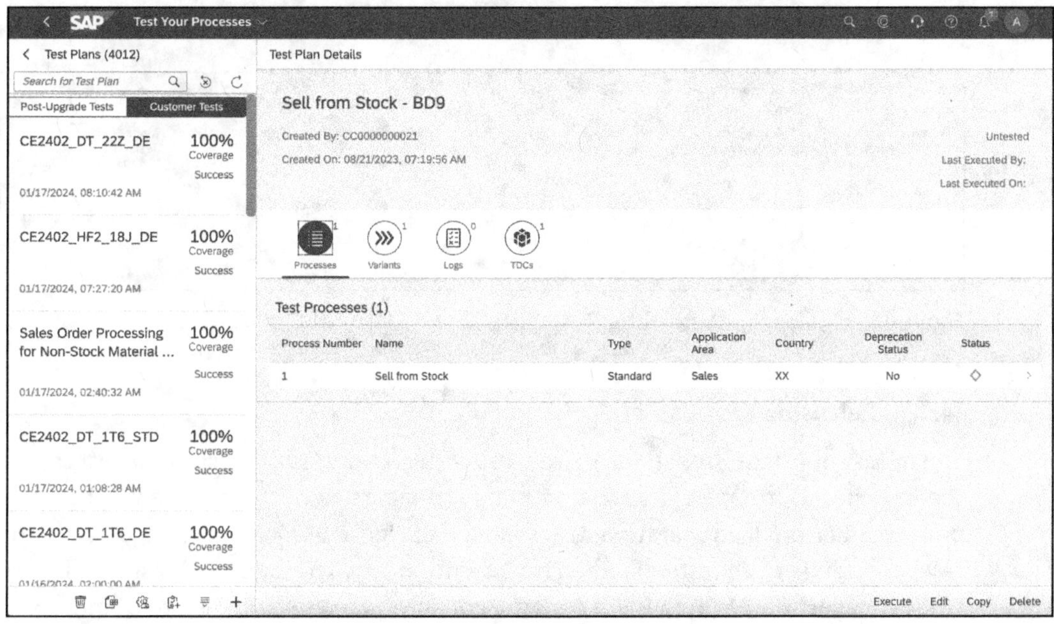

Figure 14.17 Test Processes in an Automated Test Case

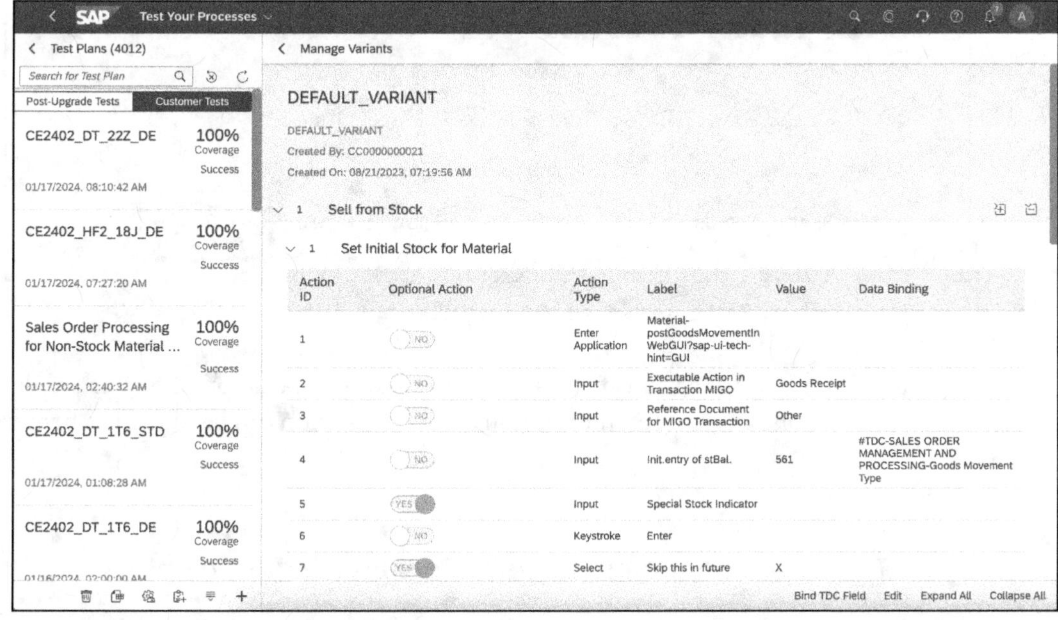

Figure 14.18 Variants of an Automated Test Case

The SAP S/4HANA Cloud test automation tool also offers you apps in the **Automatic Testing** app group for creating and managing automated test cases (see Figure 14.19). The entry point via SAP Cloud ALM is exclusively via the Test Preparation and Test Execution apps.

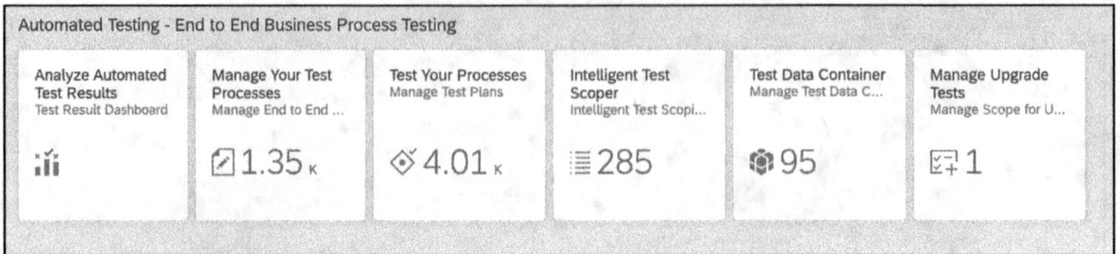

Figure 14.19 Managing Automated Test Cases in SAP S/4HANA Cloud

14.2.5 Releasing Test Cases

The release of test cases is status-driven and regulates the further use of the test case in the test execution. Open your test case and change its status for release via the dropdown menu from **In Preparation** to **Prepared** (see Figure 14.20). This change makes the test case visible to all testers in the Test Execution app. The dedicated test case is now ready for execution. Whether it is a manual or automated test case is irrelevant.

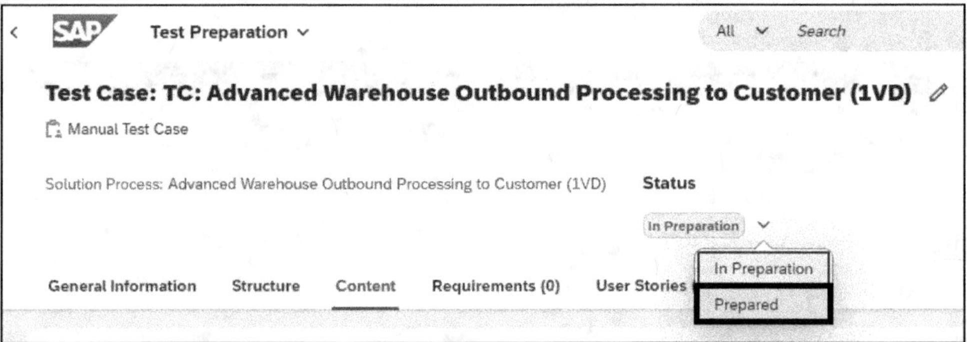

Figure 14.20 Release Test Case to Ready for Execution

The current and future options for test planning and the allocation of testers are described in Section 14.4. Further information on test execution can be found in Chapter 15.

14.3 Test Case Assignment to User Story and Requirement

After the creation of the user stories and the complete implementation of the requirements of your project, you need to check them for quality via a solution walkthrough or a test cycle. SAP Cloud ALM enables a direct link between test case and requirement or test case and user story during test case creation.

Link your test case to a user story to test the change described in it, e.g., the implementation of a new function, in a single functional test (SFT). For a user acceptance test (UAT) or functional integration test (FIT), link your test case to a requirement. You can also view or assign a linked test case at any time in the process itself on the **Test Case** tab.

To create such a link, open the test case of your choice (which is ready for testing) and, as shown in Figure 14.21, navigate to the **Requirements** tab or the **User Stories** tab, depending on the assignment. Click the **Assign** button.

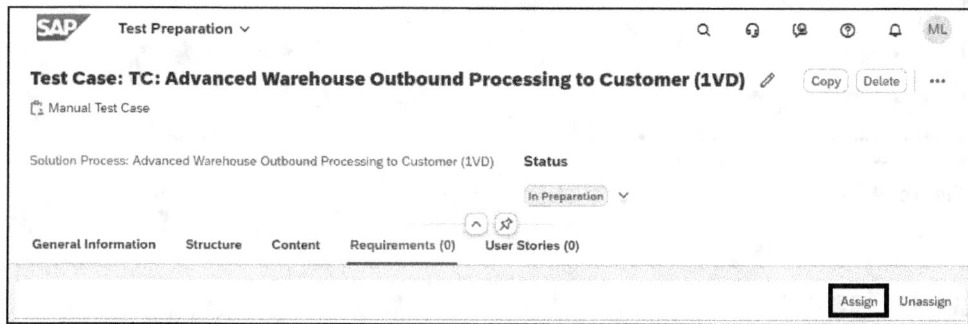

Figure 14.21 Set Test Case as Ready for Testing

In the window that opens, you can select and assign one or more requirements or user stories. As shown in Figure 14.22 in this example, the requirement **REQ - Pick Warehouse Orders: RF masks for picking** is assigned to the test case **Advanced Warehouse Outbound Processing to Customer (1VD)**.

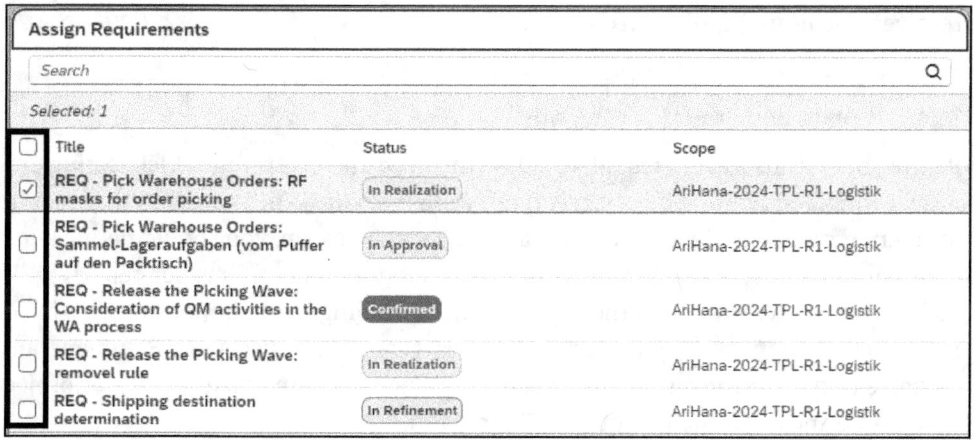

Figure 14.22 Test Case Tab to Assign Requirements

The assigned requirements or user stories are then listed on your tab and, after saving your test case, are also visible in the assigned requirement or user story on the **Test Case** tab. If you want to remove this link again, you can do so by clicking the **Unassign** button (see Figure 14.23).

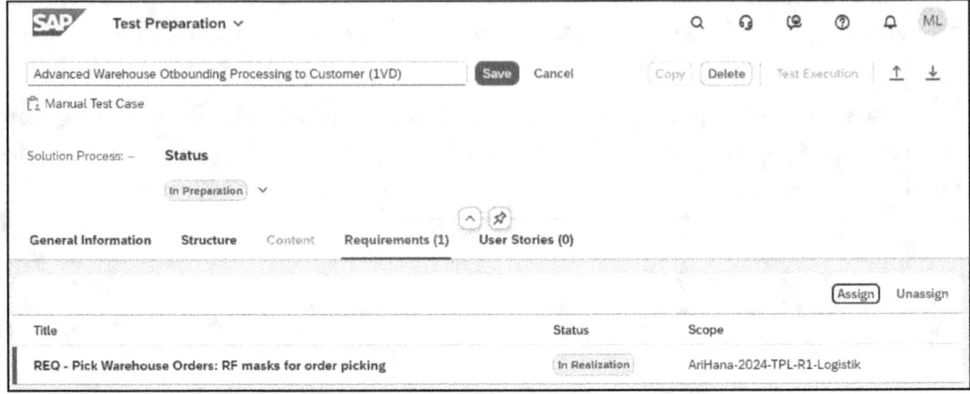

Figure 14.23 Assigned Requirement within a Test Case

14.4 Test Plan

Test plans are sets of test cases that are executed in specific test cycles, helping you run your testing activities in iterations.

While you can also carry out your testing activities without test plans for a leaner approach, in larger implementation and transformation projects, it might be necessary to reuse the same or a similar set of test cases in multiple test phases.

This way, each test case occurrence can be executed within the defined time frame and receives a dedicated execution context.

14.4.1 Create and Prepare Test Plans

Before you start to create a test plan, define the need and determine which testing purpose it should provide. Different test types were introduced in Chapter 12 that require different structure and content, for example. For this reason, you should define a test plan naming convention to ensure the right assignments and proper reporting. In addition to the description of the test plan, its scheduling is also important. This information can be retrieved by project timeboxes. Open the Projects and Setup app and navigate to the **Timeboxes** tab. Two user acceptance tests are scheduled at our project for sprint 6 and sprint 12, as shown in Figure 14.24.

In the follow example, we create a test plan for user acceptance test which bundle test case from several requirements. Navigate in the **SAP Cloud ALM for Implementation** app group and open the Test Plans app (see Figure 14.25).

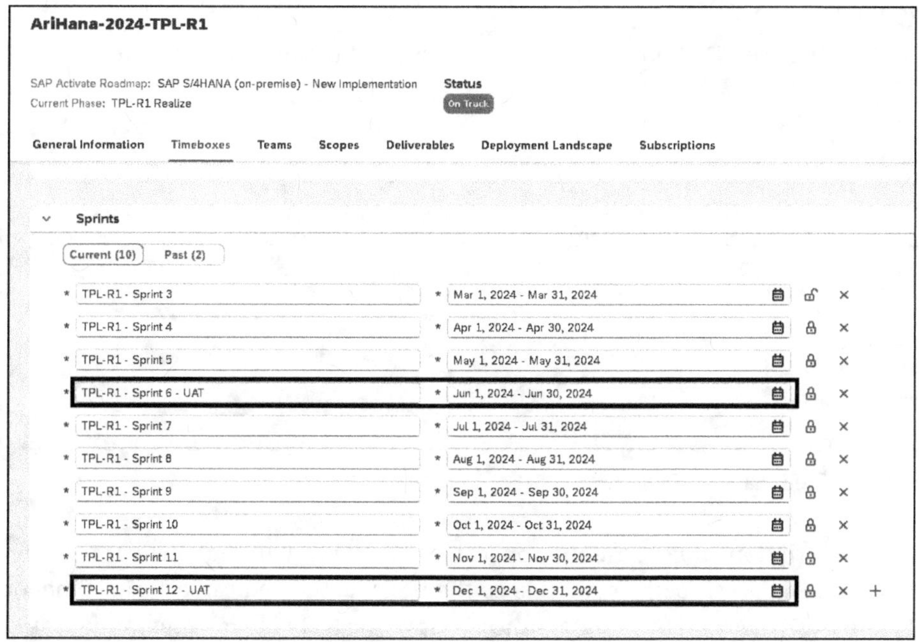

Figure 14.24 Scheduled User Acceptance Tests

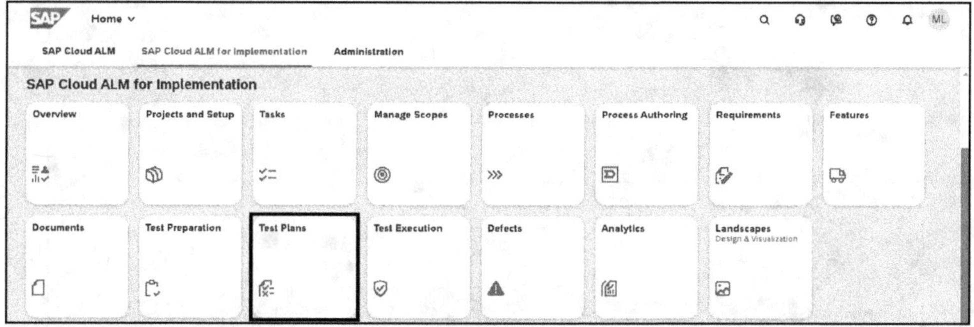

Figure 14.25 App to Set Up and Manage Test Plans

In the test plan overview, select the project and click on the **Create** button to set up a new test plan (see Figure 14.26).

On the **General Information** tab, enter a title with a predefined naming convention. The predefined naming convention helps to categorize the test plan based on certain attributes, which in turn helps you manage and report on your test plans. An example of a naming convention is as follows: *<Test type>_<Solution>_<Release>_<Sprint>_<User ID>*, where:

- *Test type* is user acceptance test (UAT)
- *Solution* is SAP S/4HANA (S4)
- *Release* is release 1 (R1)

- *Sprint* is sprint 6 (S6)
- *User ID* is the user ID of one of the dedicated users of the managed system

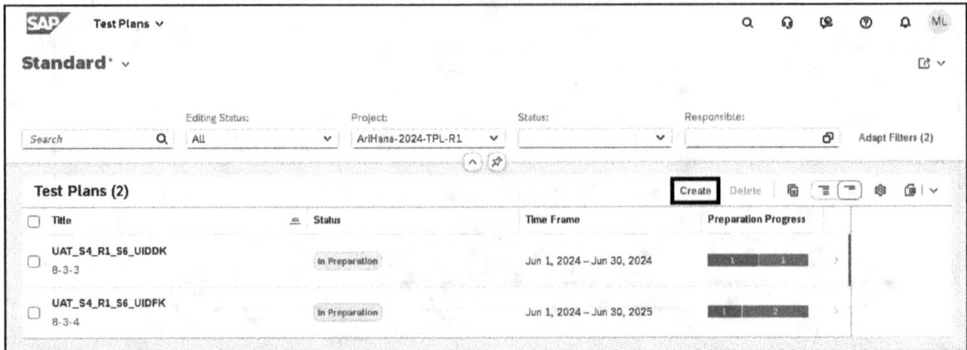

Figure 14.26 Create a New Test Plan

Enter the description and choose the applicable test plan. Fill in the calendar fields for the **Start** and **End** dates, being sure to keep them inside the dedicated sprint timebox (see Figure 14.27). Click on the **Create** button to save the test plan in **Preparation** status.

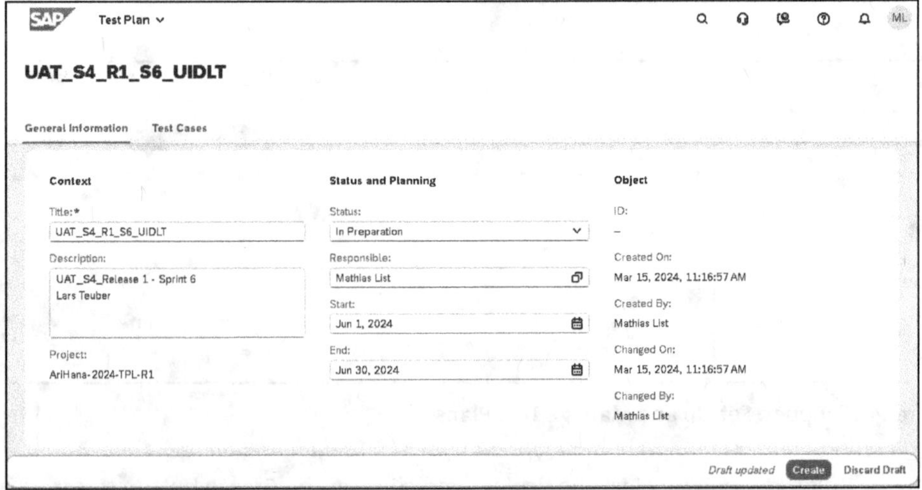

Figure 14.27 General Information of Test Plan

Navigate into the **Test Cases** tab, make sure you're in **Edit** mode, and start to assign existing test cases. Click on the **Assign Test Cases** button to open the popup window **Select: Test Cases**. Here, you have the option to filter by test case type (**Manual** or **Automated**) or by test case status (**Prepared** or **In Preparation**). You can also search for available test cases based on scope. In our example, we are searching for test cases related to the scope **AriHana-2024-TPL-R1-Logistic**. Select the dedicated test cases and confirm by clicking the **OK** button (see Figure 14.28). A combination of manual and automated test cases is possible.

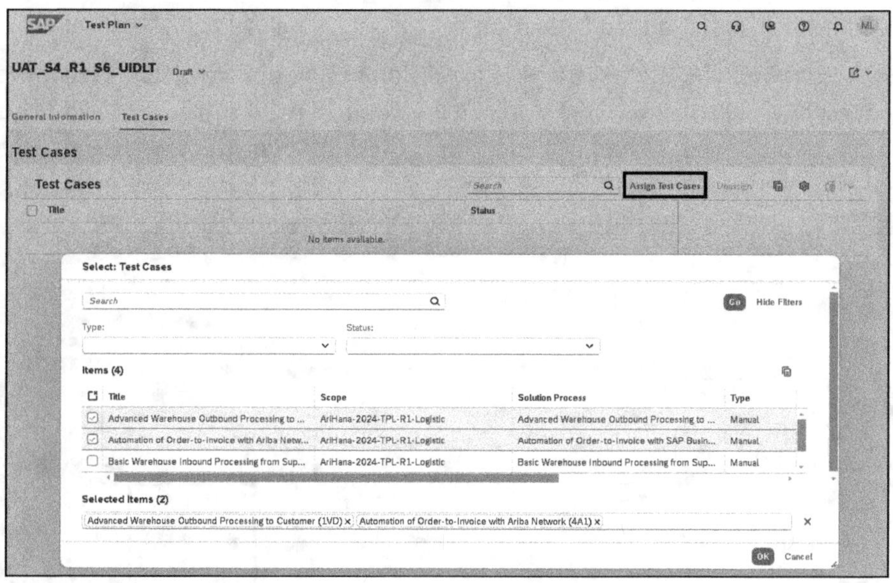

Figure 14.28 Assign Test Cases to Test Plan

We recommend that you activate all available columns in the **Test Cases** tab using the **Settings** button. In the **Test Cases** tab, switch into edit mode and fill in the **Tester** field by clicking on the field and then pressing F4. This opens the popup window **Select: Users**, where you can search by username. Select the user and confirm your settings by clicking the **Save** button (see Figure 14.29).

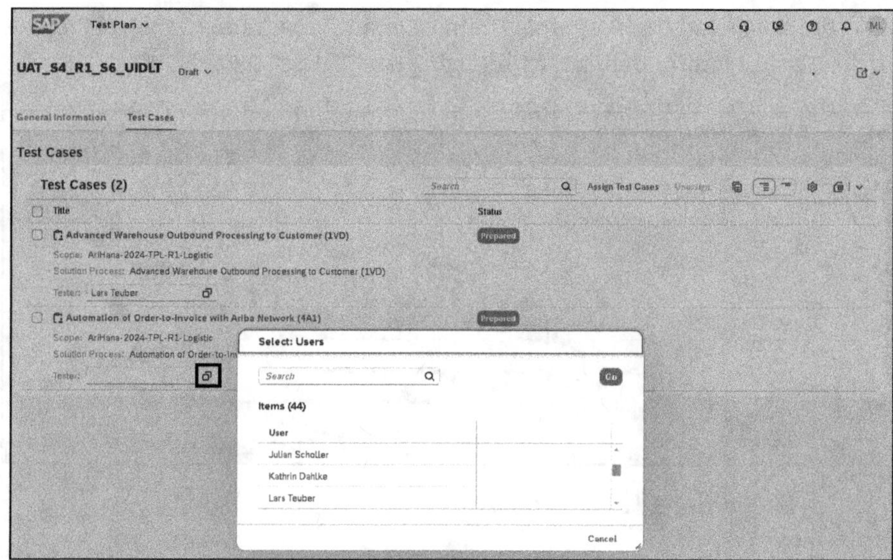

Figure 14.29 Assign Tester to Test Case

You have now completed the preparation of test plan.

14.4.2 Handover Test Plans to Execution

Before you can handover the test plans to execution for the upcoming test phase, the assigned test cases need to have the status of **Prepared**. In the test plan overview, you can see the number of assigned test cases with their different-colored statuses (see Figure 14.30).

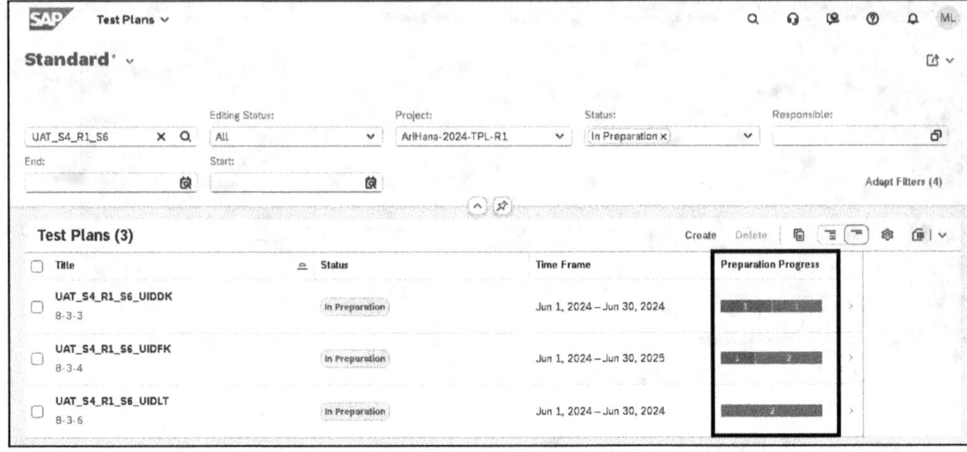

Figure 14.30 Preparation Progress of Assigned Test Cases

For test plan filtering, you can use your naming convention in the **Search** field to see relevant content only. All test plans that have test cases with **In Preparation** status (colored as a blue bar) are not ready for handover to execution. Ensure the readiness of test cases by setting their status to **Prepared**. To finalize the preparation, set the test plan's status in the **General Information** tab to **In Testing** (see Figure 14.31).

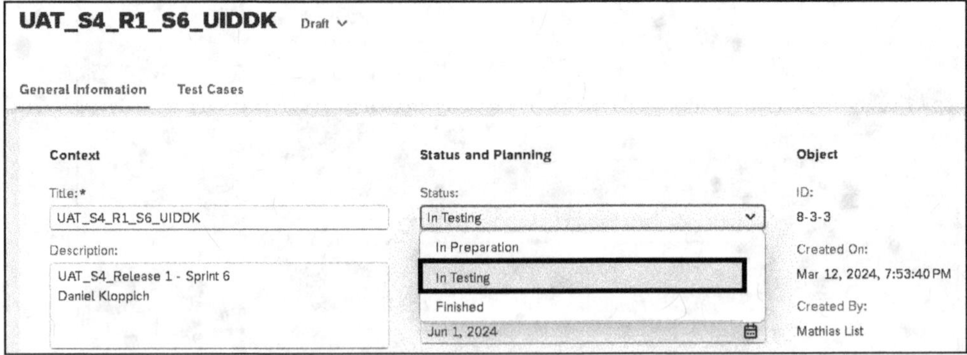

Figure 14.31 Handover Test Plan to Test Execution

14.5 Summary

The processing of requirements and user stories within the realize phase also includes the ongoing creation of associated test cases and test plans to be able to adequately test the developed function or run a test phase like an acceptance test or an integration test. This chapter has given you an overview of the creation of manual and automated test cases and the interaction between test cases and requirements as well as user stories. We have shown you how to link requirements and user stories to test cases. You also gained an insight into test planning with SAP Cloud ALM and saw how you can extend it to include project tasks. Now that you have learned how to release test cases for processing them, in the next chapter we will explain how to execute your prepared test cases and monitor the test phase. We will also show you how to document errors during testing using defects, with or without transports.

Chapter 15
Test Execution

Test execution checks the implemented SAP solution for compliance with your requirements. The test results are documented, and failed tests are logged and further processed using defect handling. A comprehensive overview of the test progress is important here. You should trace the progress of your tests and use successful test completion as milestones for the transition to the deployment phase.

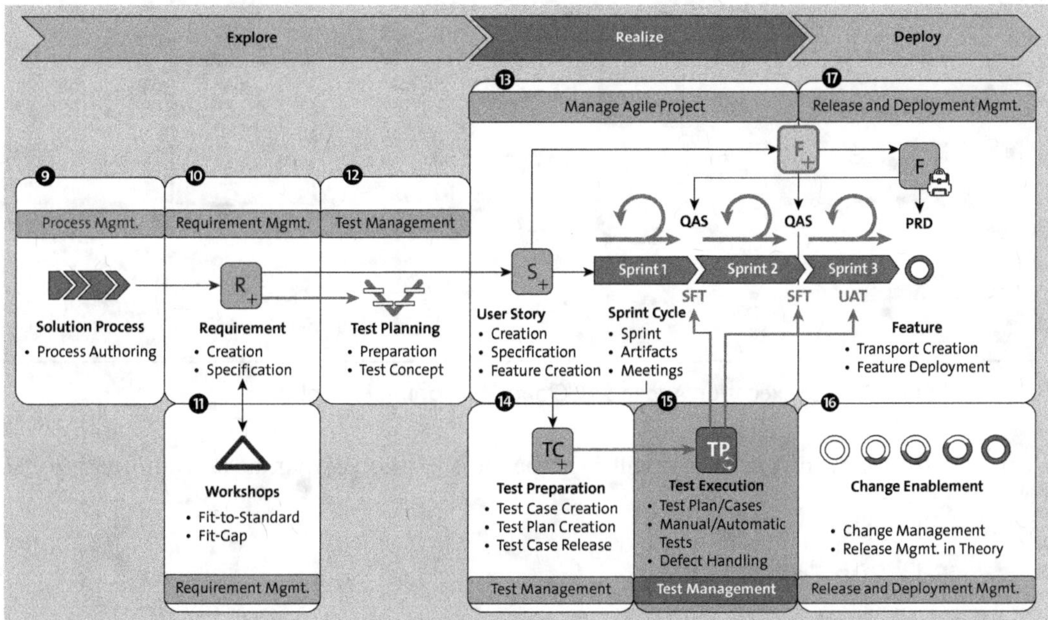

Once the user stories and/or requirements have been created, their quality and applicability must be checked in various test cycles. A single run of a test cycle during a test phase is not adequate. The execution of test cycles and the elimination of errors discovered during the tests is therefore an essential and recurring activity in the realization phase of your project.

The preparation of such a test cycle was explained in the previous chapter. A functional and clear framework is required to be able to adequately document and monitor the execution of the previously created test cases.

In this chapter, we will introduce you to the integrated execution of test cases and their documentation in SAP Cloud ALM. You will also gain insight into the troubleshooting process and the importance of correction transports into the test system as part of test management. In the role of test manager, you need to always have an overview of the current test phase and be able to evaluate it. We will show you which tools SAP Cloud ALM provides for this and give you tips on how you can document your results and transfer them to the deploy phase.

The entry point for test execution is the Test Execution app in the **SAP Cloud ALM for Implementation** app group (see Figure 15.1).

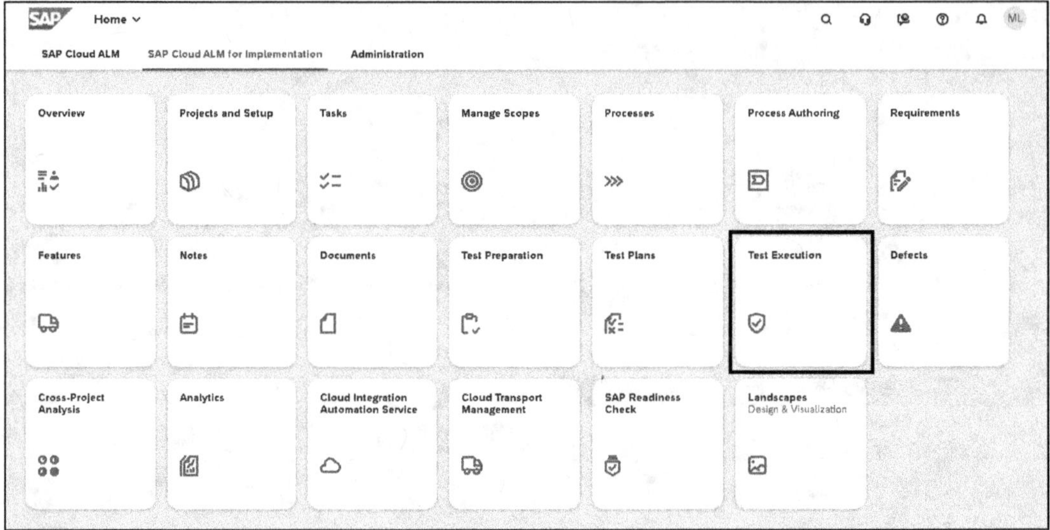

Figure 15.1 Test Execution App in SAP Cloud ALM for Implementation

In the following section, we will describe the execution of manual and automated tests.

15.1 Run Test Cycle

Once a test case and test plans are prepared, all test cases of a project are visible to the tester in the Test Execution app without an additional filtering. A common activity of a tester is to build up the current test worklist. This can be done by filtering by **Tester** and **Scope** (see Figure 15.2).

Additionally, we recommend you utilize the tagging function within the test case to optimize the structuring of test content and enhance your filtering options. For example, you can define a tag that points to an interface required to run the test case, and then filter in the Test Execution app for all test cases relevant to that interface. Another use case in which to use tags could be the need to classify a certain sprint or wave.

When the test plan is finalized and the test cases are assigned to dedicated testers, the inclusion of the test plan and tester as criteria in the filtering options is necessary. The option for a tester to save this predefined search as a separate tile is unfortunately not available. In the test case selection below the filter area, you can switch between being able to view all existing test cases, all manual test cases, or all automated test cases.

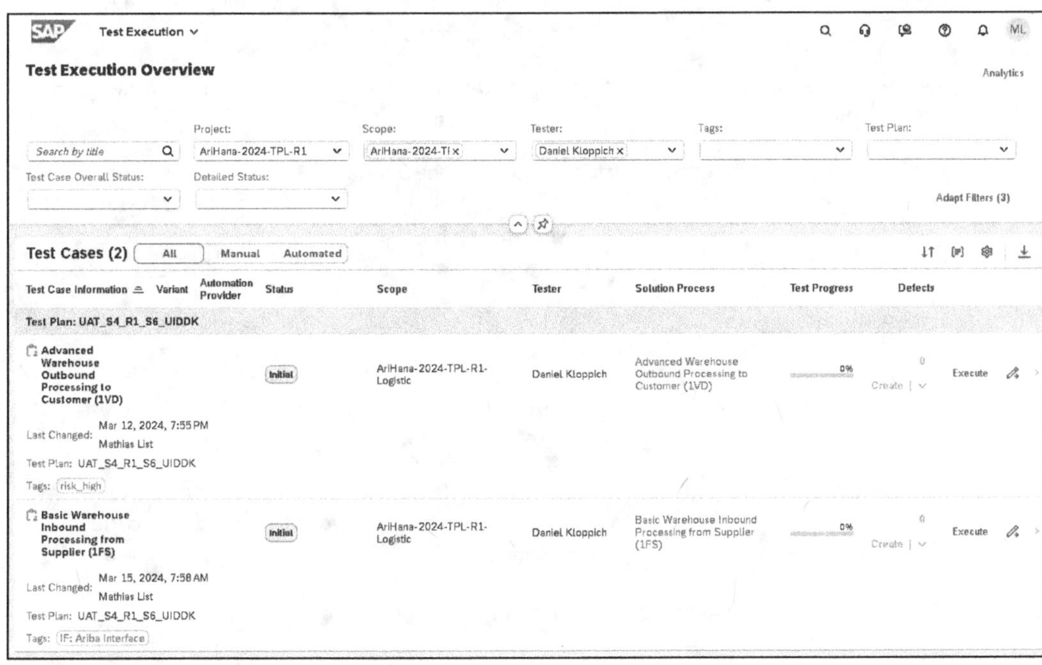

Figure 15.2 Tester Worklist in Test Execution App

Furthermore, the tester needs to be able to filter the current worklist by test case status. This status is found in the **Test Case Overall Status** field and documents the progress of the test through the statuses **Initial, In Progress, Passed**, and **Failed** (see Table 15.1).

Overall Status Value	Description
Initial	No statuses have yet been assigned to any test actions.
In Progress	The test run is open, and no test actions have failed so far.
Passed	The test run has been closed, and all test actions have passed.
Failed	At least one action has failed. This status is given as soon as one action has failed, even if the rest of the actions have passed or some actions haven't been executed yet.

Table 15.1 Overview of Test Case Overall Status

The test case's overall status (**Passed** or **Failed**) is determined by the test step status that was set during test execution (see Table 15.2). In the search, you can also filter the test by the **Detailed Status** field.

Detailed Status	Description
Pass	No errors have been detected. The application behavior matches the expected result.
Pass with Conditions	Minor issues that only slightly impair the use of the application have been detected.
Fail	The application behavior doesn't match the expected result, or the test case description is unclear. Select this status if you're unsure if it's an application issue or if the test case is inaccurate or out of date.
Fail - Application Issue	An issue with the application prevents you from executing the test action.
Fail - Test Case Correctness	There's an issue with the test case that needs to be corrected before testing can resume.
Fail - Retest Required	The previously detected issue has been addressed. You can resume the test execution and validate that the issue is indeed fixed.

Table 15.2 Overview of Detailed Test Case Status

In the next section, we explain the execution and documentation of manual and automated test cases.

15.1.1 Execution of Manual Test Cases

To run the test case, click on the **Execute** button to open a test case's content (see Figure 15.3). The overall **Status** is changed from **Initial** to **In Progress**.

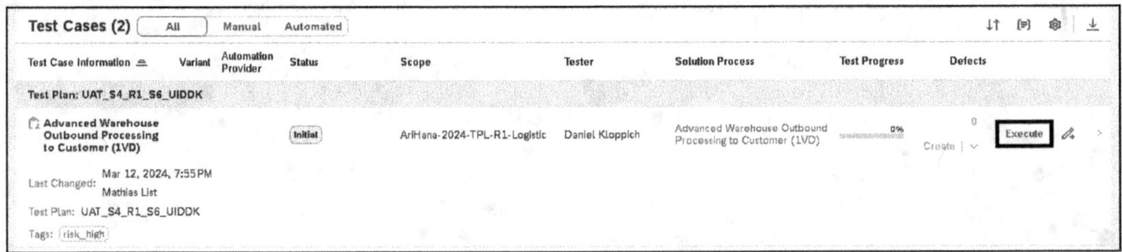

Figure 15.3 Start Execution of Manual Test Case

The test data is structured into several test steps that describe **Status, Title, Instructions, Expected Result**, and **Actions**. In Figure 15.4, you see a **Navigation** dropdown that contains 32 application test areas. Every application test area contains its own test steps. The **Navigation** dropdown allows you to navigate to a certain application test area directly by choosing an assigned test step.

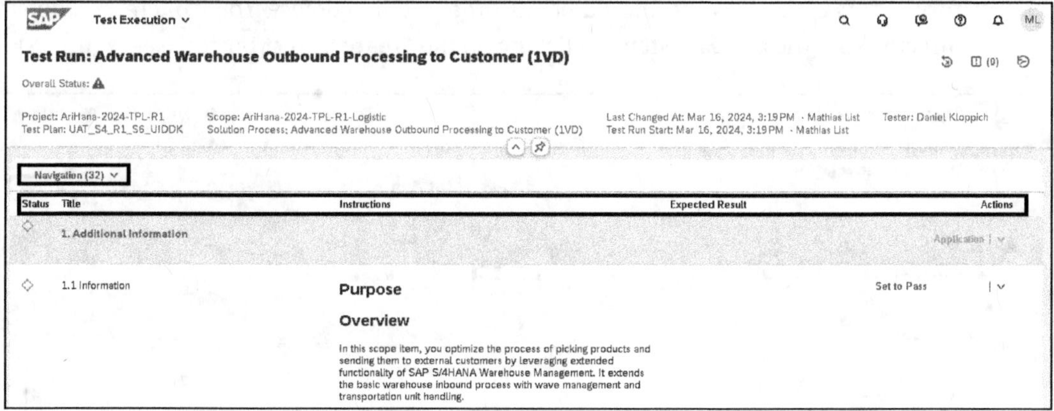

Figure 15.4 Test Case Data and Structure

As an example, we want to run testing of an application area called **Create Delivery**. The tester read the instructions and performed the test in the test system. The test result from the test system is compared with the **Expected Result** of the test case. The test step status is documented by the **Action** dropdown (see Figure 15.5).

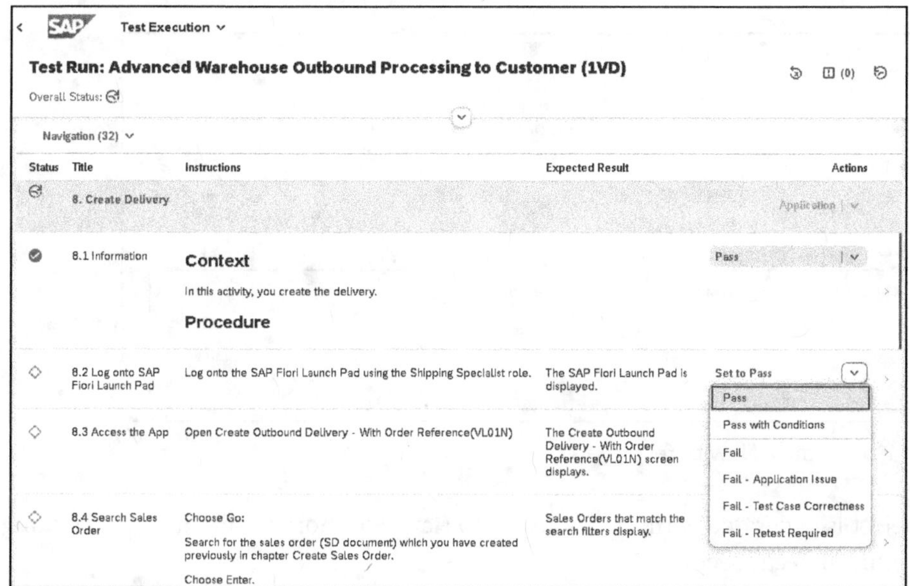

Figure 15.5 Report Test Step Result by Status

Click on the **Set to Pass** button to confirm the successful test of the selected test step. If a test step is selected, you can add comments to the document test results. As with test case creation, there are various options for text formatting. You can even insert screenshots.

If the status is set to **Fail**, the otherwise optional **Comments** text field becomes mandatory. Errors during testing should be documented in a defect. You can create a defect directly from the test case step by clicking on the **Create Defect** button (see Figure 15.6).

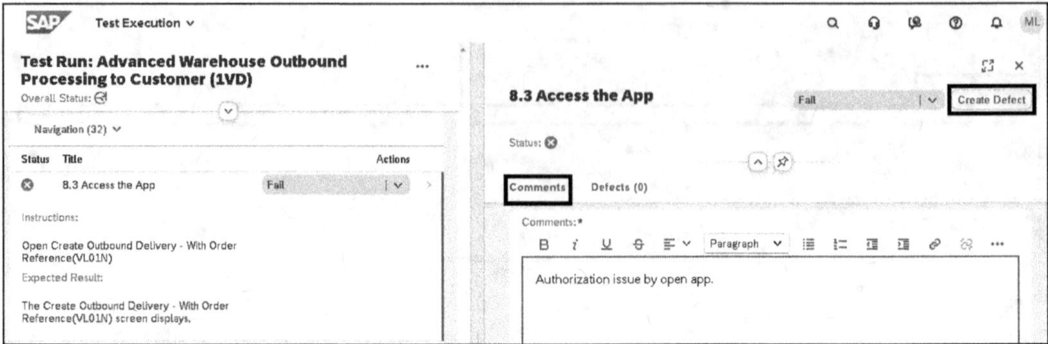

Figure 15.6 Failed Test Case Step Documented with Comments and Creation of Defect

During the defect creation, the test case name is used as the **Title** of the defect. Describe the failed test that needs to be reproduced and corrected (see Figure 15.7). When you save the new defect, an object ID is generated, and the defect and test case are automatically linked together.

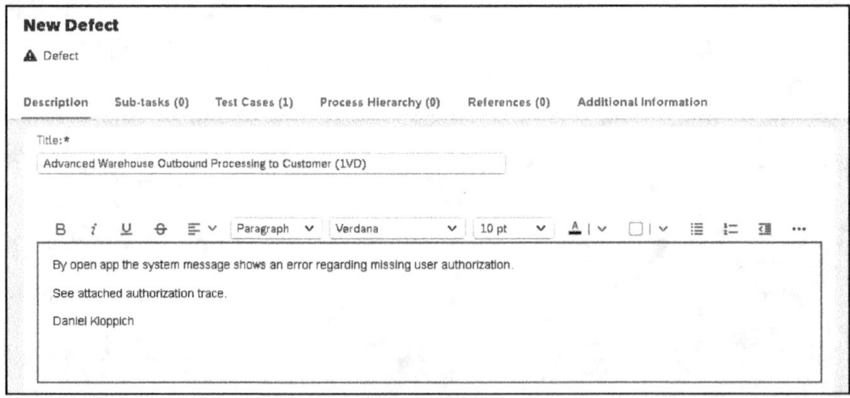

Figure 15.7 Create a New Defect

The defect is generated with the status set to **New**. For more information on creating and editing defects, Section 15.2.

[»]

Test Execution after Incorrect Test Case Step

If a test case step has been set to the **Fail** status, this *does not* block the execution of further test steps. If you have discovered an error during your test that does not hinder the execution of further test steps, you can continue with the test and continue to report back your test results step by step.

You can save your test case at any time, exit it, and continue testing at another time. In the Test Execution app, you can display the execution history of a test case by selecting it. As shown in Figure 15.8, you can view the progress of the respective test case, the last step tested, the time of the test, and the testers of all test runs carried out. You can continue working on open test runs by clicking the **Resume** button. Select the **Save and Close** button in the test case to end a test run. Results of previous test runs can be displayed by clicking on the test case to open the **Executed Test Runs** pane on the right hand-side. There, you see the history of execution with date and tester name (see Figure 15.8).

Figure 15.8 Test Progress and Execution History

The entire test case history is accessible during test case execution by clicking the **History** button, as shown in Figure 15.9. This history shows all changes during a test run.

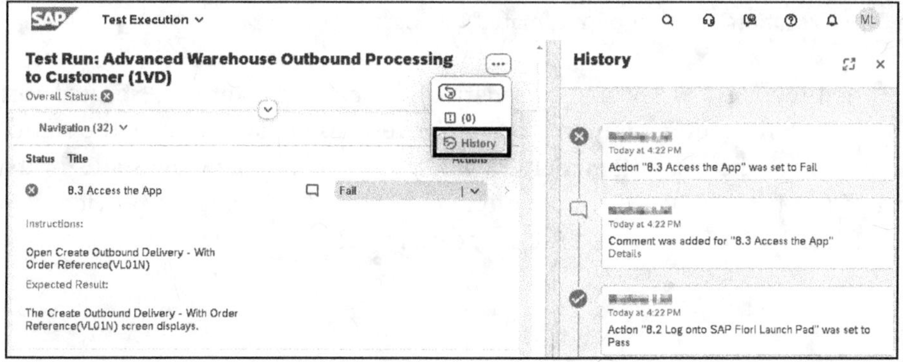

Figure 15.9 Change History of Test Case

15.1.2 Execution of Automated Test Cases

The following example of an automated test case demonstrates an integration of test automation with SAP S/4HANA Cloud Public Edition. You can find information regarding setup scenarios of test automation in Chapter 12, Section 12.3.

The execution of an automated test case is started in the Test Execution app by clicking on the **Execute** button. There, you search on the test case as described in the previous section. Alternatively, you can use the **Automation Provider** filter to show automated test cases only.

[»]

> **Scheduling Automated Tests Cases**
>
> SAP Cloud ALM does not currently (as of April 2024) offer scheduling for the automatic execution of test cases.

As soon as the test execution has been initiated, the status of the test case is displayed in the **Executed Test Runs** area (see Figure 15.10).

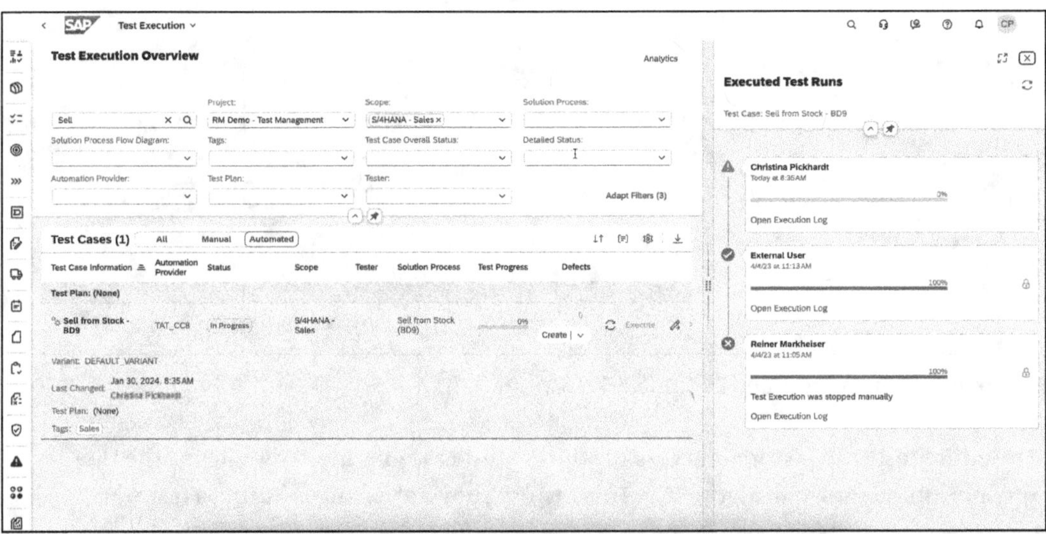

Figure 15.10 Execution Feedback of Automated Test Case

The reported status is not updated automatically. Click on the **Refresh Executed Test Runs** button for the automation provider of the test case to synchronize the status-related data from the test automation provider. You can also view the status of the automated test runs in the test automation provider itself. The **Log** tab provides you with a detailed overview of the test status of each test activity (see Figure 15.11).

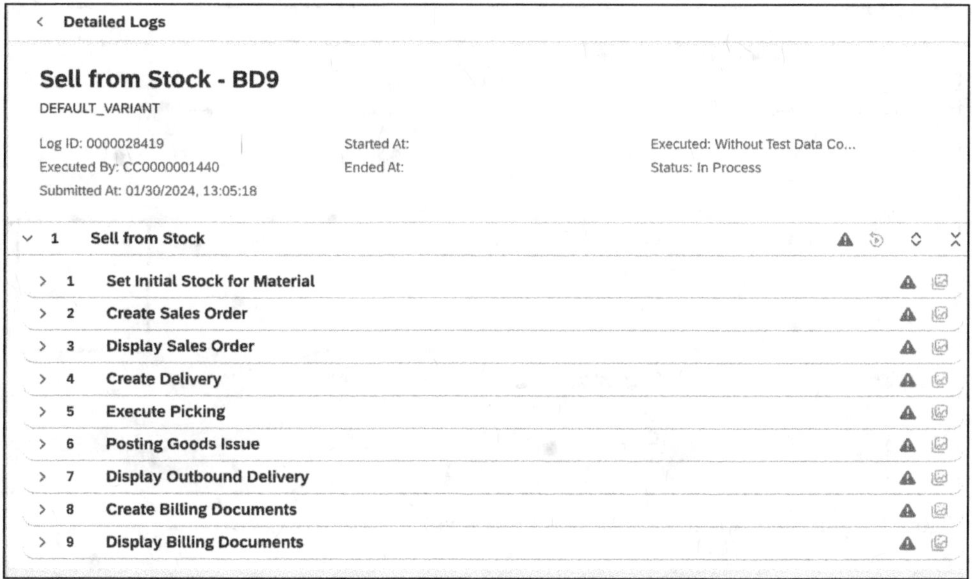

Figure 15.11 Protocol of an Automated Test Case within Test Automation

[+]

Test Execution via a Test Automation Provider

You can also execute automated test cases from a test automation provider, but the test results are not synchronized back to SAP Cloud ALM.

15.2 Defect Management

The aim of a test cycle or a solution walkthrough is to test the implanted solution process to ensure that the linked requirement or the user story is implemented error-free. Errors during testing should be reported by a defect to ensure traceability of correction(s) and retesting. All defects are managed by the Defects app in the **SAP Cloud ALM for Implementation** app group (see Figure 15.12). A defect is a task type and can therefore also be searched for in the Tasks app.

Using the Defects app, you can see all reported errors and the progress of processing. You can also limit your overview using the available filter criteria. The defect is created as a follow up to a failed test case step, as explained in Section 15.1.1. A scenario in which you would create a defect as standalone might be to report an error in a production system. Here you can create the defect in the app by clicking the **Create** button. Optionally, you could define a template to have a well-structured document for this situation (see Figure 15.13).

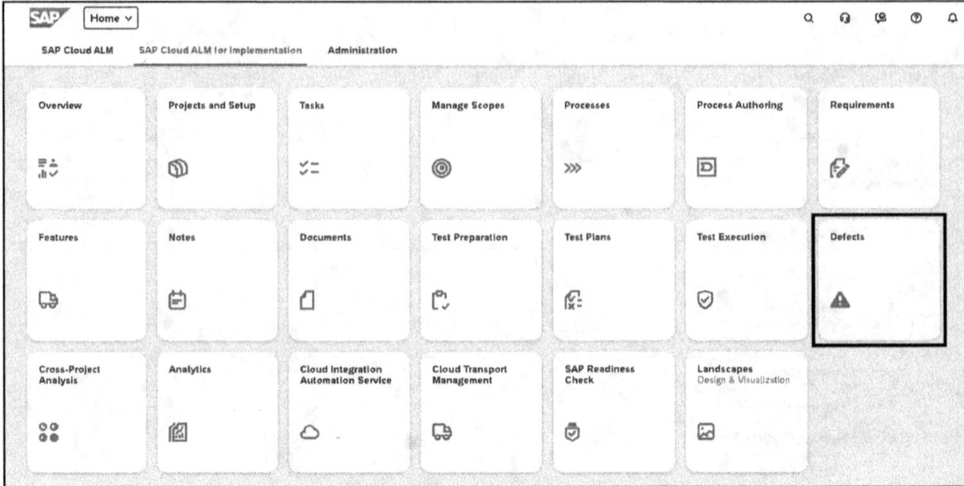

Figure 15.12 Defects App in the SAP Fiori Launchpad

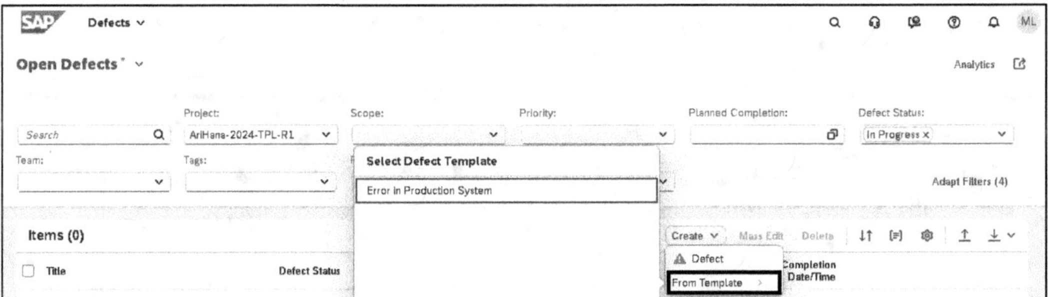

Figure 15.13 Create Defect as Standalone

The defect process chart in Figure 15.14 shows the workflow during test execution.

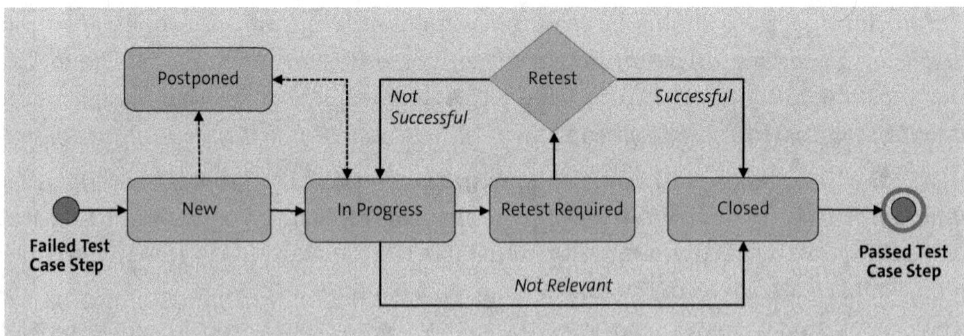

Figure 15.14 Defect Process

As the workflow begins, the initial status for the defect is **New**. The team member or responsible person needs to assign the first (dispatching) based scope, a description of the reported area, and potential solutions. A defect status of **Postponed** means that the processing is currently not possible—for example, due to missing resources or missing technical prerequisites. After a responsible person is assigned and the status is changed to **In Process**, and a deeper error analysis starts. The defect's responsible person provides a fix, which needs to be imported into the test system. Additional information can be added as comment for the tester. The defect is then set to **Retest Required** status, and the tester is assigned as the responsible person. If the retest is not successful, the tester changes the defect status to **In Process**. For a successful retest, the tester changes the defect's status to **Closed**, and the related test case step status is set to **Pass**.

15.2.1 Defect Dispatching

For new defects, the documented error needs to be investigated by the responsible person or process expert. To begin, choose the **Project** and **Scope** filters and then add information to the filters for **Defect Status** (**New**) and **Responsible/Team** (**Not Assigned** for both), as shown in Figure 15.15.

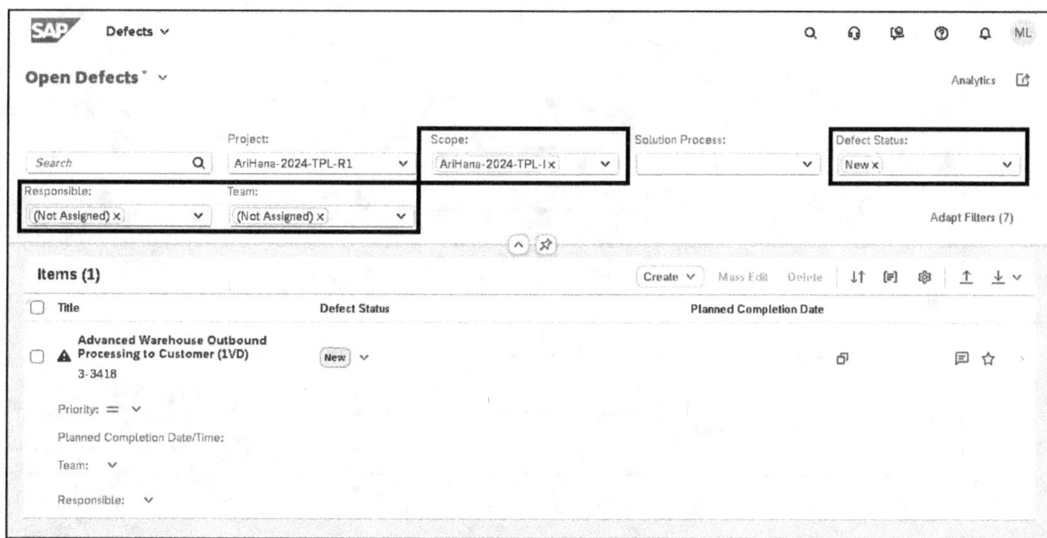

Figure 15.15 Undispatched Defects

Open the undispatched defect and fill in the required information to hand it over for processing. You will put most of this information in the **Additional Information** tab. Choose the **Release** to ensure the correction(s) will be delivered before the go-live date for the project. Assign the **Team**—or, if possible, assign a **Responsible** person who can perform the processing of the defect. Fill in the **Involved Parties** field to document the tester and to reassign them as responsible for the test if the test results in a status of **Retest Required**. Optionally, you can maintain the solution process if the reported error is related to it (see Figure 15.16).

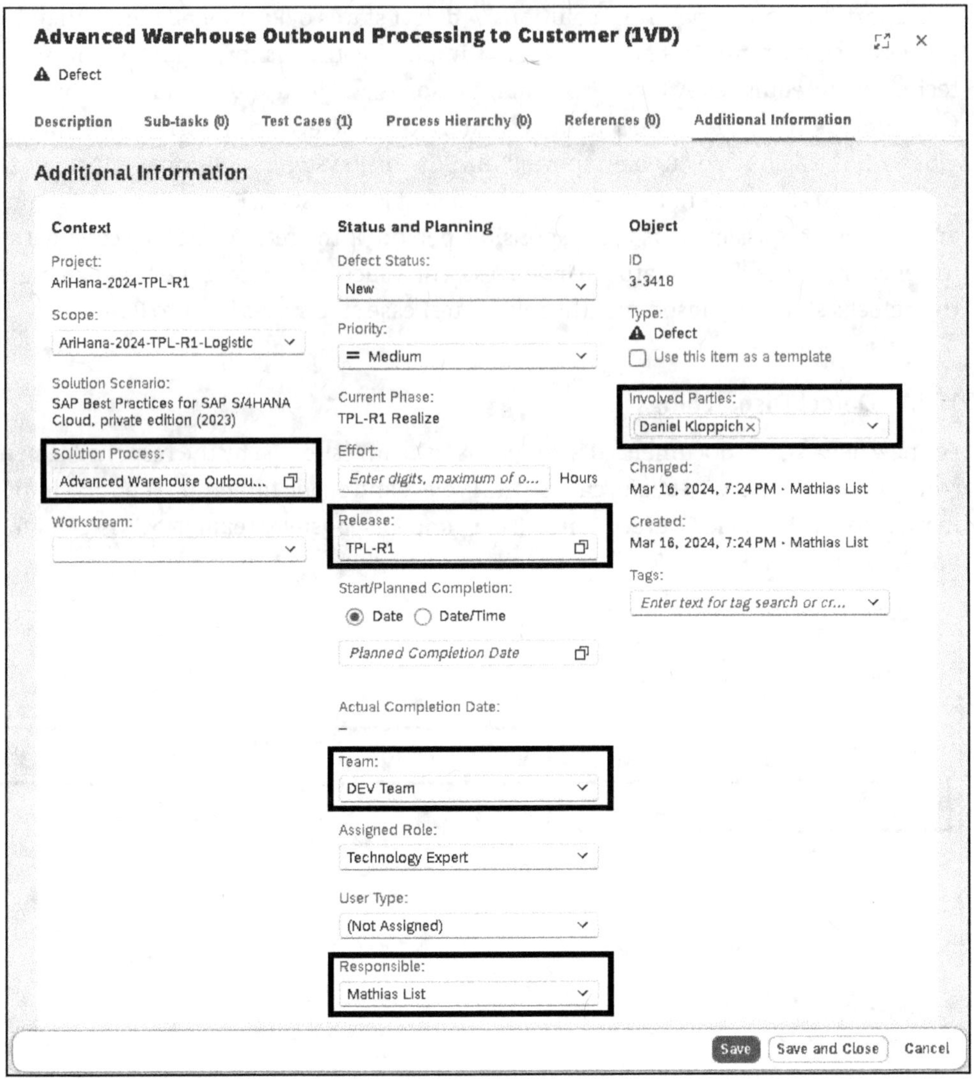

Advanced Warehouse Outbound Processing to Customer (1VD)
Defect
Description | Sub-tasks (0) | Test Cases (1) | Process Hierarchy (0) | References (0) | Additional Information
Additional Information
Context
Project:
AriHana-2024-TPL-R1
Scope:
AriHana-2024-TPL-R1-Logistic
Solution Scenario:
SAP Best Practices for SAP S/4HANA Cloud, private edition (2023)
Solution Process:
Advanced Warehouse Outbou...
Workstream:
Status and Planning
Defect Status:
New
Priority:
= Medium
Current Phase:
TPL-R1 Realize
Effort:
Enter digits, maximum of o... Hours
Release:
TPL-R1
Start/Planned Completion:
Date Date/Time
Planned Completion Date
Actual Completion Date:
–
Team:
DEV Team
Assigned Role:
Technology Expert
User Type:
(Not Assigned)
Responsible:
Mathias List
Object
ID
3-3418
Type:
Defect
Use this item as a template
Involved Parties:
Daniel Kloppich
Changed:
Mar 16, 2024, 7:24 PM · Mathias List
Created:
Mar 16, 2024, 7:24 PM · Mathias List
Tags:
Enter text for tag search or cr...
Save | Save and Close | Cancel

Figure 15.16 Enter Information to Dispatch Defect

Keep the defect status set to **New** and save your changes by clicking the **Save and Close** button. The person assigned in the **Responsible** dropdown is informed via menu notifications regarding dispatched defects, as shown in Figure 15.17.

Defect Dispatcher

The use of a defect dispatcher as a testing authority between the tester and developer is particularly useful in larger test phases or projects. In such a case, the defect dispatcher performs an initial check on the defect and assigns a responsible person for solving the defect.

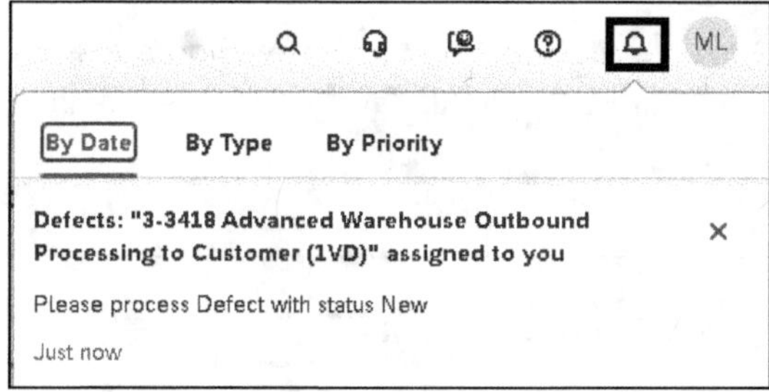

Figure 15.17 Defect Responsible Person Is Notified Regarding Status of New

15.2.2 Defect Processing

The person responsible can open the defect directly by clicking on the notification (see Figure 15.18) or filtering for the defect in the Defects app by entering their name in the **Responsible** field and setting the **Defect Status** to **New**. In the **Additional Information** tab, they then need to set the defect to **In Progress** status and start analysis and correction of the reported issue(s).

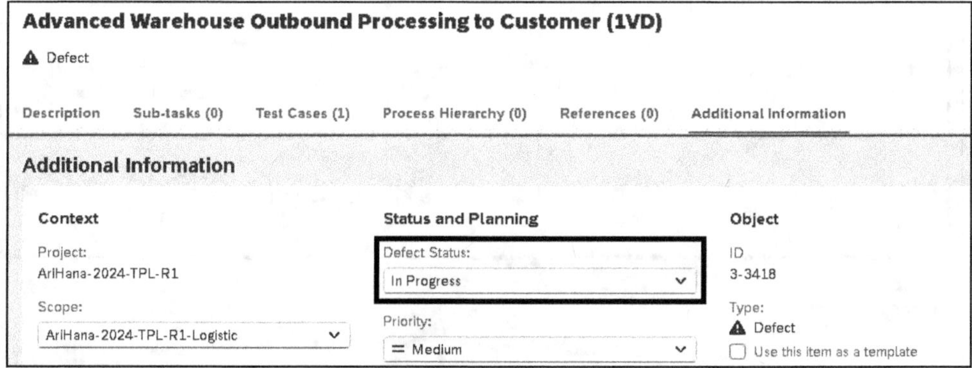

Figure 15.18 Signal Starting of Processing the Defect

For analysis, you can read the error description or jump to the linked test case to try to reproduce the error. Within the defect, you can create subtasks and assign test cases or references. The creation of subtasks is usually not carried out by the tester themselves, but by a defect dispatcher or responsible person. You might create subtasks if the defect needs to be solved with the cooperation of several experts (see Figure 15.19).

The progress of subtasks is shown in the defect header. Usually, all subtasks need to be complete upfront, before the defect can handed over for retesting.

[+]

Defect Worklist as Own Tile

During a test phase, many defects may occur to which you are assigned as the responsible person. Use the filter options in the Defects app to regularly view all defects assigned to you with the statuses of **New** and **In Progress**. As a tester, you are also responsible for defects with the status of **Retest Required**. Save your search in the Defects app as its own tile to have a quick access to the worklist.

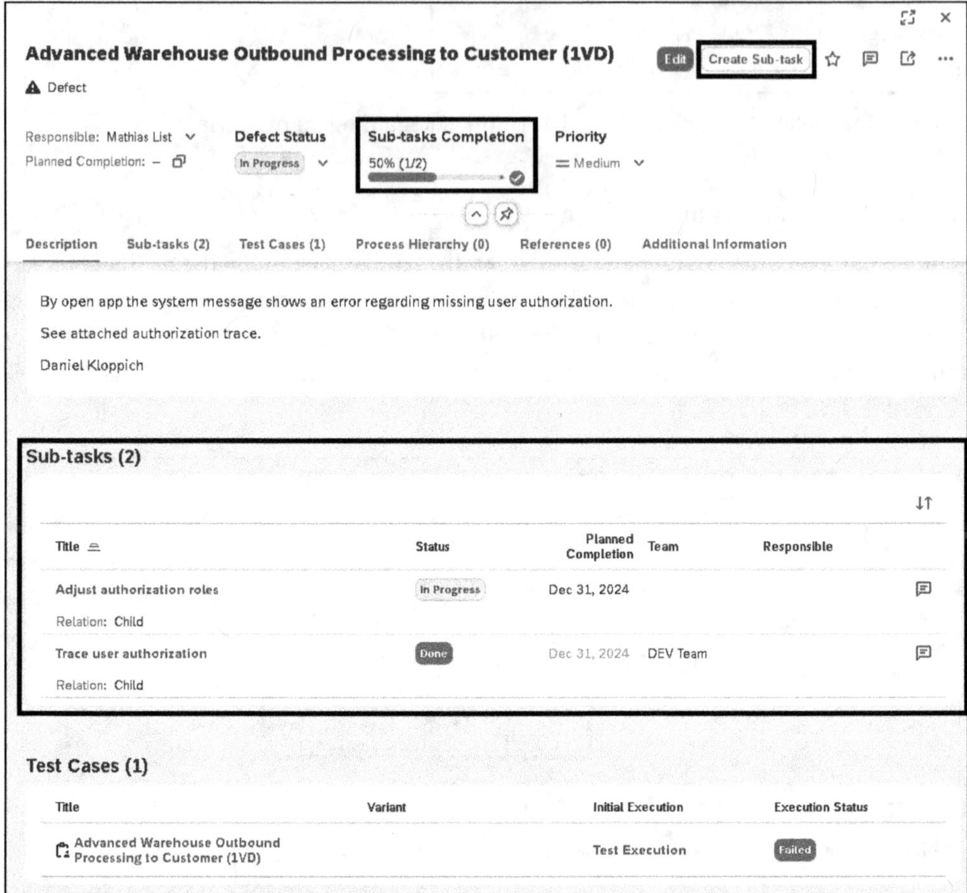

Figure 15.19 Split Defect into Subtasks to Delegate Activity

If a transport is required to resolve a defect, it cannot be linked to the defect itself. Features are the only process in SAP Cloud ALM that can be linked to a transport. If a transport is required to resolve a defect, we recommend using a feature. The entire process of features is explained in Chapter 17, Section 17.3. Figure 15.20 shows a feature that was created to manage the correction transport. Note that in the title of the feature, the defect object ID is maintained. The same description is used for the correction transport. Together with the assigned release, this helps ensures traceability.

Additionally, you can assign the feature URL as a reference within the defect to be able to track its progress. You can do this in the **References** tab by clicking on **Add**, as shown in Figure 15.21.

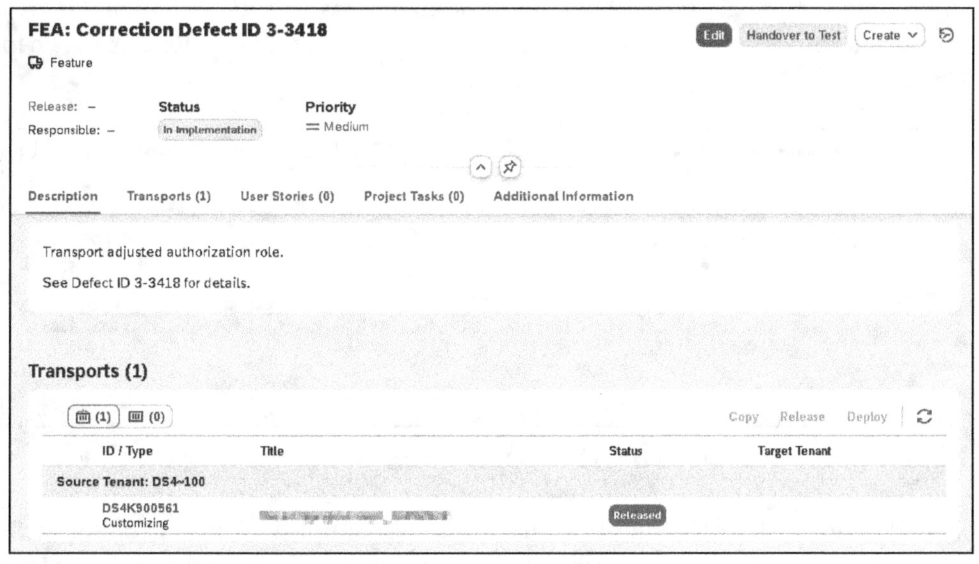

Figure 15.20 Use Feature to Manage Correction Transport

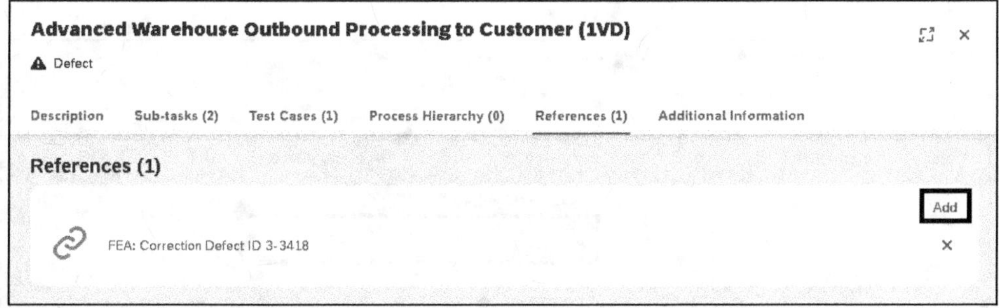

Figure 15.21 Link the Feature within Defect

After the linked subtasks are completed and correction transport is imported into test system, you can leave additional information for the tester as comment (see Figure 15.22).

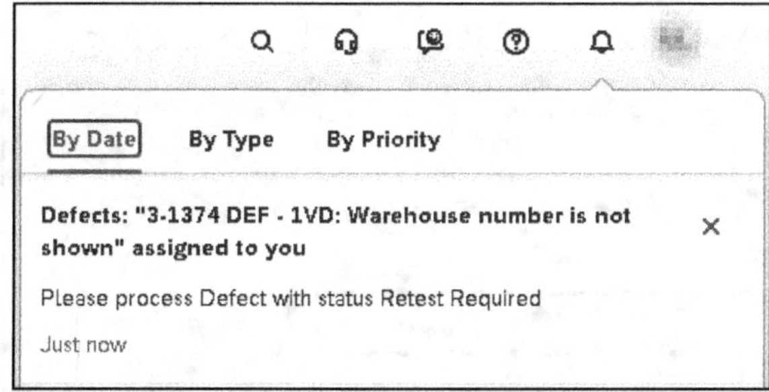

Figure 15.22 Defect Notification for Testers

In read mode of the defect, click on the **Show Comments** button The **Comments** area opens, and you can enter a new comment by clicking on the **Create** button, as shown in Figure 15.23.

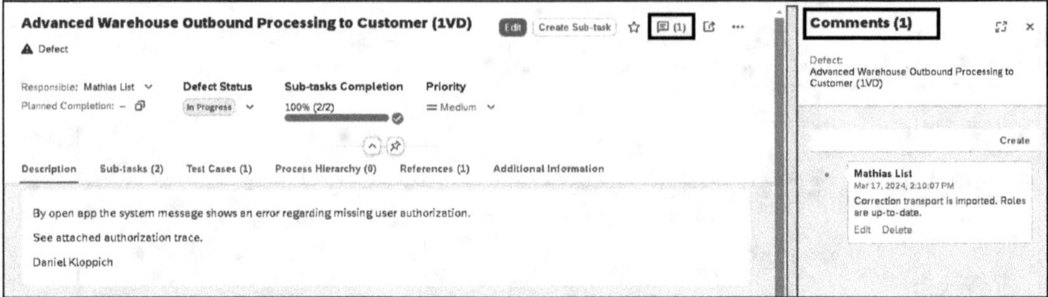

Figure 15.23 Comments in the Defect

To hand over the defect for retesting, change the **Responsible** information on the **Additional Information** tab to the tester and switch the **Defect Status** from **In Progress** into **Retest Required** (see Figure 15.24). The system will notify the tester as the new responsible person, and they can start the retest.

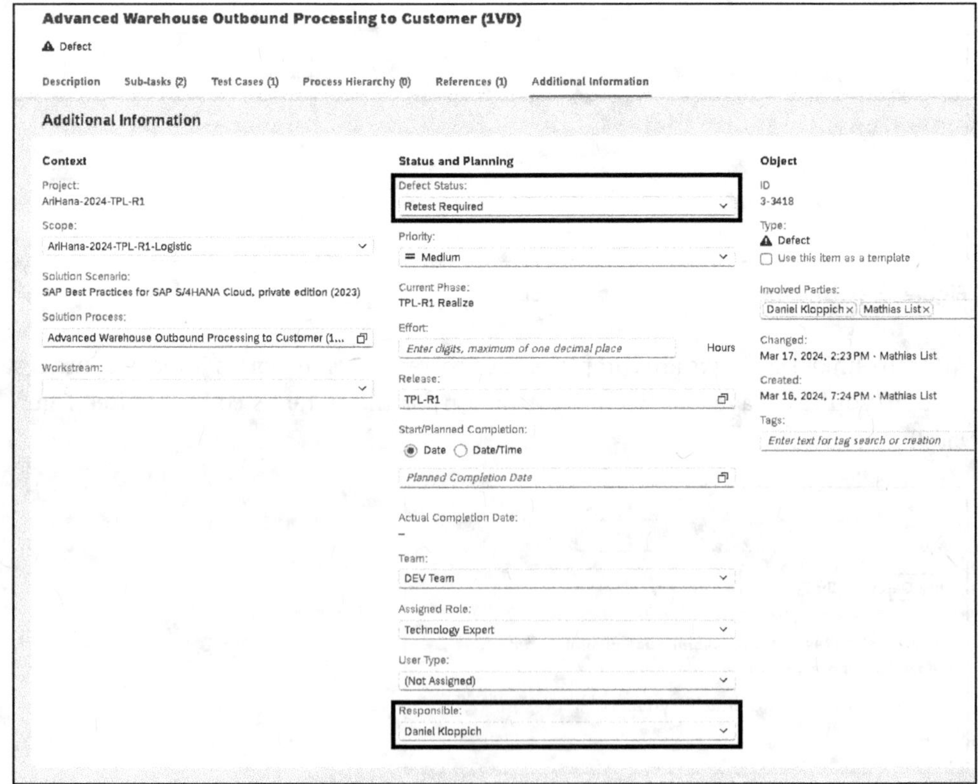

Figure 15.24 Hand Over Defect to Retest

15.2.3 Defect Closing

If a defect is corrected and set to the **Retest Required** status, the responsible tester can navigate from the notification into the defect or use the search function in the Defects app. Search criteria are the fields **Project** and **Scope** together with the **Responsible** and **Defect Status** fields. In the defect, navigate to the **Test Cases** tab and open it to resume the test. If the failed test step is not successfully retested, document this as a comment in the current defect and then change the **Responsible** person and switch the **Defect Status** to **In Progress** again. If the retest is successful, set the test case step to **Pass.** You can also leave a comment in the defect and confirm the defect is resolved by setting the status to **Closed**. Another way to switch the **Defect Status** is available in the defect header, where you can use the dropdown, as shown in Figure 15.25.

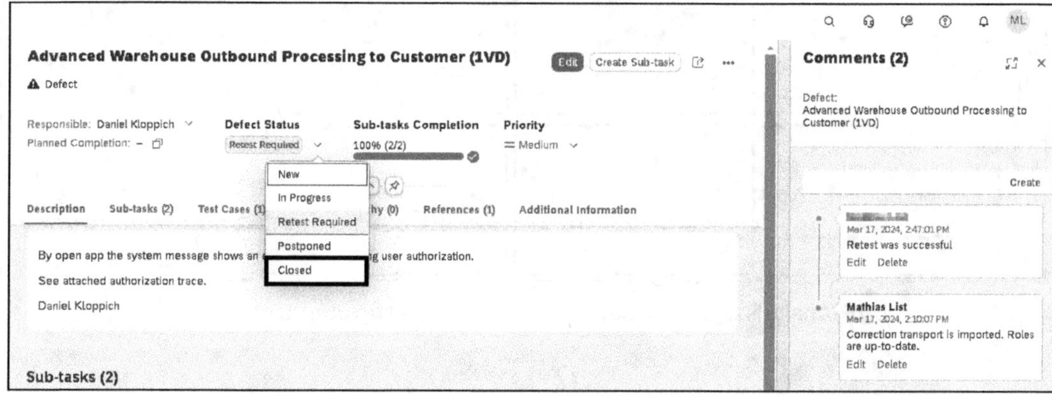

Figure 15.25 Closing Defect by Tester

15.3 Test Progress and Defect Monitoring

To promote a successful test cycle within the specified time period, SAP Cloud ALM offers you various analysis options for test progress monitoring. The options shown in Figure 15.26 serve as the entry point for the test and defect monitoring carried out by the test manager and, if applicable, the defect dispatcher. The following sections take a closer look at the analysis options and their relevance for test management:

- **Test Execution Analysis**
- **Defects Reporting**
- **Solution Processes Traceability**
- **Requirement Traceability**
- **User Story Traceability**

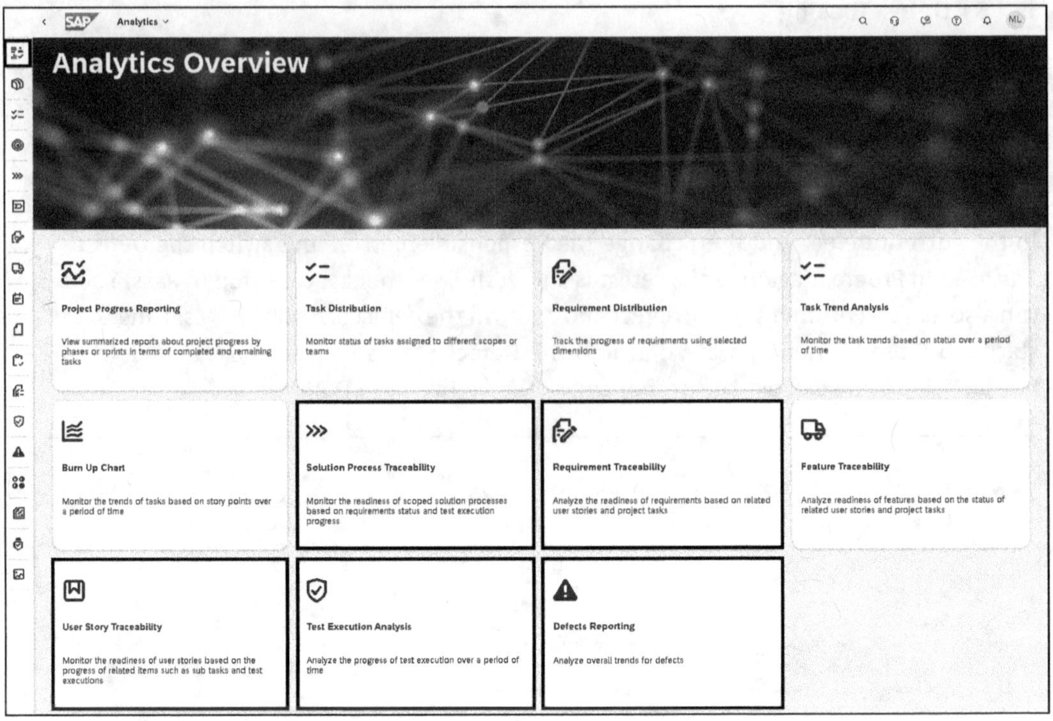

Figure 15.26 Analysis Options Relevant for Test Management

15.3.1 Test Execution Analysis

With its test execution analysis, SAP Cloud ALM provides an overview of the test cases released for testing. The dashboard serves as a tool for test management to track the history of test execution progress based on the test case status and testing speed. Note that the analytics functionality does not offer a display of test case titles or a direct jump to the respective test cases. For a closer look at specific test cases, you need to use the filter options in the Test Execution app.

In the filter options at the top of the screen, you can filter the test case display by **Project**, **Scope**, **Type (Automated/Manual)**, **Status**, **Time Frame**, and **Test Plan**. It is also possible to add the **Tags** filter as search criteria.

An overview of the progress of all test runs in the selected project is displayed on the **Execution by Status** tab. In Figure 15.27, this is illustrated using the project **AriHana-2024-TPL-R1**. The number of test runs over a certain timeframe is compared with the overall status.

When looking at the **Execution by Statuses** tab, you can switch between chart and table view using the chart icons at the top right. Using the diagram symbol on the right hand side, you can switch between the defined diagram types (see Figure 15.28).

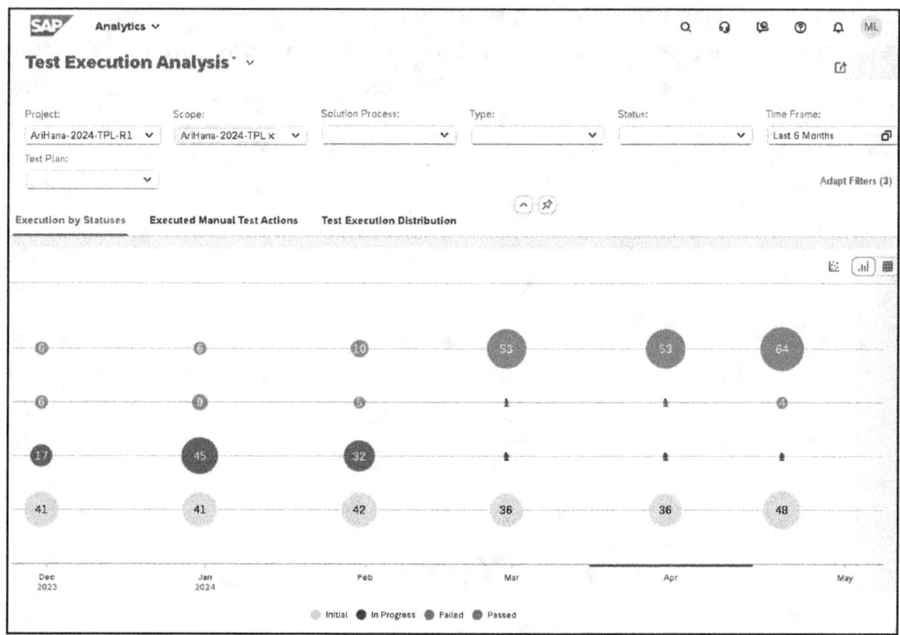

Figure 15.27 Test Execution Analysis: Execution by Statuses

Figure 15.28 Test Excecution Analytics: Changing Chart Types

On the **Executed Manual Test Actions** tab, you can track the total number of test actions. The number shown in Figure 15.29 on the bottom line (colored green) named **Executed Actions** shows the number of test actions executed daily in the test case. It provides the test manager with information about the daily speed of the testers within the test execution. The brown upper line, **Total Remaining Actions**, provides information on the total number of test actions still to be executed in the manual test cases.

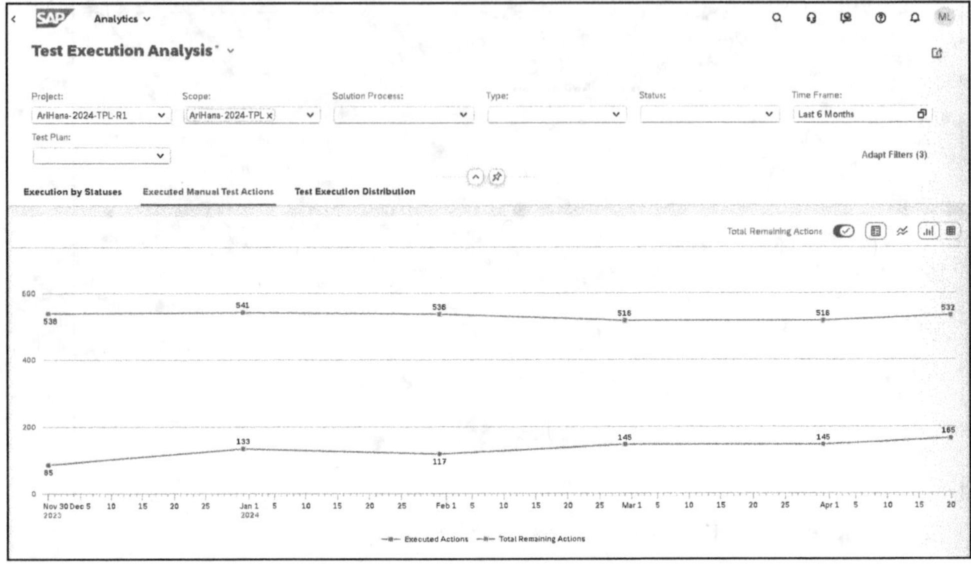

Figure 15.29 Test Execution Analysis: Executed Manual Test Actions

On the **Test Execution Distribution** tab, you can track the progress of your test executions across various scopes and test plans. You can filter your results by selecting **By Scope** or **By Test Plan** and review the status in a comparison chart (see Figure 15.30).

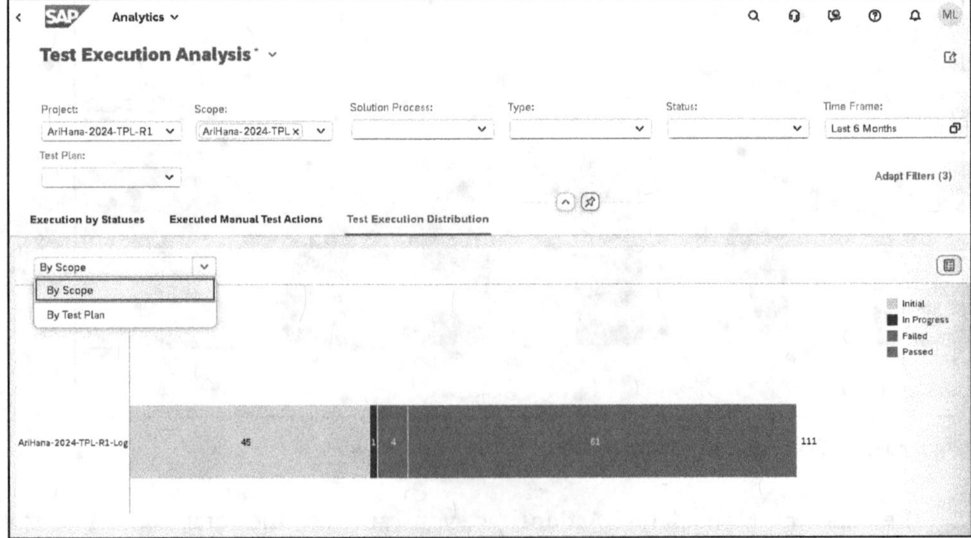

Figure 15.30 Test Execution Analysis: Test Execution Distribution

15.3.2 Defects Reporting

Defects reporting is used to analyze the overall trends of defects during the project. For the test manager and in particular the defect dispatcher, it provides an overview for tracking and monitoring the processing of defects. If a new defect has not been processed for some time or has been closed, it will show up that way in the defects reporting and addressed to the relevant person responsible. This analytics functionality does not offer a display of the defect details or a direct jump to individual defects, however. For a closer look at specific defects, use the filter options in the Defects app or the filter options described in Section 15.3.3.

As in the test execution analysis, you can use various filters in the defects reporting to search for defects and display different chart types. The **History** tab gives you an overview of the status of the defects within the selected timeframe. In Figure 15.31, this is demonstrated using our project **AriHana-2024-TPL-R1**.

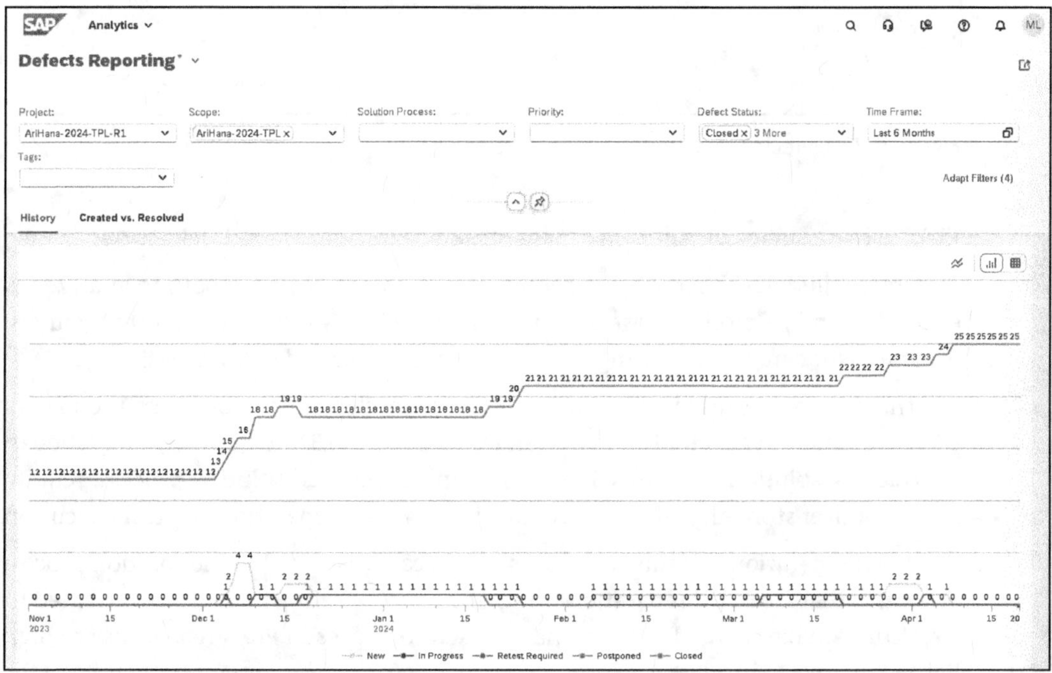

Figure 15.31 Defect Reporting: History View

The **Created vs. Resolved** tab focuses on creating and resolving defects. As shown in Figure 15.32, you can see an overview of the number of defects created and resolved over a day. You can switch the view between a chart and a table.

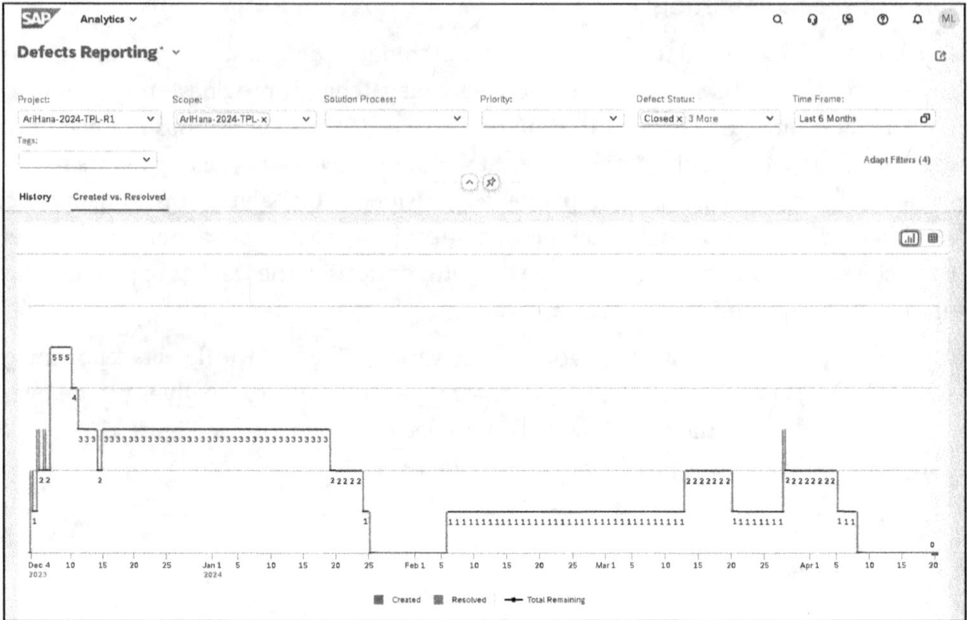

Figure 15.32 Defects Reporting: Created vs. Resolved

15.3.3 Traceability of Solution Processes, Requirements, and User Stories

The traceability dashboards described in this section can be used both to analyze test execution and to monitor test preparation. Therefore, we recommend that you also use the dashboards listed during the test planning described in Chapter 12.

The three overviews discussed in this chapter are similar in terms of their structure and their informative value for test management. In Figure 15.33, you can see the dashboard for tracking solution processes. Depending on the selected solution process, requirements, or user story, the analyses provide data on test preparation and test execution.

The **Test Preparation** column visualizes the test cases assigned to the solution process. Test cases that are in **In Preparation** status and have therefore not yet been released for a test run are highlighted in blue, and released test cases in **Prepared** status are highlighted in green. This data serves as a source of information for you in the role of test manager to check and advance the creation of test cases before a test phase or to initiate them in the event of missing test cases. The **Test Execution** column refers to the status of a test case within a test run. You can view the individual execution statuses of the released test cases.

In addition to the test execution analysis, you can get a direct overview of the test status in relation to a solution process, a requirement, or a user story. As already described, you can use the filter options to restrict the overview depending on the situation.

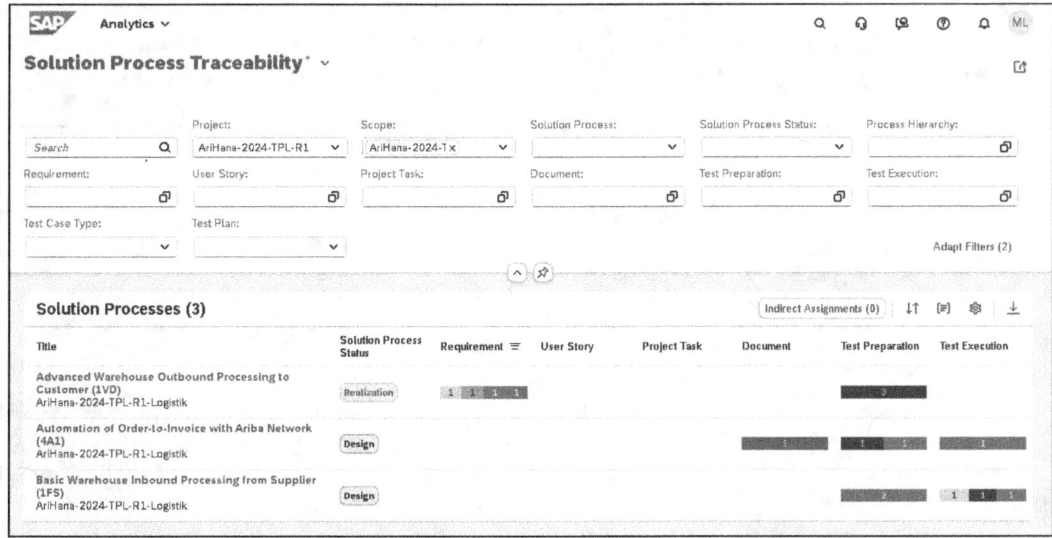

Figure 15.33 Solution Process Traceability with Test Information

You can use the **Indirect Assignment** button to display the defects resulting from a linked test case in the case of a requirement or a user story. In the case of a requirement as shown in Figure 15.34, it is also possible to display test cases or defects with an indirect assignment to the requirement. Test cases are also displayed that were not directly linked to the requirement in the Test Preparation app but have a connection to a user story that is linked to the requirement.

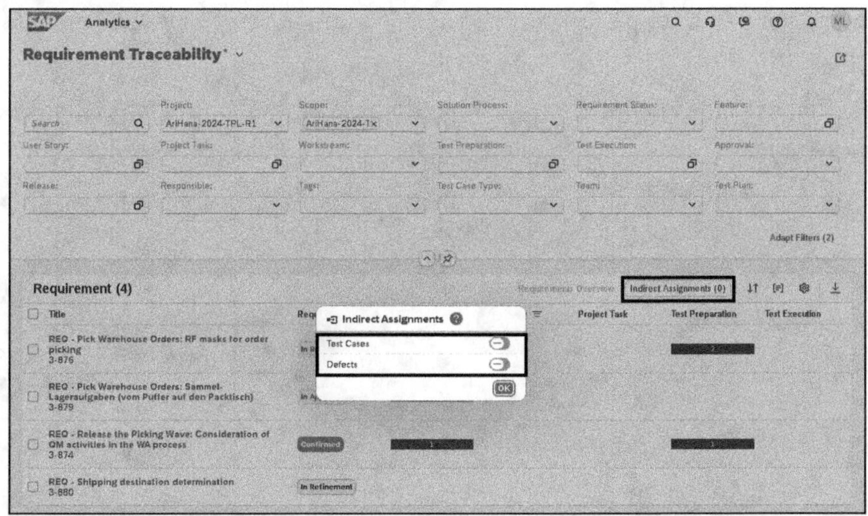

Figure 15.34 Display of Indirect Assignments in Requirement Traceability

By clicking on a colored bar in the **Test Preparation** or **Test Execution** column, a window opens, as shown in Figure 15.35, with an overview of the associated solution processes in the respective status. By selecting the linked test case, you can jump to the Test Preparation app or Test Execution app, depending on the selected column. This also applies to associated defects if you have displayed them as described earlier.

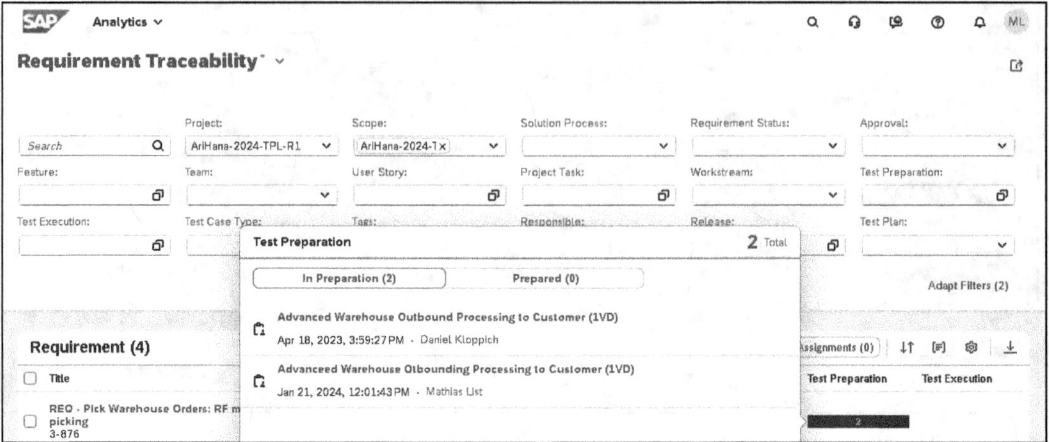

Figure 15.35 Navigate to Test Case or Defect from Chosen Requirement

15.3.4 Project Overview

In addition to the aforementioned dashboards available in the Analytics app, SAP Cloud ALM offers a project overview that can be accessed from any app via the **Overview** button. This provides, as shown in Figure 15.36, a visual overview of **Test Case Preparation**, **Test Execution**, **Failed Manual Test Runs**, and **Defects** for the selected project. As a monitoring tool for project management, the project overview also provides test management with a project-related overview.

Figure 15.36 Project Overview of Test Management

In comparison to the previously considered dashboards, it is not possible to switch between chart types or a graphical and tabular view in the project overview. However, you are offered a limited filter option via the available dropdown menus. You can jump directly to the corresponding application by clicking on the colored bars.

15.4 Documentation of Results and Handover for Deployment

Use the dashboards described in the previous sections to document your results following a test cycle or a solution walkthrough. As part of this documentation, evaluate the success of your test regarding the previously defined test strategy, test preparation, test execution, and defect processing. Include feedback from your test team and implementation team. Based on this, draw conclusions about possible improvements for your next test cycle or solution walkthrough.

The analysis options described in the previous sections help you to report the successful completion of a test phase example for a UAT or FIT test cycle. These test results are also proof of the completion of the realization phase (realize) for the tested release. The release is then handed over to the deploy phase for going live.

It is often desirable to have an overall test protocol that records the successful test cycle. For this purpose, the test results are summarized and presented in a structured manner according to defined areas. If you require this type of functionality, we would also like to draw your attention to the SAP Cloud ALM extension *X-ProCo* (Cross Project Cockpit) from NTT Data Business Solutions. This extension offers you context-related report generation, including for test management data from SAP Cloud ALM. Further information regarding X-ProCo can found in Chapter 5, Section 5.4.3.

An example of what X-ProCo can provide is comprehensive test reporting that you can present to the project management team which contains information about the associated release of the tested requirement, user story, or solution processes which is now ready for deployment to your productive system.

15.5 Summary

In this chapter, we described test execution and troubleshooting in SAP Cloud ALM using our example scenario. You now understand how testers run tests and report the status and how this affects the various dashboards for monitoring your test phase. During a defined test phase or in a realize phase, several test cycles with defined timeboxes need to be managed. For structuring these test cycles, we used test plans and test cases.

It was illustrated here how defects process is initiated in SAP Cloud ALM and how the testers and responsible persons process them. Required correction transports are managed with the help of features and by using dedicated defects. By closing the successful test phase, we move from the realize phase to the deploy phase.

In Chapter 16, we introduce the deploy phase with a theoretical consideration of the topic of change enablement. We will take a detailed look at release management in SAP Cloud ALM and hybrid system landscapes using our example scenario.

PART IV

Deploy

Chapter 16
Change Enablement

This chapter deals with the topic of change enablement, i.e., the ability to implement and manage changes. Management is viewed from a broader perspective with the bundling of related changes. This is also where the topic of release management comes into play, which is taught both theoretically and practically with reference to SAP Cloud ALM.

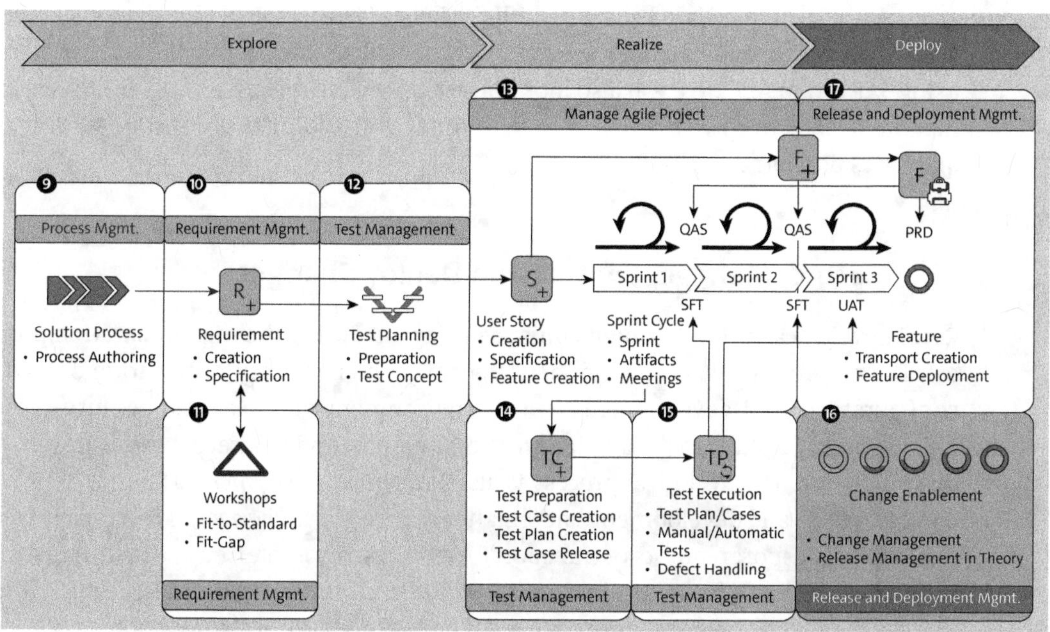

When the realize phase is completed, the deploy phase begins. We recommend that you get an overview of all the necessary roadmap tasks relating to the current deploy project phase. If you take our SAP S/4HANA project landscape as a basis, you will find all the roadmap tasks belonging to the deploy phase that are provided by SAP Activate. For example, the roadmap task "Production Cutover (New Implementation)" describes the activities for delivering the solution to the productive environment. To do this, the settings and developments tested in the realize phase must be recorded and assigned to a joint release so that a joint deployment can take place. This approach is not only used for the first rollout of the solution, but also for subsequent rollouts—due to new locations or new business areas, for example. First read how changes, such as settings or developments to be carried out, should generally be handled in Chapter 1. Information on release management—i.e., the merging of changes in preparation for release deployment—can be found in the following sections.

In this book, the term *change enablement*, mainly popularized by ITIL 4, refers to software changes. It should be possible to initiate and implement changes from every business area along the line of business (LoB). Each area should be able to define their requirements and have the experts for implementation. Certain change guidelines must be observed throughout the company—on one hand, to ensure the stability of the business application, and on the other to ensure possible audit-compliant specifications for the implementation of changes. The implementation, testing, and going live of the changes are specified by a defined change process. In the following section, we first answer the question of why changes are recorded and managed. We then look at the need to merge related changes with the help of releases. In addition to a theoretical look at the topic of release management, you will also find a practical explanation using SAP Cloud ALM. We will also look at the perspective for corrections (i.e., changes that are needed to solve a problem in the productive system or productive service as quickly as possible).

16.1 Record, Manage, Control, and Deploy Changes

Configurations and developments must be recorded during the implementation project. The reason for this is twofold: first, to ensure the necessary distribution of these configurations and developments in test environments, and second, because these records must be put into production in a certain order and at a certain time after successful testing. This technical process is also known as *deployment*. The recording of configurations and developments, hereinafter referred to as *changes*, is technically carried out in transport requests or transport tasks, hereinafter referred to as *transport*. An infrastructure must be set up in the SAP solution to be implemented to enable the management and control of changes. For an on-premise SAP S/4HANA solution, this is the *transport management system (TMS)*. An example of this can be found in Chapter 4, Section 4.2.2, as part of the project. The same applies to cloud-specific SAP solutions with the infrastructure of SAP Cloud Transport Management.

In addition to the technical basis for recording and managing changes, a predefined change process is often also defined to document the progress of changes as well as responsibilities. This is where SAP Cloud ALM uses features. This process enables the change to be recorded by the associated transport and forms the basis for its documentation, and the feature status is used to identify the current progress of the change. In Chapter 17, Section 17.2, you can find an introduction to the feature process. An approval process is often activated before a change is made; you can take this step in SAP Cloud ALM by using a request (see Chapter 10).

Before it is possible to control and deploy the changes using the feature, you have to create an integration with the managed SAP solution. See Chapter 4 to find out which transport tools are supported by the feature. When integrating an SAP S/4HANA on-premise system, for example, the transport can be created and released directly from the feature. The feature can then be used to deploy the change to the subsequent system, such as the test system and production system. The feature also offers the option of combining several related changes. To do this, the existing transports are added to the designated feature. In other words, the feature bundles these changes and thus enables joint control within the project landscape. Detailed explanations of the individual feature steps can be found in Chapter 17, Section 17.3.

16.2 Release Management in Theory

In theory, *release management* can be seen as a structured and disciplined method for managing software changes. This makes it a very important part of any successful IT project and deals with the planning, preparation, implementation, and monitoring of software releases. In this chapter, we explain the theoretical foundations of release management and present various methods and procedures that need to be considered when implementing an effective release strategy. The focus is particularly on the use of SAP Cloud ALM as a control tool and the template rollout approach as a release strategy.

Proper planning and execution of the release process ultimately helps to ensure that new functions and adaptations go into production quickly and effectively without creating unnecessary risks.

16.3 Importance of Release Management in Software Development

Release management is a critical aspect of software development that ensures changes to an application are released effectively and efficiently. This process is important to ensure that new features and bug fixes can go into production quickly, error-free, and without disruption.

Release management must interact with other important areas and teams in IT projects to coordinate an efficient change process:

- **Project management**
  This area is responsible for planning, monitoring, and controlling the project.
- **Requirements management**
  This area is responsible for recording, managing, and prioritizing requirements.

- **Quality assurance**
  This area is responsible for checking the implemented solutions for errors and defects and for the completeness of the documentation.
- **Development teams**
  Release management works closely with development teams to ensure that the software code is developed and tested according to the defined standards and guidelines and that problems can be identified during the development processes.
- **Test teams**
  Collaboration enables test periods to be planned, test plans to be developed, and defects to be identified effectively.
- **Operations teams**
  Collaboration includes planning and scheduling deployment activities, publishing to production environments, and implementing configuration requirements.
- **Change management teams**
  Collaboration includes planning, evaluation and approval, communication, monitoring, and control of changes.

16.4 Release Management with SAP Cloud ALM

This section deals with organizational tasks within release management with the aim of making release processes more efficient and effective. SAP Cloud ALM provides interesting functions and approaches for successfully managing the planning, coordination, and execution of the entire release lifecycle.

16.4.1 Developing a Sprint Concept

A sensible sprint concept always depends on the requirements and objectives of the project. As a rule, a sprint concept should enable the rapid and continuous provision of changes in the production system without jeopardizing the quality and stability of the system. Delivery after several sprints is possible.

A sensible sprint concept should also define a clear sequence of activities and goals to be achieved within a set time frame (the *sprint duration*). The goal of a sprint can be a rapid deployment of changes in the production system, but it can also be to combine several sprints to bundle larger functional changes or to implement them using a template rollout approach, for example. In both cases, the sprint concept should define a process for collaboration between developers, testers, and stakeholders to ensure a smooth and efficient transfer of requirements and changes to a production system.

In SAP Cloud ALM, the project is not the most important unit. Instead, the *import plan* (deployment plan) can be regarded as the decisive hierarchy level. An import plan can be assigned to several projects, but a project can have only one import plan. In our example (see Section 16.5), all rollout projects use the same import plan.

A release can only be assigned to a project using an attached import plan (via the Projects and Setup app). The import plan can in turn comprise several releases. A request is assigned directly to a release and does not need to be assigned to an import plan due to the restriction described above (see Figure 16.1). In the Gantt chart available in SAP Cloud ALM, the release and sprint periods can be viewed in relation to each other. Further information on this topic can be found in Chapter 17, Section 17.1.3.

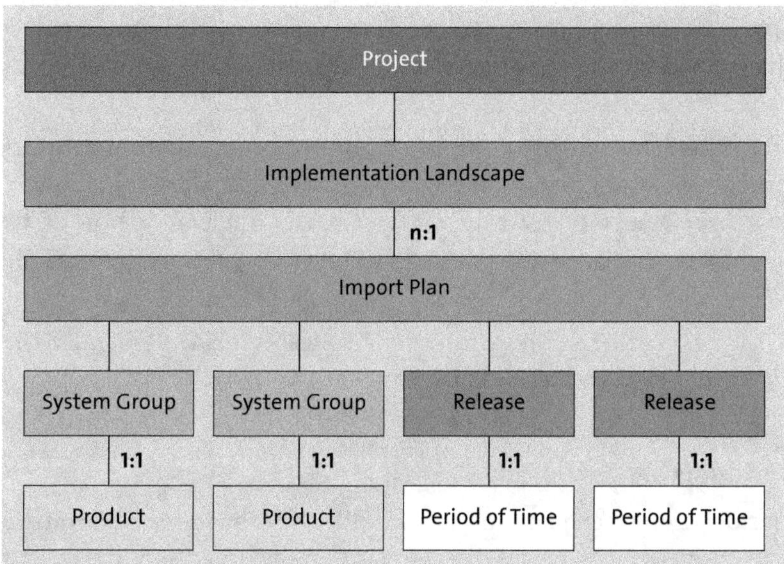

Figure 16.1 Classification of Releases in SAP Cloud ALM

A process can follow the following steps, for example, from the requirement to its implementation within a sprint period:

- **Creating a project**
  First, a project must be created in which the requirements that are to be processed in a sprint are collected and managed.
- **Definition of requirements**
  This involves collecting user stories that are to be processed in a sprint.
- **Planning the sprint**
  This is where the user stories are reviewed and the tasks to be completed in each sprint are defined.
- **Assignment of work tasks**
  All members of the development team are assigned work tasks with a clear target date.

- **Monitoring and controlling the sprint**
  This is where effective monitoring and control of the sprint progress takes place. It makes sense to regularly check the status of each work task.
- **Sprint completion**
  All work tasks completed in a sprint must be finally checked and approved.

16.4.2 Defining User Stories and Features

In SAP Cloud ALM, requirements are rolled out using features. Both entities are linked with each other for this purpose. A *feature* is therefore a function that is implemented as part of a software project to fulfill a specific requirement and can comprise several user stories.

A *user story*, on the other hand, is often considered part of a feature and describes the functionality from the end user's perspective. Furthermore, a user story contains various development components that are necessary for the implementation of the requirement. A user story can also be used as the basis for testing and acceptance criteria.

User stories and features are therefore two important concepts in SAP Cloud ALM that are used as part of agile project and release management.

16.4.3 Planning Sprints and Resources

An important consideration when planning sprints with SAP Cloud ALM is the ratio of requirements to features, as it can have a direct impact on the progress and efficiency of the project. The big advantage of one feature for each requirement (1:1 relationship) is that the project is broken down into smaller, more manageable chunks. This allows the team to progress quickly and implement the requirements incrementally. It can also help to increase transparency, as any progress can be traced back to a specific requirement.

However, the major disadvantage of this method is that it can lead to an overload of features that might not be useful or have not been sufficiently tested. This can lead to the project becoming overly complex, making it difficult to maintain an overview.

On the other hand, a single large feature with many requirements (1:*n* relationship) can help reduce the complexity of the project by summarizing the requirements in a larger context. This can help to better structure the project overall and give resources or teams a good big-picture overview of its progress. One disadvantage of this method can be that it becomes more difficult to make progress, as the team may have to work on and keep track of many requirements at the same time. In addition, it can be more difficult to attribute progress to specific requirements, which affects transparency.

Ultimately, the choice between having a large feature with *n* transport requests, many user stories, and many requirements or having one feature (each with a few transport requests) per requirement (each with 1–5 user stories) depends on your specific project requirements and circumstances. It is advisable to try out both approaches in smaller, upstream projects and see which works best.

Take, for example, the requirements for test execution and the organization of the test periods (see Figure 16.2). Both are important factors that should be considered when deciding on the ratio of requirements to features. If testing takes place after every sprint, it's better to implement one feature per requirement, as this allows you to organize a targeted import into the quality assurance system. This scenario allows you to test in a more coordinated manner, as the changes can be tested in smaller parts. Furthermore, possible errors can be quickly identified and subsequently rectified.

If the changes within a release are only transported after several sprints, it tends to point toward using a large feature that covers several requirements. Conversely, this can make it more difficult to identify and rectify possible errors, as the changes are more extensive.

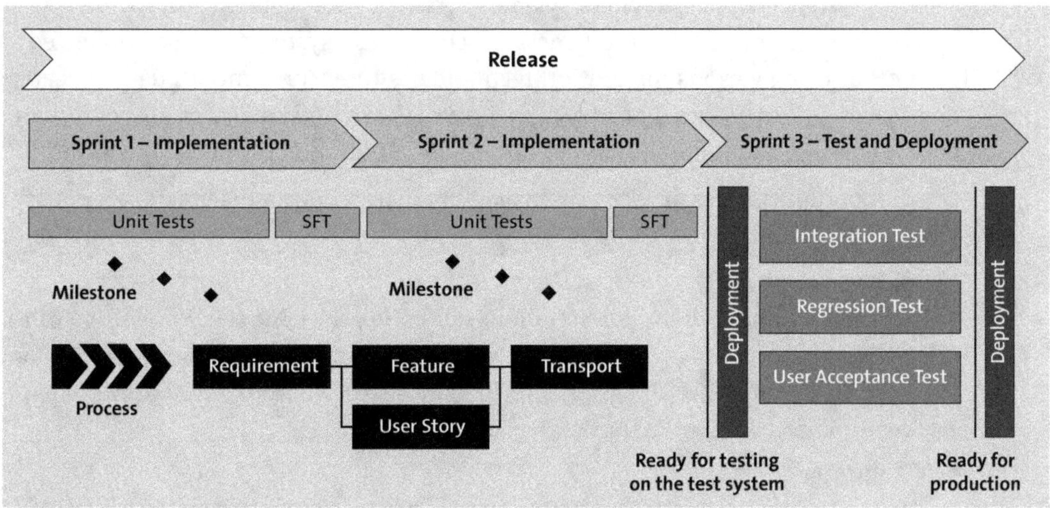

Figure 16.2 Test Execution After Several Sprints Within a Release

16.4.4 Carrying Out Sprints and Monitoring Progress

SAP Cloud ALM allows you to monitor the progress of sprints in real time. As described in the previous sections, a *sprint* is an iteration cycle in which the team works on specific user stories or features that have been previously selected. In SAP Cloud ALM, sprints can be created with the associated user stories and features. These are then assigned to the respective members of the development team.

During sprint execution, developers generally update their assigned user stories or features and use them to document the progress of their work. For example, they change the status of a user story or feature, enter the estimated effort, or indicate the progress as a percentage. This information is updated in real time and can be viewed by other team members, the release manager, and the project manager.

SAP Cloud ALM also enables the tracking of dependencies, as well as the use of dashboards and the creation of burndown charts to visualize the progress of the sprint. With these functions, the team can keep an eye on progress and adjust if necessary to achieve the sprint goal on time.

16.5 Release Management in Hybrid System Landscapes

Release management in hybrid system landscapes refers to the coordination and implementation of software releases across multiple systems, which can be located both in the public cloud and on-premise. The exact configuration depends on the company's requirements, but usually the business processes are executed in the on-premise systems, while the cloud systems serve as a supplement to provide certain functions or optimize certain business processes. The systems can be connected to each other using various integration technologies to ensure smooth data exchange. The challenge is to ensure that all components of the hybrid landscape are coordinated during a release and that release planning is synchronized in the different systems.

Some important aspects of release management in hybrid system landscapes are as follows:

- **Release planning**
  Release planning in hybrid system landscapes requires detailed knowledge of the existing systems, their dependencies, and the possible effects of changes. Good planning ensures that all systems in the landscape are updated on time to avoid downtime and guarantee smooth release processes.
- **Versioning**
  Versioning is an important aspect of software lifecycle management. In a hybrid SAP system landscape, there are different systems that interact with each other, whereby the systems can have different versions of SAP solutions. If, for example, a hybrid system landscape contains an on-premise system landscape and a public cloud system landscape, each system may contain different release statuses, making it difficult to implement changes and updates. This is where a clear concept for versioning the software in different systems helps to ensure compatibility between the different versions. Such a concept should ensure that all systems are up to date and that changes are properly documented.

- **Test and deployment processes**
  Clear test and deployment processes should be defined for the software in the various systems. By carefully planning and coordinating tests and deployment, potential problems can be avoided or identified at an early stage before they occur in production.
- **Automation**
  Especially in SAP system landscapes that contain different versions of SAP software in different systems, the automation of release processes can be challenging, which requires careful planning and implementation as well as the use of suitable tools and technologies. For example, CI/CD tools such as Jenkins, GitLab CI/CD, or Azure DevOps can be used to automatically check, test, and deploy changes to the software.
- **Monitoring**
  The monitoring of hybrid system landscapes is intended to ensure high system availability, performance, and security. It ensures that all systems and components function smoothly and communicate with each other. For example, availability monitoring can enable a rapid response with the help of defined alarms that are automatically triggered in the event of failures and malfunctions. Furthermore, system monitoring for important parameters such as CPU load, memory utilization, and database utilization as well as the monitoring of business processes has proven its worth.

In a hybrid system landscape, companies use both local IT systems and cloud-based systems. To manage such a landscape in SAP Cloud ALM, a few steps need to be followed:

- Integration: Check whether the local and cloud-based systems can be integrated with each other. SAP Cloud ALM offers API-based integration options to synchronize data and processes.
- Data management: Check how you can transfer data between systems securely and efficiently.
- Process adaptations: Check whether the processes and workflows in SAP Cloud ALM need to be adapted to the requirements of your hybrid landscape.
- Monitoring and verification: Monitor and verify the integrity and efficiency of data transfer between systems and ensure that all processes run smoothly.
- Documentation: Document the integration processes and procedures to ensure that everyone involved is informed about the hybrid system landscape and can access the documentation at any time.

API-based integration options allow companies to integrate systems in a hybrid landscape by using APIs (application programming interfaces). Using API integration, data and functions can be transferred from one system to another without the systems having to be directly connected to each other.

These integration options are particularly useful for companies that use different systems and applications but still want to maintain a uniform database and business processes. API-based integration options in SAP Cloud ALM can lead to improved data quality, better data availability, and greater efficiency in the handling of business processes.

An important aspect of the API-based integration concept is the use of REST APIs that communicate via standardized HTTP protocols. This makes integration easier, as many systems and applications already have REST APIs and can therefore be easily connected.

One example of the application of an API-based integration concept in SAP Cloud ALM can be the transfer of data between an SAP S/4HANA system and SAP SuccessFactors. REST APIs can be used here to automate the transfer of personnel and organizational data.

[»]

REST APIs

REST stands for *representational state transfer*. *REST APIs* are a standard for transferring data on the internet. It is a client-server concept in which the client (e.g. a web application or a mobile app) sends requests to its server and receives responses from the server. REST APIs use the HTTP protocol to send and receive data. The data is transferred via *resources* that are provided on the server. A resource can be identified by a unique URL. The use of HTTP methods (e.g., GET, POST, PUT, and DELETE) allows access and interaction.

REST APIs play a role in SAP Cloud ALM as they provide a simple and efficient way to exchange data and functionalities between different systems. This makes it possible to provide and receive data in a standardized format, which simplifies the integration between SAP Cloud ALM and other systems.

It is important to note that implementing an API integration in a hybrid landscape requires thorough planning and consideration to ensure that the integration solution is secure, reliable, and effective. It may also be helpful to engage an experienced consultant or service provider to support the implementation.

If you want to implement an API-based integration option for your hybrid system landscape in SAP Cloud ALM, there are some additional important considerations you should bear in mind:

- Integration requirements: Think about which data and functions you would like to integrate between the systems and which requirements this integration must fulfill.
- Security: Ensure that the integration solution is secure and that your data is protected. Consider how you can implement access controls and authentication.

- Data quality: Consider how you can ensure data quality, especially when data is transferred between systems.
- Scalability: Consider how you can scale the integration solution as your requirements grow.
- Monitoring and troubleshooting: Consider how you can monitor the integration solution and troubleshoot if problems occur.
- Documentation: Think about how you will document the integration solution and how you can ensure that everyone involved is informed about the integration processes.

[«]

Scalability

Scalability describes the ability of an IT solution to adapt to changing requirements by expanding or shrinking without compromising performance or reliability. This can mean that the solution can support more users, process more data, or automate more processes while remaining stable. A scalable solution is important to meet demand over time without requiring an overhaul of the entire IT infrastructure.

16.5.1 Template Rollout Approach

The *template rollout approach* is also a common method for hybrid system architectures to introduce SAP system solutions at several locations or in several companies. This involves creating a standard solution, known as a *template*, which serves as the basis for all further installations. This solution is then subsequently adapted and implemented at the various locations or companies by making local changes. Further information on this topic can be found in Chapter 5, Section 5.3.4.

[«]

Rollout vs. Roll-In

Rollout and roll-in are terms that are often used in project management to describe the process of introducing changes or new systems.

A *rollout* refers to the process of introducing a change or new system from a central source to multiple locations or users. An example of this could be the introduction of new software and therefore new business processes in a company, where the software is distributed to all employees in different departments or at different locations.

A *roll-in*, on the other hand, refers to the process of introducing changes or a new system from multiple sources to a central location. An example of this could be the integration of data from different departments or locations into a central system.

To summarize, a rollout describes the process of introducing changes from a central source to multiple locations or users, while a roll-in describes the process of introducing changes from multiple sources to a central location.

To implement the template rollout approach in an SAP project with SAP Cloud ALM, the following steps, among others, should be taken:

- Planning and preparation: In this step, the requirements of the various locations or companies are analyzed and documented. The template can be developed and planned based on this information.
- Template creation: The template is created by modelling and standardizing the specific business processes that are the same for all locations or companies.
- Adaptation of the template: The template is adapted to the specific requirements of each location or company. Local requirements, such as tax laws or cultural differences, are considered and implemented.
- Testing and training: The customized template is tested to ensure that it meets the requirements of the different locations or companies. It is also important to train employees to ensure that the new business processes can be understood and used.
- Implementation: The template is implemented at each location or in each company. It is important that all systems can communicate with each other so that the transition runs smoothly.
- Monitoring and maintenance: The system and software versions must be regularly monitored and maintained to ensure that they continue to function smoothly and meet the requirements.

During a rollout in a hybrid system landscape, configuration changes can relate to various systems, including applications, databases, and infrastructure components. The changes can include, for example, adjustments to functionalities, updates to software versions, new installations of systems, or changes to configuration parameters. It is important that both comprehensive planning and test execution of the changes take place before they are released in a productive system.

[+]

Configuration Parameters

For a better understanding, consider these examples of changes to configuration parameters:

- Changes to the user profile, such as adding new fields or changing the visibility of existing fields
- Changes to the access concept, such as adding access authorizations for users or changing existing authorizations
- Changes to business processes, such as the addition of new steps in an approval process or changes to an existing approval process

An example of a template rollout approach could look like this: In addition to an on-premise SAP S/4HANA solution, an organization that also uses SAP SuccessFactors as its HR system can standardize and simplify its HR management processes. By using SAP Cloud ALM as a template management solution, the processes can be standardized and codified in template projects so that they can be repeated and changed as needed. This speeds up and simplifies the HR management process by centrally defining, managing, and providing the necessary steps and data. The organization can ensure that all locations or departments follow the same processes and procedures, which can lead to improved data quality and better decisions.

[«]

SAP SuccessFactors

SAP SuccessFactors is a cloud-based HR management solution from SAP. This solution can be integrated with SAP Cloud ALM to enable efficient management of HR-related data and processes.

Another example could be the definition of approval processes for the management of employee data. These processes can also be defined in SAP Cloud ALM and applied to the instances of SAP SuccessFactors to ensure uniform management of approvals.

For example, in terms of managing employee data, the template could define certain fields that must be recorded for all employees, such as name, address, contact details, job description, and compensation details. In addition, the template also defines rules on how to make changes to employee data—e.g., which people are allowed to make changes, when approval is required, how conflicts are resolved, etc.

The template could also additionally provide certain reports and dashboards to facilitate the monitoring and review of employee data, such as staff turnover overviews, compensation analysis, turnover rates, etc.

As part of the template rollout approach with SAP Cloud ALM, the management of employee data can include the administration of HR master data, organizational structures, job and role profiles, salary data, time management, personnel development, and training as well as travel expenses. It is important to record and manage this data in a uniform and standardized format to achieve an optimal overview and efficiency in business processes.

It might be necessary to define specific rules and processes for data maintenance in order to guarantee high data quality and ensure the integrity of the data. It may also be necessary to define an authorization concept for access to the data to ensure that only authorized persons have access to certain data.

It is also important to ensure effective monitoring and review of the data through audits and testing. A regular review and update of the template might also be required to ensure that it meets the requirements and business processes.

16.5.2 Infrastructure and Components in a Template Release

The hybrid system landscape involved and to be supplied by a template approach can look very different, depending on the specific requirements of an organization. In our example, the SAP SuccessFactors system landscape could consist of the following components:

- Core HR system: This is the heart of the SAP SuccessFactors system and manages basic personnel file information such as names, addresses, Social Security numbers, etc.
- SAP SuccessFactors Talent Management: This module can provide functions such as performance management, development planning, succession management, and career planning.
- SAP SuccessFactors Recruiting Management: This module supports the entire recruitment process, from job posting to managing applications and awarding job offers.
- SAP SuccessFactors Learning: SAP SuccessFactors Learning supports training and development management for employees.
- SAP SuccessFactors Analytics: SAP SuccessFactors offers an integrated analytics platform that provides comprehensive insights into HR and business data.

It is common for public cloud solutions and system landscapes such as SAP SuccessFactors to have separate development, test, and production systems. This allows you to test changes and updates in a separate environment before they are implemented in the productive environment. It is important to ensure control over the transfer of changes between systems to ensure a stable and reliable production environment. Transfers from one system to another can be performed manually or automated by SAP Cloud ALM. With the help of SAP Cloud ALM, important aspects of release management are covered, such as testing changes and validating changes before they are made available in productive systems.

16.5.3 Planning and Implementation of Template Rollouts

When planning an extension or updating a used template, you should consider the following points:

- Extension or update: Ensure that the reasons for an extension or update are transparent to all stakeholders and communicated to all stakeholders. This can be triggered by changes to business processes, legal requirements, or technological advances.
- Check the effects: Check how the changes to the template will affect the existing processes and data.

- Definition of changes: Define the changes you want to make and make sure they meet the requirements. Monitor the implementation of the new template using modern and proven methods.
- Testing: Test the new template thoroughly to ensure that it works as expected and covers all requirements.
- Documentation: Document all changes to the template, including the reasons, the effects, and the tests carried out.
- Rollout planning: Plan the detailed process of rolling out the new template, including communication with users and monitoring the implementation.
- Review the results: Review the results and findings after the rollout of the new template and ensure that all previously defined requirements are met in operation.

16.5.4 Process Example for a Template Rollout Approach

In this example, we define the template rollout approach as an IT strategy in the hybrid system landscape described earlier. In this approach, a centralized template system is used to create and roll out new versions of applications and processes.

The aim is to shorten the time-to-market for new applications and systems while ensuring high quality, stability and standardization of all systems involved. An important advantage of this approach is the ability to improve the efficiency and consistency of the system landscape(s), as all systems are built on a common basis. In addition, changes to the systems can be made more easily and quickly as they use a uniform infrastructure.

[«]

Time-to-Market

Time-to-market (TTM) is an economic term that refers to the time it takes to bring a new product or service to market. It therefore refers to the period of time after which the product or service is actually available.

A short time-to-market is usually an important competitive advantage and depends on various factors, such as the efficiency of the development and/or production processes, the availability of resources, and the complexity of the product. Companies often rely on agile development methods to achieve the shortest possible time-to-market.

The recipients of a template, who we'll call the consumers, use their own system landscapes in addition to the template system landscape. One advantage of this architecture is that business units or consumers can use their own environment to better meet their specific requirements and needs. This can increase the agility of the company, as local adjustments can be made quickly and efficiently without affecting the template system landscape. In addition, local system landscapes can be used to test specific business processes or applications before they are transferred to the template system landscapes and delivered from there to all consumers. However, a disadvantage can be that managing and maintaining multiple system landscapes can become very complex, resulting in higher costs and labor. In addition, it is more difficult to ensure the integrity and consistency of data and processes between the different system landscapes. It is therefore necessary to consider the specific requirements and needs of a company and carry out a careful cost-benefit analysis before deciding on the future architecture infrastructure.

Figure 16.3 shows the development process of our example as part of our SAP Cloud ALM project with a template rollout approach (see Chapter 5, Section 5.3.4). The process begins with the creation of a central template that serves as the starting point for all further developments ❶.

In the next phase, the requirements for the specific locations/plants/company units are determined and the template is expanded accordingly with local content or adapted at certain points in accordance with specifications ❷? This means that all subsequent local developments or specifications and settings are made on the basis of the delivered template and ultimately result in a company-specific solution for the Dresden business unit ❸. The rollout for the Madrid business unit then begins sequentially. In addition to the template content, further local versions will also be added for this division 4. During the course of the project, various lessons learned can be identified that will help to make future rollouts successful.

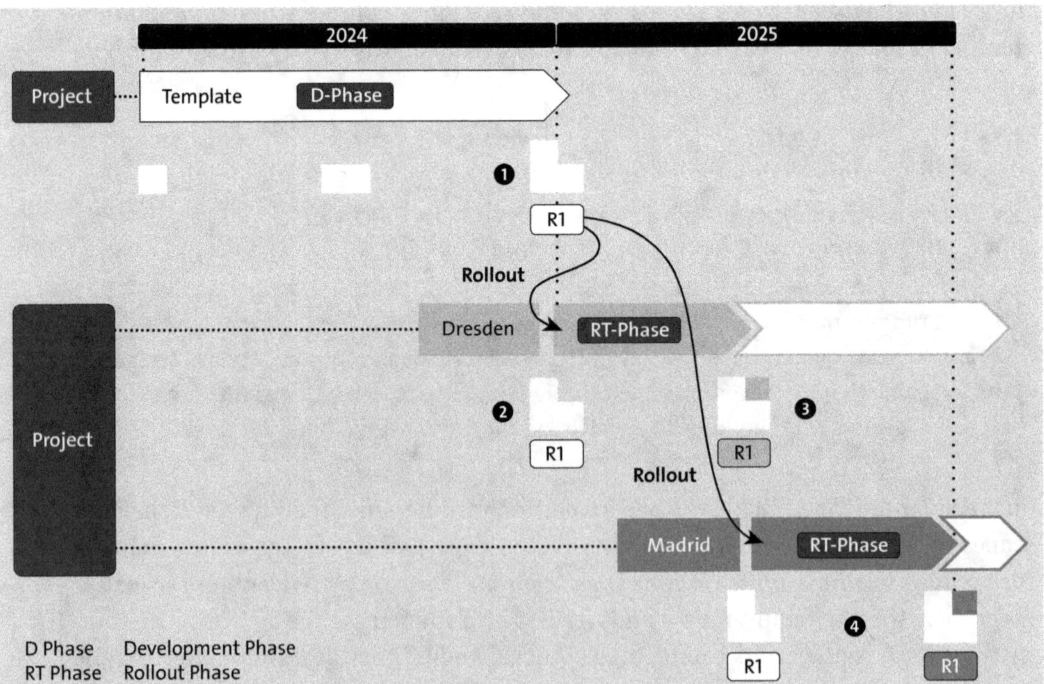

Figure 16.3 Release Management in the Template Approach: Rollout

After a successful template rollout, new requirements and changes may arise that require continuous further development of the template. Here, it is important to plan and coordinate the changes to minimize the impact on the existing systems.

To implement changes to a template, you can use *minor releases* to implement changes to a template. A minor release is a new version of the template that contains small changes or additions. It is a common method for making continuous improvements to a template while ensuring system stability at the same time. Comprehensive tests must also be carried out when implementing minor releases, whereby dependencies and effects on other business areas must be taken into account.

In our example (see Figure 16.4), the template is being further developed and outstanding requirements from the product backlog are being processed in minor release 2.1. The completion date is still within the ongoing rollout of the Dresden business unit. In this case, the recommendation is not to import the new template content during this time, but to wait for the rollout to be completed ❺.

The procedure described and used here illustrates the careful completion of the underlying project after the initial development of a template. This archives the project to fulfill possible audit purposes and to keep the documentation for future reference. It is recommended that the documentation is stored in a secure and accessible location to enable quick and easy referencing if required ❻.

[«]

Major and Minor Releases

The terms "major release" and "minor release," commonly used in software development, refer to different versions of software.

A *major release* is a new version of software that contains significant changes or enhancements and usually has a new major version number. These changes can include new functions, interfaces, or architectural changes, for example. A major release usually requires extensive testing, as the significant changes can have an impact on stability or compatibility.

A minor release is a smaller version of software that usually contains minor changes or bug fixes and has a new minor version number. These changes can include bug fixes, minor functional enhancements, or improvements to the user interface, for example.

It is common to use a version number convention to differentiate between major and minor releases. A typical convention is that the major version number (e.g., 1.0) is used for major releases and the minor version number (e.g., 1.1) is used for minor releases.

During productive operation, but also during the development phase of local changes, new requirements may arise that require the template to be adapted. It is important to evaluate these requirements carefully and decide whether they are local or global and whether they need to be implemented.

Local requirements are specific to a particular business unit and have no impact on other areas of the company. These requirements can be implemented locally without jeopardizing the stability of the entire template.

Global requirements, on the other hand, have an impact on several areas of the company or on the entire company. Here it is important to carefully evaluate the requirements and decide whether they should be integrated into the template. In our example, the global requirements validated in this way are implemented within minor releases ❼.

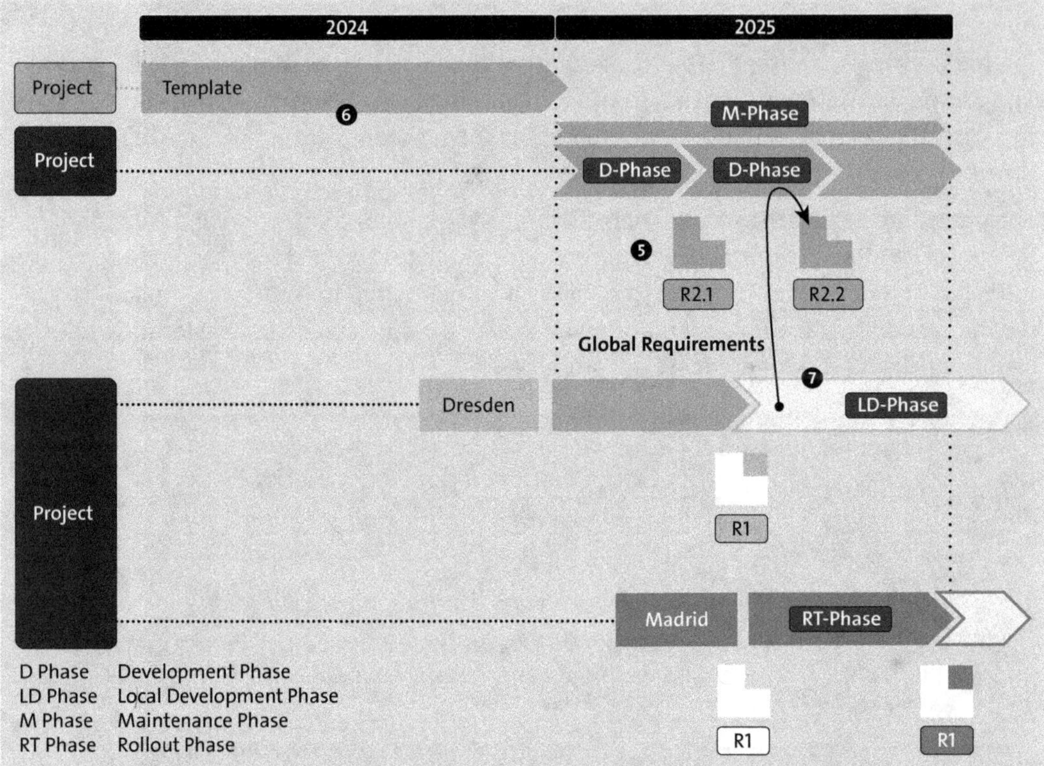

Figure 16.4 Release Management in the Template Approach: Further Development

The deployment of minor releases from the template environment to the environment of the Dresden business unit during the ongoing local development phase is an important step in implementing continuous improvements to the template (see Figure 16.5). In this case, the first delivery bundles the minor releases 2.1 and 2.2 ❽. These releases contain new functions and improvements that were requested by the Dresden business unit itself. The development of such changes in the template environment and the subsequent implementation in the environment of the business units makes it possible to use these functions and improvements while ensuring the stability of the entire system network.

When deciding whether to roll out local features after the individual sprints or in the last sprint before template deployment (sprint 3, in this example), there are various factors to consider: If features are implemented locally after each sprint, they can be used quickly, and local adjustments can be made continuously. This can lead to increased complexity and difficulties in coordinating tests, especially if several teams are working on the project. A consolidated implementation in the last sprint before the template deployment offers the advantage that there is no interference with current local developments by importing the template minor releases. In addition, the new minor releases and the local adaptations can be tested together to ensure that all changes are complete and compatible ❾.

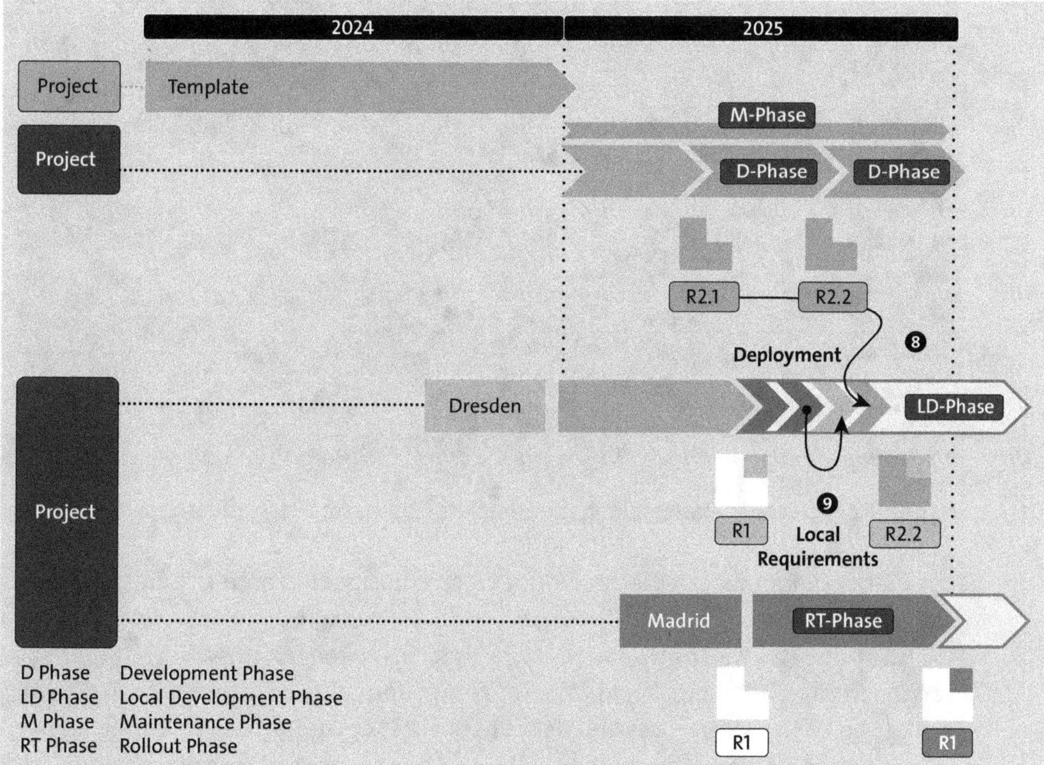

Figure 16.5 Release Management in the Template Approach: Template Deployment

As the template serves as the basis for all business units, global requirements can be submitted to the template team at any time. The template team implements them and delivers them to the consumers (see Figure 16.6). This means that changes or enhancements to the software only need to be implemented once and can then be made available globally ❿.

By using a template, local adjustments and enhancements can also be made without affecting other business units. In this way, a customized solution can be created for a unit's specific requirements while ensuring overall consistency and efficiency through the use of the template.

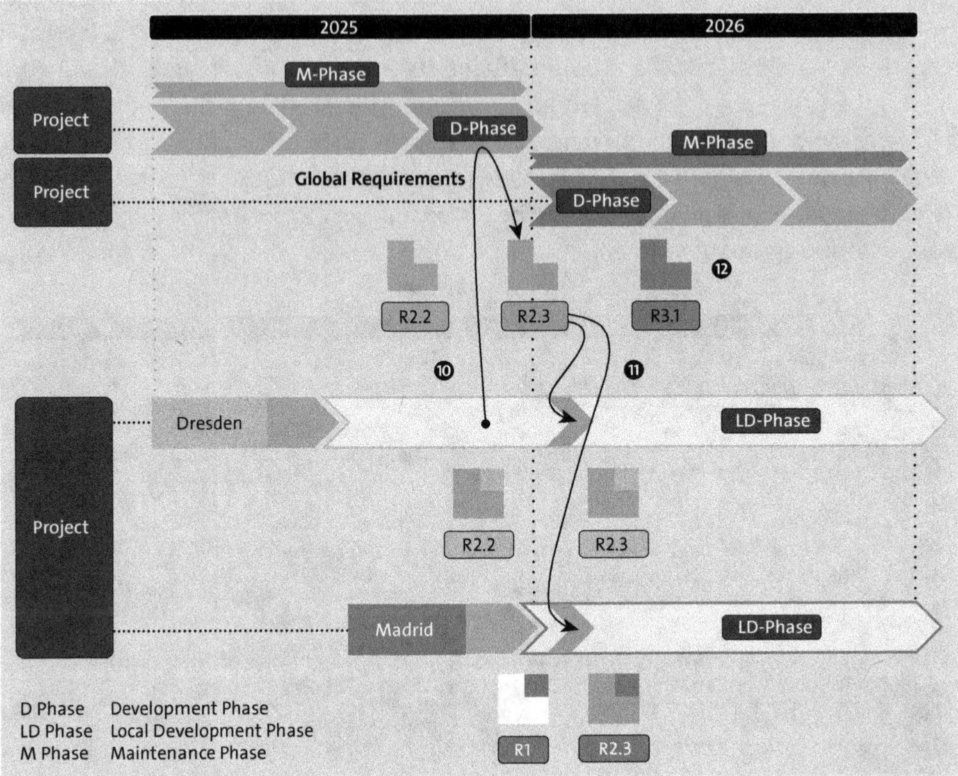

Figure 16.6 Release Management in the Template Approach: Deployment Synchronization

The delivery of minor release 2.3 in our example means that both rollouts to the Dresden and Madrid divisions are synchronized to the current template rollout status for the first time. This means that all business units using the template will have the same version from now on and can use the same functions and improvements ⓫. Synchronizing the rollouts improves the overall consistency and efficiency of the software solution used, as all business units will now be working on the same version. It also makes it easier to deliver changes, enhancements, and bug fixes globally, as all units have the same basis.

Following completion of the current minor release (R2.3), agile development of the next major release (R3) is already underway. This is a new project in SAP Cloud ALM that focuses on the further development of the template and the implementation of new functions and improvements ⓬.

In addition to the content from the delivered template releases, process adaptations and process enhancements are also carried out within the software solution at the local level of a business unit in this example (see Figure 16.7). Individual requirements are considered to optimally adapt the solution to the needs. The local development phase takes place in parallel to the global development phase, which is aimed at developing and adapting templates and global requirements.

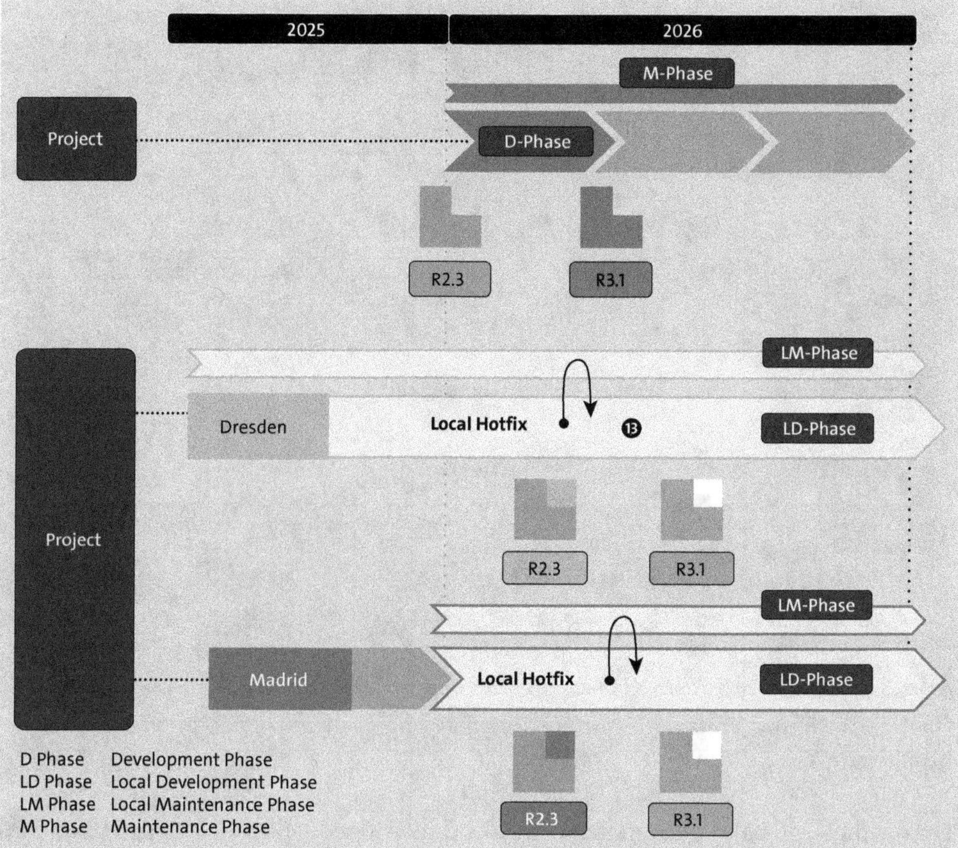

Figure 16.7 Release Management in the Template Approach: Handling Local Errors

The documentation and management of local hotfixes are an important part of the local development phase to ensure the quality and consistency of the solution. Their implementation in the projects of the business units enables effective incident management, as the local implementation teams can quickly identify and rectify any faults and errors. The developers have a clear overview of the necessary adjustments and can work on the solution in a targeted manner. We recommend documenting the errors and error corrections in a separate local maintenance phase for each rollout project. This makes it easier to separate them from the development of local requirements ⓭.

In addition to local adaptations and extensions, additional requirements can also be identified by the individual divisions (see Figure 16.8).

16

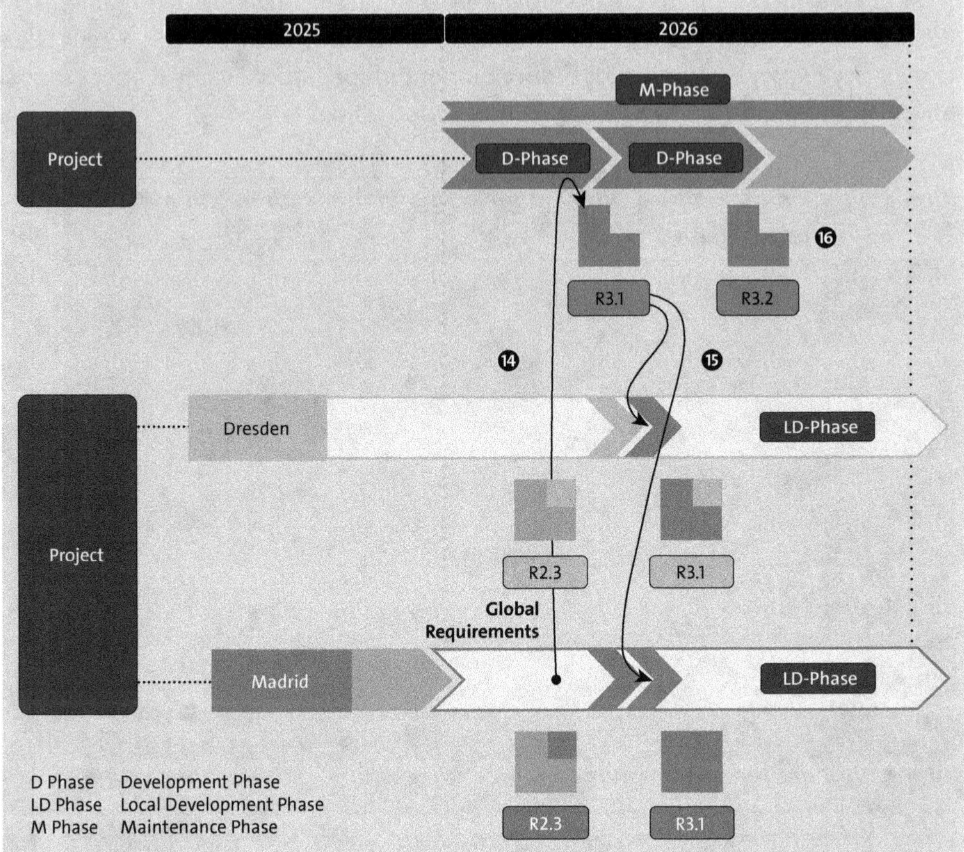

Figure 16.8 Release Management in the Template Approach: Implementation of Global Requirements

For example, global requirements can be commissioned by only one business unit ⓮ but then delivered to all template consumers ⓯. This may be the case, for example, if a specific business process or requirement needs to be implemented in all business units to improve overall efficiency or meet compliance requirements.

In such a scenario, the business unit making the global requirements should work with the template team to ensure that the requirements can be integrated into the template ⓰. Once the template is updated, all template consumers can benefit from the new features and processes. In general, when introducing global requirements, it is important to assess and consider the impact on local customizations and developments in the business units to ensure that there is no undesirable impact on business processes.

Even with the template rollout approach, errors can of course occur during productive operation despite all due care. It is important to identify, prioritize, and quickly rectify these errors to minimize the impact on business operations (see Figure 16.9).

If it is a local error that only affects a specific area of the company, local hotfixes can be implemented. These hotfixes are documented and managed in local projects, which ensures that the adjustments can be taken into account in future local development phases.

If it is a global error that affects all template consumers, this error must be passed to the template team to find a global solution. The template team should prioritize this error and provide a fix as soon as possible. Once the fix has been tested and validated, it is integrated into the template and delivered to all template consumers in advance as a global hotfix, independent of the release cycle (R3.2) ⓱. At this point, it should be mentioned that the handling of template errors is a continuous process. The continuous monitoring and improvement of the template increases the quality and efficiency of business processes for all areas of the company.

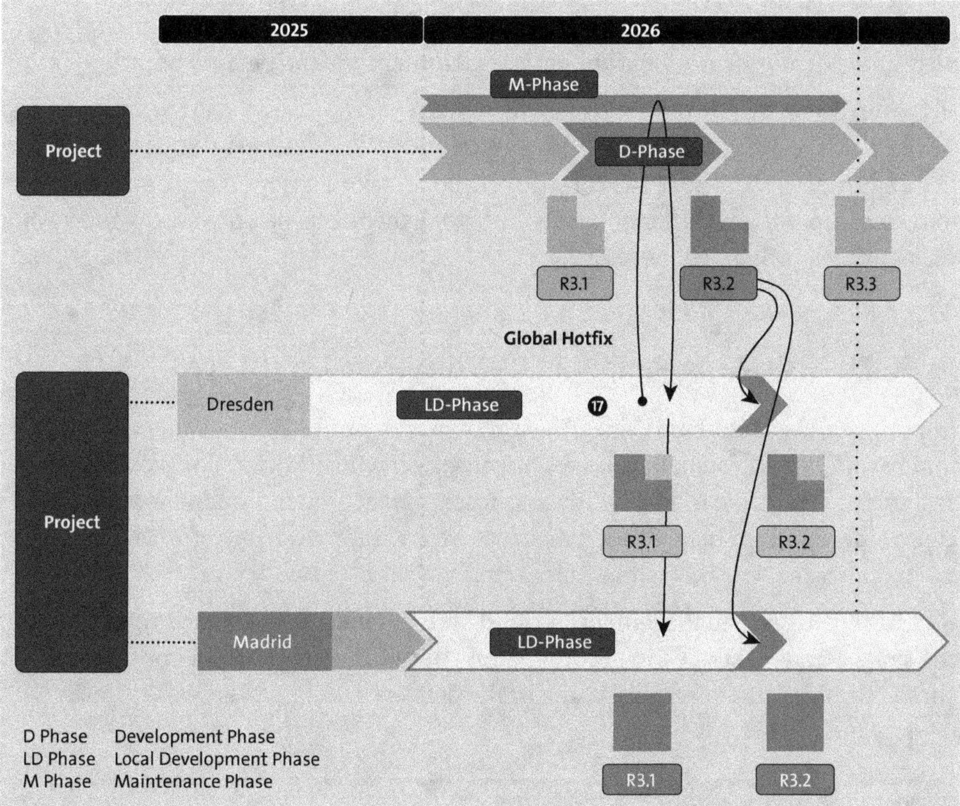

Figure 16.9 Release Management in the Template Approach: Handling of Template Errors

In summary, the separation between template and consumer can be technically implemented in various ways, depending on the needs and requirements of the company. One possibility is the use of different system landscapes. Here, the template is implemented on its own landscape, and each consumer uses its own environment. Transport routes and other technical options such as XLD (Cross-Landscape Distribution) can then be used for deployments, error corrections, or transfers of local changes to the template.

XLD

Cross-Landscape Distribution (XLD) is an SAP function that supports you in the distribution and synchronization of software changes in different SAP system landscapes. It makes it possible to distribute software components and changes that have been made in a development landscape to one or more other production landscapes. Both manual and automated processes can be defined and set up.

However, companies that develop on only one landscape for templates and consumers must find other ways to separate the various local customizations and protect the template. Exit strategies, different clients (e.g., development only takes place in the template client), namespaces, or authorization delimitations can be used here.

However, there are also scenarios in which the separation between template and consumer is less relevant, as the aim is to develop a common business solution that is used in its entirety by all areas of the company. This can save costs by avoiding different system landscapes and redundant architectures. Furthermore, balance sheets and key figures can be consolidated more easily.

16.6 Release Management in Cloud System Landscapes

In pure cloud system landscapes, release management differs in some aspects from that in traditional system landscapes. Because of the flexibility and scalability of cloud systems, it is possible to deploy releases faster and at a higher frequency. In addition, cloud systems often have a modular structure and offer a variety of APIs and integrations, which enables the seamless integration of third-party software.

In principle, release management in cloud system landscapes can be implemented in the same way as described in Section 16.5. At this point, however, an important aspect should be highlighted that exists due to the delivery of release versions through SAP's own rollout plan.

Cloud System Landscape

A *cloud system landscape* consists of a variety of IT resources, such as virtual machines, storage, and network services as well as cloud applications and platforms. These can either be operated or hosted in a public, private, or hybrid cloud.

However, to ensure effective release management in cloud system landscapes, several challenges need to be overcome. These include automating tests and deployment processes, ensuring the compatibility of updates and integration options, and monitoring systems in real time.

Implementing DevOps practices also plays an important role. These practices enable close collaboration between development and operations teams and support the rapid deployment of software updates and patches. In addition, the use of container technologies (such as Docker) enables better portability of applications between different cloud systems and environments.

The term *cloud-centric solution* describes a solution in which all applications, data, and resources are hosted in a cloud infrastructure. This means that a company does not have to operate its own physical servers and can instead use the services of a cloud provider.

Cloud-centric solutions have several advantages over locally hosted solutions, such as:

- Flexibility: Users can access the solution from anywhere, increasing both mobility and flexibility.
- Scalability: Cloud-centric solutions offer the option of adding or reducing resources quickly and easily, depending on the company's needs. This makes it possible to react flexibly to changes and save costs at the same time.
- Cost efficiency: Cloud-centric solutions are often more cost-efficient because they do not require local hardware and software investments and reduce maintenance and administration costs. Instead, you only pay for the resources you actually use.
- Higher availability: Cloud-centric solutions are often highly available because they are executed redundantly across multiple locations and systems.
- Maintenance and security: Cloud-centric solutions are often more secure because they are regularly updated by the providers with enhanced and current security standards. As a result, the system landscape always remains up to date, and there is less need for manual updates and patches.

One challenge when implementing in cloud system landscapes is when managing an ongoing project with customer-specific processes, requirements, user stories, features, and transports based on a specific release, while at the same time a new release is delivered through the SAP rollout plan. This delivery or deployment happens before the implementation project is completed.

In this case, the new functions and changes that have been included in the new release must be integrated into the current project scope. This requires a careful delta alignment to ensure that all new functions and changes are properly integrated into the project scope and that all affected processes and adjustments are taken into account.

It might also be necessary for the project teams to carry out deconstruction work to ensure that all new functions and changes can be seamlessly combined with ongoing developments. Especially if fundamental functional changes have also been delivered in the new release, it must be checked whether these changes have an impact on the customer's own processes and functions.

One approach to solving this problem could be to set up regular coordination meetings between the implementation teams and the SAP rollout team. Changes in the upcoming SAP releases and their impact on the implementation project should be discussed.

As far as is feasible, you should concentrate on separating the customer's own processes and customizations from the standard SAP functionality as far as possible. This will help to avoid potential conflicts during the upgrade.

Another approach could be to implement continuous integration and delivery (CI/CD) in the cloud system landscape. This would allow the implementation teams to react quickly to new SAP releases and make changes to the customer's own processes efficiently and securely.

CI/CD

CI/CD stands for *continuous integration/continuous delivery* and is a concept in software development that aims to automate and accelerate the provision of software. Software development and delivery are continuously integrated and delivered instead of being carried out in separate steps.

Continuous integration refers to the process in which each developer frequently and automatically transfers their own code changes to a shared code base to identify and resolve problems at an early stage. Automated tests ensure code quality and identify conflicts at an early stage.

Continuous delivery refers to the process in which the integrated software is automatically provided in a production environment. This significantly shortens the time from development to deployment of the software and reduces the susceptibility to errors.

Tools such as versioning systems (e.g., Git), build systems (e.g., Jenkins), container orchestration (e.g., Kubernetes), and testing tools (e.g., Selenium) play an important role in CI/CD. They enable the automation of tests, builds, and deployments to be optimized and errors to be detected and rectified at an early stage, thus ensuring higher-quality software development and deployment.

In tools such as SAP Solution Manager or Focused Build for SAP Solution Manager, which have already been in use for some time, seamless integration with change and deployment management is one of the most important advantages for release management. SAP Cloud ALM optimizes this integration by improving the architecture and reducing complexity specifically for cloud solutions. An important part of this optimization is the business object feature, which not only contains all landscape information of the systems involved, but also serves as a container for all transportable objects and technical change documentation. Features in SAP Cloud ALM cover deployment management for a wide range of SAP solutions, such as Change and Transport System (CTS);

SAP S/4HANA Cloud Private Edition; and SAP NetWeaver Application Server for ABAP. Solutions such as SAP BTP, Cloud Foundry environment, and SAP Cloud Transport Management are also covered. This means that all changes can be monitored, controlled, and tracked from the start of implementation in the development environment, through deployment in the test environment, and to deployment in the production system by linking requirements, user stories, and features.

SAP Cloud ALM is particularly useful for all companies that use SAP cloud solutions, such as SAP S/4HANA Cloud, SAP SuccessFactors, or SAP Ariba, and are looking for a comprehensive and integrated solution for application lifecycle management. The more complex an SAP system landscape and its applications become, the greater the need for an effective and transparent release management process.

16.7 Corrections during Operation

In addition to changes due to new requirements, changes will also be necessary due to existing production problems. This refers to corrections to faulty software that are intended to eliminate the problem in the productive environment. Here too, the guidelines for implementing the necessary change should be observed, even if a high degree of urgency is required. The verifiability of a change also remains necessary. The feature can be used to manage and deploy the change as well. Whether a process needs to be activated before the correction feature depends on the scope, criticality, and urgency of the change. The request could conceivably be an upstream process for the feature. In practice, the urgency of the change outweighs the need to fix the productive problem as quickly as possible.

We therefore recommend that you use the feature for corrections as a standalone process, but with a separate marking in the title of the correction. This allows you to evaluate such changes yourself during operation. Here, for example, it is sufficient to start with the abbreviation "KOR" (indicating corrections) to describe the title of the feature. If you want to carry out the correction in the same project, it is advisable to define a separate scope for it. A suitable name would be "maintenance" or "operation." An assignment of timeboxes would not necessarily be required.

16.8 Summary

This chapter has shown you how important it is to record and manage changes within an implementation project and during operation. Managing changes within a project also includes the need to manage and control related changes. To this end, detailed information on the topic of release management was also provided. This information provided not only a theoretical but also a practical approach using SAP Cloud ALM.

In Chapter 17, you will find detailed information on how to handle changes using the feature. You will also be shown a detailed example based on our SAP S/4HANA project landscape.

Chapter 17
Deployment with SAP Cloud ALM

In this chapter, we use our example scenario to show you how to use SAP Cloud ALM features to roll out your developments, including the associated transports, along the system line. In the technical preparations for change deployment, we look at the consolidation of systems into system groups and present the options for planning releases in a project's deployment plan.

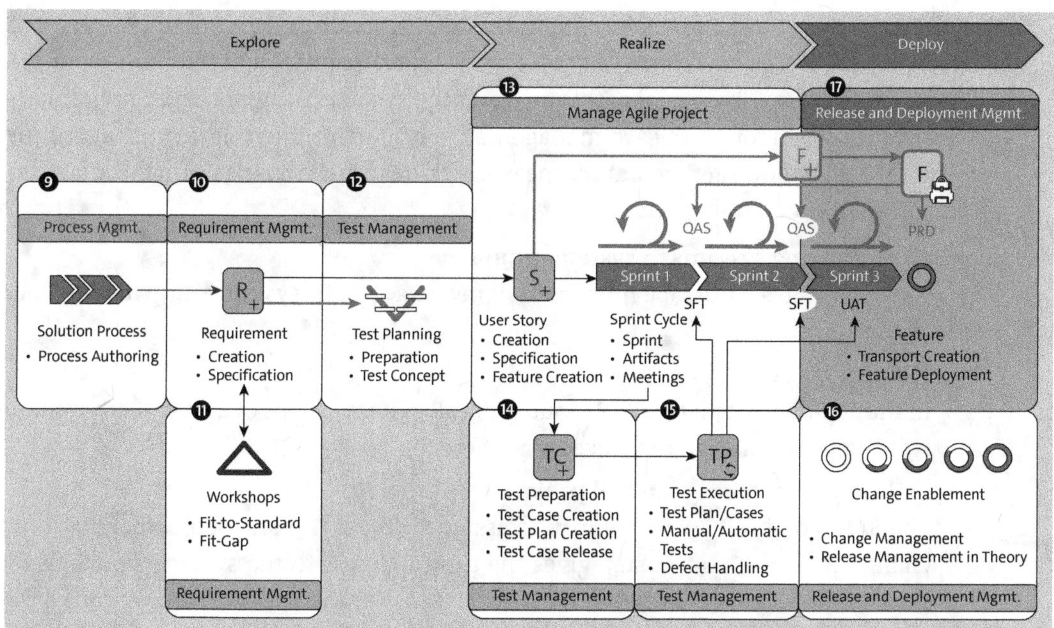

To better understand delivery in SAP implementation projects, we looked at the theoretical foundations of change enablement and release management in the last chapter. We went into detail about the release concept for the hybrid project approach in our example scenario, including the strategy and planning for the initial template creation, the plant-specific rollout projects, and the further development of the template in the form of minor releases with subsequent deployment.

In this chapter, we will focus on the technical delivery of a release with the help of features and assigned transports. You will learn how to create and specify corresponding features as a release manager, link them to requirements and user stories, populate them with transports, and import them into the systems. The deployment of functions is assigned to the deploy phase (import) within the SAP Activate methodology. In an agile project approach, however, the import and testing of developed functions already takes place incrementally in the realize phase. This mainly concerns the provision of content on the test system. If the productive environment is already available in the realize phase, you can also roll out the content after successful testing. In addition to the import to the production system, the deploy phase according to the SAP Activate methodology also includes other measures that you should consider. These include, for example, conducting end user training and handover to operations.

17.1 Technical Preparations

The change deployment and the associated change control comprise the delivery or import of developed functions of a system landscape into the subsequent systems, such as test and production systems. The technical requirements for this were provided in Chapter 4. They deal with setting up the system landscape and providing the transport tools. From a release management perspective, the release manager must now take the following preparatory measures to ensure the change control of features and the associated transports:

- **Consolidating systems in system groups**
  All systems in a landscape must be assigned to a specific system group in SAP Cloud ALM.
- **Planning the deployment**
  In the deployment plan of SAP Cloud ALM, all system groups of a project are summarized. Release planning also takes place here.
- **Linking deployment plan and project**
  The SAP Cloud ALM project must be assigned to exactly one deployment plan for change control. This last step gives the project access to the system groups of the deployment plan with the systems in it.

17.1.1 Consolidation of Systems into System Groups

To ensure that the connected systems are available for change deployment, they are consolidated in specific system groups and assigned to a project-specific deployment plan.

To do this, open the project configuration via the Projects and Setup app from the **SAP Cloud ALM for Implementation** app group (see Figure 17.1).

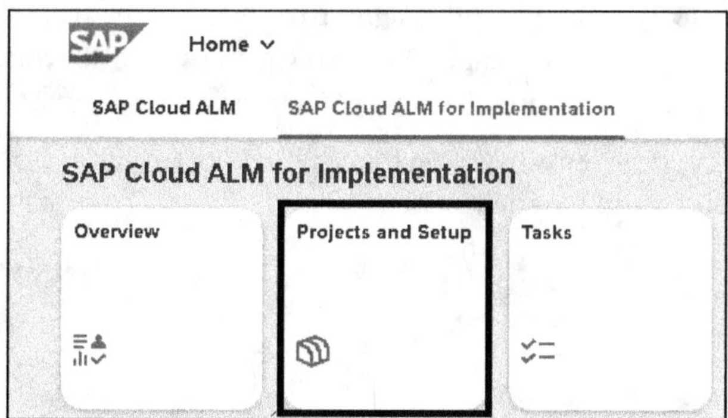

Figure 17.1 SAP Cloud ALM for Implementation: Projects and Setup App

Here you will find all projects that have been created in your SAP Cloud ALM tenant. In addition to configuring the projects themselves, you can also configure cross-project functions here, such as the roles in the project or the team setup. The **System Groups** and **Deployment Plans** tabs ❶ are relevant for change enablement at this point (see Figure 17.2).

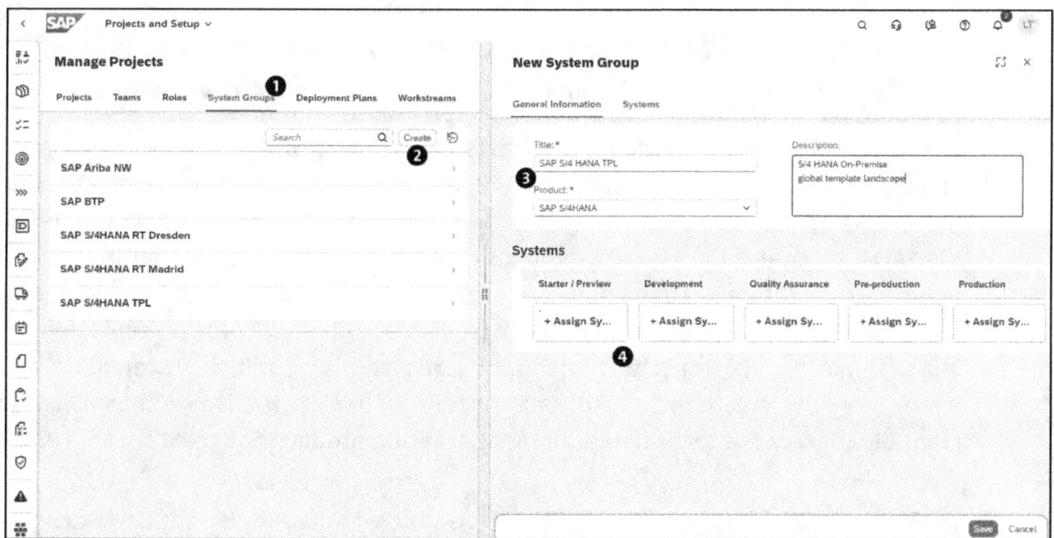

Figure 17.2 Projects and Setup: Creation and Specification of System Groups

In the first step, we consolidate the systems in the system groups. To do this, click on the tab of the same name on the left-hand side of the screen (see ❶ in Figure 17.2). Now click on the **Create** button ❷. An empty form opens on the right-hand side. To create a system group, you must assign a title and assign the group to a product. You can also add a description of the system group. In this case, we select the **SAP S/4HANA** option in the **Product** field as an example ❸. We recommend assigning the name of the system group analogous to the respective system landscape. Optionally, you could also work with the system IDs at this point to better distinguish and assign the system groups.

In the **Systems** area ❹, the associated systems from the management of the landscape (see Chapter 3) are now assigned according to their system roles:

- Starter/preview
- Development
- Quality assurance
- Production

In our example, various system groups have been created here that match our example scenario. Each of the following system groups represents a separate product landscape with a development, test, and production system in which functions can be developed, tested, and used productively:

- SAP Ariba network landscape (**SAP Ariba NW**)
- SAP Business Technology Platform (**SAP BTP**)
- SAP S/4HANA on premise, global template landscape (**SAP S/4HANA TPL**)
- SAP S/4HANA on premise, rollout landscape Dresden (**SAP S/4HANA RT Dresden**)
- SAP S/4HANA on premise, rollout landscape Madrid (**SAP S/4HANA RT Madrid**)

The system group for **SAP Ariba NW** comprises changes to SAP Ariba for the exchange of information with SAP S/4HANA. The **SAP BTP** system group is used for the development of SAP BTP applications that interact with the SAP S/4HANA system. In the **SAP S/4HANA TPL** system group, the template releases are developed, which in turn are imported into the development system of each plant and tested together with plant-specific local functions on the test system and rolled out on the production system. In our example, the plant-specific adaptations take place in separate landscapes. For this reason, we have created two additional system groups for the Dresden and Madrid plants. Each plant has its own development, test, and production system.

17.1.2 Planning the Deployment

The planning of the deployment or the release takes place in the *deployment plan*. A deployment plan is assigned to a project in SAP Cloud ALM and groups all associated system groups that together form a release. Deployment plans are created and maintained in the same way as the system groups within the Projects and Setup app are created and maintained. Switch to the **Deployment Plans** tab (see ❶ in Figure 17.3).

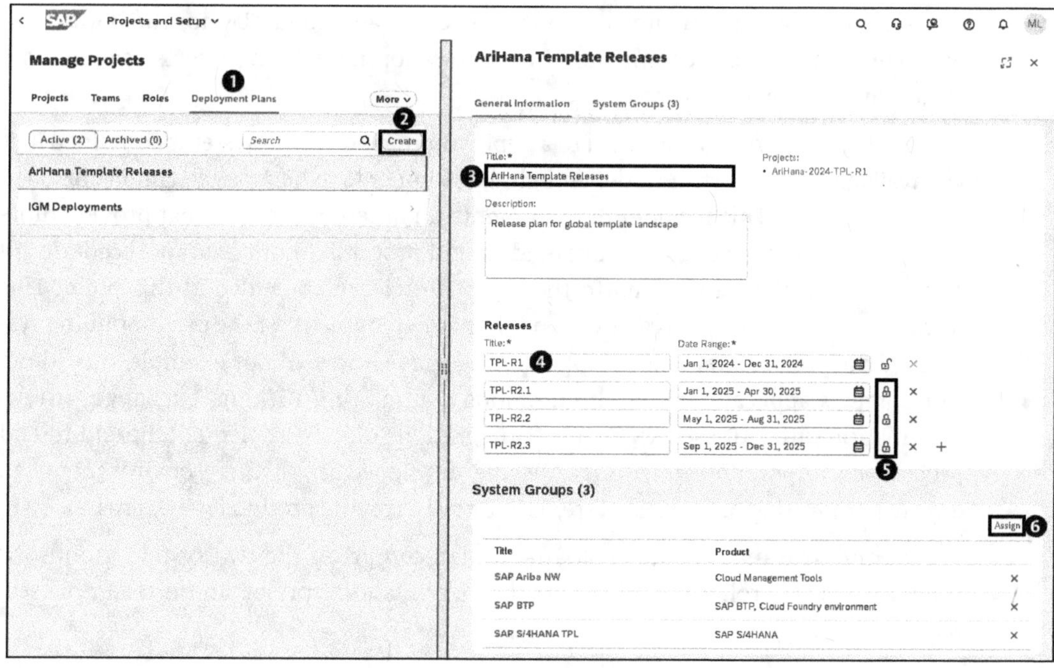

Figure 17.3 Projects and Setup: Creation and Specification of Deployment Plans

Click on **Create** to create a new deployment plan. As an example, we will create a new deployment plan for the entire template release ❷. The configuration interface opens on the right-hand side of the screen. In the first step, assign a **Title** and optionally a **Description** of the plan ❸. The title should relate to the project to reflect the intended use. Compare the naming we have chosen for this. The actual release planning takes place in the **Releases** section below. Here you can define the respective major or minor releases with details of the implementation periods ❹. We recommend that you lock future releases by clicking on the lock symbol next to the release ❺. This means that these releases are no longer available in later selection options, e.g., within a requirement, user story, or feature. In this way, you can prevent a release from being incorrectly assigned before its turn has come. Release management is responsible for gradually deactivating old releases and activating subsequent releases. Changes to the release planning are possible at any time. In the last step, the relevant system groups of a release are assigned to the import plan in the **system groups** area, which we described in Section 17.1.1 To do this, click on the **Assign** button ❻. You can select and confirm the desired groups in a selection popup window.

In our example scenario, we use the common deployment plan **AriHana Template Releases** for the template release R1 and the template follow-up releases R2.1 to R2.3. This deployment plan is later assigned to the two template projects (see Section 17.1.3).

As a reminder, we have created one project each in SAP Cloud ALM for the initial template creation and for the template's further development. Both projects use the same underlying system landscape.

If several system groups are added to a deployment plan, all the systems contained in it are available. This allows interdependent developments to be managed and controlled in a joint project. This includes developments on different systems that provide functions in a holistic business context (end-to-end process). In our example scenario, all system groups that are related to the template release are added at this point. This setting makes it possible to carry out the change deployment for the corresponding systems of the project later when implementing the features. In our example, this sometimes affects applications that are developed within SAP BTP and exchange various information with SAP S/4HANA via SAP Integration Suite. To implement the end-to-end process, corresponding functions must be developed in all three environments to ensure joint testing on the test systems and rollout on the productive systems.

The second AriHana rollouts deployment plan comprises the two system groups for the plant-specific rollouts, which are managed in a joint project in our example scenario.

17.1.3 Linking the Deployment Plan and Project

In the next step, the deployment plan and the system group it contains are assigned to a project. A deployment plan can be assigned to several projects, but this only makes sense if the deployment plan includes cross-project release planning. You can see the complete dependencies in Figure 17.4.

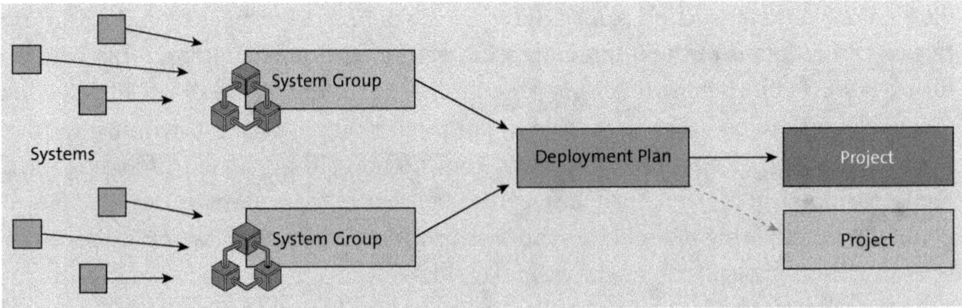

Figure 17.4 Relationships Among Systems, System Groups, Deployment Plans, and Project

To make the assignment to a project, switch to the **Projects** tab in the left-hand navigation bar (❶ in Figure 17.5) and select the desired project. In our example, we select the **AriHana-2024-TPL-R1** project ❷. The project opens on the right-hand side of the screen. Here, switch to the **Deployment Landscape** tab ❸. Select the appropriate **Deployment**

Plan from the dropdown menu and add it ❹. You can assign exactly one deployment plan to the project at this point. Once you have selected a deployment plan, the system displays the system groups assigned to the plan and the respective product.

In our example scenario, we assign the **AriHana Template Releases** deployment plan to the two template projects R1 and R2, as it uses the same landscape on the system side. Furthermore, the release planning for both releases is anchored in this deployment plan. The rollout project **AriHana-2024-TPL-R1** must in turn contain its own deployment plan, which includes the system groups for the plant-specific landscapes (one each for Dresden and Madrid). The system groups for overarching products such as SAP BTP or SAP Integration Suite also use their own system group.

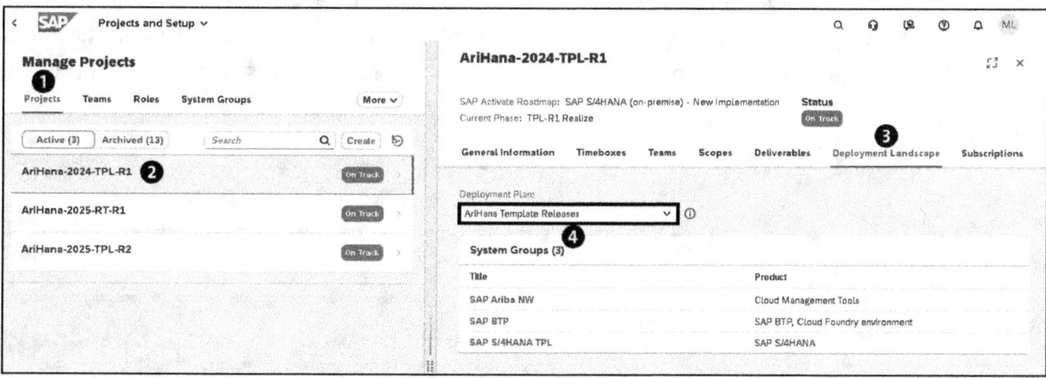

Figure 17.5 Projects and Setup: Assignment of a Deployment Plan to the Project

17.2 Features Overview

In SAP Cloud ALM, a *feature* is an object that you use to deploy changes across your entire system landscape. The feature is the central element of the deploy phase (import). It is already created in the realize phase by the product owner or release management for the implementation of a requirement and equipped with transports for the relevant development systems. The information on which system of the system group must be developed is stored in the specification of the requirement or the assigned user stories. During the processing of the user stories, the developers document their changes on the transports of the development systems. The feature is now used to transfer these changes to the assigned transports to the test and production system in a bundle. The feature contains the following important information:

- The landscape information of the project (deployment plan, system groups, and systems)
- The container for the transportable objects (transports)

- The assigned requirements or user stories
- The technical change documentation (e.g., description of the feature, description of the linked requirement or user story)

The SAP transport solutions supported by a feature are explained in more detail in Chapter 4.

17.2.1 Status Overview

Various status values can be selected for features in SAP Cloud ALM. Figure 17.6 illustrates the available status values or the states of a feature that can be set via executable actions. It also shows the options for controlling the associated transports on the system line in the status of a feature. For example, if the feature has been switched to **In Testing**, the feature's transports must be imported to the test system. This allows the developments to be tested on this system.

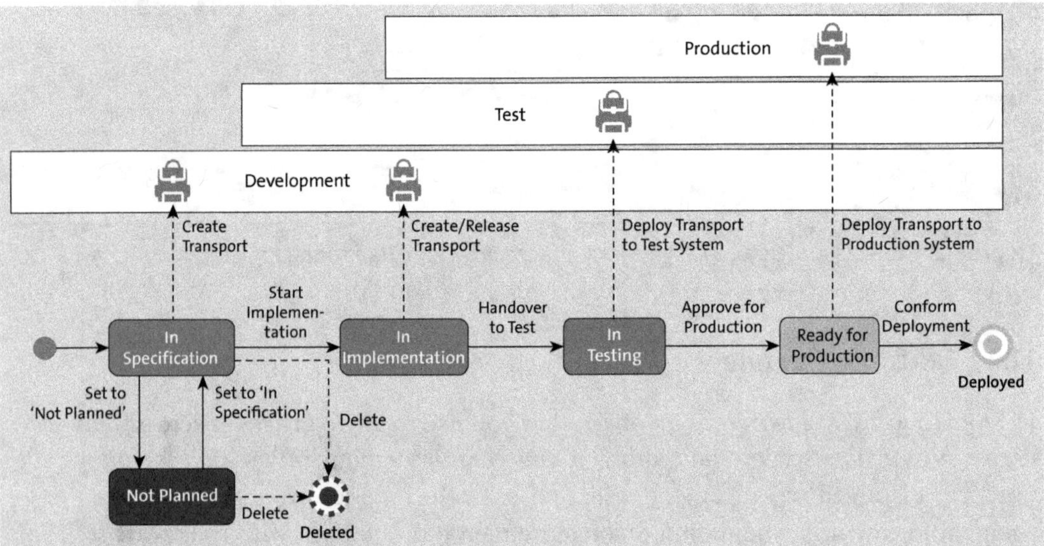

Figure 17.6 Status Overview of a Feature

The actions of a status offer different options for creating, assigning, and removing the assignment of project tasks and user stories. Furthermore, depending on the status, you can access various subsequent statuses and create, assign, release, import, or cancel the assignment of transports. A detailed explanation of the status values and all available actions can be found in Table 17.1.

Status	Activity	Possible Actions
In specification	This status represents the initial status after the feature was created. The feature is prepared for implementation and enriched with corresponding information. If necessary, transports for the developers are already created and assigned here from a release management perspective.	■ Create project task ■ Create user story ■ Assign user story ■ Cancel user story assignment
		Follow-up status: ■ Start implementation ■ Set to **Not planned** ■ Delete
		Transportation... ■ Create ■ Allocate ■ Cancel assignment
Not planned	The feature is set to the status **Not planned** if the implementation is to be postponed indefinitely.	■ Create project task ■ Create user story ■ Cancel user story assignment
		Follow-up status ■ Set to **To be specified** ■ Delete
		Transportation ■ Cancel assignment
Will be implemented	The actual development of the feature takes place in this status (i.e., the developers document their changes on the feature transports assigned to them). In an agile project approach, the transports can also be created independently by the members of the development team from within the feature.	■ Create project task ■ Create user story ■ Assign user story ■ Cancel user story assignment
		Follow-up status: ■ Handover to test
		Transportation... ■ Create ■ Allocate ■ Cancel assignment ■ Release ■ Import

Table 17.1 Description of the Status Values and Actions of a Feature

17

Status	Activity	Possible Actions
In the test	The feature transports are released and imported into the test system. This is where the feature, requirement, or user story is tested after development.	■ Create project task ■ Create user story ■ Assign user story ■ Cancel user story assignment
		Follow-up status: ■ Approve for production
		Transportation... ■ Create ■ Allocate ■ Cancel assignment ■ Release ■ Import
Ready for production	This status indicates that the feature has been successfully tested. Approval for delivery to the production system has been granted.	■ Create project task ■ Create user story ■ Assign user story ■ Cancel user story assignment
		Follow-up status: ■ Confirm import
		Transportation... ■ Allocate ■ Cancel assignment ■ Release ■ Import
Deployed	The feature has been rolled out and can be used. This status represents the final status (changes are no longer possible).	-

Table 17.1 Description of the Status Values and Actions of a Feature (Cont.)

[»]

Delete Action

Please remember that you already have the option of creating transports on the development system in the **To be specified** status. If the feature is now deleted, you will have to think about housekeeping the created transports. The feature is no longer available for reactivation after deletion.

17.2.2 Create Features

There are several ways to create features. For example, it can be done directly by the responsible product owner during sprint planning or separately by a release management team that wants to have organizational control over it. Regardless, the creation of a feature cannot be controlled via authorizations and must be defined as a work instruction for the project participants.

It is also possible to create features with or without reference to requirements or user stories. In principle, SAP Cloud ALM gives you the option of not using requirements or user stories at all. This can be particularly useful in small projects with few project participants, as the administrative effort is lower. However, to roll out changes to the system landscapes, the use of features is essential. You must weigh which approach you choose, considering your individual needs.

In our scenario, we describe below the variant of creating a feature for confirmed requirements from the perspective of release management.

Open Requirements Backlog

In our case, the responsible release management team should create corresponding features as containers for the confirmed requirements. The best way to do this is to create a requirements backlog using the requirements search. To do this, start the Tasks app within the **SAP Cloud ALM for Implementation** app group (see Figure 17.7).

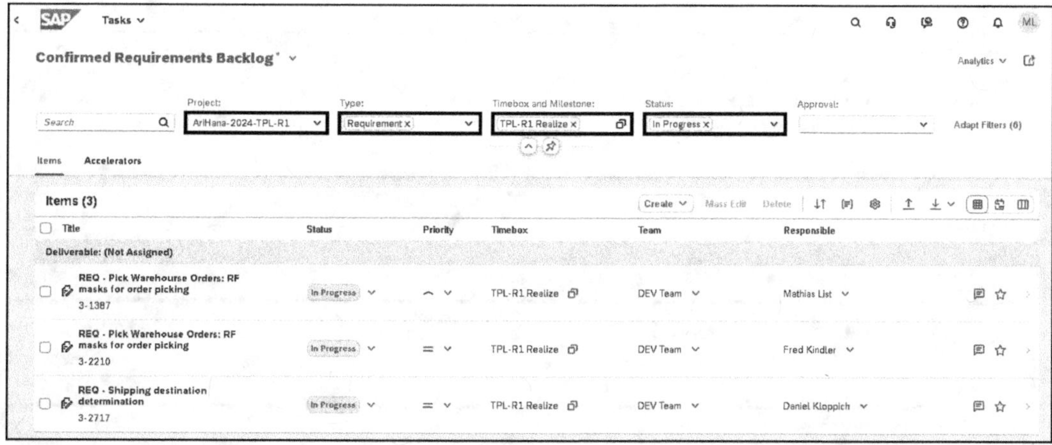

Figure 17.7 Requirement Search for the Feature System

As a rule, all requirements that are newly added or not yet confirmed are assigned to the prepare phase. We recommend assigning all requirements that are to be implemented to the realize phase. These also represent the baseline for the product owner to build the product backlog using user stories (see Chapter 5, Section 5.2.1). For the presentation of the requirements backlog, we are therefore interested in all requirements that are assigned to the realize phase. In Table 17.2 we summarize the relevant filter criteria and value pairs.

Filter	Filter Value
Project	Filter by the desired project, e.g., **AriHana-2024-TPL-R1**
Type	**Requirement**
Timebox	Realize phase (realize), e.g., **TPL-R1 Realize**
Status	**In progress:** Requirements from which user stories have already been created and thus transferred to the product backlog should be switched to the status **In progress**. These are also relevant for the feature creation.

Table 17.2 Filter Criteria and Filter Values for Requirements Backlog

We recommend that you save the defined search for later use (e.g., **Confirmed requirements backlog** in Figure 17.7). This gives you the opportunity to continuously create new features as containers for all relevant requirements quickly and efficiently during the project.

Create Feature for a Request

We open the requirement **REQ - Pick Warehouse Orders: RF masks for picking** of our example scenario from the requirement backlog (see Figure 17.8).

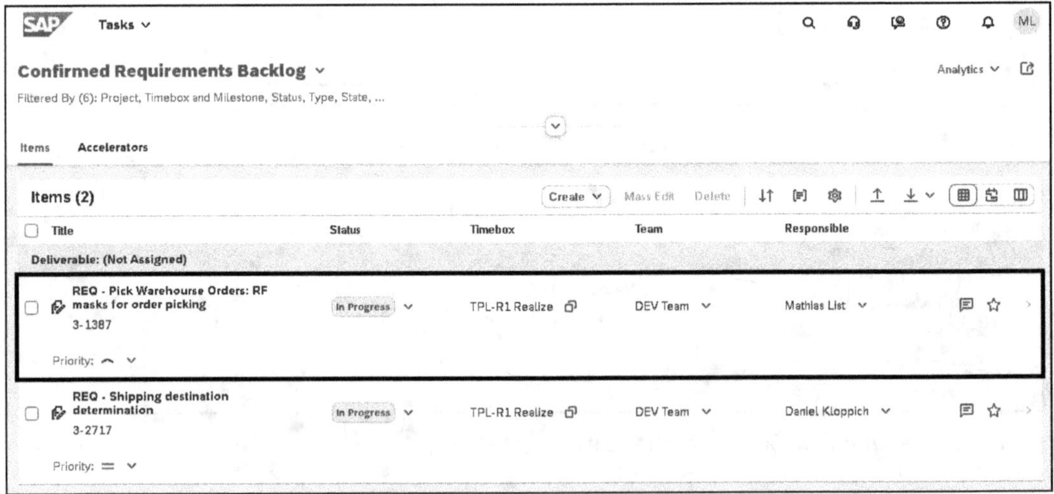

Figure 17.8 Open Request from the Requirements Backlog

Once you have selected a request, the detailed view of the request opens. All request information is displayed in the detailed view (see Figure 17.9). In the header area, you can view configuration settings such as the **Processor**, the **Approval**, or the **Request Status**. The lower section contains the assignments to **User Stories**, **Project Tasks**, **Features**, **Test Cases**, **Process Hierarchy**, **References**, and all **Additional Information** on their respective tabs.

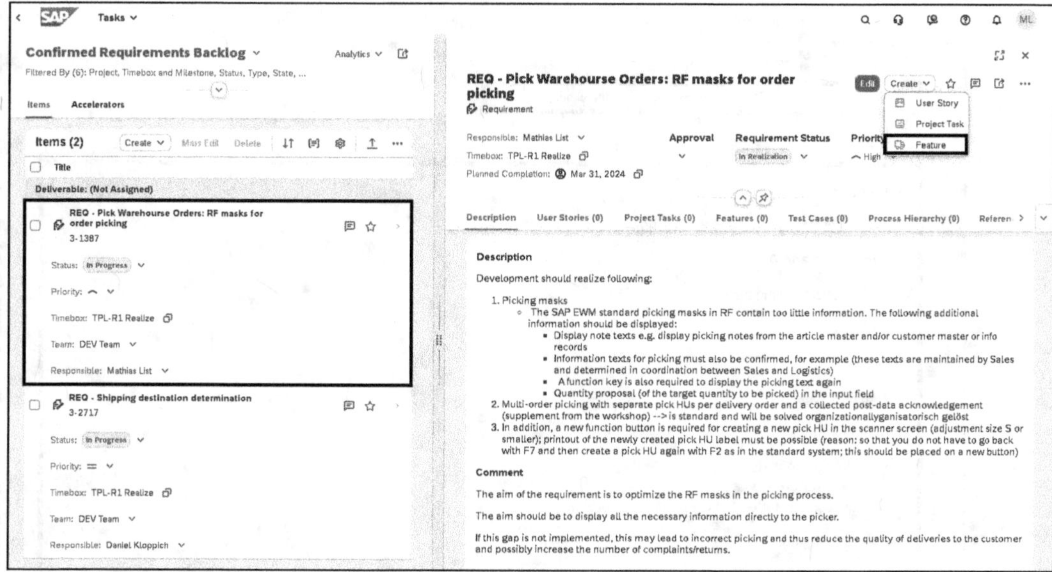

Figure 17.9 Requirements Overview and Feature Layout

The **Create** button in the selection list can now be used to create a feature as a follow-up document to the open request. Depending on the size of the requirement or necessity, several features can also be combined into one requirement (1:*n* relationship). In turn, a feature can only be assigned to a single requirement. Creating several features for a requirement is necessary if associated user stories are implemented in different sprints. This also involves different times for importing the transports to the test and production systems.

Specify Feature

After you have selected **Create • Feature**, you can enter the **Title** of the feature (mandatory) and optionally a detailed **Description** in the next step (see Figure 17.10).

As with requirements and user stories, you should establish a uniform procedure or documentation standard for the specification of features. We use the following conventions for the title to better distinguish the features according to their purpose:

- Prefix "FEA" for general features
- Prefix "COR" to document error corrections during maintenance (run phase, execution)

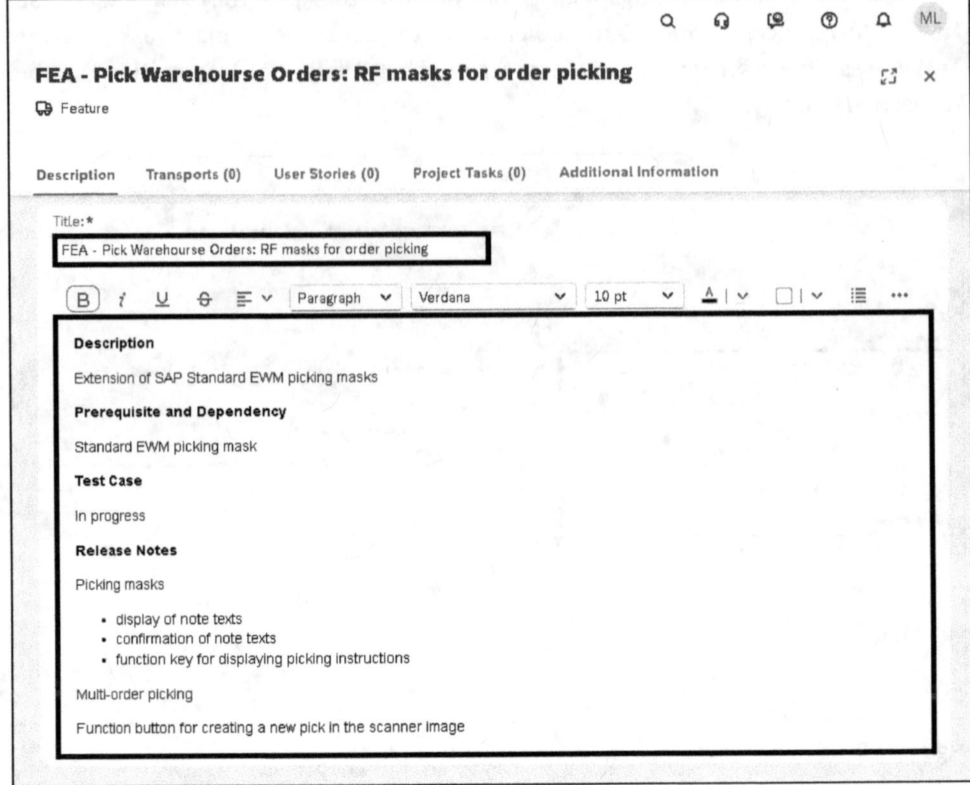

Figure 17.10 Enter Title and Description of the Feature

It is also advantageous to transfer title components from the preceding element. In our example, we transfer the text component "Pick Warehouse Orders" from the request to the feature. This originally comes from the business process from which the request was created. This makes it possible to find associated elements such as solution processes, requirements, user stories, or features easily and efficiently using a generic search at text level.

For the technical description of the features, we use the following categorization in the form of text modules:

- **Description**
  Here you can enter a short description that reflects the business context of the feature.
- **Prerequisites and dependencies**
  List all dependent features here that form a functional area and must be imported together. Sometimes the transport sequence also plays an important role.
- **Test case**
  Currently, no test cases can be physically linked to features. You can list your functional test cases for the changes here.

- **Release notes**
  Document a rough overview of the functions to be delivered for a feature here, which you can take from the associated requirements or user stories. This allows you quickly and easily create release notes for the release across all features.

[«]

Release Notes

Release notes are version notes in the form of descriptions that are distributed in addition to the delivery of developments, system changes, and services. They provide a rough overview of all changes and improvements as well as bug fixes, which are usually delivered in the form of releases. If you are continuously delivering several features or bundled feature packs, you should think about creating and publishing release notes in advance. It is a good idea to document these in bullet points within the features. In this way, you can later create the release notes for a release from the associated features without much effort.

On the **Additional Information** tab, you should, as shown in Figure 17.11, enter further important details. Important fields here are the **Workstream**, **Release**, and person **Responsible** for the feature, such as the authorized release or transport manager. It is also possible to assign topic-related tags to a feature.

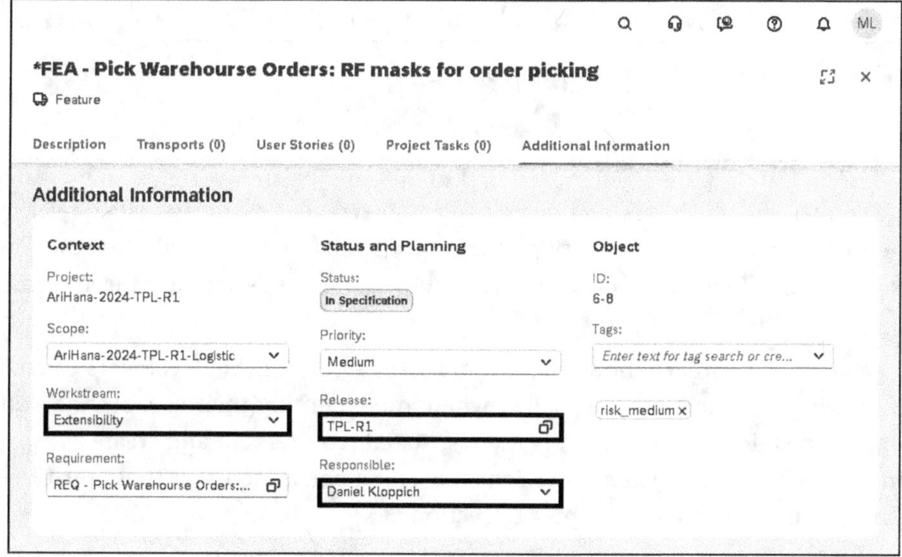

Figure 17.11 Feature: Specify Additional Information

17.3 Change Control with the Help of Features

Once the feature has been specified (see Figure 17.12), it can be enriched with transports. Depending on the SAP solution, different transport tools are integrated into SAP Cloud

ALM for this purpose (see Chapter 4). Transport tools enable change control, such as the creation, release, and import of the referenced transports into target systems such as the test and production systems. In the following example, we are using the on-premise SAP S/4HANA system landscape (explained in Chapter 3, Section 3.1.2) as a reference example. The transport tool integration is explained in Chapter 4, Section 4.2. In the following sections we run through the various status values using the feature created in advance as an example. In doing so, we will look at the technical options for controlling the transports (see Table 17.1).

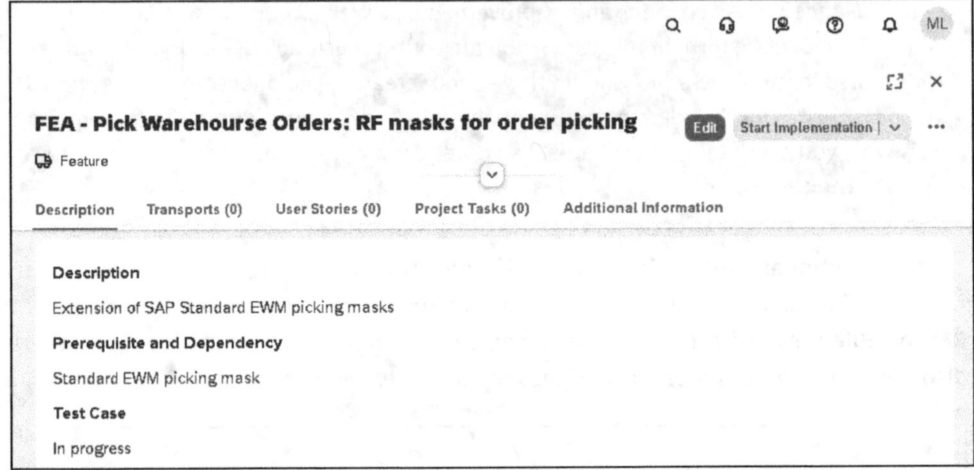

Figure 17.12 Feature: Ready for Implementation

17.3.1 Transportation Creation

In some companies, it is intended that transports are created by a responsible release management team. As already mentioned, transports can also be created when the feature has the status In Specification. To do this, click on **Create • Transport** (see Figure 17.13).

In an agile project approach, however, it also makes sense for the developers to take care of creating the transport themselves when implementing their user stories. You can jump to the referenced feature via the associated requirement and create a transport on the development system or assign an existing one if required. These relationships are described in detail in Chapter 13, Section 13.5.

When you create a new transport, a popup window opens in which you can maintain the transport information (see Figure 17.14).

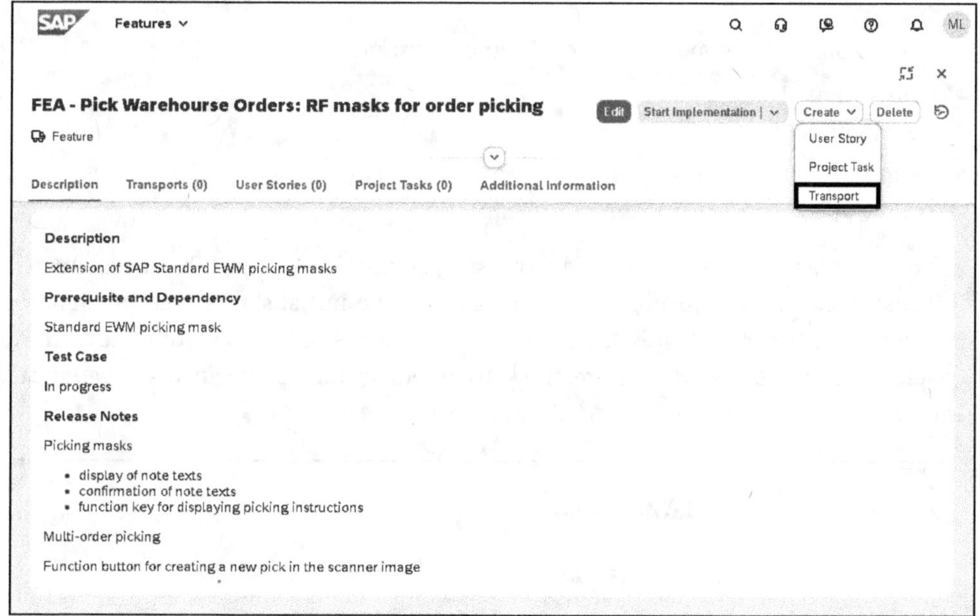

Figure 17.13 Feature: Create Transport

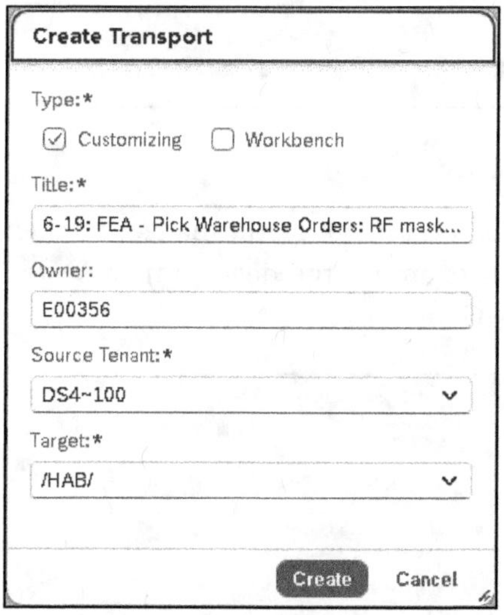

Create Transport
Type:*
Customizing Workbench
Title:*
6-19: FEA - Pick Warehouse Orders: RF mask...
Owner:
E00356
Source Tenant:*
DS4~100
Target:*
/HAB/
Create Cancel

Figure 17.14 Feature: Create Popup Window for Transport

In the popup window for creating the transport, the **Title** field is prefilled by the feature object ID and short description. This ensures the traceability of transport in the managed system landscape. Following is the information you need to enter:

- **Type**
  Workbench request/**Customizing** request

- **Owner**
  User ID of the developer in the development system
- **Source Tenant**
  Development system
- **Target**
  Follow-up system/test system or target group

Click on the **Create** button to save the transport. Once the transport has been created, it is displayed in the transport list of the feature. The initial status **Creation Pending** indicates that the transport is still being created via a specific background job on the development system. Furthermore, the subsequent system is already displayed in the **Target Tenant** column (see Figure 17.15).

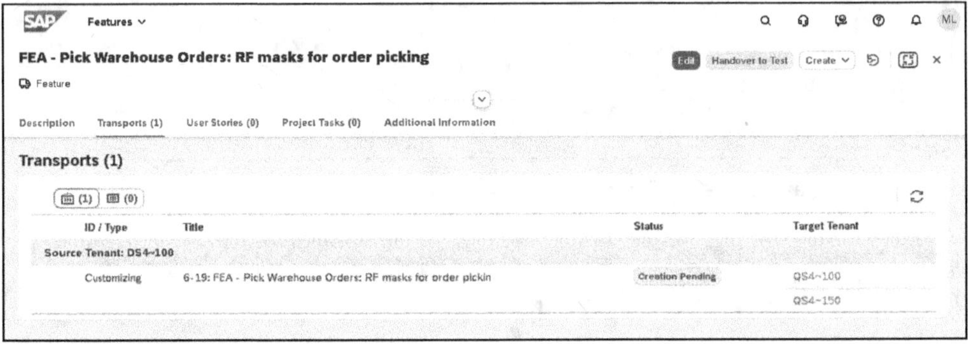

Figure 17.15 Feature: Transport List

The periodic background job /SDF/CALM_CDM_TR_POC_CL_DEP-100 on the managed system is responsible for creating the transport (in this example, the development system DS4). Figure 17.16 shows the log of the background job.

Job Log Overview for Job: /SDF/CALM_CDM_TR_PROC_CL_DEP-100 / 17084501

Date	Time	Message text	Message class	Message no.	Message type
08.03.2024	17:08:45	Job /SDF/CALM_CDM_TR_PROC_CL_DEP-100 17084501 started	00	516	S
08.03.2024	17:08:45	Step 001 started (program /SDF/CALM_CDM_TSK_PROCESS, variant &0000000155233, user ID BATCH_USER)	00	550	S
08.03.2024	17:08:45	Program /SDF/CALM_CDM_TSK_PROCESS started with destination CLOUD_ALM	/SDF/CALM_CDM_MSG	020	I
08.03.2024	17:08:45	Call post request to URL: 'https://eu20.alm.cloud.sap/api/featuresService/v1/ odata/v4/FeatureService/readTransportsForProcessin gInSourceSystemV2 '	/SDF/CALM_CDM_MSG	001	I
08.03.2024	17:08:45	Post request returned with status 200	/SDF/CALM_CDM_MSG	002	I
08.03.2024	17:08:45	Creating 1 transports for destination CLOUD_ALM	/SDF/CALM_CDM_MSG	031	I
08.03.2024	17:08:45	Transport request 'DS4K909464' created for feature id '77697788-24a9-4759-bc1a-dab784f13c30'	/SDF/CALM_CDM_MSG	023	I
08.03.2024	17:08:45	Releasing 0 transports for destination CLOUD_ALM	/SDF/CALM_CDM_MSG	032	I
08.03.2024	17:08:45	Creating 0 transports of copy for destination CLOUD_ALM	/SDF/CALM_CDM_MSG	033	I
08.03.2024	17:08:45	Call post request to URL: 'https://eu20.alm.cloud.sap/api/featuresService/v1/ odata/v4/FeatureService/handleTransportsFromDevSys temV2 '	/SDF/CALM_CDM_MSG	001	I
08.03.2024	17:08:46	Post request returned with status 204	/SDF/CALM_CDM_MSG	002	I
08.03.2024	17:08:46	Program /SDF/CALM_CDM_TSK_PROCESS ended	/SDF/CALM_CDM_MSG	021	I
08.03.2024	17:08:46	Job finished	00	517	S

Figure 17.16 Job Log for Transport System

After the transport has been created on the development system, the status of the transport in the transport list changes to **Modifiable** (see Figure 17.17). The assigned developers can now start their work and write their changes to the transport.

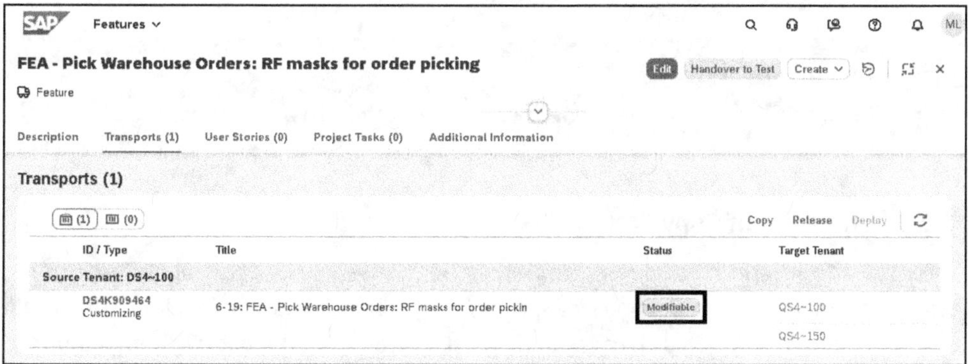

Figure 17.17 Feature: Transport List with Created Transport

17.3.2 Start Feature Implementation

Once a transport has been created and the developers have been able to document their changes, the feature should also receive the appropriate status. To do this, click on the **Start Implementation** button within the feature and confirm with **OK** (see ❶ and ❷ in Figure 17.18). The status now changes to **Implemented**.

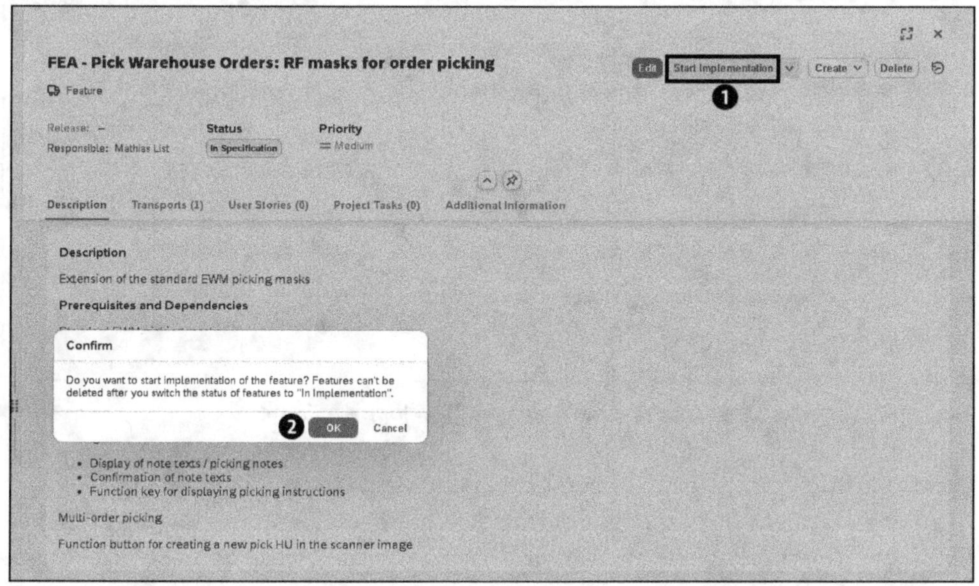

Figure 17.18 Feature: Start Implementation

If, in addition to the pure business context, the technical development components also play a role for the delivery, the relevant user stories can also be assigned to a feature. This allows you to later control and document the respective developments down to transport level for the release compilation (see Section 17.6).

Start of Implementation

Once the implementation has been started, it cannot be stopped or reversed.

17.3.3 Transport of Copy

After the development has been completed, the change can be provided to the test system for a functional test using a *transport of copy*. This allows the developer or a designated tester to make the change in the test system in advance without having to release and import the corresponding transport. Even partially completed changes can be tested in the test system in advance. A transport of copy can be created multiple times for the same transport. This procedure has a strong advantage, as it significantly reduces the number of transports that are transferred to a release. Only error-free transports that have been checked using the transport copy in the test system are released. The transport of copy can be created directly via the feature. The prerequisite is that at least one transport task has been released in advance (see Figure 17.19).

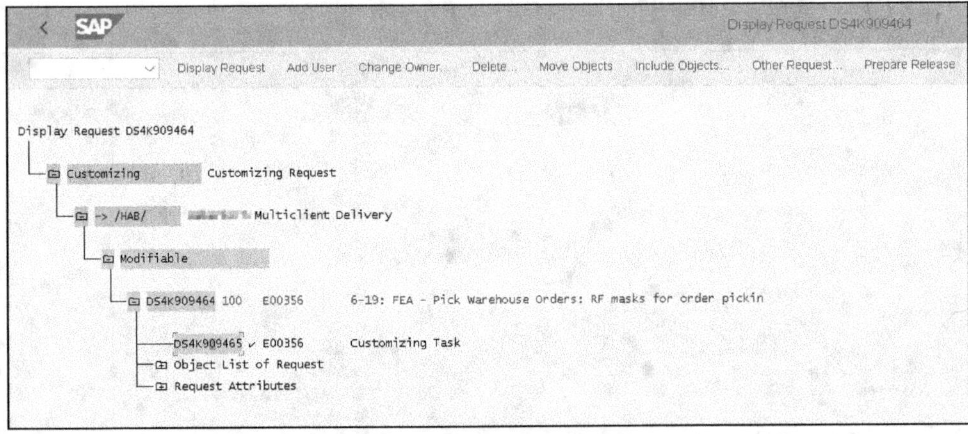

Figure 17.19 Release Transport Task in Managed Development System

Next, open the feature in SAP Cloud ALM and navigate to the **Transports** tab. Click on the **Copy** button to open the popup window **Create and Deploy Transport of Copies.** Chose the **Transport** where you want to create a transport of copy. To start the transport of copy, click the **Copy** button (see Figure 17.20).

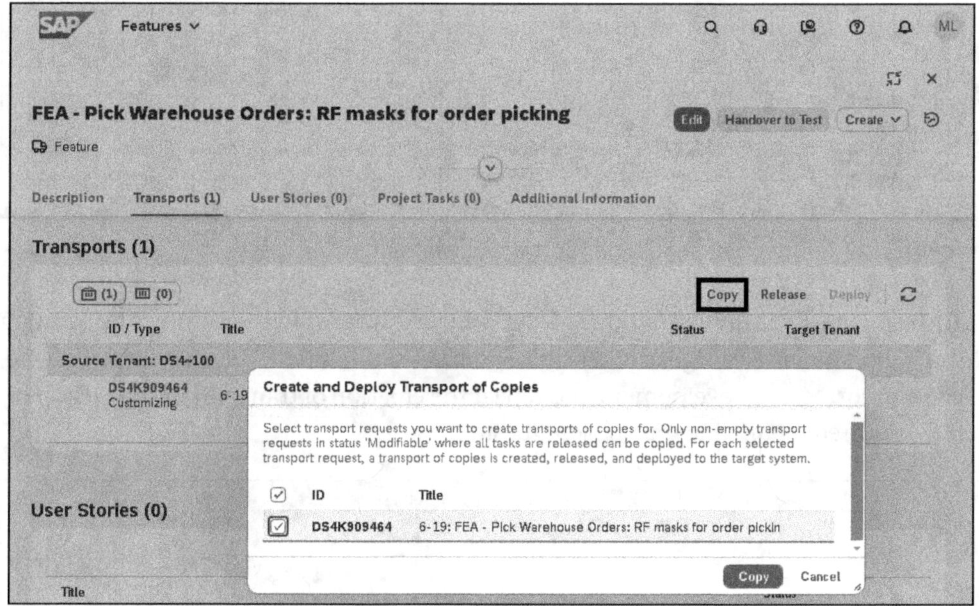

Figure 17.20 Trigger Creation of Transport of Copy in Feature

In the managed system, a job is now scheduled that will create the transport of copy with the released task content of the original transport (see Figure 17.21).

Refresh Long text Change Layout More

Job Log Overview for Job: /SDF/CALM_CDM_TR_PROC_CL_DEP-100 / 19384501

Date	Time	Message text	Message class	Message no.	Message type
08.03.2024	19:38:45	Job /SDF/CALM_CDM_TR_PROC_CL_DEP-100 19384501 started	00	516	S
08.03.2024	19:38:45	Step 001 started (program /SDF/CALM_CDM_TSK_PROCESS, variant &0000000155286, user ID BATCH_USER)	00	550	S
08.03.2024	19:38:45	Program /SDF/CALM_CDM_TSK_PROCESS started with destination CLOUD_ALM	/SDF/CALM_CDM_MSG	020	I
08.03.2024	19:38:45	Call post request to URL: 'https://eu20.alm.cloud.sap/api/featuresService/v1/ odata/v4/FeatureService/readTransportsForProcessin gInSourceSystemV2 '	/SDF/CALM_CDM_MSG	001	I
08.03.2024	19:38:45	Post request returned with status 200	/SDF/CALM_CDM_MSG	002	I
08.03.2024	19:38:45	Creating 0 transports for destination CLOUD_ALM	/SDF/CALM_CDM_MSG	031	I
08.03.2024	19:38:45	Releasing 0 transports for destination CLOUD_ALM	/SDF/CALM_CDM_MSG	032	I
08.03.2024	19:38:45	Creating 1 transports of copy for destination CLOUD_ALM	/SDF/CALM_CDM_MSG	033	I
08.03.2024	19:38:45	The object entries from DS4K909464 were added to object list DS4K909466	TK	257	S
08.03.2024	19:38:50	Transport of copies 'DS4K909466' created for transport request 'DS4K909464'	/SDF/CALM_CDM_MSG	036	I
08.03.2024	19:38:50	Call post request to URL: 'https://eu20.alm.cloud.sap/api/featuresService/v1/ odata/v4/FeatureService/handleTransportsFromDevSys temV2 '	/SDF/CALM_CDM_MSG	001	I
08.03.2024	19:38:50	Post request returned with status 204	/SDF/CALM_CDM_MSG	002	I
08.03.2024	19:38:50	Program /SDF/CALM_CDM_TSK_PROCESS ended	/SDF/CALM_CDM_MSG	021	I
08.03.2024	19:38:50	Job finished	00	517	S

Figure 17.21 Creation Transport of Copy in Managed Development System

Transport of copy details can be seen in the feature by clicking the **Status** icon in the **Transports** tab in the left-hand side of the transport request **ID** (see Figure 17.22). The resulting popup window provides information about transport of copy ID and deployment status in the dedicated target system and clients.

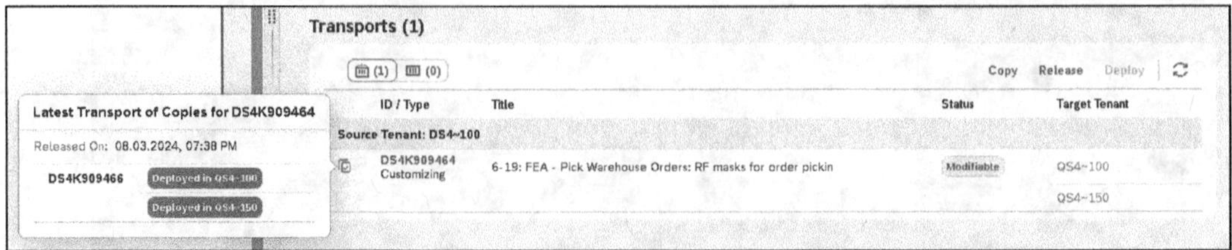

Figure 17.22 Information about Transport of Copy in Feature

You can review the import of transport of copy in the managed test system by opening the import queue (see Figure 17.23). If the proceeded functional test was not successful, you can correct the development and repeat the creation of transport of copy based on the dedicated modifiable transport request.

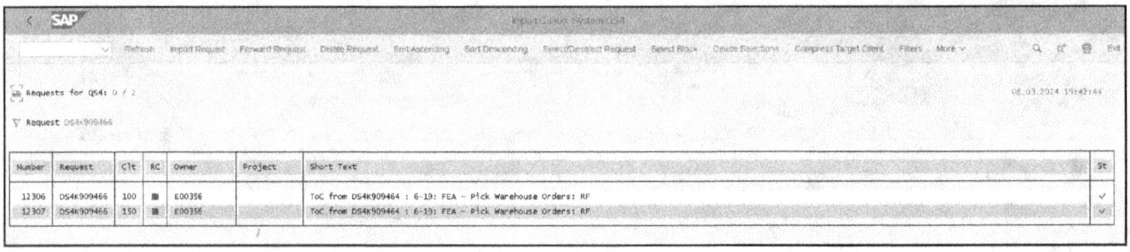

Requests for QS4: 0 / 2 08.03.2024 19:42:44

Request DS4K909466

Number	Request	Clt	RC	Owner	Project	Short Text	St
12306	DS4K909466	100	■	E00356		ToC from DS4K909464 : 6-19: FEA – Pick Warehouse Orders: RF	✓
12307	DS4K909466	150	■	E00356		ToC from DS4K909464 : 6-19: FEA – Pick Warehouse Orders: RF	✓

Figure 17.23 Import Transport of Copy in Managed Quality System Clients

17.3.4 Release Transport

After a successful functional test with help of transport of copy, the modifiable transports are released with the **Release** action (see ❶ in Figure 17.24) on the development system after which you click on **Release** in the popup ❷. After this, no more changes can be written to these transports, and they are ready for import into the subsequent system (here, the test system QS4).

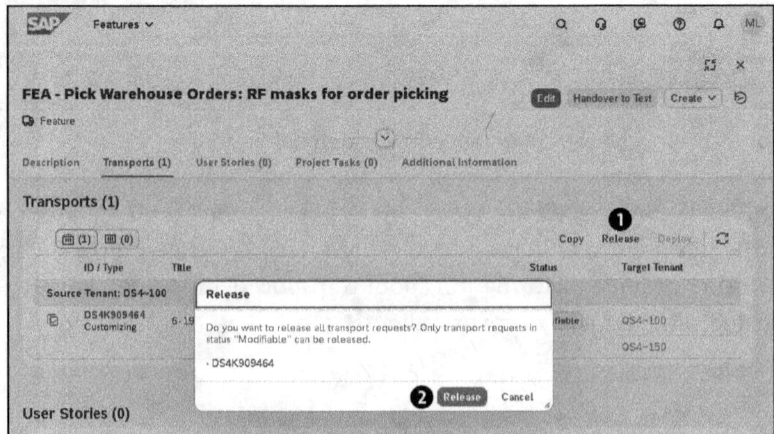

Figure 17.24 Feature: Release Transport Request

Figure 17.25 shows the updated transport list of the feature with the transport in **Released** status.

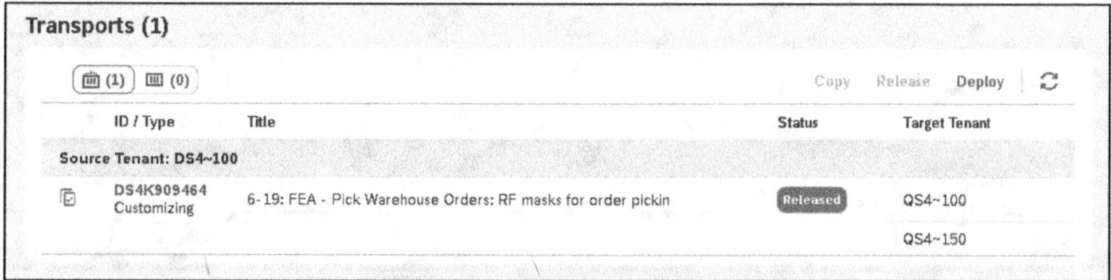

Figure 17.25 Feature: Transport List with Released Transport

17.3.5 Import Transport into the Test System

The released transports must be imported for testing in the test system. You can see the test system in the **Target Tenant** column of the transport list of the feature (see ❶ in Figure 17.26). Click on the **Import** button to initiate the import of all transports ❷ and confirm the import in the notification popup window ❸.

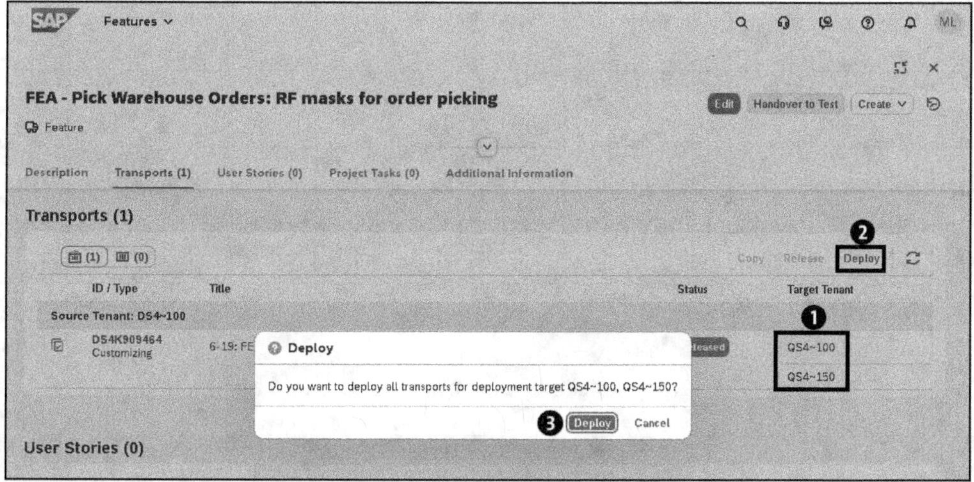

Figure 17.26 Feature: Import Transport into the Test System

The import process is carried out in a similar way to creating and releasing via a corresponding background job on the managed system. In Figure 17.27, you can see an excerpt from the processing log.

Job Log Entries for /SDF/CALM_CDM_IMPORT_TRANSPORTS / 20400400

Refresh Long text Change Layout More

Job Log Overview for Job: /SDF/CALM_CDM_IMPORT_TRANSPORTS / 20400400

Date	Time	Message text	Message class	Message no.	Message type
10.03.2024	20:40:04	Job /SDF/CALM_CDM_IMPORT_TRANSPORTS 20400400 started	00	516	S
10.03.2024	20:40:04	Step 001 started (program /SDF/CALM_CDM_TSK_IMPORT, variant &0000000000005, user ID BATCH_USER)	00	550	S
10.03.2024	20:40:04	Program /SDF/CALM_CDM_TSK_IMPORT started with destination CALM	/SDF/CALM_CDM_MSG	014	I
10.03.2024	20:40:04	Diagnostics Job already scheduled. Not scheduling again	/SDF/CALM_CDM_MSG	026	I
10.03.2024	20:40:04	Call post request to URL: 'https://eu20.alm.cloud.sap/api/featuresService/v1/ odata/v4/FeatureService/readTransportsToBeImported '	/SDF/CALM_CDM_MSG	001	I
10.03.2024	20:40:04	Post request returned with status 200	/SDF/CALM_CDM_MSG	002	I
10.03.2024	20:40:04	Read 2 transports for import from CALM	/SDF/CALM_CDM_MSG	008	I
10.03.2024	20:40:04	Starting import for 2 transports	/SDF/CALM_CDM_MSG	009	I
10.03.2024	20:40:04	Importing transport DS4K909464 QS4.100	/SDF/CALM_CDM_MSG	011	I
10.03.2024	20:40:04	Importing transport DS4K909464 QS4.150	/SDF/CALM_CDM_MSG	011	I
10.03.2024	20:40:04	Call post request to URL: 'https://eu20.alm.cloud.sap/api/featuresService/v1/ odata/v4/FeatureService/updateImportTransportsStat us '	/SDF/CALM_CDM_MSG	001	I
10.03.2024	20:40:04	Post request returned with status 204	/SDF/CALM_CDM_MSG	002	I
10.03.2024	20:40:21	Start deploying transport requests (#number 1) to target QS4.100	/SDF/CALM_CDM_MSG	012	I
10.03.2024	20:40:26	Laufzeit: 5.100 [msec]	XT	000	S
10.03.2024	20:40:31	Laufzeit: 4.943 [msec]	XT	000	S
10.03.2024	20:40:31	Job finished	00	517	S

Figure 17.27 Job Log for the Test System Import

After a successful import, the current status is displayed in the transport list of the feature. In addition, the identifier of the follow-up system now changes in the **Target Tenant** column. In our example, the transport for the import into the **PS4** production system is pending (see Figure 17.28). If necessary, refresh the view of the transport list by clicking on the Refresh icon.

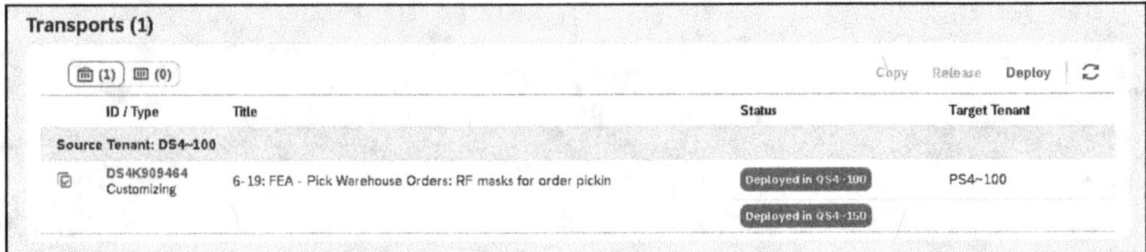

Figure 17.28 Feature: Transport List with Imported Transport into the Test System

Figure 17.29 shows an example of the import queue of the QS4 test system.

Import Queue: System QS4

Refresh Import Request Forward Request Delete Request Sort Ascending Sort Descending Select/Deselect Request Select Block Delete Selections Compress Target Client Filters More Exit

Requests for QS4: 0 / 2 10.03.2024 20:54:47

Request DS4K909464

Number	Request	Clt	RC	Owner	Project	Short Text	St
12310	DS4K909464	100	■	E00356		6-19: FEA - Pick warehouse Orders: RF masks for order pickin	✓
12311	DS4K909464	150	■	E00356		6-19: FEA - Pick warehouse Orders: RF masks for order pickin	✓

Figure 17.29 Import Queue of the Test System

17.3.6 Handover for Testing

To transfer the feature to the test management, click on the **Handover to Test** button. The status then changes to **In Testing** (see Figure 17.30).

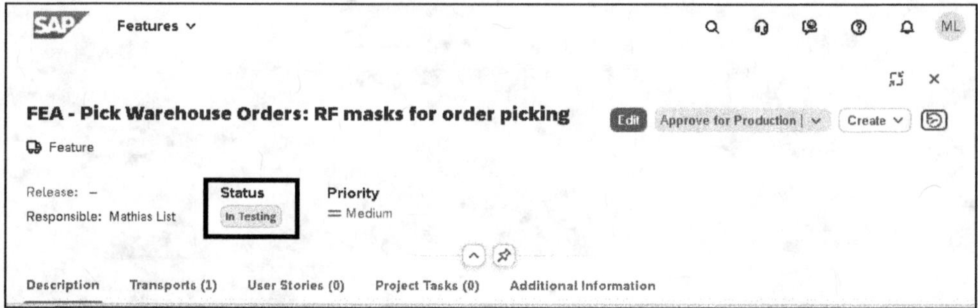

Figure 17.30 Feature: Handover to Test

17.3.7 Approve for Production

As soon as testing has been completed (see Chapter 15) and all criteria for going live have been met, the feature can be approved for production using the **Approve for Production** button (see Figure 17.31) to change the status to **Ready for Production**.

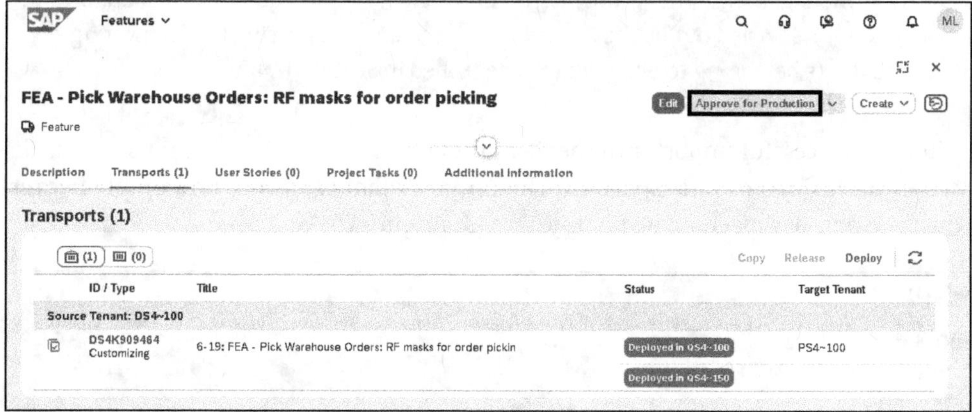

Figure 17.31 Feature: Approve for Production

17.3.8 Import Transport into the Production System

Once the feature has been approved, the transport must be imported into the production system. Follow the procedure for transporting to the test system (see Section 17.3.5). The corresponding activities for the import process are shown in Figure 17.32.

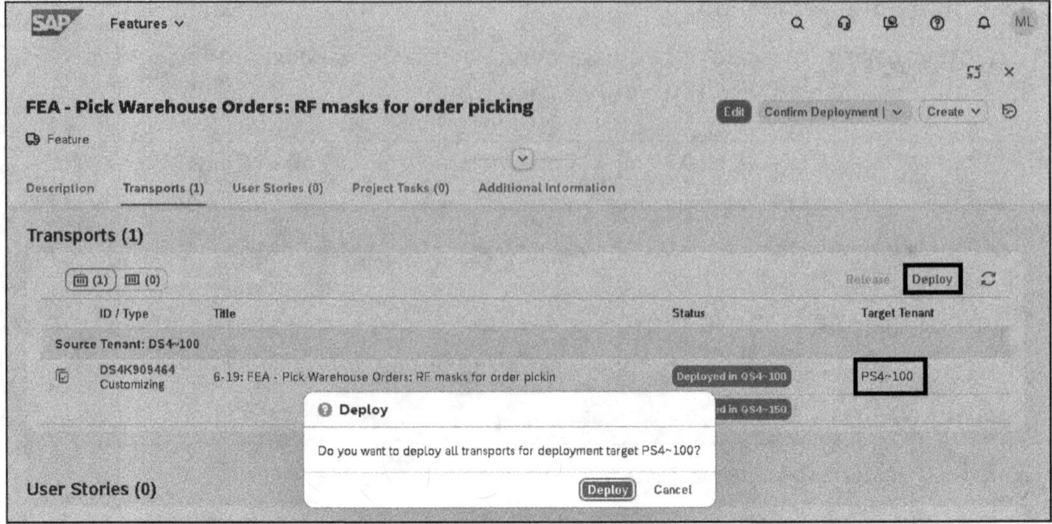

Figure 17.32 Feature: Importing the Transport into the Production System

[»]

Going Live Through Release Management

The release for production and the associated go-live are usually the responsibility of release management. The import is therefore anchored in the release calendar and should not be carried out by the individual developers. The possibility of delivering several features belonging to one release is described in Section 17.4.

After the successful import into the PS4 production system, the current status is displayed in the feature's transport list. The **Target Tenant** column is now empty because the transport has reached the final system (see Figure 17.33).

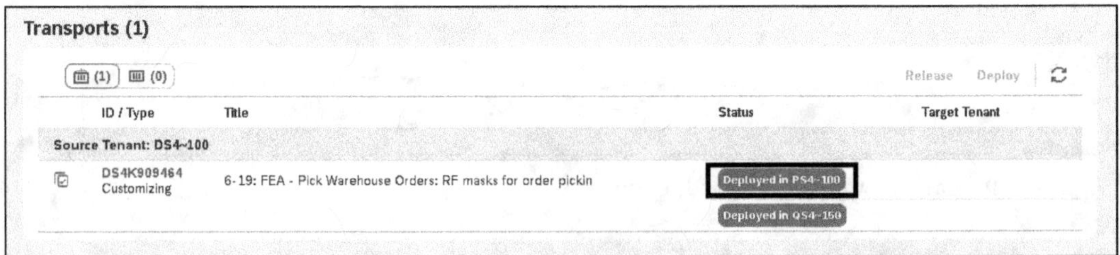

Figure 17.33 Feature: Transport List with Imported Transport into the Production System

17.3.9 Confirm and Complete Feature Import

Once all transports have been successfully set to productive and all organizational follow-up work has been completed, the import is confirmed for each feature. To do this, click on the **Confirm Deployment** button and confirm with **OK**, as shown in Figure 17.34.

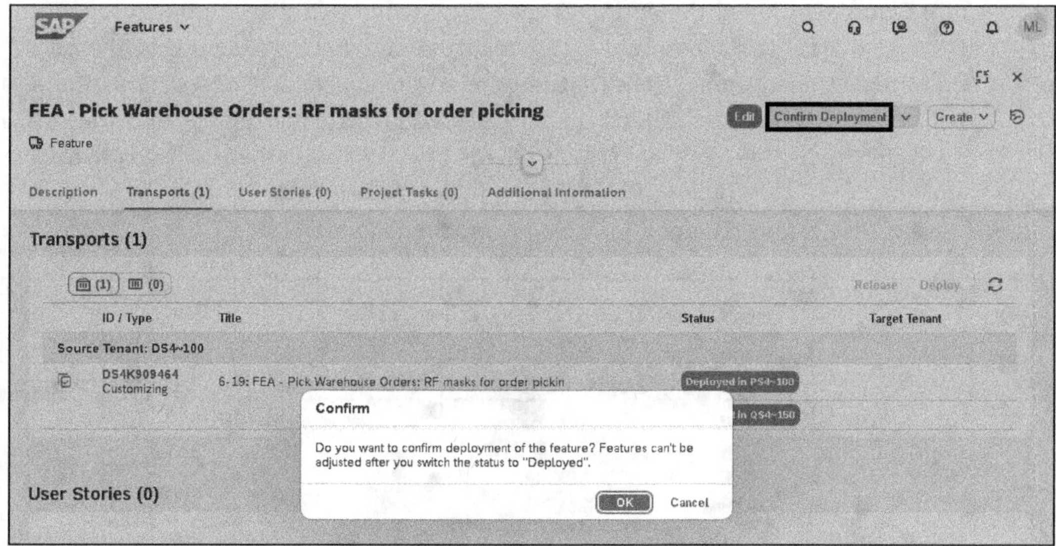

Figure 17.34 Feature: Confirm Deployment

The feature is then considered completed, which is represented by the final status **Deployed**. The **Edit** button is no longer available, and you can no longer make any changes (see Figure 17.35).

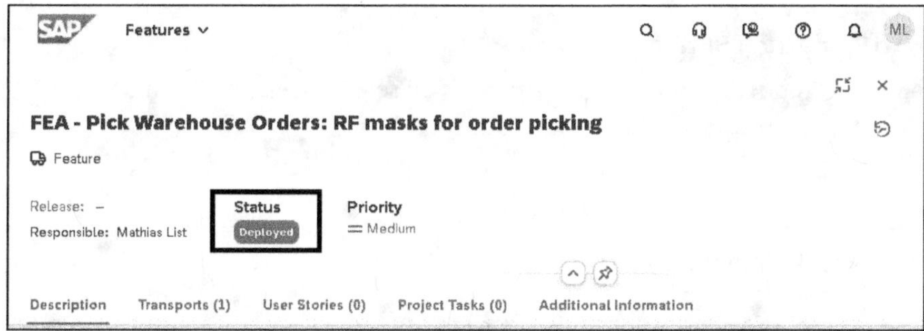

Figure 17.35 Feature with Final Status Deployed

17.4 Mass Deployment of Features into Production

The release for production or the associated go-live is usually the responsibility of release management. The import into production is determined by the deployment plan or the release plan (see Section 17.1.2).

The deployment of a release should be planned and communicated in advance. This can be done, for example, by announcing it in a newsletter and publishing the release calendar. It is also important to hand over to operations so that they can ensure support for the new function.

SAP Cloud ALM offers the option of mass deployment so that the transports of the respective features do not have to be imported individually. To do this, the release management team can use the **Features** app to search for all features relevant to the release. Important filter criteria are the desired project, the release, and the status **Ready for Production** (❶ in Figure 17.36). In the next step, all elements found can be consolidated and selected using the checkbox in the list header ❷. To start the import for the entire selection, click on the **Deploy to Production** button ❸.

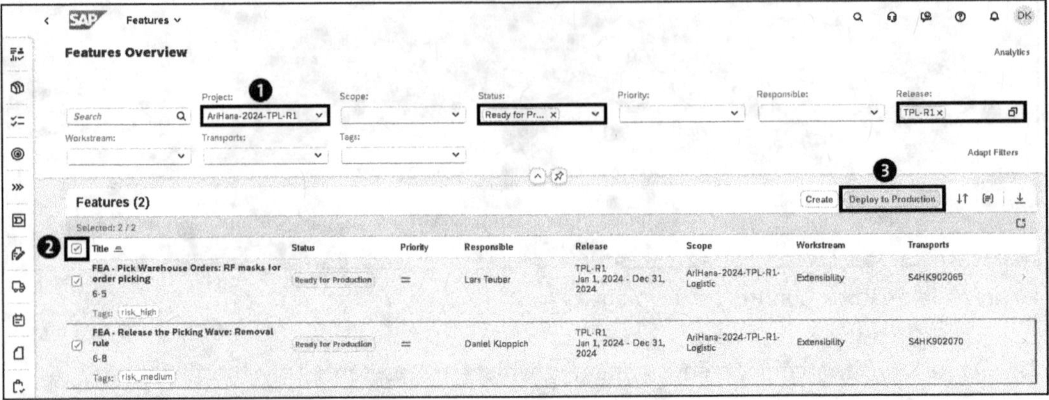

Figure 17.36 Feature List: Mass Deployment of Features into Production

[»]

Deploy to Production Button

The **Deploy to Production** button is only active if all selected features have the status **Ready For Production** and also have added transports.

17.5 Transport Protocols

The change history of a feature provides traceability and is required for change audit purposed. In addition to the individual changes to the feature itself, the transport log is also displayed here, which documents the creation, release, and import of the transports in the change history. For each change, the name of the logged-in user who made the change(s), the time, and the change itself, including the previous status, are displayed (see Figure 17.37). The transport log allows you to track the transports and the associated changes to the systems. The history can be opened in the header area of the feature at the top right via the clock symbol. The **History** window is now displayed on the right-hand side of the screen.

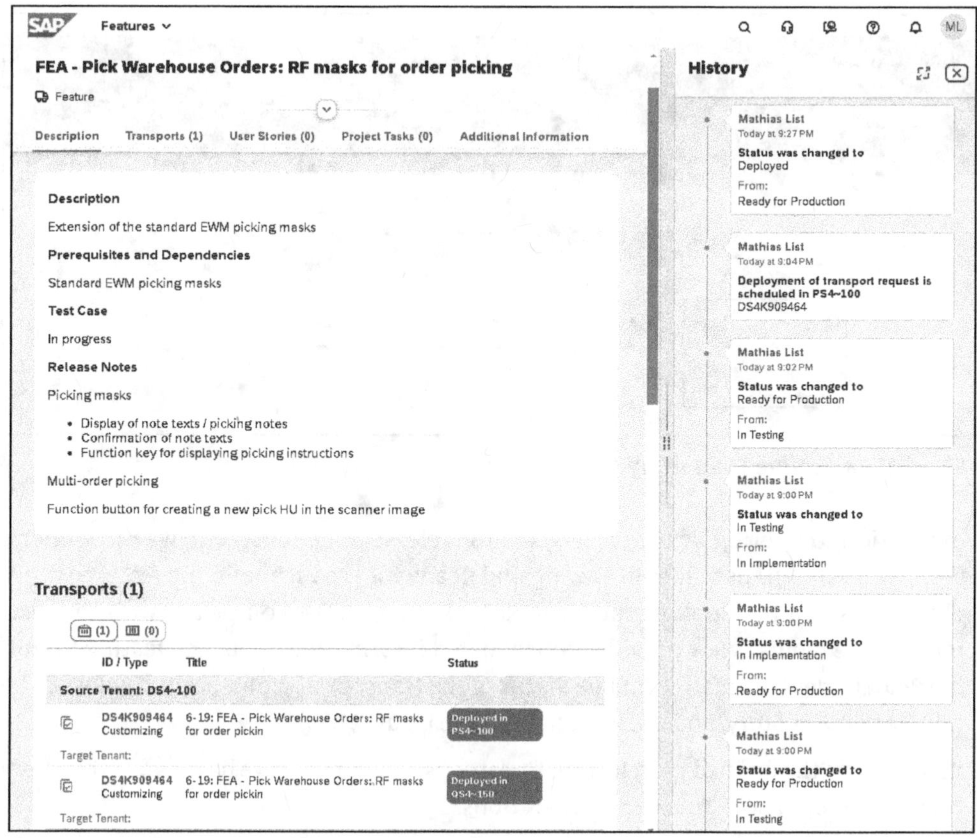

Figure 17.37 Feature History

17.6 Milestone for Handover to Operations

After deployment or after the release has gone live, the run phase begins (execution). In this phase, the implemented functions are used in live operation for the first time. Those responsible for release management should, together with the project team, enable the responsible IT support operations team to operate the new solution securely and rectify errors. The following requirements should be met:

- The release notes are documented and communicated.
- The system documentation is complete and available.
- The most important problems and priority errors identified during the test phase have either been solved or have a documented workaround available.

SAP Cloud ALM offers various functions that support you during the handover. The Feature Traceability app from the **Analytics** app group is particularly helpful (see Figure 17.38).

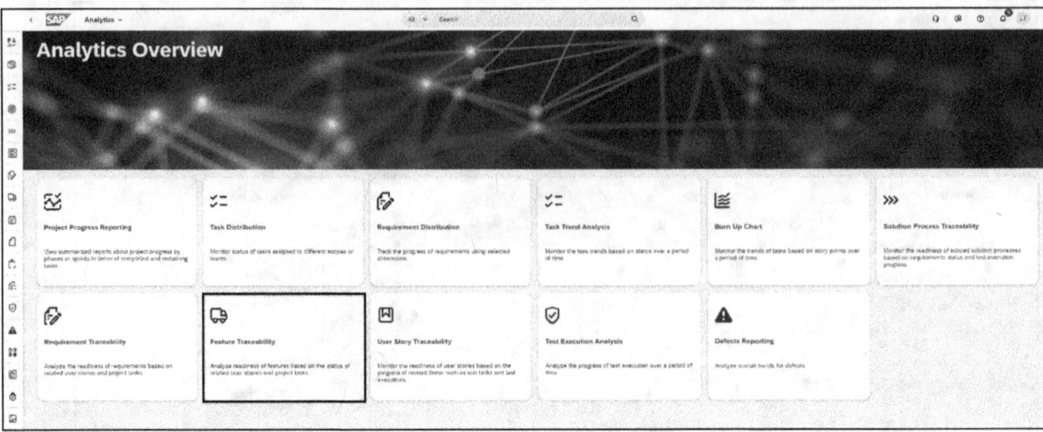

Figure 17.38 Feature Traceability App

In addition to filtering by the project and the release to be rolled out, we recommend filtering by the status values **In Testing** and **Ready for Production** in the feature status. This allows you to list all features that are expected to be completed with the next release. The tabular view provides consolidated information about the feature **Status**, the **Requirement Status**, the **Import Status** of the transports, the associated **Release**, the referenced **User Story**, and the **Project Task** (see Figure 17.39).

You can save the defined preselection for later use. In our case, we have called the query **Features - Preliminary Transfer to Operations**.

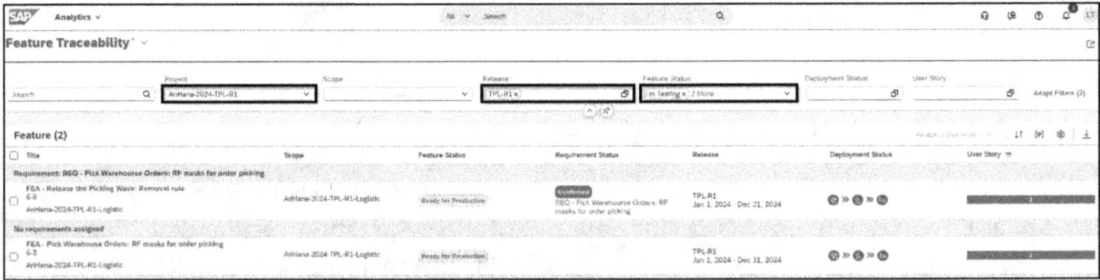

Figure 17.39 Feature Traceability: Handover to the Company

Depending on the entities in which you have maintained your information, you can use this query to quickly access all relevant sources and compile them for the IT support operations team. The following are some options:

- Jump to the feature for listing the release notes
- Jump to the requirement for the presentation of the need from a business perspective and the higher-level test cases
- Compilation of the transports via the import status (see Figure 17.40)

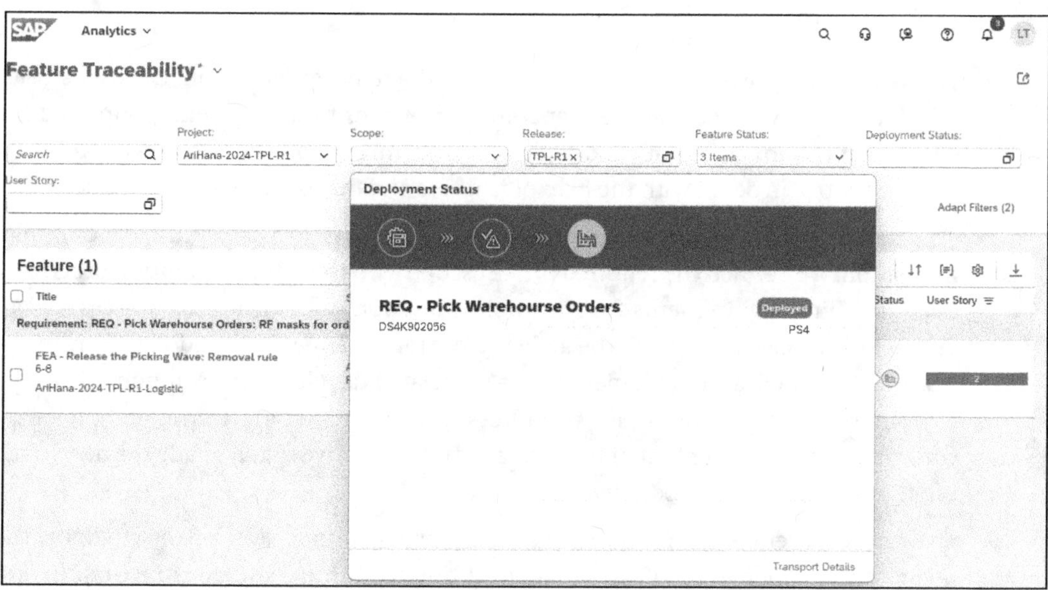

Figure 17.40 Feature Traceability: Import Status

- Jump to the user stories for the presentation of the functional requirements, test cases, development activities, and the implementation and user documentation provided (see Figure 17.41)

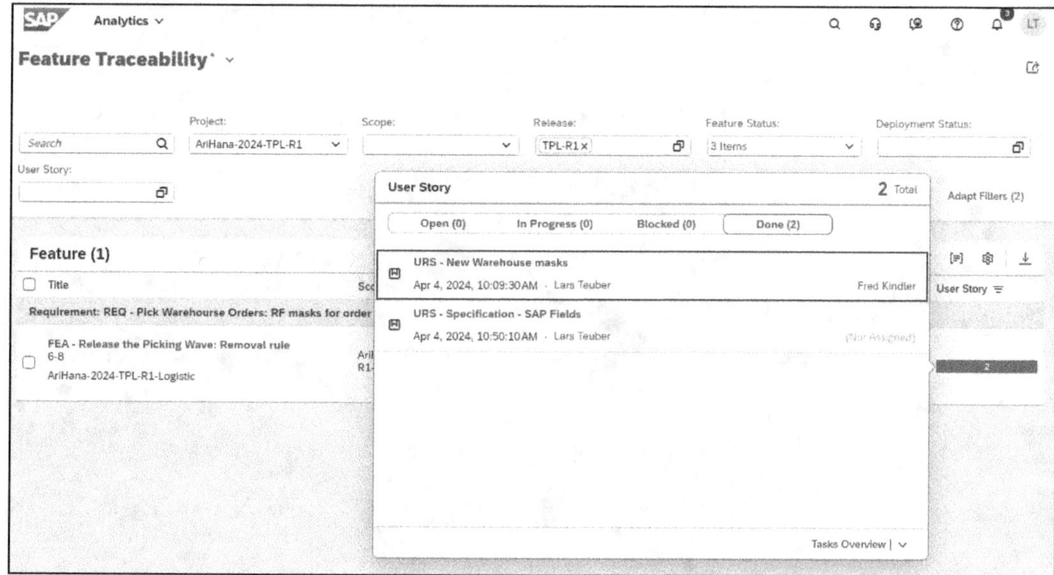

Figure 17.41 Feature Traceability: Assigned User Stories

17.7 Summary

This chapter dealt with the implementation of the requirements from the user stories using features. We looked at the preparatory measures for the system groups and the deployment plan. Using our example scenario in the logistics area, we explained change control in detail with the help of features and the associated transports.

This chapter concludes our example scenario for the part on SAP Cloud ALM for implementation. We defined the required project scope with the relevant solution processes. Corresponding requirements for the implementation of the business processes were drawn up and specified. Using the agile project management methodology Scrum, we planned, specified, and implemented the associated development components in the form of user stories. In doing so, we addressed the creation of test cases and their execution. With the deployment of features and transports, we finally made the developed functions available for productive use.

The next chapter deals with SAP Cloud ALM for operations. You will get to know the means of SAP Cloud ALM to ensure the execution of the implemented functions in productive operation.

PART V

Operate

Chapter 18

Overview of SAP Cloud ALM for Operations

Once your applications have reached the operational phase, it is important to ensure that they function adequately in terms of supporting core business processes. The options that SAP Cloud ALM offers for this phase are described in this chapter.

The need for a tool-supported operating phase (including monitoring) arises because your applications ensure the provision and processing of information for various business processes in the operational business. Malfunctions and, in the worst case, complete application failures can occur, whereby the respective causes are of course manifold. To proactively identify these situations and, ideally, prevent them before they occur, monitoring is essential. Monitoring is supported by the SAP Activate deliverable "Operate Solution." You can find more information on SAP Activate in Chapter 1, Section 1.3.

18

But before monitoring and the associated activities/actions can be implemented, they must be carefully planned. By "planning" in this context, we mean the decision-making process as to how exactly the setup should be implemented. The question of suitable software plays an important role here, with SAP Cloud ALM representing a possible solution. The following questions are central to this:

- Which monitoring scenarios does the software solution offer, and which monitoring scenario can be configured to meet your own requirements?
- How should identified disruptive situations be communicated?
- Are there ways to automate the solution to the event?
- Are there any business services that need to be monitored?
- Which cloud services or systems should be monitored?

As a first step in the planning, it makes sense to look at your own system landscape and your own business processes. This will determine the monitoring requirements. You should then familiarize yourself with the functional scope of the various software solutions that could potentially be considered. This could be a selection of different software solutions, or you might have already decided on a solution, such as SAP Cloud ALM. The main focus here should be on deciding which monitoring scenarios meet your respective requirements. Further requirements may also result from the experience you have gained in the past during the operation of the various applications, as well as from the monitoring options of the respective software.

The second step is to consider the form in which the identified disruptions should be communicated to the right recipients to initiate a solution. The final step is the targeted rectification of the event and, if possible, the attempt to prevent a recurrence of the event in the future.

The result of the planning process is a concept that defines how the configuration in SAP Cloud ALM should be carried out. Since such concepts are individual and depend on the individual requirements, we will not go into further detail in this book, but instead focus on looking at the various options that SAP Cloud ALM offers for operations. This is intended to provide you with an overview for your planning phase, which will support you in deciding on the right monitoring scenarios and in selecting additional functionalities.

In the following sections, we provide a basic introduction to the various options and monitoring scenarios in SAP Cloud ALM that support the detection, forwarding, and resolution of events and should be considered as part of the planning process. We will introduce you to business service management in Section 18.1. In Section 18.2, you will find out to what extent SAP Cloud ALM supports the event management process. Section 18.2.1 presents the individual monitoring scenarios that are provided by SAP Cloud ALM to detect events in the cloud services and in on-premise systems. The aim here is to give you an understanding of which of the available monitoring scenarios is best suited to which use case. Once an event has been detected, it must be addressed in the form of a forwarding so that you can tackle the causes of the event. The options available, as well as functionalities that support the automation of operations, are described in Section 18.2.2.

18.1 Business Service Management

Business services are cloud services and on-premise systems that support a business process and thus form a new logical unit. This allows the partly abstract technical level to be abandoned. Furthermore, information is forwarded directly from the *Cloud Availability Center* to business service management, where it is processed and displayed. This means that information on the availability and maintenance of cloud services is displayed directly at the business service level, allowing the respective service level to be mapped in different time horizons.

Business service management also offers you the option of combining your own operating events with those from the Cloud Availability Center. SAP's expert portal mentions the possibility of defining planned downtimes of systems. SAP Cloud ALM therefore offers a functionality that you are probably already familiar with from SAP Solution Manager 7.2 and SAP Focused Run, in the form of the IT calendar.

Business service management is accessed via the tile in the **SAP Cloud ALM for Operations** group, shown in Figure 18.1.

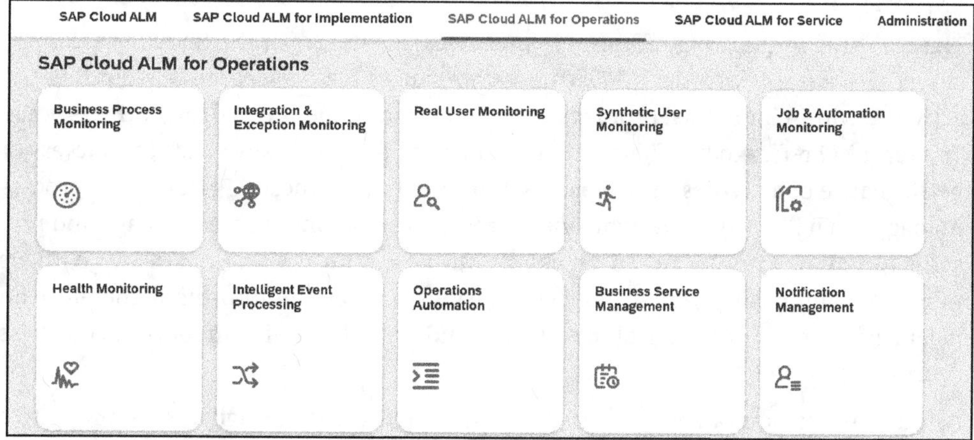

Figure 18.1 Business Service Management in the Launchpad

18.2 Monitoring and Event Management

In this chapter, the process of event detection via monitoring, event forwarding, and event resolution is referred to as *event management*. We use the term here merely as a bracket for a clear structuring. We are focusing on options that SAP Cloud ALM offers for operations in terms of monitoring and beyond.

18.2.1 Recognize Events

The detection of events that occur during the application runtime is ensured in SAP Cloud ALM by six different monitoring scenarios, which we present individually in the following sections. These monitoring scenarios enable data collection and preparation, both visually and for further processing. *Metrics* that measure certain actual states in an application or system are involved in the event detection. *Events* define data collected from monitoring as a situation, that deviates from the normal state. Events are triggered when *threshold values* are reached or exceeded, which can be defined. The defined threshold values represent the evaluation of the measured values.

The detailed information on the respective monitoring scenarios listed here as well as further information can be found in SAP's expert portal for SAP Cloud ALM for operations: *https://support.sap.com/en/alm/sap-cloud-alm/operations/expert-portal.html.*

Business Process Monitoring

The *business process monitoring* scenario is used to monitor business processes. The aim of this scenario is to ensure smooth and efficient operation without disruptions or unexpected interruptions to a company's business processes and to increase the quality and performance of business activities.

The real-time monitoring of business processes ensures, with the help of predefined *key performance indicators (KPIs)*, the greatest possible transparency regarding the activities within individual process steps and therefore the status of an entire process. The term *KPI* refers to key figures that can be used to empirically check the success or performance of activities in companies. With the help of these KPIs, business process managers can identify current business process disruptions at an early stage and proactively initiate countermeasures before business-critical situations arise. With its functional and business-oriented focus, business process monitoring is therefore an ideal complement to technical monitoring and provides useful support for business process managers.

As the composition of customer system landscapes is increasingly developing into a multi-services landscape, business process monitoring addresses precisely this transformation. Entire business processes are viewed from end to end, even if they are distributed across different IT systems.

The main focus of the monitoring scenario is on the monitoring of business processes based on hybrid landscapes, whereby the on-premise edition of SAP S/4HANA and the SAP Business Suite can be regarded as on-premise solutions. The currently supported products are listed alphabetically in the following:

- *SAP Business Suite*
- *SAP Cloud for Customer*
- *SAP Entitlement Management*
- *SAP S/4HANA* (on-premise)
- *SAP S/4HANA Cloud Private Edition*
- *SAP S/4HANA Cloud Public Edition*
- *SAP SuccessFactors*

As part of SAP Cloud ALM, business process monitoring supports the end-to-end business processes design-to-operate, lead-to-cash, recruit-to-retire, and source-to-pay. These business processes are typical, standardized core business processes of a company. The *design-to-operate* process deals with the mapping of all necessary process steps from the design to the operation of a system or the use of a product. The *lead-to-cash process* encompasses all subprocesses from initial contact and the customer's interest in buying through to the company's revenue generation and therefore affects many different departments such as marketing, sales, logistics, and finance. *Recruit-to-retire* is a human resources process in which all activities relating to the lifecycle of current and potential employees are considered. This includes recruiting, hiring, support within the company, possible skills development, and also offboarding in the event of an employee leaving the company. The *source-to-pay* process is a procurement process and includes activities relating to the search for and selection of suppliers, contract negotiations, orders, deliveries, and payment.

These processes each involve predefined substeps. The number of substeps for which KPIs are provided in business process monitoring can be seen from the arrow symbol in the **Business Process Monitoring** overview (see Figure 18.2).

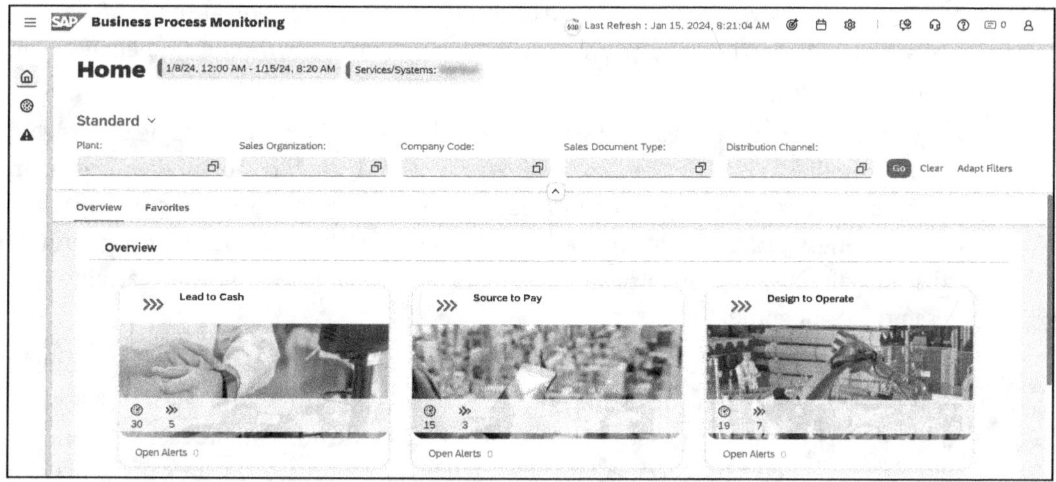

Figure 18.2 Overview of Business Process Monitoring

Directly to the left you can also see the number of active KPIs. The overview also shows the current number of alerts per process (**Open Alerts**).

You can analyze each individual process in more detail. You can open the **Process Step** view by clicking on the desired process and then see the status of all active KPIs of the assigned substeps at a glance (see Figure 18.3).

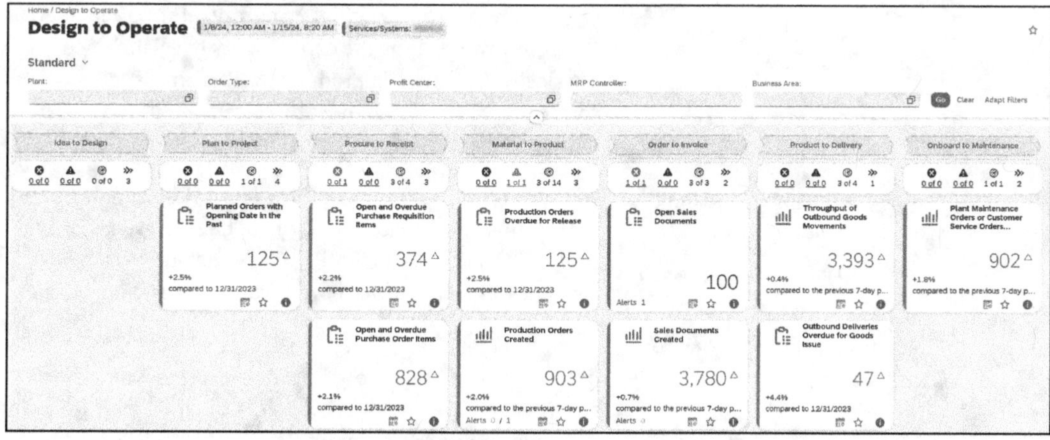

Figure 18.3 Process View in Business Process Monitoring

All active business process KPIs and the associated business process documents are displayed in an overview. A distinction is made between the KPI types backlog, throughput, and lead time. *Backlog KPIs* describe a certain number of backlog documents that have not been processed or have only been processed incompletely. *Throughput KPIs* are key figures that describe the throughput per unit of time. Lead times of certain business documents can also be monitored more specifically, which is made possible with the help of *lead time KPIs*. These different types of KPIs look at the business process with a specific focus to capture the entire process and ensure comprehensive monitoring.

As soon as the measured value deviates from the defined threshold value of a key figure, the intelligent processing of processes is activated. Via this, business process owners or executors can be informed of deviations and process anomalies. Dashboard-based business process monitoring of KPIs is particularly useful for investigating the cause of a disruption. It enables a context-sensitive drilldown into the underlying business process documents.

To make the configuration of business process monitoring in SAP Cloud ALM as simple as possible, relevant business process KPIs can be recognized automatically and the modeling of business processes can be avoided using the process, process step, and KPI selection provided. The KPIs available for each business process include those listed in Table 18.1, but this is only a small selection of all available KPIs.

KPI	Description
Design-to-Operate	
Failed Goods Movements During Process Order Confirmation	Records the number of failed goods movement postings (consumption of components and/or goods receipt of manufactured goods) during the production order confirmation. Failed goods movement postings represent a discrepancy between the actual inventory information and the book inventory, as the components have already been physically consumed in production. This error should be corrected promptly and must be corrected during period-end closing at the latest due to legal requirements.
Planned Orders with Opening Date in the Past	Records the number of planned orders whose opening date is in the past and which have not yet been converted into a production order. The identified planned orders can indicate current delays in your production process.
Production Orders Overdue for Delivery Completion	Records all released production orders whose planned end date is overdue and whose final delivery indicator is not set. The identified production orders can indicate current delays in your production process.

Table 18.1 Selection of Available KPIs

KPI	Description
Lead-To-Cash	
Deliveries Overdue for Billing	Records the number of deliveries for which the actual goods issue posting has taken place but no billing document has been created, even though the billing date is already in the past (delivered, not billed).
Incomplete Sales Documents	Records the number of sales documents that are incomplete (e.g. incomplete pricing) and are not processed until the missing fields have been completed. Delivery and/or invoicing may be delayed.
Outbound Deliveries Overdue for Goods Issue	Records the number of deliveries whose planned goods issue date is in the past and for which a complete goods issue posting has not yet been made. The identified deliveries may indicate that the delivery date cannot be met and delivery to the customer may be delayed.
Recruit-to-Retire	
Job Offers Pending	Number of vacancies awaiting approval.
Job Requisition to Hiring (Time to Hire)	Determines the average lead time (in days) between the creation and completion of a job request for which a person was hired and whose completion date is within the defined period.
Job Requisitions Without Applications	Counts the job requirements that were created in the defined period or for which no application exists.
Source-to-Pay	
Blocked Purchase Order Items	Records the number of order items that are currently blocked and cannot be processed further. This may indicate possible delays in the purchasing process.
Open and Overdue Purchase Order Items	Records the number of order items that have not yet been delivered in full and whose planned delivery date in the schedule line is in the past. They can indicate current delays in your purchasing process as well as in subsequent processes (sales, production, maintenance, or repair).
Supplier Invoice Items Blocked for Payment	Records the number of supplier invoices for which the payment block is still set for blocking reasons (e.g. price, quantity, date, manual block, or stochastic block). The identified invoices may indicate that a full or partial cash discount loss has occurred, as the payment block prevents the payment of the invoices by the payment program.

Table 18.1 Selection of Available KPIs (Cont.)

18

[»]

Complete Business Process Monitoring KPIs

A detailed list of all currently available KPIs for the individual business processes can be found in the SAP Cloud ALM for operations expert portal at *https://support.sap.com/en/alm/sap-cloud-alm/operations/expert-portal/business-process-monitoring/bpmon-content.html*

Integration and Exception Monitoring

The focus of the integration and exception monitoring scenario focuses on monitoring integration scenarios and the individual components and interfaces involved in data exchange between the various applications in these scenarios. The main focus of the monitoring scenario is on monitoring hybrid scenarios with a focus on cloud services, although SAP S/4HANA and SAP Business Suite can also be monitored as on-premise solutions. The individual products that can be monitored are listed in the following (this list will be expanded considerably in the future):

- Data ingestion for industry cloud solutions
- Integration Hub for Life Sciences
- SAP Analytics Cloud
- SAP Ariba
- SAP Batch Release Hub for Life Sciences
- SAP BTP, ABAP environment (custom-built applications)
- SAP BTP, Cloud Foundry environment (custom-built applications)
- SAP BTP, Kyma runtime (custom-built applications)
- SAP BTP, Neo environment (custom -built applications)
- SAP Business ByDesign
- SAP Business Network for Logistics
- SAP Business Suite
- SAP Cell and Gene Therapy Orchestration
- SAP Cloud for Customer
- SAP Concur
- SAP Data Intelligence
- SAP Digital Vehicle Hub
- SAP Entitlement Management
- SAP Event Broker for SAP cloud applications
- SAP Field Service Management
- SAP Fieldglass

- SAP Information Collaboration Hub for Life Sciences
- SAP Integrated Business Planning for Supply Chain
- SAP Integration Suite (Cloud Integration)
- SAP Intelligent Asset Management
- SAP Intelligent Product Recommendation
- SAP Marketing Cloud
- SAP Master Data Governance, cloud edition
- SAP Master Data Integration
- SAP Mobile Services
- SAP NetWeaver Application Server for ABAP (higher than 7.40)
- SAP Omnichannel Promotion Pricing
- SAP Order Management foundation
- SAP Predictive Replenishment
- SAP Process Integration (Java)
- SAP S/4HANA
- SAP S/4HANA Cloud Private Edition
- SAP S/4HANA Cloud Public Edition
- SAP S/4HANA Cloud Public Edition, project management, resource
- SAP Subscription Billing
- SAP SuccessFactors
- SAP SuccessFactors Employee Central Payroll

The specific content that can be monitored for each application using integration and exception monitoring varies. In general, however, monitoring is divided into two categories: messages and exceptions. *Messages* are objects that are exchanged between applications via interfaces and processed in them. This includes, among others:

- IDocs
- Web service messages
- SAP Integration Suite, managed gateway for spend management and SAP Business Network transactions
- SAP Business Network messages
- AIF messages
- SAP Concur objects
- Asset central foundation messages
- Event distributor messages
- SAP Master Data Integration messages

- SAP Mobile Services messages
- SAP Cloud for Customer messages
- SAP SuccessFactors messages

Exceptions are events that occur during the execution of an application. These events represent a disruption to the expected data processing procedure. Exceptions that can be monitored using SAP Cloud ALM include:

- ABAP runtime error
- ABAP update error
- Exceptions in the ABAP gateway
- Canceled ABAP jobs
- ABAP system log error
- Web service error for outgoing web service calls
- Exceptions in SAP SuccessFactors Employee Central Payroll
- Exceptions in extensibility transport management
- Collected exceptions that occur in the data replication framework (DFR)
- Exceptions that occur during integration with SAP Intelligent Asset Management
- Web service errors for incoming web service calls
- Messages sent or received via web services and web service proxies

[»]

Products and Content That Can Be Monitored

A detailed and up-to-date list of all products and content that can be monitored with integration and exception monitoring can be found in the SAP Cloud ALM for operations expert portal at *https://support.sap.com/en/alm/sap-cloud-alm/operations/expert-portal.html*

Monitoring can be configured for individual applications, but also for business services. Business services offer the possibility of mapping the message flows between different applications. As an SAP Cloud ALM user, the visualization in the form of a business service gives you a very good idea of which applications communicate with each other via interfaces and which data is exchanged between them. This is a very helpful, especially when it comes to complex scenarios that are sometimes very difficult to understand without a visual representation.

In addition to the pure visualization of the correlations, the errors are also displayed at a glance and highlighted by red markers, and it is possible to display message details for the single messages via drilldown functionality. The visualization of the monitoring of an individual component is shown in Figure 18.4.

Figure 18.5 shows an example of a business service consisting of an on-premise system and a cloud service.

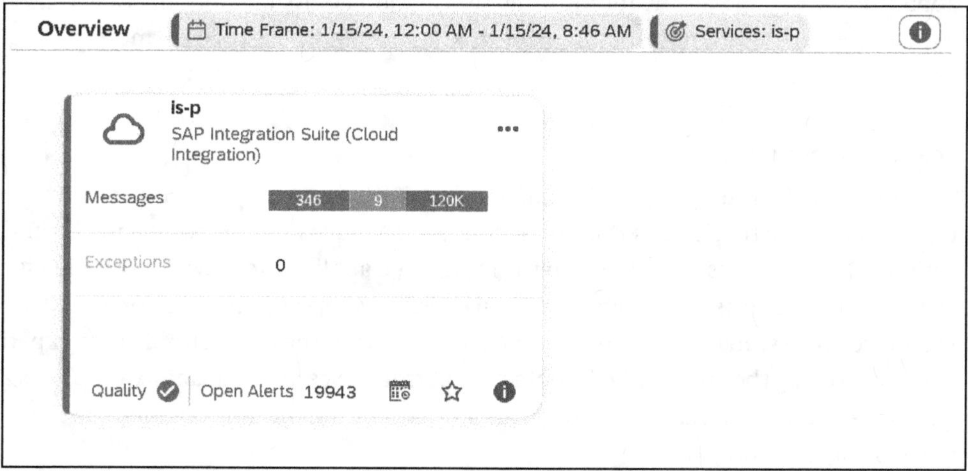

Figure 18.4 Monitoring a Single Component

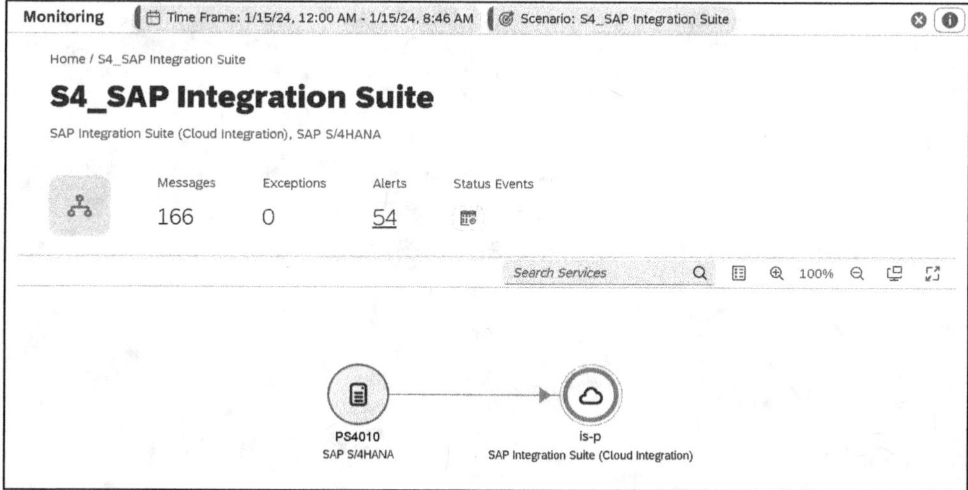

Figure 18.5 Monitoring a Business Service

Based on the monitoring data obtained, it is possible to define events and alerts that are triggered when a threshold value is reached or exceeded. Examples include messages that are in the **ERROR** status or those that have the **SCHEDULED** status for a longer period of time and remain in this status for more than 10 minutes.

In addition to events that occur in the context of message processing, it is also possible to display business information for individual messages. This relates to business context attributes, such as order or invoice numbers. If this information is transmitted to

SAP Cloud ALM by the monitored application together with log information for the notification, the end user is able to track the status of the notification across multiple applications. This makes the monitoring information accessible not only to end users with a deep technical understanding, but also to people whose focus is more on business processes.

Real User Monitoring

The aim of the monitoring scenario *real user monitoring* is to enable technical monitoring of end user activities and their performance. The focus of the monitoring is on all actions that the end users perform in the monitored application, as well as on the associated server requests (the requests from a client to a server). Both are correlated and monitored. The scenario helps to provide end users with the best possible user experience when using the application by detecting performance and availability problems at an early stage. An impression of the overview page of the monitoring scenario is provided in Figure 18.6.

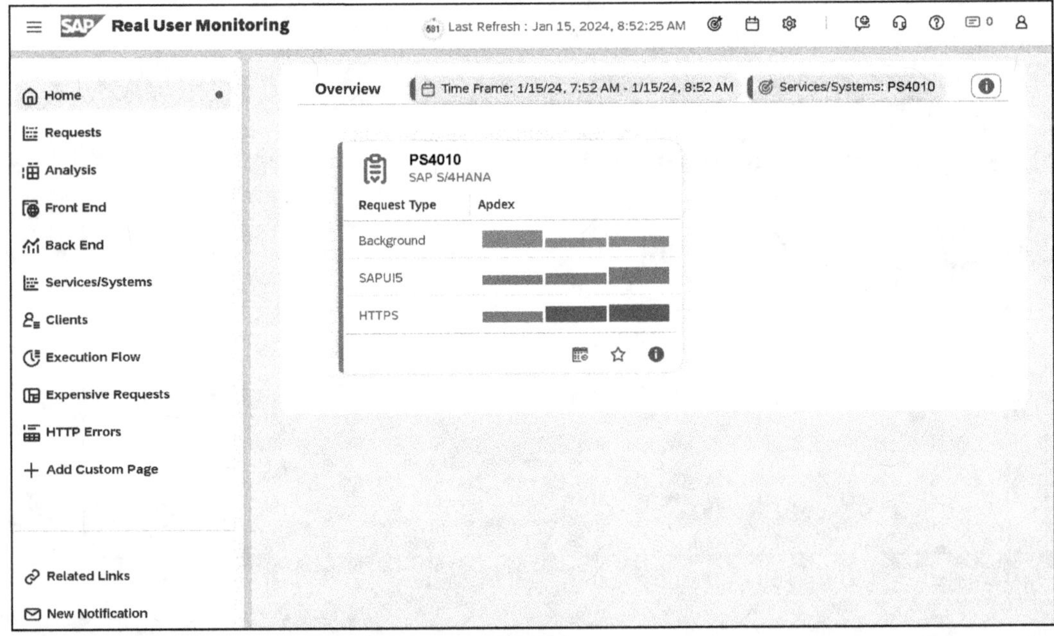

Figure 18.6 Overview Page of Real User Monitoring

Individual requests to the server are visible as a list, shown in Figure 18.7. The **Status Summary** column, which uses colored bars to draw attention to events, is very helpful.

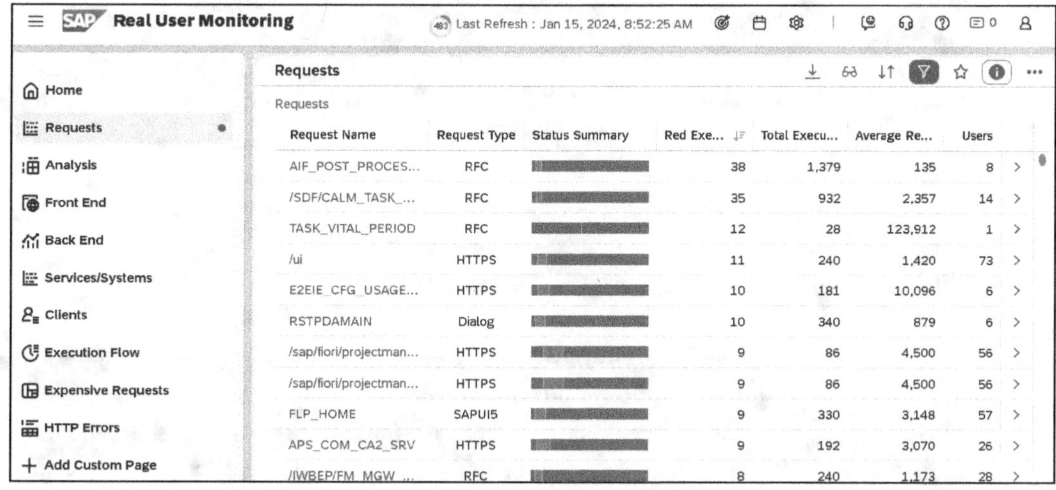

Figure 18.7 Status Overview in Real User Monitoring

The applications that can be monitored using real user monitoring are as follows:

- SAP BTP, ABAP environment
- SAP BTP, Cloud Foundry environment
- SAP BTP, Kyma runtime
- SAP BTP, Neo environment
- SAP Business Suite
- SAP Integrated Business Planning for Supply Chain
- SAP Marketing Cloud
- SAP NetWeaver Application Server for ABAP (7.40 or higher)
- SAP S/4HANA
- SAP S/4HANA Cloud Public Edition
- SAP S/4HANA Cloud Private Edition
- SAP SuccessFactors Employee Central Payroll
- SAP Cloud for Customer
- SAP Data Quality Management, microservices for location data
- SAP Ariba Category Management
- SAP Batch Release Hub for Life Sciences
- SAP Broker Reconciliation for Commodity Derivatives
- Integration Hub for Life Sciences

Both frontend and backend requests can be monitored, whereby the metrics E2E response time, backend time, network time, and execution time are available. Table 18.2 lists the respective components.

Frontend Request	Backend Request
SAPUI5/SAP Fiori	HTTP(S)
Web Dynpro (available since SAP_UI 7.53)	RFC(S) (only on premise)
SAP GUI for HTML (available since SAP_UI 7.53)	Dialog (only on premise)
	Web service

Table 18.2 Monitorable Request Types in Real User Monitoring

[»]

Products and Content That Can Currently Be Monitored

A detailed and up-to-date list of all products and content that can be monitored with Real User Monitoring can be found in the SAP Cloud ALM for operations expert portal at the following URL: *https://support.sap.com/en/alm/sap-cloud-alm/operations/expert-portal/real-user-monitoring.html.*

Synthetic User Monitoring

The *synthetic user monitoring* scenario enables the performance and availability of web applications to be monitored from the end user's perspective, using automatically executed scripts that simulate user activities. This makes it possible to detect availability and performance disruptions that affect end users, even when no end user is performing actions in the applications. In the best-case scenario, this allows you to identify and rectify events that would affect end users when using the application, even before they restrict them in their work—in other words, proactive monitoring. Theoretically, this scenario makes it possible to monitor end user activities around the clock. Figure 18.8 provides an impression of the overview page of the scenario.

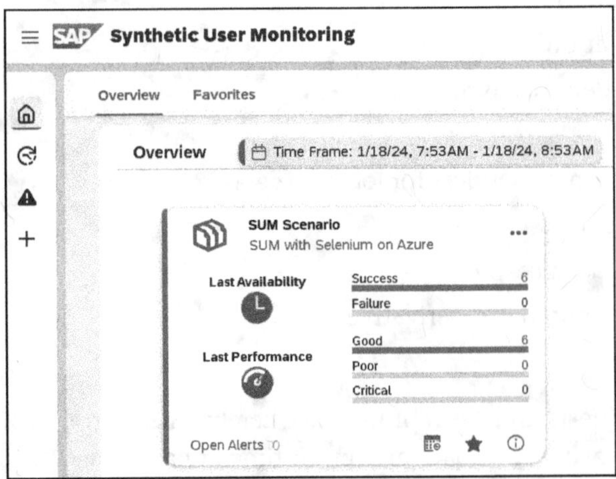

Figure 18.8 Overview Page of the Synthetic User Monitoring

The visualization of individual script executions is shown in Figure 18.9. The script execution can be analyzed using various diagrams.

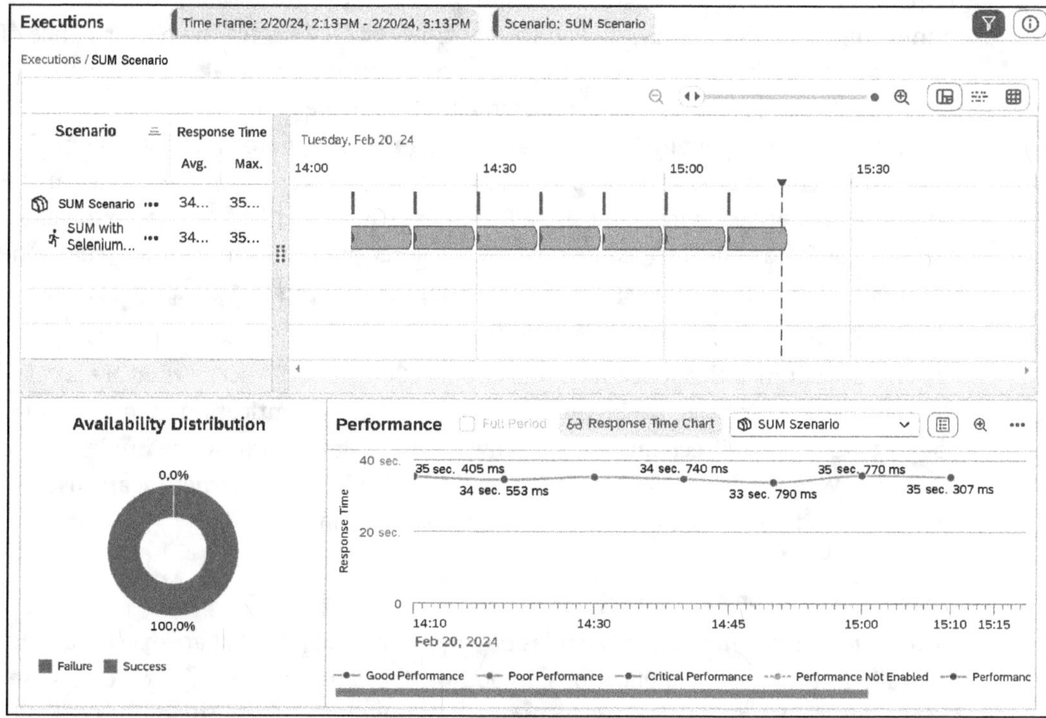

Figure 18.9 Script Executions in Synthetic User Monitoring

The special thing about synthetic user monitoring is that SAP Cloud ALM alone is not enough to enable monitoring. In addition to SAP Cloud ALM, a Selenium Remote WebDriver infrastructure and at least one .side file is required. Both are provided by Selenium and without further license costs.

What Is Selenium?

Selenium is a framework that is used for automated software testing of web applications. Selenium makes it possible to record interactions in a web application so that they can then be played back automatically. In the context of synthetic user monitoring, IDE scripts from Selenium are used to simulate the activities of end users.

In summary, synthetic user monitoring is based on the following components:

- Runner
- Resources
- Scenarios

The *runner* or SUM runner is the component that executes the script. A *resource* is a .side file that is imported into SAP Cloud ALM by uploading and that contains the scripts to be executed. The *scenarios* contain the scripts and the monitoring settings for monitoring the script execution. You can set threshold values for the performance of the script execution (in relation to the entire script or to individual script steps), with the script steps representing the end user activities or the commands. The threshold values can be set manually or automatically. In the automated variant, the performance values of the first successful script execution are taken as the basis. The threshold values are derived from there and are set for a status change from green to yellow (warning) or from yellow to red (error). If, during script execution, a step exceeds the threshold value that is typical for its execution, it is given a corresponding status, which can be seen in the UI of the scenario.

By using Selenium, web applications that support Selenium IDE scripts and are accessible from the SUM Runner can be monitored. However, authentication management is not provided—only direct authentication via user and password can be implemented in a script. Web applications that are based on the shadow DOM structure are also not supported, and the range of supported commands is limited.

Job and Automation Monitoring

Job and automation monitoring enables the monitoring of jobs that are executed automatically in the various applications. In addition to pure monitoring, trend analyses are also possible, which allow conclusions about impending non-availabilities of applications so that they can be counteracted at an early stage. The products that can be monitored using job and automation monitoring are as follows:

- SAP S/4HANA Cloud
- SAP S/4HANA and SAP Business Suite
- SAP Build Process Automation
- SAP Intelligent Robotic Process Automation
- SAP Marketing Cloud
- SAP Integrated Business Planning for Supply Chain
- SAP SuccessFactors Employee Central Payroll
- SAP BTP, ABAP environment
- SAP BTP, Cloud Foundry (customer-built application)

The job types that can be monitored include:

- SAP application job
- ABAP job

- SAP BW process chain
- SAP Build Process Automation job
- SAP Intelligent RPA job
- SAP Business Workflow
- SAP Job Scheduling service job (of an SAP BTP customer-built application)

[«]

Products and Content That Can Be Monitored

A detailed and up-to-date list of all products and job types that can be monitored with Job and Automation Monitoring can be found in the SAP Cloud ALM for operations expert portal at *https://support.sap.com/en/alm/sap-cloud-alm/operations/expert-portal/job-monitoring/job-automation-monitoring-content.html*.

The monitoring scenario makes it possible to obtain information on whether a job has been successfully executed in an application, whether it has processed the data correctly, whether it has started as planned, and whether its runtime was within the usual range. It is also possible to monitor if SAP application jobs and ABAP jobs have left a critical entry in the application log. All of this detailed information is clearly displayed in the UI of the scenario. If jobs have a status that deviates from the norm, an event action can be triggered. The initial dashboard that you see when you call up the monitoring scenario provides an overview of the job monitoring status of the individual monitored applications (see Figure 18.10).

Figure 18.10 Overview of Job and Automation Monitoring

The details of individual monitored jobs are shown in Figure 18.11. An overview of all jobs and various status information is provided in tabular form.

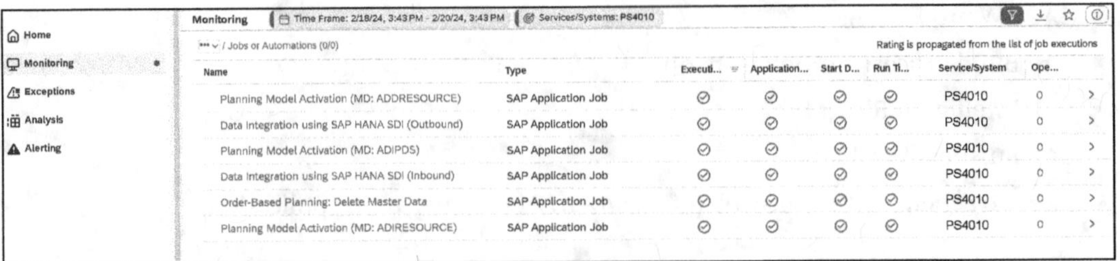

Figure 18.11 Detailed View of the Jobs in Job and Automation Monitoring

Health Monitoring

The *health monitoring* scenario provides technical monitoring that shows the health status of the systems and cloud services from the customer's perspective at all times. To determine the health status of the applications, technical information is collected at regular intervals, whereby the metrics provided depend on the application. Health monitoring is currently available for the following applications, among others:

- SAP S/4HANA Cloud Public Edition
- SAP Marketing Cloud
- SAP S/4HANA
- SAP S/4HANA Cloud Private Edition
- SAP Business Suite
- SAP SuccessFactors Employee Central Payroll
- SAP Intelligent Robotic Process Automation
- SAP Integration Suite – API Management
- SAP Integration Suite – Cloud Integration
- SAP Integrated Business Planning
- SAP HANA Cloud
- SAP Cloud ALM
- SAP Build Process Automation
- SAP BTP, Neo environment
- SAP BTP, Cloud Foundry environment
- Cloud connector

For SAP BTP, Neo environment, for example, the response time, CPU utilization, and hard disk space used can be monitored. For the SAP HANA Cloud database, the number of connections, host disk usage, and column store merge requests can be monitored, among other things.

Total Monitoring Content of Health Monitoring

The expert portal for SAP Cloud ALM for operations provides a list of all available monitoring content provided by health monitoring. The expert portal can be found at *https://support.sap.com/en/alm/sap-cloud-alm/operations/expert-portal/health-monitoring/health-mon-content.html*.

As with the other monitoring scenarios, the information is also displayed in the UI in the form of dashboards in health monitoring. Information about malfunction situations is also provided in the form of alerts, email notifications, etc. The status overview, which is the first dashboard in **Health Monitoring**, is shown in Figure 18.12.

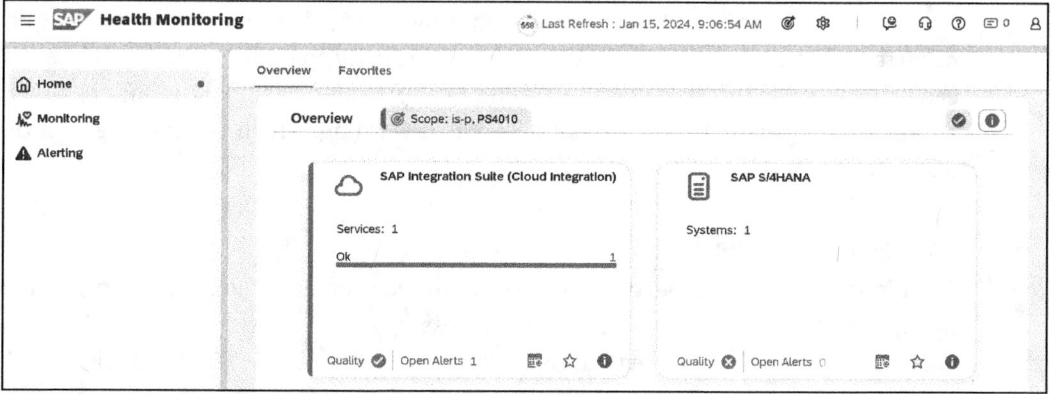

Figure 18.12 Status Overview in Health Monitoring

Figure 18.13 shows an example of the detailed view of the health status of individual components. This provides an initial overview of the status of the components, with the color-coded visualization greatly simplifying orientation.

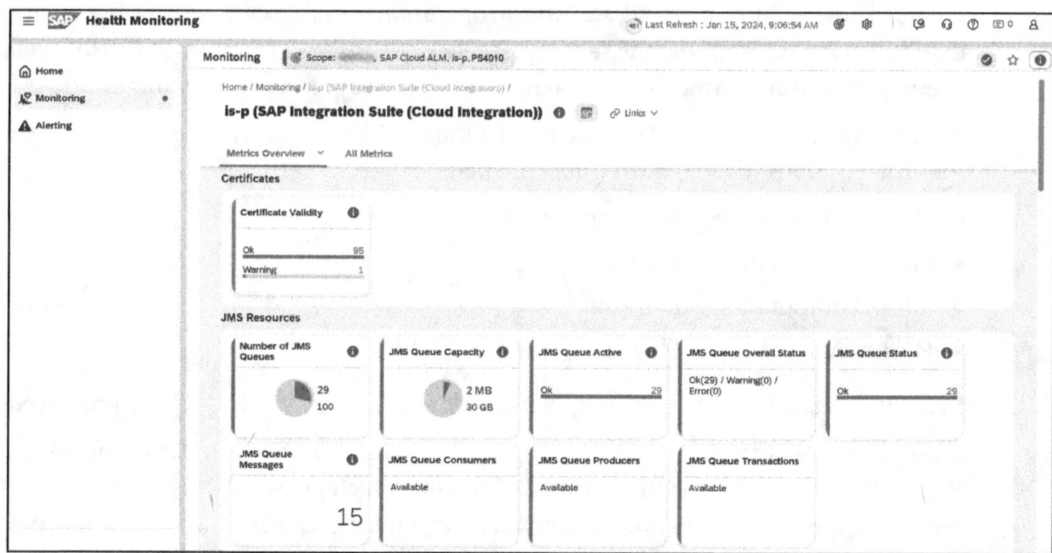

Figure 18.13 Detailed View of the Health Status of Individual Components

18.2.2 Forwarding and Resolving Events

In Section 18.2.1, we showed how events in the applications can be detected using the monitoring scenarios. The determined data is displayed visually. Besides that, alerts can also be generated.

The aim must be to initiate the troubleshooting process based on the event that has occurred; i.e., the event must be forwarded accordingly so that the cause of the event can be resolved in the next step. SAP Cloud ALM provides intelligent event processing and event management functions for event forwarding, which we describe in more detail below.

Intelligent Processing of Events

SAP Cloud ALM enables the forwarding of identified event situations by means of intelligent event processing. The intelligent processing of events in SAP Cloud ALM standardizes incoming events, regardless of whether they originate from SAP Cloud ALM itself, another SAP application, or a non-SAP application. This standardization is continued in the configuration of event forwarding and forwarding rules. It is important to differentiate between alerts and events, which are considered independently of each other by the intelligent processing of events. While an event signals a situation that deviates from the normal state, an alarm marks the event as critical. In this case, an alarm is defined as a possible action following the occurrence of an event. Other actions included are sending an e-mail, starting an *operation flow* (the creation of a ticket in an external ticket system), and the sending of chat notifications. To summarize: Not every event results in an alarm, but every alarm is the consequence of a previous event.

The intelligent processing of events in SAP Cloud ALM is currently fed by data from the following monitoring scenarios (more are planned for the future):

- Integration and exception monitoring
- Business process monitoring
- Job and automation monitoring
- Health monitoring

In addition, data from the SAP Alert Notification service for SAP BTP can be processed.

The event actions are activated directly in the individual monitoring scenarios. Within the configuration, there is the **Events** menu item, which gives you access to the event settings. Here you can activate the individual event actions using buttons. Figure 18.14 shows how alerting has been activated for the **Erroneous Messages(IDOC) Detected (Grouped)** event in the integration and exception monitoring scenario.

If you have defined e-mail notification as the event action, an e-mail is sent when the corresponding event occurs, which has the format shown in Figure 18.15. This provides the recipient with an initial indication of which event has occurred and a link that enables direct access to SAP Cloud ALM.

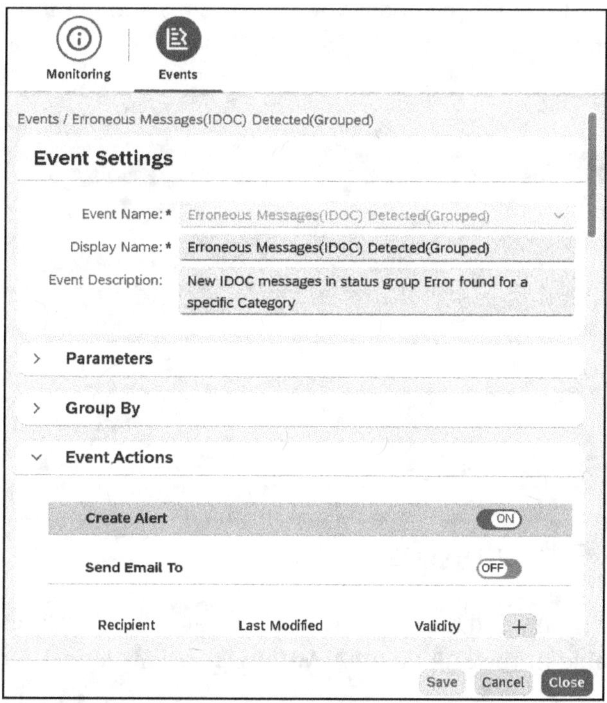

Figure 18.14 Activation of the Alarm for an Event

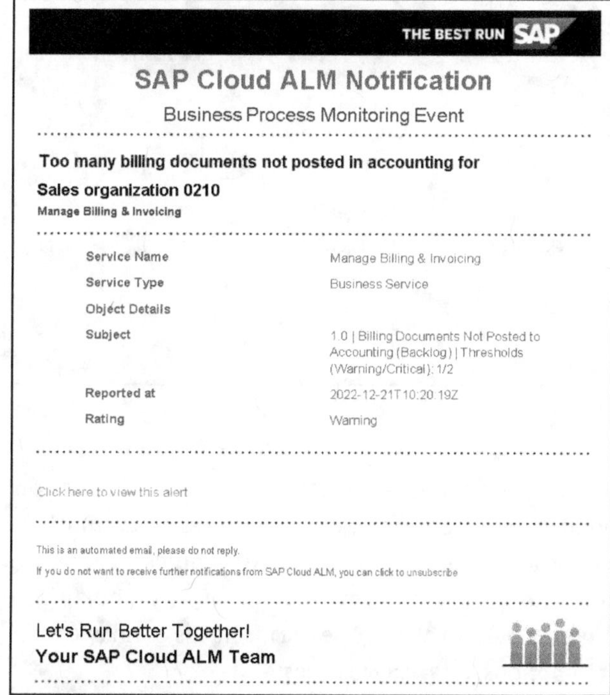

Figure 18.15 Email Notification for an Event

If you configure the creation of a ticket, e.g., in ServiceNow, the ticket is automatically created based on the information in the event (see Figure 18.16).

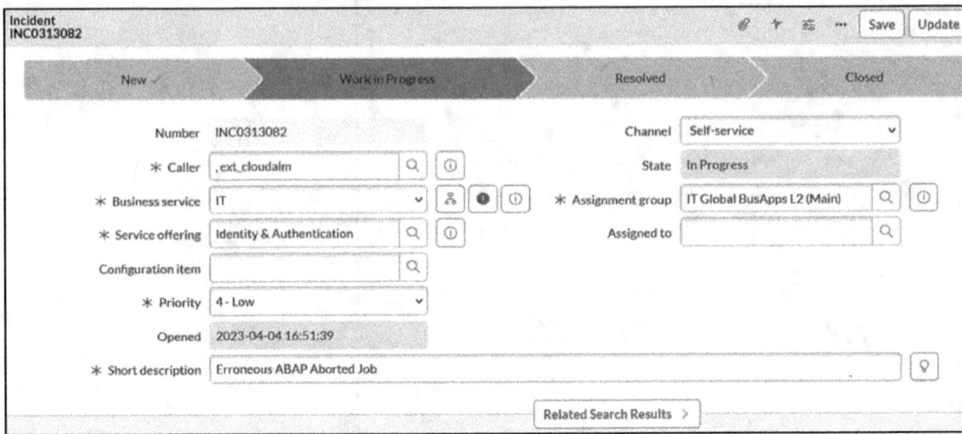

Figure 18.16 Creating a ServiceNow Ticket for an Event

For central access to all triggered events and the event rules, a separate tile called **Intelligent Event Processing** is available in the **SAP Cloud ALM for Operations** launchpad group (see Figure 18.17).

Figure 18.17 Intelligent Processing of Events

The **Overview** page in the left-hand menu provides information about event situations (see Figure 18.18). The information displayed here depends on the filters that may have been set. The **Event Situation Overview** area uses tiles to provide a quick overview of the events and is also used for navigation. A list of the event situations in tabular form can be found in the corresponding area below the overview. The status of an event situation can be **Open, Closed, Error in Actions**, or **With No Rules.**

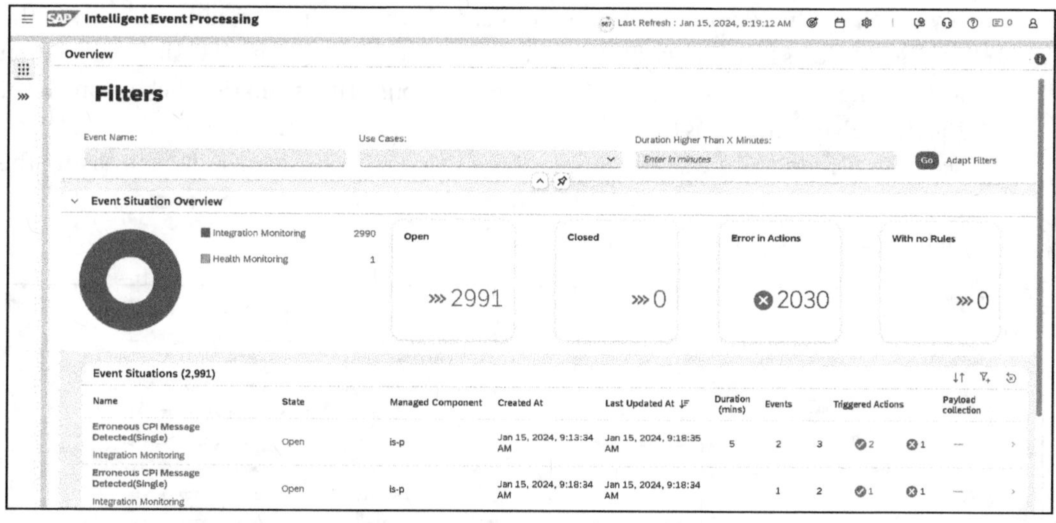

Figure 18.18 Intelligent Processing of Events: Overview Page

If you click on one of the event situations, you navigate to the overview of the actions that were triggered by the respective event. In the example in Figure 18.19, 177 alerts were triggered for the **Erroneous CPI Messages Detected (Grouped)** event in the period from 12/18/2023 to 01/15/2024.

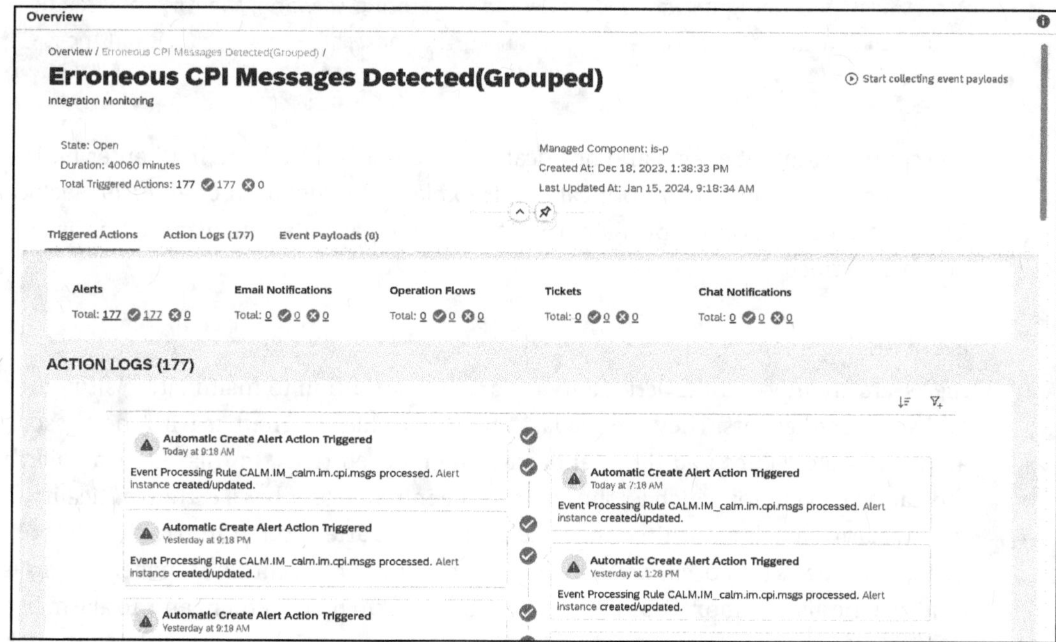

Figure 18.19 Triggered Actions

If you call up the **Event Processing Rules** page via the left-hand menu, you will see the configured rules listed, regardless of the selected focus (see Figure 18.20). Here you can also see the rules that were created directly in a monitoring scenario. Clicking on one of the rules gives you access to rule details, such as the name of the rule and the associated scenario **(Origin)**. You can create rules here using the plus icon, edit them using the pencil icon, and delete them with the trash can icon. These buttons are located to the right of the search field.

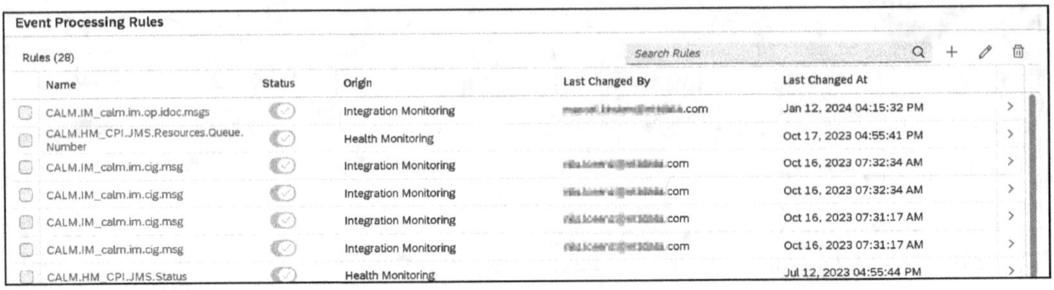

Event Processing Rules

Rules (28) Search Rules

Name	Status	Origin	Last Changed By	Last Changed At
CALM.IM_calm.im.op.idoc.msgs		Integration Monitoring	[illegible].com	Jan 12, 2024 04:15:32 PM
CALM.HM_CPI.JMS.Resources.Queue.Number		Health Monitoring		Oct 17, 2023 04:55:41 PM
CALM.IM_calm.im.cig.msg		Integration Monitoring	[illegible].com	Oct 16, 2023 07:32:34 AM
CALM.IM_calm.im.cig.msg		Integration Monitoring	[illegible].com	Oct 16, 2023 07:32:34 AM
CALM.IM_calm.im.cig.msg		Integration Monitoring	[illegible].com	Oct 16, 2023 07:31:17 AM
CALM.IM_calm.im.cig.msg		Integration Monitoring	[illegible].com	Oct 16, 2023 07:31:17 AM
CALM.HM_CPI.JMS.Status		Health Monitoring		Jul 12, 2023 04:55:44 PM

Figure 18.20 Event Processing Rules

[!]

Editing Options for Event Processing Rules

You can only edit and delete rules that were created directly via intelligent event processing. If the configuration was made in a monitoring scenario, the rule can only be deleted and edited at this original location. Furthermore, alarm activation is only possible there.

The configuration of events and notifications is described in detail in Chapter 19. The configuration of the e-mail notifications is explained in detail in Chapter 19, Section 19.8, where the activation of the automatic ticket creation as an event action is also briefly described.

Event Management

Alerts are an important functionality for drawing attention to malfunction situations and associated events. They are the last link in the chain of metrics with their associated threshold values and the events with their stored rules. In this chain, an alarm indicates an event for which a solution must be initiated as it has a certain criticality. A yellow alarm signals a warning, and a red alarm draws attention to an error in the application. Any alerts that occur are displayed centrally. In SAP Solution Manager 7.2, this is done via the alarm inbox, in SAP Focused Run via alarm management and an alarm list, and in SAP Cloud ALM via alarm lists in the respective monitoring scenarios, although it is not possible to create alerts in every scenario.

Alerting is available for the following monitoring scenarios (at the time of writing); it is planned for other scenarios in the future:

- Business process monitoring
- Integration and exception monitoring
- Job and automation monitoring
- Health monitoring

If alerting has been activated, the alerts are generated for each monitoring scenario according to the configuration. To call up the alerts, use the **Alerting** page that is available in the monitoring scenarios in the left-hand menu (see Figure 18.21).

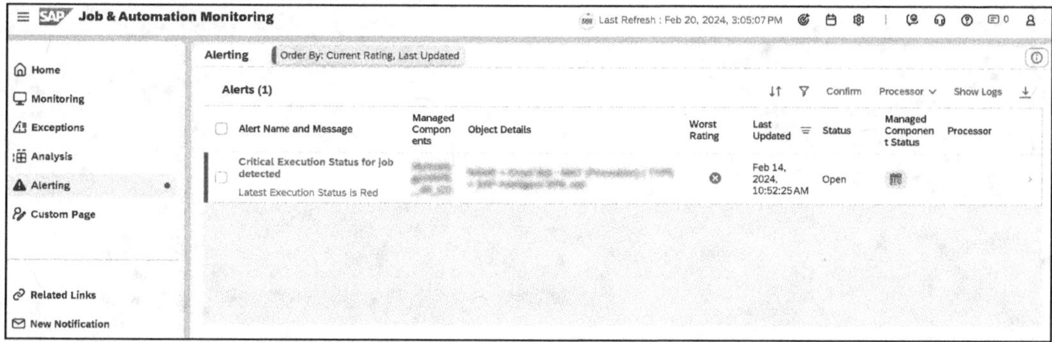

Figure 18.21 Alarm List in Alarm Management

This initial view provides an overview of all the alerts generated. The list shows information on the alert name and the associated message (**Alert Name and Message**), the monitored component (**Managed Components**), and detailed information on the object that led to the alert in the respective component (**Object Details**). In addition, the **Worst Rating** of the alert is displayed, information on the last update (**Last Updated**), the status of the alert (**Status**), the status of the managed component (**Managed Component Status**), and whether a **Processor** is already dealing with the alert (**In progress**).

Various functions are available directly for this list. The sorting can be changed, filter options are available (see Figure 18.22), and agents can be assigned to or removed from the alerts. You can also view a log for each alarm (**Show Logs**) and export the complete alarm list as an Excel file.

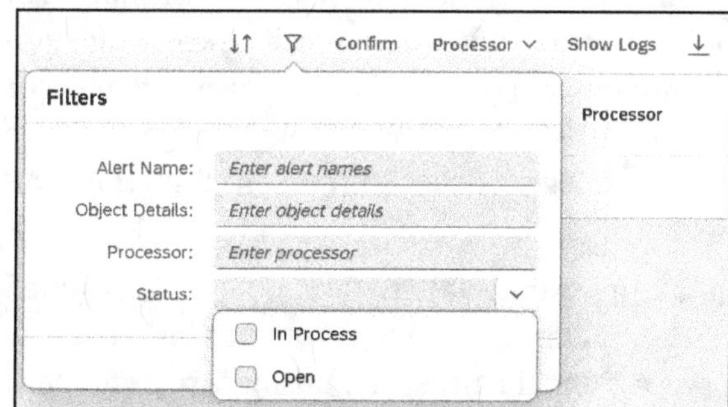

Figure 18.22 Filter Options in the Alarm List

Further detailed information on the respective alarm and additional actions, such as adding comments to the alarm (**Add Comments**), can be called up by clicking on the individual alarm (see Figure 18.23).

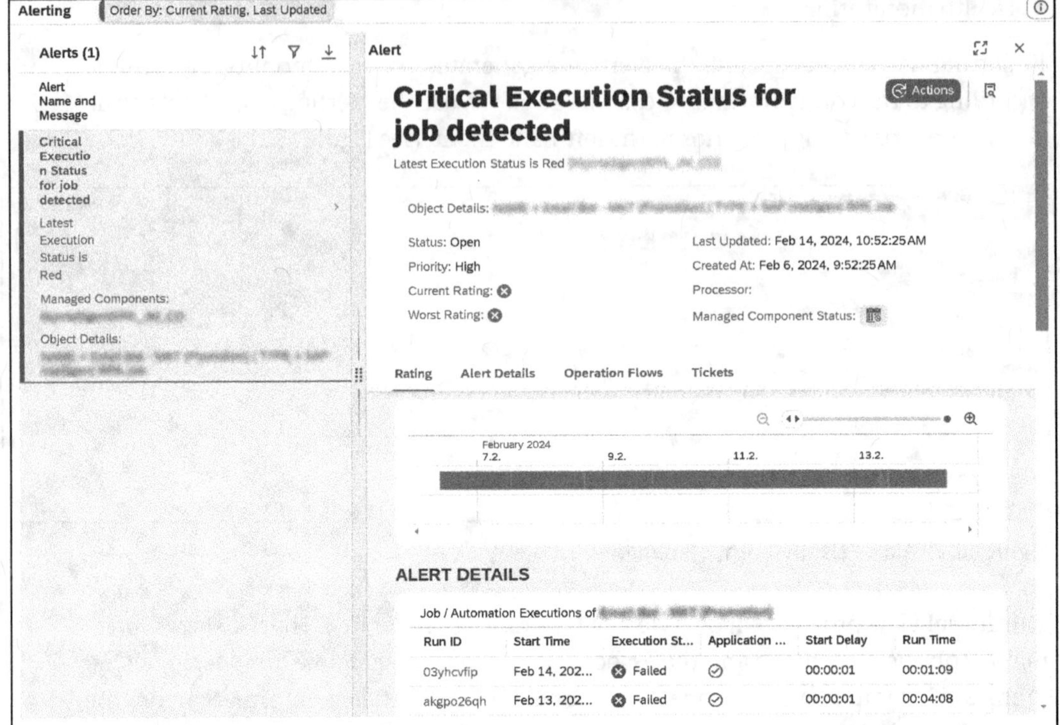

Figure 18.23 Alarm Details with More Detailed Information About the Alarm

Automation of the Operation

Once data on the various applications and systems has been collected by means of monitoring and situations that deviate from the normal state have been addressed by events and alerts, the event situation must be resolved. SAP Cloud ALM supports automation mechanisms, which we will look at briefly in this section.

With the operation automation functionality, SAP Cloud ALM offers you the option of reacting automatically to malfunction situations. By automatically executing standardized processes (also known as *operation flows*), you can react to events that occur and rectify the underlying cause. SAP Cloud ALM and SAP Workflow Management are integrated for this purpose, as well as SAP Intelligent Robotic Process Automation (SAP Intelligent RPA) and SAP Automation Pilot.

SAP Workflow Management is executed in a separate SAP BTP account. It is best suited for the automation of tasks that involve user interactions and automated tasks because it allows existing requirements for a process and the necessary business rules for decisions to be quickly and easily mapped digitally.

SAP Intelligent RPA is also executed in a separate SAP BTP account. It is particularly suitable for automating repetitive UI interactions.

SAP Automation Pilot is best suited for automating DevOps tasks. It gives you access to a comprehensive range of fully automated processes, including database and application lifecycle management, monitoring, HTTP requests, and more. You can model and implement your individual scenario by either using already integrated content or creating custom content.

The services are integrated manually via the administration of the landscape. Once you have configured an endpoint for the service, you can register the operation flows directly via the scenario. They are then stored in a processing rule for an event and executed automatically as soon as the event occurs. You can also trigger the operation flows manually via the alarm list.

18.3 Summary

This chapter has given you an overview of the options you have with SAP Cloud ALM in the operational phase. The next step is to put these options into practice. Chapter 19 focuses on the implementation of monitoring scenarios.

Chapter 19
Implementation

Once you have familiarized yourself with the possibilities that SAP Cloud ALM offers for the operational phase and you have created your concept, you can start with the implementation, which is described in this chapter.

Once you have completed the planning of the future setup (see Chapter 18), and the concept is complete, you can start implementing the monitoring and the associated functions. In SAP Cloud ALM, a basic prerequisite for the implementation is that the connection between the applications to be monitored and SAP Cloud ALM has been successfully set up to collect or provide the data via the push or pull method.

To provide the connection and data, preparatory activities must be carried out in the application to be monitored and, in some cases, also in the administration of the SAP Cloud ALM landscape.

The examples in this chapter are based on the practical example in Chapter 1, Section 1.4 but do not follow it 100%. We added some applications in this chapter to illustrate the possibilities of SAP Cloud ALM for operations even more extensively.

For the connection of SAP Business Suite and SAP S/4HANA with SAP Cloud ALM, which are used as an example, the push data provider must first be configured in the relevant client. When this happens, the use cases (monitoring scenarios) for which data is to be provided are also activated. This means that monitoring using SAP Cloud ALM is activated directly. In addition, the service key from SAP Cloud ALM is also stored. You can find further details regarding the service key in Chapter 3, Section 3.4.

For the cloud service, SAP Integrated Business Planning for Supply Chain is another application that serves as an example in this chapter. A communication system and communication arrangement must be set up to enable the push of the data to SAP Cloud ALM.

For the fourth example application, SAP Integration Suite (Cloud Integration), a service instance must be created in the SAP BTP cockpit. This enables OAuth authentication between SAP Cloud ALM and Cloud Integration. A service key is also generated. This service key is necessary for the configuration in the administration of the SAP Cloud ALM landscape, as an HTTP endpoint must be configured to enable pulling data. You can find details on this in Section 19.3.2. Further information on managing the landscape can be found in Chapter 3.

For detailed descriptions of the system preparation for the individual monitorable products, including illustrative videos, refer to the expert portal for SAP Cloud ALM for operations at *https://support.sap.com/en/alm/sap-cloud-alm/operations/expert-portal/integration-monitoring/int-mon-videos.html.*

In this chapter, we describe in Section 19.1 how to set up business service management. In Section 19.2 to Section 19.7, we present the configuration of the individual monitoring scenarios in the order that you already know from Chapter 18, Section 18.2.1. The implementation of the individual monitoring scenarios is generally divided into the following points:

- Checking the prerequisites
- Initial configuration
- Specific configuration

The initial configuration as well as the specific configuration is limited to activities in SAP Cloud ALM itself. This is only different in the synthetic user monitoring scenario, as you will see below. Furthermore, for systems and cloud services that transmit data to SAP Cloud ALM via push, no activities are described in the initial configuration that serve to activate monitoring. In these cases, it is shown which monitoring information is initially available. The reason for this is that data collection and monitoring are activated automatically as soon as such an application is connected to SAP Cloud ALM.

Section 19.8 describes the configuration of email notifications and the automatic creation of tickets in ServiceNow. In Section 19.9, we give you tips for optimizing the configuration and briefly discuss the use of SAP Cloud ALM as a reverse proxy for SAP Focused Run, as well as give a bit of information regarding the usage of OpenTelemetry data by SAP Cloud ALM.

19.1 Business Service Management

You access the Business Service Management app in the launchpad via a tile in the **SAP Cloud ALM for Operations** group (see Figure 19.1).

To start the configuration, click on the gear icon at the top (see Figure 19.2). This button is always the access point to the configuration, and you will encounter it in all monitoring scenarios. If you expand the first menu item, **Infrastructure**, the status for the **Intelligent Event Processing** job is displayed. This job is the basic requirement for business service management. The **SAP Focused Run System Configuration** area provides the opportunity to activate a forwarding of cloud service events to an SAP Focused Run system, if the respective system has a configured HTTP endpoint in the landscape management and an active landscape synchronization.

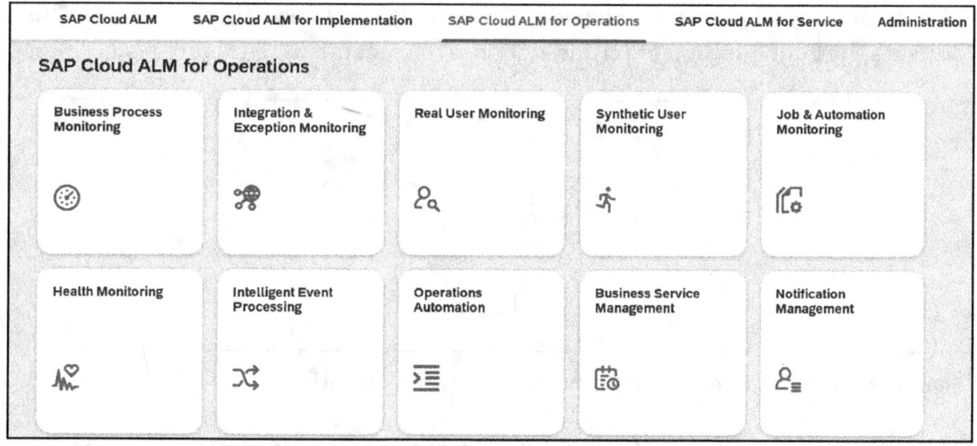

Figure 19.1 Access to Business Service Management

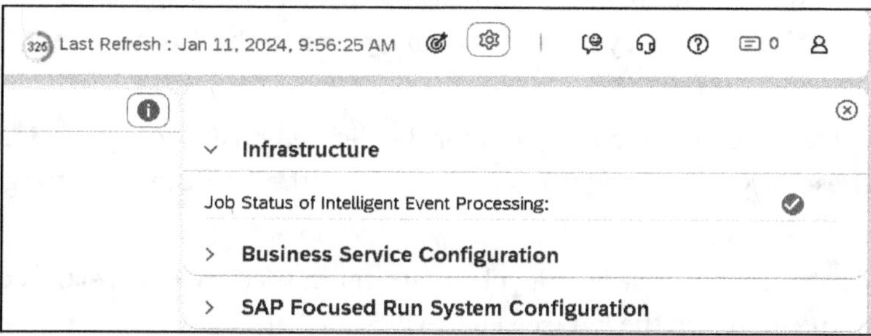

Figure 19.2 Access to the Configuration

If you open the **Business Service Configuration** menu item, you will see the already configured business services. You can enter the setup via the pencil icon to customize/create them (see Figure 19.3).

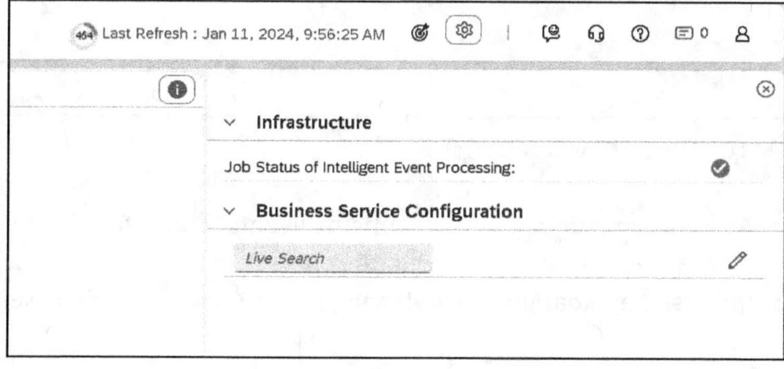

Figure 19.3 Accessing the Event Settings

In the window that opens, use the **Add** button to define a new business service. This is done via the **Configuration of Business Service Management** (see Figure 19.4).

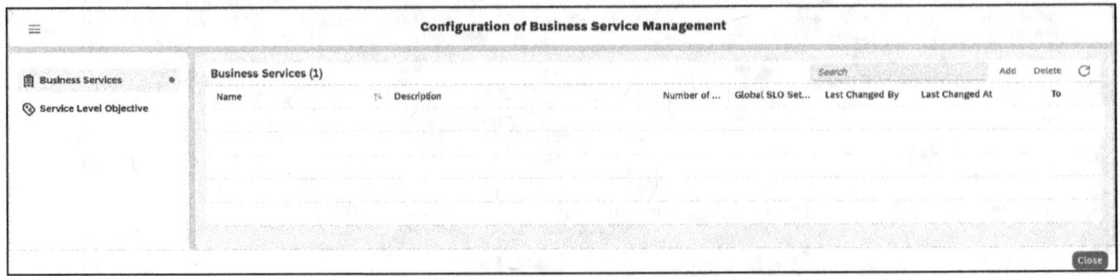

Figure 19.4 Configuration of the Business Service Administration

In this menu, you assign a name and a description for the business service. You can then use the **Add Cloud Services** and **Add Technical Systems** buttons to add the cloud services and the on-premise systems that belong to the respective business service. Once you have selected the services/systems, they will appear in the **Business Service Configuration** section.

Initial **Service Level Objectives** are preconfigured for the defined business service; it is possible to adjust these targets after you have deactivated the **Use Global Settings** checkbox.

To change the global settings for the service level targets, select the **Service Level Objective** menu item from the menu on the left (see Figure 19.5).

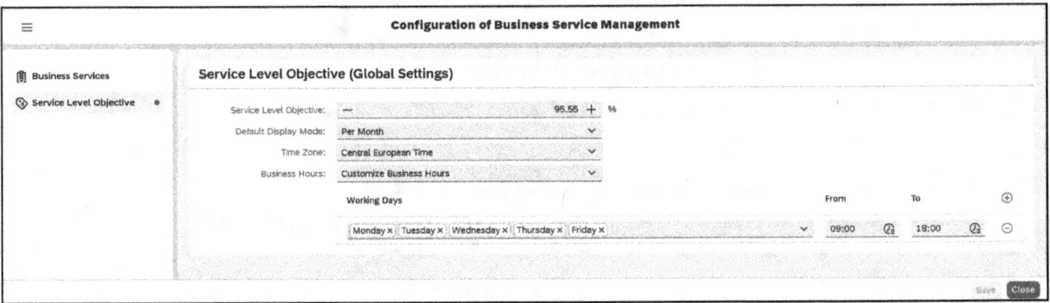

Figure 19.5 Service Level Objective (Global Settings)

If you open the **Event Actions** menu item, you can configure events actions for planned and unplanned operational events that can occur in the context of an SAP cloud service. These are **Disruptions, Degradations, Maintenance**, and **Communication** (see Figure 19.6).

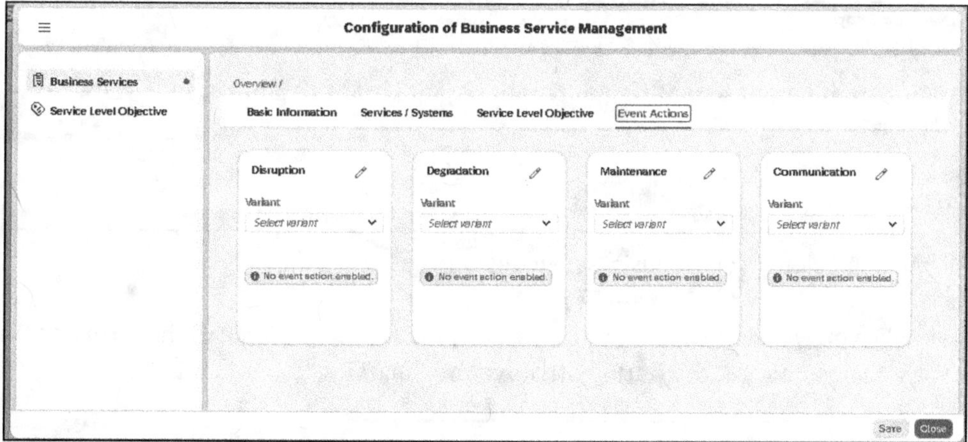

Figure 19.6 Event Actions for Business Services

For each event, you can decide whether you want to have an email notification and/or an automated created ticket. You can also decide the condition for the event action. In Figure 19.7, you can see the setup for an email notification and an automated created ticket whenever a disruption is detected for the business service.

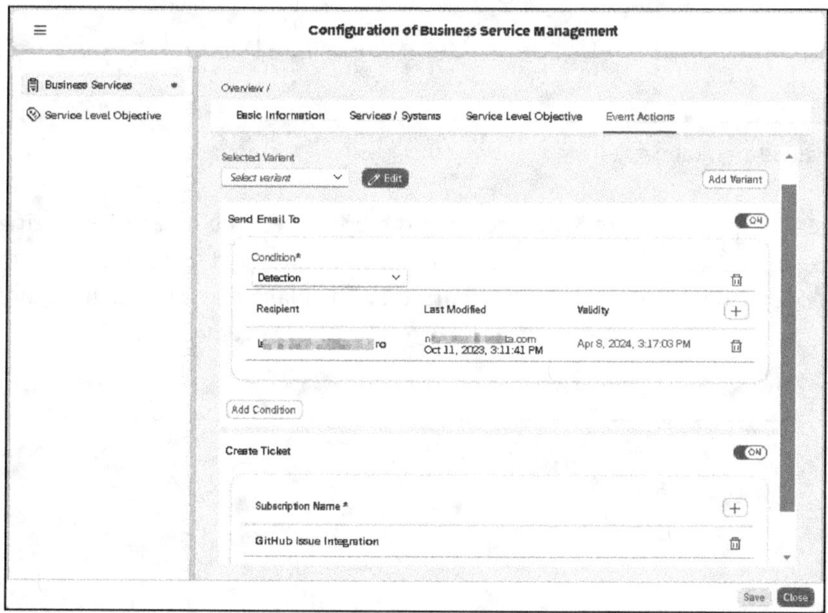

Figure 19.7 Configuration of Event Action for Business Service

Once the configuration of the business service has been completed, you can find it in the scope selection, as shown in Figure 19.8.

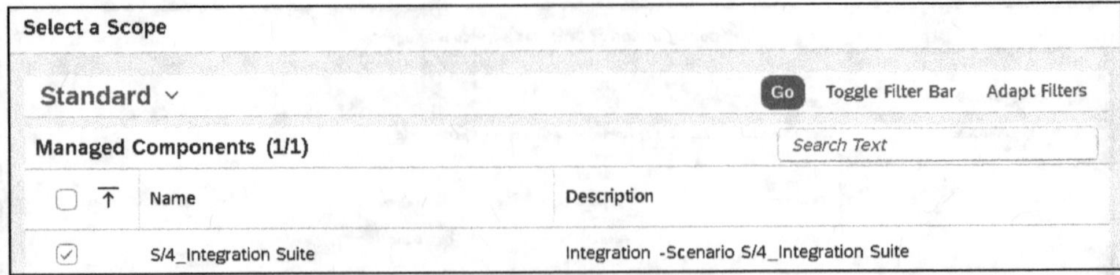

Figure 19.8 Selection of the Business Service

Initially, you will see an overview of the business service in **Home** in the form of a tile that visually indicates the status with its colored marking (see Figure 19.9).

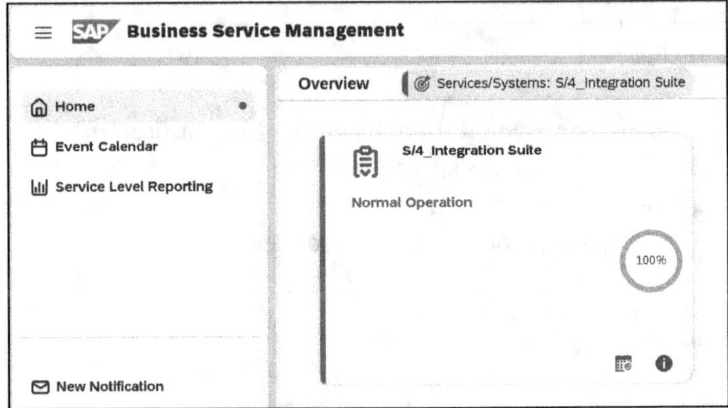

Figure 19.9 Overview on the Start Page

In the **Event Calendar**, which you can access via the button on the left, the operating events are displayed for the business service and for the individual systems and cloud services. The operating events that have occurred are displayed as an icon. In Figure 19.10, you can see there was communication from SAP for several events in several months.

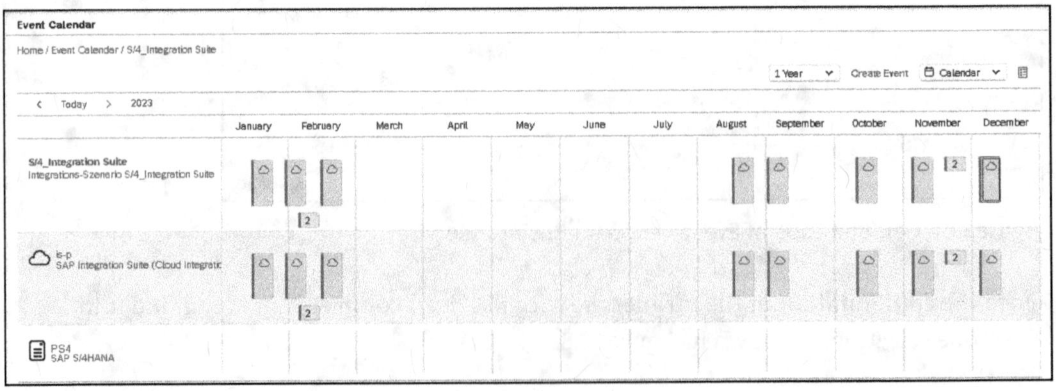

Figure 19.10 Event Calendar

By clicking on the icon for the event, you can call up the event details. Via the **Action Log** tab, which can be seen in Figure 19.11, you can access the contents of the communication directly.

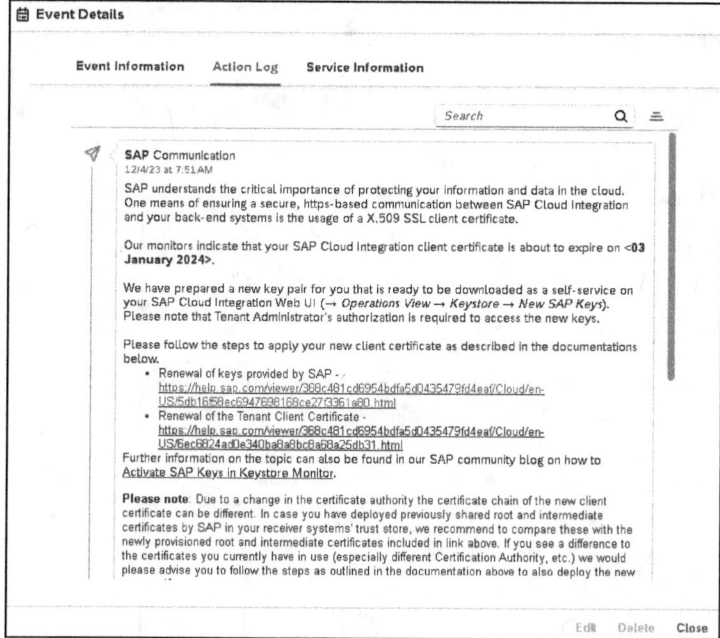

Figure 19.11 Action Log for an Event

Via **Service Level Reporting** in the left-hand sidebar, you can access reporting information for the defined business service (see Figure 19.12).

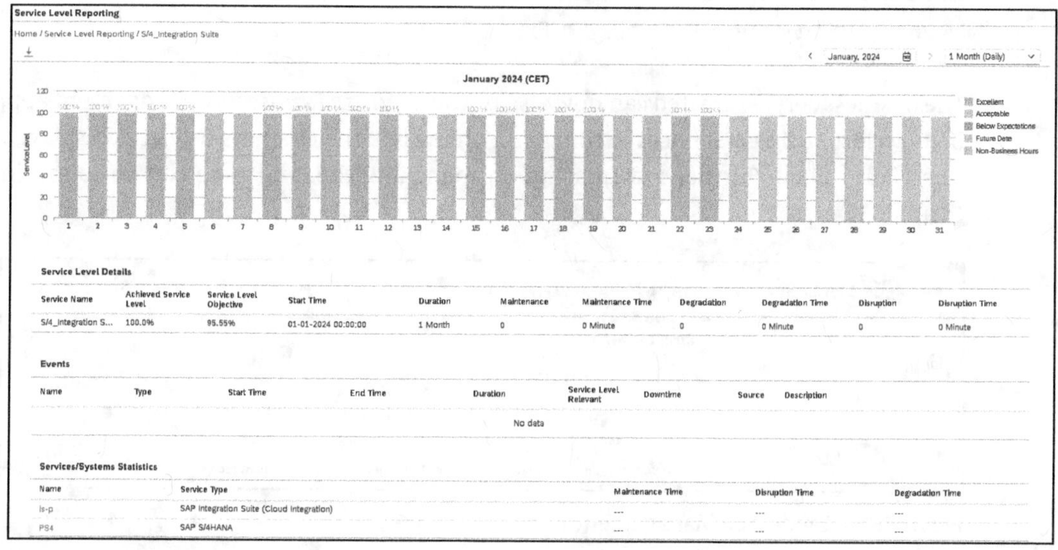

Figure 19.12 Service Level Reporting

In addition to filter options with regard to the period under review, a download of the reporting information is also available, which automatically generates a report in the form of a PDF (see Figure 19.13).

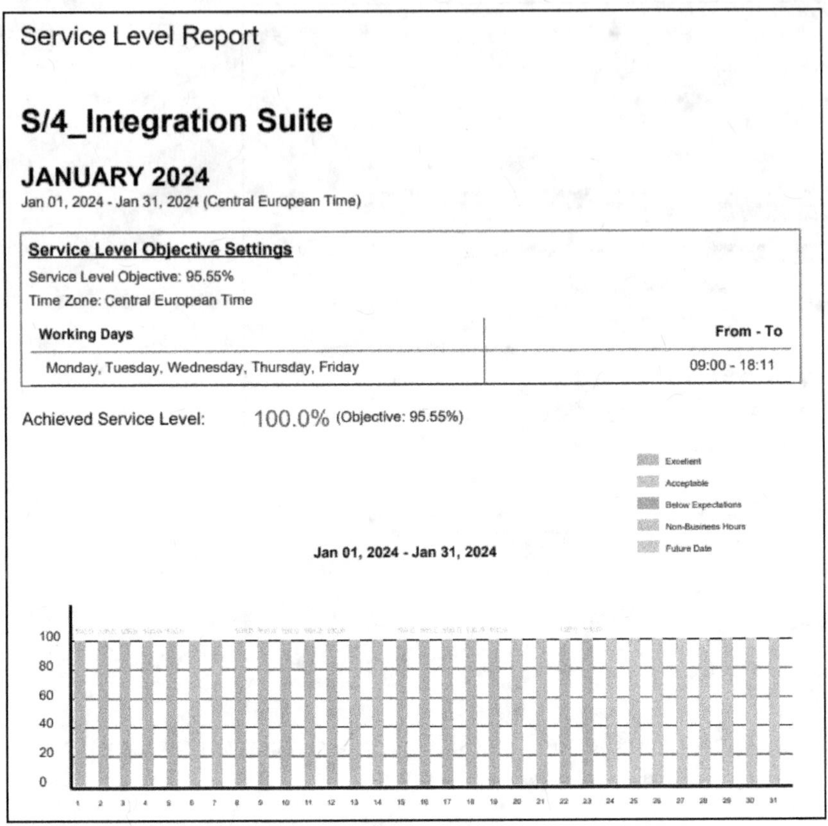

Figure 19.13 Service Level Report as PDF Export

The business service created can now also be found in the Landscape Management app. If you select **Business Services** in the menu on the left, the business service is displayed (see Figure 19.14). Further information on managing the landscape can be found in Chapter 3.

SAP Landscape Management — Last Refresh : Jan 11, 2024, 11:04:49 AM

Home
Services & Systems
Business Services

Business Services

Business Services (18) — Live Search

Name	Description	Use Case	Number of ...
Schedule Product...	Design to Operate -> Material to Product -> ...	Business Process...	0
S4_SAP Integrati...	Integration -Scenario S4-Integration Suite	Integration Monit...	2

Figure 19.14 Landscape Management

19.2 Business Process Monitoring

To illustrate the implementation of the business process monitoring scenario, in this example we use an SAP Business Suite system and set up monitoring for the lead-to-cash process.

19.2.1 Checking the Prerequisites

To be able to start implementing business process monitoring with regard to your own requirements, you must first check whether data collection has been activated for the relevant ABAP system and is running successfully. Take a look at the Landscape Management app in the **Administration** launchpad group (see Figure 19.15).

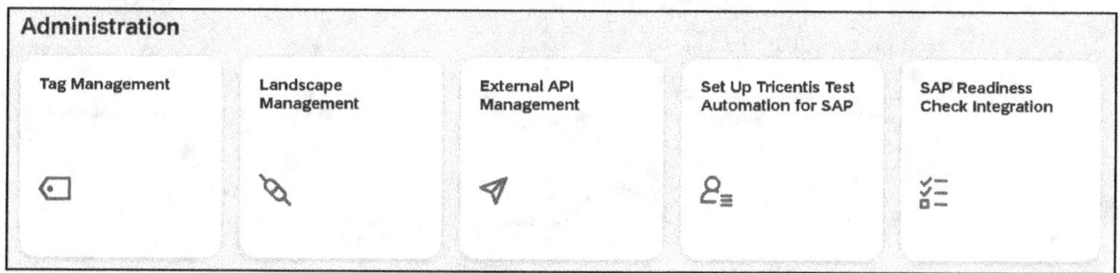

Figure 19.15 Access to the Landscape Management

In the left-hand menu of the Landscape Management app (see Figure 19.16), you will find the **Services & Systems** page.

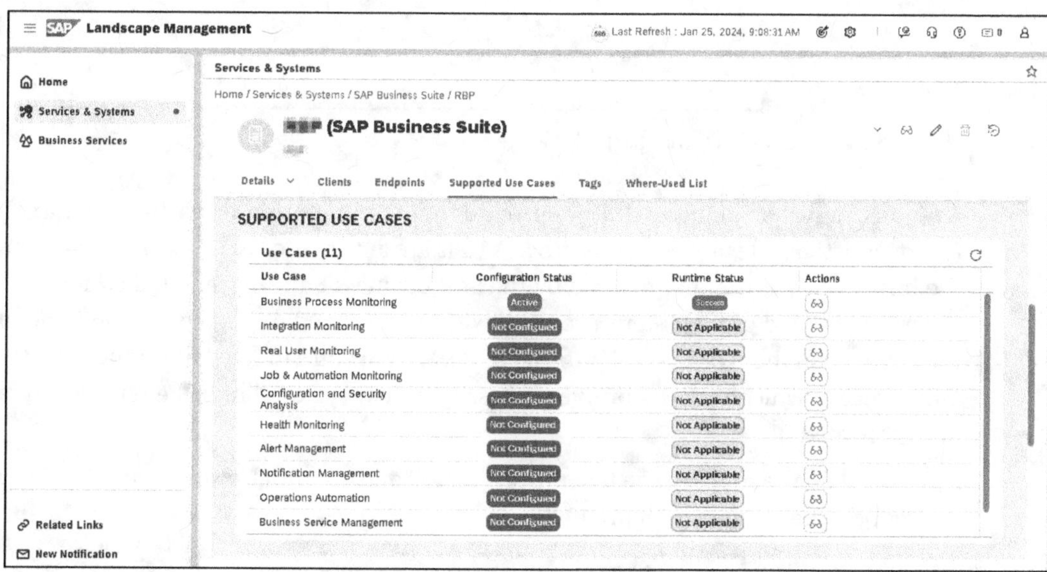

Figure 19.16 Checking the Prerequisites

After you select **Services & Systems** and select the system type in the scope selection, you will see the on-premise systems of this type that are available in SAP Cloud ALM. After selecting our example system and the **Supported Use Cases** tab, you can see that the **Business Process Monitoring** scenario, including data collection via the push method, has already been activated for the system (**Configuration Status Active**). Furthermore, the data collection is successful (**Runtime Status Success**). Due to the previous configuration of the push data provider in the system to be monitored, both were activated automatically.

19.2.2 Initial Configuration

To start configuring process monitoring, select the Business Process Monitoring app in the SAP Cloud ALM launchpad. This is easier if you select **SAP Cloud ALM for Operations** in the top menu (see Figure 19.17).

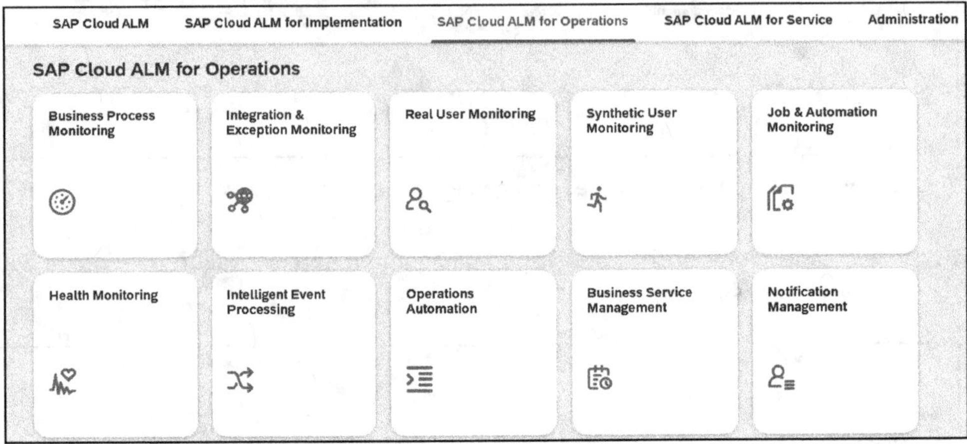

Figure 19.17 Access Via Launchpad Tile

As soon as a system is successfully registered for the monitoring scenario and data collection is active, relevant business process KPIs are automatically recognized based on the business information provided by the SAP Business Suite system and assigned to the processes and process steps provided within SAP Cloud ALM. The first opening of the application then looks similar to Figure 19.18. Here you can see that for the **Lead-to-Cash** process, which consists of five process sections a total of 24 key figures have been activated.

By clicking on the process, the five process stages and an initial assignment of the key figures becomes clear (see Figure 19.19).

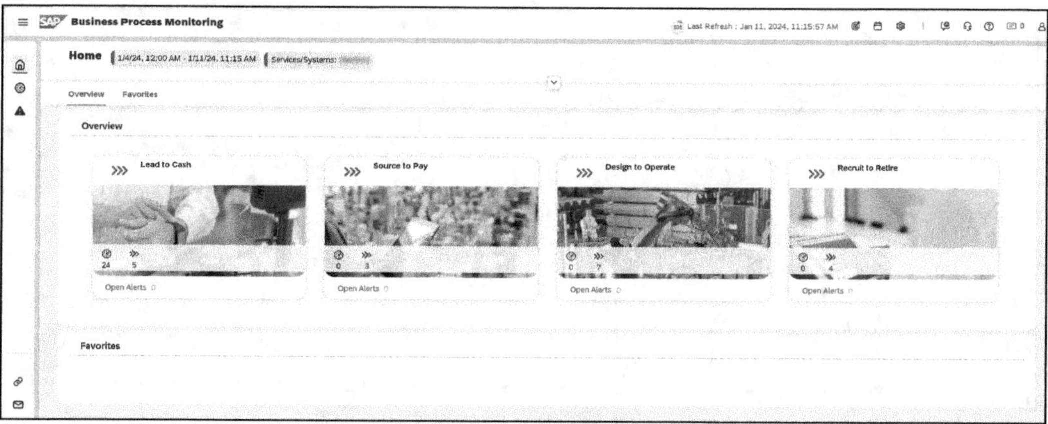

Figure 19.18 Initial Situation for Business Process Monitoring

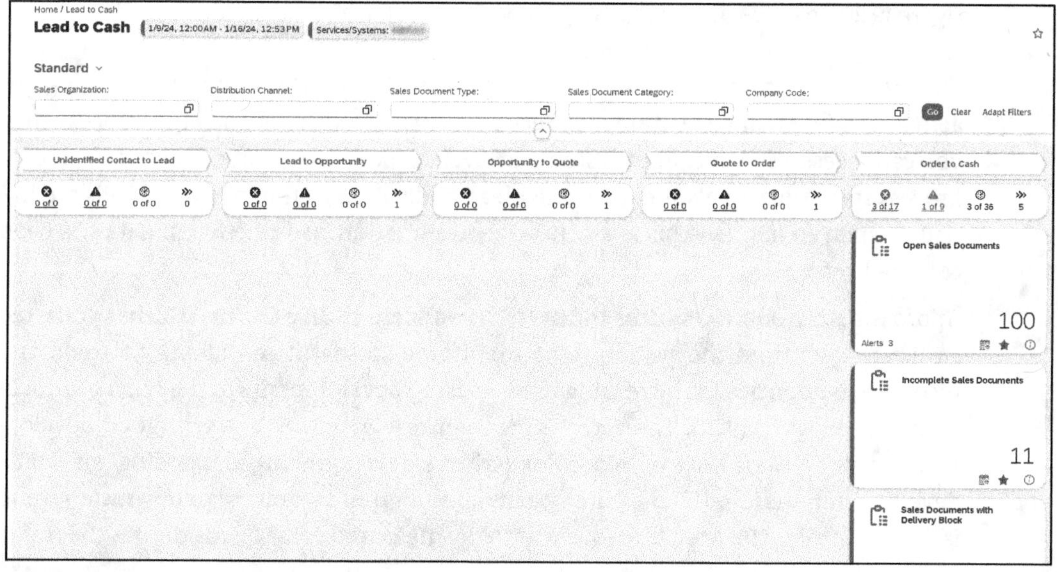

Figure 19.19 Process Section View for Lead-to-Cash

As you can see, not all of the 24 active KPIs can be displayed. If you look at the last process section, **Order to Cash**, it becomes clear that this consists of five further substeps. You can drill down into this specific process section again by clicking on its name. After opening the **Order to Cash** process, the other five detailed process steps become visible (see Figure 19.20).

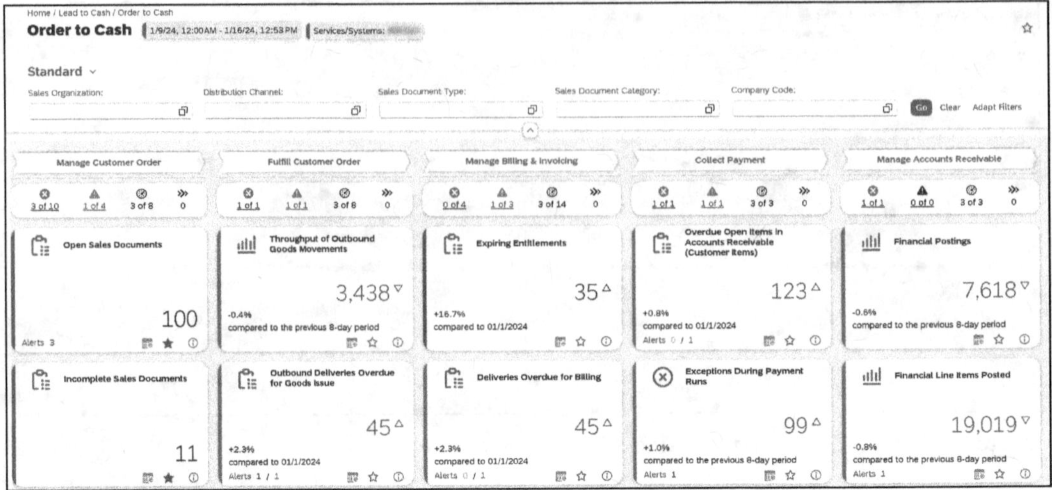

Figure 19.20 Lead-to-cash Process Step View "Order to Cash"

This overview also only lists a selection of the available KPIs (e.g., only three of the eight active KPIs for **Manage Customer Order**). To display all KPIs assigned to a process step, you can access a specific process step one level deeper. To do this, click on the relevant process step. You can also see that the measured values of the individual KPIs are initially displayed in gray. This is a neutral representation, as not enough data has been collected yet.

Approximately one week after the system connection, SAP Cloud ALM has collected sufficient data from the system to be monitored so that it can identify a trend and recognize tendencies for increasing or decreasing developments in the business documents (see Figure 19.21). The active KPIs then receive either a green (good development), yellow (warning), or red color (critical development), depending on which development was identified by the system and whether an upward or downward trend is desired for the KPI. A small arrow next to the measured value also indicates the trend.

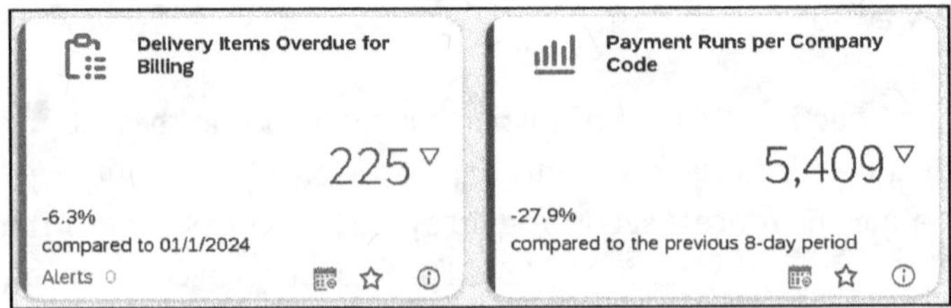

Figure 19.21 Automated Evaluation of KPI Developments

19.2.3 Specific Configuration

For the specific configurations, switch to the **Configuration** screen via the gear wheel symbol [gear icon] (see Figure 19.22).

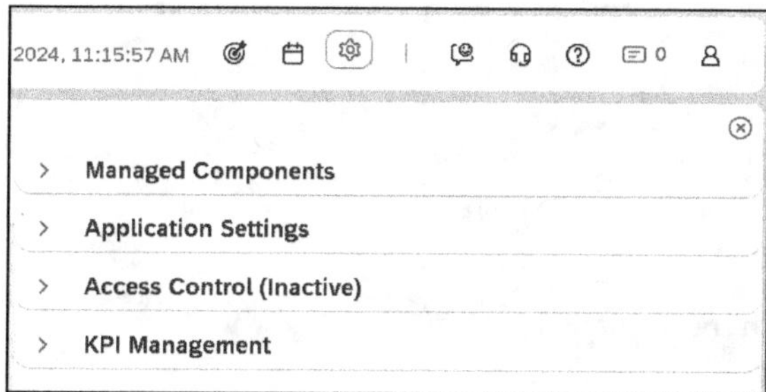

Figure 19.22 Configuration Options

Here you can make adjustments to the following categories. **Managed Components**, **Application Settings**, **Access Control** (inactive in our case), and **KPI Management**. In the **Application Settings**, the current memory utilization is displayed first (see Figure 19.23). As a process monitoring administrator, you can make settings here regarding the retention time of individual business documents and aggregated data. The retention time for individual executions defines the period in days for which data for individual executions (detailed business documents) is retained for throughput KPIs. Enter a value between 1 and 30, which is up to half the retention time of the aggregated data. The retention time for aggregated executions defines the period in days for which data is retained in the data history. After this retention period has expired, the data is deleted. Enter a value between 2 and 365 that is at least twice as long as the retention period for the raw data.

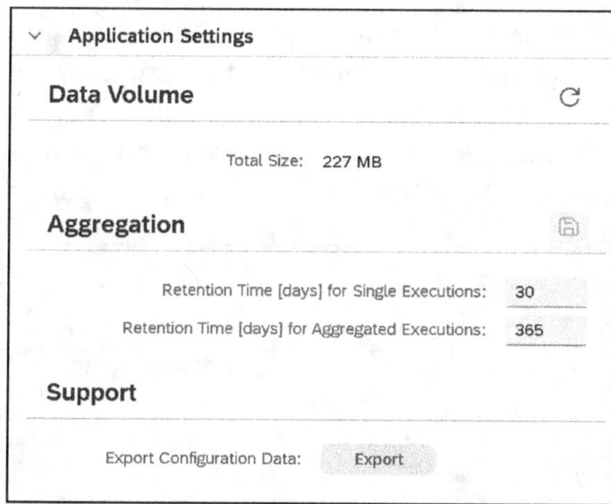

Figure 19.23 Global Settings

The specified retention times apply to all monitored systems and services. You can see for which services and systems the business process monitoring scenario is active in the **Managed Components** section (see Figure 19.24). In our example, the scenario is activated for the entire SAP Business Suite.

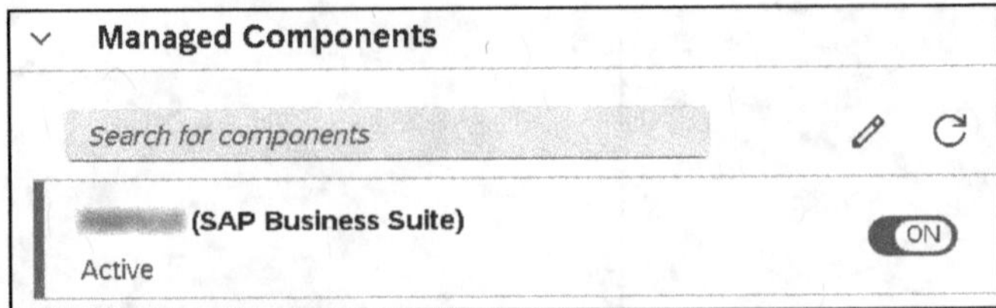

Figure 19.24 Services/Systems for Which the Scenario is Active

In the **KPI Management** section (see Figure 19.25), all available KPIs from SAP Cloud ALM are displayed. Active KPIs can be recognized by the green marking. Inactive KPIs are not highlighted. You can use the search field to filter the KPI catalog according to the KPI you are looking for.

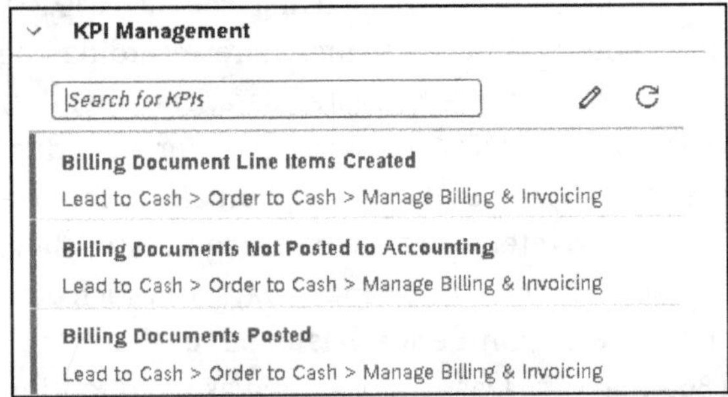

Figure 19.25 KPI Management

In our example, the key figure **Billing Documents Posted** is to be configured as part of the lead-to-cash process. To do this, click on the name of the KPI, and a new window will open. The KPI management overview (see Figure 19.26) includes the following initial information for each KPI: **Activation Status**, **Last Configuration Change**, and **Current Global Value**. You will find a description and the KPI category under each key figure.

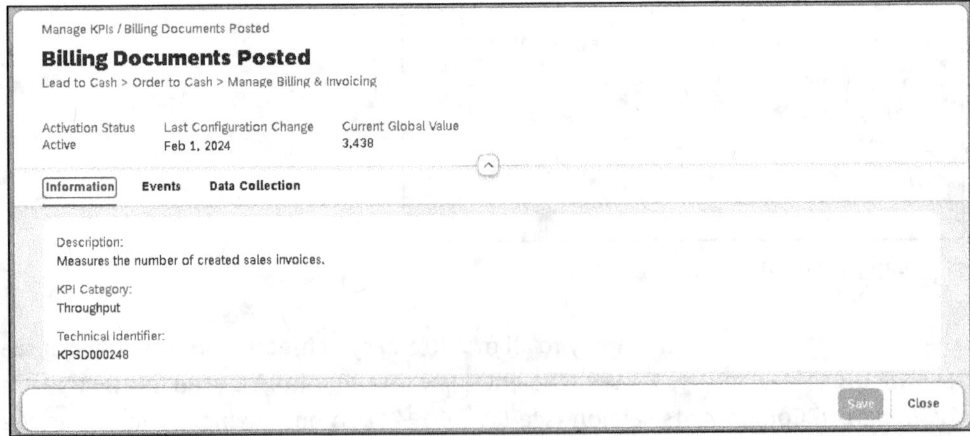

Figure 19.26 KPI Management Using the Example of Billing Documents Posted

To interpret the measured value, define events for each KPI. Within **Event Configuration** (see Figure 19.27), you can define the event or alert name and threshold values for a **Warning** and the **Critical** status. You can also add **Filters** if not all documents in the monitored system should be in focus for this event. In the event definition details, use the **ON/OFF** switch to define whether an alarm should be generated, an email notification sent, an operation flow started, or a ticket created.

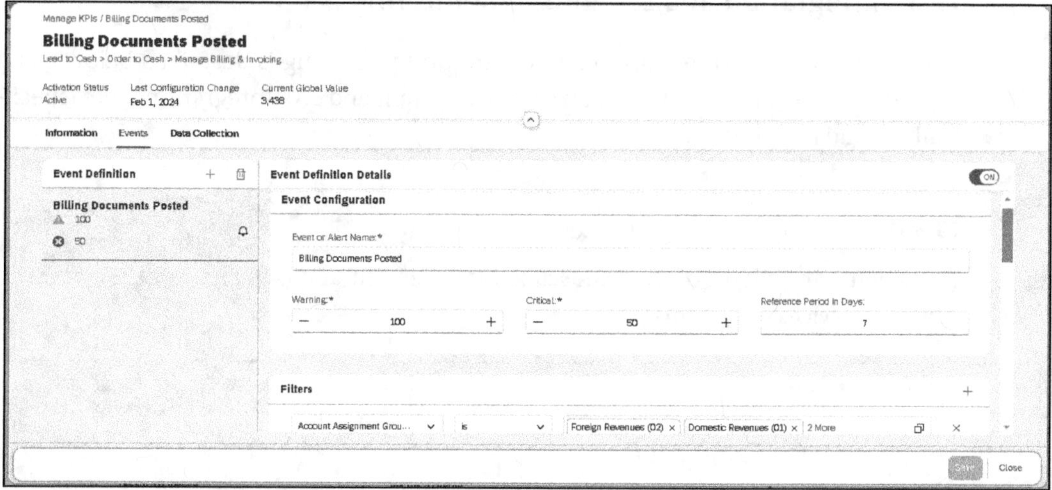

Figure 19.27 Creating Events

If you have configured an alarm for this event and the defined threshold value is exceeded, an alarm appears in your integrated alert inbox in the Business Process Monitoring application (see Figure 19.28).

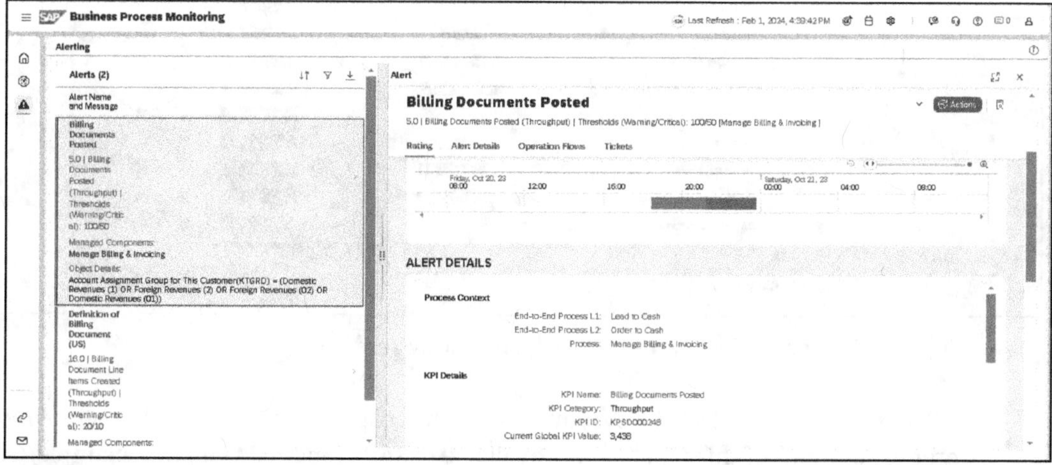

Figure 19.28 Display of an Alarm in the Alert Inbox

19

The alarm is then also visible on the overview page of the KPIs and can therefore also be integrated directly on the start page if the KPI has been marked as a favorite using the star symbol (**Add to Favorites**). The symbol can be found in the KPI tile on the overview page.

19.3 Integration and Exception Monitoring

The configuration of integration and exception monitoring is described below using the example of an SAP S/4HANA on-premise system and SAP Integration Suite (specifically, Cloud Integration).

19.3.1 Checking the Prerequisites

In the left-hand menu of the Landscape Management app (see Figure 19.29), you will find the **Services & Systems** page.

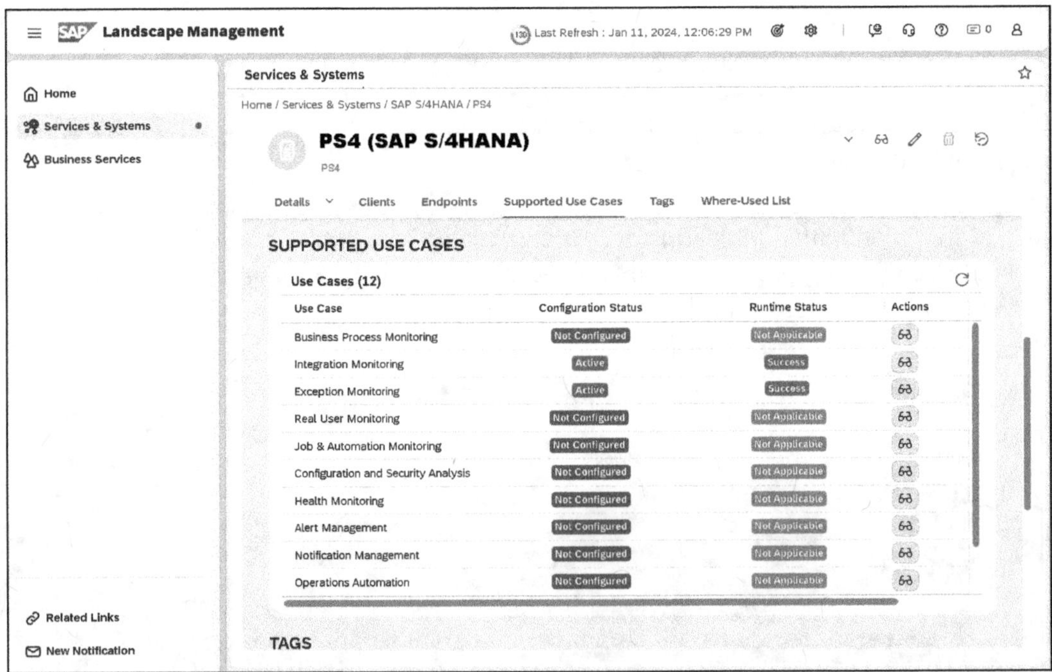

Figure 19.29 Connection Status in the On-Premise System

If you select it and select the system type in the scope selection, you will see the on-premise systems of this type that are available in SAP Cloud ALM—in our example, we want to set up monitoring for the system with the SID **PS4**. The **Supported Use Cases** tab shows the monitoring scenarios for which the on-premise system provides data. You can see that monitoring for integration and exception monitoring is already active

(**Integration Monitoring** and **Exception Monitoring** use cases), including data collection via the push method. Furthermore, the data collection is successful (**Runtime Status** of **Success**). This has been activated automatically by the previous configuration of the push data provider in the system to be monitored.

If you select **Services & Systems** in the menu on the left (see Figure 19.30), you can see whether the cloud service for which you want to set up monitoring is available in SAP Cloud ALM—in our example, this is the **is-p** service. This check is sufficient at this stage.

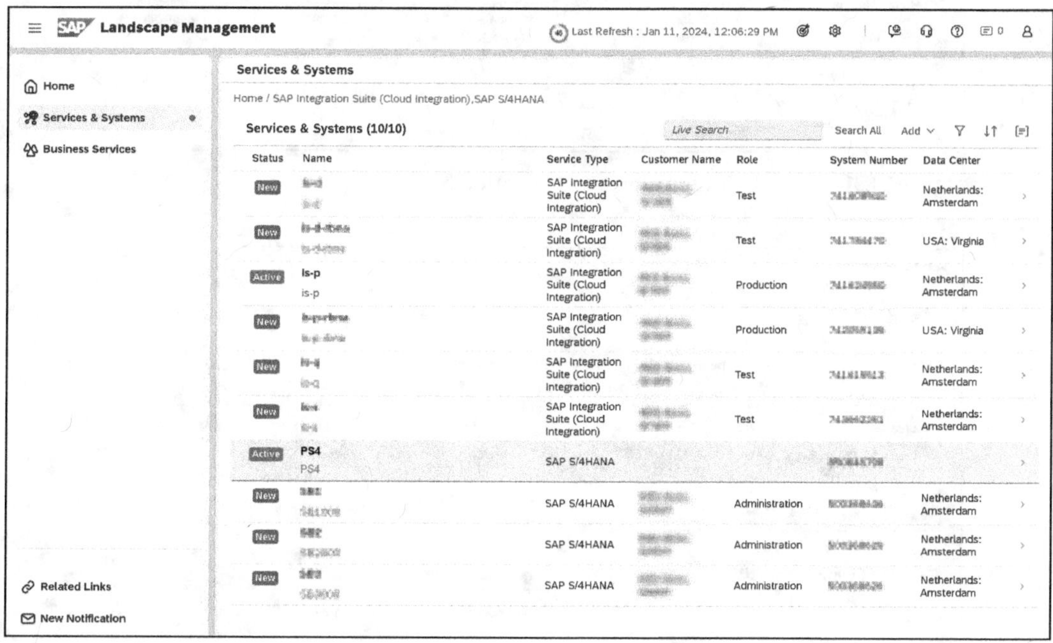

Figure 19.30 Connection Status for the Cloud Service

19.3.2 Initial Configuration

If the systems to be monitored are successfully connected to SAP Cloud ALM, you can start configuring the monitoring. In the launchpad, navigate to the **SAP Cloud ALM for Operations** group and click on the **Integration & Exception Monitoring** tile (see Figure 19.31).

The window you see is used to select the scope. This is where you define the system for which you want to display monitoring data. In our example, we would like to focus on the data for the PS4 system. To do this, we select the **Systems** under **Managed Components**. We select **PS4010** as the **Name**, where 010 stands for the client for which we want to view the monitoring information. After clicking on **Go**, the search begins and the results are displayed. The selection of the scope is completed by ticking the checkbox and clicking on the **Apply** button. The procedure described is shown in Figure 19.32.

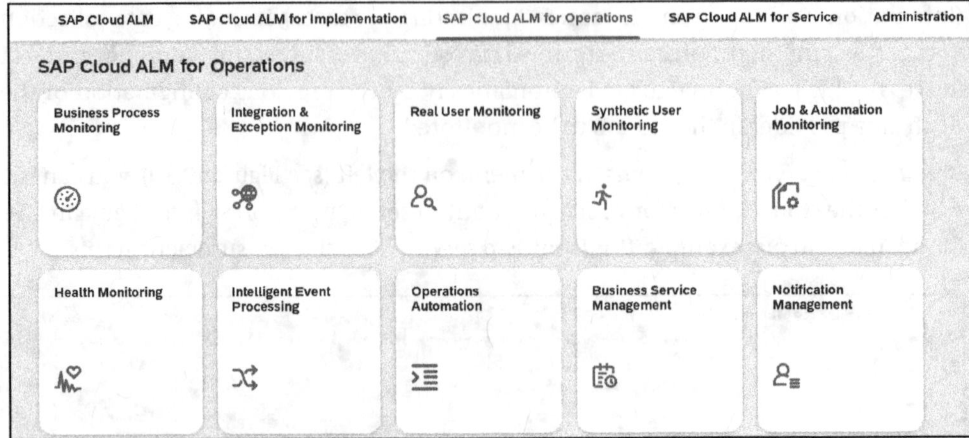

Figure 19.31 Access to Integration and Exception Monitoring

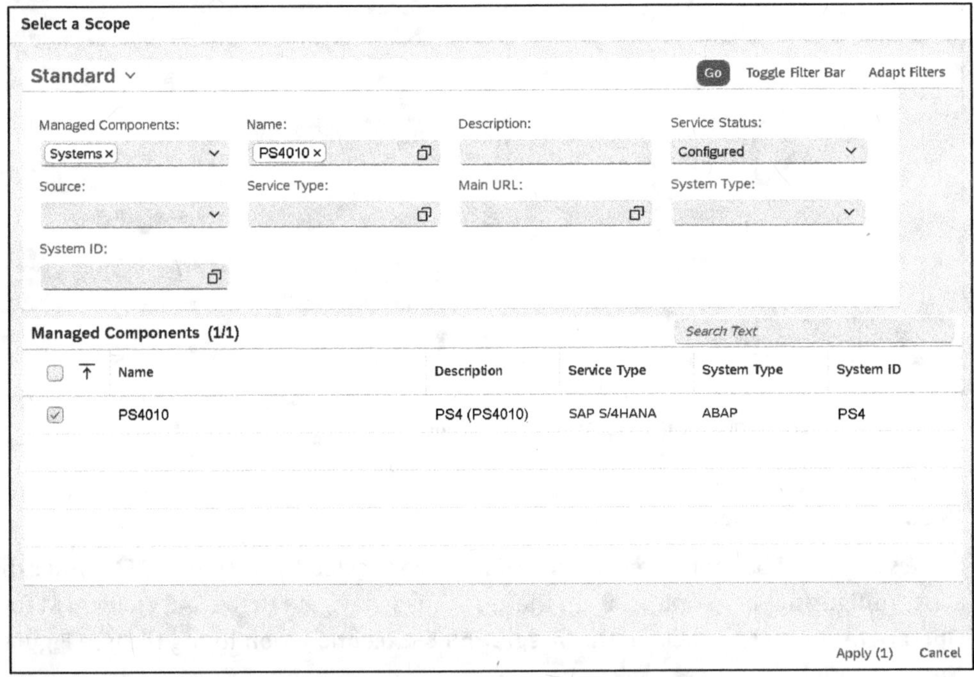

Figure 19.32 Selection of the Scope

You will then find yourself on the **Home** page of the monitoring scenario (see Figure 19.33), visible in the left-hand menu, where the monitoring data for the **Messages**, **Exceptions**, data quality (**Quality**), and the open alerts (**Open Alerts**) for the on-premise system. It is not necessary to activate monitoring separately; this is done automatically via the push method. Information on the messages and exceptions is collected and displayed in the form of a tile.

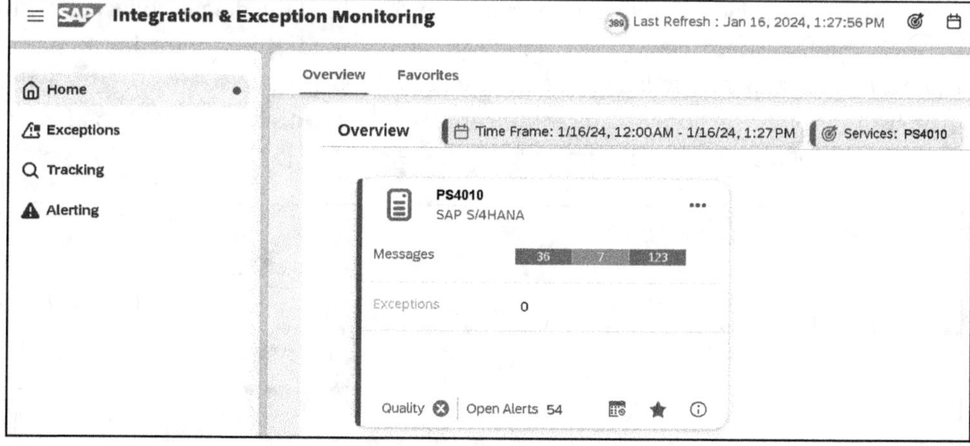

Figure 19.33 Start Page for Monitoring an On-Premise System

The initial configuration of the cloud service in our example involves defining an HTTP endpoint in the administration of the landscape after the preparatory work in the cloud service itself has been successfully completed.

The addition of such an HTTP endpoint is shown in Figure 19.34. The data that must be entered in the respective fields is derived from the information in the service key that was created as a preparatory measure in the cloud service. Once the HTTP endpoint has been successfully added, you enter the integration and exception monitoring scenario again.

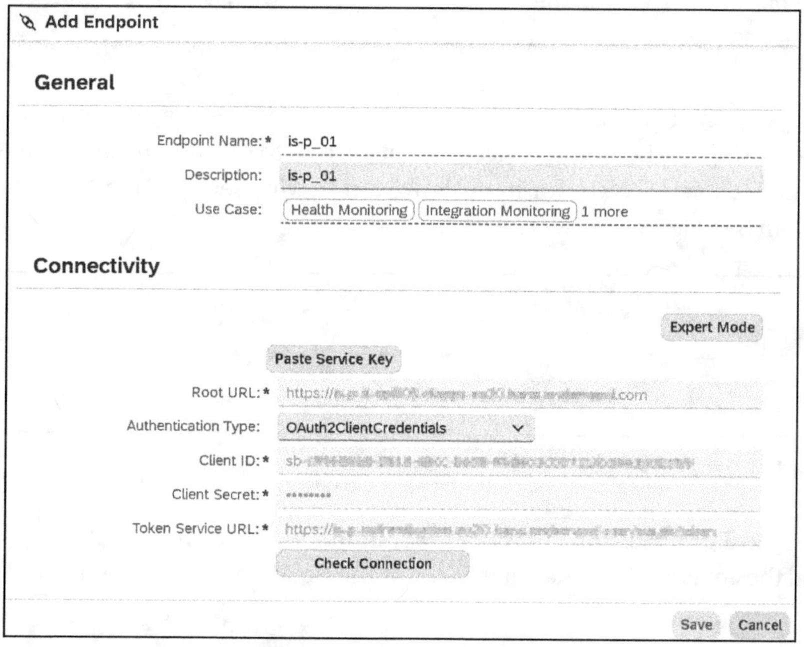

Add Endpoint
General
Endpoint Name:* is-p_01
Description: is-p_01
Use Case: Health Monitoring | Integration Monitoring | 1 more
Connectivity
Expert Mode
Paste Service Key
Root URL:*
Authentication Type: OAuth2ClientCredentials
Client ID:*
Client Secret:* ********
Token Service URL:*
Check Connection
Save Cancel

Figure 19.34 Adding an HTTP Endpoint

Monitoring for the cloud service in the example, for which the data is collected using the PULL method is currently not automatically activated for this monitoring scenario, which becomes apparent when you select the cloud service in the scope selection—the displayed tile does not contain any monitoring information, and the status of the service is not displayed in color, as shown in Figure 19.35.

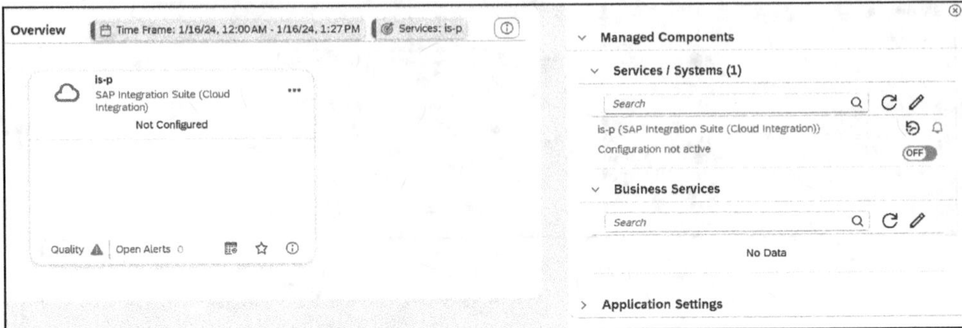

Figure 19.35 Configuration for Monitoring a Cloud Service

Monitoring of Applications: Pull Method

If you want to monitor an application for which the data is provided using the pull method, after creating an HTTP endpoint in the administration of the landscape, the monitoring must currently still be activated initially in the respective monitoring scenario—it is generally not activated automatically. Automatic activation only takes place for the business process monitoring and health monitoring scenarios. In the future, this will also be the case for other scenarios.

Open the configuration via the gear icon, and you will see the selected cloud service in blue. By clicking on the cloud service, you gain access to the monitoring configuration. The initial data collection for monitoring can now be set to **ON** in the **Data Collection** column (see Figure 19.36).

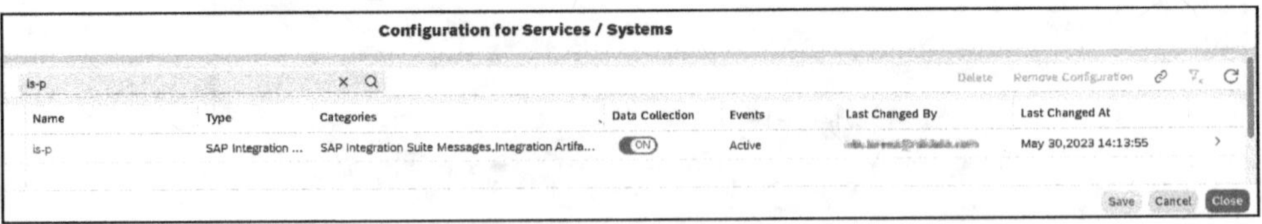

Figure 19.36 Start of Data Collection

After activating the data collection, it takes a few minutes until data for the **Messages**, **Exceptions**, and the generated alerts are available for the cloud service (see Figure 19.37).

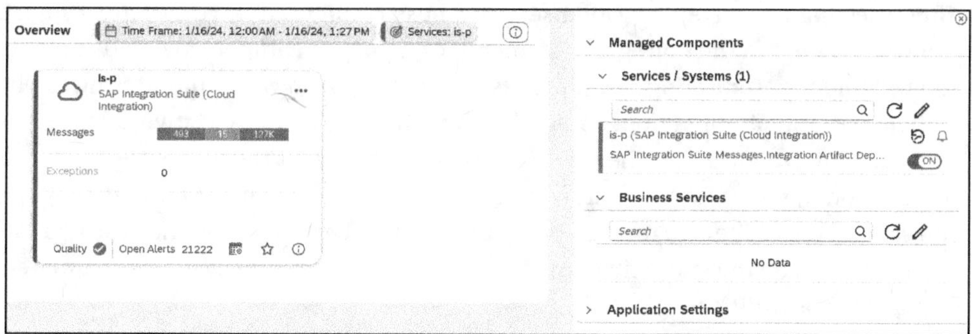

Figure 19.37 Monitoring Data for a Cloud Service

19.3.3 Specific Configuration

Once the initial configuration has been completed, you can continue with the specification of the monitoring, i.e., you configure the monitoring so that exactly the data that is of interest to you is collected. You can also create the events and resulting alerts.

To continue with the configuration, click on the gear icon again. It is important for the following configuration that you select exactly the systems and cloud services for which you want to make configurations in the scope selection, as only these are listed under the **Managed Components** menu item (see Figure 19.38).

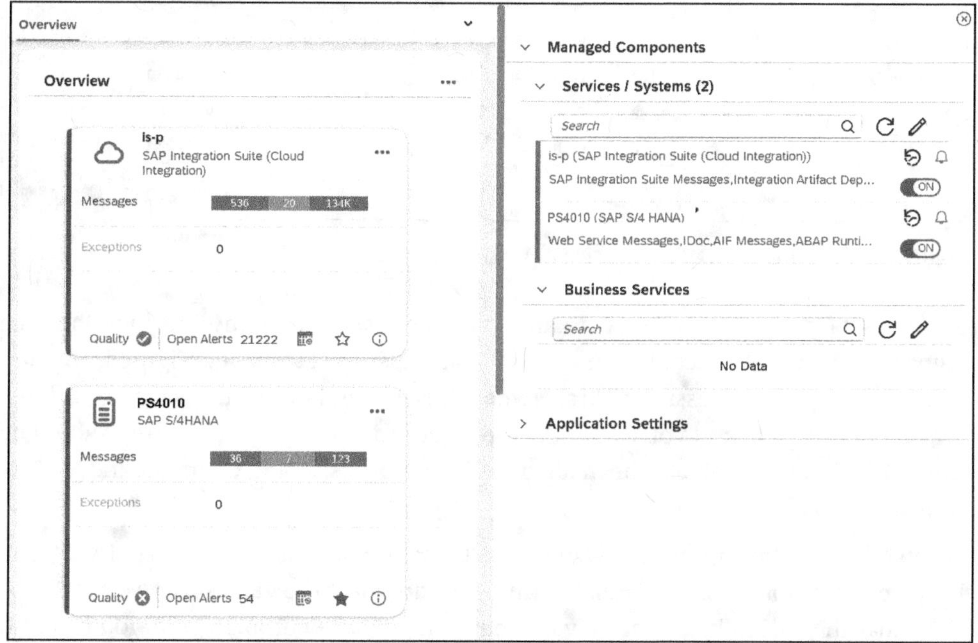

Figure 19.38 Monitoring Data for a Cloud Service

After selecting the system or cloud service whose configuration you want to adjust, you will be taken to the Configuration menu for managed components. In this menu, you will find the **Monitoring** menu item (see Figure 19.39), in which monitoring can be switched on and off for various categories using the button in the **Active** column. For each category, you will find the arrow button on the right at the end of the line [>]. If you click on it, you will be forwarded to the filter options for the respective category. You can use these filters to restrict monitoring to a specific category. The available categories depend on the respective system or cloud service, and the available parameters for the filters also differ.

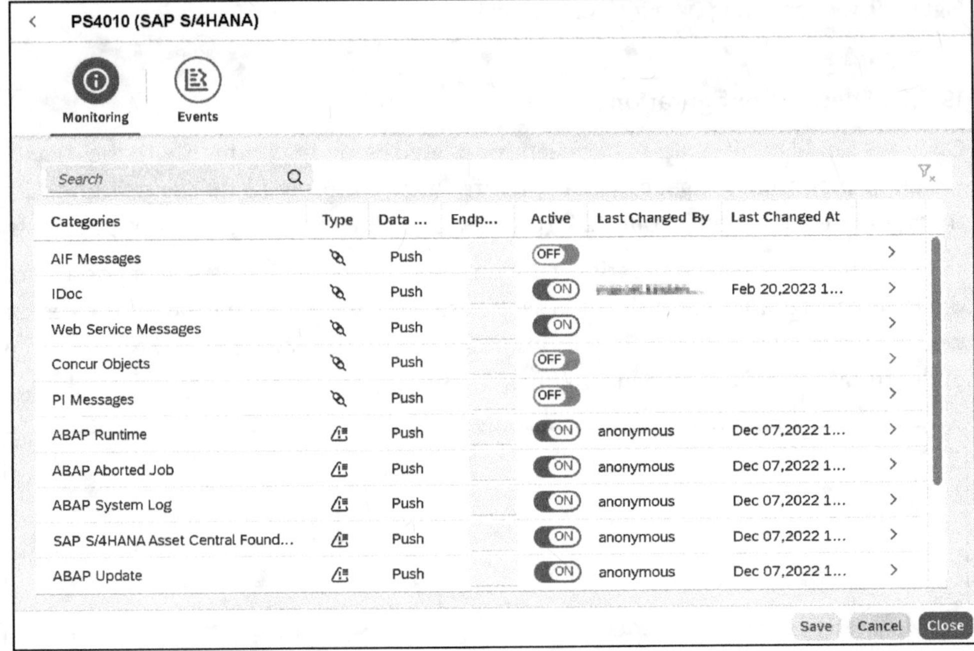

Figure 19.39 Monitoring Setup for Each Category

You can set filters via **Add**—in our example, a filter is created for monitoring IDocs (see Figure 19.40). A new area opens in which the filter parameters are configured. The first thing you should do is assign a **Filter Name**—this is mandatory. The available parameters are preselected, and you can select the values to which the monitoring should be limited. Restrictions can also be made using operators. **Is**, **Is Not**, **Contains**, and **No Pattern** are available for this purpose.

When selecting the appropriate values, a value help, which lists the values already collected by SAP Cloud ALM, is available, and you can select the values using a checkbox. Note that the values listed in the value help may not be complete, but you can also make entries directly.

In addition to the parameters, further attributes can be collected, which you can add in the **Collect Additional Attributes**, below the area for the filter parameters. Once you have completed the configuration, save it using the **Save** button. You will now see the filter you have just created in the table for the filters.

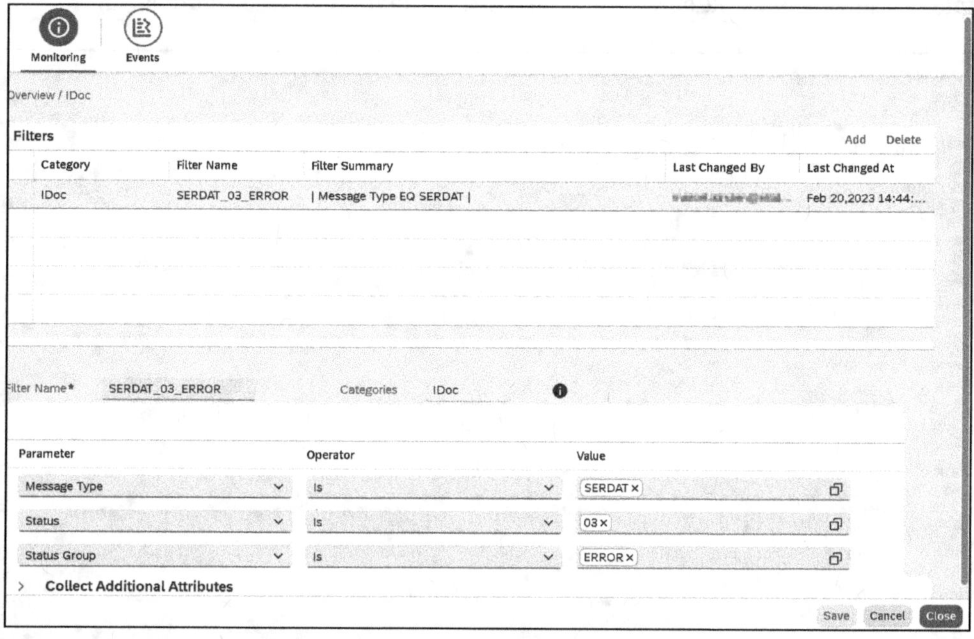

Figure 19.40 Adding Monitoring Filters

19

To the right of the **Monitoring** menu item, you will find the **Events** menu item, which is used to configure events, alerts, and other event actions. If you select the menu item, you will see predefined events, some of which are already activated. You can select which of these events should be activated using the switch in the **Active** column (see Figure 19.41).

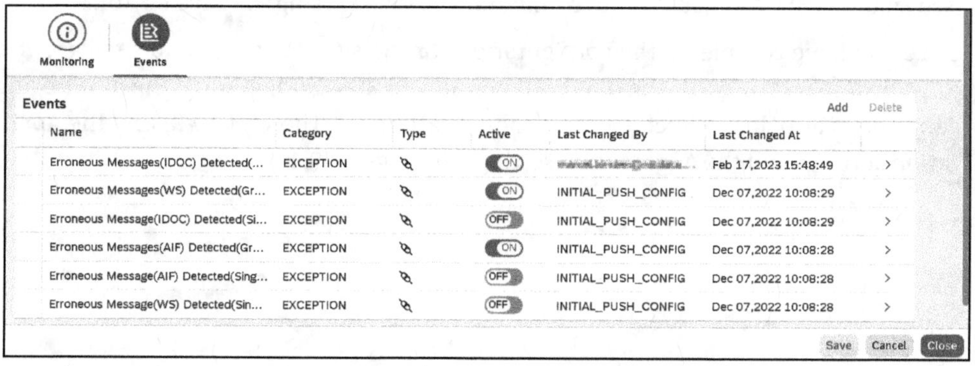

Monitoring | Events

Events — Add | Delete

Name	Category	Type	Active	Last Changed By	Last Changed At
Erroneous Messages(IDOC) Detected(...	EXCEPTION		ON	[illegible]	Feb 17,2023 15:48:49
Erroneous Messages(WS) Detected(Gr...	EXCEPTION		ON	INITIAL_PUSH_CONFIG	Dec 07,2022 10:08:29
Erroneous Message(IDOC) Detected(Si...	EXCEPTION		OFF	INITIAL_PUSH_CONFIG	Dec 07,2022 10:08:29
Erroneous Messages(AIF) Detected(Gr...	EXCEPTION		ON	INITIAL_PUSH_CONFIG	Dec 07,2022 10:08:28
Erroneous Message(AIF) Detected(Sing...	EXCEPTION		OFF	INITIAL_PUSH_CONFIG	Dec 07,2022 10:08:28
Erroneous Message(WS) Detected(Sin...	EXCEPTION		OFF	INITIAL_PUSH_CONFIG	Dec 07,2022 10:08:28

Save | Cancel | Close

Figure 19.41 Calling Up the Events

Use the arrow button [>] at the right-hand end of each event line, which takes you to a submenu that allows you to edit the events. The display name of the event and the event description are freely customizable here. Furthermore, parameters for the respective event and the event actions can be configured, including the activation of alerts. Alerting is activated via the slider next to **Create Alert** (see Figure 19.42).

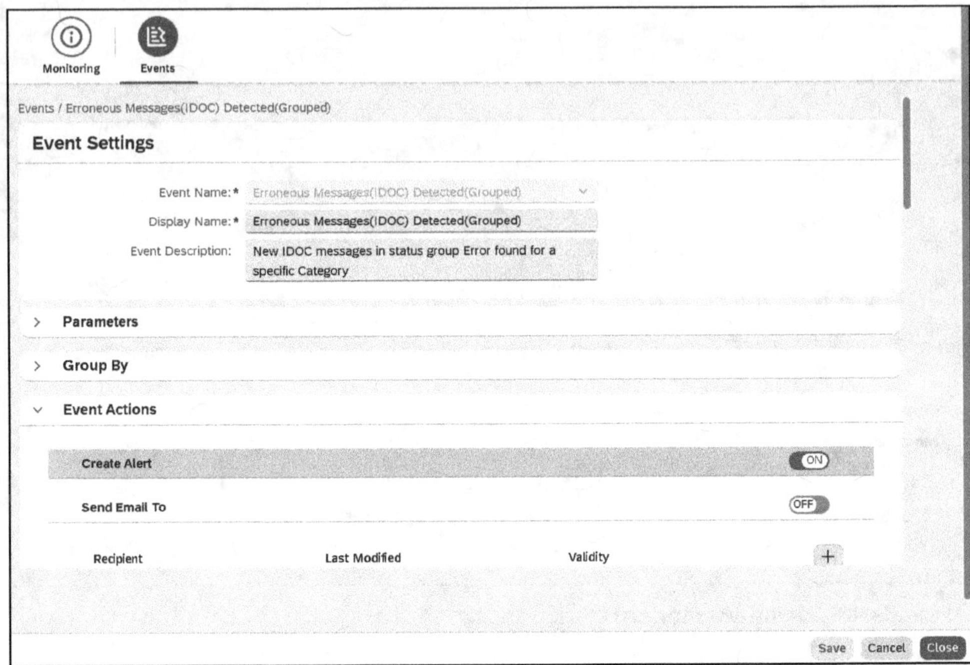

Figure 19.42 Activation of the Alarm for an Event

An important area for the configuration of events can be found at the end of the submenu. This is where you will find the **Filter Configuration**, which allows you to enter parameters for a very granular configuration of events for each monitored category (see Figure 19.43). After successful configuration, save it using the **Save** button.

Once you have completed the configuration for the systems and cloud services, you can adjust the settings for the retention period and the aggregation of the monitoring data. This is also done directly via the configuration [⚙]. There you will find the **Application Settings** and the **Aggregation** settings (see Figure 19.44).

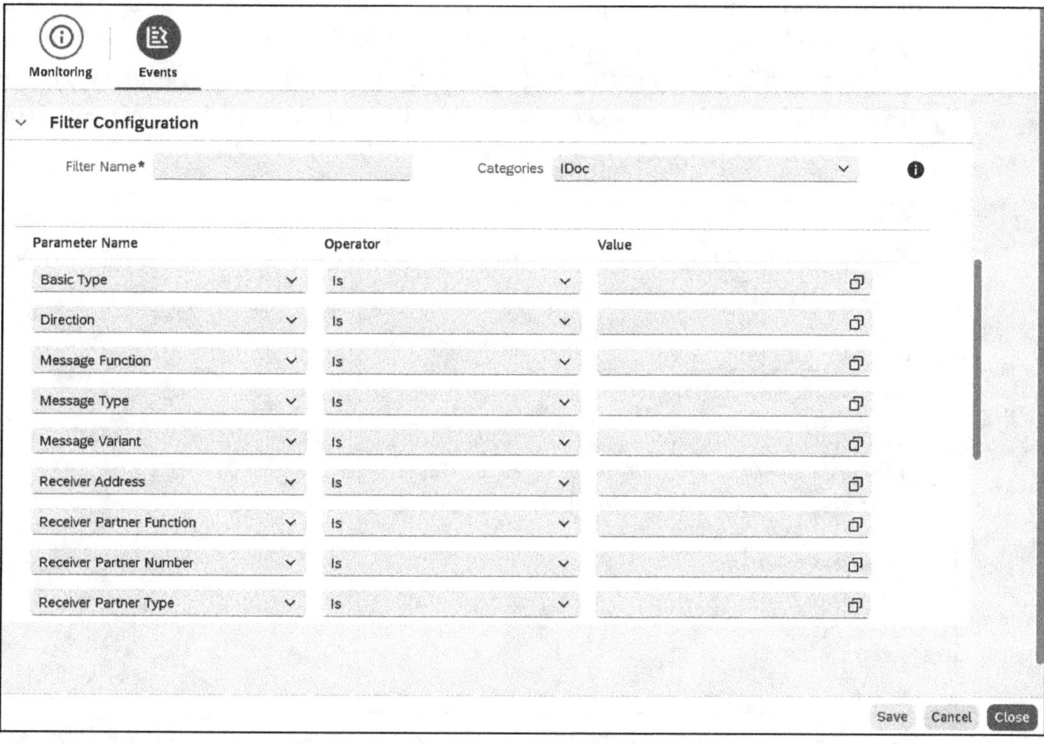

Figure 19.43 Configuring Filters for Events

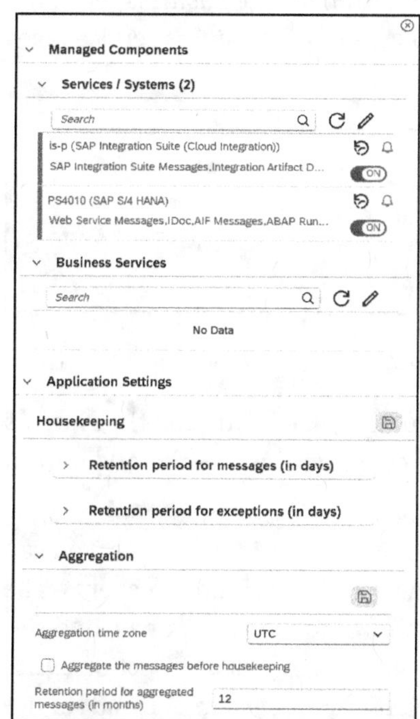

Figure 19.44 Housekeeping in Integration and Exception Monitoring

In addition to monitoring on-premise systems and cloud services, integration and exception monitoring can also be used to monitor business services. To start setting up the monitoring, it is necessary to call up the configuration again. The **Business Services** item is located below the managed components (see Figure 19.45). Use the pencil icon to access the submenu where you will create the business service.

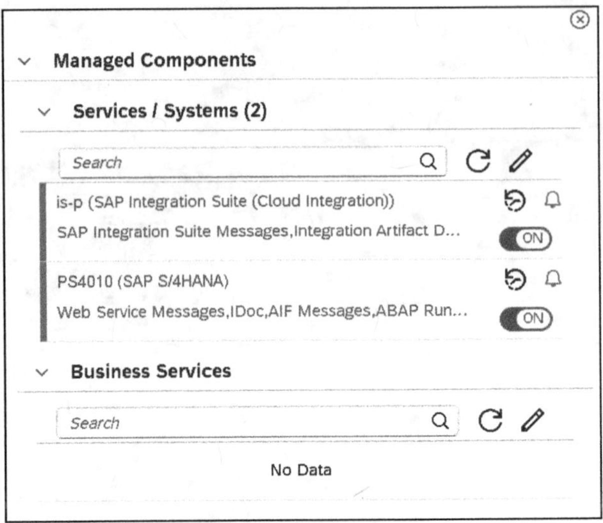

Figure 19.45 Access to the Configuration of the Business Service Monitoring

After clicking on **Add** and assigning a name and description for the business service, it is created, as shown in Figure 19.46. Clicking on the name of the business service takes you to the configuration.

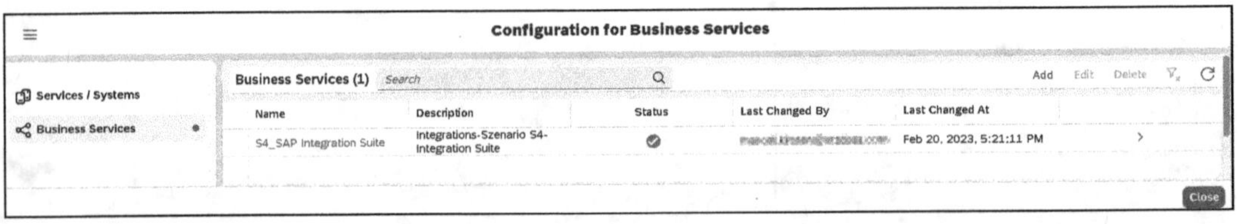

Figure 19.46 Created Business Service

The first step takes you to the **Service/Systems** menu item, where you add the systems and cloud services that belong to the business service. To do this, use the **Add** button again (see Figure 19.47).

Once you have added the systems and cloud services for the business service, you can continue with the configuration of the filters for the monitoring and the events. The procedure is identical to the configuration already shown for a system or a cloud service as a single managed component (see Figure 19.48), as described in Section 19.3.2. The configuration of the filters and events always relates to the system or cloud service selected in the **Services/System** table.

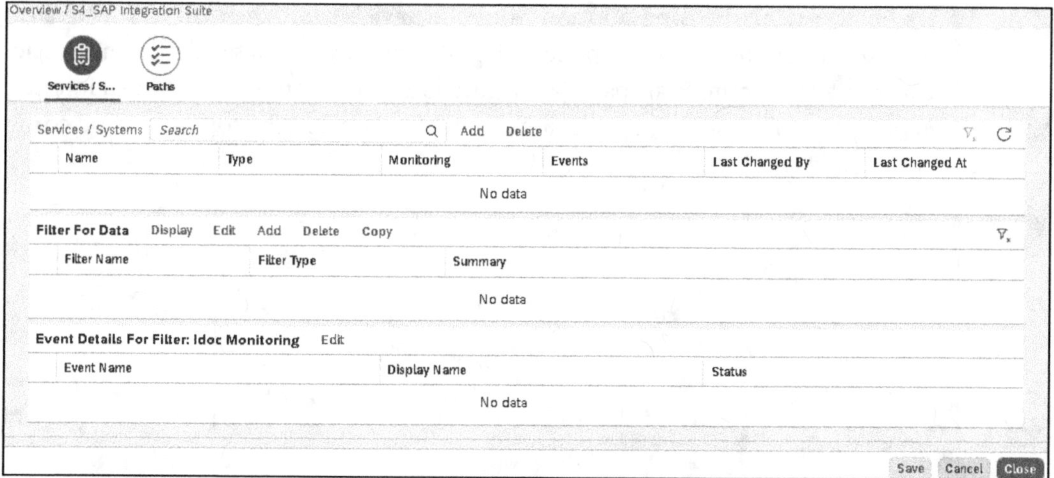

Figure 19.47 Initial View of the Business Service Configuration

Figure 19.48 Filters and Events for Business Service

Once you have configured the filters and events, switch to the **Paths** menu item. Here, the connection paths between the individual systems/cloud services are created by activating the checkboxes (shown in Figure 19.49). In the example, the **PS4010** system is the source system, and the **is-p** cloud service is the target system. A path corresponds to the direction in which data is exchanged between the components of a business service.

After finally saving the configuration using the **Save** button, the business service can be found and selected in the scope selection. It is displayed as a tile on the **Integration & Exception Monitoring** start page (see Figure 19.50). You can obtain detailed information on the business service by clicking on the three dots in the top right-hand corner of the tile and clicking on **Go to Details**.

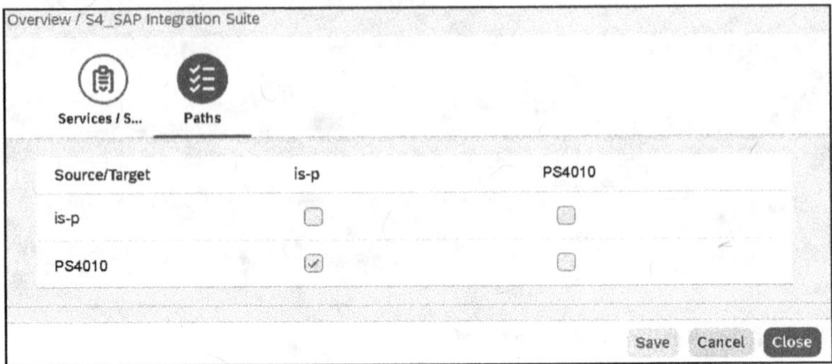

Figure 19.49 Create Paths

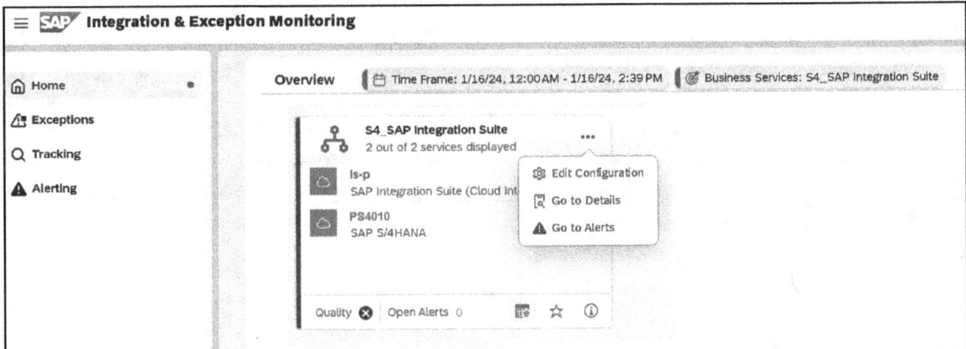

Figure 19.50 Business Service, Displayed as a Tile

The following view shows the individual components of the business service including the paths (see Figure 19.51). In addition to the general information of the **Messages**, **Exceptions**, and **Alerts** of the business service, which you can see below the name of the business service, you can call up the details of the individual components by clicking on them. An area opens on the right with the specific information on the messages and exceptions for the component.

Starting from the detailed information, you can click on the number of messages or exceptions to navigate to a submenu that provides an insight. Figure 19.52 shows an example of this for the exceptions that have occurred for the **is-p** service. The information is displayed both in the form of diagrams and in a table. You can see that one resolved exception is available.

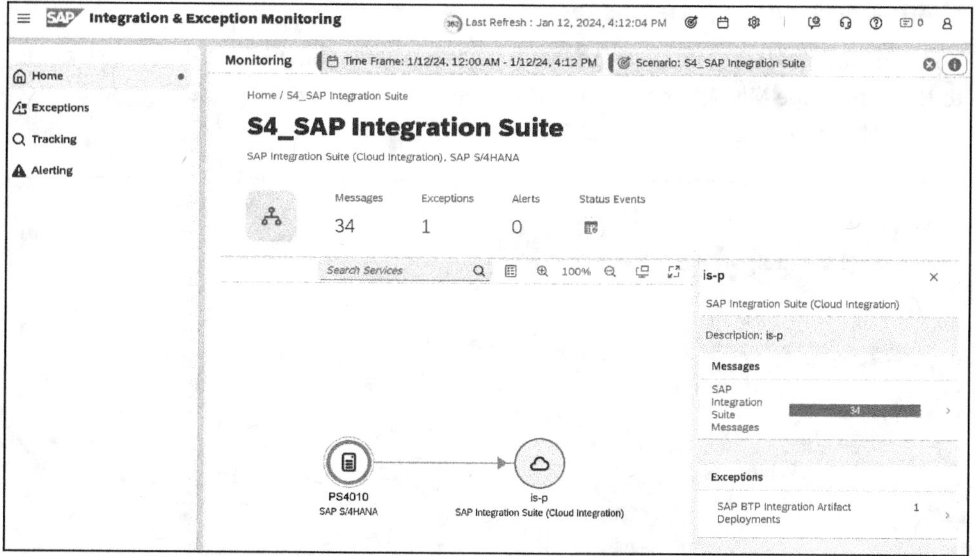

Figure 19.51 Business Service with Path and Detailed Information

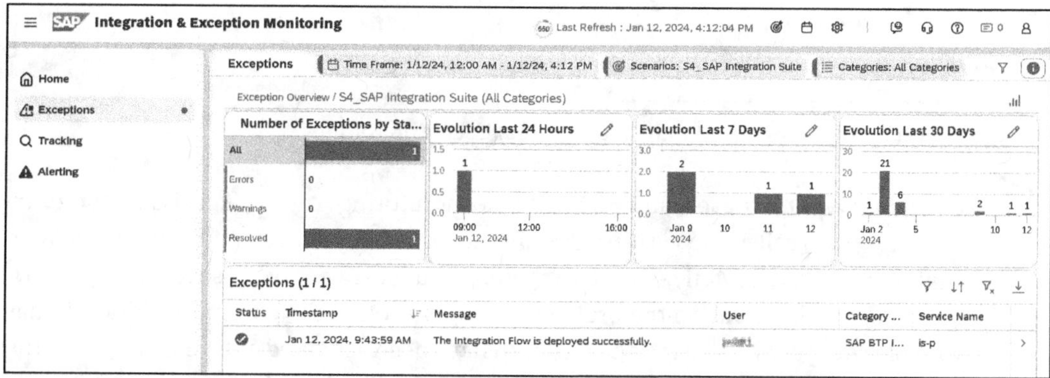

Figure 19.52 Detailed Information on Exceptions

19.4 Real User Monitoring

The monitoring scenario *real user monitoring* is shown in this section using SAP Integrated Business Planning (SAP IBP) as an example. As no events, alerts, thresholds, etc. can currently be configured for this monitoring scenario, the section for the specific configuration is omitted.

19.4.1 Checking the Prerequisites

You can check whether real user monitoring is available for this cloud service by going to the Landscape Management app, calling up the **Services & Systems** page, and selecting the cloud service (see Figure 19.53).

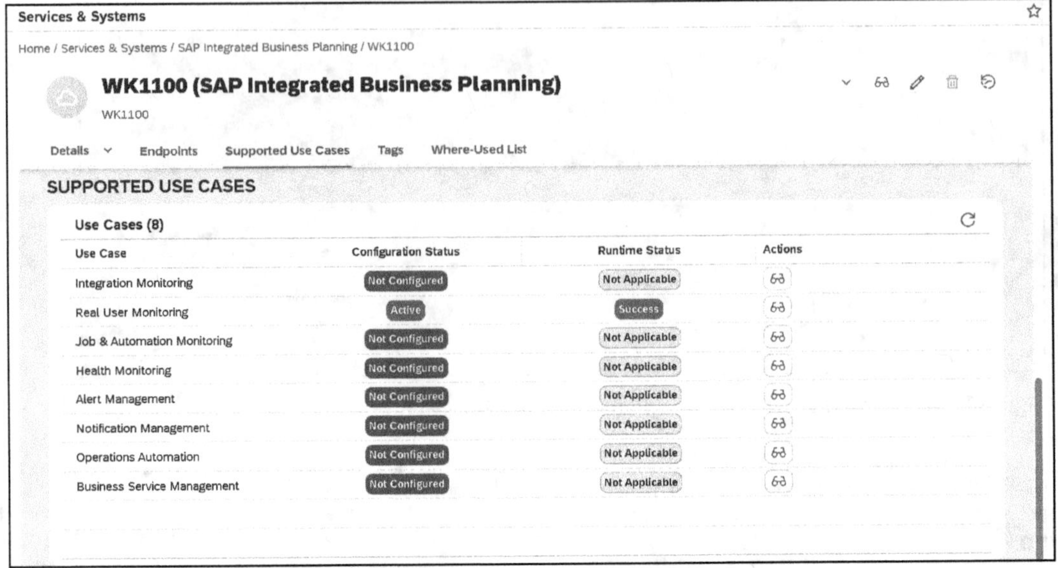

Figure 19.53 Landscape Management in Real User Monitoring

The **Supported Use Cases** tab shows that the monitoring scenario **Real User Monitoring** including data collection via the PUSH method is already activated for the system (**Configuration Status** of **Active**). Furthermore, the data collection is successful (**Runtime Status** of **Success**). Due to the previous configuration of the communication arrangement in the system to be monitored, both have been activated automatically. Whether an HTTP endpoint is defined is not relevant for the cloud service in the example.

19.4.2 Initial Configuration

The monitoring scenario is called up via the launchpad and the corresponding tile in the **SAP Cloud ALM for Operations** group, as shown in Figure 19.54.

Once you have selected the system or cloud service in the scope selection for which you want to view the monitoring, the system/cloud service is displayed as a tile on the start page. Data collection and monitoring are already active. If you call up the configuration via the gear icon [⚙], you can also deactivate the data collection and thus the monitoring in the **Managed Components** area by clicking on the **ON** slider. The button then changes to **OFF** (see Figure 19.55).

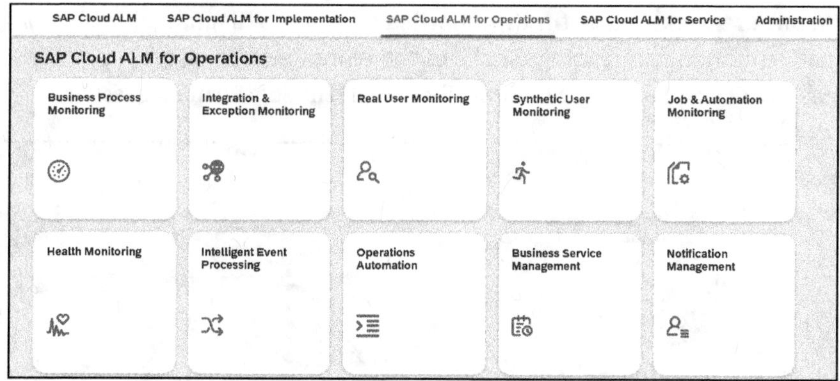

Figure 19.54 Access to the Real User Monitoring

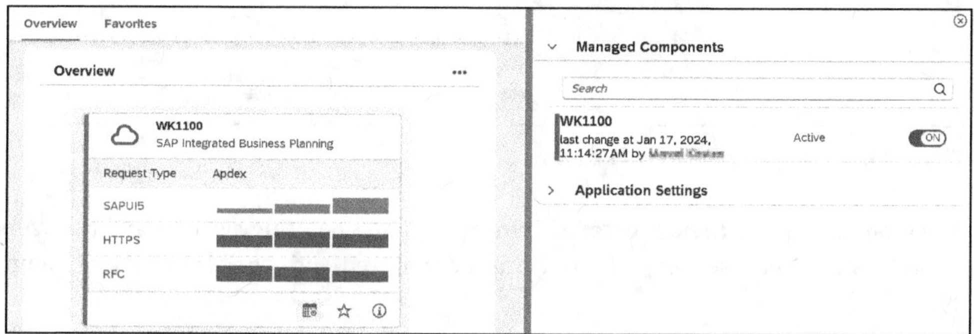

Figure 19.55 Activation of the Monitoring

If you switch to the **Requests** page in the left-hand menu, you will get an insight into the individual requests, where you can see detailed information on an individual **Execution Time** (see Figure 19.56).

Real User Monitoring — Last Refresh : Jan 17, 2024, 11:13:46AM

Home · Requests · Analysis · Front End · Back End · Services/Systems · Clients · Execution Flow · Expensive Requests · HTTP Errors · Add Custom Page

Requests — Time Frame: 1/17/24, 10:13AM - 1/17/24, 11:13AM — Services/Systems: WK1100

Requests / /IBP/PLANNING_AREA_READ / POST-batch

Execution Time	Type	Exec...	Status Summary	Response Ti...	Service/Syst...	Net Time Per...	Net Time
11:05:00AM		1		397	MyIBP_01	100 %	397
11:04:32AM		1		838	MyIBP_01	100 %	838
11:03:10AM		1		133	MyIBP_01	100 %	133
11:01:04AM		1		952	MyIBP_01	100 %	952
10:56:01AM		1		443	MyIBP_01	100 %	443
10:54:53AM		1		115	MyIBP_01	100 %	115
10:51:40AM		1		118	MyIBP_01	100 %	118
10:51:40AM		1		120	MyIBP_01	100 %	120
10:50:43AM		1		861	MyIBP_01	100 %	861
10:49:03AM		1		148	MyIBP_01	100 %	148

Figure 19.56 Display of Requests

If you click on one of the execution times, you will be taken to a diagram view showing the individual components, including details of the components and response times (see Figure 19.57). This display is very helpful when analyzing performance issues.

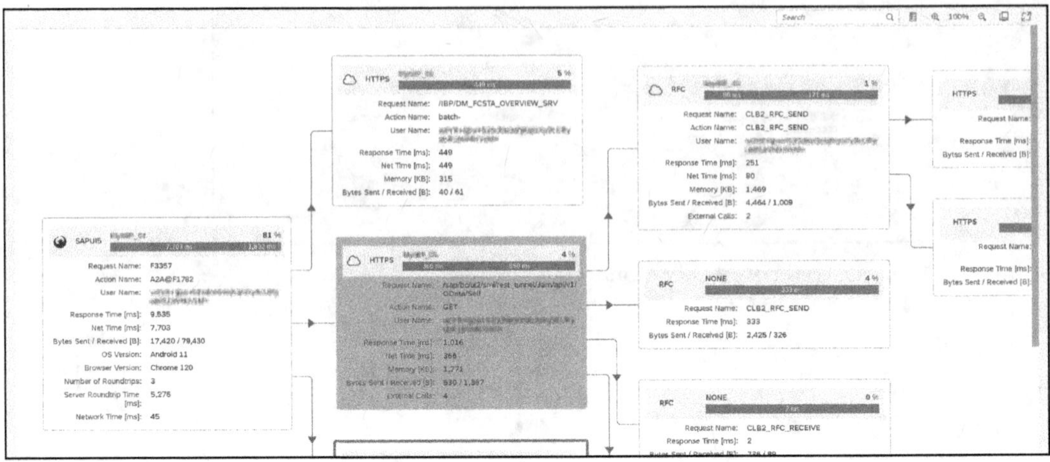

Figure 19.57 Component View of Requests

The **Front End** page provides detailed information on operating systems and browsers as well as the end user and network times associated with the requests (see Figure 19.58).

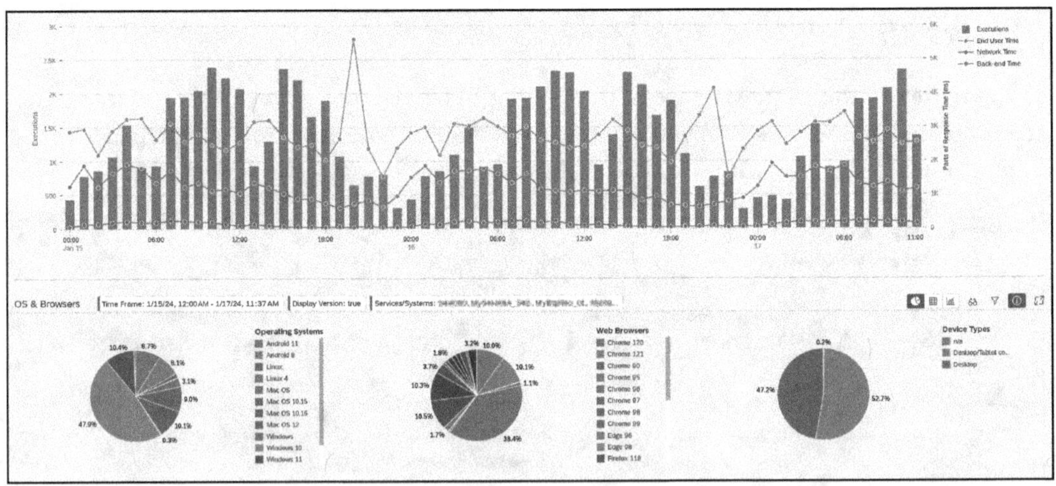

Figure 19.58 Information on the Front End

The **Back End** page gives you an insight into the most important performance and usage metrics for backend requests. The number of executions of the respective request type is displayed (see Figure 19.59).

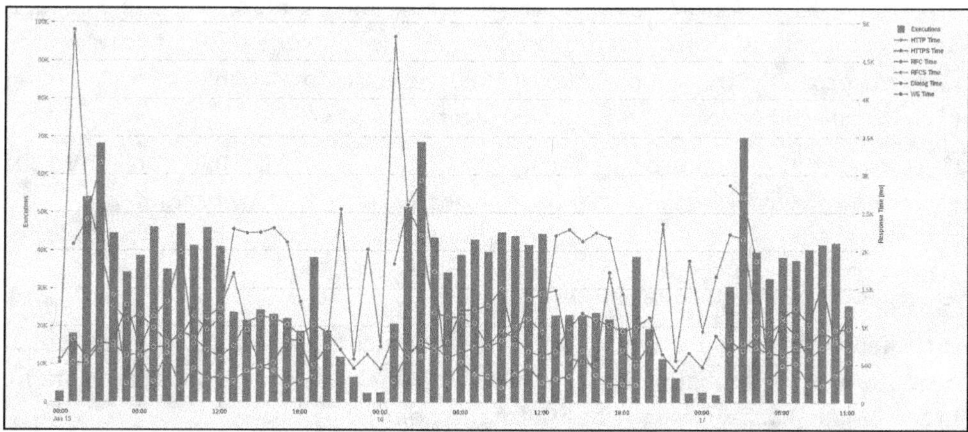

Figure 19.59 Usage Metrics for Backend Requests

The last page to be mentioned here is the **Expensive Requests** page. It provides you with information in the form of a heat map, with the scale ranging from green (requests with a short execution time) to red (requests with a long execution time). Figure 19.60 shows such a heat map.

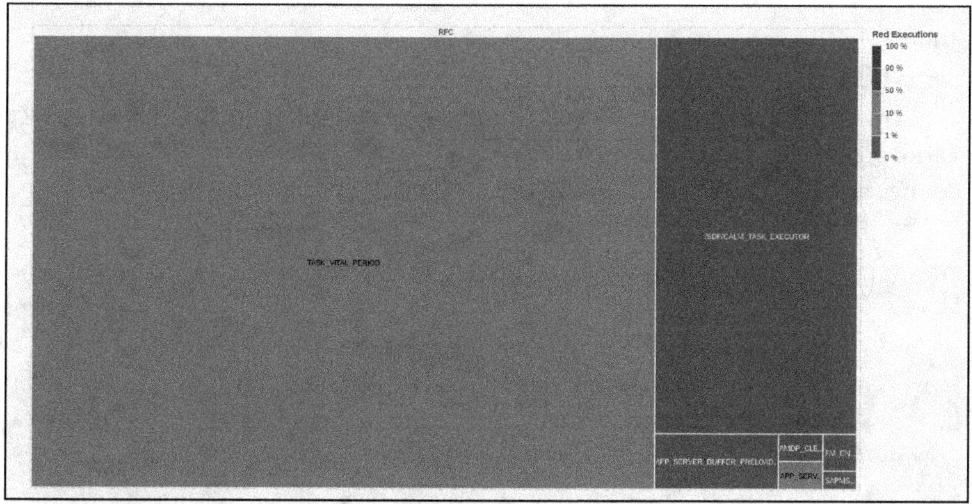

Figure 19.60 Heat Map for Real User Monitoring

19.5 Synthetic User Monitoring

As explained in Chapter 18, Section 18.2.1, for the synthetic user monitoring scenario, in addition to SAP Cloud ALM, you need a Selenium—remote WebDriver infrastructure—server via which the scripts are executed.

In Section 19.5.1, we will look at the most important components of synthetic user monitoring with a focus on scripting. In Section 19.5.2, we describe the configuration that takes place directly in SAP Cloud ALM. It is not necessary or possible to check the connection/data collection via landscape management.

If you require information on setting up a Selenium server, refer to the Selenium documentation, which is available at *https://www.selenium.dev/documentation/*.

19.5.1 Initial Configuration

In the following section, we will explain the components that you need before you can continue with the configuration steps in SAP Cloud ALM. As this is not a configuration in SAP Cloud ALM itself, we will be brief at this point.

Selenium Remote WebDriver infrastructure

As already mentioned, you need a Selenium server. You also need a cloud connector, which enables secure communication from your internal network to SAP Cloud ALM via an HTTPS tunnel. The cloud connector is set up in the same network as the Selenium infrastructure.

Runner

The Selenium server has a URL that is stored in the synthetic user monitoring of SAP Cloud ALM in the runner as an endpoint. The *runner* defines where the monitoring scenarios are executed. This URL must be a secure HTTPS address; HTTP or IP addresses are not permitted at this point. You must then assign a name and a password. The runner is now fully configured.

Scripts

Scripts are created using the Selenium IDE browser extension, which you can download free of charge from the browser extension stores. After starting the plug-in (usually in browsers on the right-hand side under extensions), it opens in a new window. You can now start creating a new project or open an existing one (see Figure 19.61).

The menu for creating scripts is intuitive and simple, as shown in Figure 19.62. Above the display of all scripts on the left-hand side, you can see the assigned project. In the middle, you will find the individual script instructions of the selected script. Above that is the main URL from which the script is started, as well as the various commands for execution itself (execute all or the current script, jump to specific script instructions, speed). At the top right you will find buttons for saving and starting/stopping a new script. Clicking on the red **Record** button opens a new window with the stored URL. Now clicks and user input can be recreated. Once you have completed all the steps, click on the Record button again to end the recording and save it in Selenium. After recording a new script, you can edit individual steps using the fields in the middle. These fields can be used to adjust the **Command**, **Target** object, and **Value** of a step. A **Description** can also be stored, which is then displayed in the sequence table instead of the commands. Below you will find a **Log** in which status messages of the current script steps, including a timestamp, are displayed.

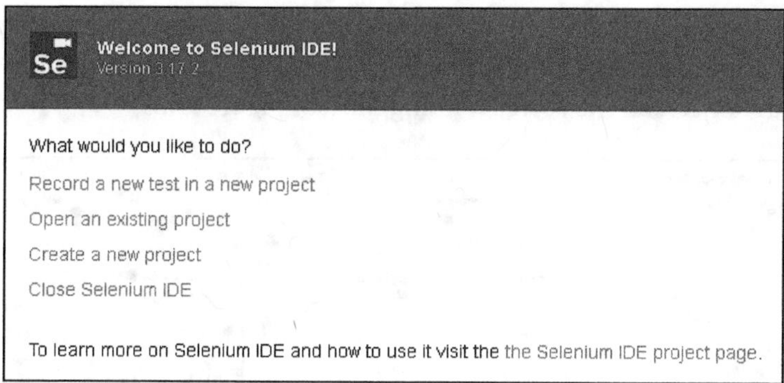

Figure 19.61 Selenium IDE

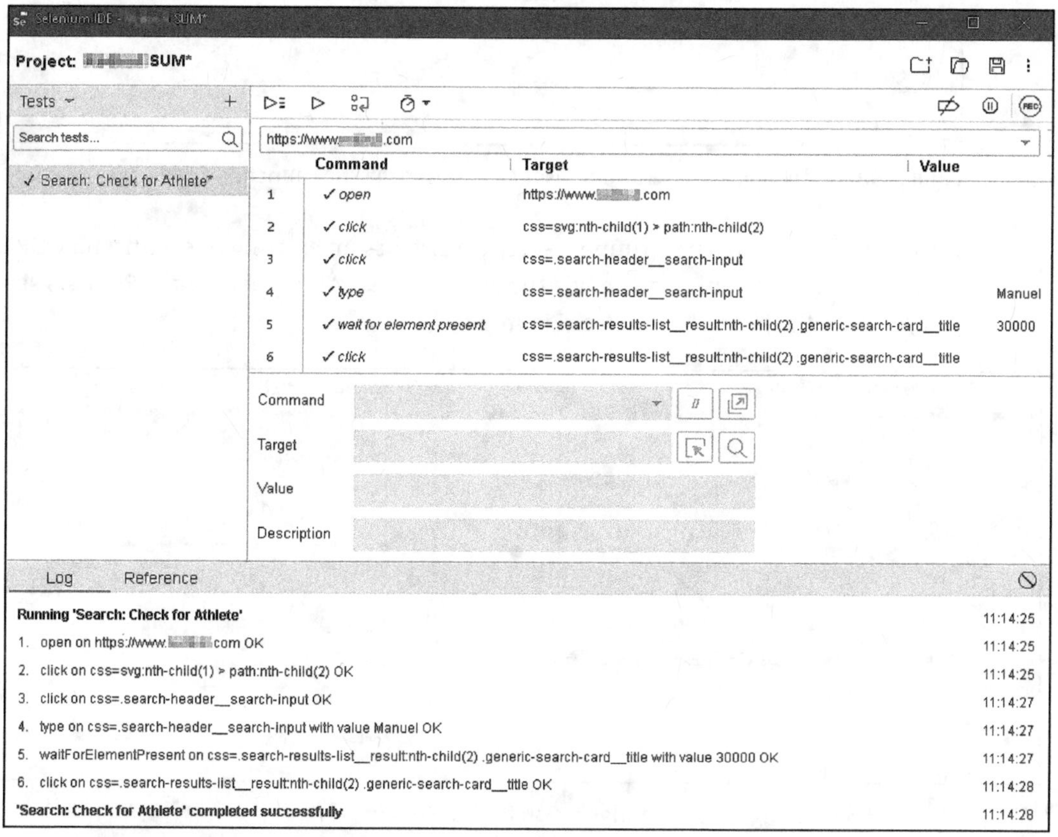

Figure 19.62 Menu for Creating Scripts

19.5.2 Specific Configuration

This section explains the configuration steps that you need to perform in SAP Cloud ALM to put synthetic user monitoring (sometimes abbreviated as SUM) into operation.

Add Runner

Once you have successfully set up the Selenium environment and created a script, you can start with the configuration in SAP Cloud ALM. To do this, first access the monitoring scenario via the launchpad and the corresponding tile (see Figure 19.63).

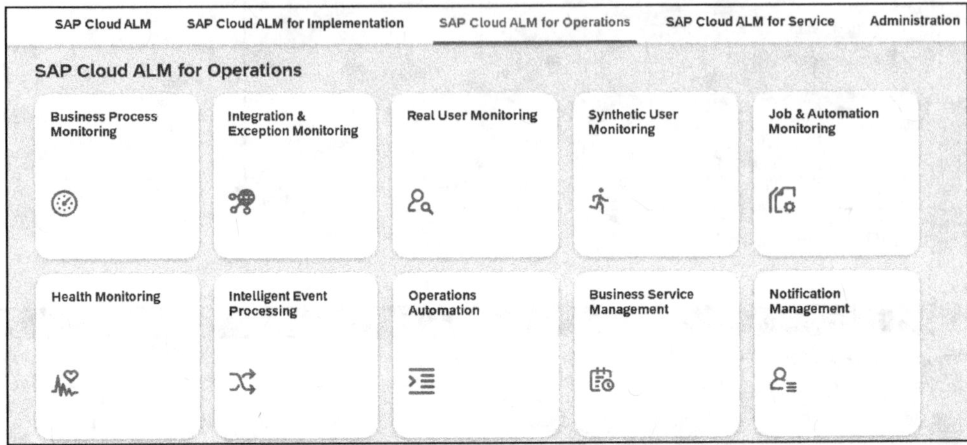

Figure 19.63 Calling up the Synthetic User Monitoring via the Launchpad Tile

First you need to add a new runner—you can also use an existing one with a new URL. To add it, open the configuration via the gear icon. After opening the **Runners** section, click on the pencil button (see Figure 19.64).

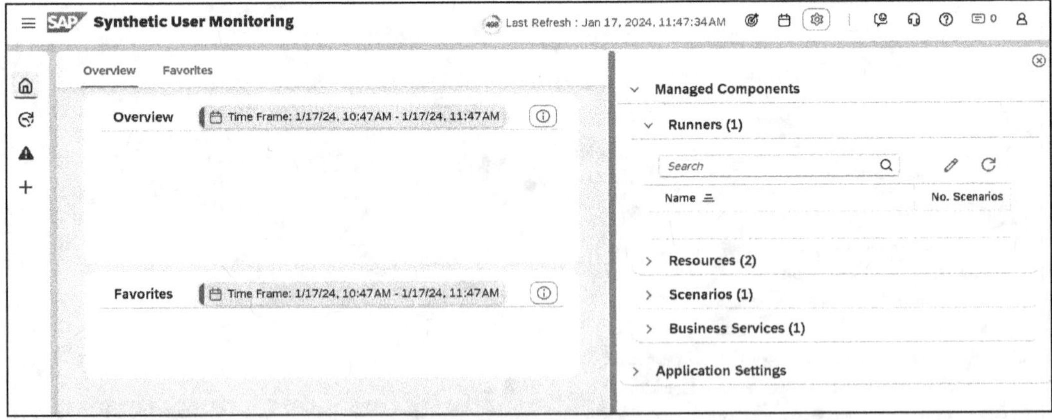

Figure 19.64 Calling Up the Configuration for the Runner

You can now add a new runner using the **Add** button, in which you store the configuration information (see Figure 19.65). The fields **Name**, **Type**, **URL**, **Username**, and **Password** are mandatory entries. Optionally, you can also add a description.

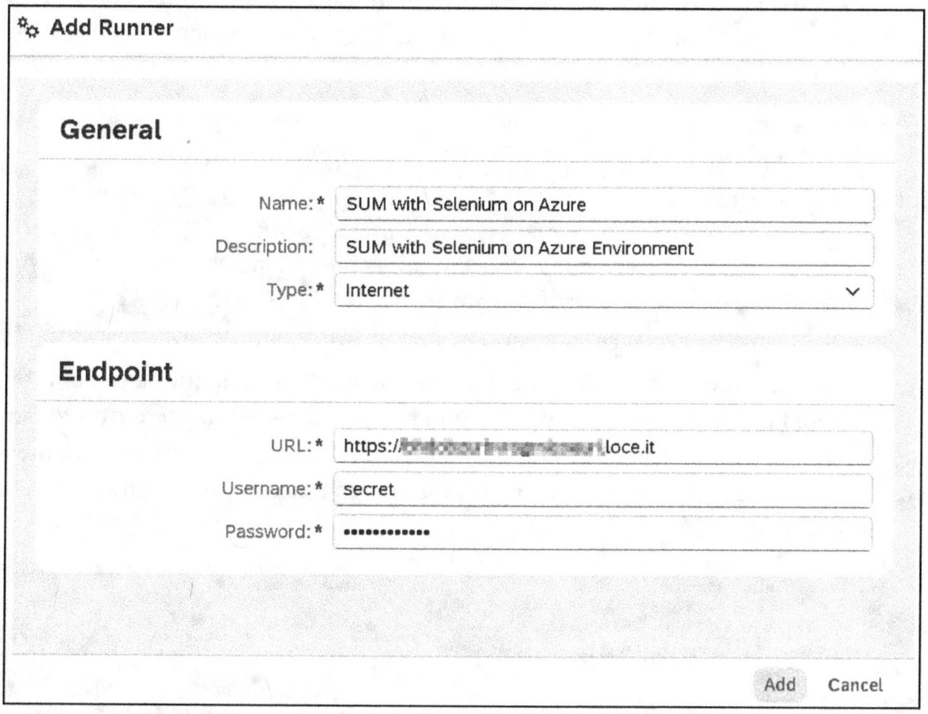

Figure 19.65 Add Runner

Manage Resources

The scripts created using the Selenium IDE add-on are entered via the **Resources** menu item. You can upload them using the **Upload** button in the **Resources** menu item, as shown in Figure 19.66. If you upload a file with the same name to SAP Cloud ALM, the old file is automatically updated.

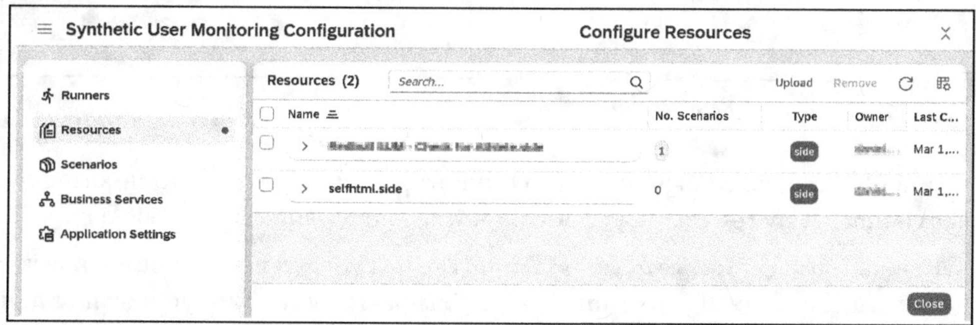

Figure 19.66 Manage Resources

Create Scenarios

Scenarios represent the various cases under which the scripts are executed on the Selenium server. You can create and edit scenarios via the **Scenarios** menu item. Here, the resources are assigned to the respective runners, and the runner is activated. Additional parameters such as threshold values or the execution frequency and descriptions can also be added.

[»]

Synthetic User Monitoring Scenarios

From a given Selenium IDE file, several monitoring scenarios can be generated. Also multiple scenarios can be linked to one or more runners. Each scenario is linked to at least one business service, but it can also be linked to more business services, and each business service is linked to at least one or more cloud services or systems.

Once you have assigned a name and a description for the scenario in the **General** tab and have stored the script for the scenario, you can add the scenario to a runner (see Figure 19.67). To do this, add the desired runner in the following window under **Runners** and displayed in the lower list by clicking on the **Assign** button. Use the **ON** button to activate the runner.

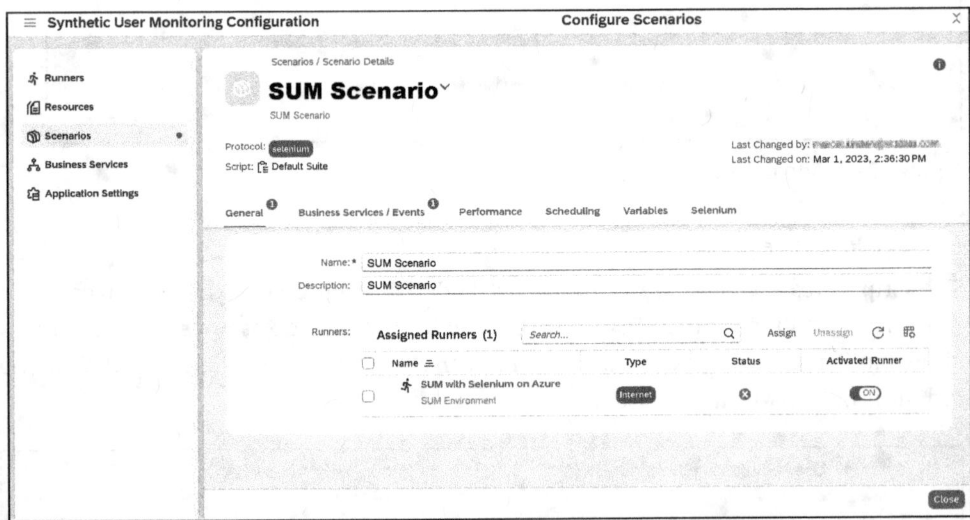

Figure 19.67 Create Scenario

Within the scenario configuration, you can now also make further settings using the tabs **Business Services/Events, Performance, Scheduling, Variables,** and **Selenium.**

Within the **Business Services/Events** tab, you can link your scenario to the appropriate SUM business service. If you're unsure which business service to use, you can link your scenario to the **Default SUM Business Service** (see Figure 19.68).

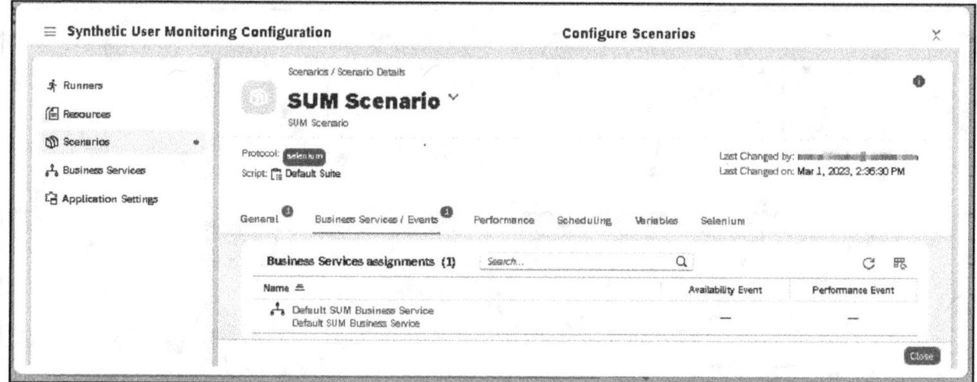

Figure 19.68 Business Services/Events

After clicking on the business service, you can configure the business service itself—assign a scenario to it, add systems/services, and define event actions for the **Availability Event** and/or **Performance Event** (see Figure 19.69).

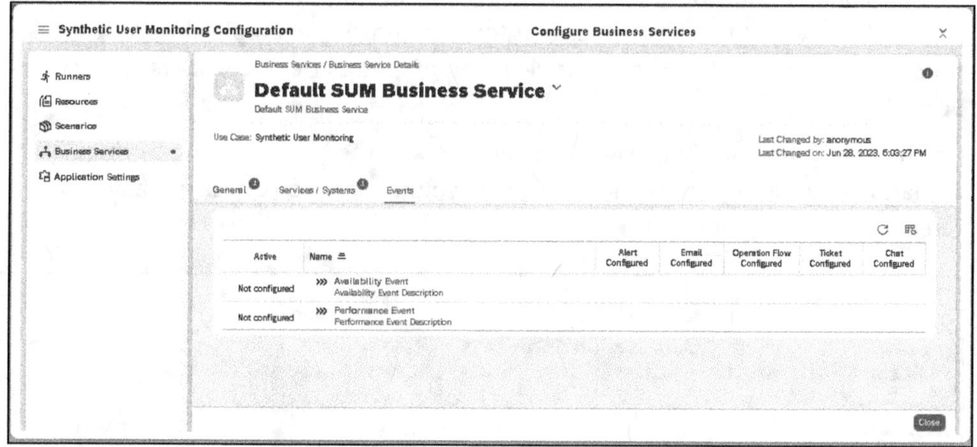

Figure 19.69 Definition of Event Actions

When you click on the event name, you can define which event action will be triggered (alert, email, etc.).

When you go back to the **Scenarios** page in the left-hand menu, you can use the **Performance** tab to define threshold values for the script (see Figure 19.70). If the script consists of several steps, you can also define threshold values for them. In the section **Global Performance Ratings**, you can decide, by selecting dropdown entry **Enabled**, if the performance shall be calculated using dynamic thresholds. This means, that with each run, AI now automatically calculates thresholds based on the historical data of successful executions.

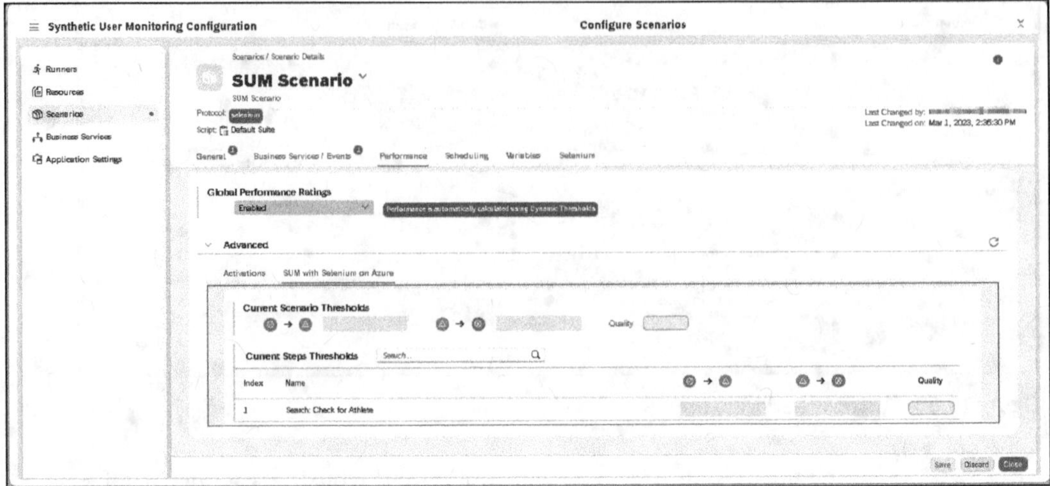

Figure 19.70 Define Threshold Values

Further configuration options for the scenario are the definition of the execution period of the scripts and specific settings for Selenium. For example, the browser selection, WebDriver arguments, and the selection of automatic screenshots are available here.

On the **Application Settings** page, you get access to the housekeeping settings of the scenario. Here you get an insight to used data volume, and you can also change the data retention time (see Figure 19.71).

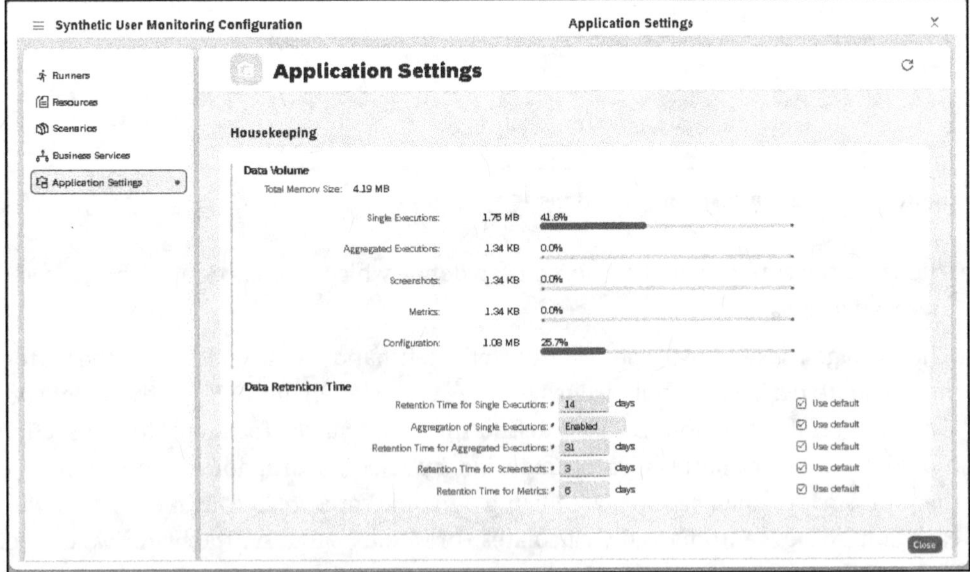

Figure 19.71 Application Settings

As soon as the runner has been activated in the scenario, you must define the scope. Select the desired scenario and click on the **Apply** button (see Figure 19.72). The monitoring data is then displayed.

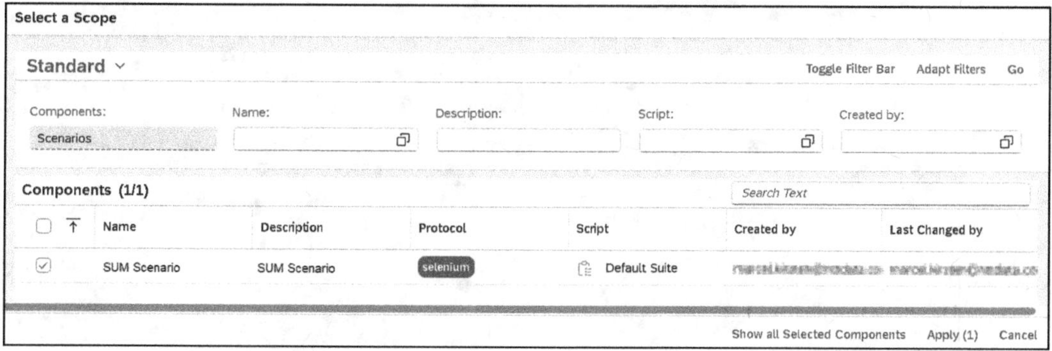

Figure 19.72 Definition of the Scope

An overview of all running scenarios is displayed on the **Home** page, with a brief presentation of the results—this is particularly useful if many scenarios are running in parallel and is therefore good for an initial overview (see Figure 19.73).

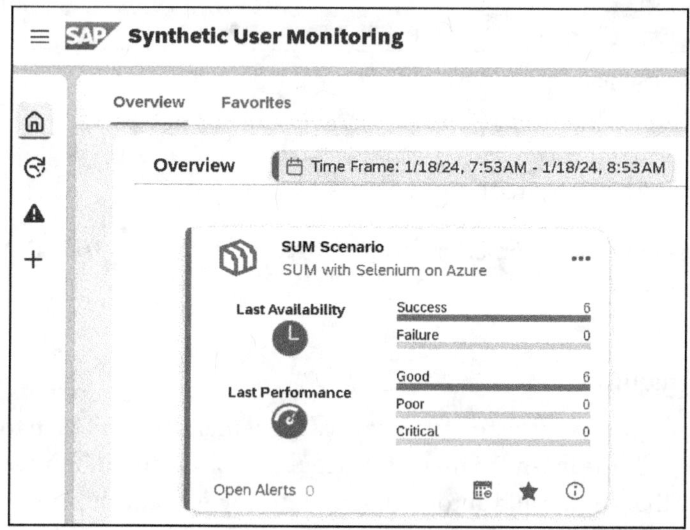

Figure 19.73 Home Page with Active Scenarios

Click on the tile of a scenario to open the detailed overview page (see Figure 19.74). This page shows the scenario with the associated runner and more detailed data on the execution of the script. The lower diagrams show the availability distribution (could the script be executed without errors?) and the response times (including additional parameters—poor and critical performance values, if set up). The times of the individual executions are also displayed.

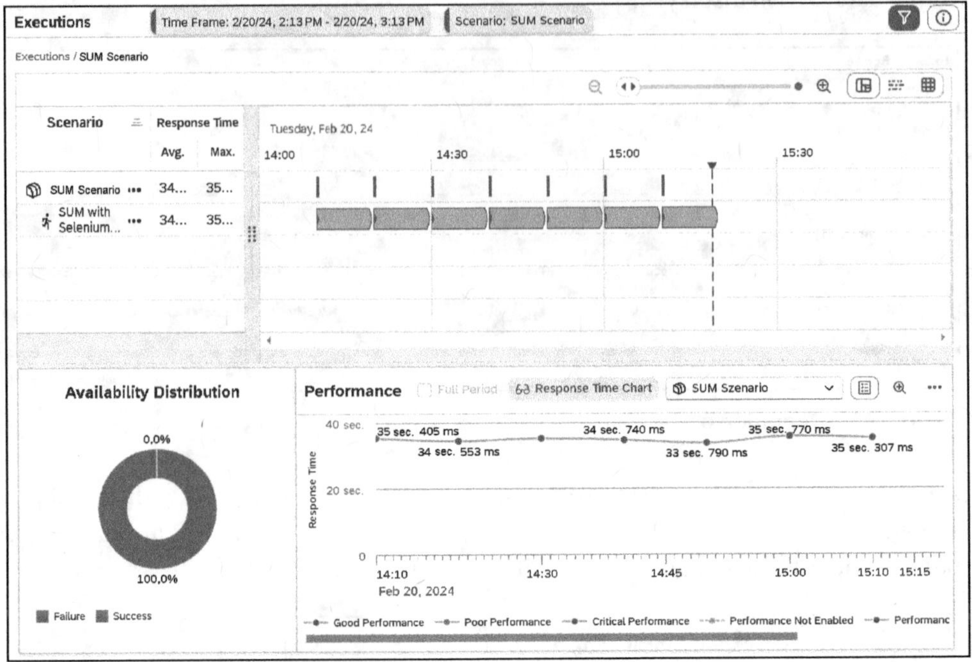

Figure 19.74 Script Execution Details

19.6 Job and Automation Monitoring

To show the configuration of job and automation monitoring, we'll use an SAP S/4HANA on-premise system as an example.

19.6.1 Checking the Prerequisites

The first step of the configuration leads to the Landscape Management app and then to the **Services & Systems** page in the left-hand menu (see Figure 19.75). On the **Supported Use Cases** tab, you can see that the job and automation monitoring scenario, including data collection via the PUSH method, has already been activated for the system (**Job & Automation Monitoring Use Case**, **Configuration Status** of **Active**). Furthermore, the data collection is successful (**Runtime Status** of **Success**). Due to the previous configuration of the push data provider in the system to be monitored, both were activated automatically.

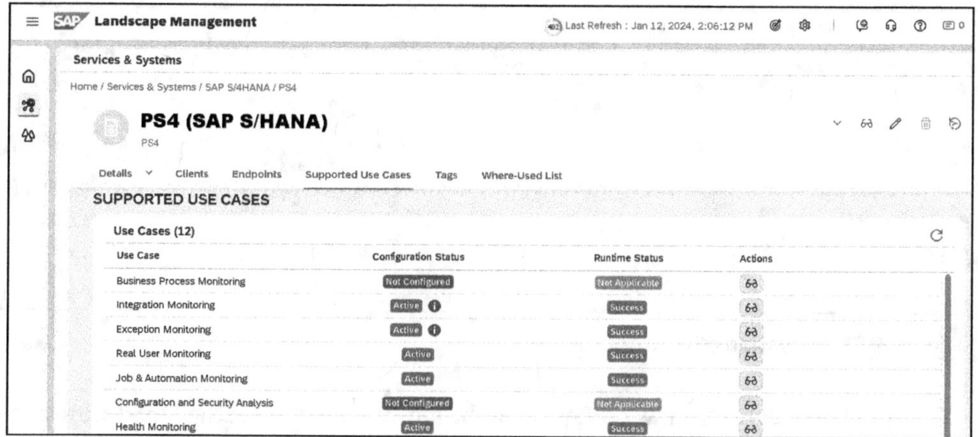

Figure 19.75 Connection Status in Job and Automation Monitoring

19.6.2 Initial Configuration

First, you need to access the monitoring scenario via the launchpad. This is again done via the **SAP Cloud ALM for Operations** group and the corresponding Job & Automation Monitoring app (see Figure 19.76).

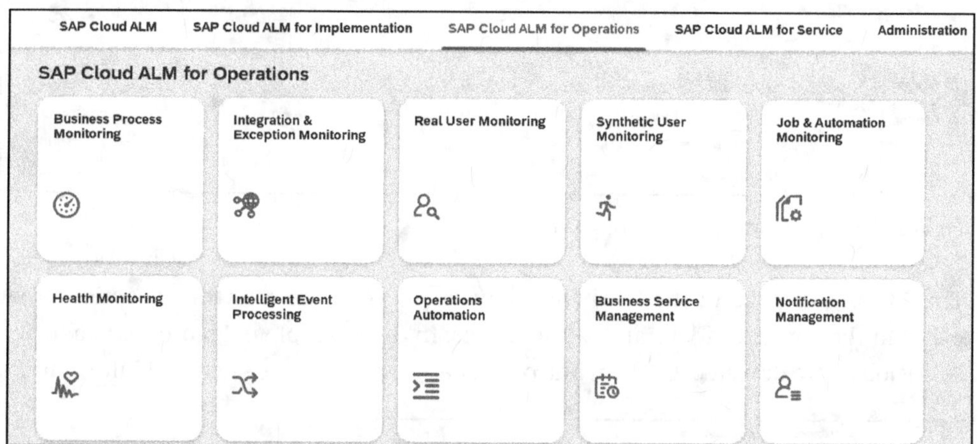

Figure 19.76 Access to Job and Automation Monitoring via the Launchpad Tile

After selecting the system via the scope selection, you can view the monitoring data directly; separate activation is not necessary, as this is a system that transmits the data using the push method. If you open the configuration via the gear icon and open the **Managed Components** item, you will see the system you have selected. Click on the **ON** button (see Figure 19.77), you can deactivate the data collection and thus the monitoring (the button then changes to **OFF**).

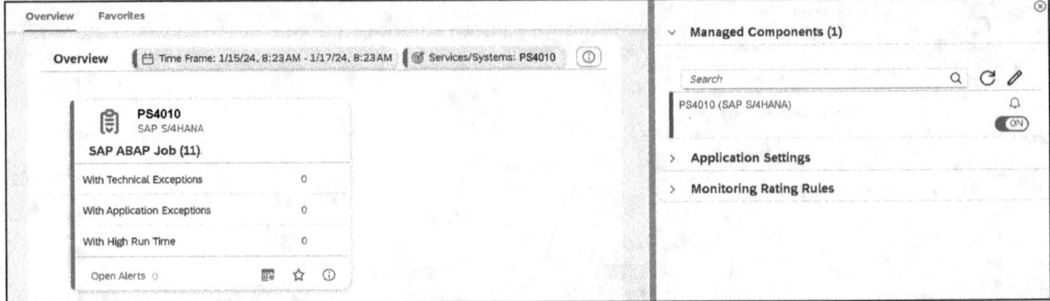

Figure 19.77 Activation of Monitoring

The monitoring data can now be viewed in the left-hand menu via the **Monitoring** page (see Figure 19.78). This view lists all jobs that are determined by SAP Cloud ALM for the selected component. Information on the **Execution Status**, **Application Status**, **Start Delay**, and **Runtime** is displayed with an icon. You can view information on the various points in time at which a job was executed by clicking on the job name.

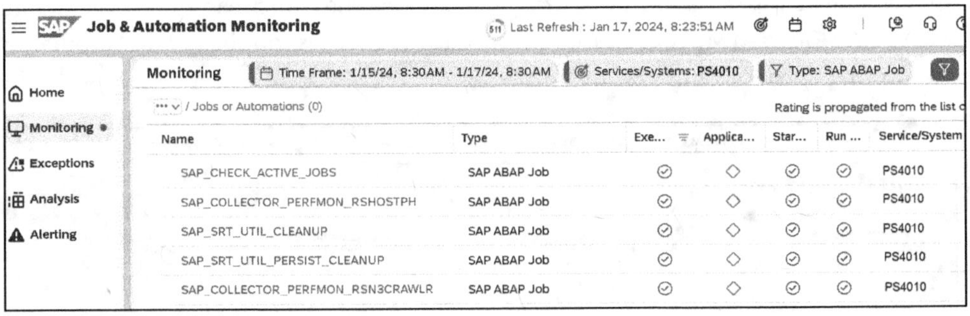

Job & Automation Monitoring — Last Refresh : Jan 17, 2024, 8:23:51 AM

Home | Monitoring | Exceptions | Analysis | Alerting

Monitoring — Time Frame: 1/15/24, 8:30AM - 1/17/24, 8:30AM — Services/Systems: PS4010 — Type: SAP ABAP Job

/ Jobs or Automations (0) — Rating is propagated from the list

Name	Type	Exe...	Applica...	Star...	Run ...	Service/System
SAP_CHECK_ACTIVE_JOBS	SAP ABAP Job					PS4010
SAP_COLLECTOR_PERFMON_RSHOSTPH	SAP ABAP Job					PS4010
SAP_SRT_UTIL_CLEANUP	SAP ABAP Job					PS4010
SAP_SRT_UTIL_PERSIST_CLEANUP	SAP ABAP Job					PS4010
SAP_COLLECTOR_PERFMON_RSN3CRAWLR	SAP ABAP Job					PS4010

Figure 19.78 View of the Monitoring Data

The **Exceptions** page in the left-hand menu gives you access to the exceptions that occur in the context of jobs and automations. In the example in Figure 19.79, aborted ABAP jobs were detected. Click on one of the **Categories** to access detailed information.

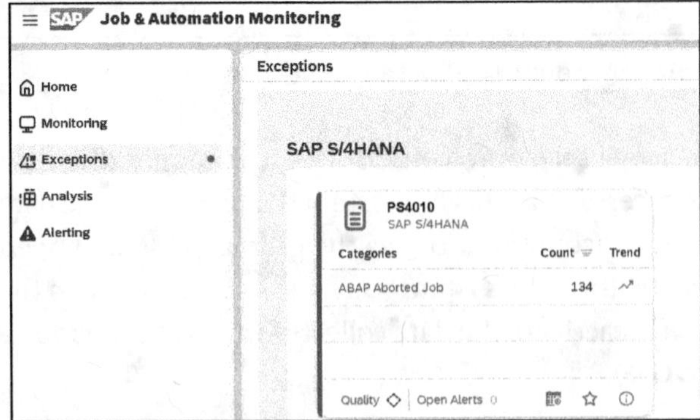

Figure 19.79 Access to Exceptions

The **Analysis** page, which can also be accessed from the left-hand menu, provides values for various metrics for jobs or automations. Here you can view trends for various status values and obtain information on the average and total runtimes of jobs (see Figure 19.80).

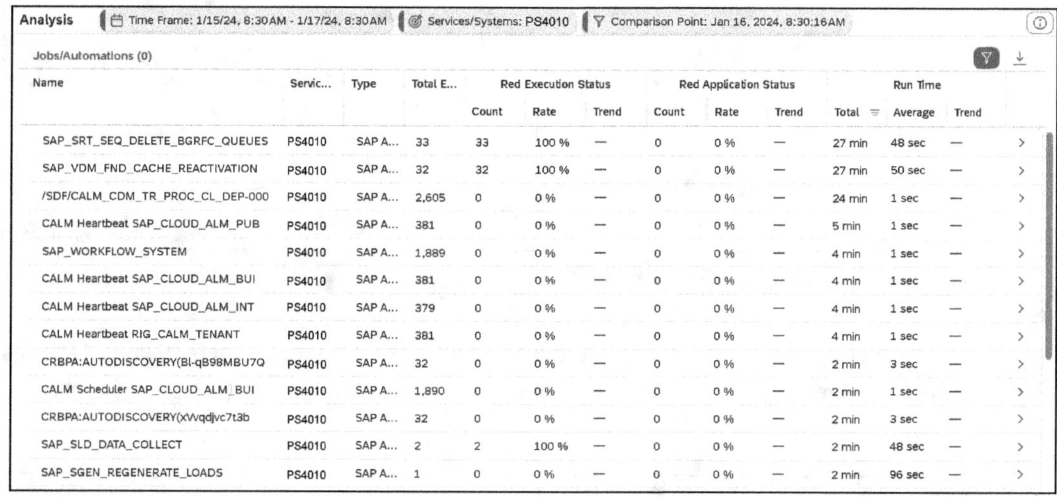

Analysis | Time Frame: 1/15/24, 8:30AM - 1/17/24, 8:30AM | Services/Systems: PS4010 | Comparison Point: Jan 16, 2024, 8:30:16AM

Jobs/Automations (0)

Name	Servic...	Type	Total E...	Red Execution Status			Red Application Status			Run Time			
				Count	Rate	Trend	Count	Rate	Trend	Total	Average	Trend	
SAP_SRT_SEQ_DELETE_BGRFC_QUEUES	PS4010	SAP A...	33	33	100 %	—	0	0 %	—	27 min	48 sec	—	>
SAP_VDM_FND_CACHE_REACTIVATION	PS4010	SAP A...	32	32	100 %	—	0	0 %	—	27 min	50 sec	—	>
/SDF/CALM_CDM_TR_PROC_CL_DEP-000	PS4010	SAP A...	2,605	0	0 %	—	0	0 %	—	24 min	1 sec	—	>
CALM Heartbeat SAP_CLOUD_ALM_PUB	PS4010	SAP A...	381	0	0 %	—	0	0 %	—	5 min	1 sec	—	>
SAP_WORKFLOW_SYSTEM	PS4010	SAP A...	1,889	0	0 %	—	0	0 %	—	4 min	1 sec	—	>
CALM Heartbeat SAP_CLOUD_ALM_BUI	PS4010	SAP A...	381	0	0 %	—	0	0 %	—	4 min	1 sec	—	>
CALM Heartbeat SAP_CLOUD_ALM_INT	PS4010	SAP A...	379	0	0 %	—	0	0 %	—	4 min	1 sec	—	>
CALM Heartbeat RIG_CALM_TENANT	PS4010	SAP A...	381	0	0 %	—	0	0 %	—	4 min	1 sec	—	>
CRBPA:AUTODISCOVERY(BI-qB98MBU7Q	PS4010	SAP A...	32	0	0 %	—	0	0 %	—	2 min	3 sec	—	>
CALM Scheduler SAP_CLOUD_ALM_BUI	PS4010	SAP A...	1,890	0	0 %	—	0	0 %	—	2 min	1 sec	—	>
CRBPA:AUTODISCOVERY(xWqdjvc7t3b	PS4010	SAP A...	32	0	0 %	—	0	0 %	—	2 min	3 sec	—	>
SAP_SLD_DATA_COLLECT	PS4010	SAP A...	2	2	100 %	—	0	0 %	—	2 min	48 sec	—	>
SAP_SGEN_REGENERATE_LOADS	PS4010	SAP A...	1	0	0 %	—	0	0 %	—	2 min	96 sec	—	>

Figure 19.80 Analysis of Trends and Maturities

19.6.3 Specific Configuration

Use the pencil button to access the configuration of events. In the **Configuration for Services/Systems** menu, click on the system name (see Figure 19.81) to access the menu for the actual creation of events.

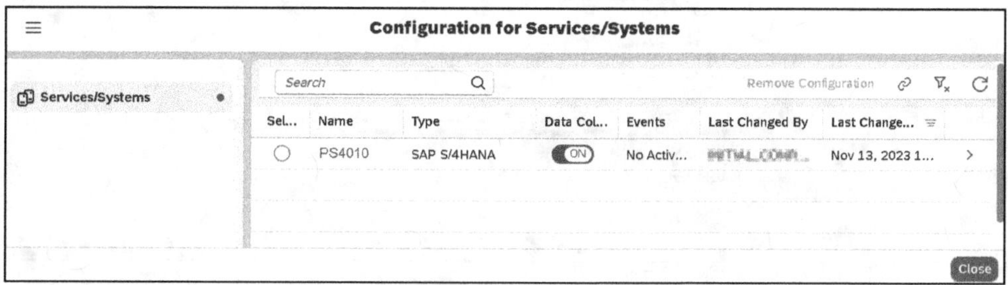

Figure 19.81 System Selection for Events

No events are listed initially. You must add the events that are already preconfigured using the plus symbol [+] (see Figure 19.82).

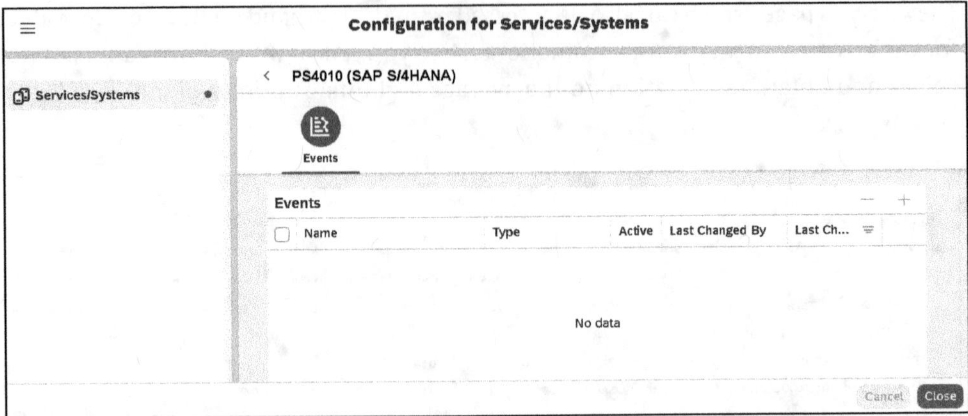

Figure 19.82 Initial View of the Events

In the menu that opens, select the event type in the first area labeled **Event Settings** select the **Event Type**—a dropdown menu with the available events opens (see Figure 19.83).

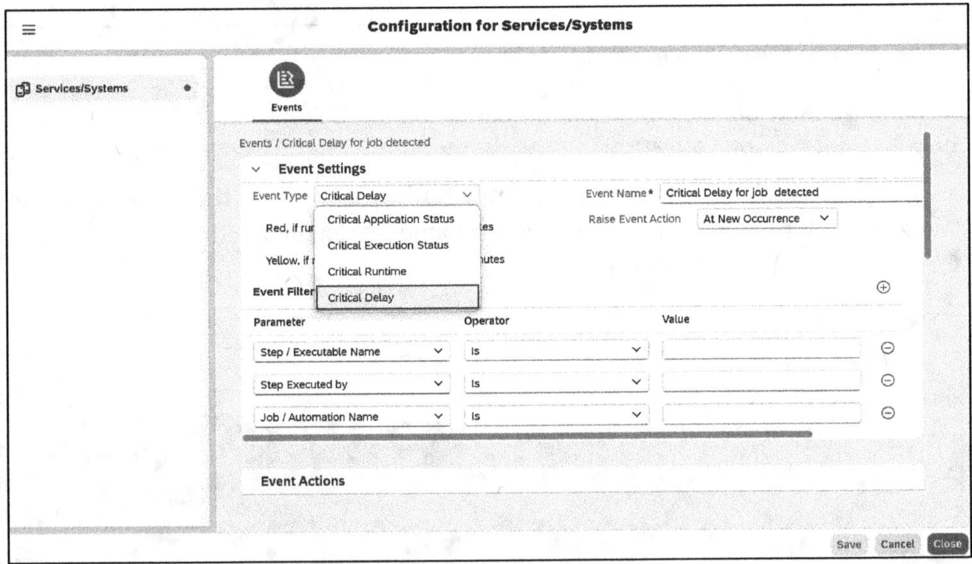

Figure 19.83 Selecting the Event Type

Threshold values can also be defined in the **Event Settings**. In the example in Figure 19.84, a delay for a job of longer than 1 minute is given a yellow rating, while a delay of longer than 2 minutes is given a red rating. In the **Event Actions** area, you can finally activate the alert for the event by setting the switch in the **Create Alert** line to **ON**.

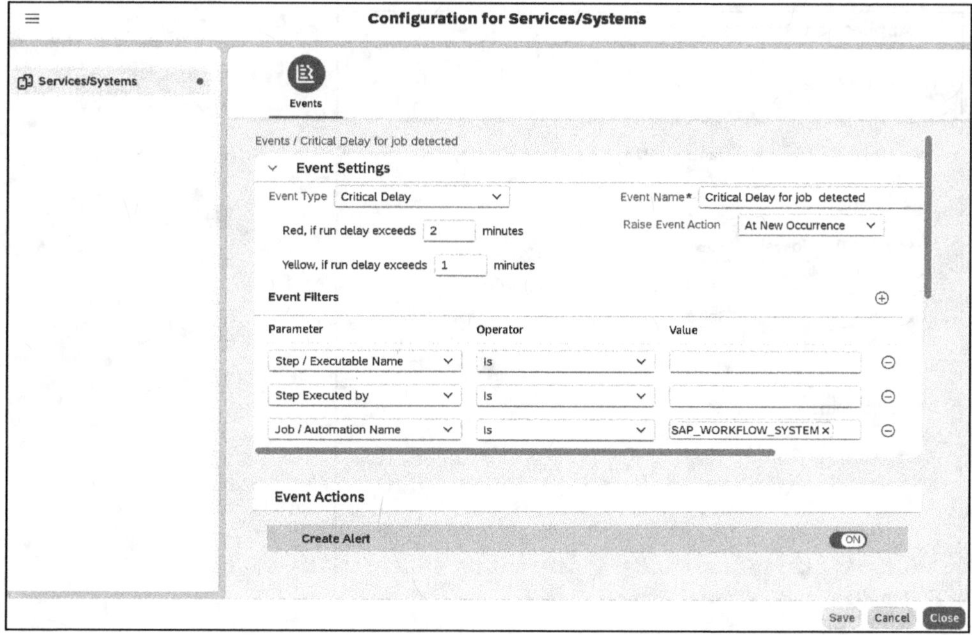

Figure 19.84 Configuration of an Event

After you save the configuration using the **Save** button, the created event is displayed in a list, as shown in Figure 19.85. If you want to delete the event, you must select it and click on the minus [—] icon.

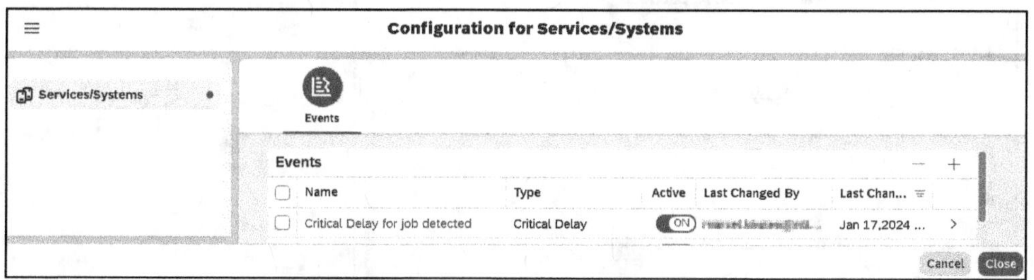

Figure 19.85 View of a Created Event

Via the gear icon [⚙], you can view **Application Settings** (**Housekeeping**, **Aggregation**, **Alert**, **Data Volume**) and **Monitoring Rating Rules** (**Execution Status**, **Application Status**, **Start Delay**, and **Run Time**) and adjust some of them to suit your needs (see Figure 19.86).

19

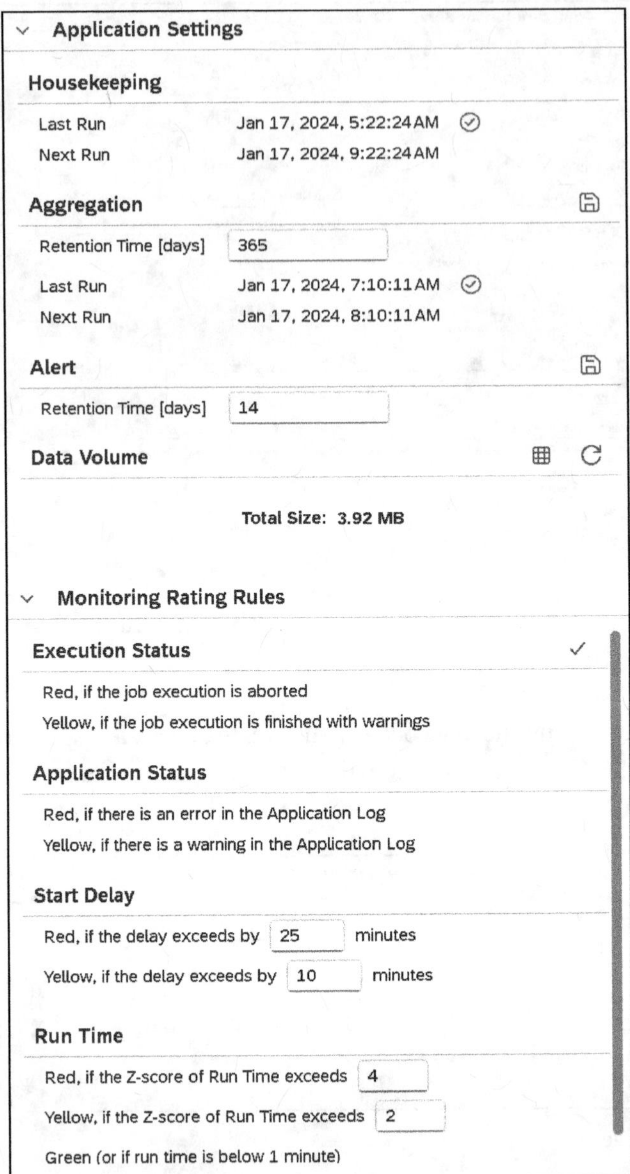

Figure 19.86 Application Settings for Job and Automation Monitoring

19.7 Health Monitoring

We will demonstrate the configuration of health monitoring again using the example of an SAP S/4HANA on-premise system and SAP Integration Suite (Cloud Integration) in Cloud Foundry.

19.7.1 Checking the Prerequisites

The connection is checked via the Landscape Management app. There you will find the **Services & Systems** page on the left. If you select this and select the system type in the scope selection, you will see the on-premise systems of this type that are available in SAP Cloud ALM—in our example, we want to set up monitoring for the system with the SID **PS4**, as shown in Figure 19.87. On the **Supported Use Cases** tab, you can see that the monitoring scenario **Health Monitoring** including data collection via the push method has been activated for the system (**Configuration Status** of **Active**). The data collection is also successful (**Runtime Status** of **Success**). Due to the previous configuration of the push data provider in the system to be monitored, both were activated automatically.

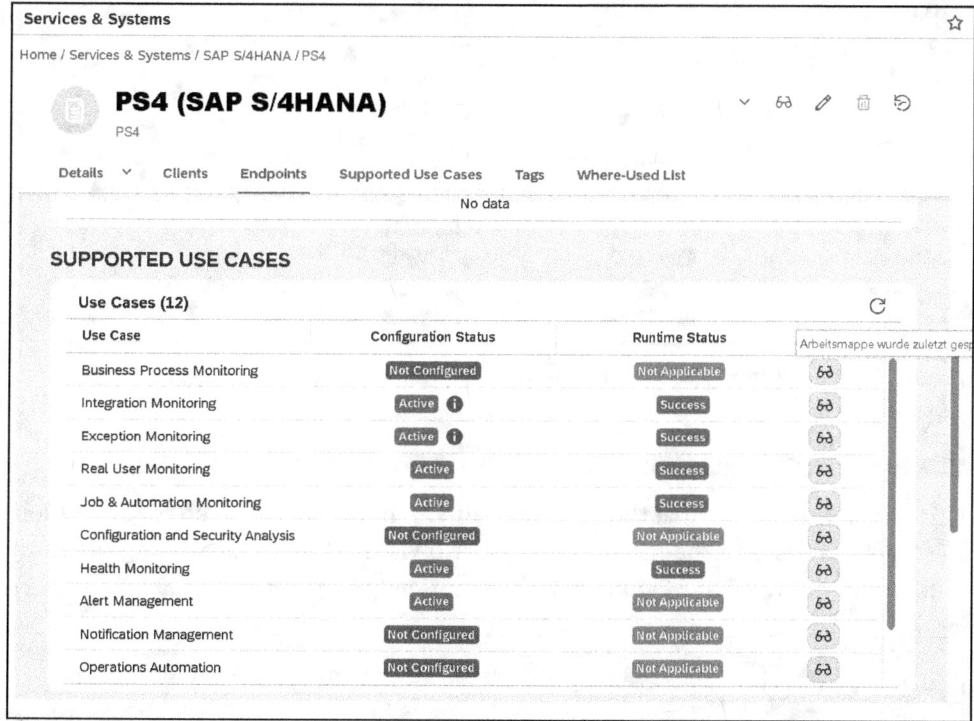

Figure 19.87 Checking the Connection of the On-Premise System

To integrate the cloud service in the example on health monitoring, you must search for the cloud service via the **Services & Systems** menu item. On the **Endpoints** tab, an entry with a green status must be visible (see Figure 19.88). Furthermore, **Health Monitoring** must be listed on the **Supported Use Cases** tab, while the configuration status must be set to **Active** and the runtime status to **Success**. If no endpoint is defined, you must create one (see Section 19.3.2).

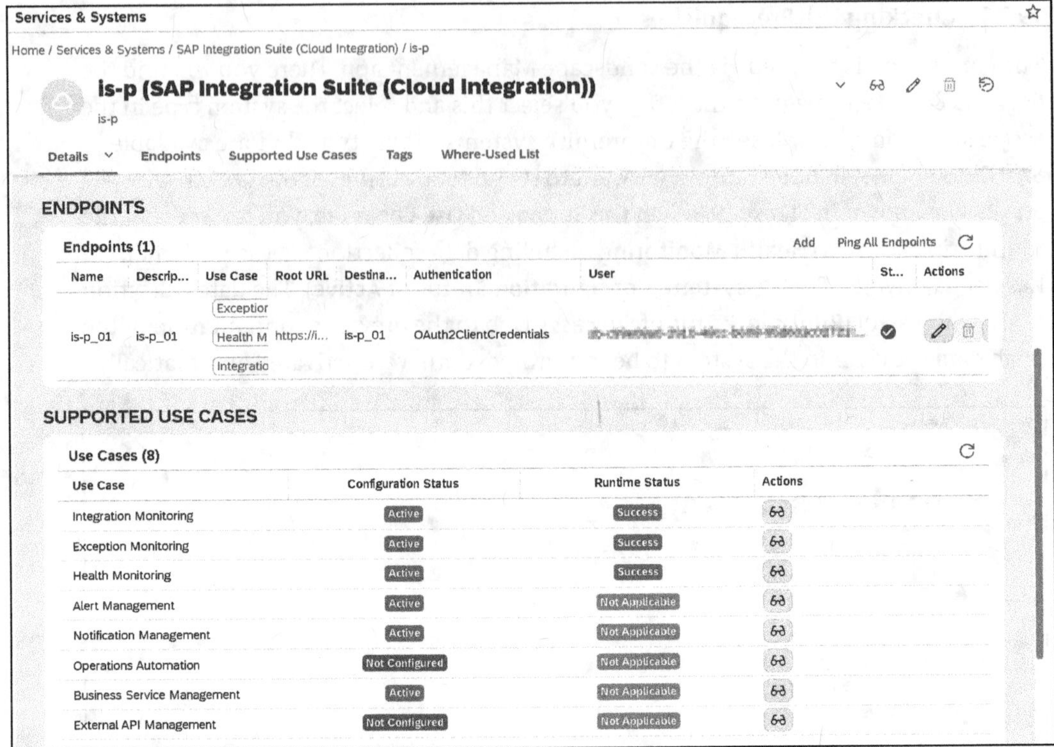

Figure 19.88 Checking the Connection to the Cloud Service

19.7.2 Initial Configuration

Once you have ensured that the data for health monitoring is transferred to SAP Cloud ALM, you navigate to the monitoring scenario via the **Health Monitoring** tile in the launchpad within the **SAP Cloud ALM for Operations** group (see Figure 19.89).

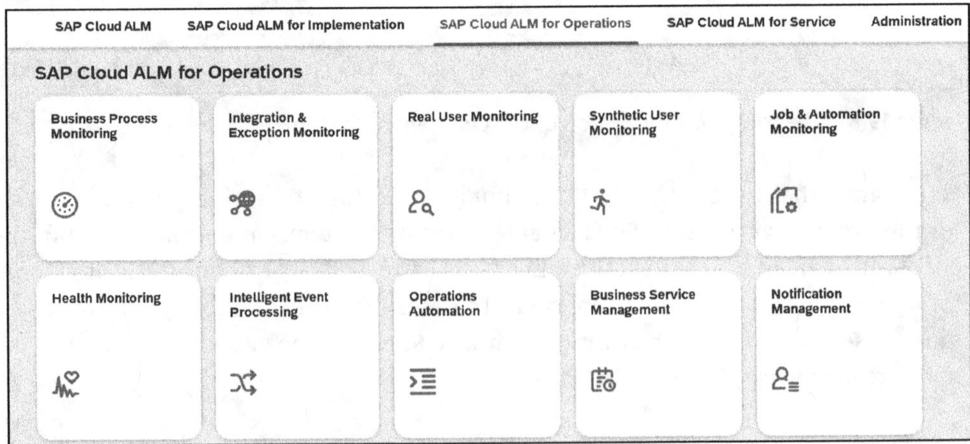

Figure 19.89 Calling up Health Monitoring via the Launchpad Tile

You will return to the **Home** page. There you will initially see a list of all cloud services and on-premise systems that have been recognized by SAP Cloud ALM (see Figure 19.90). Systems and cloud services for which data collection is active are already initially integrated into health monitoring without the need for configuration—this can be seen in Figure 19.90 on the first two tiles.

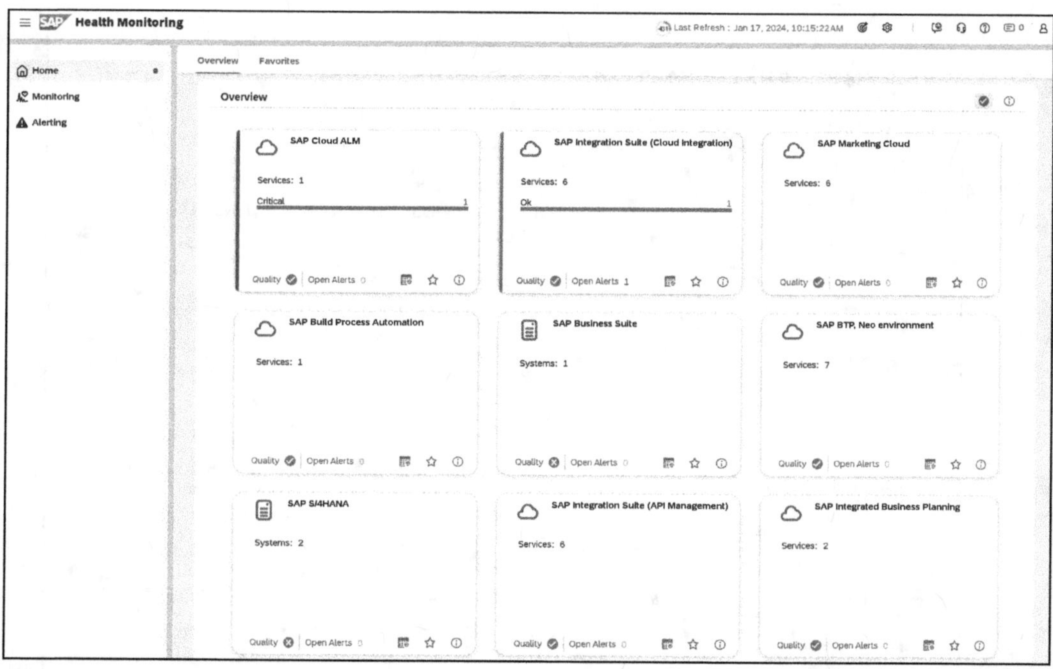

Figure 19.90 Start Page of the Health Monitoring

If you click on one of the tiles that already contains status information, you switch to the **Monitoring** page; this corresponds to the second menu item in the left-hand menu. In the example, shown in Figure 19.91, all SAP S/4HANA systems or their clients are listed, which are of type SAP S/4HANA and are connected to SAP Cloud ALM. The symbol in the **Data Collection** column indicates for which of the two clients it is active.

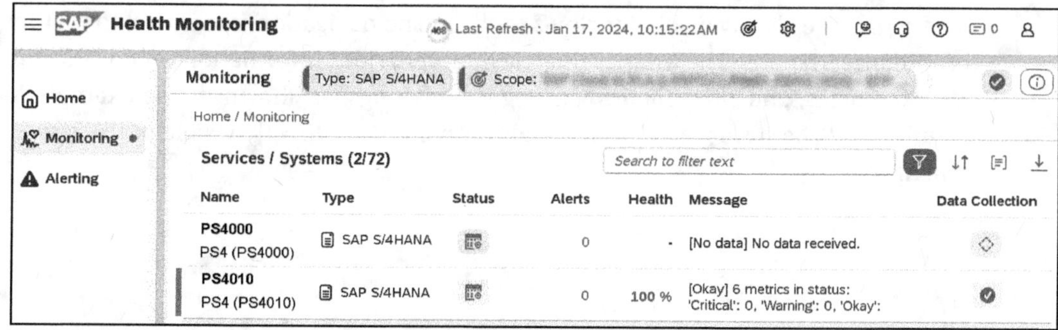

Figure 19.91 Monitoring of the On-Premise System

When you click on a system for which data is being collected, you get access to various key figures (to be found on the **Metrics Overview** tab, shown in Figure 19.92). Monitoring information on the ABAP system, the SAP HANA database, ICM, and many other components is available immediately and without any further configuration steps.

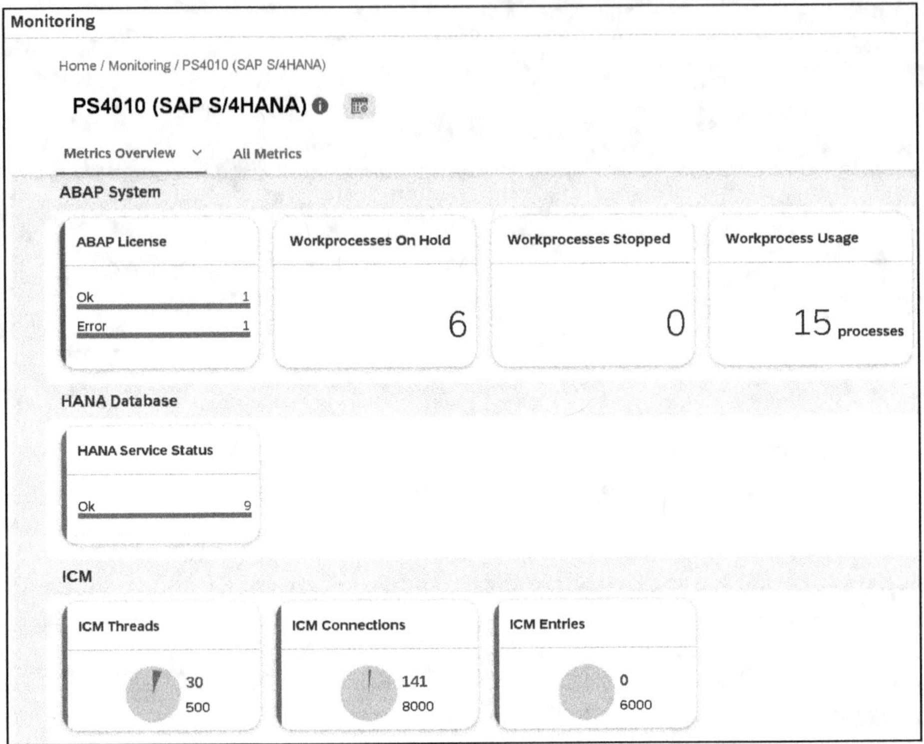

Figure 19.92 Overview of Key Figures for the On-Premise System

If you select the **All Metrics** tab, you can access a table view that provides detailed information on all key figures, whereby clicking on the info icon in the **Details** column allows you to access additional information that supplements the information in the table (see Figure 19.93).

Key figures are also available for cloud services. The navigation is identical to that for on-premise systems. The **Metrics Overview** tab provides information on **Certificates**, **JMS Resources**, and **JMS Queues**, among other things, as shown in the example in Figure 19.94. The status of the individual components is directly visible in the respective tiles.

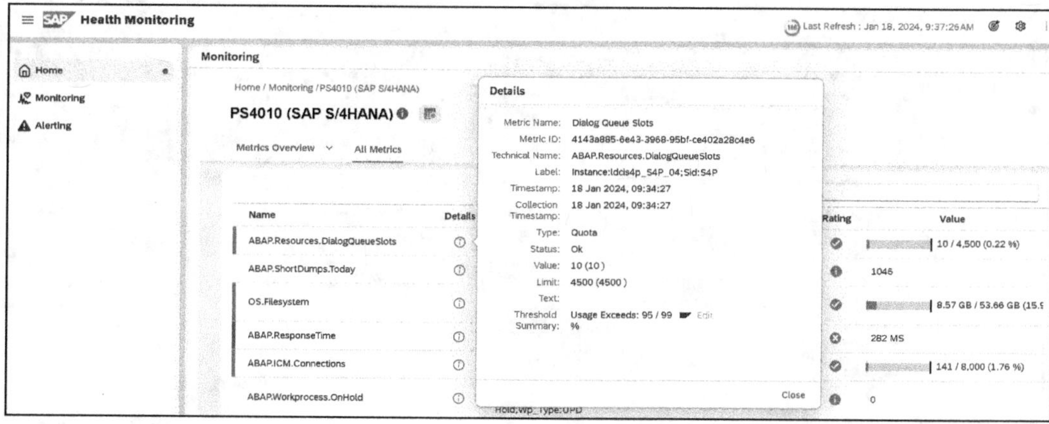

Figure 19.93 Key Figure Details for the On-Premise System

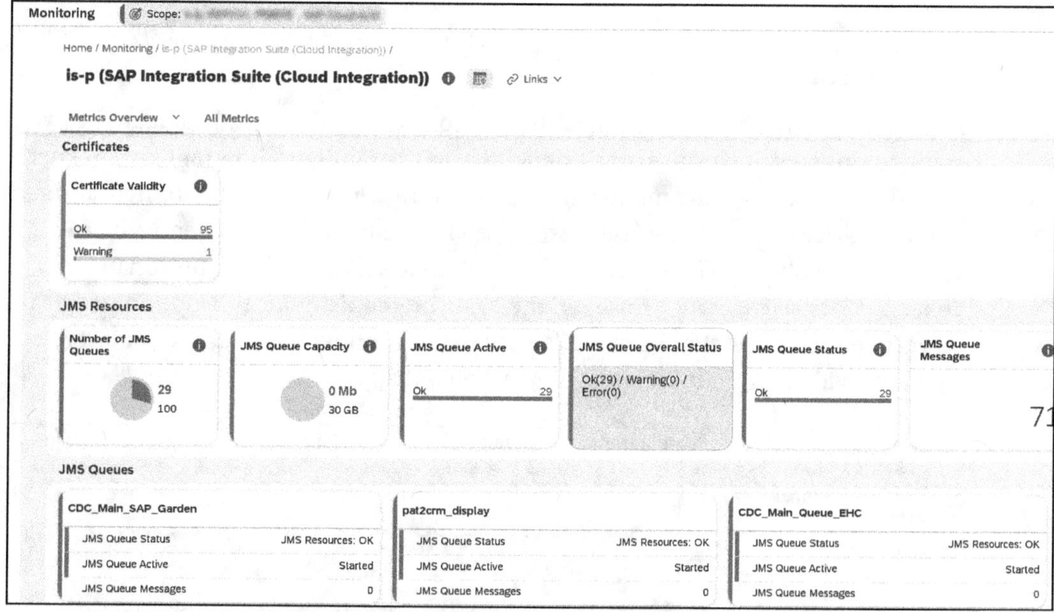

Figure 19.94 Overview of Key Figures for the Cloud Service

You can call up detailed information for an individual tile by clicking on it. In the example in Figure 19.95, details regarding the **Certificate Validity** (the remaining days of validity) are immediately visible. Further details can be called up via the familiar ⓘ icon. Historical data can also be viewed via the icon in the **History** column.

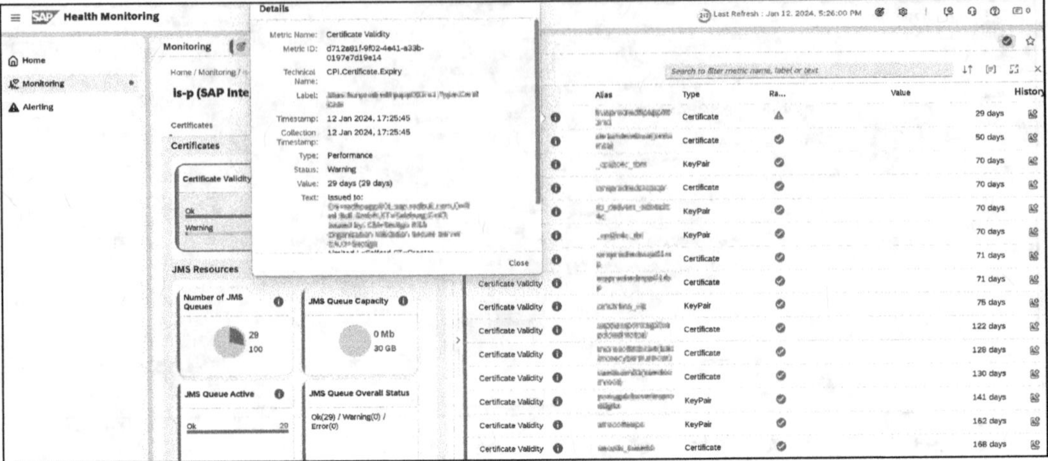

Figure 19.95 Key Figure Details for the Cloud Service

19.7.3 Specific Configuration

The initial configuration of the health monitoring can of course be adapted to your own requirements (via the gear symbol [gear icon]). In the **Configuration for Systems** menu, first select in the left-hand menu whether you want to customize the configuration for a cloud service or an on-premise system (see Figure 19.96). In this menu, you can also activate data collection for other systems and cloud services (via the button in the **Data Collection Active** column). However, the basic prerequisite for this option is that the respective system or cloud service is successfully connected to SAP Cloud ALM and data is provided—see the Landscape Management app.

Configuration for Systems

Services
Systems

Systems (2/2)

Live search

System Name	System Type	Active Al...	Data Col...	Last Data Collection	Data Collector St...
[blurred]	SAP Business Suite	0	OFF	18 Jan 2024, 10:32:13	
SP4010	SAP S/4HANA	0	ON	18 Jan 2024, 10:32:13	

Figure 19.96 Data Collection for Health Monitoring

If you select a system/cloud service for which data collection is active, you can define threshold values for individual metrics, as shown in Figure 19.97. There is a separate menu item for this, which you can use to call up a list of the available **Metrics**.

If you click directly on a metric name, you will be taken to a menu that allows you to define threshold values for the selected metric. In the example in Figure 19.98, it is defined for the metric **Batch Jobs Aborted** that a warning is issued as soon as more than five batch jobs are aborted and that a critical situation exists as soon as more than 10 batch jobs are aborted.

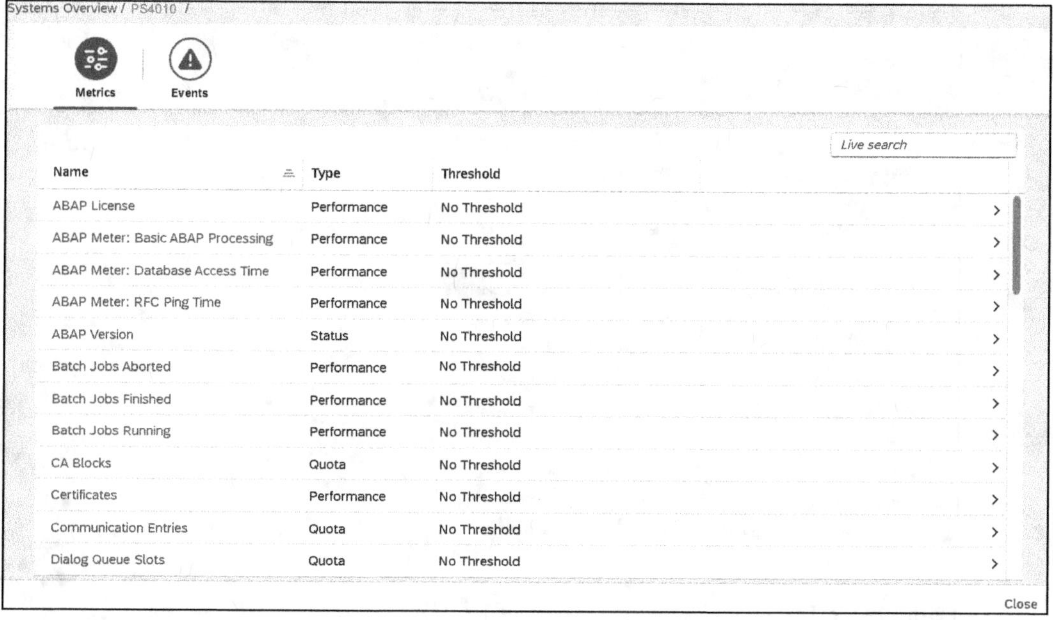

Figure 19.97 Metrics for Health Monitoring

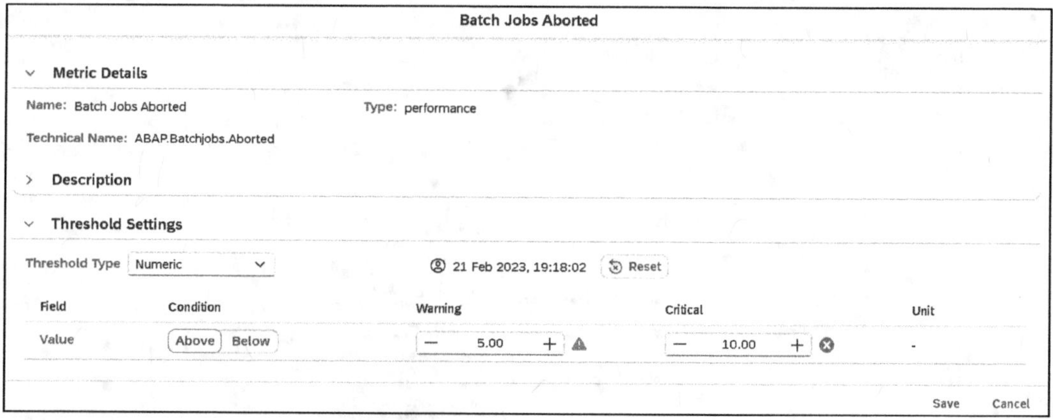

Figure 19.98 Define Threshold Values

After you save the configuration of the threshold value using the **Save** button, the threshold value also appears in the metrics menu. For the example configured here, this is shown in Figure 19.99.

The second menu item, which allows adjustments to be made to the initial configuration, is located directly to the right of the **Metrics** menu item and is called **Events**, as shown in Figure 19.100. Here you can again configure alerts and further actions for events.

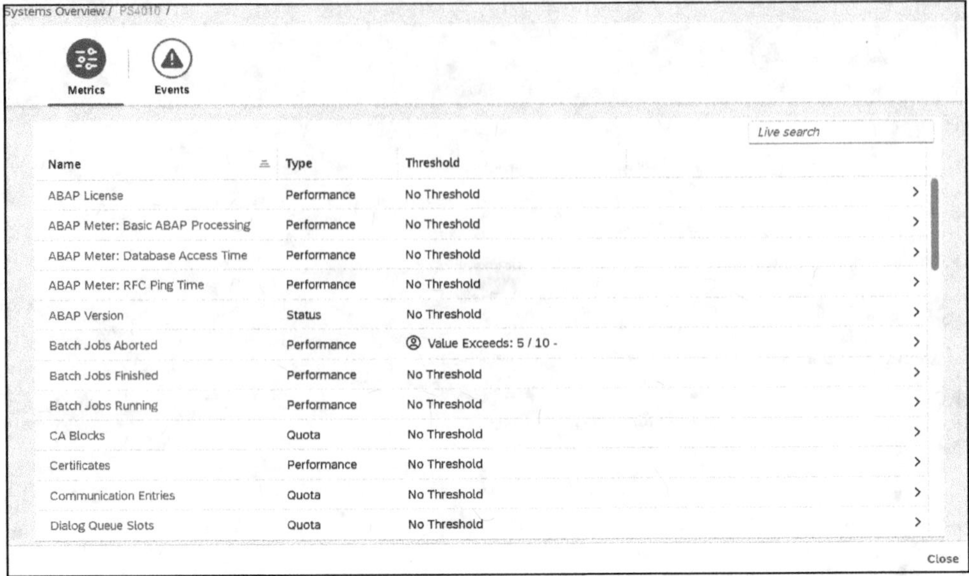

Figure 19.99 Threshold Value Set in the Metrics

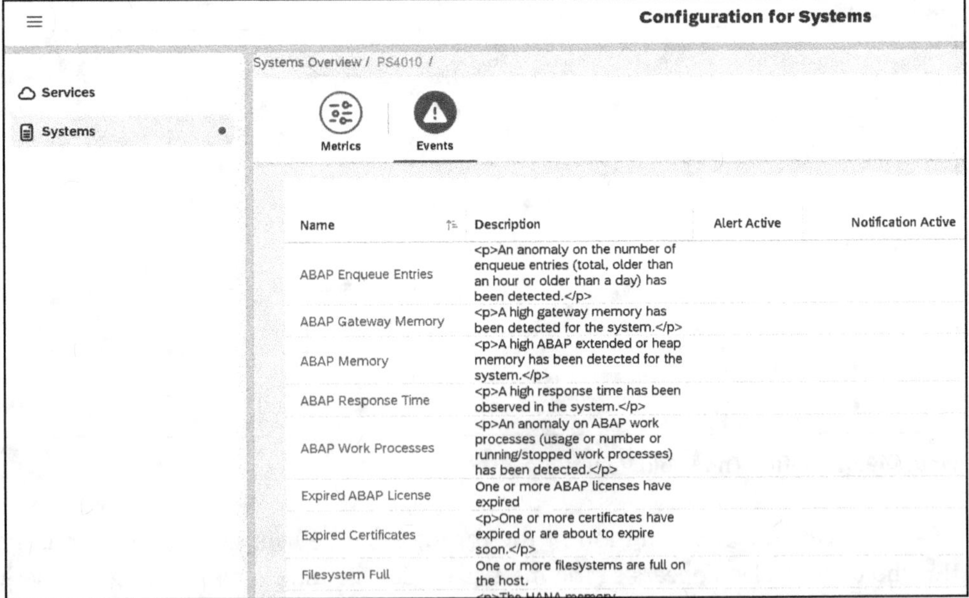

Figure 19.100 Configuring Events and Alerts

[!]

Configuration of Events

The preconfigured events cannot be customized as in the other monitoring scenarios—the event name and event description cannot be changed, and no parameters or filters are available.

Once you have selected one of the events, you can set the alarm to **ON** using the familiar switch and thus activate it, as shown in Figure 19.101. Once the settings have been successfully saved using the **Save** button, the alarm for the respective event is active.

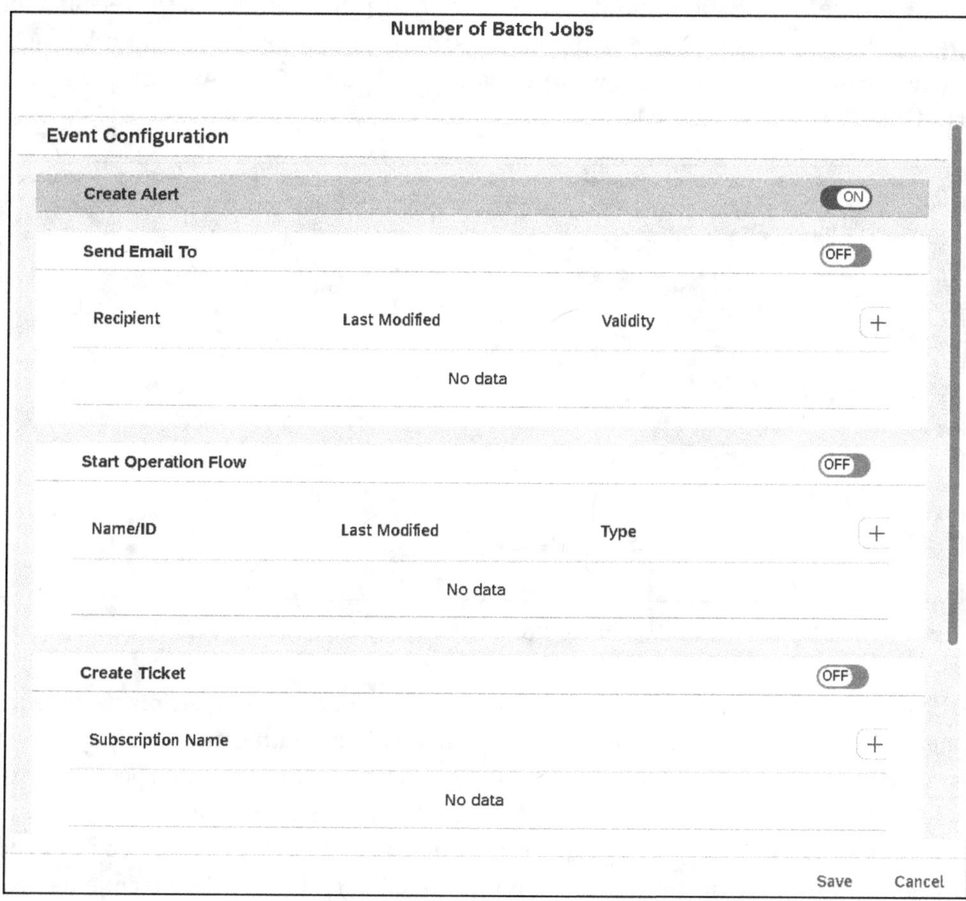

Figure 19.101 Activation of the Alarm for an Event

19.8 Email Notifications and Ticket Creation

Here we will show you how to configure email notifications as an event action. We will also briefly explain how to activate the automatic creation of tickets.

19.8.1 Set Up Email Event Action

SAP Cloud ALM ensures that only confirmed users receive emails and that the regulations of the European General Data Protection Regulation (GDPR) are complied with. As part of this, every newly created user in SAP Cloud ALM receives an email notification. This serves to verify the email address and to ensure that the person also wants to receive notifications. Furthermore, the email address is automatically added to the possible recipients of notifications in SAP Cloud ALM.

In Chapter 18, Section 18.2.2, email notifications were mentioned as event actions for the first time. Before such a notification to a person can be activated, the recipient of the emails must be stored in SAP Cloud ALM. This is done in the Notification Management app, which you can access via a separate tile in the **SAP Cloud ALM for Operations** launchpad group (see Figure 19.102).

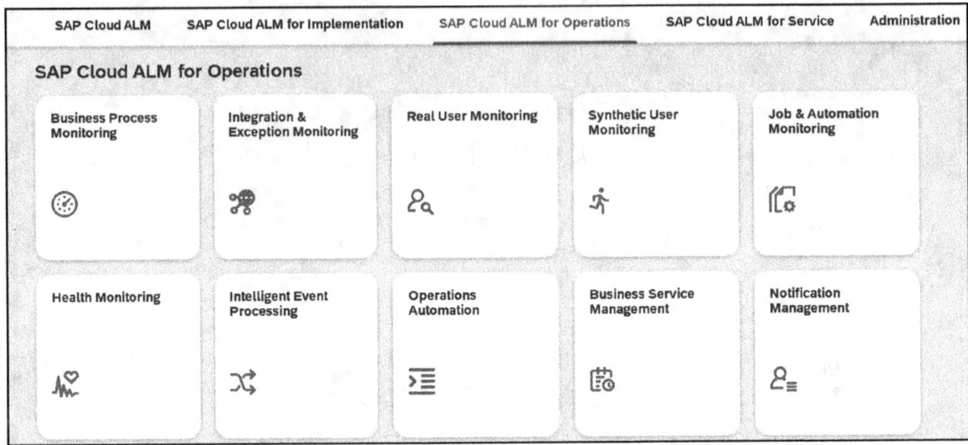

Figure 19.102 Access Tile for Managing Notifications

There is exactly one page available to you in the left-hand menu: **Recipient Management**. On this page, you can view the recipients of email notifications already stored and information on their respective status. This includes whether the respective email address or person has already been verified or whether this is still pending, and you can view the **Validity** of the recipients (shown in Figure 19.103). At this point, you can add recipients using the buttons in the top right-hand corner [+], delete recipients [ ], extend their validity [ ], and send another invitation to verify the email address [ ].

Notification Management

Recipient Management

Recipients (4)

Recipient	Status	Validity	Last Modified By	Last Modified At
[illegible]	Verified	Apr 8, 2024, 3:17:03 PM	[illegible]	Oct 11, 2023, 3:11:41 PM
[illegible]	Verified	Apr 8, 2024, 3:12:05 PM	[illegible]	Oct 11, 2023, 3:11:26 PM

Figure 19.103 Recipient Management for Email Notifications

If you would like to add a new recipient, you must click on the [+] button and enter the recipient's email address in the following window and then save it using the **Save** button (see Figure 19.104). If you want to add several recipients, you can do it by separating the individual email addresses with a comma.

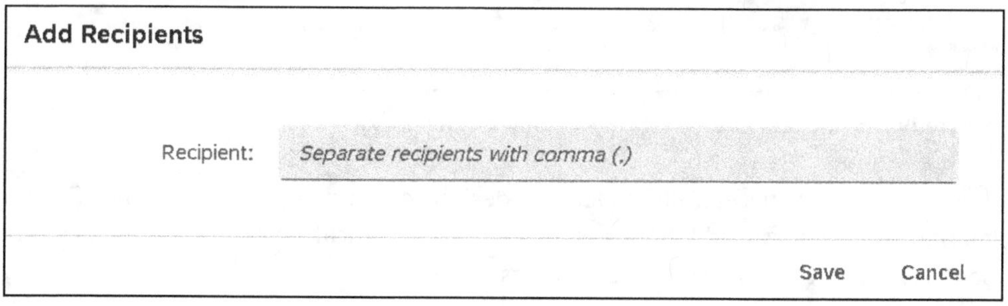

Add Recipients

Recipient: *Separate recipients with comma (,)*

Save Cancel

Figure 19.104 Add Receiver

Once the recipient has been created, he/she/they receives an email informing them that the email address has been stored as the recipient address in SAP Cloud ALM (shown in Figure 19.105).

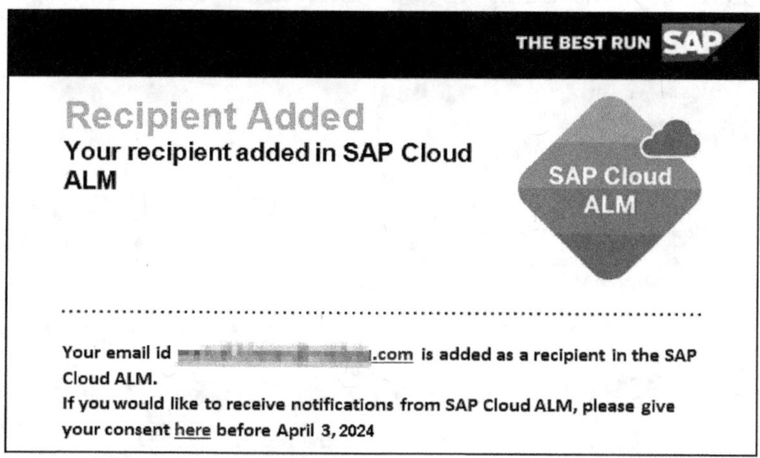

Figure 19.105 Welcome Email

This email also contains a link that the recipient must use to give their consent to receive email notifications. This consent must be given within a limited period of time, or the offer expired.

Once the user has given their consent, they are listed as a verified email recipient in the recipient management list, recognizable by the entry in the **Status** column. The recipient created in the example is listed in the first line in Figure 19.106. If the potential recipient does not give their consent within the specified period, you can send a new invitation using the aforementioned paper flyer button, the outer button to the right of the search field. All you have to do is select the recipient using the checkbox and then click on the button.

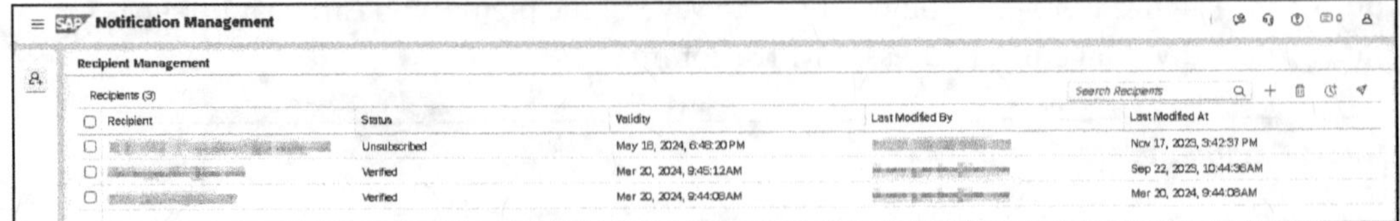

Figure 19.106 Created Users in the Recipient Administration

If you have now successfully added the desired email recipient via the notification management, you can add them as a recipient of a notification email. This email will be sent as soon as the respective event occurs.

For the configuration, you must access the event actions, which you can find in the respective monitoring scenarios, among other things. In the **Event Actions** you will find the **Send Email To** item, which can be seen in Figure 19.107. After clicking on the plus symbol [+], you can now select the recipient from a list. This list is based on the recipients that have been stored in the management of notifications. Once successfully added, this person is visible in the list. Use the **ON** button to activate the sending of emails for the event.

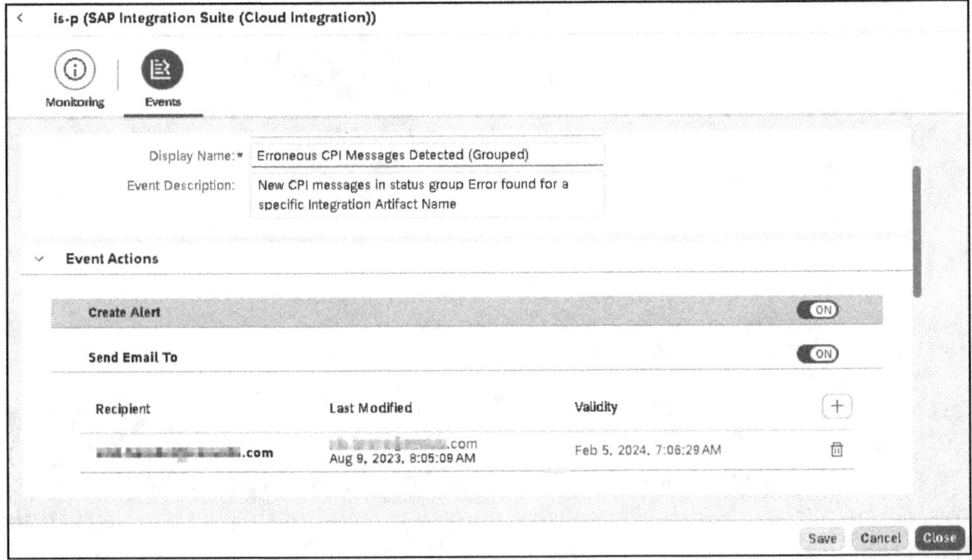

Figure 19.107 Configuration of the Event Action

19.8.2 Set Up Ticket Event Action

Here we briefly describe how to activate the automatic creation of ServiceNow tickets as an event action. The automatic creation of tickets in ServiceNow is configured in the same place in SAP Cloud ALM where you can also set up the email notifications: in the **Events** menu and in the **Event Actions** area. Here you will find the **Create Ticket** item, shown in Figure 19.108. The plus symbol [+] gives you access to the subscriptions that you can manage or create via the External API Management app. Once you have selected the subscription to be used for ticket creation in ServiceNow, it will appear under **Subscription Name**. Use the familiar **ON** button to activate the automatic ticket creation in ServiceNow as an event action for the event.

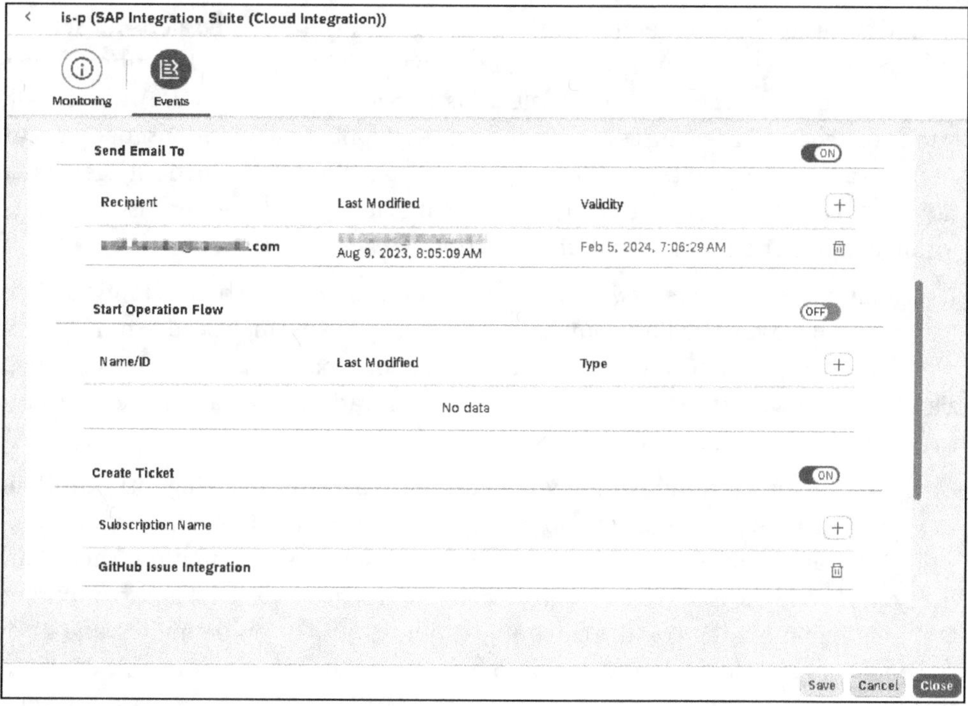

Figure 19.108 Create Ticket Event Action

19.9 Notes for the Operating Phase

In this section, we give you some tips on how to use SAP Cloud ALM for the operational phase and explain how SAP Cloud ALM and SAP Focused Run can be used in combination.

19.9.1 Optimization the Configuration

If you have initially configured SAP Cloud ALM to detect events to ensure and support the operation of your applications in the best possible way, you will find that the existing configuration should be continuously optimized.

If your experience with a particular application is still limited and you find it difficult to distinguish between critical and non-critical events, your threshold values are likely set too sensitively. This has the effect of generating too many events, alerts, and ultimately tickets that cannot be processed adequately. In the flood of alerts and tickets, those that actually mark a critical situation are then overlooked. To avoid this, it is essential that you observe the monitoring behavior; check the events, alerts, and tickets generated for their meaningfulness; and, if necessary, adjust the configuration of the monitoring from time to time. With regard to the threshold values, you need to question whether they are set too sensitively or perhaps too optimistically. You must examine the configured events to see whether they are relevant for you. The activated alerts must be set in such a way that they actually identify events that are to be regarded as critical. It does not make sense to create as many alerts as possible so that you are theoretically informed about every event, as this means you lose sight of the situations that are actually critical—at this point, less is more. The same applies to automatically generated tickets. Tickets should only be created as a result of an event if they are also processed within a short period of time in order to rectify the cause of the event.

The result of the critical review may be an optimization of the existing configuration. This may require you to adjust some threshold values. Events may need to be redefined or deactivated completely. For some of the events, it may make more sense to only alert them or forward them to experts by email notification, whereas other events need to lead directly to a ticket.

Furthermore, you must integrate new applications into the monitoring with SAP Cloud ALM as soon as they are put into operation. As with the initial planning, you should write down all monitoring requirements and the resulting configuration. This means that you expand the initial concept to meet the new requirements. It is also important that you document the completed configuration so that it can be understood by all stakeholders. This documentation is also important if events occur in an application that have not been identified and addressed by SAP Cloud ALM. In this case, the documentation provides a quick overview of the extent to which the configuration must or can be adjusted.

Changes to the existing monitoring configuration should not be made arbitrarily, but it should be planned—this can be done on a quarterly basis. In particularly critical cases, when it is certain that critical events will not be detected on a regular basis, action should of course be taken immediately. However, you must always prevent the configuration from changing several times at short intervals. To avoid a certain arbitrariness, it is also advisable to implement the adjustments in several steps. The first step is to define the threshold values and activate events and alerts. After a validation phase, in which it is checked whether the existing setup meets the set requirements, it is rolled out, i.e., the stakeholder groups are defined in the form of recipients for email notifications, and the automatic generation of tickets is activated.

In summary, it can be said that monitoring is an ongoing process that needs to be scrutinized regularly and requires continuous adjustments to the configuration. Good planning and documentation are helpful here.

19.9.2 SAP Focused Run Reverse Proxy

SAP offers with the *SAP Focused Run reverse proxy* a way to implement push data collection between SAP cloud services and SAP Focused Run without exposing SAP Focused Run to a security risk through this method. We will briefly explain the functionality here. The specific configuration steps can be accessed via the expert portal for SAP Cloud ALM for operations, which can be found at *https://support.sap.com/en/alm/sap-focused-run/expert-portal/managed-systems-maintenance-guide/frun-reverse-proxy.html.*

The functionality is currently available for the following SAP cloud services:

- SAP S/4HANA Cloud
- SAP Integrated Business Planning for Supply Chain
- SAP Marketing Cloud
- SAP S/4HANA Cloud Private Edition
- SAP SuccessFactors

You can use SAP Focused Run to monitor your hybrid landscapes. One challenge is that SAP Focused Run is operated separately from the internet and protected by a firewall. However, many of the SAP cloud services offer *push data collection*. This means that the cloud service itself triggers the data collection and automatically transmits the collected data to the respective monitoring platform. The advantage of this is that data is only transmitted when it is actually generated. Setting up data collection and transmission is much easier than with pull data collection, for example, where the monitoring platform initiates the data collection itself.

If you want to establish a push data collection between an SAP cloud service and SAP Focused Run, you would have to open the firewall to the internet, which is a risk in terms of security. To avoid this security risk, SAP Cloud ALM can be used together with a cloud connector. Both components together serve as an SAP Focused Run reverse proxy and enable push data collection from the SAP cloud service to SAP Focused Run without compromising on security. In this scenario, SAP Cloud ALM acts as the component that forwards the data from the SAP cloud service to the cloud connector using the SAP Connectivity service. The cloud connector transmits the data to SAP Focused Run. SAP Cloud ALM and the SAP cloud service are located in front of the firewall on the internet. The cloud connector and SAP Focused Run are located behind the firewall, outside the Internet. The interaction of the individual components is shown in Figure 19.109.

To be able to use the SAP Focused Run reverse proxy functionality, you must have at least one SAP Focused Run 3.0 with FP 03 available.

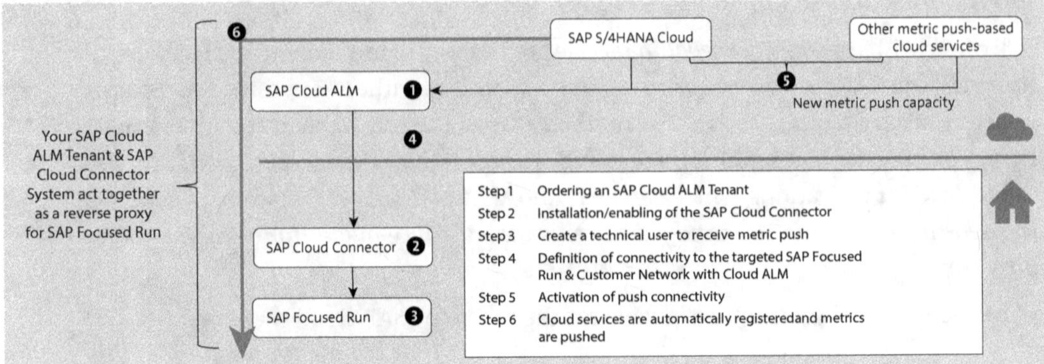

Figure 19.109 SAP Focused Run Reverse Proxy Infrastructure

To implement the SAP Focused Run reverse proxy, configuration steps must be carried out in the cloud connector, in SAP Focused Run, SAP Cloud ALM, and in the respective SAP cloud service. In Table 19.1, you will find a list of the steps to be carried out for each component involved.

Component	Configuration Steps
Cloud connector	■ Create subaccount for SAP Cloud ALM ■ Add connection to SAP Focused Run ■ Check SAP Web Dispatcher rules (if SAP Web Dispatcher is used)
SAP Focused Run	■ Determine admin request parameter for the customer networks that are to receive push data ■ Create technical users ■ Register SAP Focused Run in SAP Cloud ALM
SAP Cloud ALM	■ Manual creation of the SAP Focused Run via landscape management (only if not registered) ■ Create proxy endpoint to SAP Focused Run (for each relevant customer network)
SAP cloud service	■ Activate push data transmission ■ Configuration is individual, per SAP cloud service

Table 19.1 . Necessary Configuration Steps for the SAP Focused Run Reverse Proxy

19.9.3 Next Generation Data Collection Infrastructure with OpenTelemetry

SAP is providing a new way to collect data from SAP-built SaaS applications as well as customer-built SAP PaaS applications, by using the standard OpenTelemetry. This is possible because both are using the same data collection instrumentation libraries. Besides that, SAP has also developed own OpenTelemetry extensions.

OpenTelemetry is an open source observability framework. It provides provider-independent or neutral APIs, software development kits, and other tools for capturing telemetry data from cloud-native applications and their underlying infrastructure, which makes it possible to get insights to their performance and status by using logs, metrics, and traces.

Using OpenTelemetry together with SAP Cloud ALM makes it possible to get insights to data from non-SAP data providers as well as from non-SAP data consumers, directly in SAP Cloud ALM for operations. It's important to note that to make use of the OpenTelemetry data, some development skills are needed.

Currently, the following monitoring scenarios of SAP Cloud ALM can make use of OpenTelemetry data:

- Real user monitoring (supported for Java and Node.js)
- Health monitoring (supported for Java and Node.js)
- Exception monitoring (only supported for Java)
- Job and automation monitoring (SAP Job Scheduler service jobs in Java and Node.js)

The supported environments are currently:

- SAP BTP, Cloud Foundry environment
- SAP BTP, Kyma runtime

In total, the configuration consists of the following three major steps:

- Enablement of the frontend data collection (only needed if end user response times will be available on the real user monitoring scenario)
- Enablement of the backend data collection
- Make authentication possible (for the export of the data to SAP Cloud ALM)

All necessary prerequisites and a detailed documentation of the three major configuration steps can be found in the expert portal for SAP Cloud ALM for operations, which you can access at *http://s-prs.co/v589703*.

19.10 Summary

This chapter has provided you with an overview of how to setup most of the monitoring scenarios and functionalities that SAP Cloud ALM for operations provides. Now that you have successfully implemented the individual scenarios in SAP Cloud ALM, you have a tool at your disposal that provides you with the best possible support for operating your applications. It's up to you to make the best use of the tool and to constantly adapt the monitoring scenarios to your own needs and also to make sure that new applications are implemented in SAP Cloud ALM for operations as soon as they are used productively.

PART VI

Extend and Integrate

Chapter 20
Extension Basics

After describing the use of SAP Cloud ALM in the previous chapters, we will look at the extension concepts in this chapter. Using practical examples, you will learn about the options for extending SAP Cloud ALM.

In the previous chapters, you have become familiar with the numerous functions of SAP Cloud ALM. Nevertheless, it may be that the functions offered by the solution are not sufficient to cover all your requirements. For this reason, SAP Cloud ALM offers the option of adding functions through various extensions. In this chapter, we will give you an overview of the different types of enhancements and discuss the functions of SAP Business Accelerator Hub and SAP Integration Suite. An insight into the basic usage of and best practices for extensions round off this chapter.

20.1 Extension Types

In addition to the functional enhancement types, which we will discuss in more detail in a moment, the infrastructural enhancements of SAP Cloud ALM should also be mentioned in this context. In the text box, you will find details on the framework conditions that the fair use principle specifies, which is relevant because the functional enhancements are the specifications on data transfer limitations. Further information can also be found in Chapter 2, Section 2.4. At this point, we would like to point out once again that these framework conditions will certainly change during further product development on the part of SAP.

[!]

Restrictions When Using SAP Cloud ALM

SAP provides every authorized customer with a productive SAP Cloud ALM tenant. The current fair use principle includes the use of an SAP HANA database with 8 GB (see Chapter 2, Section 2.4). If this is not sufficient, it can be expanded a further 4 GB using SAP Cloud ALM, memory extension. In addition, the volume for outgoing data transfer is limited to 8 GB per month. A corresponding service to extend that limit is not yet available.

20.1.1 SAP BTP Services

SAP Business Technology Platform (SAP BTP) is SAP's standard extension platform and offers many different services that you can subscribe to as a customer. Many of these services are also useful when using SAP Cloud ALM and can be incorporated into the ALM solution via preconfigured integrations. In the following section, we will introduce you to two services that are automatically provided with SAP Cloud ALM and therefore do not incur any additional costs:

- **SAP Cloud Transport Management**
  This is a service for the transport management of cloud solutions. The service includes up to 4GB of bandwidth (defined as the amount of traffic transmitted and received by the cloud service). For further details on configuration, refer to Chapter 4.
- **Cloud Integration Automation**
  The service provides an infrastructure for the automatic description and execution of integration scenarios.

Both services are provided directly in the SAP Cloud ALM subaccount, which you can manage in the SAP BTP cockpit.

Another group of SAP BTP services is provided as a free tier offering with a limited scope of use. This allows you to assess the benefits of each solution and check whether it is worth using in your company. If you would like to use the services beyond the free tier, you can switch to a standard service subscription with a full range of functions. The billing conditions may vary depending on the service. Bookings are made via the SAP BTP cockpit. These types of services include, for example

- **SAP Document Management service, integration option**
  This service is used to save documents (e.g. Word, Adobe, process images).
- **SAP Build Process Automation**
  The service supports customers in the automation of process flows.
- **SAP Automation Pilot**
  With this service, workflows can be created and SAP Intelligent Robotic Process Automation can be used.

20.1.2 SAP Cloud ALM APIs

Another category of extensions can be found in SAP Business Accelerator Hub (formerly SAP API Business Hub). Here you will find all public application programming interfaces (APIs) for SAP Cloud ALM. Many of the services mentioned above are based on existing APIs that have been developed for various SAP Cloud ALM deployment scenarios. SAP Business Accelerator Hub pursues the approach of offering open APIs that can be used flexibly by customers and partners to connect third-party tools to an existing ALM process and, in particular, to extend functions in the process that are not provided by SAP Cloud ALM itself. In the following sections, we take a closer look at the use of SAP Business Accelerator Hub and the various interfaces (APIs).

20.1.3 SAP Solution Extensions

SAP Solution Extensions are a category of solution extensions that are provided by SAP or OEM partners and are deeply integrated into the SAP Cloud ALM processes. They are based on standalone tools but only offer a reduced range of functions compared to those tools. The extensions and functional enhancements are covered by the fair use license. The range of functions should be sufficient for most customer scenarios and offer added value to the existing SAP Cloud ALM functions. You can also extend the functionality to get even more value from these solutions by purchasing additional licenses for the full or selected scope. After purchasing the full license, the entire range of functions is available.

SAP Solution Extensions currently include these three solutions:

- **SAP Signavio Process Modeler**
  With SAP Signavio Process Modeler, business processes can be recorded, edited and modeled.
- **SAP Test Automation by Tricentis**
  SAP Test Automation by Tricentis is a solution for executing manual and automatic tests. The tool supports various user interfaces.
- **SAP Business Transformation Center**
  SAP Business Transformation Center enables optimized and guided data migration and transformation, accelerating migration and transformation projects by reducing the complexity of planning and execution.

20.1.4 Your Applications

The last extension category is the creation of your own cloud application. SAP BTP offers a range of services and functions designed for writing modern cloud applications. The development of a cloud application usually follows the following steps:

1. Design of the application: This includes the creation of a model of the data structure and the business logic of the application.
2. Development of the application: This is where the actual application is created using the programming languages and tools supported by SAP BTP. This is usually carried out using SAP Cloud Application Programming Model and Java or Node.js.
3. Testing the application: After development, the application is tested to ensure that it works as expected.
4. Deployment of the application: Once the tests have been successfully completed, the application can be deployed on SAP BTP.

The data is used and SAP Cloud ALM is extended via the aforementioned APIs, and tests can also be carried out via SAP Cloud ALM. Another aspect is the use of the Next Generation – Data Collection Infrastructure (NG–DCI). Libraries for data collection in the frontend and backend are integrated into the company's own applications and can then be evaluated in real user or health monitoring, for example.

[»]

Next Generation – Data Collection Infrastructure

The *Next Generation – Data Collection Infrastructure* consists of the following components:

- Data collection instrumentation (DCI)
- DCR (SAP BTP service)
- SAP Cloud ALM (including the data routing infrastructure [DRI])

The actual instrumentation is carried out with DCI. Further information can be found in the SAP Cloud for operations expert portal: *https://support.sap.com/en/alm/sap-cloud-alm/operations/expert-portal/data-collection-infrastructure.html*

20.2 Introduction to SAP Business Accelerator Hub

SAP Business Accelerator Hub (formerly SAP API Business Hub) is a central online platform through which companies can access an extensive collection of APIs provided by SAP and SAP partners. The platform offers a standardized interface through which developers, consultants, and architects can access, test, and integrate the APIs in order to develop their own applications and solutions. The SAP Business Accelerator Hub is divided into different categories that cover a wide range of business areas, including finance, procurement, human resources, sales, marketing, and many more. Each category contains a variety of APIs specifically designed to support the business processes in these areas. The APIs on the platform are fully documented and have sample code and applications to speed up the development process. In addition, SAP provides tools that facilitate the development and integration of applications, such as Open Connectors, which facilitates the integration of APIs with other cloud-based applications.

SAP Business Accelerator Hub is therefore a central point of contact for companies to quickly and easily access a wide range of APIs to improve and optimize their applications and business processes. The start page of SAP Business Accelerator Hub can be accessed via *https://api.sap.com*, as shown in Figure 20.1.

Select the **Categories** tab to display the various interfaces or interface packages by category (see Figure 20.2).

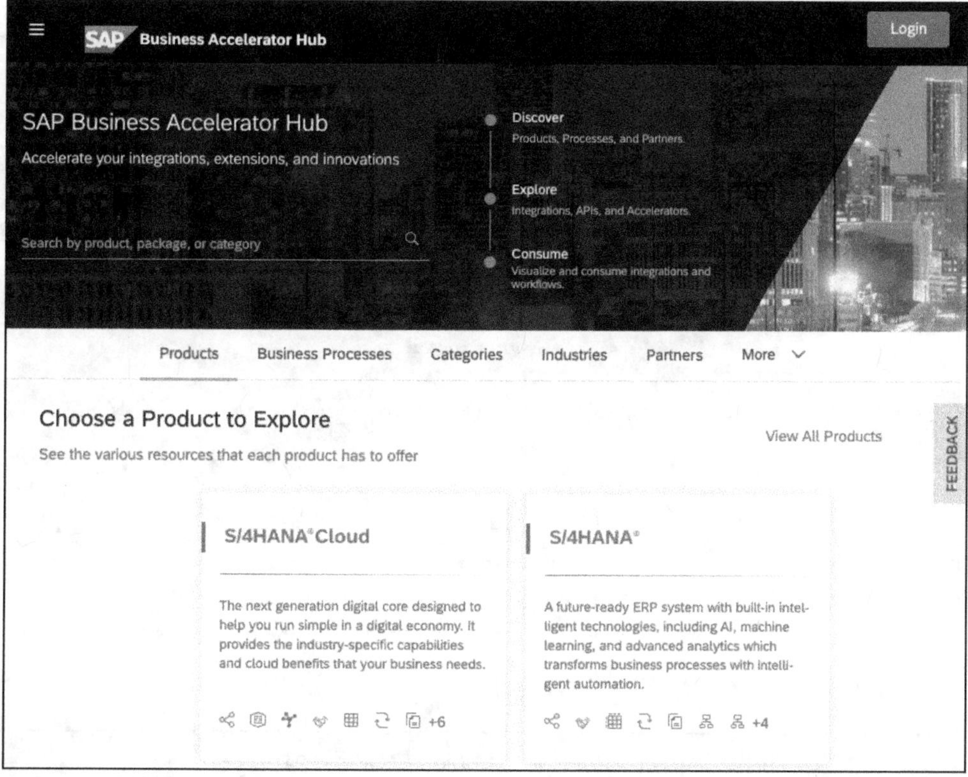

Figure 20.1 SAP Business Accelerator Hub

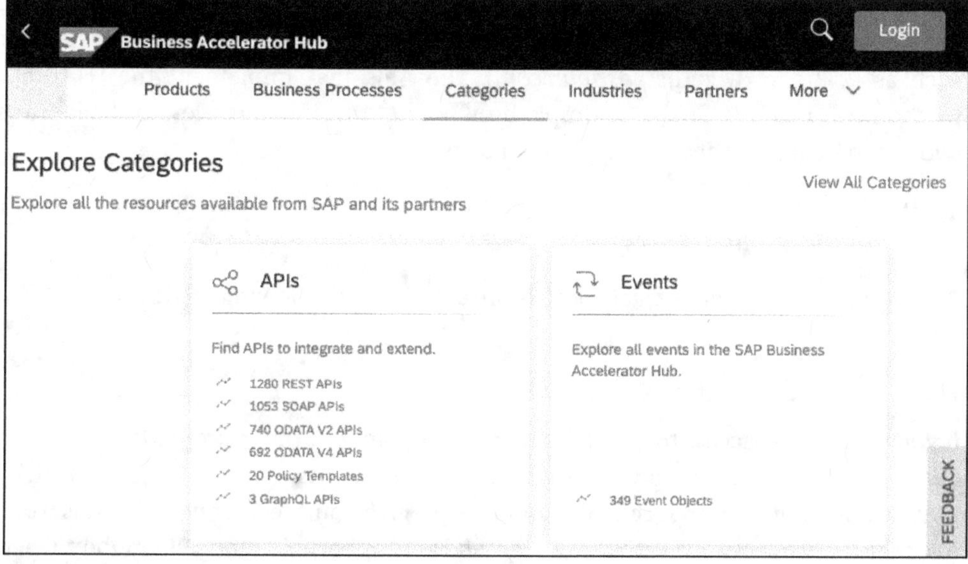

Figure 20.2 SAP Business Accelerator Hub: Categories

If you click on the **APIs** tile, you will be taken to an overview page on which the various API packages are listed (see Figure 20.3). This list can be sorted according to the technical type in which the interface was published, e.g. according to the types SOAP, OData, or REST. Select the **SAP Cloud ALM** interface package here.

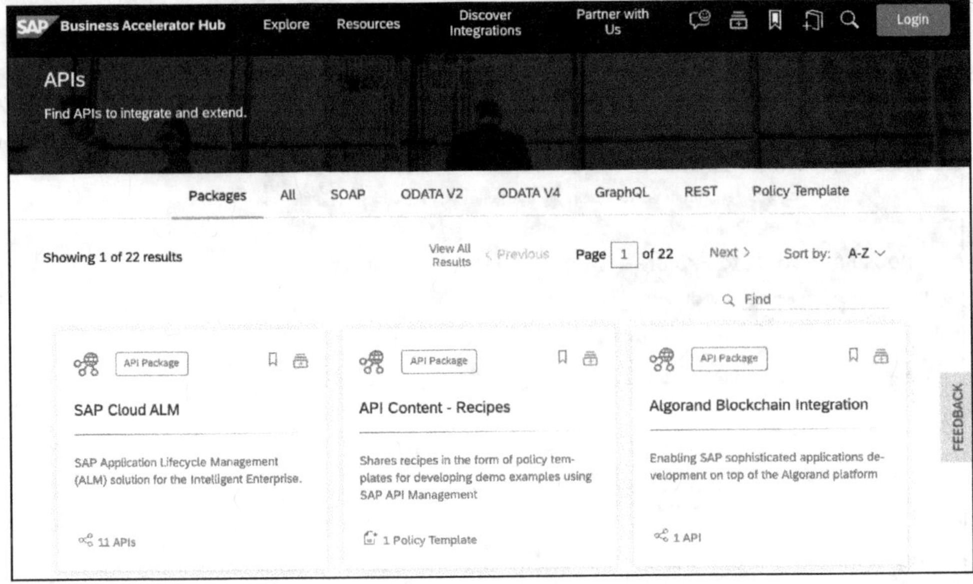

Figure 20.3 SAP Business Accelerator Hub: SAP Cloud ALM API Package

In Figure 20.3, note how you can explore the different APIs. In the following section, we will focus on the **SAP Cloud ALM** package. With SAP Business Accelerator Hub, you can test APIs directly without implementing a line of code or using a third-party REST client such as Postman (*www.getpostman.com*). The APIs that implement the HTTP GET method can also be tested in a simple browser. SAP Business Accelerator Hub offers two different approaches for carrying out tests:

- The API sandbox
- An Cloud Integration tenant

Both test approaches are explained in more detail in the following sections.

20.2.1 Testing with the API Sandbox

If you do not have access to a Cloud Integration tenant or other third-party software to test the interfaces, you can always use the *API sandbox*, a test environment provided by SAP. It is filled with test data and is a good way to give you a feel for how the APIs work. Note that only operations with the GET method are supported in the API sandbox. Operations with the POST method that require write access require a login to the API sandbox.

API Key

Each user of SAP Business Accelerator Hub receives a personalized *API key*. This API key authenticates the user to the sandbox environment. When executing the APIs from the SAP Business Accelerator Hub, this API key is forwarded internally, and you receive a corresponding response. To test the APIs in an external environment, an API key must be generated and used for access.

For example, let's use the project API to illustrate testing with the API sandbox approach: Navigate to the menu shown in Figure 20.3 and then to the **SAP Cloud ALM Projects** API. Then select the **API Reference** tab on the next page (see Figure 20.4). All operations of the API are listed on this page. Within the list, click on the GET field of the API /projects. The task of this operation is to list all projects with the ID.

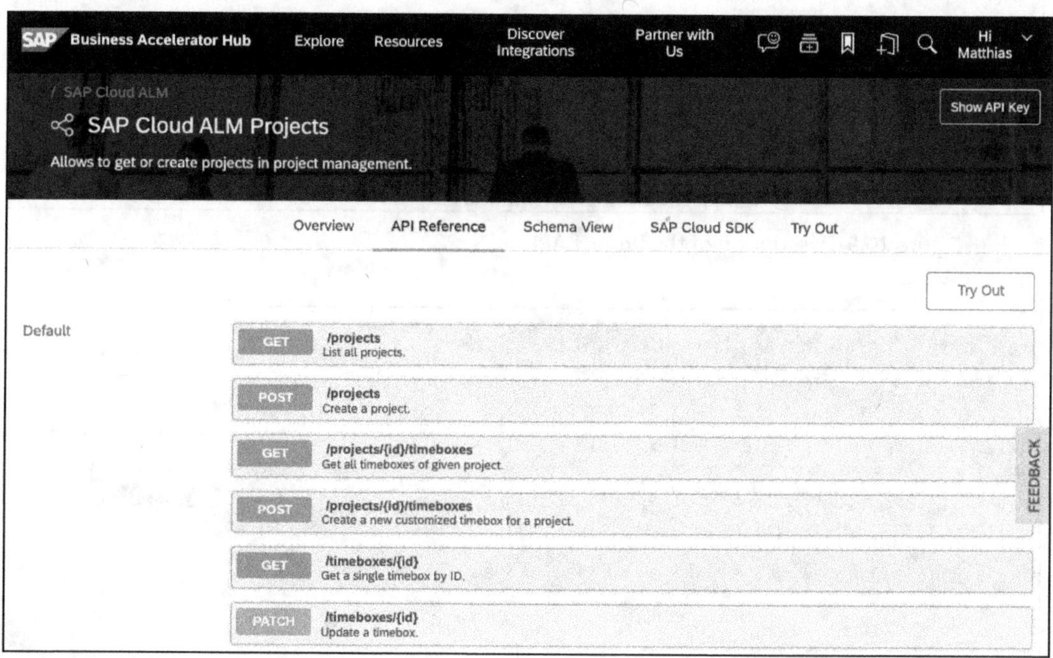

Figure 20.4 List of Project API Operations

The detailed view opens. Here you can see an exact description of the operation, the necessary parameters, and an example of a positive and a negative response from the API. Note that if parameters are required for the initial call, they must be filled in accordingly. In our example, no further input parameters are required to query the project list on SAP Cloud ALM (see Figure 20.5).

Click on the **Try Out** button to test the API. You will then be taken to a test environment in which you can store the necessary parameters and header information (see Figure 20.6).

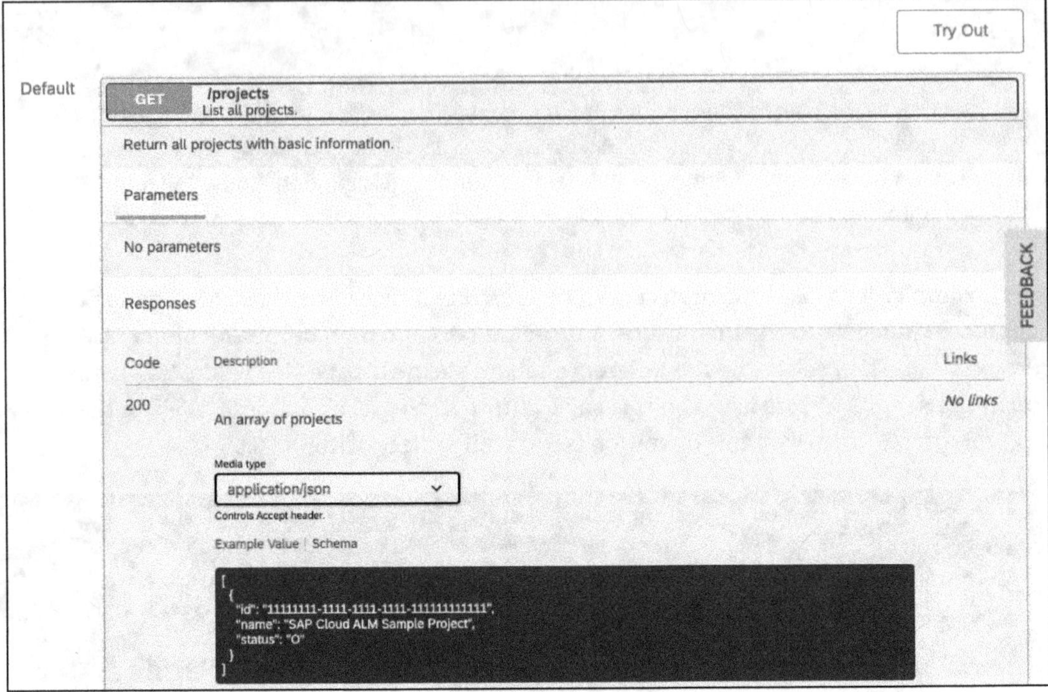

Figure 20.5 Description of the Project API

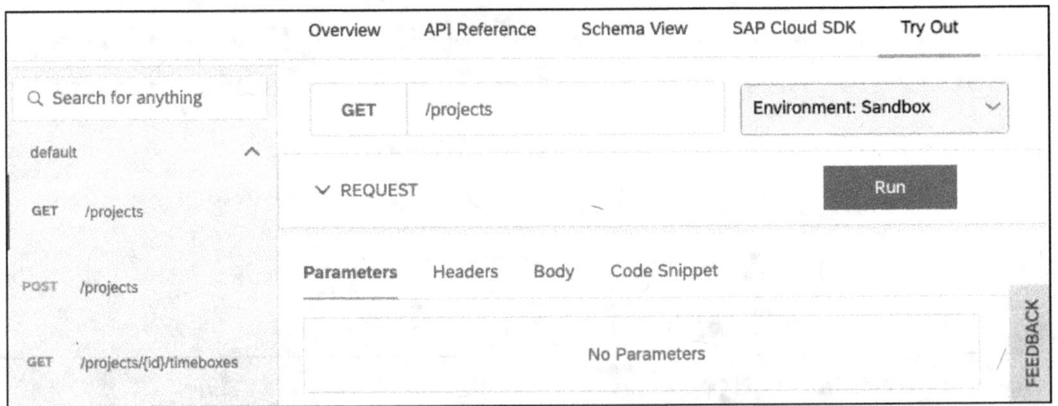

Figure 20.6 SAP Business Accelerator Hub: Empty Test Environment

Another useful piece of information are the *code snippets*, which are programming examples for calling the API. They are available in the common programming languages Java, JavaScript, and Swift, but also ABAP and SAP BTP, ABAP environment (see Table 20.1).

Functionality	Description
Code snippets	Access to documentation and various programming examples for calling the API, which are provided in the common programming languages Java, JavaScript, and Swift, but also ABAP and SAP BTP, ABAP environment.
Show API key	Simple call of the API key in the sandbox environment to be able to try out the APIs in an external environment if necessary.
Download SDK	Download a preconfigured Java Software Development Kit (SDK) for the API. Note that this download contains a complete Java project that you can use.

Table 20.1 Functions of the SAP Business Accelerator Hub on the Try Out Tab

If all the information required to call the API is stored in our example, no further information or parameters need to be stored for a successful call. If you can execute the API by pressing the **Run** button, you will receive a corresponding response (response body). In our example, this is the complete project list from our SAP Cloud ALM sandbox tenant (see Figure 20.7).

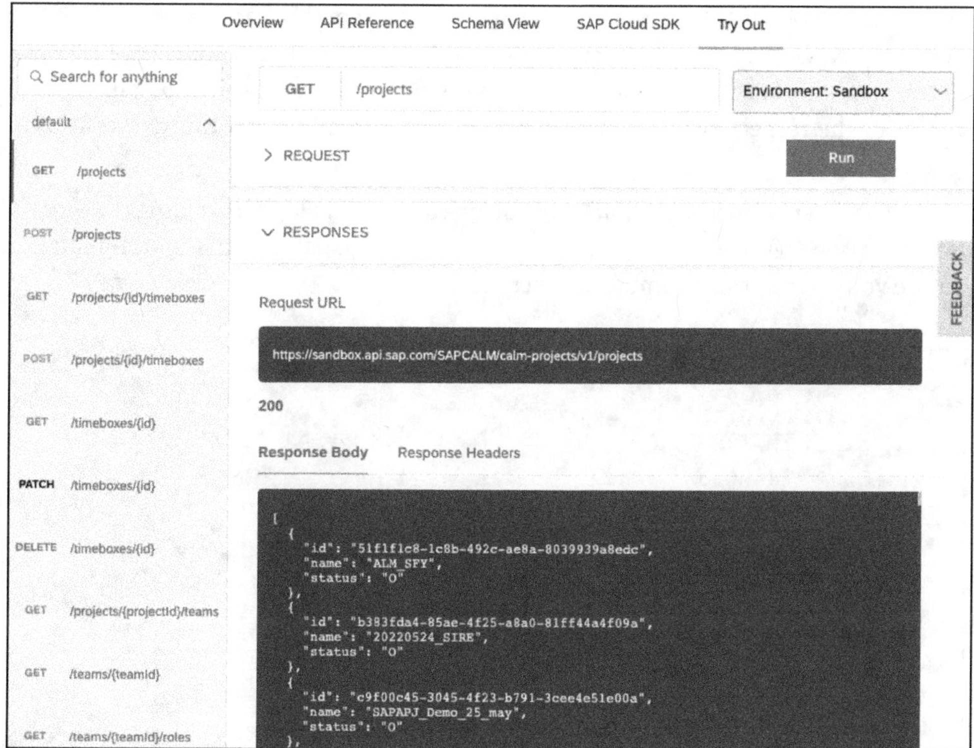

Figure 20.7 Calling the Project API and Test Result

20.2.2 Testing with Cloud Integration

An alternative to the sandbox for testing the API is to integrate the API into your Cloud Integration instance. To do this, first navigate to the Integration Advisor of your Cloud Integration instance and make the SAP Cloud ALM package available (see Figure 20.8).

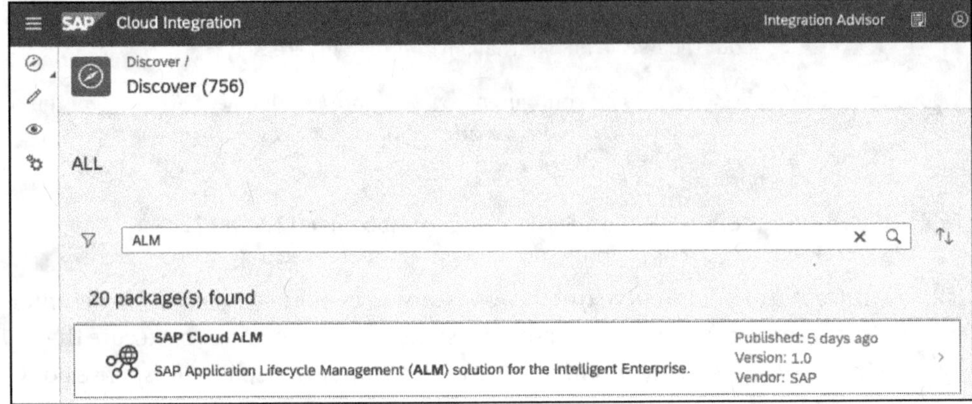

Figure 20.8 API Package in Cloud Integration

Select the **SAP Cloud ALM** package. You will now see a general overview and description of the API. Navigate to the **Artifacts** and select the **SAP Cloud ALM Projects** API. Similar to the steps you take in SAP Business Accelerator Hub, you will then be taken to the details of the respective API, including sample calls and feedback in the event of an error.

Click on the **Generate Integration Flow** button to open a dialog for creating the integration flow (see Figure 20.9). Note that you must create a package in Cloud Integration before you can run this example in practice.

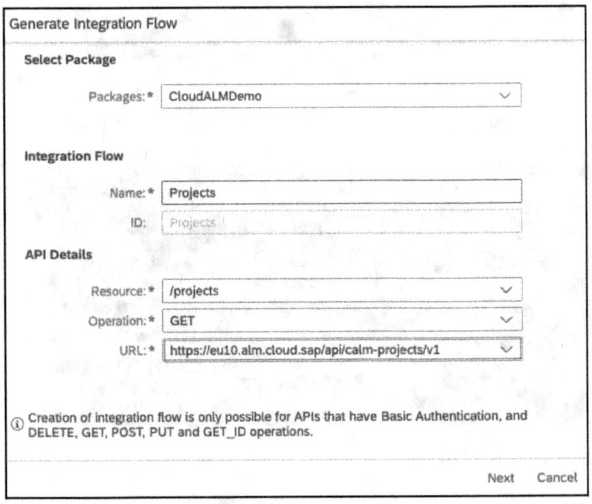

Generate Integration Flow

Select Package

Packages:* CloudALMDemo

Integration Flow

Name:* Projects

ID: Projects

API Details

Resource:* /projects

Operation:* GET

URL:* https://eu10.alm.cloud.sap/api/calm-projects/v1

Creation of integration flow is only possible for APIs that have Basic Authentication, and DELETE, GET, POST, PUT and GET_ID operations.

Next Cancel

Figure 20.9 Create Integration Flow

Now maintain the relevant mandatory fields with the information from Table 20.2.

Fields	Description
Packages	Package in which the integration flow is to be stored; in the example: "CloudALMDemo"
Name	Name of the integration flow; in the example: "Projects"
ID	Unique ID; it is assigned automatically when created
Resource	Call or parameterization of the API, in the example `/projects`
Operation	`GET` or `POST method`, in the example `GET`
URL	URL of the called API
Receiver Address	URL of the API to be called
Sender Address	URL of the sending API
Authentication	In the example: `None`
SAP Product Profile	In the example: `iflmap`

Table 20.2 Information on the Integration Flow

Once you have entered the relevant information, click on the **Next** button to access a second input screen in which you can select the remaining fields, **Receiver Address**, **Sender Address**, **Authentication**, and **SAP Product Profile.** When you click on the **Finish** button, the integration flow is created. An integration is created in Cloud Integration as an integration flow and represents the sequence of a message flow. The terms "integration flow" and "integration" are used synonymously in the following. Once you have created the integration, it is displayed as a list entry in your package.

20

Click on the pencil icon in the left-hand toolbar takes you to the **design view**, where you can edit the previously created integration (see Figure 20.10). Navigate to the package you have created and then switch to the **Artifacts** tab to customize the previously generated integration flow. The work interface of the integration opens in read mode. You can now model the integration flow by clicking **Edit** in the top right-hand corner. For our example, navigate to the corresponding connections in the **Sender** and **Receiver** components and store the relevant authentication features via the **Connection** tab, as shown in our example in Figure 20.10 (lower part of the figure). When you click on **Save**, you have created a functional integration. To test this, now select **Deploy.** Deployment publishes the integration and makes it available for access.

We have made a Postman collection available via GitHub so that the deployment of the integration can be tested. The execution of the Postman collection is shown in Figure 20.11.

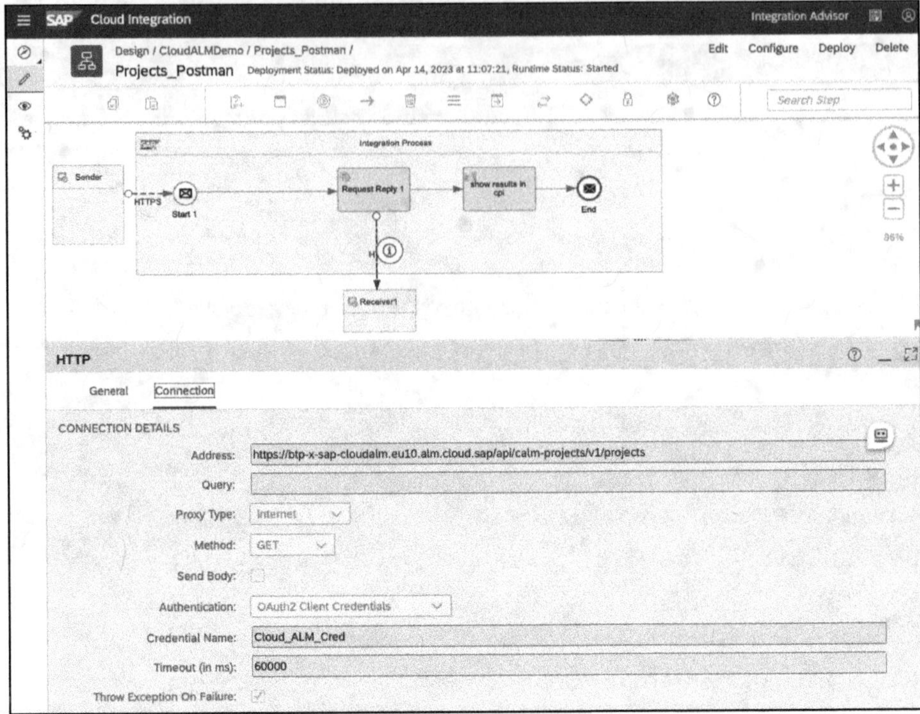

Figure 20.10 Customize Integration Flow

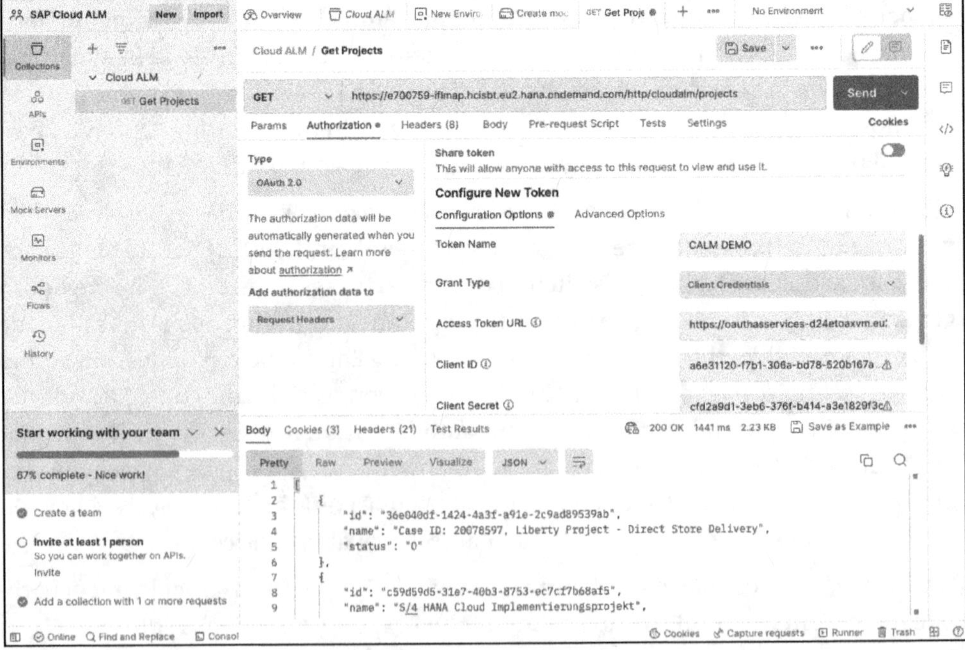

Figure 20.11 Successful Call of the Projects GET Operation Using Postman via Cloud Integration

20.3 SAP Integration Suite

SAP Integration Suite is a cloud-based integration platform developed by SAP to help companies seamlessly integrate different systems, applications, and data sources. The suite offers a comprehensive range of tools and functions for effective integration, including connectors, APIs, data mapping, process automation, business-to-business communication, and more. In the following sections, we will briefly introduce you to the tools of SAP Integration Suite. At the heart of SAP Integration Suite is the Cloud Integration function as a middleware solution for integrating different systems. We will then use this tool to build our configuration examples.

20.3.1 Cloud Integration

Cloud Integration is an integration platform as a service (iPaaS) solution that enables companies to network data, applications, processes, and systems in a clean, standardized, and clear way. IT and business processes can be seamlessly linked and operated without interruption. Cloud Integration enables the seamless connection of cloud applications with other cloud and on-premise applications, both from SAP and third-party providers. It offers prebuilt connectors for popular enterprise applications such as Salesforce, Workday, SAP SuccessFactors, and more. Various integration patterns such as point-to-point, hub-and-spoke and publish-subscribe are supported. Cloud Integration provides a unified user interface for designing and monitoring integrations, making it easier for end users to manage their integration workflows.

Integration Designer

20

The central component of cloud integration is this integration workflow (also known as integration flow or iFlow for short), which defines the protocols and transformation steps of a message between the applications involved. The data from application A can be accessed or received via various internet protocols and adapters. The data is processed according to a defined processing logic and made available to application B via supported protocols and adapters.

These integration workflows are modeled using the integration designer, a web-based interface. Depending on the scenario, several recipient systems can be addressed, and comprehensive routing and processing logics can be defined.

Clicking on the pencil icon in the toolbar on the left-hand side takes you to the design view, where you can create your own integration. There you can use the **Create** button to create a new package that can contain several integrations.

To do this, proceed as follows (see Figure 20.12):

1. Create a new integration package and enter a package name and description.

2. The **Technical Name** field is filled in automatically; special characters and spaces are standardized. If necessary, you can edit the technical name here—but subsequent changes are not possible.
3. Complete the creation of the package with **Save**.

Switch to the **Artifacts** tab to create a new integration workflow. A list of existing integrations is displayed here. An integration is referred to as an integration flow in Cloud Integration and represents the sequence of a message flow.

1. To do this, select the integration package you have just created and click on **Add • Integration Flow**.
2. Assign a descriptive name for the integration workflow; a unique ID is assigned automatically.
3. Confirm the entries with **OK**.

A more detailed example of creating an integration workflow can be found in Section 20.5.2.

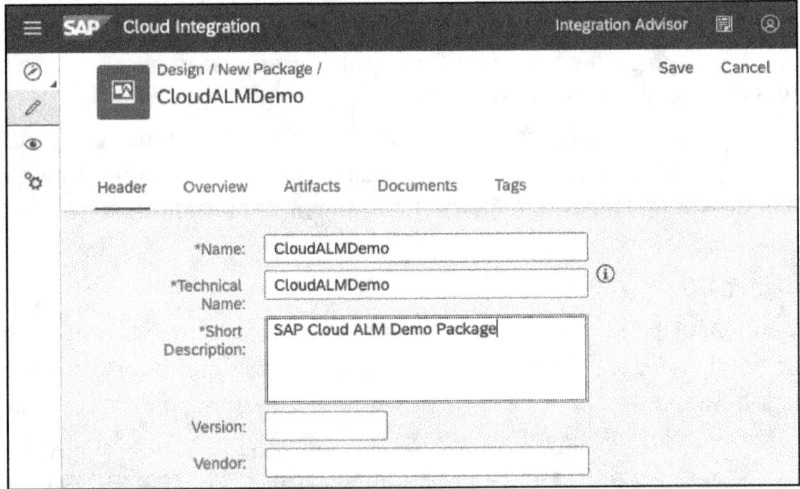

Figure 20.12 Cloud Integration: Create Package

Discovery

In addition to the option of freely modeling such integration workflows in the integration designer and determining the necessary integration steps and transformations yourself, Cloud Integration also offers access to many predefined integration packages and templates for the integration of processes and data between SAP and non-SAP applications. For example, you can access these predefined mappings of data structures and entire integration workflows when integrating SAP S/4HANA Cloud, SAP SuccessFactors, SAP Customer Experience, and many other applications. This minimizes the integration effort and saves time and costs during implementation.

There are two ways to access the contents of the catalog:

- Via the publicly accessible SAP Business Accelerator Hub (Section 20.2)
- Via your private instance of Cloud Integration

You can access the same content in both ways.

Monitoring

In addition to the configuration of integration content in the integration designer, Cloud Integration offers its own monitoring environment for monitoring integration artifacts and message flows. The various monitoring areas provide information on the status of a message, whether there are errors, or whether communication certificates are about to expire and need to be renewed.

Further Documentation

SAP offers various further documentation on the Cloud Integration web interface. You can access these in the following ways: In the top right-hand corner, you will find the symbol of a person . Click on this icon to open the documentation shown in Table 20.3.

Element	Description
About	Displays the version of Cloud Integration and the tenant ID.
Online Guide	Opens the Cloud Integration documentation.
Design Guidelines	Opens the design guidelines, which are part of the documentation and show example scenarios for integrations.
Troubleshooting Guide	Opens an interactive guide with problem-solving suggestions.

Table 20.3 Further Documentation of Cloud Integration

Click on the newspaper symbol in the top right-hand corner, you will be taken to the **What's New** area. All new features are displayed in this area and links are provided to further documentation, for example on SAP Help Portal.

20.3.2 Other Tools

In addition to the Cloud Integration function, there are other tools that are part of SAP Integration Suite. The individual functions and tools are only briefly outlined here to give you a complete overview and an idea of the other possibilities.

API Management

Alongside cloud integration, API Management is an important component of SAP Integration Suite. The API Management service supports you in your digital transformation by providing simple, scalable and secure access to your company's digital resources via APIs. API Management enables your customers and partners to use and test your company's APIs. The more such APIs a company makes available, the greater the administrative effort. Their utilization must be monitored, usage by customers must be billed, and the systems must be protected against unauthorized access or external attacks. The most important functions of API Management include the following:

- An API designer based on the OpenAPI specification standard
- An API catalog with a large selection of ready-made APIs
- Billing of API accesses according to calls or according to a defined quota
- Comprehensive analyses of access over the last six months, including API response times and processing latencies

Integration Advisor

Integration Advisor provides a new procedure for establishing communication between different systems on the basis of templates. Communication is set up in five steps:

1. In the first step, a communication type is selected from a library of industry standards. Each standard contains detailed documentation of the communication.
2. Message implementation guidelines (MIG) are provided for message exchange. In this step, the structure of the data to be exchanged, such as the field length, can be changed.
3. Mapping guidelines (MAG) are provided for mapping the data structures in the source and target systems. In this step, adjustments can be made to the mapping and then simulated.
4. The data generated for communication in the previous steps is downloaded to an archive in this step.
5. In the fifth and final step, the data from the archive is uploaded to Cloud Integration and published. The integration can then be monitored there.

SAP Alert Notification Service for SAP BTP

The SAP Alert Notification service can be used to control messages for monitoring system and service availability as well as other warning signals. The service is a kind of intermediary that does not generate its own warning signals but receives them from the services (e.g., SAP BTP or another hyperscaler) and applications (including customer-specific ones) that are connected to it. In addition to the warning signals generated by the connected services, their availability is also checked, and a warning message is issued if they are unavailable.

Test SAP Alert Notification Service

The SAP Alert Notification service can be tested in the SAP BTP, Cloud Foundry environment trial account. SAP Automation Pilot is an alternative for implementing simpler notification scenarios.

SAP Cloud Transport Management

SAP Cloud Transport Management, which we mentioned at the beginning of this chapter, has the task of transporting applications and other content within the SAP BTP landscapes, e.g., from a development environment to a production environment. Source and target systems are modeled as nodes and transport paths as routes. The following types of objects can be transported with SAP Cloud Transport Management:

- Multitarget applications (MTA)
- Integration packages from Cloud Integration
- Development units from SAP HANA
- SAP Cloud ALM

You can find further information on the setup in Chapter 4, Section 4.1.2.

20.3.3 Versions

SAP Integration Suite is available in several versions. You can find a current overview of the versions in the SAP Help Portal. The overview in Table 20.4 is based on the document status as of December 2023 and is intended to provide an overview of which functions are included in which version. The term "tenant" corresponds to an account via which the customer can access the rented resources. For example, a landscape with a test system and a productive system consists of two tenants.

Billing According to Messages

The use of SAP Integration Suite is billed according to the number of messages processed. Each message is counted with a size of 250 KB.

Version	Description
SAP Integration Suite, Basic Edition	▪ One tenant ▪ Cloud Integration without support for asynchronous message queues ▪ Billing is based on messages.
SAP Integration Suite, Standard Edition	▪ One tenant ▪ Billing is based on messages. ▪ Cloud Integration with support for asynchronous message queues ▪ API Management ▪ Open Connectors ▪ Integration Advisor
SAP Integration Suite, Premium Edition	▪ Three tenants ▪ 1,000,000 messages per month ▪ Cloud Integration with support for asynchronous message queues ▪ API Management ▪ Open Connectors ▪ Integration Advisor ▪ SAP Alert Notification service ▪ SAP Cloud Transport Management

Table 20.4 Versions of SAP Integration Suite

20.4 External API Management

This section gives you an overview of the SAP Cloud ALM application External API Management, as this application is crucial for the usage of push APIs but may also be relevant for specific scenarios when using Pull APIs.

You can reach the External API Management app within the **Administration** group. The app offers the following menu items, which we will explain in more detail.

- **Subscription Management**
- **Webhooks Management**
- **Mappings Management**
- **Console**

Figure 20.13 shows the correlation of subscriptions, webhooks, and mappings. The subscription is the central object to combine the other artifacts into an integration scenario. So it's possible to reuse webhooks and mappings in multiple subscriptions, so that similar integrations to different tools, or different integrations to the same tool, may be possible.

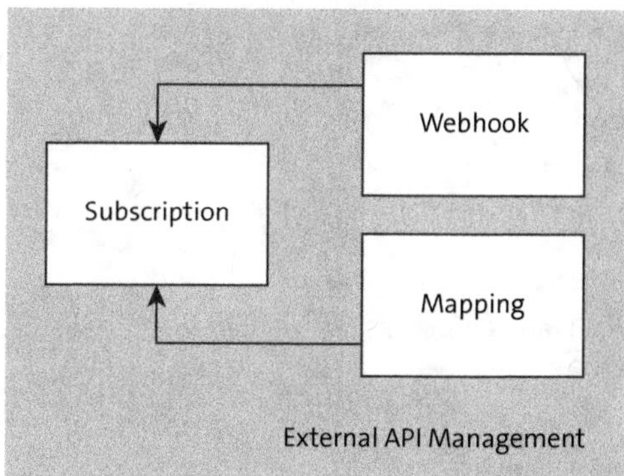

Figure 20.13 Correlation of Artifacts in External API Management

We will go through each menu item and explain its use in detail.

Within **Subscriptions Management**, which is shown in Figure 20.14, you can define the integration scenarios. As already explained, an integration scenario combines a webhook and mapping to integrate a use case of SAP Cloud ALM into an external application. It is complemented by additional metadata, as shown in Table 20.5.

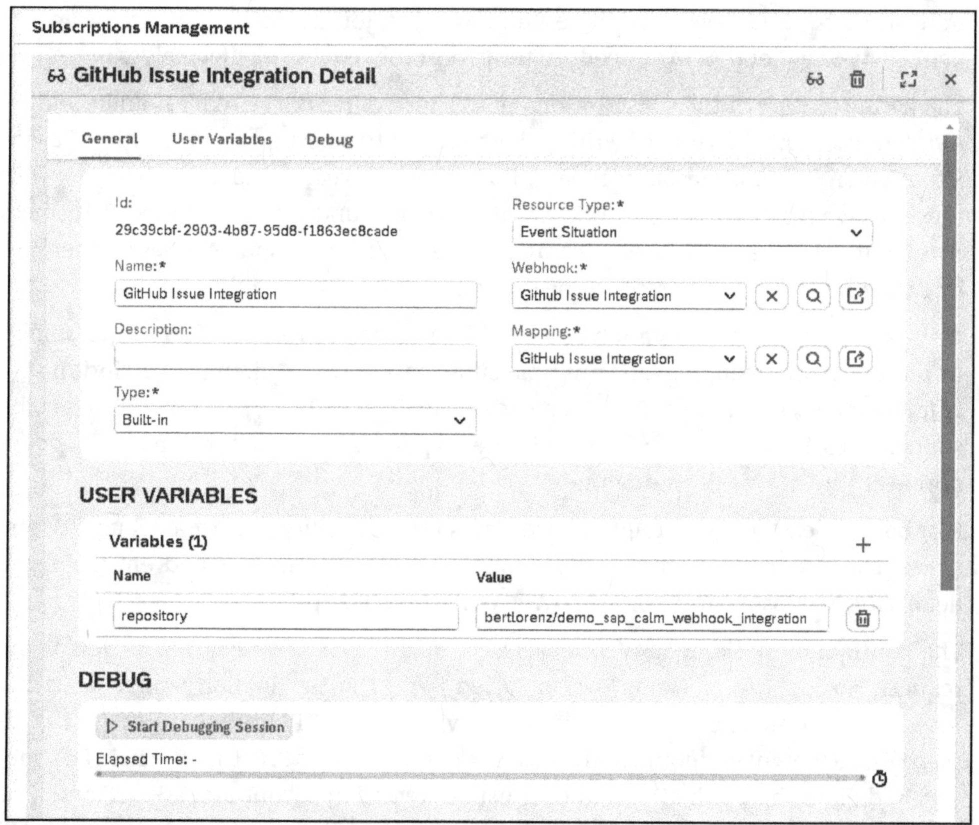

Figure 20.14 External API Management Subscription

20

Field	Usage
Id	Automatically generated identification of the subscription
Name	Descriptive name
Description	Short explanation
Type	Differentiation between built-in and external
Resource Type	Use case of the integration
Webhook	Referenced target of the external application
Mapping	Referenced request manipulation to external application

Table 20.5 Fields of a Subscription

The **Type** of the subscription defines how the request to the external application will be transformed. When using the type **Built-in**, a mapping can be referenced so that the request to the external application is transformed by SAP Cloud ALM. When using the type **External**, the request to the external application is sent without any transformation, so that a middleware or the external application needs to provide the API as defined by SAP Cloud ALM or needs to transform the request by themselves.

The **Resource Type** of the subscription defines the source of the request and the used API definition. At the time of writing, it's possible to select a resource type of **Event Situation** or **Task**. Both resource types follow the same basic concept of integration, but they do have their own API definition, which can be found in SAP help or SAP Business Accelerator Hub. In the future, SAP may provide additional resource types to offer new integration use cases or to streamline the configuration of the existing ones.

It's possible to define **User Variables** for a subscription. A variable offers the possibility to reuse existing mappings for multiple subscription while adding subscription specific flavors to the mapping. The given variable in Figure 20.14 can be referenced via the syntax `#{@USER_VAR(repository)}` within the mapping. Note that not all properties support user variables at the moment.

Last but not least, the subscription also offers the possibility to start a debugging session for it. A debugging session lasts for maximum 30 minutes and enables more details in the console of the External API Management app.

The configuration element **Webhook** (see Figure 20.15) describes the target of the request. An identifying **Name** is also stored here. The **Destination Source** describes whether the connection is to be established via an SAP BTP destination or via a landscape management endpoint. In the case of a connection via SAP BTP, the field **Destination ID** defines the destination, which must be created in advance in SAP BTP and thus describes the basic accessibility of the external application for the External API Management app. Depending on the requirements, the external application can be operated publicly via the internet or only in the local intranet. In the case of operation on the local intranet, it is necessary to use the cloud connector, which must also be configured and connected in the corresponding SAP BTP subaccount.

If the LMS endpoint is selected as the destination source, a service and an endpoint registered in the SAP Cloud ALM landscape management must be selected. The landscape management of SAP Cloud ALM is described in more detail in Chapter 3. While the destination or the landscape management endpoint defines the full name of the domain (FQDN) and any system-relevant paths, the application-relevant path is defined via the **Path** attribute of the webhook. For example, the destination *https://webhook.site* and the path `ced4d06f-37ac-482d-8bda-1cb733a9ed2f` are used to create the complete target address of the webhook to be configured: *https://webhook.site/ced4d06f-37ac-482d-8bda-1cb733a9ed2f*.

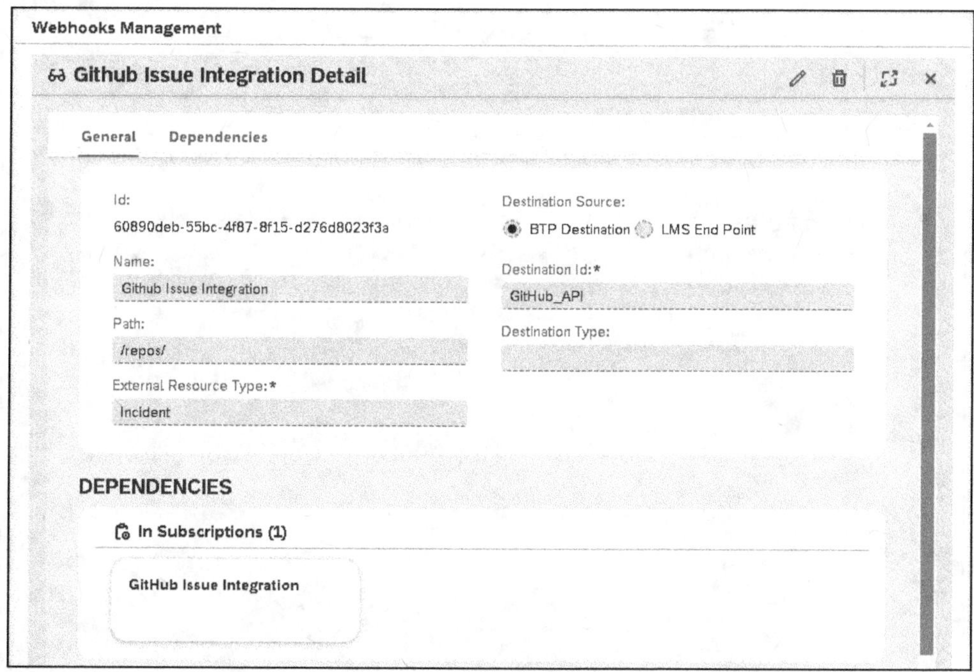

Figure 20.15 External API Management Webhook

The property **External Resource Type** defines the type of object that is created within the external application. You can choose between **Incident**, **Chat**, or **Task**, whereby the field acts as a free text field so you can also enter your own values. The selection has no influence on the technical functionality of the webhook, but it leads to a more precise specification of it so that it can be filtered according to the corresponding integration scenario in the configuration of the various applications. The same applies to the **Destination** type attribute, where ServiceNow, Microsoft Teams, or another product can be selected, or a free text can be entered.

The mapping configuration element (see Figure 20.16) enables the definition or linking of data that the external application expects or sends back. The fields **Description**, **Version**, **Source**, and **Destination** are used to describe the mapping. The field **MAPPING RULES** is used to define the desired logic in JSON format. At minimum, the objects `_information`, `structure` (in the case of payload-based mapping), `request` and `response` (in the case of rule-based mapping), and `eventTypeMapping` are defined, as shown in Listing 20.1. We will go into the mapping types in more detail later in this section.

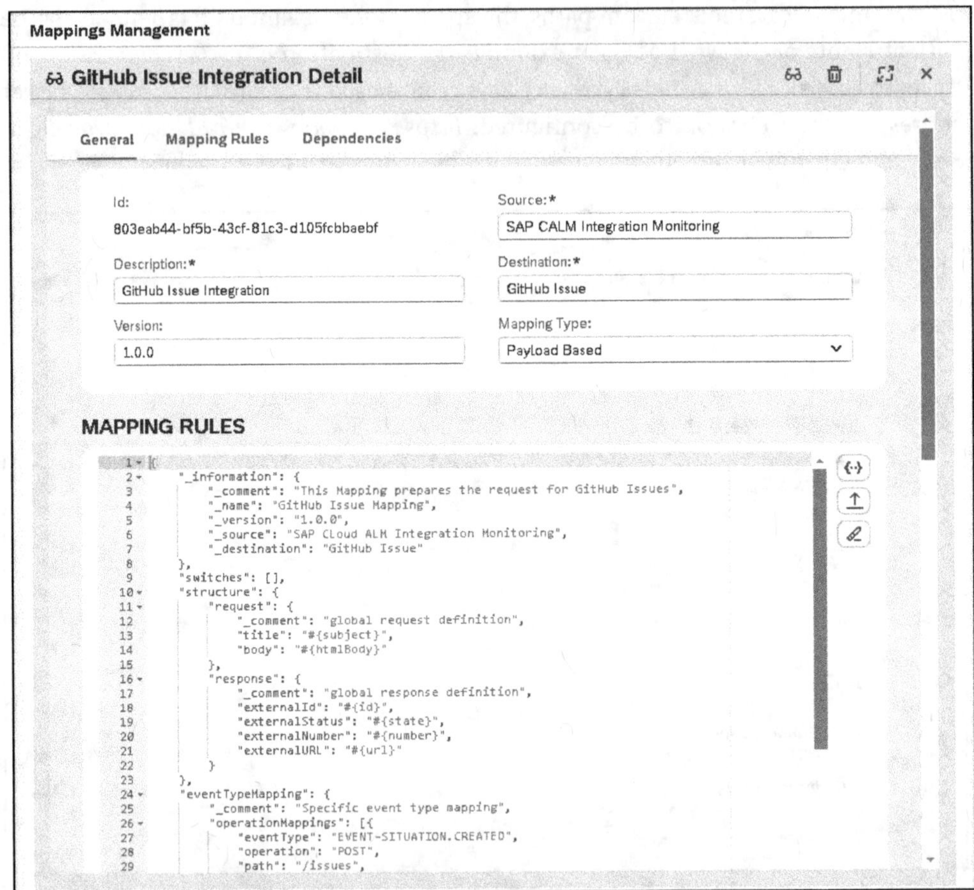

Figure 20.16 External API Management Mapping

```
{
    "_information": {},
    "request": {},
    "response": {},
    "eventTypeMapping": {}
}
```

Listing 20.1 Objects of the Mapping Rule

The object _information contains general information on the mapping definition. It requires the properties _version, _source, _destination, and _name and can optionally also contain the property _comment. The object eventTypeMapping enables the event type dependent mapping definition. The events EVENT-SITUATION.CREATED, EVENT-SITUATION.UPDATED, EVENT-SITUATION.CLOSED, and EVENT-SITUATION.PING are available for the Resource Change API for event situations and defined in the array OperationMappings. Each event type can have the values listed in Table 20.6 or override the associated global configurations.

Feature	Meaning
operation	HTTP operation that is to be executed, e.g., GET
responseRoot	Root node to be considered for analyzing the response
path	Parameter to be appended to the URL of the external application
request	Overriding the global request object in the case of rule-based mapping
response	Overriding the global response object in the case of rule-based mapping
structure	Overriding the global structure object in the case of payload-based mapping
displayURL	Rules for defining the URL of the created object in the external system

Table 20.6 Properties of an Operation Mapping Object

The objects request, response, and structure can be defined both in the eventTypeMapping object and globally, depending on the mapping types defined in the External API Management app. A distinction is made between rule-based and payload-based mapping types. The rule-based mapping enables the transformation of the sent data so that, for example, different technical status values can be translated accordingly. For this purpose, the mapping types listed in Table 20.7 are available for this purpose.

Rule	Meaning
maps	Direct definition of properties
expressions	Definition of expressions and use of placeholders
constants	Definition of constants
choices	Conditional assignment of properties

Table 20.7 Available Rules for Rule-Based Assignment

Listing 20.2 illustrates the structure of the rule-based assignment.

```
{
    "_information": {},
    "request": {
        "_comment": "Mapping Rules on Request Level",
        "maps": [],
        "constants": [],
        "choices": [],
        "expressions": []
    },
    "response": {
        "_comment": "Mapping Rules on Response Level",
        "maps": []
    },
    "eventTypeMapping": {
        "operationMappings": [
            {
                "eventType": "EVENT-SITUATION.CREATED",
                "response": {
                    "_comment": "Event specific Mappings",
                    "maps": []
                }
            }
        ]
    }
}
```

Listing 20.2 Structure of the Rule-Based Assignment

While the rule-based mapping manipulates the original HTTP request and expected response from SAP Cloud ALM so that it can be processed by the external application, the *payload-based mapping* defines the HTTP request exactly as the external application expects or sends it back. Within the mapping definition, the object `structure` is defined with the fields `request` and `response`. The field `request` of the assignment definition must describe the structure as the external application expects it as a request. The field `response` of the mapping definition corresponds to the expected response from SAP Cloud ALM. Payload-based mapping `switches` can be used to transform data. These are comparable to the rule `choices` within the rule-based mapping. Listing 20.3 illustrates the structure of payload-based mapping.

```
{
    "_information": {},
    "switches": [],
    "structure": {
        "request": {
            "systemName": "#{serviceName}",
```

```
            "description": "#{subject}"
        },
        "response": {
            "externalId": "#{ticket_id}"
        }
    },
    "eventTypeMapping": {}
}
```

Listing 20.3 Structure of the Payload-Based Assignment

As the last menu item offered by the External API Management app, we will explain the *console*. As shown in Figure 20.17, the console gives an overview of all API calls made via the External API Management app. This overview is created regardless of a debugging session activated for a subscription. Beside the **Timestamp** and relevant **Subscription**, the triggering **Event Type**, the **Delivery Status**, **Partner Response Time**, **Method**, and **Partner HTTP Status** (HTTP return code) are displayed in a tabular view sorted by latest calls first. The console offers a good overview, whether the subscriptions are generally working fine or not.

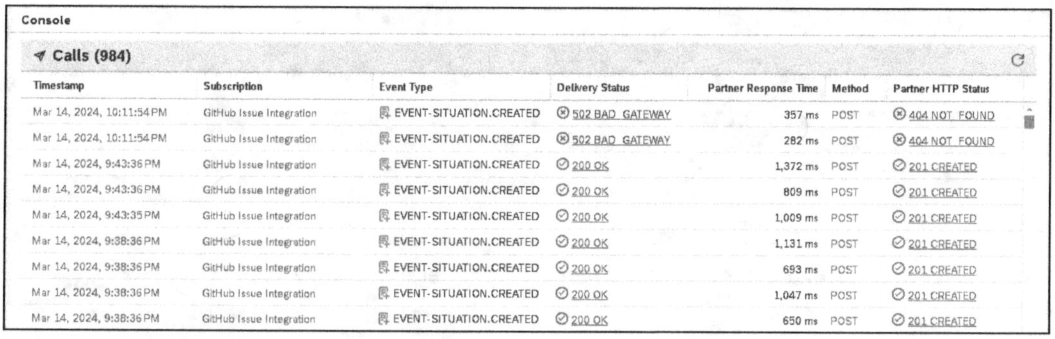

Console

Calls (984)

Timestamp	Subscription	Event Type	Delivery Status	Partner Response Time	Method	Partner HTTP Status
Mar 14, 2024, 10:11:54 PM	GitHub Issue Integration	EVENT-SITUATION.CREATED	502 BAD GATEWAY	357 ms	POST	404 NOT FOUND
Mar 14, 2024, 10:11:54 PM	GitHub Issue Integration	EVENT-SITUATION.CREATED	502 BAD GATEWAY	282 ms	POST	404 NOT FOUND
Mar 14, 2024, 9:43:36 PM	GitHub Issue Integration	EVENT-SITUATION.CREATED	200 OK	1,372 ms	POST	201 CREATED
Mar 14, 2024, 9:43:36 PM	GitHub Issue Integration	EVENT-SITUATION.CREATED	200 OK	809 ms	POST	201 CREATED
Mar 14, 2024, 9:43:35 PM	GitHub Issue Integration	EVENT-SITUATION.CREATED	200 OK	1,009 ms	POST	201 CREATED
Mar 14, 2024, 9:38:36 PM	GitHub Issue Integration	EVENT-SITUATION.CREATED	200 OK	1,131 ms	POST	201 CREATED
Mar 14, 2024, 9:38:36 PM	GitHub Issue Integration	EVENT-SITUATION.CREATED	200 OK	693 ms	POST	201 CREATED
Mar 14, 2024, 9:38:36 PM	GitHub Issue Integration	EVENT-SITUATION.CREATED	200 OK	1,047 ms	POST	201 CREATED
Mar 14, 2024, 9:38:36 PM	GitHub Issue Integration	EVENT-SITUATION.CREATED	200 OK	650 ms	POST	201 CREATED

Figure 20.17 External API Management Console Overview

You can drill down to the actual call by clicking on one of the linked fields. This will show a detailed view of the call made to the external application, as shown in Figure 20.18. Beside the related subscription, request and response data for the **Delivery** and the **Partner** is being shown. *Delivery* describes the connection between the SAP Cloud ALM source application and the External API Management app, whereas the *partner* describes the connection between the External API Management app and the external application. Note that the data is cut, so that it does not provide all details for the call made to the external application. When the debugging session for a subscription is activated, more and sensitive data is displayed in the request and response block for the delivery and partner. Even when the debugging session is activated, the data will be cut.

Figure 20.18 External API Management Console Detail View

20.5 Basic Use of APIs

SAP Cloud ALM offers two basic APIs for extending the applications and processes supplied:

- Pull APIs (inbound interfaces)
- Push APIs (outbound interfaces)

Data is retrieved from SAP Cloud ALM via *pull APIs*. This includes, for example, information on projects, tasks, processes, and events. *Push APIs* are also referred to as *webhooks*. In SAP Cloud ALM, *webhooks* offer the option of responding to events, such as exceptional situations in monitoring, with a query to a connected system.

In this section, we will first look at the general use of inbound interfaces before going into more detail on outbound interfaces in the rest of this section.

20.5.1 Pull APIs

To use the pull APIs, an additional configuration is required that is not part of the standard SAP Cloud ALM delivery. One reason for this is that not every customer has a need for integration, and another is that the *pull APIs* use a concept that supports flexible configuration scenarios. The concept of integration is shown in Figure 20.19. It shows that access to SAP Cloud ALM is granted by means of a separate *service instance* as well as a corresponding *service key*. In the rest of this section, we will take a closer look at the configuration of the various components.

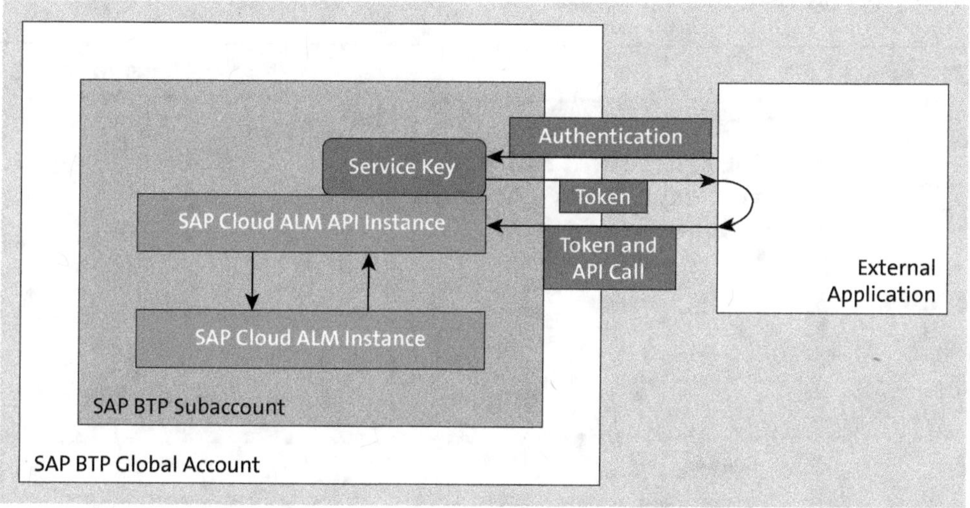

Figure 20.19 Integration Concept for Incoming Interfaces

As SAP Cloud ALM is a software as a service (SaaS) offering within a private cloud from SAP, a separate global account and an associated subaccount are usually created during provisioning. SAP Cloud ALM is then made available by a subscription in the subaccount, enabling subaccount-specific access to the central SAP SaaS solution. The encapsulation of the individual SAP Cloud ALM installations in customer-specific accounts enables separation between the customers, with the subscription in turn ensuring the central supply of SAP Cloud ALM with corrections and features.

The concept of multitenancy is described in detail in the SAP Help Portal of SAP BTP (*https://help.sap.com/docs/btp/sap-business-technology-platform*) within the **Developing Multitenant Applications in the Cloud Foundry Environment** section. To be able to use interfaces to SAP Cloud ALM, a *service instance* must be created in the same subaccount in which SAP Cloud ALM is operated. This service instance has access to SAP Cloud ALM and provides access to endpoints that can be used to query the data. The authentication of requests is carried out using the *OAuth 2.0* authorization protocol, which is the de facto standard for modern authentication. One or more service keys can be generated for the SAP Cloud ALM API instance. In addition to the `clientID` (comparable to the username) and the `clientSecret` (comparable to the password), a service key contains further details for authentication in JSON format. Before an external application can access an endpoint, a short-lived token must be requested from the authentication authority using the service key, which can then be used to prove authorized access to the endpoint. The practical use of authentication and calling the interfaces is explained in Chapter 22, Section 22.2.

To create the service instance, use the SAP BTP cockpit. First navigate to the SAP Cloud ALM global account (see Figure 20.20).

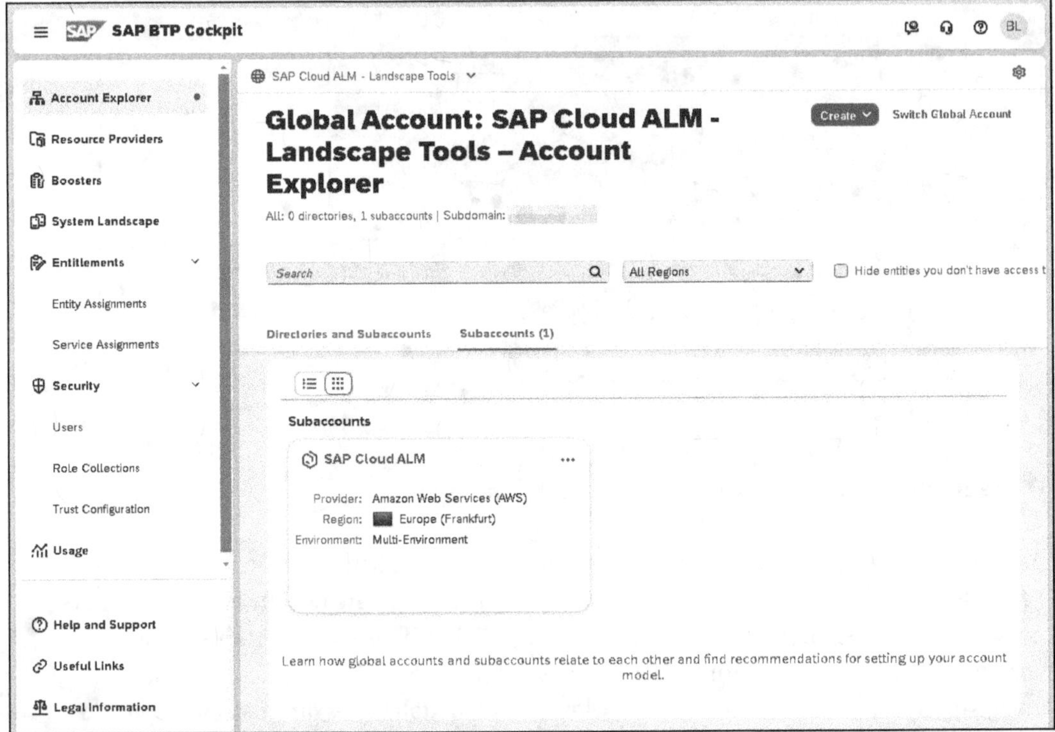

Figure 20.20 Overview of SAP BTP Accounts for SAP Cloud ALM

Note that initially only the user who has requested the provisioning of SAP Cloud ALM has access to the SAP BTP global account of SAP Cloud ALM. If other users are to administer the SAP BTP account, the corresponding access must be set up manually. Further details on this can be found in the SAP Help Portal for SAP BTP (*https://help.sap.com/docs/btp/sap-business-technology-platform*) within the section **Security Administration: Managing Authentication and Authorization**.

You can add additional licensed or license-free services from SAP to the subaccount via the **Entitlements • Entity Assignments** menu. The **SAP Cloud ALM API** service, which is required for pull APIs, is already licensed in the global account. The **SAP Cloud ALM** subaccount can be selected in the **Entity Assignments** overview, whereupon the existing assignments are displayed, as shown in Figure 20.21.

SAP Cloud ALM — Edit

Service	Service Technical Name	Plan	Set Quota Limit	Quota Assignment	Remaining Quota	Actions
Application Autoscaler	autoscaler	standard		Assigned (1 unit)		
Application Logging Service	application-logs	lite		Assigned (1 unit)		
Audit Log Management Service	auditlog-management	default		Assigned (1 unit)		
Authorization and Trust Management Service	xsuaa	application		Assigned (1 unit)		
		space		Assigned (1 unit)		
		apiaccess		Assigned (1 unit)		
Cloud Identity Services	identity	application		Assigned (1 unit)		
Cloud Integration Automation	cias	oauth2		Assigned (1 unit)		
		standard (Application)		Assigned (2 units)		
Cloud Transport Management	alm-ts	standard (Application)		Assigned (1 unit)		
Connectivity Service	connectivity	lite		Assigned (1 unit)		
Content Agent Service	content-agent-ui	free (Application)		Assigned (1 unit)		

Figure 20.21 Overview of the Entity Assignments for the SAP Cloud ALM Subaccount

The edit mode is activated via the **Edit** button. Within the popup, you can assign the service **SAP Cloud ALM API** with the service plan **standard** to the subaccount by activating the corresponding checkbox and clicking on the button **Add 1 Service Plan**, as shown in Figure 20.22.

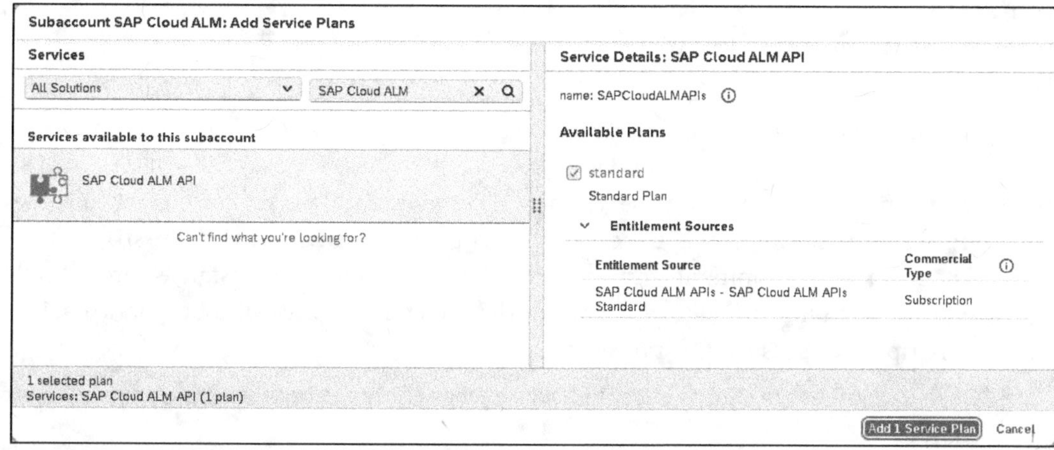

Figure 20.22 Assigning the SAP Cloud ALM API Service to the Subaccount

Once you have saved the changes to the assignments, the service **SAP Cloud ALM API** is available alongside the **SAP Cloud ALM** service (see Figure 20.23).

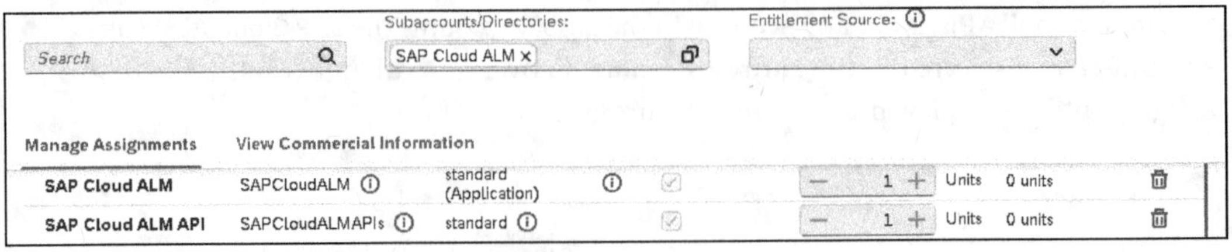

Figure 20.23 Updated Overview of the Entity Assignments

By assigning the service, the subaccount can operate an instance of the service **SAP Cloud ALM API**. To create such an instance, first navigate to the subaccount via the Account Explorer. You can then create a new instance of **SAP Cloud ALM API** via the menu item **Services • Service Marketplace** by selecting the relevant service and clicking on the **Create** button (see Figure 20.24).

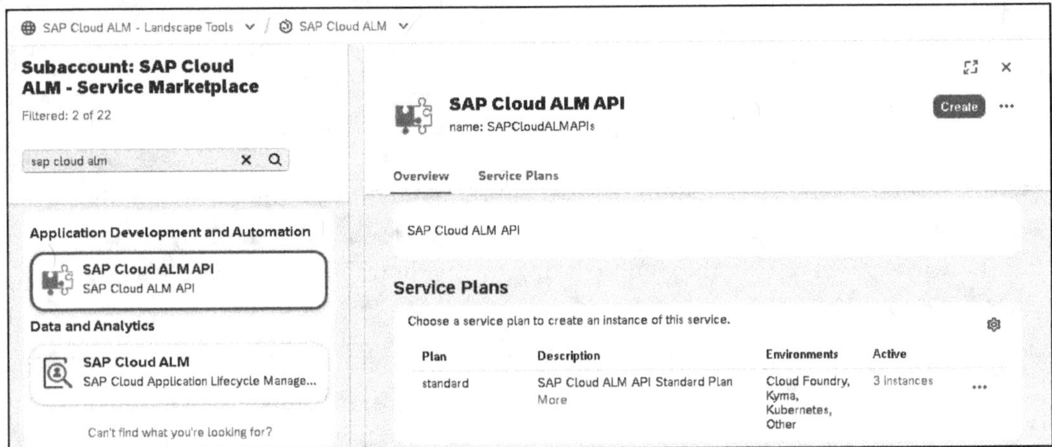

Figure 20.24 Service Marketplace in the Subaccount

In the first step of creating a service instance, general data is requested. This includes the service and the associated service plan, the runtime environment, the space within the runtime environment, and a technical instance name. The instance name should be command-line friendly, which means that a technical name without spaces that is as descriptive as possible must be used.

Possible values are shown in Table 20.8 for where only the instance name can be freely defined.

Parameters	Value
Service	SAP Cloud ALM API
Plan	Standard

Table 20.8 Basic Information on Creating the SAP Cloud ALM API Instance

Parameters	Value
Runtime Environment	Cloud Foundry
Space	SAP Cloud ALM
Instance name	sap-calm-api

Table 20.8 Basic Information on Creating the SAP Cloud ALM API Instance (Cont.)

Once the basic information about the instance has been defined, you must specify the scope of the instance in more detail in the second step of the creation. The *scope of the instance* defines which areas of SAP Cloud ALM this service instance can access and with which authorities. It is possible to define the broadest possible scope for the instance so that it can access all the interfaces provided by SAP Cloud ALM. However, as several service instances can also be created and operated in parallel, it is also possible to configure the instances more specifically so that only the scope that is actually required for the planned integration is defined. Listing 20.4 shows the configuration required to define the entire scope of the interface. The `xsappname` field must match the instance name of the basic information. The `authorities` field describes the list of available API authorizations for the integration.

```
{
    "xs-security": {
        "xsappname": "<Please enter your instance name here>",
        "authorities": [
            "$XSMASTERAPPNAME.calm-api.analytics.read",
            "$XSMASTERAPPNAME.calm-api.analytics.providers.read",
            "$XSMASTERAPPNAME.calm-api.demo-tasks.read",
            "$XSMASTERAPPNAME.calm-api.projects.read",
            "$XSMASTERAPPNAME.calm-api.projects.write",
            "$XSMASTERAPPNAME.calm-api.tasks.read",
            "$XSMASTERAPPNAME.calm-api.tasks.write",
            "$XSMASTERAPPNAME.calm-api.tests.read",
            "$XSMASTERAPPNAME.calm-api.integration-monitoring.read",
            "$XSMASTERAPPNAME.calm-api.bsm.read",
            "$XSMASTERAPPNAME.calm-api.defects.read",
            "$XSMASTERAPPNAME.calm-api.requirements.read",
            "$XSMASTERAPPNAME.calm-api.jobs.read",
            "$XSMASTERAPPNAME.calm-api.processmanagement.read",
            "$XSMASTERAPPNAME.calm-api.processmanagement.write",
            "$XSMASTERAPPNAME.calm-api.processmanagement.delete",
            "$XSMASTERAPPNAME.calm-api.processauthoring.read",
            "$XSMASTERAPPNAME.calm-api.processauthoring.write",
            "$XSMASTERAPPNAME.calm-api.processauthoring.delete",
```

20

```
            "$XSMASTERAPPNAME.calm-api.event-situations.write",
            "$XSMASTERAPPNAME.calm-api.rum.read",
            "$XSMASTERAPPNAME.calm-api.bm.read",
            "$XSMASTERAPPNAME.calm-api.hm.read",
            "$XSMASTERAPPNAME.calm-api.metrics.read",
            "$XSMASTERAPPNAME.calm-api.metrics.write"
        ]
    }
}
```

Listing 20.4 Instance Parameters for Configuring the Entire Interface Scope

As SAP has placed the focus on integration in SAP Cloud ALM, it is possible that further APIs will be made available in the future. In this case, the existing service instance must be updated accordingly. The current configuration of the interface scope can be found in SAP Help Portal for SAP Cloud ALM (see *https://help.sap.com/docs/cloud-alm*). To find this, navigate to the section **API Guide for SAP Cloud ALM**. In the menu item **Getting Started with APIs • Pull API • Managing Scopes** you will find a sample configuration. Please note that SAP Help Portal is updated regularly, so the structure of the website may differ from the one shown here.

To use the cloud-based transport system, which is described in Chapter 4, you must also create a service instance with the configuration of the corresponding scope. This scope is shown in Listing 20.5, where you can also find the current configuration in SAP Help Portal. To find this, open the *Setup and Administration Guide* (see *https://help.sap.com/docs/cloud-alm/setup-administration*) and navigate to the menu item **Integration and configuration options • Activation of the SAP Cloud ALM API • Procedure • Maintain Instance.**

```
{
    "xs-security": {
        "xsappname": "<Please enter your instance name here>",
        "authorities": [
            "$XSMASTERAPPNAME.imp-cdm-feature-display-ui",
            "$XSMASTERAPPNAME.imp-cdm-feature-manage-ui"
        ]
    }
}
```

Listing 20.5 Instance Parameters for Configuring the Cloud-Based Transport System

Once the service instance has been successfully created, a service key can be generated for it. To do this, navigate to the service instance you have just created via the menu item **Services • Instances and Subscriptions** and click on the button **Create** in the area **Service Keys**, as shown in Figure 20.25.

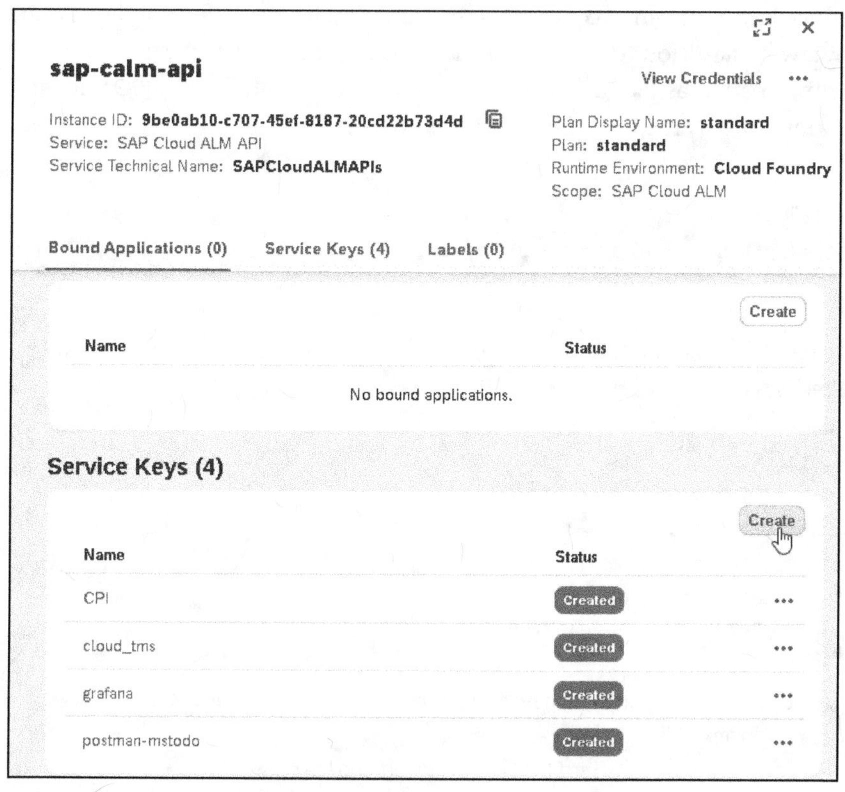

Figure 20.25 Newly Created SAP Cloud ALM API Service Instance

In the following window, you only need to enter a name that is as descriptive as possible. As several service keys can be created for a service instance, it makes sense to create a separate service key for each integration. As with the service instance, an extended configuration can also be defined for the service key via JSON, but this is not used in the SAP Cloud ALM API, which is why you can leave the **Configure Binding Parameters** field empty, as shown in Figure 20.26.

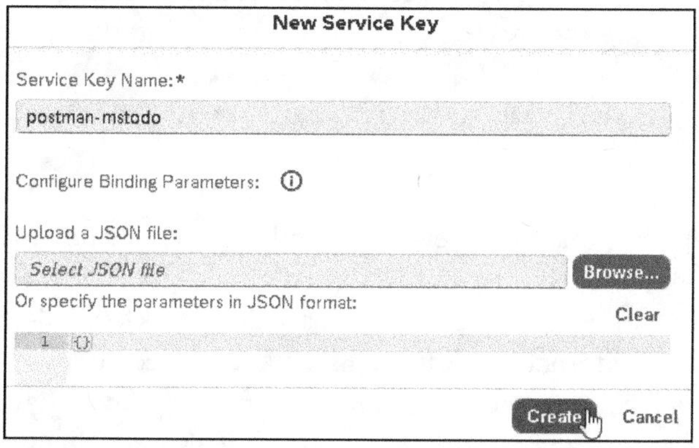

Figure 20.26 Creation of a Service Key

Once the service key has been successfully created, it can be displayed via the context menu item **View**. In addition to a form-based display, the service key can also be displayed and downloaded as JSON. Listing 20.6 shows a complete key in JSON format, with a small amount of sensitive information removed.


```
{
    "endpoints": {
        "api": "https://eu10.alm.cloud.sap/api"
    },
    "uaa": {
        "clientid": "sb-sap-calm-api!b160695|sapcloudalm!b16907",
        "clientsecret": "2e0f9150-35d2-45e9-baa5-518d8ecf7421$h32_
wWpYOH4TergQxcVOcjXkC8c_2LceO_W6FoyOCb4=",
        "url": "https://***.authentication.eu10.hana.ondemand.com",
        "identityzone": "***",
        "identityzoneid": "***",
        "tenantid": "***",
        "tenantmode": "shared",
        "sburl": "https://internal-xsuaa.authentication.eu10.hana.
ondemand.com",
        "apiurl": "https://api.authentication.eu10.hana.ondemand.com",
        "verificationkey": "----",
        "xsappname": "sap-calm-api!b160695|sapcloudalm!b16907",
        "sub-accountid": "***",
        "uaadomain": "authentication.eu10.hana.ondemand.com",
        "zoneid": "***",
        "credential-type": "binding-secret"
    }
}
```


Listing 20.6 Service Key in JSON Format

Once the service key is available, SAP Cloud ALM has been successfully configured for pull APIs. Further details on using the service key and calling the interfaces are described in Chapter 22, Section 22.2.

20.5.2 Push APIs

Beside endpoints for external applications to retrieve and manipulate data from SAP Cloud ALM, there are also possibilities to trigger calls to external APIs. In general, this is called a *webhook*. It means that specific events, such as the creation of a task or an alert, will trigger a communication to an external application. As a standardized approach, SAP Cloud ALM provides the integration via External API Management, which we have analyzed in Section 20.4 already. At the time of writing this book, the resource changes for event situations and resource changes for tasks, defects, and requirements are supported by the External API Management. Additionally, also very specific integrations that are not handled within External API Management are available. Those integrations are set up directly in the corresponding application of SAP Cloud ALM itself and offer only very limited possibilities of transformation, so that either a middleware has to be used or the external application needs to implement the API exactly as specified. The Test Automation API is such an integration. Ideally, SAP should harmonize the configuration of the different push APIs, so that there's a common approach for how webhooks to external applications are implemented.

Independent to the integration scenario, both options share the same approach on how to connect to the external application. The general connection is set up via a destination in SAP BTP or via a landscape management endpoint. This defines the target of the integration, as well as possible authentications. Especially for external applications hosted in the local intranet, it may be required to use an SAP BTP destination, which is properly connected with a cloud connector for cloud to on-premise connections. Beside of this, no further preparation is required to make use of push APIs. As already mentioned, the Test Automation API needs to be implemented in the corresponding application, whereas the Resource Changes API is implemented within the External API Management app. We'll show a full configuration of an integration scenario in Chapter 22.

20.6 Best Practices

In this section, we will discuss best practices for extending SAP Cloud ALM. Our best practices should be seen as recommendations and not as guidelines—after all, there is a use case for every implementation method.

First, let's look at the topic of microservices. With SAP Cloud ALM as an SaaS offering, SAP has also changed the release concept. Whereas in the past only a few support and feature packages were published each year, SAP Cloud ALM has a release cycle of a few weeks. Corrections can be imported into the offering on a daily basis and thus made available to all customers. In addition to the shorter release cycle for new functions, customers also have no choice as to whether or when a new release is imported. It is therefore necessary to monitor change notices and react accordingly. This release concept has implications for how enhancements should be implemented. Instead of large, complex monoliths that combine all the functions of an enhancement in a single application, the use of microservices enables a more flexible response to adaptations by SAP, as well as faster delivery of corrections or new functions of your own enhancement. When using microservices, the aim is to operate the individual components of an application separately from each other.

A common example to illustrate the concept of using microservices is an application consisting of a frontend, a backend, and a database. The database is usually operated separately in production environments, while the frontend and backend of a monolith application are often operated together in one environment. Changes to the frontend or backend can only be updated together. With the concept of microservices, the backend and frontend are operated separately from each other so that the frontend usually accesses the backend via web services. On the one hand, this makes the backend scalable so that computing power can be added for more extensive use; on the other hand, the frontend and backend can also be operated independently of each other and can therefore be exchanged flexibly. By using microservices to extend SAP Cloud ALM, you are able to change individual components of the application. In addition to the testing effort, this also reduces the maintenance effort and increases the stability of the extension.

One quality feature of an extension or development is maintainability. After all, a development should not cause any additional costs in the economic area but should ultimately reduce costs or increase competitiveness through optimization. During the development of an enhancement, you should always question whether the enhancement actually brings the desired benefits or whether it should simply be implemented because the processes that existed before the use of SAP Cloud ALM "have always been like this." Particularly when it comes to less specific use cases, such as integration into common products, especially those offered by SAP, it makes sense to approach SAP as part of the Customer Influence program and express the need. Due to the aforementioned release concept of SAP Cloud ALM, it is quite possible that SAP will integrate the required customization in the standard if needed.

If an extension is nevertheless required, it makes sense to keep the scope of customer-specific development as small as possible. One way of reducing the scope is to outsource parts of the application to existing services. The simplest example of this is the database. Of course, a database can be operated in-house. But this causes implementation and maintenance costs that can be avoided by using a cloud service. However, smaller components of an application, such as an enterprise service bus, workflows or sets of rules, can also be outsourced to services. All that needs to be done is to integrate the use of the corresponding service into the application and configure the service according to requirements.

Even when using new technologies, such as chatbots, the Internet of Things, artificial intelligence, or blockchain, it makes sense to rely on existing offerings, as most use cases for an extension revolve around using a technology and not about developing the technology further. But please be careful when using existing services, as you always become dependent when using them. The services are updated regularly, just like all other offers, which in turn can result in changes to the configuration or connection of the service. Services may also be discontinued by the provider or replaced by new offerings. If existing services are used excessively, this can ultimately lead to the costs being significantly higher than a small in-house development.

In addition to outsourcing individual components of an application, it may even be possible to outsource the entire application to existing services. This concept is called *low-code* or *no-code*. In this case, a development is no longer developed with code, but is configured as far as possible and supplemented with the help of smaller code snippets. This type of implementation has the advantage that only the processes that actually need to be implemented need to be considered during implementation and maintenance. At the same time, however, this places you in a position of maximum dependency on the provider, as the offerings differ to such an extent that migration from one provider to another cannot usually be automated.

We have deliberately not mentioned SAP's offerings in this section but would now like to present another best practice: SAP BTP. SAP BTP combines general services such as an enterprise service bus or a workflow engine as well as innovative services, such as artificial intelligence or blockchain services, in such a way that almost any requirement, including those relating to the operation of an application, can be implemented. For example, it is possible to implement the operation of an application using your own continuous integration/continuous delivery pipelines, or you can use project "Piper" in conjunction with the DevOps service from SAP BTP. A further level of integration in this context is the cloud transport management system, which applies the proven concept of transport requests to SAP BTP. The *multitarget application model* forms the basis of the customer-specific application on SAP BTP. In a nutshell, the MTA combines the application with its dependencies (such as event bus, database, workflow engine, etc.) and the desired scaling in a common archive. This archive can then be implemented on the various accounts within SAP BTP, whereupon the application and its dependencies are provided automatically. In summary, SAP BTP offers both common and innovative services in such a way that users and developers can easily use, configure and integrate them. However, it is also important to note that SAP BTP offers can be terminated or replaced if necessary. As a rule, however, SAP offers a migration path that reduces the effort on the customer side as much as possible.

Authentication and secure access to company data have become an important aspect when using existing services, whether on SAP BTP or from other providers. After all, applications are no longer only provided on the intranet, but are in principle available to every internet user. One standard in this area is Open Authorization (OAuth) 2.0. This means that the authentication of a user is no longer carried out by the application itself but is outsourced in the spirit of microservices. If a user attempts to log in to an application, they are forwarded to the authentication authority for authentication. After successful authentication, this is communicated to the application, and the user is forwarded accordingly. This concept of authentication enables central user administration on the one hand and secures the application in such a way that a key aspect of securing publicly available applications can be centralized on the other. The application itself now only has to regulate the assignment of authorizations for authenticated users. Whenever you provide an application for end users, you should therefore consider OAuth as an authentication method.

Let us now return to SAP's release concept with regard to SAP Cloud ALM. As already mentioned, SAP Cloud ALM regularly receives new functions and corrections. It may prove difficult to maintain an overview and assess the effects accordingly. While SAP Help Portal (*https://help.sap.com*) used to be rather brief and concise, particularly with regard to application lifecycle management, which led to the need for external consulting, it has now become the first port of call for the implementation, use, and expansion of SAP Cloud ALM. To be fair, it should be mentioned that the SAP Help Portal has also improved significantly in the area of SAP Solution Manager. For complex issues, however, SAP Help Portal is still unsuitable as the sole medium, meaning that other sources of information or external advice must be accessed. As already described, however, you should avoid complex use cases.

Finally, we would now like to discuss the topic of development operations (DevOps). DevOps is often reduced to the topics of operation in Kubernetes, Docker, and continuous integration, and continuous delivery/deployment. Of course, Kubernetes and Docker infrastructures, which are often used in DevOps, are an important aspect, but that alone is not enough. Almost more important than the technical foundation is the corresponding organization and mentality behind it.

Practical Example

The lack of clarity in relation to DevOps can be illustrated by an example from our customer. They wanted to improve the operation of their internal applications through DevOps by running as many applications as possible in a Docker Swarm. This also included the installation of GitLab, a GitHub-like platform for managing source code. While running his own applications on the Docker Swarm proved to be advantageous, using the productive GitLab instance had significant disadvantages. The administration effort increased, as not only GitLab but also the corresponding Docker configuration had to be managed. In addition to the effort involved, this also created uncertainty, as the configuration of the network, storage location, backup and update raised new questions. At the same time, on closer inspection, there was no need for the flexibility that Docker brings. In the context of the GitLab instance, Docker was only used as a replacement for a dedicated server. Running the GitLab instance on a physical or virtual server would have been the better choice for the customer in terms of stability, security, and administration effort.

When applying DevOps as a methodology, it should therefore be noted that technology alone is not enough. The organization and mentality in the company are an essential component for the success or failure of DevOps.

20.7 Summary

To summarize, common development practices for cloud-native developments apply to the extension of SAP Cloud ALM. On the one hand, there are ready-to-run integrations possible; on the other hand, extensive custom scenarios may be implemented using the Push and Pull APIs of SAP Cloud ALM. For those integrations, we have explained some of the resources provided by SAP, such as SAP Business Accelerator Hub, SAP Integration Suite, and the External API Management app. Finally, we also provided some best practices for your custom extension of SAP Cloud ALM. Consider a microservice architecture to reduce the dependencies of the application's components. Use services from cloud providers to reduce unnecessary in-house development. Use automation to minimize manual activities such as testing and delivery while increasing the stability, maintainability and delivery speed of corrections and functions. Only implement enhancements that offer sustainable added value. And finally: Please note that all measures will only be effective if the organization and the mindset of those involved are aligned accordingly.

Chapter 21
Overview of APIs

Following on from the previous chapter, which provided an overview of the enhancement concepts and tools, we will now take a closer look at the SAP Cloud ALM APIs and explain the basics of using them.

This chapter provides an overview of the different types of interfaces in SAP Cloud ALM, how they work, data flows, and examples of how to use them. Figure 21.1 shows a first impression of the general data flow of incoming and outgoing interfaces. The end users authenticate themselves on the SAP BTP subaccount and thus gain access to the data granted by authorization. In the case of incoming interfaces, this process is regulated using the API key. Outgoing interfaces are authorized accordingly via the SAP Connectivity service.

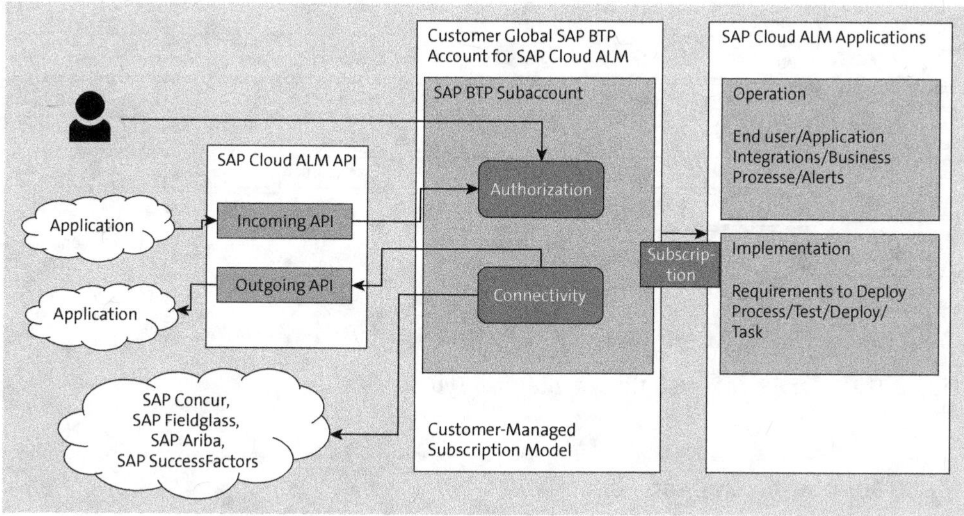

Figure 21.1 Calling the Incoming and Outgoing Interfaces

In addition to the type of interface (incoming or outgoing), a distinction is also made between the following interface types:

- **Transactional APIs**
  Describes interfaces for modifying SAP Cloud ALM resources (usually a REST API for reading, modifying, and deleting an object [e.g., projects or tasks]).

- **Resource changes APIs**
  Describes interfaces that pass event-based changes to SAP Cloud ALM resources to the outside world; these are usually REST APIs (e.g., `ALERT-SITUATION.CREATED` or `TASK.CREATED`).
- **Analytics APIs**
  Refers to REST/OData interfaces for reading aggregated data in time series or in table format (e.g., all tasks, closed for the last five days or last five weeks).
- **Raw data APIs**
  Refers to inbound and outbound APIs for reading/writing raw data using the OpenTelemetry Protocol (e.g., health monitoring metrics for an SAP S/4HANA system; see Section 21.2.5).

An overview of the classification of these four interface types can be found in Figure 21.2. In the next section, we will go into the various interfaces in more detail.

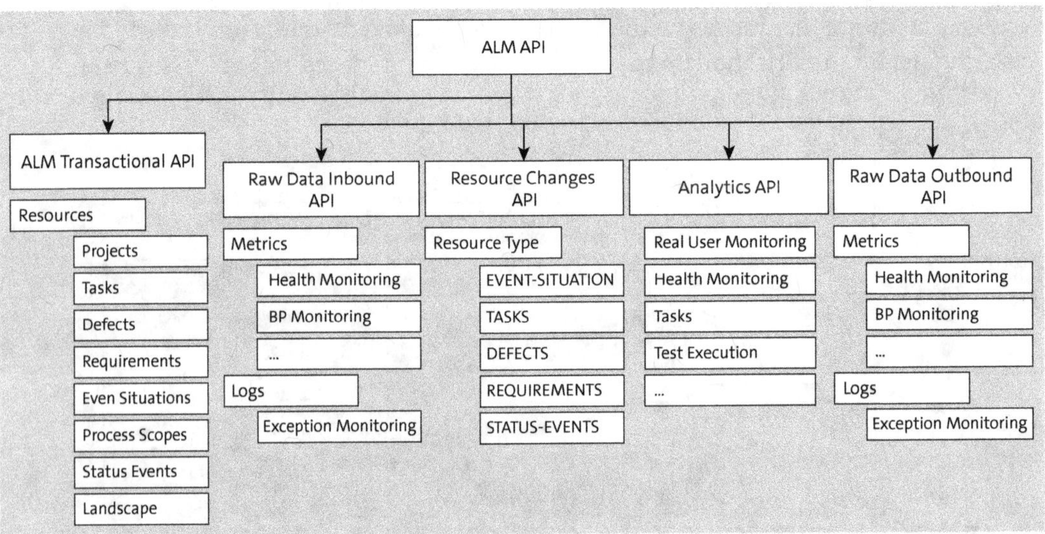

Figure 21.2 Overview of the Different Interface Types

[»]

SAP Business Accelerator Hub and GitHub

The SAP Cloud ALM APIs are documented in SAP Business Accelerator Hub. You can find an overview at *https://api.sap.com/package/SAPCloudALM/overview*.

SAP also provides examples and best practices, which are available on GitHub. For example, you can find a Postman collection that you can use to create a simple test scenario at:

https://github.com/SAP-samples/cloud-alm-api-examples

21.1 APIs in the Implementation Area: Build

Different types of interfaces are offered in the SAP Cloud ALM build and implementation scenarios. The following types are currently supported:

- Analytical APIs
- Transactional APIs

In the following section, we will take a closer look at the various interfaces, list their general properties, and provide you with some practical examples.

21.1.1 Analytics API

The SAP Cloud ALM Analytics API can be used to create dashboards and reports that aggregate all types of data managed by SAP Cloud ALM in the implementation and operation phase. The SAP Cloud ALM *Analytics API* provides an interface with both OData and REST endpoints for this purpose (see Figure 21.3).

Figure 21.3 Endpoints of the SAP Cloud ALM Analytics API

The following points are covered by the Analytics API:

- Uniform access to all managed data from SAP Cloud ALM using a generic API (single entry point)
- Merging data from different use cases and scenarios with a generic API
- Access to time series and generic table information
- Access to SAP Cloud ALM data within SAP Analytics Cloud or other consumers (e.g., SAP Datasphere) via OData
- Access to SAP Cloud ALM data from other applications via the REST API

Figure 21.4 shows the general sequence of a call to the Analytics API. Let us explain some of the key terms:

- **Data provider**
  Data providers are analysis data sources that correspond to the various objects or scenarios managed in SAP Cloud ALM. These can be tasks, projects, or alarms, for example.
- **Dimension**
  The *dimension* corresponds to the various fields of the SAP Cloud ALM entities (e.g., the name of the projects, the status for tasks, etc.).

- **Metrics**
  Metrics are quantitative measurements from different data sources (e.g., number of tasks, average response time); different aggregation methods are supported for each metric (e.g., average, sum, maximum, minimum, etc.).

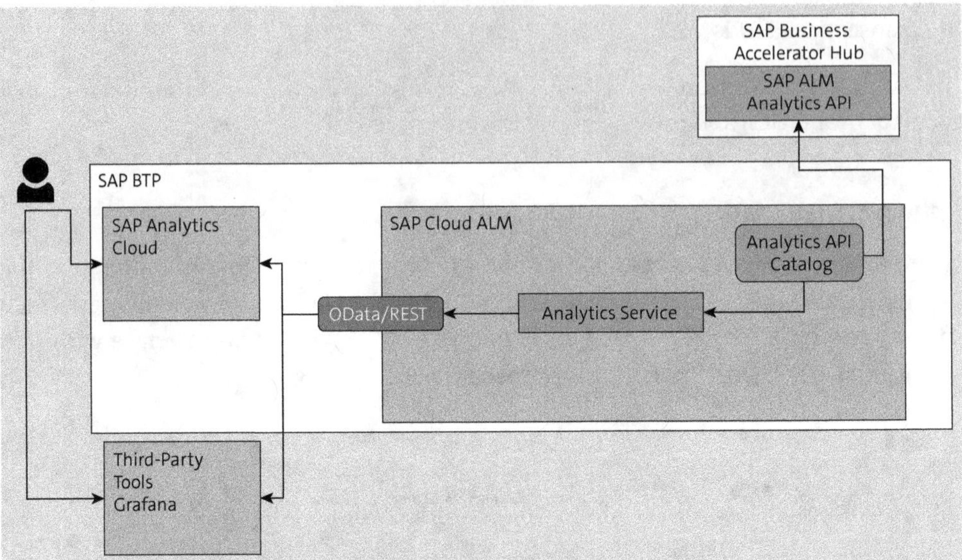

Figure 21.4 General Sequence of a Call to the Analytics API

The data from the SAP Cloud ALM Analytics API is retrieved and formatted or structured using a query. A corresponding query is responsible for returning either a set of key figures (series) or a row-and-column-structured data format (table) for a specific data provider and a specific time period based on filters and columns that are structured by selected dimensions of the data sources.

You have already learned how to test the web service calls in SAP Business Accelerator Hub. We recommend that you test the corresponding web service that you want to use later in advance.

You can find an overview of all data providers at *https://sandbox.api.sap.com/SAPCALM/calm-analytics/v1/analytics/providers*. You will receive a list of the data providers on your SAP Cloud ALM instance (see Listing 21.1).

```
{
    "name": "DP_TASKS",
    "shortName": null,
    "description": "ALM Tasks",
    "service": "imp-ca-ana-srv",
    "domain": "Analytics",
    "scope": "calm-api.tasks.read",
    "writeScope": null,
```

```
    "version": [
        "v1"
    ],
    "plan": "standard",
    "calmMarket": "BUILD"
}
```

Listing 21.1 Return of the Data Provider List (Excerpt)

You can now use additional parameters to customize the result of the query. For example, you can restrict the data provider and specify the return values accordingly. An overview of the various areas (data providers) in the implementation scenario is shown in Table 21.1.

Parameters	Description
Kind	Incoming interface
Scenario	Implementation: build
Type	Analytical interface
Analytics operations	`GET /analytics/providers`: List of all data providers with the supported metrics and dimensions `POST /analytics/providers/data`: List `POST /analytics/providers/filters`: List

Authorization	`calm-api.analytics.read`
	`calm-api.analytics.providers.read`
	`calm-api.projects.read`
	`calm-api.tasks.read`
	`calm-api.monitoring-integration.read`

Areas	`Tasks Analytics`
	`Project Analytics`
	`Test Execution Analytics`
	`Defect Analytics`
	`Requirements Analytics`

Table 21.1 Properties of the Analytics API in the Implementation Scenario

In the following example, we want to evaluate the `DEMO_TASKS` data provider. In the first example, we want to retrieve the data from the last few days. The call looks like this:

```
GET https://<tenant name>/api/calm-analytics/v1/odata/v4/analytics/DataSet?
$filter=provider eq 'DEMO_TASKS' and period eq 'L2D' and resolution eq 'D'&
$select=timestamp,period,resolution,date,week,dayOfWeek,week,provider,d2k,
d2v,m1k,m1v
```

The result of this call is shown in Listing 21.2.

```
{
    "@context" : "$metadata#DataSet(ID,period,resolution,provider,timestamp,
date,week,dayOfWeek,d2k,d2v,d3k,d3v,m1k,m1v)",
    "@metadataEtag" : "W/\"97a60111ccc922023de9898ba8d0046388fb97e729f2825b01f
5a0f134cd1203\""
    ,
    "value" : [
    {
        "@id" : "DataSet(fe63d9ad-53be-470b-b12b-b88d9cea9463)"
        ,
        "ID" : "fe63d9ad-53be-470b-b12b-b88d9cea9463" ,
        "period" : "L1D"
        ,
        "resolution" : "D" ,
        "provider" : "DEMO_TASKS"
        ,
        "timestamp" : "20210527000000"
        ,
        "date" : "20210527" ,
        "week" : "202121" ,
        "dayOfWeek" : "4" ,
        "d2k" : "project" ,
        "d2v" : "CRM" ,
        "d3k" : "name" ,
        "d3v" : "Design the feature" ,
        "m1k" : "Count" ,
        "m1v" : "1"
    }
```

Listing 21.2 Result of the Analytics API Call: Demo_Tasks (Example 1)

In the second example, we retrieve the changes of the last seven days in the data provider DEMO_TASKS as a time series. The call for this looks as follows:

```
GET https://<tenant name>/api/calm-analytics/v1/odata/v4/CalmExtApiService/
DataSet?$filter=provider eq 'DEMO_TASKS'&$select=
provider,d1k,d1v,d2k,d2v, m1k,m1v
```

The result of this call is shown in Listing 21.3.

```
{
    "@context" : "$metadata#DataSet(ID,provider,d1k,d1v,d2k,d2v,m1k,m1v)"
    ,
    "value" : [
```

```
        {
            "@id" : "DataSet(9d1fe147-6d4f-4b78-bb90-1f7e80a8fdcb)"
            ,
            "ID" : "9d1fe147-6d4f-4b78-bb90-1f7e80a8fdcb" ,
            "provider" : "DEMO_TASKS"
            ,
            "d1k" : "PROJECT" ,
            "d1v" : "Implementation Project" ,
            "d2k" : "STATUS" ,
            "d2v" : "Open" ,
            "m1k" : "TASK_STATUS_COUNTER" ,
            "m1v" : "333" },
        {
            "@id" : "DataSet(a80bfe5e-21ad-4018-b698-f09b42c14b9d)"
            ,
            "ID" : "a80bfe5e-21ad-4018-b698-f09b42c14b9d" ,
            "provider" : "DEMO_TASKS"
            ,
            "d1k" : "PROJECT" ,
            "d1v" : "test project 1" ,
            "d2k" : "STATUS" ,
            "d2v" : "Open" ,
            "m1k" : "TASK_STATUS_COUNTER" ,
            "m1v" : "11" },
        {
            "@id": "DataSet(744a1929-e370-40ec-b823-cd91f0b2342e)",
            "ID": "744a1929-e370-40ec-b823-cd91f0b2342e",
            "provider": "DEMO_TASKS",
            "d1k": "PROJECT",
            "d1v": "Implementation Project",
            "d2k": "STATUS",
            "d2v": "In Progress",
            "m1k": "TASK_STATUS_COUNTER",
            "m1v": "3"
        }
    ]
}
```


Listing 21.3 Result of the Analytics API Call: Demo_Tasks (Example 2)

21.1.2 Project API

The *Project API* processes operations for reading project information or creating a project. Table 21.2 provides an overview of its functions.

Parameters	Description
Kind	Incoming interface
Scenario	Implementation: build
Type	Transactional interface
URL format	*https://<tenant name>.<region>.alm.cloud.sap/api/calm-projects/v1/projects*
Project operations	`GET /projects`: List of all projects `POST /projects`: Create a new project
Authorization	`calm-api.projects.read`: Read authorization projects `calm-api.projects.write`: Write authorization for projects

Table 21.2 Properties of the Project API

The following call returns the list of all projects:

```
GET https://<tenant name>.<region>.alm.cloud.sap/api/calm-projects/v1/projects
```

The result of this call is shown in Listing 21.4.

```
{
    "projectId" : "56381be7-f1d7-4de2-97a2-54990391b6e7"
    ,
    "title" : "0db16c53-418b-43d6-a34a-3cf7f819845f",
    "type": "Review Overview Documentation" ,
    "status": "Review Overview Documentation",
    "externalId" : "CIPTKOPEN",
    "dueDate" : "2021-02-26",
}
```

Listing 21.4 Result List of the Project API Call

21.1.3 Task API

The *Task API* processes operations for reading or creating tasks. Table 21.3 provides an overview of the functions.

Parameters	Description
Kind	Incoming interface
Scenario	Implementation: build

Table 21.3 Properties of the Task API

Parameters	Description
Type	Transactional interface
URL format	*https://<tenant name>.<region>.alm.cloud.sap/api/calm-tasks/v1/tasks*
Task operations	GET /tasks: List of all tasks POST /tasks: Create a new task PATCH /tasks: Change the task DELETE /tasks: Delete the task
Authorization	calm-api.tasks.read calm-api.tasks.write

Table 21.3 Properties of the Task API (Cont.)

As an example, we would like to read out the various tasks for a specific project. The call looks like this:

```
GET: https://eu10.alm.cloud.sap/api/calm-tasks/v1/tasks?projectId=35b22fcc-cd63-
4d6f-97ba-3a2be95a7367
```

The result of this call looks like Listing 21.5.


```
[
    {
        "id": "0034f9f2-b187-486b-8b97-92b349b19520",
        "projectId": "35b22fcc-cd63-4d6f-97ba-3a2be95a7367",
        "title": "Setup Project",
        "type": "CALMTMPL",
        "status": "CIPTKOPEN",
        "externalId": null,
        "dueDate": null
    },
    {
        "id": "660a3ad4-5c2d-45ea-97d7-94f016502670",
        "projectId": "35b22fcc-cd63-4d6f-97ba-3a2be95a7367",
        "title": "Setup Central Business Configuration Integration (Optional)",
        "type": "CALMTMPL",
        "status": "CIPTKOPEN",
        "externalId": null,
        "dueDate": null
    }
]
```


Listing 21.5 Result of the Task API Call

21.1.4 Custom Process API

The *Custom Process API* allows you to upload, modify, and activate SAP Cloud ALM custom solution processes. Table 21.4 provides an overview of the functions. The calls of the individual operations for the business processes, solution processes, assets, etc., are the same or similar, which is why we have omitted the labeling in the lower part of the table. We then provide two examples of how to call the API: The first example shows how to read business processes, while the second example shows how to read solution processes.

Parameters	Description
Kind	Incoming interface
Scenario	Implementation: build
Type	Transactional interface
URL format	*https://{tenant}.{region}.alm.cloud.sap/api/calm- processauthoring/v1*
Business process operations	`GET/businessProcesses`: List of business processes `POST/businessProcesses`: Creating a business process `GET/businessProcesses/$count`: Number of business processes `GET/businessProcesses/{businessProcessId}`: Read a business process by ID `PATCH/businessProcesses/{businessProcessId}`: Change a business process by ID `DELETE/businessProcesses/{businessProcessId}`: Delete a business process by ID
Solution process operations	`GET/solutionProcesses`: List of a solution process `POST/solutionProcesses`: Creating a solution process `GET/solutionProcesses/$count`: Number of solution processes `GET/solutionProcesses/{solutionProcessId}`: Reading a solution process `PATCH/solutionProcesses/{solutionProcessId}`: Edit a solution process `DELETE/solutionProcesses/{solutionProcessId}`: Delete a solution process `POST /publishSolutionProcess`: Publish a solution process `POST /createDraftSolutionProcess`: Creating a (draft) solution process

Table 21.4 Properties of the Custom Process API

Parameters	Description
Asset operations	GET/solutionProcesses/{solutionProcessId}/assets: List of assigned assets POST/solutionProcesses/{solutionProcessId}/assets: Creating and assigning an asset GET/solutionProcesses/{solutionProcessId}/assets/$count: Number of assigned assets GET/assets: List of all assets GET/assets/{assetId}: Reading an asset by ID DELETE/assets/{assetId}: Deletion of an asset PATCH/assets/{assetId}: Change an asset GET/assets/$count: Number of assets
Solution process flow operations	GET/solutionProcesses/{solutionProcessId}/solutionProcessFlows GET/solutionProcesses/{solutionProcessId}/solutionProcessFlows/$count GET/solutionProcessFlows GET/solutionProcessFlows/$count GET/solutionProcessFlows/{solutionProcessFlowId} PATCH/solutionProcessFlows/{solutionProcessFlowId}
Solution flow diagram operations	GET/solutionProcesses/{solutionProcessId}/solutionValueFlowDiagram GET/solutionValueFlowDiagrams GET/solutionValueFlowDiagrams/$count GET/solutionValueFlowDiagrams/{solutionValueFlowDiagramId} PATCH/solutionValueFlowDiagrams/{solutionValueFlowDiagramId}
Solution process flowchart operations	GET/solutionProcessFlows/{solutionProcessFlowId}/solutionProcessFlowDiagrams GET /solutionProcessFlows/{solutionProcessFlowId}/solutionProcessFlowDiagrams/$count GET/solutionProcessFlowDiagrams GET/solutionProcessFlowDiagrams/$count GET/solutionProcessFlowDiagrams/{solutionProcessFlowDiagramId} POST/solutionProcessFlowDiagrams/{solutionProcessFlowDiagramId} PATCH/solutionProcessFlowDiagrams/{solutionProcessFlowDiagramId} DELETE/solutionProcessFlowDiagrams/{solutionProcessFlowDiagramId}
Authorization	calm-api.processauthoring.read calm-api.processauthoring.write

Table 21.4 Properties of the Custom Process API (Cont.)

In our first example, we call up the first five business processes. The call to the Custom Process API to read the business processes looks like this:

```
GET: https://<tenant name>.<region>.alm.cloud.sap/SAPCALM/calm-
processauthoring/v1/businessProcesses?$skip=10&$top=6&$count=true&$orderby=asc
```

As a result of the request, you will receive a response in the format shown in Listing 21.6.

```
{
    "count": "5",
    "value":
        [
            {
                "name": "I2P_A456_AUTO",
                "description": null,
                "id": "55294fcd-911d-4b06-800e-445950c560bf"
            },
            {
                "name": "new Business Process 01 ssk",
                "description": null,
                "id": "4fe050ac-4f09-4df7-9ff9-e61f1b2d642b"
            },
            {
                "name": "AO Testing",
                "description": null,
                "id": "55c29644-62f9-4058-a595-cf96bac24e9e"
            },
            {
                "name": "My Process (Variant 1)",
                "description": null,
                "id": "88f6a363-2828-460b-b98a-6e9e5a34225c"
            },
            {
                "name": "Subscription Billing Integration",
                "description": null,
                "id": "52ad93cc-ed1a-46dc-869d-7341bf13ec68"
            }
        ]
]
```

Listing 21.6 Result of the Call to the Custom Process API: Business Processes

With the following call to the Custom Process API, we only receive the details of a specific business process:

```
GET https://sandbox.api.sap.com/SAPCALM/calm-processauthoring/v1/
businessProcesses/4fe050ac-4f09-4df7-9ff9-e61f1b2d642b
```

The return of our response looks like in Listing 21.7.

```
{
    "name": "new Business Process 01 ssk",
    "description": null,
    "id": "4fe050ac-4f09-4df7-9ff9-e61f1b2d642b"
}
```

Listing 21.7 Result of the Custom Process API Call: Concrete Business Process

21.1.5 Process Scopes API

The *Process Scopes API* allows you to read and modify SAP Cloud ALM custom solution scopes. Table 21.5 provides an overview of the functions.

Parameters	Description
Kind	Incoming interface
Scenario	Implementation: build
Type	Transactional interface
URL format	*https://{tenant}.{region}.alm.cloud.sap/api/calm-processmanagement/v1*
Scope operations	`GET /scopes`: Lists of all scopes `POST /scopes`: Create a scope `GET /scopes/$count`: Number of scopes `GET /scopes/{id}`: Read a scope by ID `PATCH /scopes/{id}`: Change a scope `DELETE /scopes/{id}`: Deletion of a scope

Table 21.5 Properties of the Process Scopes API

21

Parameters	Description
Solution scenario operations	GET /scopes/{id}/solutionScenarioVersions: Read a solution scenario by scope POST /scopes/{id}/solutionScenarioVersions: Assign one or more solution scenario(s) per scope GET /scopes/{id}/solutionScenarioVersions/$count: Number of solution scenario versions per scope GET /solutionScenarioVersions: Retrieve a list of solution scenario versions, filtered by the specified query parameters GET /solutionScenarioVersions/$count: Determine the total number of versions of the solution scenario GET /solutionScenarioVersions/{id}: Retrieve a specific version of the solution scenario, identified by the ID
Solution process operations	GET /solutionProcesses: Retrieve a list of solution process scopes and their scope status GET /solutionProcesses/$count: Retrieve the number of solution processes with the given filter criteria PATCH /solutionProcesses/scopeAssignments: Changing a scope process assignment
Authorization	calm-api.processmanagement.read calm-api.processmanagement.write

Table 21.5 Properties of the Process Scopes API (Cont.)

The Process Scopes API is called using the following command:

```
GET https://<tenant name>.<region>.alm.cloud.sap/api/calm-pm/v1/scopes/
{scopeUuid}
```

Listing 21.8 shows the result of the call to the Process Scopes API.

```
{
    "id": "b82ec4ec-6225-4cc4-a758-c1ff3ef25c9b",
    "projectId": "11111111-1111-1111-1111-111111111111",
    "name": "Scope name",
    "description": "Scope description"
}
```

Listing 21.8 Result of the Call to the Process Scopes API

21.2 APIs in the Operation Area: Run

Different types of interfaces are also used in the run or operating phase of SAP Cloud ALM. The following interface types are currently supported (see Figure 21.2):

- Analytical APIs
- Transactional APIs
- Integration scenarios
- Raw data APIs

Examples of further integration scenarios can be found in Chapter 22.

[«]

External API Management: Operating Phase

Further information on the interfaces in the run scenario can be found in the SAP Cloud ALM for operations expert portal in the SAP Service Marketplace under the keyword **External API Management**: *https://support.sap.com/en/alm/sap-cloud-alm/operations/expert-portal/calm-apis-for-operations.html*

21.2.1 Analytics API

Functionally, the same options and features apply in the operation scenario as in the implementation part. Particularly in the area of operating (cloud) applications, requirements are often condensed into a list of KPIs that are displayed within a dashboard. Here, SAP Cloud ALM offers a comprehensive tool using the Analytics API, which can be integrated into SAP Analytics Cloud or Grafana, for example. In Figure 21.4 shows schematically how data is provided in the various use cases. The details of the interfaces described here can be found in SAP Business Accelerator Hub. The following use cases are currently supported specifically in the operation of SAP Cloud ALM:

- Business process monitoring
- Integration and exception monitoring
- Real user monitoring
- Job and automation monitoring
- Health monitoring
- Business service management

Table 21.6 provides an overview of the functions.

Parameters	Description
Kind	Incoming interface
Scenario	Operation: run

Table 21.6 Properties of the Analytics API in the Operating Phase

Parameters	Description
Type	Analytical interface
Analytics operations	GET /analytics/providers: List of all data providers with the supported metrics and dimensions POST /analytics/providers/data: Retrieve analytics data in table format or in time series format POST /analytics/providers/filters: List of all filters from data providers
Authorization	calm-api.analytics.read calm-api.analytics.providers.read calm-api.bm.read calm-api.monitoring-integration.read
Areas	Business Process Monitoring Analytics Integration & Exception Monitoring Analytics Real User Monitoring Analytics Job & Automation Monitoring Analytics Health Monitoring Analytics Business Service Management Analytics

Table 21.6 Properties of the Analytics API in the Operating Phase (Cont.)

21.2.2 Status Event API

The Status Event API allows you to extract status events from SAP Cloud ALM. These status events are disruptions, degradations, or maintenances. Table 21.7 provides an overview of the functions.

Parameters	Description
Kind	Incoming interface
Scenario	Operation: run
Type	Transactional interface
URL format	*https://<tenant name>.<region>.alm.cloud.sap/api/bsm-service/v1*
Task operations	GET /eventsDATA: OData V4 API for returning all events GET /events: REST API for returning all events
Authorization	calm-api.bsm.read

Table 21.7 Properties of the Status Event API

You can use the Status Event API to query the current events and their status. During the query, you can select filter settings, specify services, or sort the results list. In the following example, we would like to list two events and the associated details:

```
GET: https://sandbox.api.sap.com/SAPCALM/bsm-service/v1/eventsData?$top=2
```

The result is a list of the selected events (see Listing 21.9).


```
{
    "@context": "$metadata#eventsData",
    "@metadataEtag": "W/\"8923fc8e33b892a7ac2052c6d7e32d88133227c08b8e58dd71e3e8
d1fac52811\"",
    "value":
    [
        {
            "eventId": "90976d3d-bdea-43f7-9b1d-6a1d5cd14c1a",
            "eventName": "EV17401796",
            "eventDescription": "",
            "eventType": "Maintenance",
            "eventSource": "SAP",
            "eventStartTime": "2023-01-22T04:00",
            "eventEndTime": "2023-01-22T11:00",
            "eventDuration": 420,
            "eventStatus": "Closed",
            "serviceName": "cdh",
            "serviceDescription": "cdh (OM1P1D4)",
            "serviceId": "911bbf7c-dc8b-4698-9a65-6238dcfc1dc0",
            "systemNumber": "740066724",
            "serviceType": "SAP SuccessFactors HXM Suite",
            "type": "CloudService"
        },
        {
            "eventId": "e9641430-be51-4d35-98b0-2d716c112c4d",
            "eventName": "EV13071758",
            "eventDescription": "",
            "eventType": "Maintenance",
            "eventSource": "SAP",
            "eventStartTime": "2022-10-02T04:01",
            "eventEndTime": "2022-10-02T08:27",
            "eventDuration": 266,
            "eventStatus": "Closed",
            "serviceName": "cdh",
            "serviceDescription": "cdh (OM1P1D4)",
            "serviceId": "911bbf7c-dc8b-4698-9a65-6238dcfc1dc0",
            "systemNumber": "740066724",
```

```
            "serviceType": "SAP SuccessFactors HXM Suite",
            "type": "CloudService"
        }
    ]
}
```

Listing 21.9 Result of the Call to the Status Event API

21.2.3 Event Situations API

You can use the *Event Situations API* to close an event situation and all open alarms and external tickets that are assigned to the event situation. Table 21.8 provides an overview of the functions.

Parameters	Description
Kind	Incoming interface
Scenario	Operation: run
Type	Transactional interface
URL format	*https://<tenant name>.<region>.alm.cloud.sap/api/calm-event-situations/v1*
Task operations	`PUT /event-situations/{id}`: Close an event situation, alerts, and external tickets with ID `PUT /event-situations/externalid/{id}`: Close an event situation, from alerts with external ID
Authorization	`calm-api.event-situations.write` scope

Table 21.8 Properties of the Event Situations API

In our example, we want to close a specific event situation. To do this, we need an ID and a body that we can pass to the API call. The call then looks like in Listing 21.10.

```
PUT https://<tenant name>.<region>.alm.cloud.sap/api/calm-event-situations/v1/
event-situations/04b4bb08-d142-3eff-bfff-42043fe6aad4
{
"state": "CLOSED"
}
```

Listing 21.10 Calling the Event Situations API

21.2.4 Resource Changes API

You can use the *Resource Changes API* to monitor changes to SAP Cloud ALM objects and be informed of any changes. The API is implemented as a webhook and reacts to value changes in SAP Cloud ALM entities. The information sent contains data on which resources have been changed and how. Resources are entities that are described in the SAP Cloud ALM API catalog and that are accessible via the SAP Cloud ALM REST API (e.g., EVENT-SITUATION.CREATED). Table 21.9 provides an overview of the functions.

The Resource Changes API contains the following information:

- Event type that describes the type of change event (e.g., EVENT-SITUATION.CREATED)
- Type of resource (e.g., EVENT SITUATION)
- Payload of the resource, which is described in SAP Business Accelerator Hub

Parameters	Description
Kind	Outgoing interface
Scenario	Operation: run
Type	Transactional interface
URL format	*https://<tenant name>.<region>.alm.cloud.sap/api/calm-event-situations/v1*
Resource change operations	POST /resource-change Supported resource types: EVENT SITUATION Supported event types: EVENT-SITUATION.CREATED EVENT-SITUATION.UPDATED EVENT-SITUATION.CLOSED EVENT-SITUATION.PING
Authorization	The user needs access to the SAP BTP destination from the SAP Cloud ALM tenant to the webhook provider.

Table 21.9 Properties of the Resource Changes API

The Resource Changes API is called via a destination; details on configuration can be found in Chapter 22, Section 22.1.1. In addition to the resource types already presented here, there will be others in the near future, such as TASK, FEATURE, and TEST-CASE. These will allow for a more flexible use of the Resource Changes API.

21.2.5 Raw Data Outbound Metrics API

The *Raw Data Outbound Metrics API* can be used to read metrics in the OpenTelemetry Metrics Exporter format, e.g. via a third-party tool using the *OpenTelemetry Protocol* to integrate. The number of data points returned by the API depends on the time period parameters and time resolution used in the API calls.

[»]

OpenTelemetry Metrics Exporter Format

The OpenTelemetry Metrics Exporter format refers to the specific structure and sequence used to export metrics from applications to external monitoring systems. This format ensures consistency and interoperability so that different OpenTelemetry-compatible applications and monitoring backends can communicate seamlessly with each other.

Table 21.10 provides an overview of the functions.

Parameters	Description
Kind	Outgoing interface
Scenario	Operation: run
Type	Raw data interface
URL format	*https://<tenant name>.<provider name>/api/calm-metrics/v1/metrics*
Resource change operations	`GET /metrics` Supported scenarios: `hm`: Health monitoring (see Chapter 19) `rum`: Real user monitoring (see Chapter 19)

Authorization	`calm-api.metrics.read`
	`calm-api.analytics.read`
	`calm-api.hm.read`
	`calm-api.rum.read`

Table 21.10 Properties of the Raw Data Outbound Metrics API

The example shows the call of the Raw Data Outbound Metrics API in the health monitoring scenario. The call looks like this:

```
GET https://sandbox.api.sap.com/SAPCALM/calm-metrics/v1/metrics?format=text&
provider=hm&resolution=5Mi&shift=1&metrics=measure%3Avalue&dimensions=
serviceType%2CserviceName%2CserviceId%2CmetricName%2CmetricLabel&filter1=
filter1Value&filter2=filter2Value
```

The result of the call (excerpt) is shown in Listing 21.11.

```
# TYPE value gauge
# HELP value Value of the metric
value{measure="value",serviceType="SAP_CC",metricName=
"Cloud.Connector.Certificates",metricLabel="Alias:08e4bee1-fede-46bd-92c9-
7ddba69c2131;Type:subaccount",serviceId="e5fd8c2e-4326-4813-befa-
```

```
23b22c148059",serviceName="SCC"}. 21.0
1690146000000
value{measure="value",serviceType="SCP_HCAAS",metricName=
"HDBConnectionCount",metricLabel="Host:fd6725a4-52fb-4eb1-9dd3-10ff9d51deab-
38df1;Port:30001;Servicename:nameserver",serviceId="e5fd8c2e-4326-4813-befa-
23b22c148051",serviceName="MyHanaCloud_01"} 6.0 1690146000000
value{measure="value",serviceType="SAP_CP_NEO",metricName=
"CP.NEO.Busy.Threads",metricLabel=
"Application:sysmontasksapplication;Process:2328d5a2e38222f85af42adc3ed8c4a1fc9
352dd",serviceId="5f84da21-0f35-41f3-a494-c656c535ac26",serviceName="MyBtpNeo_
01"} 0.0 1690146000000
value{measure="value",serviceType="IRPA",metricName=
"Distribution.processed",metricLabel="Environment:Tokio",serviceId="92c4b1d0-
f0be-42ac-9688-2f0655419827",serviceName="MyIRPA_01"} 0.0
1690146000000
value{measure="value",serviceType="SCP_HCAAS",metricName=
"HDBSystemCPU",metricLabel="Host:fd6725a4-52fb-4eb1-9dd3-10ff9d51deab-
38df1;Port:30001;Servicename:nameserver",serviceId="e5fd8c2e-4326-4813-befa-
23b22c148051",serviceName="MyHanaCloud_01"} 0.0 1690146000000
```

Listing 21.11 Result of the Call (Excerpt) of the Raw Data Outbound Metrics API

21.3 Summary

This chapter has given you a general overview of the different types of interfaces in SAP Cloud ALM. In the following chapter, we will give you an outlook on the upcoming enhancements in this context. In addition to the enhancements to the Resource Changes API already mentioned, there will also be enhancements in the analytics environment. It is planned to expand the synthetic user monitoring, alert management, and configuration and security scenarios as well as the features/transports. The following transactional interfaces are also to be added: landscape, features, test cases, solution documentation, and inbound raw data logs. As you can see, the possibilities for reading, adapting, or deleting data in SAP Cloud ALM will continue to increase in the coming months, and SAP Cloud ALM will grow into an attractive ecosystem with a wide range of possible applications. In the next chapter, we would like to look at some exciting application examples that can be mapped with the currently published interfaces.

Chapter 22
Practical Examples

Now that we have laid the theoretical foundations, we want to examine practical examples in detail in this chapter. Although we will focus on SAP Cloud ALM, we will also look at the integrated applications.

Everything always sounds easier in theory than it actually is in practice. We will therefore now take a closer look at the two basic integration options using practical examples.

First of all, in Section 22.1, a push API replicates the alerts generated in the integration and exception monitoring as GitHub issues. We will go into the necessary configurations of SAP Cloud ALM and GitHub in detail. Finally, we look at the practical use of the configured integration.

In Section 22.2, we use a pull API to replicate project tasks from SAP Cloud ALM as personal tasks in Microsoft To Do. Here, too, we consider both the configuration on the SAP Cloud ALM side and on the Microsoft API side. We use Postman as an intermediary between the two applications. Postman is a tool for developing integrations and offers the option of calling interfaces explicitly. See more details for Postman at *https://www.postman.com*.

Both integrations are also available in our GitHub repository (*https://github.com/SAP-Cloud-ALM-Book*).

22

[!]

Ever-Changing Interfaces

As SAP is constantly developing the SAP Cloud ALM interfaces, we can only reflect the status at the time of writing this book. It is therefore possible that individual configuration steps or interfaces may have already changed, although this should not alter the basic method of implementation.

22.1 Creating GitHub Issues for Alarms in Integration Monitoring

With the increasing establishment of DevOps as an operating concept, smaller project teams or organizations are increasingly taking on responsibility for operating an application or its components. The aim is to keep the process of operation as lean as possible and to tailor it precisely to the needs of the team. Extensive fault reporting processes take a back seat, while processes integrated into development take over. Source code management platforms such as GitHub, GitLab, or something similar are increasingly being used. This offers the great advantage that source code, feature planning, and bug reports are brought together in one place, while the processes are streamlined in that way, and that no extensive creation masks or regulations need to be adhered to. In this section, we will discuss the integration of integration and exception monitoring alerts in a GitHub repository. As soon as the integration and exception monitoring detects a faulty message, a corresponding issue is created on GitHub, which can be further processed by the developers.

Figure 22.1 illustrates the logical link between the various SAP Cloud ALM modules and the GitHub API. In our case, the Intelligent Event Processing app of SAP Cloud ALM acts as a logical link between integration monitoring and exception monitoring. When looking into other integrations, you'll see that this is not necessarily required from technical point of view.

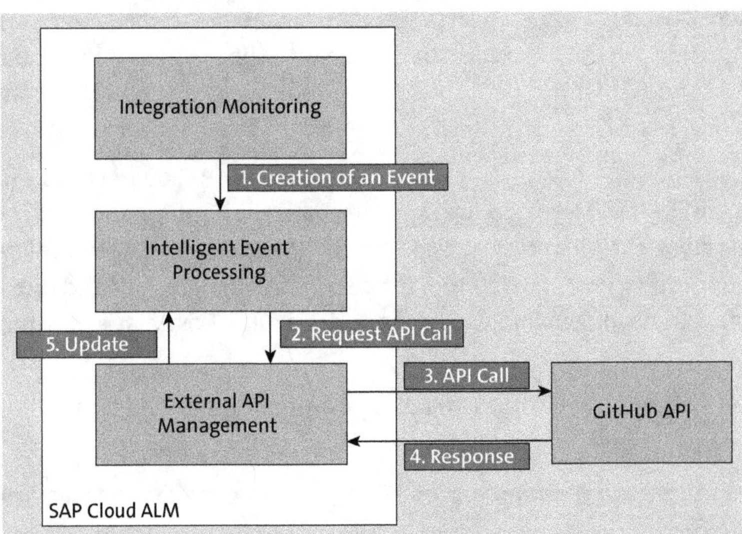

Figure 22.1 Schematic Representation of the Integration of GitHub in SAP Cloud ALM via Webhook

22.1.1 GitHub API for Issues

Before configuring a webhook in SAP Cloud ALM, you must first familiarize yourself with the interface of the target application. GitHub provides extensive interfaces in almost all areas. The basic concept of the interface, such as authentication, is the same in each case, meaning that the various interfaces in the application differ only in the data required or provided. The central endpoint of the interfaces is provided at *https://api.github.com*. The documentation of the interfaces can be found at *https://docs.github.com/rest*.

Calling an interface from GitHub requires the authentication of the user, whereby various types are available. Please note that not every authentication type is supported by SAP BTP. As our focus is on SAP Cloud ALM and not SAP BTP, we use the simplest authentication option, the *personal access token*. The personal access token is a defined access in the context of your GitHub user. The token and the permitted scope of access can be configured in the GitHub profile settings in the **Developer Settings • Personal access tokens** menu. Since our integration scenario only creates and edits issues, authorizations for reading and writing issues as well as reading the metadata of a repository, as shown in Figure 22.2, are sufficient for creating a new token.

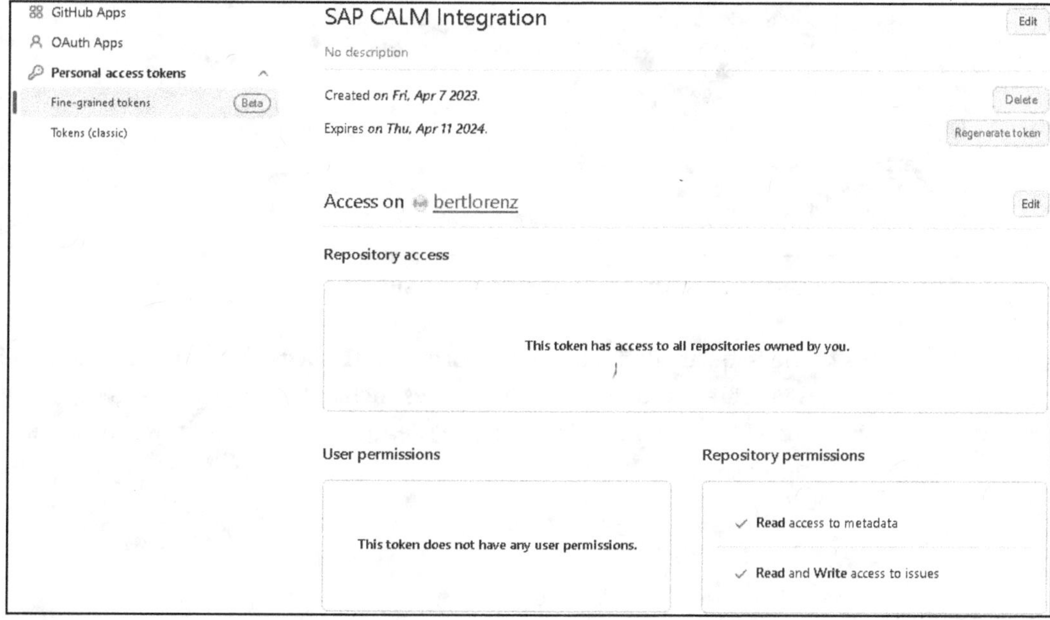

Figure 22.2 GitHub Personal Access Token

22

The generated personal access token can then be used as a password for requests to the interfaces of GitHub.

Issues are created via a repository-specific endpoint. This means that the target repository is not included in the data of a request, i.e., the `body`, but that each repository has its own endpoint. Based on an example repository with the name `demo_sap_calm_webhook_integration`, this results in the following endpoint for the issue interface: *https://api.github.com/repos/bertlorenz/demo_sap_calm_webhook_integration/issues.* A `POST` operation is required to create a new issue. The username and password are supplied via `BasicAuthentication`, as shown in Figure 22.3.

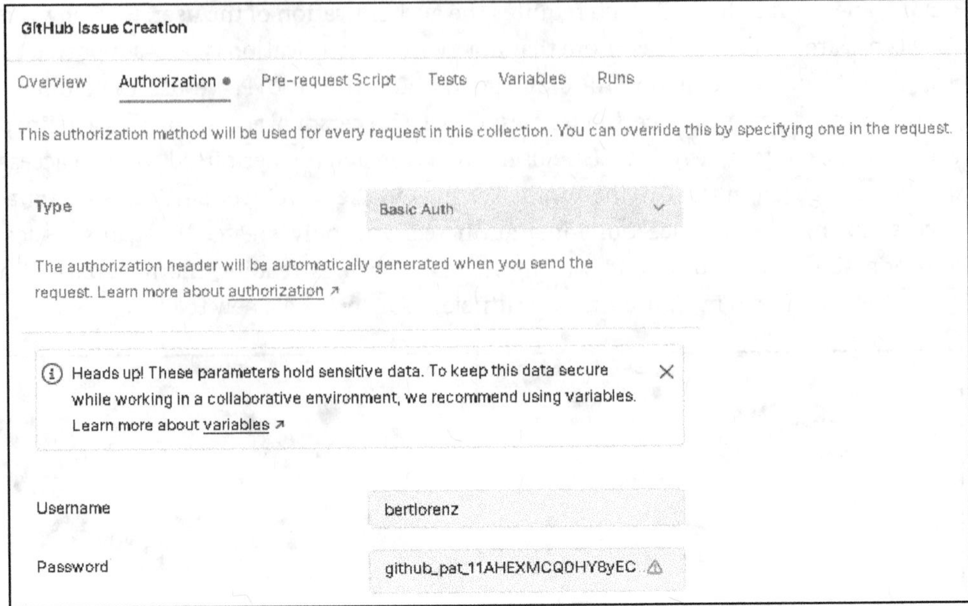

Figure 22.3 Postman Authorization for Creating an Issue

The details of the issue are defined in JSON format in the body of the request. An issue requires at least a `title`, although other properties such as the body, an `assignee` or several `assignees`, `milestones` and `labels` can also be defined. Listing 22.1 shows an example body for creating an issue.

```
{
    "title": "Found a bug",
    "body": "I'm having a problem with this."
}
```

Listing 22.1 Body of the HTTP Request to Create an Issue in GitHub

If the request is successful, the entire issue is returned as a response in JSON format. An important property of the issue for subsequent identification is not the ID, such as 1671744317, but the number, such as 169, as this is also unique in combination with the repository. The issue can be updated using patch requests to the specific endpoint *https://api.github.com/repos/bertlorenz/demo_sap_calm_webhook_integration/issues/169*.

Comments on issues are created via an interface below the issue. Using the example of issue number 169, comments can be managed via the following endpoint: *https://api.github.com/repos/bertlorenz/demo_sap_calm_webhook_integration/issues/169/comments*. To create a comment, only the body property needs to be defined. In response, GitHub also returns the entire object of the comment:

```
{
    "body": "I got the same issue!"
}
```

With the basics of the GitHub interface shown, you will be able to create new issues in a repository, update them, and add further comments.

22.1.2 External API Management

The External API Management app is used for the central administration of webhooks. However, central does not currently mean that all possible webhooks are actually configured in this functionality. For example, the integration of test automation is configured directly in the test management application and not in the External API Management app. SAP will certainly harmonize this in the future. For our integration, we need a subscription with an associated webhook and suitable mappings as already described in Chapter 20, Section 20.4.

The webhook defines in connection with a destination of the SAP BTP the endpoint of the target application. In our example, the GitHub Issue API is to be called for a defined repository. The endpoint is shown in Figure 22.4.

The SAP BTP destination contains the general access to the GitHub API. In addition to the root URL, authentication is also carried out at this level. As an example, we have used a user-specific personal access token. Here, the authentication method of the SAP BTP destination must be set to **BasicAuthentication**. The username can then be entered in the **User** field and the generated personal access token in the **Password** field. The configured SAP BTP destination is shown in Figure 22.5.

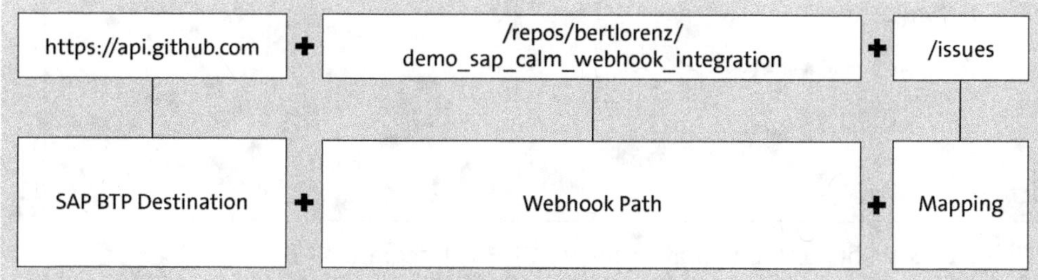

Figure 22.4 Subdivision of the GitHub API Endpoint

Once the SAP BTP destination has been successfully created and checked, it can be referenced in the webhook. The integration-specific path is stored in the **Path** field. We use the **Incident** type as the **External Resource Type**, as we want to create a type of fault message in GitHub. As already described in Chapter 21, the **Chat** and **Task** types are also available for selection in addition to the **Incident** type. The **Destination Type** field can be left empty. The configuration of the webhook is shown in Figure 22.6.

Destination Configuration

Name:*	GitHub_API	**Additional Properties**
Type:	HTTP	
Description:		Use default JDK truststore
URL:*	https://api.github.com	
Proxy Type:	Internet	
Authentication:	BasicAuthentication	
User:*	bertlorenz	
Password:	••••••••	

Figure 22.5 SAP BTP Destination for the Connection to the GitHub API

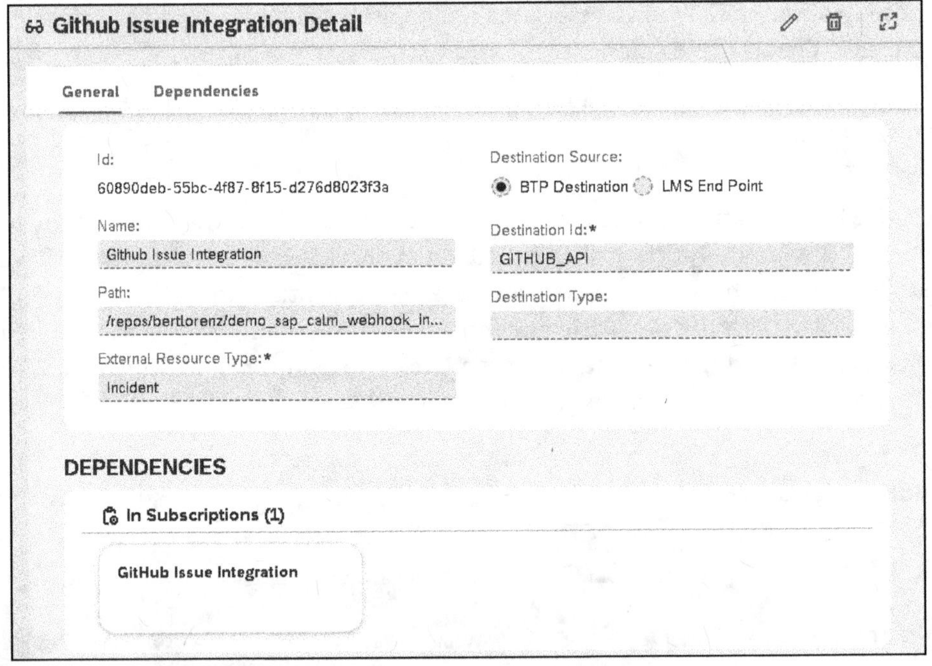

Figure 22.6 Webhook Configuration for the GitHub Integration

Once the webhook has been created, the corresponding mapping must be configured within the **Mappings Management** view. In addition to the description, version, source and destination, the mapping type of the mapping must also be defined. Based on the selected type, the mapping is either rule-based or payload-based. Further details on this distinction can be found in Chapter 20. As there is no translation of technical values when integrating GitHub, payload-based mapping is suitable in this case. This enables a simple definition of the target and response structure. First, the general information, as shown in Listing 22.2, is defined.

```
{
    "_information": {
    "_comment": "This Mapping prepares the request for GitHub Issues",
    "_name": "GitHub Issue Mapping",
    "_version": "1.0.0",
    "_source": "SAP CLoud ALM Integration Monitoring",
    "_destination": "GitHub Issue"
}
```

Listing 22.2 Object "information" for the GitHub Mapping

Switches for translating values are not required, so the object `structure` is defined in the next step. This object defines at a global level how the request to GitHub and the response from GitHub are to be processed. The request to create an issue must be stored as the object `request` and contains the properties `title` and `body`. The property `title` defines the short text, i.e. the title, and the property `body` the long text, i.e. the description of the issue to be created. As already described, you can define further properties when creating an issue. However, this is not necessary in our integration scenario. The response from GitHub contains all the properties of the issue created. For correct linking in SAP Cloud ALM, these properties must be assigned to the properties known to SAP Cloud ALM. In addition to the `externalId` and `externalNumber`, it is advisable to define the properties `externalStatus` and `displayURL` properties when creating the issue. The corresponding properties of the response from GitHub are assigned to the properties in SAP Cloud ALM. This provides SAP Cloud ALM with all relevant information about the issue created. The entire specification of the `structure` object is shown in Listing 22.3.

```
{
    "structure": {
        "request": {
            "_comment": "global request definition",
            "title": "#{subject}",
            "body": "#{htmlBody}"
        },
        "response": {
            "_comment": "global response definition",
            "externalId": "#{id}",
            "externalStatus": "#{state}",
            "externalNumber": "#{number}",
            "displayURL": "#{html_url}"
        }
    }
```

Listing 22.3 Object "structure" for the GitHub Mapping

The last object to be configured is the eventTypeMapping. Depending on the integration scenario, this can be the most extensive object, as a separate mapping can be defined for each event type. Table 22.1 defines the event types and the corresponding action in GitHub.

Event Type	Action in GitHub
EVENT-SITUATION.CREATED	Create issue
EVENT-SITUATION.UPDATED	Add comment to issue
EVENT-SITUATION.CLOSED	Close issue
EVENT-SITUATION.PING	Read issue

Table 22.1 Event Type with Associated Action in GitHub

The integration must act differently based on the different event types. With the EVENT-SITUATION.CREATED event type, a new issue is to be created using the POST operation. The structures of the request and response do not need to be defined again, as they have already been defined globally. For the EVENT-SITUATION.UPDATED event type, a new comment is to be created for the issue using a POST operation. The structure must be changed as a different interface is called. You can use the property Path to overwrite the path of the interface. As an issue should already exist at the time an event situation is updated, you can refer to the corresponding reference, resulting in the following configuration for Path: /issues/#{externalNumber}/comments.

The structure of the request and the response must also be defined, as the interface for creating comments requires or returns different data than the interface for creating issues. This data is used, for example, by intelligent event processing, to show the ID and provide a link to the created issue later on. The event type EVENT-SITUATION.PING queries the details of the created issue using a GET operation. The event type EVENT-SITUATION.CLOSED uses the PATCH operation to update the issue so that it is closed. The technical status of GitHub (closed) must be defined in the request. Listing 22.4 shows the entire specification of the eventTypeMapping object.

```
{
    "eventTypeMapping": {
        "_comment": "Specific event type mapping",
        "operationMappings": [
            {
                "eventType": "EVENT-SITUATION.CREATED",
                "operation": "POST",
                "path": "/issues",
```

```
                "responseRoot": ""
            },
            {
                "eventType": "EVENT-SITUATION.UPDATED",
                "operation": "POST",
                "path": "/issues/#{externalNumber}/comments",
                "responseRoot": "",
                "structure": {
                    "request": {
                        "body": "#{htmlBody}"
                    }
                }
            },
            {
                "eventType": "EVENT-SITUATION.PING",
                "operation": "GET",
                "path": "/issues/#{externalNumber}",
                "responseRoot": "",
                "structure": {
                    "request": {},
                    "response": {
                        "externalStatus": "#{state}"
                    }
                }
            },
            {
                "eventType": "EVENT-SITUATION.CLOSED",
                "operation": "PATCH",
                "path": "/issues/#{externalNumber}",
                "responseRoot": "",
                "structure": {
                    "request": {
                        "state": "closed"
                    },
                    "response": {}
                }
            }
        ]
    }
}
```

Listing 22.4 Object "eventTypeMapping" for the GitHub Mapping

22.1.3 Intelligent Event Processing

The intelligent processing of events is used to centrally bundle and process events in SAP Cloud ALM. As the details of this functionality have already been described in Chapter 18, Section 18.2.2, we would like to go into the specific points for our integration here. First, the desired event must be registered: It is recommended, especially for integration and exception monitoring, to create the event directly in the application itself and not in intelligent event processing. On the SAP Cloud ALM start page, first open the Integration and Exception Monitoring app and use the gear icon to navigate to the configuration of the systems. After selecting a system, you will be taken to its configuration. You have the option of creating a new event via the **Events** tab. In our example, faulty SAP Integration Suite messages are to be grouped with an issue. To do this, use the event name **Erroneous CPI Messages Detected(Grouped)**. The **Display Name** can be freely defined. Please note that in this specific case, the **Display Name** is available as a `subject` in the mapping of the External API Management app. We have described the `subject` in Section 22.1.2, as the title of the issue to be created. Within the **Event Actions**, the **Create Alert** toggle can be deactivated, and the **Create Ticket** toggle can be activated. When creating tickets, the already configured subscription must be selected. The configuration is shown in Figure 22.7 and in Figure 22.8.

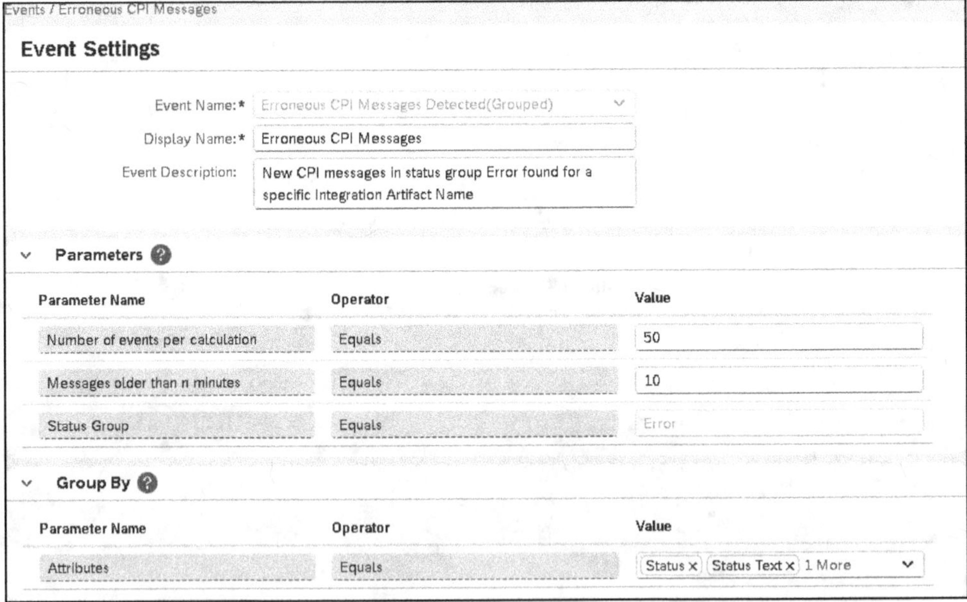

Figure 22.7 Event Configuration

After saving the configuration, the event is available in the **Event Processing Rules** menu item of the Intelligent Event Processing app, as shown in Figure 22.9.

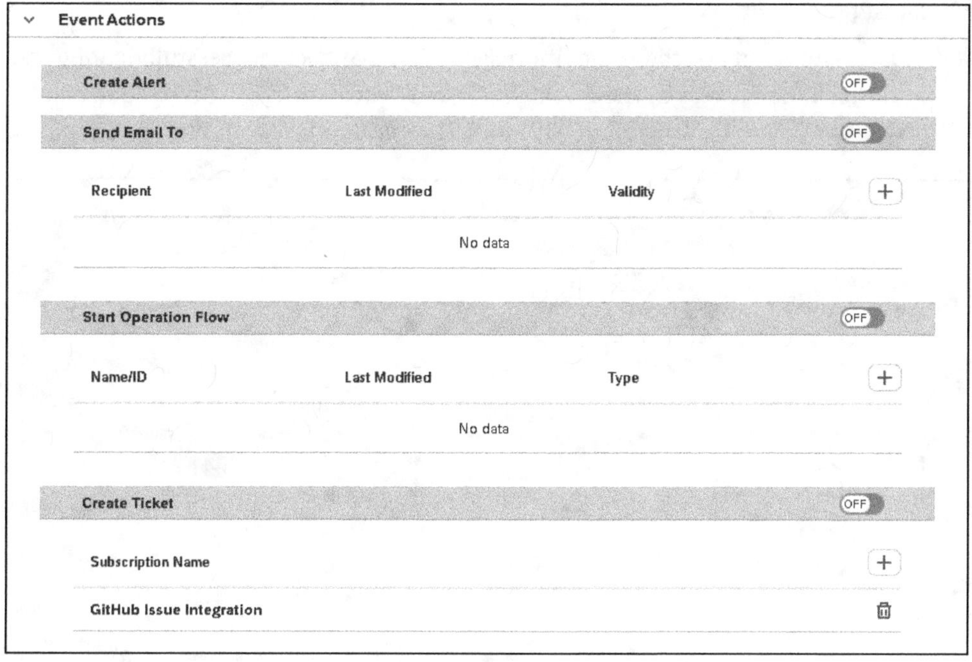

Figure 22.8 Event Action Configuration for Ticket Creation

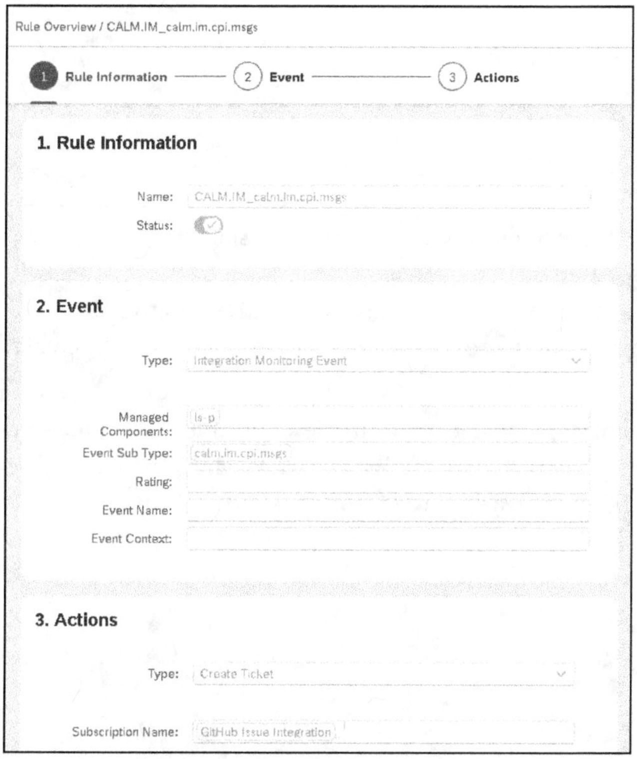

Rule Overview / CALM.IM_calm.im.cpi.msgs

1 Rule Information — 2 Event — 3 Actions

1. Rule Information

Name: CALM.IM_calm.im.cpi.msgs

Status:

2. Event

Type: Integration Monitoring Event

Managed Components: ls-p

Event Sub Type: calm.im.cpi.msgs

Rating:

Event Name:

Event Context:

3. Actions

Type: Create Ticket

Subscription Name: GitHub Issue Integration

Figure 22.9 Event Processing Rule

22.1.4 Practical Application

Once all components of the integration scenario have been successfully configured, you can track the process in the Intelligent Event Processing app. Grouped events are displayed in the overview together with the triggered actions, as shown in Figure 22.10.

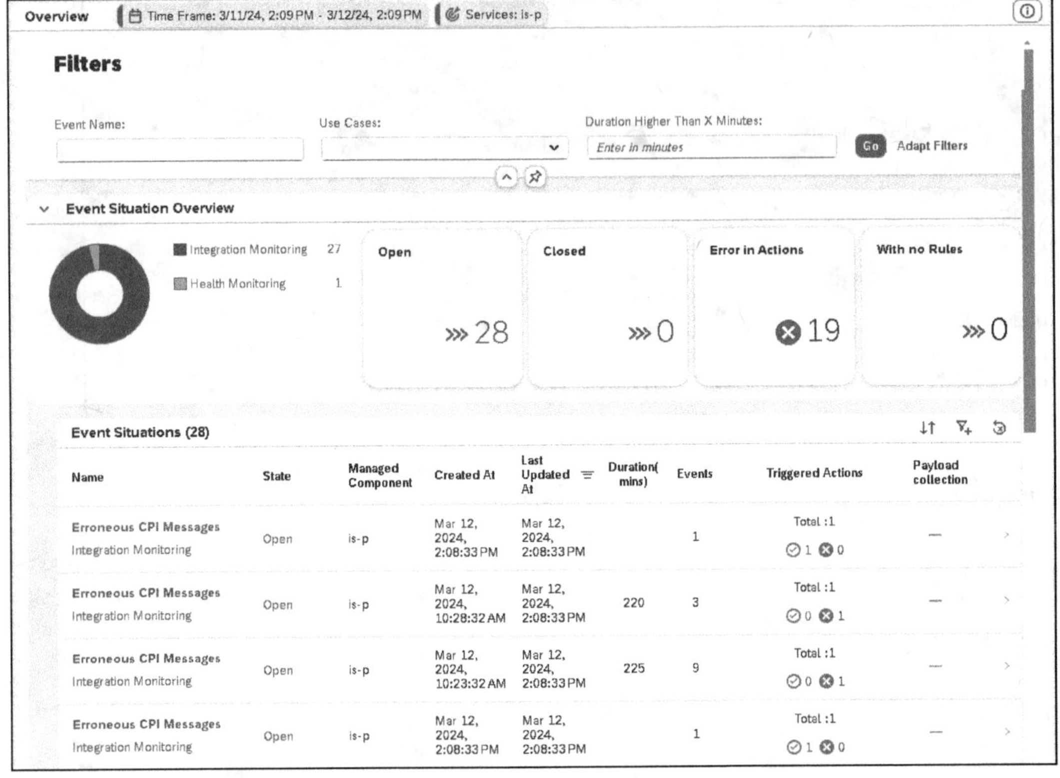

Figure 22.10 Overview of Event Situations in Intelligent Event Processing

When navigating to an event situation, further details are displayed. In the **Automatic Ticket Action Triggered** message, you can see that a ticket has been successfully created (see Figure 22.11).

In the External API Management app **Console**, you can view and track the requests of the configured subscription (see Figure 22.12). In addition to the **Event Type**, the **Partner Response Time**, the **Method**, and the **Partner HTTP Status** are also displayed.

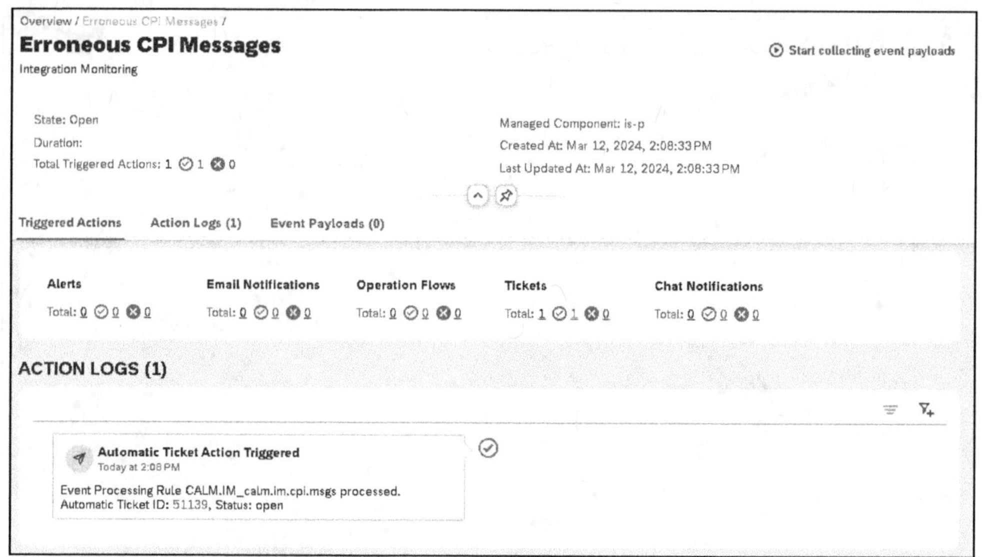

Figure 22.11 Event Situation in Intelligent Event Processing

Console

Calls (27)

Timestamp	Subscription	Event Type	Delivery Status	Partner Response Time	Method	Partner HTTP Status
Mar 12, 2024, 2:08:37 PM	GitHub Issue Integration	EVENT-SITUATION.CREAT	200 OK	876 ms	POST	201 CREATED
Mar 12, 2024, 2:08:36 PM	GitHub Issue Integration	EVENT-SITUATION.CREAT	200 OK	779 ms	POST	201 CREATED
Mar 12, 2024, 2:03:37 PM	GitHub Issue Integration	EVENT-SITUATION.CREAT	200 OK	863 ms	POST	201 CREATED
Mar 12, 2024, 2:03:37 PM	GitHub Issue Integration	EVENT-SITUATION.CREAT	200 OK	633 ms	POST	201 CREATED
Mar 12, 2024, 2:03:36 PM	GitHub Issue Integration	EVENT-SITUATION.CREAT	200 OK	1,212 ms	POST	201 CREATED
Mar 12, 2024, 1:53:36 PM	GitHub Issue Integration	EVENT-SITUATION.CREAT	200 OK	913 ms	POST	201 CREATED
Mar 12, 2024, 1:53:36 PM	GitHub Issue Integration	EVENT-SITUATION.CREAT	200 OK	863 ms	POST	201 CREATED
Mar 12, 2024, 1:53:36 PM	GitHub Issue Integration	EVENT-SITUATION.CREAT	200 OK	800 ms	POST	201 CREATED
Mar 12, 2024, 1:43:34 PM	GitHub Issue Integration	EVENT-SITUATION.CREAT	502 BAD_GATEWAY	0 ms	N/A	N/A
Mar 12, 2024, 1:33:36 PM	GitHub Issue Integration	EVENT-SITUATION.CREAT	502 BAD_GATEWAY	0 ms	N/A	N/A
Mar 12, 2024, 1:33:36 PM	GitHub Issue Integration	EVENT-SITUATION.CREAT	502 BAD_GATEWAY	0 ms	N/A	N/A
Mar 12, 2024, 1:33:35 PM	GitHub Issue Integration	EVENT-SITUATION.CREAT	502 BAD_GATEWAY	0 ms	N/A	N/A
Mar 12, 2024, 1:33:35 PM	GitHub Issue Integration	EVENT-SITUATION.CREAT	502 BAD_GATEWAY	0 ms	N/A	N/A

Figure 22.12 Console in External API Management

Finally, issues are mapped in the configured Github repository (see Figure 22.13). As the `htmlBody` of the Resource Changes API was used as the long text for the issue in the assignment configuration, the issue is displayed in the appropriate format.

22

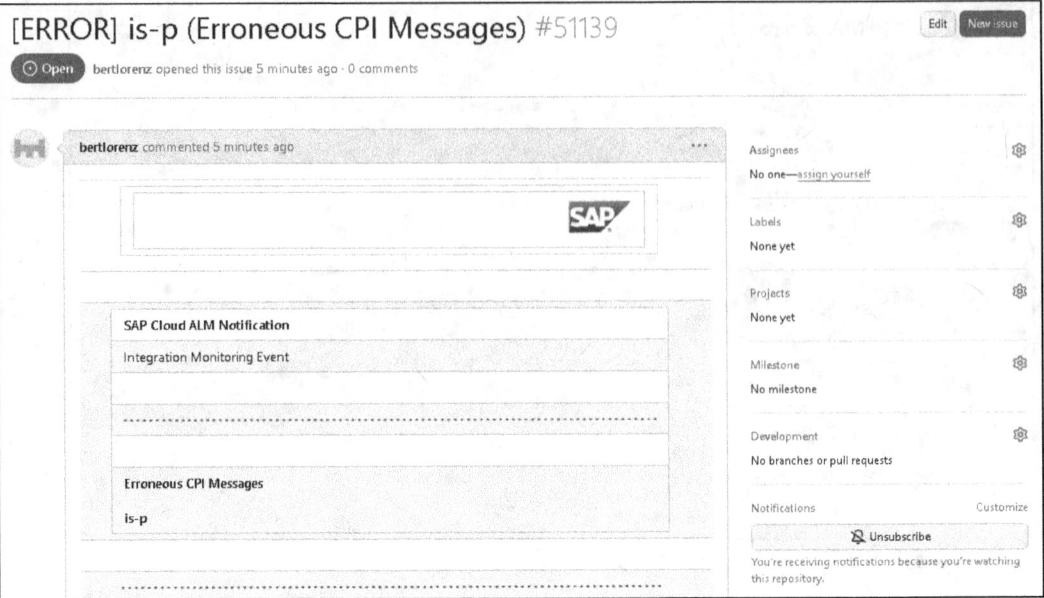

Figure 22.13 Created Issue in GitHub

22.2 Integration of Microsoft To Do

With the cloud approach of Microsoft 365, Microsoft has established a seamless integration of the included applications such as Outlook, Excel, SharePoint, and Teams, which are constantly being expanded. One of these integrations is the task planner. While tasks in previous versions of Office mostly consisted of marked emails and manually created tasks in Outlook, tasks are now collected from the various Microsoft 365 applications and can be managed centrally in the *Microsoft To Do* application and on the **Tasks** tab in Microsoft Teams. Depending on the project situation and your personal preferences in terms of personal task planning, you can increase your own productivity, as not all, but many tasks can be overviewed, and simple notifications and completion deadlines can also be defined.

This section describes a standalone solution for integrating SAP Cloud ALM into Microsoft To Do. We will focus on the steps required to configure the integration without going into detail about all the technical requirements, as this would go beyond the scope of this manual and is subject to constant change.

Figure 22.14 shows a diagram of the planned integration. In addition to the existing integration of Microsoft Outlook/Exchange and Microsoft Teams Planner, it should also be possible to manage tasks from SAP Cloud ALM in Microsoft To Do.

SAP Cloud ALM project tasks, such as defining the scope and objectives, should be bundled in a separate list, as shown in Figure 22.15.

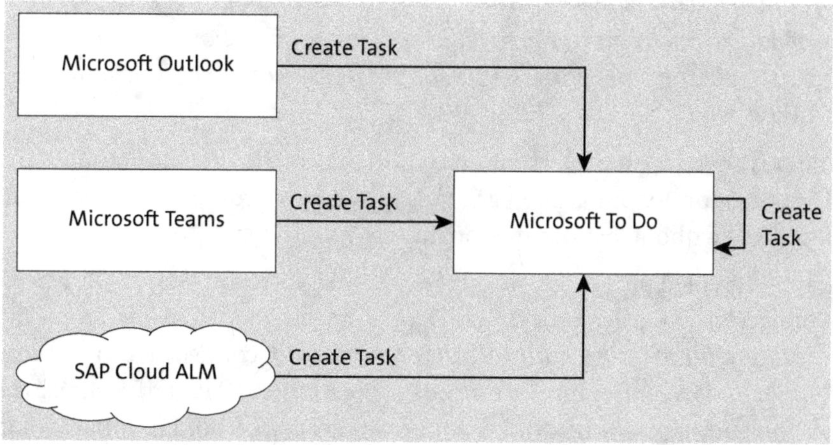

Figure 22.14 Schematic Representation of the Integration in Microsoft To Do

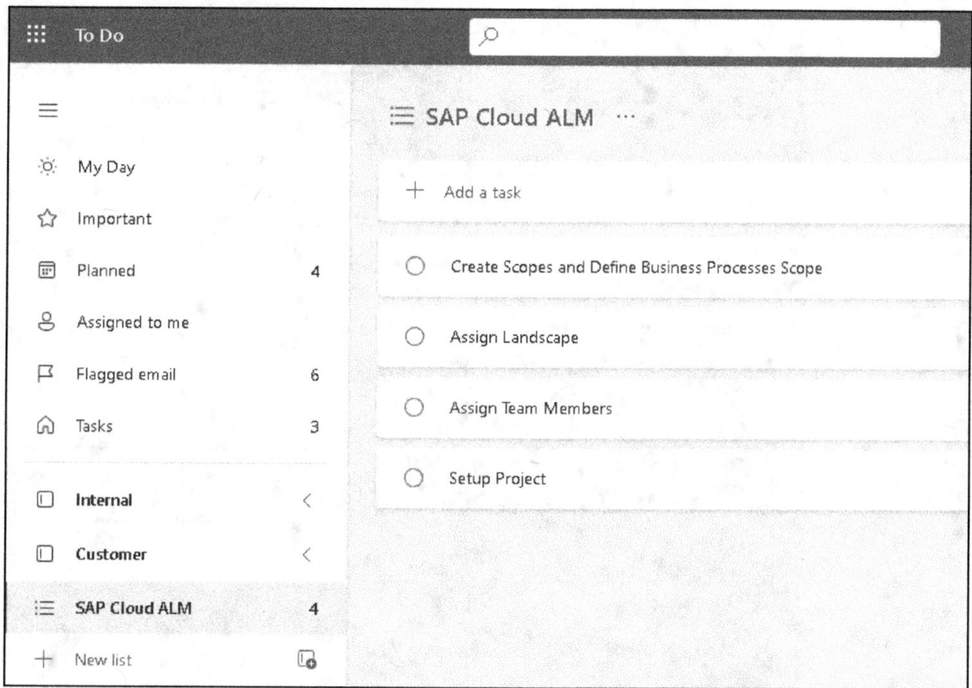

Figure 22.15 Planned Representation of SAP Cloud ALM Tasks in Microsoft To Do

22.2.1 Technical Aspects of the Integration of Microsoft To Do

Microsoft provides APIs for Microsoft To Do via *Microsoft Graph*. Microsoft itself describes Microsoft Graph as follows:

> *"Microsoft Graph is the gateway to data and intelligence in Microsoft 365. It provides a unified programmability model that you can use to access the tremendous*

amount of data in Microsoft 365, Windows, and Enterprise Mobility + Security. Use the wealth of data in Microsoft Graph to build apps for organizations and consumers that interact with millions of users. " (Source: https://learn.microsoft.com/en-US/graph/overview).

Microsoft Graph is therefore a central interface into selected Microsoft products and offers simple interaction for simple integrations, while more complex scenarios can also be implemented without loss of functionality.

Microsoft Graph is divided into three main components: Microsoft Graph API, Microsoft Graph Connectors and Microsoft Graph Data Connect. The Microsoft Graph API provides a central endpoint for executing CRUD operations for connected products, i.e. for the `create`, `read`, `update`, and `delete` operations of objects. REST APIs and SDKs can be used to access the endpoint. Microsoft Graph Connectors can be used to provide data from external data storage providers, such as Google Drive. Microsoft Graph Data Connect includes tools for providing data in Microsoft Azure, Microsoft's cloud platform, and applications connected to it.

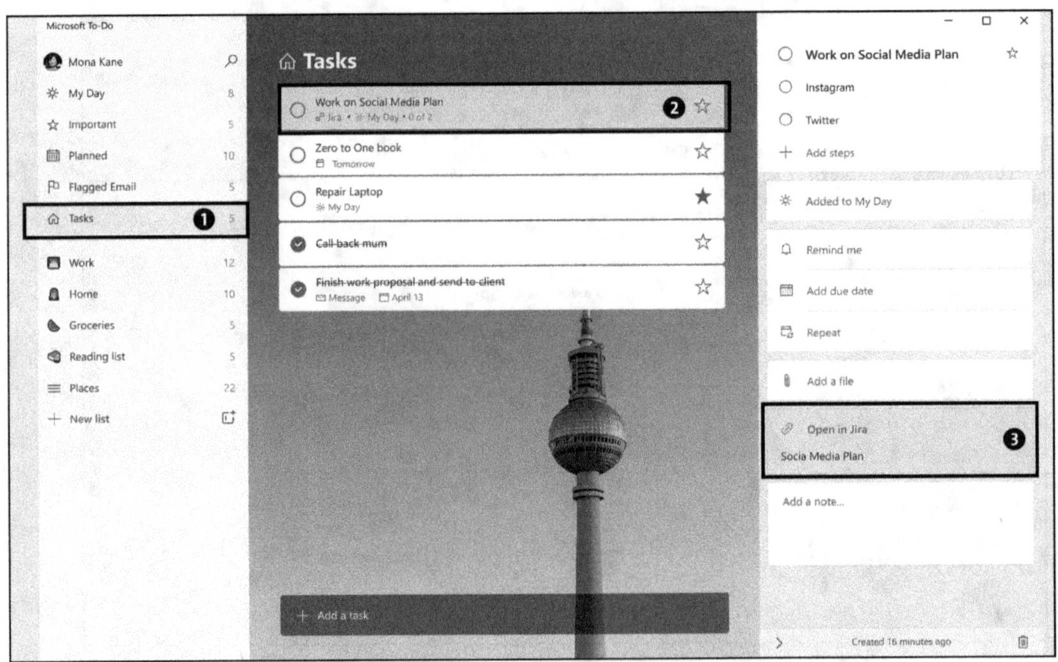

Figure 22.16 Application Components of Microsoft To Do (Source: Microsoft)

Microsoft provides interfaces to three different components of the application for Microsoft To Do. These are divided into a task list ❶, tasks ❷, and linked resources ❸, as shown in Figure 22.16. The possible operations for the components are divided as follows:

- `GET` for reading objects
- `POST` to create objects
- `PATCH` for updating objects
- `DELETE` to delete objects

22.2.2 Implementation of the Microsoft Graph API for Microsoft To Do

Authentication in Microsoft Graph takes place via an access token, i.e. a generally short-lived defined access to a resource. This access token contains information about the application and requires corresponding authorizations for the application. An application is registered as part of Microsoft Entra ID within the Microsoft Azure portal.

A new application can be registered within the **Enterprise applications** menu item. A descriptive name and basic authorizations for access to the Microsoft Entra ID can be stored for the application. The name is usually the target of the application, in this case "Microsoft To Do Integration." The authorization options for access are as follows:

- Only accounts in this organizational directory (single tenant)
- Accounts in any organizational directory (any Microsoft Entra ID tenant: multi-tenant)
- Accounts in any organizational directory (any Microsoft Entra ID tenant: multi-tenant) and personal Microsoft accounts (e.g., Skype, Xbox)
- Only personal Microsoft accounts

Microsoft provides further details on the various options to help you decide. Using this integration as an example, the application is defined as a single client for use in the current organizational directory (see Figure 22.17).

A redirect URI is required if user authentication is needed in the dialog. As the solution provides for synchronization in the background (i.e., without any interaction with the user), no redirect URI is required.

After registering the application, two important pieces of information are available to you (see Figure 22.18):

- Application ID (client)
- Directory ID (tenant)

In addition to the registration, the authentication of the application must be defined. This works via a certificate or a password for the application. The use of a certificate is recommended for productive use scenarios in particular, as it offers significantly higher security. You define the authentication within the **Properties • Application Registration • Certificates & Secrets** menu path.

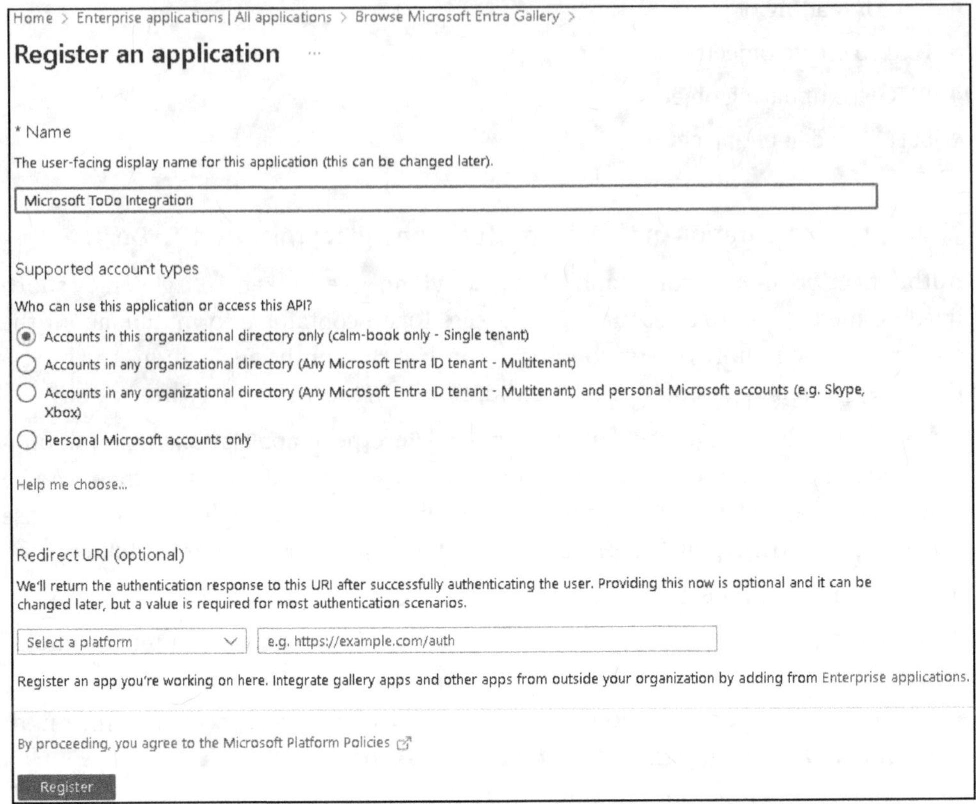

Figure 22.17 Definition of the Application in Microsoft Entra ID of Microsoft Azure

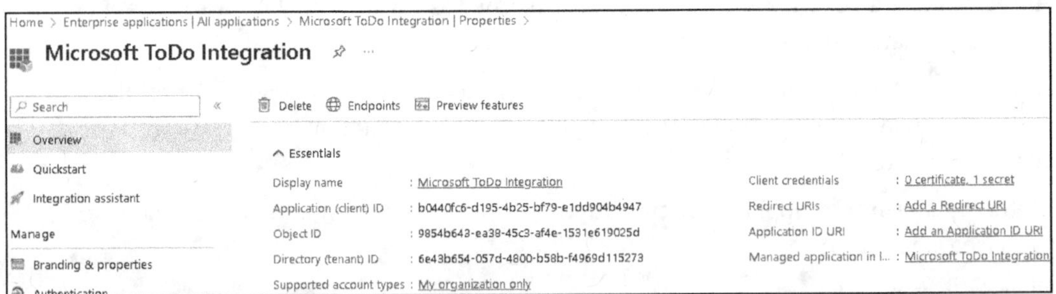

Figure 22.18 Overview of the Application Registration

In this book, we use a secret service key to illustrate the integration. The service key has a description and a validity. As several endpoints can access the same application, a reference to the source application is a good description, in this case, SAP Cloud ALM. After defining a secret, the necessary information, i.e. the **Description**, validity (**Valid Until**), the **Value**, and the **Secret ID**, is displayed (see Figure 22.19). The secret service key is only displayed immediately after creation and should therefore be saved temporarily for later use.

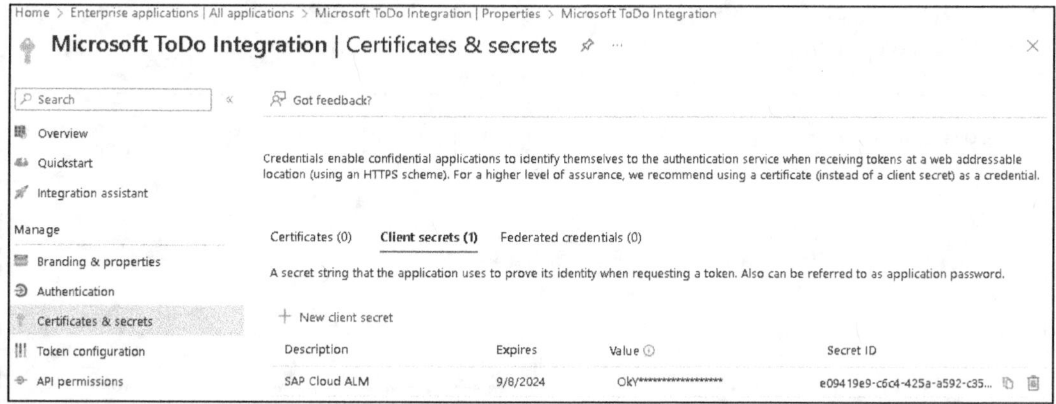

Figure 22.19 Secret Service Key of the Application

All the information required to authenticate the application is now available. Table 22.2 summarizes this again.

Feature	Value
Application ID	`b0440fc6-d195-4b25-bf79-e1dd904b4947`
Directory ID	`6e43b654-057d-4800-b58b-f4969d115273`
Secret key	`~2d8Q~mqw54-hQqBdCW3zfsJYF0z2EH0THhSRbQV`

Table 22.2 Authentication Information

Finally, the application requires authorizations for the corresponding components of the Microsoft Graph API. To do this, grant a new authorization for the Microsoft Graph API under **API permissions**. The authorization can be granted as a delegated authorization or as an application authorization. As the integration is executed in the background for all users, you should use the application authorization in this case. When granting authorization for the Microsoft Graph API, individual components can also be authorized. The principle here is that only the absolutely necessary access should be defined, so the `Tasks.ReadWrite.All` component is used in this case. This includes reading and writing user `tasks` and `tasklists`. In addition, the permission `User.ReadWrite.All` is required to be able to map SAP Cloud ALM users to users in Microsoft Entra ID. Integrations that can read and modify user data require the consent of the relevant users for data protection reasons, among others. In the case of delegated access, the integration is executed in dialog, meaning that the respective user must give their consent during the first synchronization. As this integration has received application authorization, the individual cannot give consent, so the administration team of the organizational directory must give consent via the action **Grant Admin Consent for "Default Directory."**

After authorization assignment and administrator approval, the authorizations are displayed with a positive **Status**, as shown on the right in Figure 22.20.

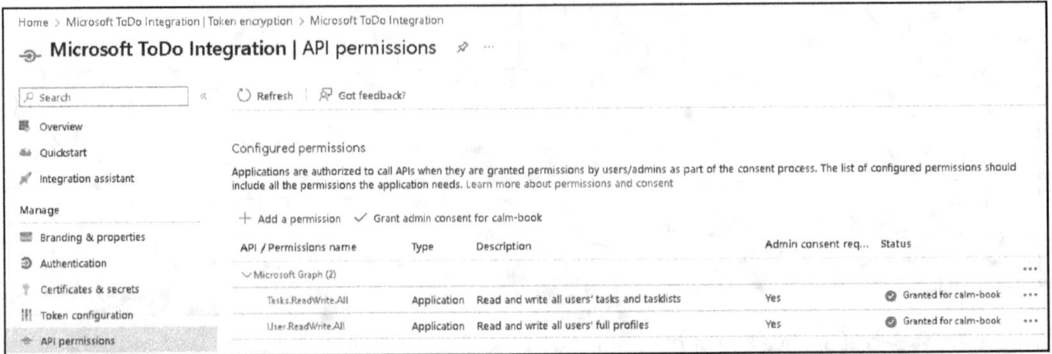

Figure 22.20 API Authorization for the Application

After application registration, the generated service key can be used to authenticate the requests in Postman or another suitable tool for testing interfaces. For this purpose, a new collection is created so that all underlying requests can access a common authentication token. On the **Authorization** tab, the data listed in Table 22.3 must be configured (see Figure 22.21).

Field	Value
Type	**OAuth 2.0**
Add auth data to	**Request Headers**
Header prefix	**Bearer**
Token name	speaking name
Access Token URL	*https://login.microsoftonline.com/{{tenant}}/oauth2/v2.0/token*
Client ID	{{client}} (This is the application ID.)
Client Secret	{{secret}} (This is the secret service key.)
Scope	https://graph.microsoft.com/.default
Client Authentication	**Send client credentials in body**

Table 22.3 Value Entries of the Authentication Configuration for Microsoft Graph

On the tab **Variables**, the variables (the values with the pattern `{{variable name}}`) must be maintained accordingly, as shown in Figure 22.22. Each variable is given a name, an **INITIAL VALUE**, which is visible to everyone who can access the collection, and a **CURRENT VALUE**, which can only be viewed by you.

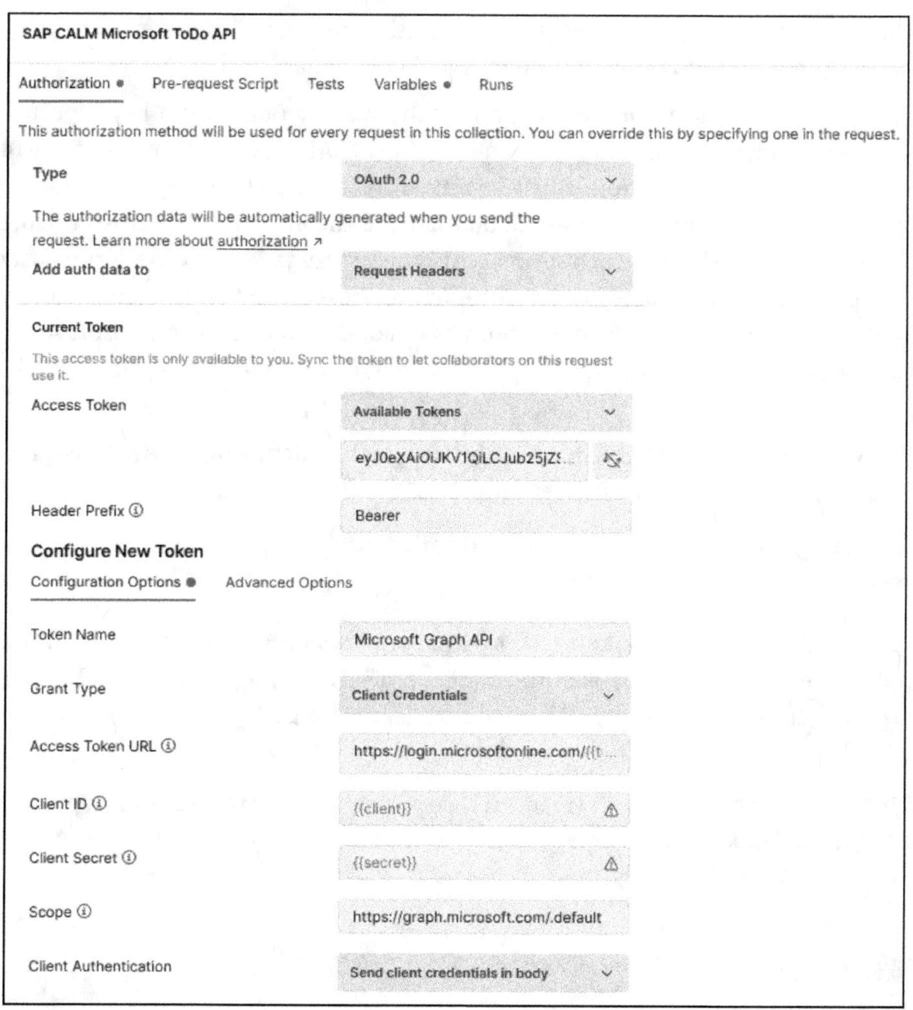

Figure 22.21 Authentication Configuration for Microsoft Graph

SAP CALM Microsoft ToDo API — Share | Fork 0 | Run | Save | •••

Authorization ● Pre-request Script Tests Variables ● Runs

These variables are specific to this collection and its requests. Learn more about collection variables. ↗

	VARIABLE	INITIAL VALUE	CURRENT VALUE	••• Persist All Reset All
☑	tenant		6e43b654-057d-4800-b58b-f4969d115273	
☑	client		b0440fc6-d195-4b25-bf79-e1dd904b4947	
☑	secret		~2d8Q~mqw54-hQqBdCW3zfsJYF0z2EHOTHhSRbQV	
	Add a new variable			

Figure 22.22 Variable Configuration for Microsoft Graph

You can then generate a token and use it for further requests. The token expires by default after one hour and must then be renewed in Postman.

In addition to authentication, the mapping of the various objects is a key aspect of interface development. In the case of the integration of SAP Cloud ALM in Microsoft To Do, the users of both applications must be linked correctly so that the correct task is created for a corresponding user. SAP Cloud ALM and the Entra ID of Microsoft Azure each have their own user administrations, so it is necessary to identify the correct user in the respective identity management. The email address is used as the main property for identification, as this is available in both SAP Cloud ALM and the Microsoft Azure Entra ID. Depending on the design of the directories, other characteristics can also be used if necessary.

The following request can be used to identify users in the Microsoft Azure Active Directory based on their email address (see Figure 22.23):

```
curl–location–globoff 'https://graph.microsoft.com/v1.0/users/{{email}}/id' \
--header 'Authorization: Bearer {{token}}'
```

The variable `{{mail}}` must be replaced by the corresponding e-mail address (in this example "peterparker@calmbook.onmicrosoft.com"). The value for the key `value` contains the user ID that is used for further requests (see Listing 22.5).

```
{
    "@odata.context": "https://graph.microsoft.com/v1.0/$metadata#users('peter
parker%40calmbook.onmicrosoft.com')/id",
    "value": "ab990765-b034-47b7-831b-bdac79af2cb2"
}
```

Listing 22.5 Response to the Request for a User ID from Microsoft Graph

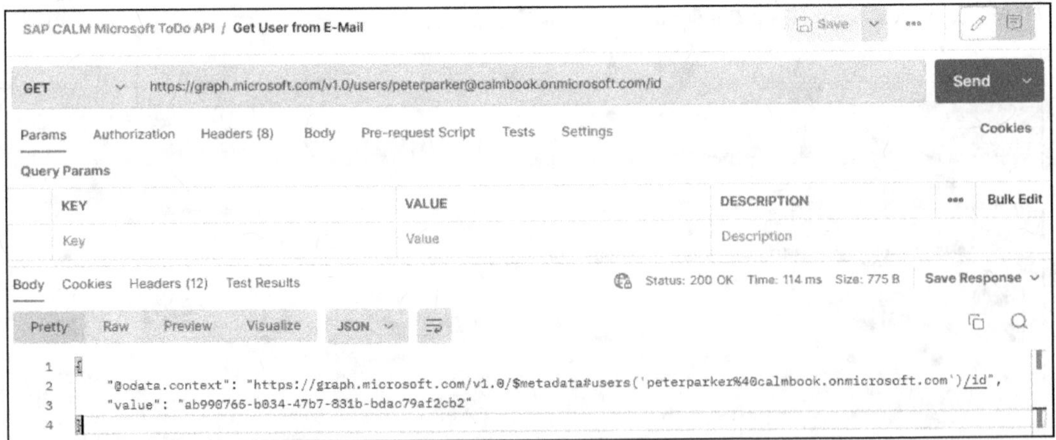

Figure 22.23 Request to Determine the User ID in the Microsoft Azure Active Directory

After assigning a user in the Microsoft Entra ID based on the user in SAP Cloud ALM, you need a task list in Microsoft To Do in which the tasks from SAP Cloud ALM are to be created. It is advisable to use a separate task list for the integration so that the tasks can be clearly separated from each other. So that the user does not have to create the task list independently, it can be created automatically via the API. The following request is necessary for this:

```
POST https://graph.microsoft.com/v1.0/users/{{userid}}/todo/lists
```

The variable `{{userid}}` must be replaced by the value you specified in Listing 22.5 have determined. In addition, the title of the task list must be entered in the `body` of the request:

```
{
    "displayName": "SAP Cloud ALM"
}
```

The entire task list object is returned in response to the request, as shown in Listing 22.6.

```
{
    "@odata.context": "https://graph.microsoft.com/v1.0/$metadata#users('ab9907
65-b034-47b7-831b-bdac79af2cb2')/todo/lists/$entity",
    "@odata.etag": "W/\|f9BsxewIoOm2RuEe06AflQAAAAAEAg==\|",
    "displayName": "SAP Cloud ALM",
    "isOwner": true,
    "isShared": false,
    "wellknownListName": "none",
    "id": "AAMkADMyY2UwNWYyLTJhYzEtNDQ1Ni05NDAwLThkOWQyZGQ3Otk2NAAuAAAAAAApdGsr
z7JOR5RYrKyGFcLaAQB-OFLF7AijSbZG4R7ToB_VAAAAAFPoAAA="
}
```

Listing 22.6 Response to the Creation of a New Task List

Figure 22.24 shows the complete request in Postman with its end point, the `body`, and the response.

Once a task list has been created, it must also be correctly identified during subsequent synchronizations. A user's task lists can be read using the following request:

```
GET https://graph.microsoft.com/v1.0/users/{{userid}}/todo/lists
```

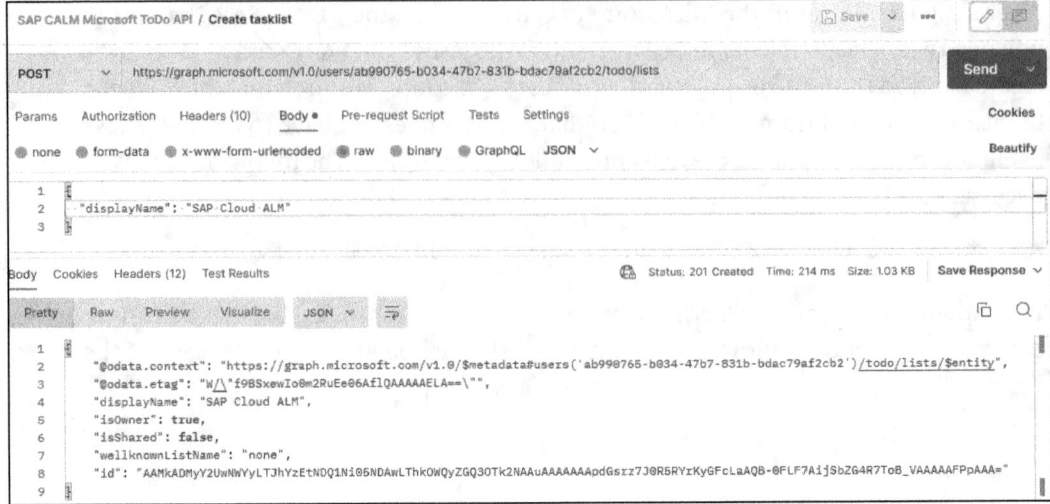

Figure 22.24 Request to Create a Task List

Here, too, the variable `{{userid}}` must be replaced by the value you entered in Listing 22.5. In this example, the user ID is `ab990765-b034-47b7-831b-bdac79af2cb2`. As a result of the request, all of the user's task lists are listed, as shown in Listing 22.7.

```
{
    "@odata.context": "https://graph.microsoft.com/v1.0/$metadata#users('ab9907
65-b034-47b7-831b-bdac79af2cb2')/todo/lists",
    "value": [
        {
            "@odata.etag": "W/\|f9BsxewIoOm2RuEe06Af1QAAAAAAoA==\|",
            "displayName": "Tasks",
            "isOwner": true,
            "isShared": false,
            "wellknownListName": "defaultList",
            "id": "AAMkADMyY2UwNWYyLTJhYzEtNDQ1Ni05NDAwLThkOWQyZGQ3Otk2NAAuAAAA
AAApdGsrz7JOR5RYrKyGFcLaAQB-0FLF7AijSbZG4R7ToB_VAAAAAAESAAA="
        },
        {
            "@odata.etag": "W/\|f9BsxewIoOm2RuEe06Af1QAAAAAELA==\|",
            "displayName": "SAP Cloud ALM",
            "isOwner": true,
            "isShared": false,
            "wellknownListName": "none",
            "id":
"AAMkADMyY2UwNWYyLTJhYzEtNDQ1Ni05NDAwLThkOWQyZGQ3Otk2NAAuAAAAAAApdGsrz7JOR5RYrK
yGFcLaAQB-0FLF7AijSbZG4R7ToB_VAAAAAFPpAAA="
        },
```

```
        {
            "@odata.etag": "W/\|f9BsxewIoOm2RuEe06AflQAAAAAAA2A==\|",
            "displayName": "Flagged Emails",
            "isOwner": true,
            "isShared": false,
            "wellknownListName": "flaggedEmails",
            "id": "AAMkADMyY2UwNWYyLTJhYzEtNDQ1Ni05NDAwLThkOWQyZGQ3Otk2NAAuAAA
AAAApdGsrz7JOR5RYrKyGFcLaAQB-OFLF7AijSbZG4R7ToB_VAAAAAAEeAAA="
        }
    ]
}
```

Listing 22.7 Listing the Task Lists of a User

Figure 22.25 shows the entire request in Postman, with its endpoint, the body, and the response.

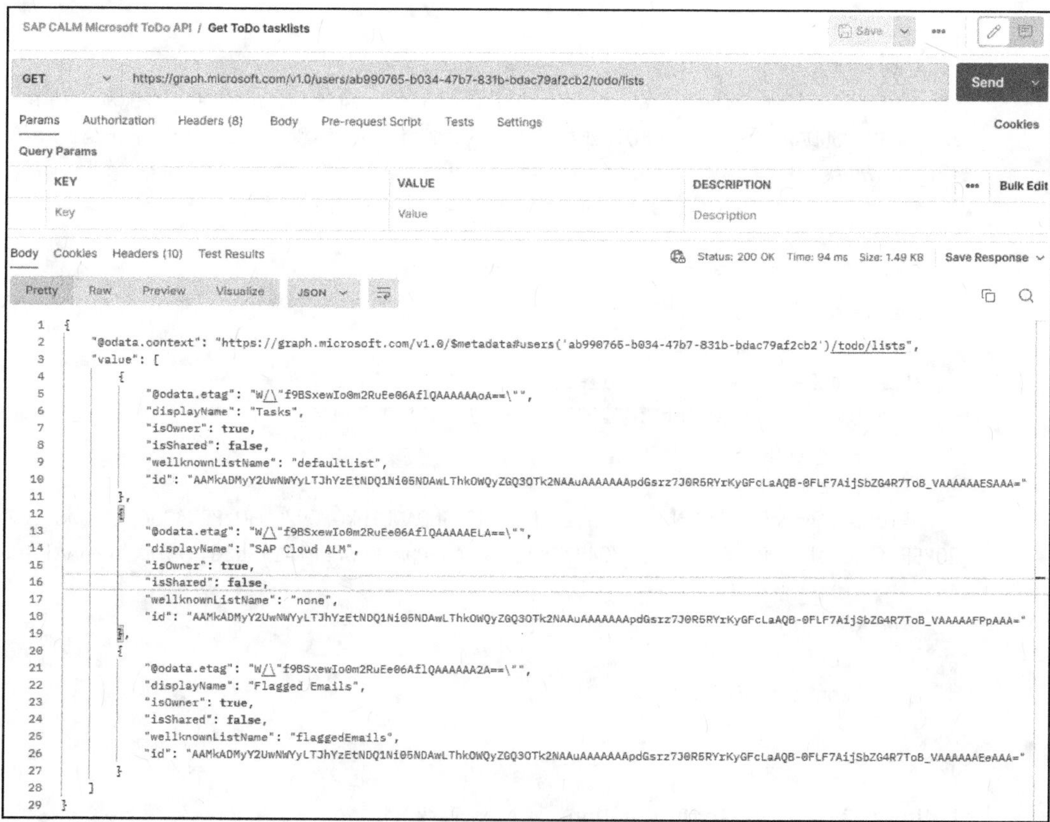

Figure 22.25 Request to Determine the Task Lists of a User

Once all the prerequisites have been met with authentication, user identification, and the creation of a task list, a new task can be created with the following request:

```
POST https://graph.microsoft.com/v1.0/users/{{userid}}/todo/lists/
{{tasklistid}}/tasks
```

The variable {{userid}} must be replaced by the value you entered in Listing 22.5, and the variable {{tasklistid}} must be replaced by the value you determined in Listing 22.6 or Listing 22.7. Further information is required to create a task. At least the title of the task must be provided, although further information, such as importance, dueDateTime, or recurrence, can also be defined during creation.

```
{
    "title": "A new task"
}
```

The entire object of the task is returned as the response, as shown in Listing 22.8.


```
{
    "@odata.context": "https://graph.microsoft.com/v1.0/$metadata#
users('ab990765-b034-47b7-831b-bdac79af2cb2')/todo/lists('AAMkADMyY2UwNWYyLTJhY
zEtNDQ1Ni05NDAwLThkOWQyZGQ3OTk2NAAuAAAAAAApdGsrz7JOR5RYrKyGFcLaAQB-OFLF7AijSbZG
4R7ToB_VAAAAAFPpAAA%3D')/tasks/$entity",
    "@odata.etag": "W/\"f9BsxewIoOm2RuEe06AflQAAAAAERw==\"",
    "importance": "normal",
    "isReminderOn": false,
    "status": "notStarted",
    "title": "A new task",
    "createdDateTime": "2022-11-16T15:02:29.6966149Z",
    "lastModifiedDateTime": "2022-11-16T15:02:29.7903458Z",
    "hasAttachments": false,
    "categories": [],
    "id": "AAMkADMyY2UwNWYyLTJhYzEtNDQ1Ni05NDAwLThkOWQyZGQ3OTk2NABGAAAAApdGsrz7
JOR5RyrKyGFcLaBwB-OFLF7AijSbZG4R7ToB_VAAAAAFPpAAB-OFLF7AijSbZG4R7ToB_VAAAAAILeA
AA=",
    "body": {
        "content": "",
        "contentType": "text"
    }
}
```


Listing 22.8 Response to the Creation of a New Task

Figure 22.26 shows the entire request in Postman.

Figure 22.26 Request to Create a Task

The tasks in a task list can be read using the following query:

```
GET https://graph.microsoft.com/v1.0/users/{{userid}} /todo/lists/
{{tasklistid}}/tasks
```

The variable `{{userid}}` must be replaced by the value you entered in Listing 22.5, and `{{tasklistid}}` must be replaced by the value you determined in Listing 22.6 or Listing 22.7, respectively. All task objects are returned as the response, as shown in Listing 22.9.

```
{
    "@odata.context": "https://graph.microsoft.com/v1.0/$metadata#users('ab9907
65-b034-47b7-831b-bdac79af2cb2')/todo/lists('AAMkADMyY2UwNWYyLTJhYzEtNDQ1NiO5ND
AwLThkOWQyZGQ3OTk2NAAuAAAAAAApdGsrz7JOR5RYrKyGFcLaAQB-0FLF7AijSbZG4R7ToB_VAAAAA
FPpAAA%3D')/tasks",
    "value": [
        {
            "@odata.etag": "W/\|f9BSxewIoOm2RuEe06AflQAAAAAERw==\|",
            "importance": "normal",
            "isReminderOn": false,
            "status": "notStarted",
            "title": "A new task",
            "createdDateTime": "2022-11-16T15:02:29.6966149Z",
            "lastModifiedDateTime": "2022-11-16T15:02:29.7903458Z",
            "hasAttachments": false,
            "categories": [],
```

```
            "id": "AAMkADMyY2UwNWYyLTJhYzEtNDQ1Ni05NDAwLThkOWQyZGQ3OTk2NABGAAAA
AAApdGsrz7JOR5RYrKyGFcLaBwB-OFLF7AijSbZG4R7ToB_VAAAAAFPpAAB-OFLF7AijSbZG4R7ToB_
VAAAAAILeAAA=",
            "body": {
                "content": "",
                "contentType": "text"
            }
        }
    ]
}
```

Listing 22.9 Response to the Request for Tasks from a Task List

Figure 22.27 shows the entire request in Postman.

A task can be updated using the following request:

```
PATCH https://graph.microsoft.com/v1.0/users/{{userid}}/todo/lists/
{{tasklistid}}/tasks/{{taskid}}
```

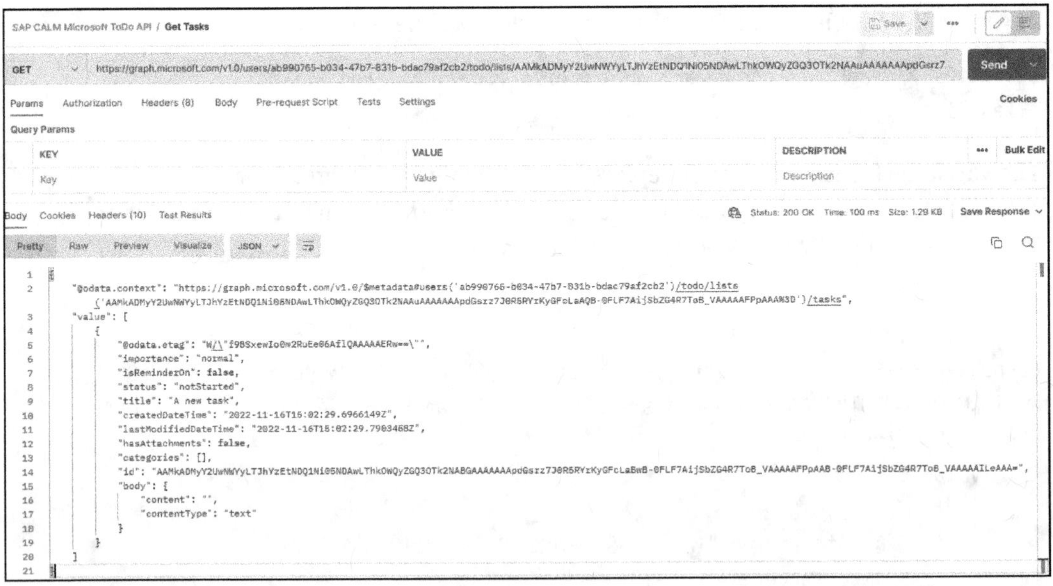

Figure 22.27 Request to Read the Tasks of a Task List

The variable {{userid}} must be replaced by the value you entered in Listing 22.5. The variable {{tasklistid}} must be replaced by the value you determined in Listing 22.6 or Listing 22.7, and the variable {{taskid}} by the value you determined in Listing 22.8 or Listing 22.9. The properties to be changed must be defined in the body of the request:

```
{
    "status": "completed"
}
```

The updated task object is returned as the result of the request, as shown in Listing 22.10.

```
{
    "@odata.context": "https://graph.microsoft.com/v1.0/$metadata#
users('ab990765-b034-47b7-831b-bdac79af2’cb2')/todo/lists('AAMkADMyY2UwNWYyLTJh
YzEtNDQ1NiO5NDAwLThkOWQyZGQ3OTk2NAAuAAAAAAApdGsrz7JOR5RYrKyGFcLaAQB-OFLF7AijSbZ
G4R7ToB_VAAAAAFPpAAA%3D')/tasks/$entity",
    "@odata.etag": "W/\|f9BSxewIoOm2RuEe06AflQAAAAAEbQ==\|",
    "importance": "normal",
    "isReminderOn": false,
    "status": "completed",
    "title": "A new task",
    "createdDateTime": "2022-11-16T15:02:29.6966149Z",
    "lastModifiedDateTime": "2022-11-16T15:11:16.6574048Z",
    "hasAttachments": false,
    "categories": [],
    "id": "AAMkADMyY2UwNWYyLTJhYzEtNDQ1NiO5NDAwLThkOWQyZGQ3OTk2NABGAAAAAAApdGsr
z7JOR5RYrKyGFcLaBwB-OFLF7AijSbZG4R7ToB_VAAAAAFPpAAB-OFLF7AijSbZG4R7ToB_VAAAAAIL
eAAA=",
    "body": {
        "content": "",
        "contentType": "text"
    },
    "completedDateTime": {
        "dateTime": "2022-11-16T00:00:00.0000000",
        "timeZone": "UTC"
    }
}
```

Listing 22.10 Response to the Update of a Task

Figure 22.28 shows the entire request in Postman.

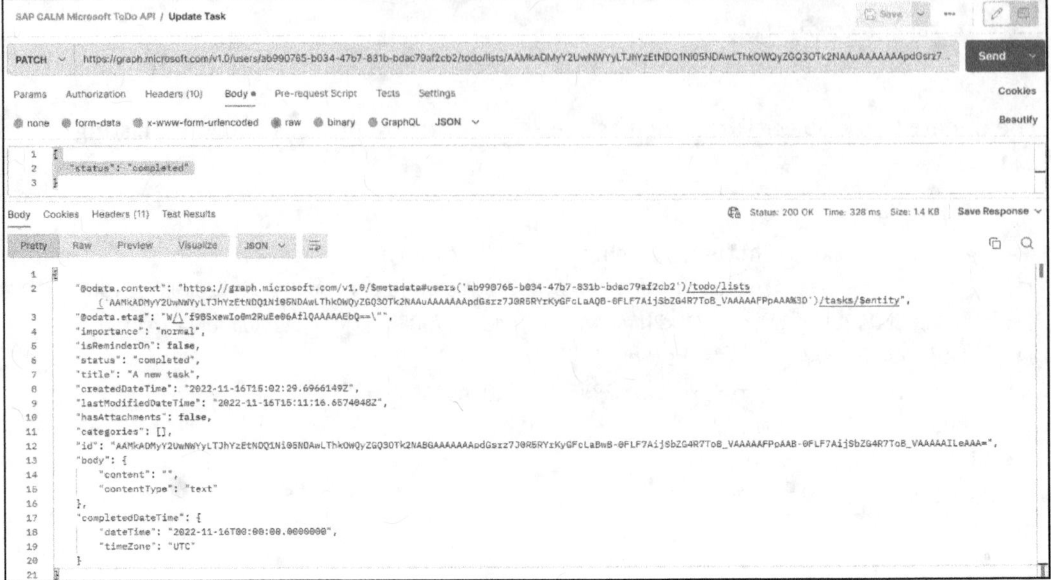

Figure 22.28 Request to Update a Task

22.2.3 Implementation of the SAP Cloud ALM API

In Chapter 20, Section XXX, we described that a service instance must be created to call pull APIs of SAP Cloud ALM. The service instance can be defined with specific authorizations so that a separate service instance can be created for each use case. To integrate Microsoft To Do, you need access to projects and tasks in SAP Cloud ALM. You can define these authorizations using the code in Listing 22.11.

```
{
    "xs-security": {
        "xsappname": "sap-calm-api",
        "authorities": [
            "$XSMASTERAPPNAME.calm-api.projects.read",
            "$XSMASTERAPPNAME.calm-api.tasks.read",
            "$XSMASTERAPPNAME.calm-api.tasks.write"
        ]
    }
}
```

Listing 22.11 Service Instance Parameters of the SAP Cloud ALM API for Microsoft To Do Integration

Once the service instance and a service key have been created, the configuration of the authentication can be carried out in Postman. It is advisable to carry out the configuration at the level of a collection so that you can reuse all underlying requests for this authentication. The configuration can be carried out on the **Authorization** tab. The **Add auth data to** field must be filled with **Request Headers**. Select **OAuth 2.0** as the **authentication type**. Table 22.4 shows the necessary configuration in the **Configure New Token** area. The variables `{{calmSubDomain}}`, `{{calmRegion}}`, `{{clientID}}`, and `{{clientSecret}}` must be supplemented with the corresponding values of the service key.

Parameters	Value to Be Set
Token name	Freely selectable, e.g., "SAP Cloud ALM"
Grant Type	**Client Credentials**
Access Token URL	https://{{calmSubDomain}}.authentication. {{calmRegion}}.hana.ondemand.com/oauth/token
Client ID	{{clientID}}
Client Secret	{{clientSecret}}
Scope	-
Client Authentication	**Send as Basic Auth Header**

Table 22.4 Token Configuration for the SAP Cloud ALM API in Postman

You can then test the access using the button **Get New Access Token**. The token is then displayed, which is a short-lived token. This means that in the case of SAP Cloud ALM, the token must be renewed after 60 minutes. Please note that the token must be activated for use after it has been created. This is done via **Use Token**. All requests that are now created within the Postman collection automatically use this token for authentication so that you can then concentrate on the requests themselves.

First, you need an overview of all projects so that you can then retrieve the tasks of the projects. The end point for the interface to the projects can be reached via the following URL: *https://{{calmRegion}}.alm.cloud.sap/api/calm-projects/v1/projects*. If you have already defined the variable `{{calmRegion}}` at the collection level, it can be used directly in the request. A `GET request` without further parameters to this endpoint returns all projects of the SAP Cloud ALM instance. The return is in JSON format, as shown in Listing 22.12.

```
[
    {
        "id": "c59d59d5-31e7-40b3-8753-ec7cf7b68af5",
        "name": "S/4 HANA Cloud Implementation Project",
        "status": "0"
    }
]
```

Listing 22.12 Response of the SAP Cloud ALM Project API

The property `id` identifies the project and is required later on to request the associated tasks. The properties `name` and `status` are self-explanatory, with the property `status` being the technical value. The technical value can be translated, as shown in Table 22.5.

Technical Value	Description
O	Active project
C	Hidden project

Table 22.5 Technical Status Values of Projects

After determining a project, you can retrieve all tasks for this project via the endpoint `https://{{calmRegion}}.alm.cloud.sap/api/calm-tasks/v1/tasks?projectId={{projectID}}`. As with reading the projects, the variable `{{calmRegion}}` is used. The desired project ID with the request-specific variable `{{projectID}}` is entered directly in the URL as a `GET` parameter. This is a mandatory parameter so that a cross-project request for tasks is not possible. In response to the request, all tasks are returned in JSON format, as shown in Listing 22.13.


```
[
    {
        "id": "99977f3d-e8cf-4b47-85f5-16e9e00d2eb9",
        "projectId": "c59d59d5-31e7-40b3-8753-ec7cf7b68af5",
        "title": "Assign Landscape",
        "type": "CALMTMPL",
        "status": "CIPTKOPEN",
        "externalId": null,
        "startDate": "2024-01-16",
        "dueDate": "2024-01-31",
        "priorityId": 30,
        "assigneeId": "peterparker@calmbook.onmicrosoft.com",
        "assigneeName": "Peter Parker",
        "parentId": null,
        "subState": null,
        "approvalState": null,
        "scopeId": null,
        "scopeName": null,
        "assignedRoleId": "d8d90840-0df4-43bc-b253-07314bebbb82",
        "assignedRoleName": "System Administrator",
        "storyPoints": null,
        "obsolete": false,
        "workstream": "WS002",
        "deliverableId": "96b73d4c-ed79-451d-9435-2b976221d4c5",
        "createdTimestamp": "2024-01-10T12:03:09.883598Z",
        "lastChangedTimestamp": "2024-03-13T07:57:15.142210Z"
```

```
    }
]
```

Listing 22.13 Response of the SAP Cloud ALM Task API

The properties `id`, `title`, and `dueDate` are required for the initial creation of tasks in Microsoft To Do, whereby the property `id` is stored as a reference in Microsoft To Do. This reference enables both navigation to SAP Cloud ALM and updating of the Microsoft To Do task.

22.2.4 Practical Application

After we have looked at the various interfaces of Microsoft and SAP Cloud ALM, requests to these must be orchestrated accordingly. Figure 22.29 shows the logical orchestration of the requests toward a functioning integration, whereby we focus on the essential requests so that authentication at Microsoft, for example, is not taken into account.

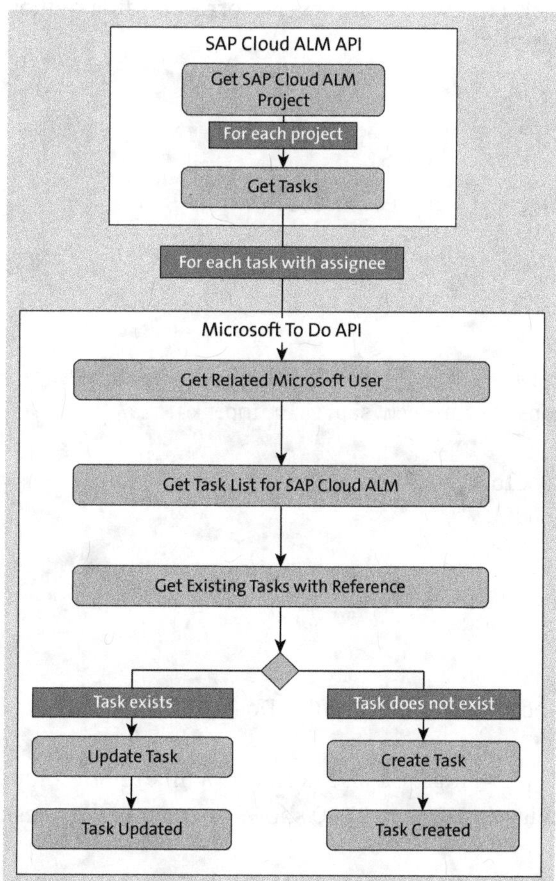

Figure 22.29 Flowchart for the Integration of SAP Cloud ALM Tasks with Microsoft To Do

First, the projects must be read from SAP Cloud ALM. You can then request the corresponding tasks for each project. The list of tasks should be filtered so that only tasks with an assigned agent are considered further. With these two steps, the processing of the SAP Cloud ALM API is already complete, as all information on the tasks is available.

Each task must then be considered individually. First, the assigned agent is determined on the Microsoft To Do layer. If this was successful, the corresponding task list for tasks within SAP Cloud ALM can be determined. It is also possible for the integration to automatically create task lists, e.g., for each SAP Cloud ALM project. Please note, however, that you are interfering with employees' personal task planning. It is therefore advisable to use an opt-in procedure, for example, by creating a corresponding task list, so that tasks are only synchronized if this task list has already been created. Once the task list has been determined, you can check whether the task is already available. The ID of the task in SAP Cloud ALM is checked as a reference to the task in Microsoft To Do. If the task already exists, it is compared with the task in SAP Cloud ALM and updated accordingly. If a task is not yet available in Microsoft To Do, it is created. By defining corresponding references, all relevant information, such as the due date and a direct jump to SAP Cloud ALM, can be stored in the task. Listing 22.14 shows the corresponding request to create a referenced task in Microsoft To Do.

```
{
    "title": "{{title}}",
    "dueDateTime": {
        "dateTime": "{{calmTaskDueDate}}T04:00:00.0000000",
        "timeZone": "UTC"
    },
    "linkedResources": [
        {
            "webUrl": "https://{{calmSubdomain}}.{{calmRegion}}.alm.cloud.sap/
launchpad#task-management?sap-ui-app-id-hint=com.sap.calm.imp.tkm.ui&/
taskDetail/{{calmTaskId}}",
            "applicationName": "SAP Cloud ALM",
            "displayName": "SAP Cloud ALM",
            "externalId": "{{calmTaskId}}"
        }
    ]
}
```

Listing 22.14 Request to Create a Referenced Task in Microsoft To Do

In addition to the due date, the corresponding task also contains a link to SAP Cloud ALM, which can be used for direct navigation. Figure 22.30 shows an example of a task.

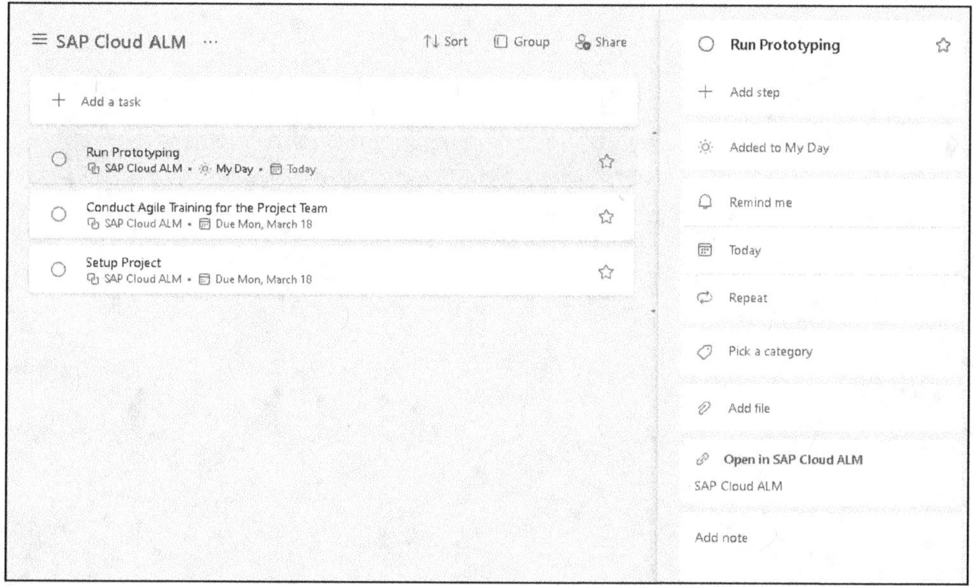

Figure 22.30 Referenced Task in Microsoft To Do

The practical implementation of integration can be realized in various environments. Within the SAP product catalog, the use of the SAP Integration Suite is obvious. However, other low-code or no-code implementations are also possible. The integration can be implemented both on a time-controlled or an event-controlled basis. In case of a time-controlled implementation, the integration needs to check all tasks every *n* minutes or hours in SAP Cloud ALM and compare with those with the data available in Microsoft To Do. Alternatively, an event-controlled implementation can be done via the usage of the Resource Changes API for tasks. With that, a push event of the task can be used to trigger the synchronization of this (and other) tasks to Microsoft To Do and back to SAP Cloud ALM. We have provided a sample implementation in our GitHub repository (*https://github.com/SAP-Cloud-ALM-Book*). You can use this to understand the details of the requests or create your own implementation on this basis.

22.3 Summary

In this chapter, we have looked at the use of inbound and outbound interfaces using practical examples. Even if the solutions shown are not to be used directly in a productive environment, they demonstrate the basic approach to integrating SAP Cloud ALM.

In our GitHub repository (*https://github.com/SAP-Cloud-ALM-Book*), you will find the examples, which we adapt to the current status of SAP Cloud ALM whenever possible.

The Authors

Matthias Friedrich has been working at NTT DATA Business Solution GBMS GmbH since 2008, where he is responsible for the global innovation and optimization service area. With a global team, he supports customers in the expansion, optimization, and development of applications and their integration into business processes. He is a business IT specialist and began his career in 1998 at SAP Systems Integration AG in Dresden. The focus of his work was on the analysis and development of IT service and change management concepts, the coordination of national and international projects, and the implementation of exciting product ideas.

Fred Kindler holds a degree in business informatics and began his career at NTT DATA Business Solution GBMS GmbH in Dresden in 2011. In addition to consulting in the field of ALM, he is currently responsible for managing an ALM team and for the application transition practice in the Application Solution Advisory division. His focus is on advising on the ALM processes required to implement hybrid landscapes in the cloud.

Bert Lorenz is employed as a development consultant at NTT DATA Business Solutions GBMS GmbH and supports customers in the areas of SAP Business Technology Platform, automation, low-code development, integration, and application lifecycle management. At NTT DATA Business Solutions GBMS, he is responsible for global development operations within the SAP and application services division and is committed to providing customers with continuous integration of the methodology into their business areas.

Mathias List has developed into an ALM expert over the years and has been working for NTT DATA Business Solutions GBMS GmbH since 2018. He began his career in 2008 as an SAP consultant for SAP Solution Manager at BIT Group GmbH. There he developed as an expert in the fields of IT service management and change and release management. In his national and international consulting and implementation projects, he also uses methodological approaches such as ITIL, Scrum and SAP Activate. He coauthored two SAP PRESS books on IT service management with SAP Solution Manager. He regularly uses his knowledge and experience in projects relating to ALM solutions such as SAP Cloud ALM. Mathias List studied business informatics at BA Bautzen.

Daniel Kloppich is one of the experts for application lifecycle management at NTT DATA Business Solutions GBMS GmbH in Dresden. He leads the global center of excellence for application lifecycle management at NTT DATA Business Solutions. He began his career in 2009 as a consultant at Comarch AG in the telecommunications sector. He studied media informatics at TU Dresden.

Marcel Kirsten has been working at NTT DATA Business Solutions GBMS GmbH in Dresden since 2016. Since completing his master's degree at Chemnitz University of Technology, he has been working as a consultant in the field of application lifecycle management with a focus on application operations. In addition to supporting implementation projects in the SAP Solution Manager 7.2 environment, he is also responsible for the development and implementation of monitoring services in the context of SAP Focused Run. Since 2020, his areas of expertise have also included SAP Cloud ALM, about which he advises customers on its introduction and implementation.

Lars Teuber has been working at NTT DATA Business Solutions GBMS GmbH in Dresden since 2008. Since 2017 he has been responsible for and headed the application lifecycle management consulting service area. He has been managing projects for the entire application lifecycle management since 2005. Since 2020, he has also been supporting the transformation of ALM processes to SAP Cloud ALM. His expertise lies in the optimization of ALM processes using various tools such as SAP Solution Manager and SAP Focused Run.

Index

G

H

I

J

K

L

M

N

O

P

Q

R

U

V

W

X

- Manage your cloud or on-premise SAP S/4HANA project
- Discover tools for configuration, data migration, extensibility, integration, and testing
- Explore hybrid scenarios across key lines of business

Sven Denecken, Jan Musil, Srivatsan Santhanam

SAP Activate

Project Management for SAP S/4HANA and SAP S/4HANA Cloud

Planning your SAP S/4HANA implementation project? This book has the expert guidance you need for every step of the journey! Learn about SAP Activate, including key concepts, the methodology, and supporting technologies. Understand your workstreams for configuration, data migration, extensibility, and more—then dive into deployment. Walk through the key phases and deliverables for your project, whether you're performing a new implementation or system conversion, and see what's possible with a two-tier ERP setup. Get ready for a successful go-live!

579 pages, 2nd edition, pub. 04/2022
E-Book: $84.99 | **Print:** $89.95 | **Bundle:** $99.99

www.sap-press.com/5463